T0259104

Return to the River
Restoring Salmon to the Columbia River

Return to the River
Restoring Salmon to the Columbia River

Edited by

Richard N. Williams

ELSEVIER
ACADEMIC
PRESS

AMSTERDAM • BOSTON • HEIDELBERG • LONDON
NEW YORK • OXFORD • PARIS • SAN DIEGO
SAN FRANCISCO • SINGAPORE • SYDNEY • TOKYO

Elsevier Academic Press
30 Corporate Drive, Suite 400, Burlington, MA 01803, USA
525 B Street, Suite 1900, San Diego, California 92101-4495, USA
84 Theobald's Road, London WC1X 8RR, UK

This book is printed on acid-free paper. ∞

Library of Congress Cataloging-in-Publication Data

Return to the river : restoring salmon to the Columbia River / edited by Richard N. Williams. – 1st ed.
    p. cm.
    Includes bibliographical references and index.
    ISBN 0-12-088414-3 (alk. paper)
    1. Pacific salmon fisheries–Columbia River Watershed–Management.
    2. Pacific salmon–Conservation–Columbia River Watershed. 3. Steel–head fisheries–Columbia River Watershed–Management. 4. Steelhead (Fish)–Conservation–Columbia River Watershed. I. Williams, Richard Nicholas.
    SH348.R43 2005
    639.2′756′09797–dc22   2005013382

British Library Cataloguing in Publication Data
A catalogue record for this book is available from the British Library

ISBN 13: 978-0-12-088414-8
ISBN 10: 0-12-088414-3

For all information on all Elsevier Academic Press publications visit our Web site at *www.books.elsevier.com*

Printed and bound by CPI Group (UK) Ltd, Croydon, CR0 4YY

Transferred to Digital Print 2011

Contents

Contributors   vii
Foreword   ix
Preface   xi
Acknowledgments   xiii

Part I
A NEW CONCEPTUAL FOUNDATION

Chapter 1.   The Problem of the Columbia River Salmon   1
Willis E. McConnaha, Richard N. Williams, and James A. Lichatowich

Chapter 2.   The Conceptual Foundation   29
James A. Lichatowich, Willis E. McConnaha, William J. Liss, Jack A. Stanford, and Richard N. Williams

Chapter 3.   A Foundation for Restoration   51
William J. Liss, Jack A. Stanford, James A. Lichatowich, Richard N. Williams, Charles C. Coutant, Phillip R. Mundy, and Richard R. Whitney

Part II
THE SALMON ECOSYSTEM AND ITS MANAGEMENT

Chapter 4.   The Diversity, Structure and Status of Populations   99
Richard N. Williams, James A. Lichatowich, and Madison S. Powell

Chapter 5.   The Status of Habitats   173
Jack A. Stanford, Christopher A. Frissell, and Charles C. Coutant

Chapter 6.   Hydroelectric Systems and Migration   249
Charles C. Coutant, and Richard R. Whitney

Chapter 7.   Sources of Mortality and Effectiveness of Mitigation   325
Richard R. Whitney, Charles C. Coutant, and Phillip R. Mundy

**Chapter 8.   Artificial Production and the Effects
of Fish Culture on Native Salmonids**   417
James A. Lichatowich, Madison S. Powell, and Richard N. Williams

Chapter 9.   Harvest Management   465
Phillip R. Mundy

Chapter 10.   The Estuary, Plume, and Marine Environments   507
Daniel L. Bottom, Brian E. Riddell, and James A. Lichatowich

Part III

A NEW VISION FOR RESTORATION

Chapter 11.   Monitoring and Evaluation   571
*Lyman L. McDonald, Charles C. Coutant, Lyle D. Calvin,
and Richard N. Williams*

**Chapter 12.   The Federal Approach to Salmon Recovery
at the Millennium**   601
*Peter A. Bisson, James A. Lichatowich, William J. Liss, Daniel Goodman,
Charles C. Coutant, Lyman L. McDonald, Dennis Lettenmeier,
Eric J. Loudenslager, and Richard N. Williams*

**Chapter 13.   Conclusions and Strategies for Salmon Restoration
in the Columbia River Basin**   629
*Richard N. Williams, Jack A. Stanford, James A. Lichatowich,
William J. Liss, Charles C. Coutant, Willis E. McConnaha,
Richard R. Whitney, Phillip R. Mundy, Peter A. Bisson,
and Madison S. Powell*

Subject Index   667

Contributors

Peter A. Bisson, Ph.D., (Chapters 12, 13) USDA Forest Service, Pacific Northwest Research Station, Olympia, Washington 98512

Daniel L. Bottom, M.S., (Chapter 10) National Marine Fisheries Service, Hatfield Marine Science Center, 2030 SE Marine Science Drive, Newport, Oregon 97365

Lyle D. Calvin, Ph.D., (Chapter 11) 3463 NW Crest Dr., Corvallis, Oregon 97330

Charles C. Coutant, Ph.D., (Chapters 3, 5, 6, 7, 11, 12, 13) Oak Ridge National Laboratory, Oak Ridge, Tennessee 37831

Christopher A. Frissell, Ph.D., (Chapter 5) Pacific Rivers Council, PMB 219, 1 Second Avenue, East, Suite C, Polson, Montana 59860

Daniel Goodman, Ph.D., (Chapter 12) Department of Ecology, Montana State University, Bozeman, Montana 59717

Dennis Lettenmeier, Ph.D., (Chapter 12) Department of Civil and Environmental Engineering, University of Washington, Seattle, Washington 98105

James A. Lichatowich, M.S., (Chapters 1, 2, 3, 4, 8, 10, 12, 13) Alder Fork Consulting, PO Box 36343, Miloris Way, Columbia City, Oregon 97018

William J. Liss, Ph.D., (Chapters 2, 3, 12, 13) Department of Fish and Wildlife, Oregon State University, Nash Hall 104, Corvallis, Oregon 97331

Eric J. Loudenslager, Ph.D., (Chapter 12) Department of Fisheries, Humboldt State University, Arcata, California 95521

Willis E. McConnaha, Ph.D., (Chapters 1, 2, 13) Jones and Stokes, 317 SW Alder, Suite 800, Portland, Oregon 97204-2583

Lyman L. McDonald, Ph.D., (Chapters 11, 12) West, Inc., 217 South First Street, Suite 5, Laramie, Wyoming 82070

Phillip R. Mundy, Ph.D., (Chapter 3, 7, 9, 13) Fisheries and Aquatic Sciences, 1019 Medfra Street, Anchorage, Alaska 99501-4013

Madison S. Powell, Ph.D., (Chapter 4, 8, 13) Center for Salmonid and Freshwater Species at Risk, University of Idaho, 3059 F National Fish Hatchery Road, Hagerman, Idaho 83332

Brian E. Riddell, Ph.D., (Chapter 10) Pacific Biological Station, Department Fisheries and Oceans, Hammond Bay Road, Nanaimo, British Columbia V9R 6N7 Canada

Jack A. Stanford, Ph.D., (Chapters 2, 3, 5, 13) Director and Bierman Professor of Ecology, Flathead Lake Biological Station, 311 BioStation Lane, Polson, Montana 58912

Richard R. Whitney, Ph.D., (Chapter 3, 6, 7, 13) 16500 River Road, Leavenworth, Washington 98826

Richard N. Williams, Ph.D., (Chapters 1, 2, 3, 4, 8, 11, 12, 13) Center for Salmonids and Freshwater Species at Risk, University of Idaho, 3059 F National Fish Hatchery Road, Hagerman, Idaho 83332

Foreword

The Columbia River is a river transformed. Over the course of a single generation, hard work, brilliant engineering, and massive federal expenditures have changed it from a vibrant biological system—home to the largest Chinook salmon run in the world—to a machine, an "organic machine" to use Richard White's term[1] (White 1995). The Columbia River is now a hydroelectric generating system providing inexpensive electricity to much of the west coast. It is a transportation system creating the nation's furthest inland seaport at Lewiston, Idaho. It supplies one of the world's largest irrigation systems transforming the arid interior Columbia Basin into an agricultural powerhouse. And it is a system to control flooding, allowing development of its floodplain into industrial parks and airports. To the original designers, the transformation of the Columbia River has been an overwhelming success allowing development of the Columbia Basin and providing benefits to local communities, the region and the nation.

The cost of this transformation has been the loss of much of the natural character of the river and its immense fish and wildlife resources. A steady decline in salmon abundance coincided with development of the river and tracked a less noticed loss of other aquatic resources and wildlife throughout the basin. Native American tribes that had depended on the return of salmon for millennia suddenly found salmon in short supply to the detriment of their traditional economies and culture. The decline in salmon culminated in the early 1990s with the listing of many populations under the Endangered Species Act.

Given the engineered transformation of the Columbia River, it is no surprise that an engineered solution was devised to address the loss of salmon. The intent of salmon restoration has been largely to support commercial fisheries in the lower river and ocean and not to restore aquatic environments or native cultures. Fish restoration strategies have been devised to protect the economic benefits afforded by development of the river. In effect, a second, artificial environment was created for salmon while allowing the river itself to be managed to meet human needs. Hatcheries replaced natural spawning and rearing areas lost to inundation by the dams and habitat degradation in the tributaries. In the Columbia and Snake rivers, the salmon's migration route was replaced by elaborate systems to collect and transport juvenile fish

---

1. White, R. 1995. The Organic Machine: the Remaking of the Columbia River. Hill and Wang, New York, NY.

around the dams, while fish ladders aided upstream migration of adults. The problem with this system, elegant and innovative as it was, is that it hasn't worked. Salmon continue to decline.

Into this setting, *Return to the River* suggested its own radical transformation of the Columbia River in regard to vision, if not in the landscape itself. It advanced the radical notion that the Columbia River is, in fact, a river, and that the solution to the loss of species, heralded by the decline in salmon, is to "return to the river"—to a vision of the Columbia River as an ecosystem rather than an economic machine. *Return to the River* is not a Luddite prescription of a "return to nature", but rather is a pragmatic vision of the Columbia River as a dynamic natural system that spans natural and cultural elements. It suggested that compromise between human and natural economies was not only possible but necessary, and that the notion that we could have it all—abundant salmon, cheap hydroelectricity and the power to transform the dessert—was not tenable—that nature would have the final say over human hubris.

It is difficult to appreciate how radical that commonsensical notion was when first advanced by *Return to the River* in 1996. When the Northwest Power Planning Council commissioned the report from its board of independent scientists, the prevailing wisdom was not that the region's approach to salmon recovery was necessarily wrong. Rather, the failure to recover salmon was due to not having gone far enough with existing programs and that we really needed more of the same—more hatcheries and more grandiose development of fish passage and transportation engineering. *Return to the River* suggested that the whole approach to salmon recovery needed to be redirected and used as evidence the fact that after almost a century of application, the engineering solution had failed to halt the decline of salmon in the Columbia River. Instead, after thoughtful review of the problem and study of the existing scientific literature, the group concluded that efforts to separate salmon from their environment were doomed to failure and that what was needed was to recouple salmon and the Columbia River and to allow the river to express a measure of its natural character, thereby creating conditions conducive to salmon and other native species.

Return to the River was the metaphorical grenade on the table of a staid decision making process that had attempted for years to nibble around the edges of the problem while preserving the *status quo* envisioned by the original developers of the river. Judge Marsh eloquently stated this in his review of the problem in 1994 quoted at the start of Chapter 1. Regional decision makers, including the Northwest Power Planning Council, were confronted with radical ideas like removing dams and allowing the river to re-establish habitats and conditions essential to native fish and wildlife. Even those that viewed the engineered infrastructure as a fundamental component of the landscape began to recognize that we were not going to be able to engineer our way out of the salmon crisis. Instead, the management of ecosystems became part of the region's thinking while the role of natural environmental cycles—in the ocean and more locally—was recognized as being fundamental to the success of restoration programs.

The sad truth is that the heady times of the late 1990s when new ideas were possible did not last. The Columbia River in 2005 does not look markedly different than it did in 1990. All the dams remain in place, fish bypass and transportation remain the prevailing method to assist downstream movement of juvenile salmon and most salmon in the Columbia River continue to return to hatcheries. A more conservative political sentiment has constrained change, while a period of good ocean conditions has led to increased returns that have let the region, at least temporarily, off the hook. There are signs that the problem has not gone away. As we write this, the region is witnessing a record low return of spring Chinook salmon to the Columbia River—much to the surprise of biologists and decision makers alike. New evidence points to a possible downturn in ocean conditions that will affect salmon abundance coast-wide. Human development of the region continues and population projections portend continued loss of aquatic and terrestrial habitats. Climate change, whether natural or human-caused, will increase challenges on natural systems and exacerbate the effects of habitat loss and degradation on salmon and other species. The lessons of *Return to the River* are that if the magnificent heritage of Columbia River salmon is to continue, we must recognize the fundamental linkages between species and their environments, view change and variation as constructive features of natural systems and accept our role as participants and stewards of the environment.

Willis E. (Chip) McConnaha, Ph.D.
Senior Fisheries Scientist, Northwest Power Planning Council (2000—2003)

Richard N. Williams, Ph.D.
Chair, Independent Scientific Group (1993–1996)
Chair, Independent Scientific Advisory Board (1997–1999)
Chair, Independent Scientific Review Panel (1997–2005)

Preface

Return to the River is the fruit of a unique 15-year collaboration by a diverse group of scientists working on the management and recovery of salmon, steelhead trout, and wildlife populations in the Pacific Northwest. *Return to the River* evolved while most of the book's authors were members of the Independent Scientific Group (ISG), which provided scientific and technical advice and peer-review for the massive program to recover Pacific salmon and steelhead in the Columbia Basin. The ISG was created and funded by the Northwest Power Planning Council[1], Bonneville Power Administration, and the National Marine Fisheries Service (now NOAA-Fisheries).

Return to the River arose directly from the Northwest Power Planning Council's request in the mid-1990s to the Independent Scientific Group to review the scientific basis for its Fish and Wildlife Program and identify critical uncertainties. The ISG delivered its final report to the Council in 2000 (*Return to the River 2000*) after taking into account constructive criticism from scientific and technical peer-review and public comment on an earlier draft.

Many of the complex issues discussed in detail in this volume, such as habitat degradation, juvenile survival through the hydrosystem, the role of artificial production, and harvest reform, have been recognized as areas of critical concern for decades. However, until the publication of *Return to the River* there has never been a comprehensive, scientific review of the programs intended to address those issues. Prior to the publication of *Return to the River*, the program focused on juvenile passage and hatcheries. Those two issues accounted for the majority (approximately 75-80%) of the Bonneville Power Administration's and the Council's Fish and Wildlife Program annual expenditures (~$450 million) to support fish and wildlife costs. *Return to the River* shifted attention to a broader suite of issues affecting salmon and steelhead production in the Columbia Basin.

While the ISG's 2000 report focused very closely on the Council's Fish and Wildlife Program, this volume is written to address regional and

---

[1]The Northwest Power Planning Council, formed in 1981 by the Northwest Power Act, was recently renamed the Northwest Power and Conservation Council (www.nwcouncil.org) to reflect the Council's dual role in protecting and conserving the region's fish and wildlife resources and providing reliable and economical hydroelectric power to the region.

international issues involved in managing fish and wildlife resources in a large regulated river ecosystem. Consequently, we treat the Council's Fish and Wildlife Program as a case study, rather than as the focus of the book.

The management and proposed recovery actions for Columbia River salmon and steelhead contain many lessons to the fisheries and conservation communities as the scale of the effort and expenditures involved in the Columbia River Basin are nearly unmatched. Recovery of salmon and steelhead is an epic undertaking, with epic consequences. We believe *Return to the River* will continue to foster regional, national, and international discussion and contribute toward the restoration of our precious salmon and steelhead resources.

Rick Williams

Acknowledgements

Many individuals and institutions helped us with the preparation of *Return to the River* directly or indirectly. A few of those, we thank explicitly below, and apologize for not mentioning all.

Foremost, we thank the Northwest Power and Conservation Council and Council staff for understanding our initial vision for *Return to the River*, urging us to proceed, and for their patience (and financial support) as the work incubated and grew to completion.

We were always willingly and promptly assisted by the staff of many institutions within the Columbia Basin. We thank Marianne McClure and James Berkson, Columbia River Inter-Tribal Fish Commission, Portland, for advice on Pacific Salmon Commission matters; John G. Williams (NMFS) for assistance on the bypass section and specifics on juvenile smolt transportation; Lyman L. McDonald (West, Inc.) for assistance on our flow-survival analysis; Tom Iverson and the staff of the Columbia Basin Fish and Wildlife Authority for various background materials and logistical support; and Dennis Dauble and David Geist of Batelle's Pacific Northwest Division for the tours and the lessons about the Hanford Reach over the course of several visits there. We also thank Chuck Peven (Chelan County PUD Number 1) for providing access to and clarification of the Habitat Conservation Plans (HCPs) of Chelan and Douglas County PUDs and Stuart Hammond (Grant County PUD Number 1) for providing unpublished information on Grant County PUD's progress in development of bypass facilities at Wanapum and Priest Rapids dams.

Return to the River was dramatically improved by comments provided by peer reviewers and by an interested public. We thank the anonymous reviewers of an early draft of *Return to the River* and of our 1999 *Fisheries* article on Columbia River Salmon Recovery[2]. Particularly insightful were comments provided by Pete Bisson and Brian Riddell (members of the NRC's panel on Pacific Salmon and later, colleagues on the Council's independent scientific advisory panels), and by Ernie Brannon (University of Idaho) and Fred Utter (NMFS). Extensive comments were also provided through the Council's public comment period by CRITFC, ODFW, John Palmisano, Bill McNeil, and Suck Cho Chyung.

---

[2]Independent Scientific Group. 1999. Scientific issues in the restoration of salmonid fishes in the Columbia River. *Fisheries* 24:10–19.

Finally, generous support for publication of *Return to the River* was provided by the following contributors (listed alphabetically): American Rivers, Kenai River Sportfishing Association, National Fish and Wildlife Foundation, Native Fish Society, Oregon Trout, Save Our Salmon, and Trout Unlimited. We greatly appreciate the assistance and support of these institutions. Their support made possible, among other things, inclusion of the many supporting color figures and photos used throughout the book. We appreciate their support and patience for the project.

Funding was provided in support of publication for *Return to the River* by the following institutions:
American Rivers
Kenai River Sportfishing Association
National Fish and Wildlife Foundation
Native Fish Society
Oregon Trout
Save Our Salmon
Trout Unlimited

"Nature, to be commended, must be obeyed."

Sir Francis Bacon, *Novum Organum*, 1620.

1

Introduction and Background of the Columbia River Salmon Problem

Willis E. McConnaha, Ph.D., Richard N. Williams, Ph.D., James A. Lichatowich, M.S.

Introduction
Background
  Salmon in the Columbia River
  Salmon Management in the Columbia River
    Northwest Power Act
    Endangered Species Act
    Federal Treaties with Indian Tribes
Summary of Our Findings
Relationship to Other Plans and Reviews
Where Do We Go From Here?
Literature Cited

> *". . .the process is seriously, significantly, flawed because it is too heavily geared towards a status quo that has allowed all forms of river activity to proceed in a deficit situation–that is, relatively small steps, minor improvements and adjustments–when the situation literally cries out for a major overhaul."*
> —Judge Marsh in his review of the 1993 National Marine Fisheries Service Biological Opinion on Columbia River mainstem operations (Idaho Department of Fish and Game v. National Marine Fisheries Service, Civil No. 92-973-MA, slip opinion at p. 36 (D. Ore. 1994)).

Introduction

On October 16, 1805, Meriwether Lewis, William Clark, and the other members of the *Corps of Discovery* reached the confluence of the Columbia and Snake Rivers following their 17-month overland trek from St. Louis, Missouri. Their arrival coincided with the arrival of vast numbers of salmon (anadromous salmon and steelhead trout, *Oncorhynchus* spp.) returning to the river on their annual spawning migration. William Clark's journal entry of October 17, 1805, noted, "The number of dead Salmon on the Shores & floating in the river is incredible (sic) to say . . ." (DeVoto 1953). They described a thriving native culture centered on the salmon: "They have only to collect the fish Split them open and dry them on their Scaffolds on which they have great numbers" (Clark: DeVoto 1953). The number of salmon returning annually to the Columbia River prior to European settlement has been estimated to be between 10 and 16 million fish (Northwest Power Planning Council [NPPC] 1987).

The arrival of the *Corps of Discovery* heralded a period of dramatic environmental change brought on by the encroachment and subsequent growth of European civilization in the Pacific Northwest. Lewis and Clark encountered a native culture in the Columbia Basin that was centered on the salmon and relied on its annual bounty for sustenance and wealth (Lichatowich 1999; Scarce 2000; Figure 1.1). Within a relatively brief time after the return of the *Corps of Discovery*, the spiritually based conceptual foundation that linked the native culture to salmon was replaced by a commodities-driven foundation based on 19th-century laissez-faire economic thought (Lichatowich 1999; Figure 1.2). In the latter part of the 19th century, salmon became the

Figure 1.1 Native American salmon fishing by dipnetting from scaffolding at Celilo Falls, prior to its inundation by The Dalles Dam in 1957. Photo from U.S. Army Corps Digital Visual Library, website at *http://images.usace.army.mil/photolib.html.*

Figure 1.2   The Dalles Dam juxtaposed with a Native American fishing scaffold. Photo credit: Mike Pinney (*www.photobymike.com*).

object of an intense commercial fishery in the Columbia River that culminated with an estimated catch of 40 million pounds of salmon in 1883 (Van Hyning 1973; Taylor 1999). At the same time, logging, agricultural development, and urbanization produced profound changes in the natural character of the Columbia Basin. Coincident with these changes, salmon abundance in the Columbia River during the 20th century showed an overall pattern of decline. The average annual return of all salmon to the Columbia River from 1996 to 2000 averaged 1.1 million fish, the majority of which were produced artificially in hatcheries. Many wild populations have been eliminated and most others are severely depressed (Nehlsen et al. 1991). Salmon and steelhead populations throughout the Columbia and Snake River basins are listed as endangered or threatened under the Endangered Species Act.

The decline in salmon has not gone unattended. The 20th century saw several salmon recovery programs, most of which were associated with mitigation for fishery impacts from development of the river's hydroelectric potential. In order to support the commercial fishery as habitat was lost or degraded through development, an extensive hatchery system was developed, which currently releases some 235 million juvenile salmon and steelhead each year (Northwest Power and Conservation Council 2003). The General Accounting Office estimated that from 1982 through 2001, federal agencies expended about 6.4 billion dollars on salmon restoration in the Columbia Basin (General Accounting Office 2002).

This book presents a review of salmon and steelhead management in the Columbia Basin and suggests a new conceptual foundation for salmon management in the 21st century. The region's failure to halt the decline in salmon populations is the result of numerous social, economic, and scientific issues, most of which are generally recognized and have been discussed extensively (e.g., National Research Council 1996). What is less recognized is the lack of an explicit and scientifically based conceptual foundation and the consequences of this on salmon management and recovery actions. A conceptual foundation or "world view" (Costanza 2000) is fundamental to how we interpret the "facts" garnered from observation or scientific investigation, and, in turn, to how we manage human interactions with the environment. The commodities-driven conceptual foundation that guided much of 20th-century fishery management was based first on the belief in nearly inexhaustible resources and later on faith in technology to replace natural functions lost as a result of human actions (Bottom 1997). The decline of salmon in the Columbia Basin over the course of the 20th century, despite massive infusions of money and technology, proves the failure of the old paradigm and the need for a new conceptual foundation for salmon management. On the basis of our review of existing salmon restoration efforts in the Columbia Basin, we present a conceptual foundation for 21st-century salmon management that stresses the role of the environment—with all its variability and complexity–in shaping species performance and persistence. This foundation is based less on the notion of engineering nature to fit the needs of human society and more on the idea that human activities can be managed to facilitate natural processes that shape the environment and ensure resiliency of species.

Development of this new conceptual foundation is based on our review of the Fish and Wildlife Program developed for the Columbia River by the Northwest Power and Conservation Council (NPCC)[1]. The Council's program directs about 137 million dollars in annual funding for fish and wildlife restoration projects in the Columbia Basin by the Bonneville Power Administration (BPA). While the Council's program addresses the impact of the hydroelectric system on both fish and wildlife, the bulk of the program is devoted to the restoration of fish, especially salmon[2]. Likewise, our review has focused on the impact of the Council's program on aquatic habitats and salmon populations. In the 1992 Fish and Wildlife Program (NPPC 1992a, 1992b), the Council brought together most of the contributors to this book to review the scientific basis for the region's efforts. In that review (Independent

---

[1]For most of its history, the Council was known as the Northwest Power Planning Council (NPPC). In 2003, the Council changed its name to the Northwest Power and Conservation Council (NPCC).

[2]Unless otherwise stated, for purposes of this report the term "salmon" will refer to anadromous species of the genus *Oncorhynchus* that includes Pacific salmon and steelhead trout.

Scientific Group 1998; 1999), we were struck by the power of the region's implied conceptual foundation to shape how scientists and managers interpreted information and designed restoration activities. We became convinced of the need to develop an explicit conceptual foundation that could be critically examined, and one that incorporated contemporary scientific thought on species and their environment. We developed the conceptual foundation presented here as a synthesis of existing scientific information to serve as a foundation for 21st-century salmon management in the Columbia River and elsewhere. Since our initial review, the Council has led the region through two revisions of its fish and wildlife program. The region is currently preparing restoration plans for each subbasin in the Columbia Basin that may be incorporated into a future fish and wildlife program. In short, the Council's program is an ever-changing reflection of the region's legal, economic, and social priorities. Because of this, our review does not address a single version of the Council's program but rather the region's collective efforts in the early part of the 21st century as embodied in the Council's programs. Further, because the region's efforts are largely based on the prevailing paradigm that guides fisheries management, our conclusions can be broadly applied to Pacific and Atlantic salmon management in general.

This book is organized into three sections:

Part I. An introduction and background to the salmon problem (Chapter 1), followed by a description of the current conceptual foundation directing salmon restoration and an analysis of the scientific basis for the assumptions and beliefs implied by measures in the Council's Fish and Wildlife Program (Chapter 2), and finally, an explicit description of an alternative ecologically based conceptual foundation for fish and wildlife management (Chapter 3).

Part II. A technical review and documentation of major scientific issues and topics supporting the conceptual foundation (Chapters 4-10).

Part III. A review of the role of monitoring and evaluation in salmon restoration (Chapter 11), the current federal approach to salmon recovery (Chapter 12), and our conclusions and strategies for restoration from the overall review (Chapter 13).

Background

Salmon in the Columbia River

The Columbia River is the 18th largest river in the world in terms of discharge and the 2nd largest river in the United States (Leopold 1994). The river drains a watershed of 258,000 square miles covering much of the states of Washington and Oregon and major portions of the Province of British Columbia and the states of Idaho, western Montana, and small parts of

Figure 1.3 Map showing the Columbia River Basin and subbasins (within the U.S. portion). The green color denotes the entire basin, while yellow indicates the portion of the basin to which anadromous salmonids presently have access. Figure provided by Columbia Basin Fish and Wildlife Authority and StreamNet.

Nevada, Utah, and Wyoming (Figure 1.3). The river has two major tributaries, the Snake River in southern Idaho and eastern Oregon, and the Willamette River in western Oregon. The basin is divided by the Cascade Mountains into a region of high rainfall and mild winters to the west and a dry region with harsher winters to the east. Historically, anadromous salmon and steelhead occurred widely throughout the basin except where blocked by natural falls, such as Shoshone Falls on the Snake River in south central Idaho (Figure 1.4).

Historically, the Columbia River produced the world's largest runs of Chinook salmon (*Oncorhynchus tshawytscha*) (Van Hyning 1973). In addition

Figure 1.4   Shoshone Falls in south central Idaho is an upstream migration barrier to anadromous salmon and steelhead in the Snake River basin. Photo credit: Mike Pinney (*www. photo bymike.com*).

to Chinook, large numbers of coho (*O. kisutch*), sockeye (*O. nerka*), and steelhead trout (*O. mykiss*) returned, as well as smaller numbers of chum (*O. keta*) and pink salmon (*O. gorbuscha*). Pre-development run size estimates of all salmon range from 6.2 million fish (Pacific Fisheries Management Council 1979) to 10–16 million fish (NPPC 1986).

This abundance of salmon has been estimated to have been composed of over 200 distinct anadromous stocks (NPPC 1986; Nehlsen et al. 1991) (Chapter 4). Today, with few exceptions, most chum, pink, and wild coho stocks are extinct, and the other species are at risk of extinction. Nehlsen et al. (1991) identified 69 extinct stocks and 75 others at risk of extinction in some areas of the basin. Only Lewis River (WA) and Hanford Reach (WA) fall Chinook, Lake Wenatchee and Lake Osoyoos (WA) sockeye, and five summer steelhead stocks in the John Day River (OR) can be classified as healthy[3] (Mullan et al. 1992; Huntington et al. 1996). Total returns of

---

[3]Huntington et al. (1996) noted that native stocks of anadromous salmonids could be considered healthy from several perspectives. Their final criteria for classifying stocks as healthy were that they be at least one-third as abundant as expected in the absence of human impacts, abundant relative to current habitat capacity, self-sustaining (not recently declining in abundance or dependent on hatchery supplementation), and not previously identified as being at substantial risk of extinction.

cultured and wild Chinook and sockeye reached an all time low in 1995 (Figure 1.5). Likewise, resident (non-anadromous) salmonid populations, such as bull trout and westslope cutthroat trout, also are increasingly isolated by habitat fragmentation and have been eliminated from many river segments. Many remaining populations are reduced in size and vulnerable to extinction. Evaluation of native salmonids in headwater reaches of the Columbia River shows that the distribution of healthy stocks are reduced to 10% to 30% of their original distribution, depending on the species (Behnke 1992; Henjum et al. 1994; Anderson et al. 1996; Quigley et al. 1996).

The rise of European civilization in the region in the latter half of the 19th century brought with it the decline in salmon abundance in the Columbia Basin. With the development of technology to preserve and can salmon in the late 1800s, an intense commercial fishery developed in the Columbia River. During the period 1875 to the 1920s, 20 to 40 million pounds of salmon were harvested and canned annually (Van Hyning 1973). The combined pack of salmon of all species in the Columbia River peaked in 1885. After this, the commercial catch, an indicator of total abundance, declined steadily throughout the 20th century (Figure 1.5).

The decline of salmon in the Columbia Basin has many causes that will be discussed in later chapters. Briefly however, salmon abundance declined initially as a result of fishing pressure (Figure 1.5; Chapter 9). At about the

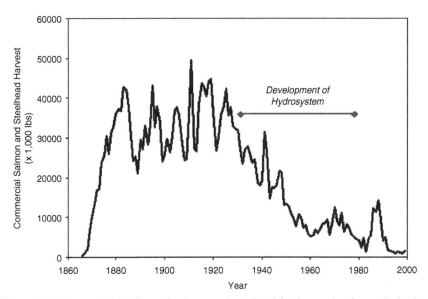

Figure 1.5 Commercial landings of salmon and steelhead in thousands of pounds in the Columbia River, 1866 to 2000 (Washington Department of Fish and Wildlife and Oregon Department of Fish and Wildlife 2001).

same, time, rapidly escalating development led to widespread modification of the landscape and the degradation of aquatic environments throughout the basin (Lichatowich 1999; Chapter 5). Riverine environments were altered or destroyed by agriculture, logging, mining, urbanization, and the construction of dams for irrigation, navigation, and power production (Figure 1.6; Lichatowich 1999). These same forces continue to alter terrestrial and aquatic habitats in the Columbia River to the present day.

As the habitat was lost and fish populations declined, the region responded with two remarkable engineering solutions: first, to replace mainstem

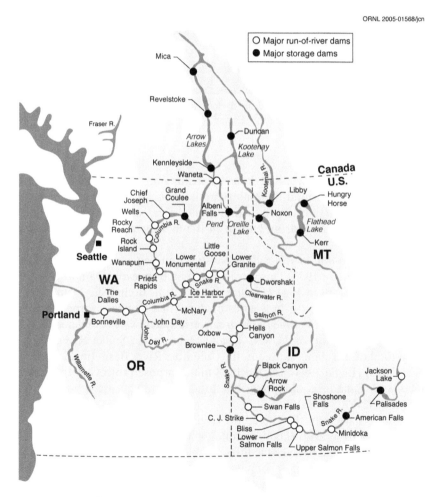

Figure 1.6 Map of the Columbia River Basin, showing major rivers, lakes, run-of-the-river dams and storage dams.

Figure 1.7   Aerial view of Bonneville Dam, the lowermost of the Columbia River Hydrosystem projects (photo from U.S. Army Corps Digital Visual Library *http://images.usace.army.mill photolib.html*).

migrational habitat affected by dams by creating an artificial migration corridor through and around the mainstem dams (Chapter 6), and second, by development of artificial production (hatcheries) on a massive scale to replace spawning and rearing habitat in the mainstem river and tributaries (Chapter 8).

The era of Columbia River dam construction began with completion of Rock Island Dam in 1933. This was followed by Bonneville Dam in 1938 and Grand Coulee Dam in 1941 (Figures 1.5 and 1.6). Bonneville Dam (Figure 1.7) is now the lowermost in a series of dams that stretches throughout the Columbia Basin. By the time the hydroelectric system was completed in 1975 with the construction of Lower Granite Dam on the Snake River (Figure 1.8), a total of 211 dams existed in the Columbia River Basin (including tributary dams) (Logie 1993), of which 83 multipurpose projects in the United States and Canada could provide a total generating capacity of 30,813 megawatts (NPPC 1986).

The dams have radically altered the riverine environment of the Columbia River and its tributaries (Dauble et al. 2003). Grand Coulee Dam was constructed without fish passage facilities and blocked access to 2,240 kilometers of highly productive salmon habitat in the upper Columbia River. Because of the lack of access to habitat above Grand Coulee, fish passage facilities were deemed unnecessary at Chief Joseph Dam, located not far below Grand Coulee. Similarly, Hells Canyon Dam blocked 338 kilometers of mainstem

Figure 1.8 Aerial view of Lower Granite Dam, the uppermost of the four dams on the lower Snake River. Salmon and steelhead returning to central Idaho pass through eight hydroelectric projects before reaching their spawning grounds in the Salmon and Clearwater Rivers (photo from U.S. Army Corps Digital Visual Library *http://images.usace.army.mil/photolib.html*).

habitat in the Snake River. Below these points that now mark the upstream limit of anadromous fish passage, 13 federal and private dams affect riverine habitats and fish survival (Figure 1.6). Presently there are only two free-flowing riverine reaches of the Columbia system above Bonneville Dam that are accessible to salmon and steelhead: the Hanford Reach in the Columbia River and Hells Canyon in the Snake River (Figure 1.9). With these two exceptions, almost all salmon production above Bonneville Dam has been pushed out of the mainstem river and now originates in hatcheries and in the various tributaries to the Columbia and Snake Rivers. The mainstem river sections below Chief Joseph and Hells Canyon dams have been converted to a series of slack water reservoirs that bear scant resemblance to the original river. The reservoirs inundated mainstem habitat that provided spawning for large numbers of Chinook salmon and other species. Overall, productive mainstem spawning areas in the Columbia Basin have been reduced to 6 percent of the original area as a result of dam construction (Dauble et al. 2003).

The impacts of the 13 dams below Chief Joseph and Hells Canyon dams have been addressed by developing a complex system of screens, transport vehicles, and project operations designed to move adult and juvenile fish around and through the hydroelectric system. All of these dams were built

Figure 1.9   Hells Canyon, along the Idaho-Oregon border, remains one of only two free-flowing sections of the main Snake and Columbia Rivers used by remnant wild salmon and steelhead populations. Photo credit: Mike Pinney (*www.photobymike.com*).

with ladders for adult upstream passage (Figure 1.10). When most of the dams were constructed, little thought was given to downstream passage of juvenile fish. However, studies by the National Marine Fisheries Service in the 1970s pointed to severe mortality to juvenile downstream migrants as they passed through turbines or over spillways (Ebel et al. 1979). Development of engineering measures to improve juvenile survival was a topic of intense research during the late 1970s and 1980s (Raymond 1988). As a result, almost all dams have been retrofitted with elaborate screening systems to divert juvenile fish away from turbines and deliver them below the dam. Many of these systems are used to collect juvenile migrants at the dams for downstream transportation. In the Snake River, the majority of downstream juvenile salmon migrants are collected at the dams and transported to below Bonneville Dam in barges or trucks.

   The use of hatcheries to replace spawning habitat has a long history in the Columbia River. The first hatchery in the Columbia River was built on the Clackamas River (a major tributary to the Willamette River) in 1877 (Stone 1879). However, through the first half of the 20th century, hatcheries made relatively minor contributions to returns, and most returning fish were progeny of wild spawners. During the 1960s, the region embarked on a massive program to develop hatcheries to replace lost habitat and to support declining

Figure 1.10 Fish ladder for returning adult salmon and steelhead at Bonneville Dam, the lowermost of the dams on the Columbia River. Photo by R. N. Williams.

fish populations as a way to maintain harvest levels (Bottom 1997). Initial success was followed by precipitous declines in abundance in the late 1970s that coincided with completion of the hydroelectric system and an ocean regime shift that reduced ocean survival for salmon populations throughout the Pacific Northwest (Pearcy 1992). Hatcheries continue to be a major

Figure 1.11   Hatchery raceways at Bonneville Dam. Covers protect the young salmon from bird predation. Photo by R. N. Williams.

component of salmon returns to the Columbia River (Figure 1.11). Currently, federal, state, and tribal managers in the basin have planned releases of juvenile fish of some 235 million juveniles from hatcheries that are concentrated in the lower portion of the basin (NPCC 2003). About 80 percent of the salmon returning to the Columbia River now originate from hatcheries (NPPC 1992a).

Despite the use of these technologies to replace habitat and to mitigate environmental change, salmon populations continued to decline throughout the latter part of the 20th century. Concern about depressed salmon and steelhead populations in the late 1970s prompted inclusion of language in the Northwest Power Act that created the NPCC. During the 1980s and 1990s the region embarked on an ambitious restoration program based around the fish and wildlife program developed by the Council. Despite expenditures of over 100 million dollars each year by the BPA in response to the Council's program, salmon and steelhead populations, especially wild populations continued to decline. While the various versions of the Council's program contained many innovative ideas and concepts, in practice most actions were expansions of existing techniques of replacing mainstem migrational habitat with new screens, transportation and flow operations, and production habitats with new and expanded hatcheries.

The perilous condition of many salmon and steelhead populations in the Columbia River during the 1980s prompted reviews for listing under the Endangered Species Act (e.g., Myers et al. 1998). At the present time, 13 groups of salmon and steelhead, termed Evolutionarily Significant Units or ESUs under the Endangered Species Act (Waples 1995), in the Columbia Basin are listed under the Endangered Species Act. Efforts to address the needs of these listed populations have become a major priority in the region and occupy much of the attention of the federal management agencies that have special obligations under the Endangered Species Act.

Salmon Management in the Columbia River

In the early part of the 21st century, fish recovery efforts in the Columbia River are structured by three legal mandates–the Northwest Power Planning and Conservation Act of 1980 (The Northwest Power Act), the Endangered Species Act, and federal treaties with Indian Tribes. The three legal mandates structuring fish and wildlife recovery in the Columbia Basin represent the states composing the NPCC, the federal government represented by the agencies responsible for implementation of the Endangered Species Act, and Indian Tribes with interest and management authority secured by federal treaties. Notwithstanding arguments that can be made regarding legal authorities, in practice most fish recovery actions in the basin are arrived at through negotiation and compromise between these three authorities, with occasional intervention of federal courts.

Part of the reason for cooperation between the three governing forces in the region is that they all largely draw on a single source of funding for fish restoration—electric ratepayer funds collected by the BPA. BPA is the federal agency created to market and transmit power created by the Columbia River hydroelectric system. The agency has annual gross revenues from power sales in excess of 2 billion dollars (Bonneville Power Administration 2002). Of this, around 240 million dollars is annually devoted to funding of fish and wildlife restoration projects in the Columbia Basin to mitigate for habitat lost to development and operation of the hydroelectric system. The Northwest Power Act created the Northwest Power Planning and Conservation Council in part to develop a regional program to direct the use of Bonneville's fish and wildlife restoration funds. BPA annually funds 135 million dollars in projects called for under the Council's program. BPA also provides funding for actions determined to be "reasonable and prudent" to protect salmon and other species listed under the Endangered Species Act. Even funding directed by other federal agencies is often reimbursed by ratepayer funds from BPA. Examples include funds allocated by the Army Corps of Engineers for construction and operation of fish passage facilities at hydroelectric projects and funding for federal hatchery programs designed to mitigate for habitat lost to hydroelectric development.

As background for our review of salmon management in the Columbia Basin, each of these three legal mandates will be discussed in more detail.

Northwest Power Act

The Pacific Northwest Electric Power Planning and Conservation Act of 1980 (the Northwest Power Act) was intended to restructure the Northwest electric power industry and especially the role of the BPA (Blumm 2002). Because most of the electricity in the region results from hydroelectric generation, the Northwest Power Act also contained important provisions regarding mitigation for the impacts of hydroelectric development on fish and wildlife in the basin. To provide a regional voice to energy planning and fish and wildlife mitigation, the act created the Northwest Electric Power Planning and Conservation Council (generally shortened to the Northwest Power Planning Council and, more recently, the Northwest Power and Conservation Council). The Council is composed of two representatives appointed by the governors from each of the states of Montana, Idaho, Washington, and Oregon. The Northwest Power Act included many groundbreaking concepts for both electric power production and fish and wildlife management. However, most important, from the standpoint of fish and wildlife, was its role in creating the concept of the Columbia River as an integrated *system* rather than as the collection of parochial interests that had characterized the river through much of the 20th century. The act also attempted to create greater parity between fish and electric power production in terms of river management. While it can be argued that the Northwest Power Act failed to live up to its full promise (Blumm 2002), it has nonetheless succeeded in creating a regional forum for discussion of fish and wildlife issues. The Council has led the region to consider new ideas such as adaptive management (Lee and Lawrence 1986; Lee 1993) and to critically examine fundamental management assumptions such as the role of artificial production (NPCC 2003) and this review of the overall conceptual foundation behind fish and wildlife restoration and management.

The Northwest Power Planning and Conservation Council was directed in the Northwest Power Act to develop a program to direct BPA funding to "protect, mitigate, and enhance" fish and wildlife as affected by development and operation of the hydroelectric system. By its design in the Northwest Power Act, the Council's program was fundamentally weakened as a vehicle for comprehensive fish and wildlife restoration. The Council's program is intended to only mitigate for the effects of hydroelectric development and many factors, such as harvest, that have contributed to the decline of salmon are outside the Council's purview. The Council also was granted little power to enforce its program, but instead must rely on often-imperfect cooperation by federal, state, and tribal management agencies. Finally, the U.S. Congress directed that the Council's program be based on recommendations from concerned parties and especially those from the region's fish and wildlife managers. Their recommendations often reflect disparate mandates and objectives and leave the Council open to the criticism that its program is simply a potpourri of good ideas without a cohesive framework (Independent Scientific Group 1998).

Congress directed the Council as its first act to prepare a fish and wildlife plan to address the loss of fish and wildlife in the Columbia River Basin resulting from the development and operation of the hydroelectric system. The first Fish and Wildlife Program was adopted in November 1982, following an extensive public process to garner ideas and projects. Since that time, the Council's program has been revised numerous times, most recently in 2000 (NPPC 2000). As part of the most recent program, the Council is leading the region in preparation of restoration plans for each of the 62 subbasins within the Columbia River Basin.

Each version of the Council's Fish and Wildlife Program has described a variety of actions to be carried out by the BPA, other federal agencies, and the region's state and tribal fish and wildlife managers. Many of these actions have focused on in-river returns and production of anadromous salmonids, while others have addressed the needs of resident fish and wildlife. Reflecting the legislated focus on the impacts of the hydroelectric system, the program emphasizes actions to increase survival of salmon and steelhead in the Lower Snake River (i.e., downstream from Hells Canyon Dam, the most upstream point currently accessible to anadromous fish), the middle and lower reaches of the mainstem Columbia River (i.e., downstream from Chief Joseph Dam, the upstream point of current accessibility in the Columbia River), and their accessible tributaries (Figure 1.6). These actions include modification of mainstem dam operations and facilities to improve passage of adults and juveniles and coordination of river operations to enhance flows for fish migration. Other actions call for reduction of predators of downstream migrating juveniles, construction and operation of hatcheries, and modification of existing artificial production operations, including supplementation of naturally reproducing populations with hatchery-raised juveniles. In recent years, the Fish and Wildlife Program has attempted to address stream habitat through implementation of "best management practices" for land use activities and protection of many tributaries from further hydroelectric development. The program also contains a variety of research and monitoring projects designed to answer critical questions. Many of these activities and projects have been funded or implemented through complementary programs overseen by the National Marine Fisheries Service and the U. S. Army Corps of Engineers.

Endangered Species Act

The Endangered Species Act of 1973 was enacted to "provide a means whereby the ecosystems on which endangered species and threatened species depend may be conserved, [and] to provide a program for the conservation of such endangered species and threatened species" (16 USC 1531). NOAA Fisheries (formerly the National Marine Fisheries Service) is charged with review and determination of the status of anadromous salmon under the Endangered Species Act. NOAA Fisheries first began to analyze the status of Snake River salmon populations in 1979 to determine if they warranted protection under the Act (ESA; 43 Fed. Reg. 45628 (1978)). Inclusion of fish and wildlife protection during the development of the Northwest Power Act, and its ultimate passage in 1980, helped forestall ESA listings for more than a decade. However, the abundance of many salmon populations continued to decline, while others were extirpated, after passage of the Power Act and creation of the Council's first Fish and Wildlife Program in1982. These declines led to petitions and listings under the Endangered Species Act. Between 1991 and 1998, first sockeye *O. nerka*, then spring-summer and fall Chinook *O. tshawytscha*, and finally steelhead *O. mykiss* from the Snake River were listed under the Endangered Species Act (Matthews and Waples 1991; Waples et al. 1991a, 1991b; Waples 1995; Busby et al. 1996). Another species of salmon known to have occurred in the Snake River Basin, the coho salmon *O. kisutch*, declined sharply in the 1980s to become extinct some time prior to 1992 (Hassemer et al. 1996).

These and other recent listings[4], and the listing of Kootenai River white sturgeon *Acipenser transmontanus* (1991) and the bull trout *Salvelinus confluentes* (1997), have added another layer of complexity and additional capital cost to the restoration effort in the Columbia River. Development of recovery plans for listed anadromous salmonid fish populations in the Columbia River are the responsibility of NOAA Fisheries, while the U.S. Fish and Wildlife Service (USFWS) has responsibility for the listed resident fish species such as bull trout, westslope cutthroat trout *O. clarki lewisi*, and sturgeon[5].

---

[4]Presently, 13 species (or "evolutionary significant units" of species) of salmon and steelhead that spawn in the Columbia River or its tributaries have now been listed as threatened or endangered under the Endangered Species Act. These include Snake River fall Chinook, Snake River spring/summer Chinook, Snake River sockeye, Snake River steelhead, upper Columbia River spring Chinook, upper Columbia River steelhead, middle Columbia River steelhead, lower Columbia River spring Chinook, lower Columbia River coho, lower Columbia River steelhead, Columbia River chum salmon, upper Willamette River spring Chinook, and upper Willamette River steelhead.

[5]Although sturgeon are anadromous fish, populations above Bonneville Dam are prevented from migrating to the ocean by the dams and are considered resident species for management purposes.

Federal Treaties with Indian Tribes

When Lewis and Clark entered the Columbia Basin in 1805, they encountered a rich native culture composed of many groups and bands only loosely organized into larger tribal groups. Over the next few decades, European civilization rapidly expanded in the region, and friction between Indian and non-Indian residents increased. In the 1850s, Washington Governor Isaac Stevens and Oregon Governor Joel Palmer negotiated treaties with many Indian groups that ceded most of the land that now comprises the states of Idaho, Montana, Oregon, and Washington to the United States. Indians were concentrated in a number of reservations that brought together many disparate tribal groups. In exchange, the tribes reserved their right to hunt and fish in their "usual and accustom" places "in common" with non-Indian residents.

In the ensuing years, the tribes fought, and generally won, numerous court battles to affirm their hunting and fishing rights. In most cases, the conflict concerned the right of the states to enforce harvest regulations on the tribes. The conflict culminated with the decision by federal district judge Robert Belloni in 1970 that the treaties clearly reserved for the tribes on the Columbia River unique rights that superceded the authority of the states, and that the regulatory authority of the states over tribal fishing was limited to conservation. He concluded that the "in common" language in the treaties entitled the tribes to a "fair share" of the harvest, and he ruled that the tribes should participate as co-managers of fish and wildlife and have a meaningful role in establishing regulations (Blumm 2002).

In the mid-1970s, federal judge George Boldt applied many of Judge Belloni's conclusions to fishing by tribes in Puget Sound. Further, Judge Boldt clarified that the language in the Stevens' treaties reserving the right of the tribes to fish "in common" with non-Indians meant that the tribes were entitled to 50 percent of the allowable harvest of salmon. Most of Judge Boldt's decision was later upheld by the Supreme Court (Blumm 2002).

The Belloni and Boldt decisions radically transformed fishery management in the Pacific Northwest. The 50/50 sharing formula made tribal commercial fishing a viable economic base for many tribes. During the 1980s, the tribes developed their own fishery programs that included harvest management, habitat protection and restoration, and the development of tribal hatcheries. The tribes established themselves as credible natural resource managers both on and off the reservations. For the most part, the tribes are now accepted in the region as co-equal fishery managers and are participants in most management decisions.

In the Columbia River, the tribes are fundamental components of fish and wildlife management. The tribes have been particularly active participants in the Council's program that has provided a forum for discussion of fish and wildlife issues among the federal, state, and tribal management agencies.

Much of the BPA funds for fish and wildlife restoration under the Council's program have gone to support tribal programs including construction of hatcheries and restoration and protection of aquatic and terrestrial habitat (Figure 1.12). The four Stevens' treaty tribes[6] have created their own fishery management and restoration program for the Columbia River, *Wy-Kan-Ush-Mi Wa-Kish-Wit* (the Spirit of the Salmon) (Columbia River Inter-Tribal Fish Commission [CRITFC] 1995).

Summary of Our Findings

Throughout our review, we attempt to identify ecological processes that require restoration. Because of our emphasis on the need to restore functioning ecosystems in order to restore salmon and other species of interest, we take a somewhat negative view of many traditional technological fixes to ecological problems. This does not mean that we are antitechnology. The Columbia River, like most rivers, exists within a human cultural system. It will continue to be managed to provide goods and services to the people of the Pacific Northwest. We do not propose or advocate turning back the clock and returning the river to some state that predates human development of the basin. Nonetheless, we stress the close coupling of species of interest, such as salmon, with their associated ecosystems. We conclude that efforts to apply technology to sustain salmon production in the absence of ecological functions (i.e., engineer substitutes for ecological functions) have generally proven unsuccessful, largely because the efforts have failed to understand and incorporate the ecological processes they wished to replace. Instead, they often created something totally artificial. Restoration of particular species implies restoration of their ecosystems to the greatest extent possible. If we are to have salmon and other species of interest, technology, either as a means to restore fish and wildlife or to benefit humans in some fashion, must operate within the context of restoring and maintaining necessary ecosystem functions. The technological fixes, where used, must mimic natural conditions, not run counter to them. Moving the Columbia River toward a healthy state of ecological functions with respect to salmon and other native species will constrain or eliminate some activities and is likely to have significant economic impact. However, we conclude that it is the only route that, in the long run, will meet the Council's goals for fish and wildlife, satisfy the requirements of the Northwest Power Act, and successfully deal with populations listed under the Endangered Species Act.

---

[6]Stevens' treaties were negotiated with groups that formed the Yakama, Warm Springs, Umatilla, and Nez Perce tribes. These four tribes formed a common fishery management resource group, the CRITFC. There are nine other tribes in the basin with similar reserved fishing and hunting rights.

Figure 1.12 Critical spring Chinook spawning habitat in the Middle Fork of the John Day in central Oregon. Much of the prime Chinook spawning habitat in the John Day system has been protected either through conservation easements or through acquisition for conservation through the fish and wildlife programs of the Northwest Power and Conservation Council and BPA. Photo by R. N. Williams.

In our review, we describe the characteristics of the Columbia River ecosystem with respect to salmon, recognizing that there are other fish and wildlife species of interest. The paradigm that has governed fisheries management for most of this century holds that destruction of ecological functions by human actions can be compensated by using technology to devise substitutes (Bottom 1997). We believe this paradigm to be false. Despite decades of effort and the expenditure of billions of dollars, the present condition of fish and wildlife in the Columbia River Basin and elsewhere demonstrates the failure of the technological paradigm. Technology provides no lasting substitutes for the benefits of ecosystem functions. Technology can only be effective in the context of functioning ecosystems where it can augment, but not replace, natural processes. In most cases, ecological restoration requires relaxation of human-imposed constraints to allow reexpression of natural physical and biological processes (Figure 1.13).

We believe that the conceptual foundation presented in Chapter 3 is consistent with the objectives of the Northwest Power Act and the broad policies expressed in the Council's Fish and Wildlife Program (e.g., NPPC 1994; 2000). It is equally applicable to recovery programs aimed at endangered species. Nevertheless, it is a departure from the overall approach to restoration that has characterized the region's efforts to date and the assumptions underlying the Council's program (see Chapters 2, 3, and 11) and in fish and wildlife management in general.

In our opinion, failure to adopt an ecologically based conceptual foundation and to change the approach to salmon restoration in the basin will lead to more listings of salmon and other species under the Endangered Species Act, continued expenditures, and little progress toward the Council's rebuilding goals. Temporary increases in some populations may occur in response to fluctuations in ocean and climatic conditions, but the overall downward trend in returns that has occurred throughout the 20th century will likely continue. To us, the continued failure of restoration to date calls for the region to question the basic premises of its fish and wildlife recovery effort and to consider alternatives. It is our task in the following chapters to describe such an alternative, an ecologically based conceptual foundation for the Columbia River.

Relationship to Other Plans and Reviews

Ours is the first scientific review that has focused on the Council's Fish and Wildlife Program as an expression of fishery management in the Columbia Basin. Other insightful historical and scientific syntheses of salmonid fisheries problems in the Columbia River and adjacent regions predate our effort, including Netboy (1980), Ebel et al. (1989), Rhodes et al. (1994), and Lichatowich et al. (1995). Also, at least seven reviews from the 1990s provide detailed action plans or recommendations to reduce mortality and increase salmonid production, in addition to reviewing the status of the fisheries and the causes and consequences of declines (Chapman et al. 1991, 1995; Henjum et al. 1994; Botkin et al. 1995; CRITFC 1995; National

Figure 1.13   Habitat restoration in the Asotin subbasin in steelhead spawning and rearing habitat supported by the fish and wildlife programs of the NPCC and BPA. Restoration involved rebuilding the riparian corridor and stream channel through active restoration, willow plantings, and removal of cattle grazing pressure. Photo by R. N. Williams.

Marine Fisheries Service (NMFS) 1995; National Research Council 1996; Quigley et al. 1996). The review by the National Research Council, in particular, addressed issues that are also addressed in our analysis. A main theme in these reviews, and ours, is that the downward trend in numbers (i.e., adult returns in anadromous species and population size in resident

species) and stock diversity is due, in large part, to human actions and institutional conflicts occurring against a backdrop of natural environmental
change. Agents of natural environmental changes are cyclic oceanic
changes such as El Niño, floods, drought, predation, competition, and disease. Examples of human-mediated environmental change are related to
habitat degradation and loss, including those lost to the effects of dams,
irrigation withdrawals, hatchery effects, harvest, and introductions of non-
native biota, including predators, as well as hydropower. Effects of human-
mediated changes may be exacerbated by conflicting sets of values and
goals and ineffective transfer of information among research scientists,
managers, and policy makers.

Where Do We Go From Here?

The conceptual foundation presented here represents a new approach to
salmon management and restoration in the Columbia River Basin. It is one
with which the region has little experience. The approach is based on the relationship between natural ecological functions and processes, including habitat diversity, complexity, and connectivity, and salmonid diversity and productivity.

Recovery actions will occur against the backdrop of regional environmental change and fluctuations that will dominate short- and long-term trends in
fish abundance. Thus, it is not possible to predict the exact relationship
between improved conditions and salmon production. A variety of responses
are possible. The underlying relationship might be linear with salmon production increasing continuously in proportion to the improvement in ecological conditions (Figure 1.14). Alternatively, the response might be nonlinear
increasing at first in small proportional increments with little or no discern-
able increase in production until significant changes accumulate which precipitate rapid increases in production (Figure 1.14). A third possibility would
be characterized by a series of thresholds and plateaus. In this case, as riverine conditions improve, little increase in salmon production might be
observed until a threshold is reached precipitating a subsequent increase in
production to a new level or plateau. The shape of the response of the ecosystem (and salmon) to restoration actions has important implications for scal-
ing the region's expectations and the amount of effort required to elicit
identifiable change.

The region does have experience with taking very small steps toward
improving ecological conditions and tinkering around the edges of the existing system of natural resource use in the basin (e.g., quote from Judge Marsh
at start of this chapter). Unfortunately, those small steps have produced little
discernible progress toward the objectives of the Northwest Power Act, the

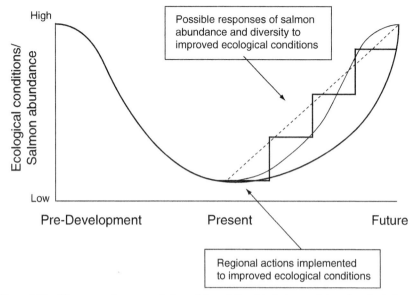

Figure 1.14   Diagrammatic representation of the relationship between status of ecological conditions and salmon abundance at historic, present, and potential future points in time. Future responses are based on the assumption that the region initiates a salmon management plan that improves ecological conditions in the future. The different-shaped future responses of the ecosystem and salmon resources to improved ecological conditions indicate our uncertainty about the nature and timing of their response. The diagram shows only four examples of many possible response curves.

Council's goals, or the condition of populations listed under the Endangered Species Act. Because of this, it is reasonable to question the underlying rationale that has guided these efforts. More substantial changes, based on a scientifically derived rationale, must be taken. At the same time, our knowledge of how to restore key attributes of an ecological system of the scope and complexity of the Columbia River is imperfect, and a rigorous program of evaluation, monitoring, and research will be required. In the following chapters, we present a scientifically rigorous framework for making these major changes.

A fish and wildlife program based on this conceptual foundation is unlikely to be socially painless or inexpensive nor is it likely to provide short-term gratification. Scientific uncertainties abound, and unforeseen events will occur. However, we believe that an approach based on the principles described in the following pages, combined with an implementation program governed by the principles of adaptive management, offers the best hope for preventing large-scale extinction of salmon in the basin and making meaningful progress toward the Council's goals.

Literature Cited

Anderson, D. A., G. Christofferson, R. Beamesderfer, B. Woodard, M. Rowe, and J. Hansen. 1996. StreamNet: Report on the Status of Salmon and Steelhead in the Columbia River Basin - 1995. Bonneville Power Administration. DOE/BP-65130-1. Portland, Oregon. 76 p.

Behnke, R. J. 1992. Native trout of western North America. American Fisheries Society, Bethesda, Maryland.

Blumm, M. C. 2002. *Sacrificing the salmon. A legal and policy history of the decline of Columbia Basin salmon.* Book World Publications, Portland, Oregon.

Bonneville Power Administration. 1996. 1996 Annual Report. Bonneville Power Administration. Portland, Oregon.

Bonneville Power Administration. 2002. 2002 Annual Report of the Bonneville Power Administration. Bonneville Power Administration, Portland, Oregon.

Botkin, D., K. Cummins, T. Dunne, H. Regier, M. Sobel, L. Talbot, and L. Simpson. 1995. Status and future of salmon of western Oregon and northern California: Findings and options. Center for the Study of the Environment. 8. Santa Barbara, California. 300 p.

Bottom, D. L. 1997. To till the water: A history of ideas in fisheries conservation. Pages 569–597 in R. J. Naiman and D. Stouder, eds. *Pacific Salmon and Their Ecosystems: Status and Future Options.* Chapman Hall, New York.

Busby, P. J., T. C. Wainwright, G. J. Bryant, L. J. Lierheimer, and R. S. Waples. 1996. Status review of west coast steelhead from Washington, Idaho, Oregon, and California. National Marine Fisheries Service Northwest Fisheries Science Center. NOAA Technical Memorandum, NMFS-NWFSC-10. Seattle, Washington.

Chapman, D., A. Giorgi, M. Hill, A. Maule, S. McCutcheon, D. Park, W. Platts, K. Pratt, J. Seeb, L. Seeb, and F. Utter. 1991. Status of Snake River Chinook salmon. Don Chapman Consultants, Boise, Idaho. 520 p.

Chapman, D., C. Peven, A. Giorgi, T. Hillman, and F. Utter. 1995. Status of spring Chinook salmon in the mid-Columbia region. Don Chapman Consultants, Boise, Idaho. 477 p.

Columbia River Inter-Tribal Fish Commission (CRITFC). 1995. *Wy-Kan-Ush-Mi Wa-Kish-Wit,* Spirit of the Salmon, The Columbia River Anadromous Fish Restoration Plan of the Nez Perce, Umatilla, Warm Springs, and Yakama Tribes. Volume I. Columbia River Inter-Tribal Fish Commission. Review Draft, Portland, Oregon.

Costanza, R. 2000. Visions of alternative (unpredictable) futures and their use in policy analysis. Page 5 in *Conservation Ecology* (online).

Dauble, D. D., T. P. Hanrahan, and D. R. Geist. 2003. Impacts of the Columbia River hydroelectric system on main-stem habitat of fall Chinook salmon. *North American Journal of Fisheries Management* 23:641–659.

DeVoto, B. 1953. *Journals of Lewis and Clark.* Houghton Mifflin, Boston, Massachusetts.

Ebel, W. J., G. K. Tanonaka, G. E. Monan, H. L. Raymond, and D. L. Park. 1979. Status report-1978: The Snake River salmon and steelhead crisis: Its relation to dams and the national energy shortage. NWAFC Processed Report 79-09, National Marine Fisheries Service, Seattle, Washington.

Ebel, W. J., C. D. Becker, J. W. Mullan, and H. L. Raymond. 1989. The Columbia River: Toward a holistic understanding. Pages 205–219 in D. P. Dodge, ed. *Proceedings of the International Large River Symposium (LARS).* Special Publication of the *Canadian Journal of Fisheries and Aquatic Sciences.*

General Accounting Office. 2002. Columbia River Basin salmon and steelhead. Federal agencies' recovery responsibilities, expenditures and actions. General Accounting Office, Washington, DC.

Hassemer, P. F., S. W. Kiefer, and C. E. Petrosky. 1996. Idaho's salmon: Can we count every one? Pages 113–125 in R. J. Naiman and D. Stouder, eds. *Pacific Salmon and Their Ecosystems: Status and Future Options*. Chapman Hall, New York.

Hayden, M. V. 1930. History of the Salmon Industry of Oregon. Master's thesis, University of Oregon, Eugene.

Henjum, M. G., J. R. Karr, D. L. Bottom, J. C. Bednarz, S. G. Wright, S. A. Beckwitt, and E. Beckwitt. 1994. Interim protection for late-successional forests, fisheries, and watersheds: National forests east of the Cascade Crest, Oregon and Washington. The Wildlife Society. Bethesda, Maryland. 245 p.

Huntington, C., W. Nehlsen, and J. Bowers. 1996. A survey of healthy native stocks of anadromous salmonids in the Pacific Northwest and California. *Fisheries* 21:6–14.

Independent Scientific Group (ISG). 1993. Critical uncertainties in the Columbia River Basin Fish and Wildlife Program. Bonneville Power Administration. Report to Policy Review Group, SRG 93–3. Portland, Oregon. 17 p.

Independent Scientific Group. 1998. Return to the river: An ecological vision for the recovery of the Columbia River salmon. *Environmental Law* 28:503–518.

Independent Scientific Group. 1999. Scientific issues in the restoration of salmonid fishes in the Columbia River. *Fisheries* 24:10–19.

Lee, K. N., and J. Lawrence. 1986. Adaptive management: Learning from the Columbia River Basin fish and wildlife program. *Environmental Law* 16:431–460.

Lee, K. N. 1993. *Compass and gyroscope: Integrating science and politics for the environment*. Island Press, Washington, DC.

Leopold, L. 1994. *A view of the river*. Harvard University Press, Cambridge, Massachusetts.

Lichatowich, J., L. Mobrand, L. Lestelle, and T. Vogel. 1995. An approach to the diagnosis and treatment of depleted Pacific salmon populations in Pacific Northwest watersheds. *Fisheries* 20:10–18.

Lichatowich, J. 1999. *Salmon without rivers*. Island Press, Washington, DC.

Logie, P. 1993. Power system coordination. A guide to the Pacific Northwest Coordination Agreement. Columbia River system operation review. Joint Project of U.S. Dept. of Energy, Bonneville Power Administration; U.S. Dept. of the Army, Corps of Engineers, North Pacific Division; and U.S. Dept. of Interior, Bureau of Reclamation, Pacific Northwest Region. 46 p.

Matthews, G. M., and R. S. Waples. 1991. Status review for Snake River spring and summer Chinook salmon. National Marine Fisheries Service, Northwest Fisheries Science Center. NOAA Technical Memorandum, NMFS F/NWC-200. Seattle, Washington. 75 p.

Mullan, J. W., A. Rockhold, and C. R. Chrisman. 1992. Life histories and precocity of Chinook salmon in the Mid-Columbia River. *Progressive Fish-Culturist* 54:25–28.

Myers, J. M., R. G. Kope, G. J. Bryant, D. Teel, L. J. Lierheimer, T. C. Wainwright, W. S. Grant, F. W. Waknitz, K. Neely, S. T. Lindley, and R. S. Waples. 1998. Status review of Chinook salmon from Washington, Idaho, Oregon and California. Technical Memo NMFS-NWFSC-35, U.S. Dept. of Commerce, Seattle, Washington.

National Marine Fisheries Service (NMFS). 1995. Proposed recovery plan for Snake River salmon. U.S. Dept. of Commerce, National Oceanic and Atmospheric Administration, Washington, DC. 387 p.

National Research Council. 1996. *Upstream: Salmon and society in the Pacific Northwest*. National Academy Press, Washington, DC.

Nehlsen, W., J. E. Williams, and J. A. Lichatowich. 1991. Pacific salmon at the crossroads: Stocks at risk from California, Oregon, Idaho, and Washington. *Fisheries* 16:4–21.

Netboy, A. 1980. *The Columbia River salmon and steelhead trout: Their fight for survival*. University of Washington Press, Seattle, Washington.

Northwest Power and Conservation Council (NPCC). 2003. Artificial Production Review and Evaluation. September 2003 Northwest Power Planning Council, Portland, Oregon.

Northwest Power Planning Council (NPPC). 1986. Council Staff Compilation of Information on Salmon and Steelhead Losses in the Columbia River Basin. Northwest Power Planning Council. Portland, Oregon.

NPPC. 1987. 1987 Columbia River Basin Fish and Wildlife Program. Northwest Power Planning Council. Portland, Oregon. 246 p.

NPPC. 1992a. Strategy for Salmon, Vol. 1. Northwest Power Planning Council. 92-21. Portland, Oregon. 43 p.

NPPC. 1992b. Strategy for Salmon, Vol. 2. Northwest Power Planning Council. 92-21. Portland, Oregon. 98 p.

NPPC. 1994. Columbia River Basin Fish and Wildlife Program. Northwest Power Planning Council. Portland, Oregon.

NPPC. 2000. Columbia River Basin Fish and Wildlife Program. Council Document 2000-19, Northwest Power Planning Council, Portland, Oregon.

Pacific Fisheries Management Council. 1979. Freshwater habitat, salmon produced, and escapement for natural spawning along the Pacific Coast of the United States. Report prepared by the Anadromous Salmonid Task Force of the Pacific Fisheries Management Council. 68 p.

Pearcy, W. G. 1992. *Ocean ecology of North Pacific salmonids*. University of Washington Press, Seattle, Washington.

Quigley, T. M., R. W. Haynes, and R. T. Graham. 1996. Integrated scientific assessment for ecosystem management in the Interior Columbia Basin and portions of the Klamath and Great Basins. U.S. Department of Agriculture, Forest Service, Pacific Northwest Research Station. General Technical Report, PNW-GTR-382. Portland, Oregon.

Raymond, H. L. 1988. Effects of hydroelectric development and fisheries enhancement on spring and summer Chinook salmon and steelhead in the Columbia River basin. *North American Journal of Fisheries Management* 8:1–24.

Rhodes, J. J., D. A. McCullough, and J. F. A. Espinosa. 1994. A coarse screening process for evaluation of the effects of land management activities on salmon spawning and rearing habitat in ESA consultations. Columbia River Inter-Tribal Fish Commission. Technical Report, 94-1. Portland, Oregon.

Scarce, R. 2000. Fishy business: Salmon, biology, and the social construction of nature. Temple University Press, Philadelphia. 236 p.

Stone, L. 1879. Report of operations at the salmon-hatching station on the Clackamas River, Oregon, in 1877. Pages Part 11 in Part 5. *Report of the Commissioner for 1877*. U.S. Commission of Fish and Fisheries, Washington, DC.

Taylor J. E., III. 1999. *Making salmon: An environmental history of the Northwest fishery crisis*. Weyerhaeuser Environmental Books, W. Cronon, ed. University of Washington Press, Seattle, Washington. 421 p.

Van Hyning, J. M. 1973. Factors affecting the abundance of fall Chinook salmon in the Columbia River. *Research Reports of the Fish Commission of Oregon* 4:1–87.

Waples, R. S. 1995. Evolutionarily significant units and the conservation of biological diversity under the Endangered Species Act. *Evolution and the aquatic ecosystem: Defining unique units in population conservation* 17:8–27.

Waples, R. S., O. W. Johnson, and R. P. Jones Jr. 1991a. Status review for Snake River sockeye salmon. National Marine Fisheries Service Northwest Fisheries Science Center. NOAA Technical Memorandum, NMFS-F/NWC-195. Seattle, Washington.

Waples, R. S., J. R. P. Jones, B. R. Beckman, and G. A. Swan. 1991b. Status review for Snake River fall Chinook salmon. National Marine Fisheries Service Northwest Fisheries Science Center. NOAA Technical Memorandum, NMFS F/NWC-201. Seattle, Washington. 71 p.

Washington Department of Fish and Wildlife, and Oregon Department of Fish and Wildlife. 2001. Status report. Columbia River fish runs and fisheries, 1938–2000. Washington Department of Fish and Wildlife and Oregon Department of Fish and Wildlife. Olympia, Washington, and Clackamas, Oregon.

2

The Existing Conceptual Foundation and the Columbia River Basin Fish and Wildlife Program

James A. Lichatowich, M.S., Willis E. McConnaha, Ph.D.,
William J. Liss, Ph.D., Jack A. Stanford, Ph.D.,
Richard N. Williams, Ph.D.

What Is a Conceptual Foundation?
A Brief History of Conceptual Foundations for Salmon Management
Evaluation of the Conceptual Foundation of Salmon Recovery
  Background on the Council's Fish and Wildlife Program
  Conceptual Framework of the Council's Fish and Wildlife Program
Literature Cited

> *"It is easy to see why the conventional wisdom resists so stoutly such change. It is a far, far better thing to have a firm anchor in nonsense than to put out on the troubled seas of thought."*
> *—John Kenneth Galbraith, The Affluent Society, 1959.*
> *"It has become equally clear that the simple solutions so often advanced to solve the ills of the fishery were, in light of our growing knowledge of the resource, only simple minded."*
> *—James A. Crutchfield and Giulio Pontecorvo, The Pacific Salmon Fisheries: A Study of Irrational Conservation, 1969.*

Restoration and management of salmon populations, like most human endeavors, reflect our training, beliefs, and experience. We term this body of prior knowledge a conceptual foundation. The conceptual foundation is the "lens" through which we interpret observations and thereby form conclusions. We contend that management of Pacific salmon has been characterized by conflicting and poorly articulated conceptual foundations. Different groups of scientists and managers interpret the same observations through differing lenses reflecting different academic disciplines, social objectives, and belief systems. As a result, restoration programs, such as the Northwest Power Planning Council's (NPPC's)[1] fish and wildlife program, are often a

---

[1]Recently, the Council was renamed the Northwest Planning and Conservation Council.

hodge-podge of ideas arrived at through compromise without a unifying scientific basis. While unification of social objectives and values is likely not possible and perhaps not even desirable, recognition of the power of conceptual foundations and their clear articulation is fundamental to a critique of management programs. Critical examination of prevailing conceptual foundations in light of current scientific knowledge leads to development of new conceptual foundations and improvement in management actions.

In this chapter we describe and critique the conceptual foundation that was the basis for salmon management throughout most of the 20th century and continues to guide many aspects of management in the 21st century. Remaining chapters will describe a new conceptual foundation based on current ecological thinking. Because it is fundamental to our discussion, the term "conceptual foundation" needs further definition.

What Is a Conceptual Foundation?

"The problems that foresters faced in the Blue Mountains flowed as much from their own scientific paradigms as from the ecological phenomena going on in the forest itself—phenomena that those paradigms rendered all too invisible. The moral of this story should be clear. Even well-intentioned management can have disastrous consequences if it is predicated on the wrong assumptions, and yet testing those assumptions is much harder than people realize."
    –W. Cronon. 1995. Forward: with the best intentions. Page ix in N. Langston. Forest Dreams, Forest Nightmares: The Paradox of Old Growth in the Inland West. University of Washington Press, Seattle, Washington

A conceptual foundation is the set of principles and assumptions that gives direction to management and research activities, including fishery restoration programs. It determines what problems (e.g., limitations on fish production) are identified, what information is collected and how it is interpreted, and as a result, establishes the range of appropriate solutions (Lichatowich et al. 1996). Because it influences the interpretation of information, the conceptual foundation can be a powerful scientific element of management and restoration plans, and it can determine the success or failure of those plans. Natural resource management carried out with the best intentions and methodological expertise can have disastrous consequences if based on incorrect assumptions (Cronon 1995). The importance of a conceptual foundation and the problems created by the failure to explicitly define it extends beyond natural resource management. For example, Heilbroner and Milberg (1995) attributed chaos in economic analysis to the lack of a central vision, or in our terminology, a conceptual foundation.

To illustrate the importance of a conceptual foundation, think of it as analogous to the picture that comes with a jigsaw puzzle. Each piece of the puzzle is a small data set containing usable information, but interpreting the relevance of that information is difficult or even impossible without referring to the picture. Salmon managers generate and review many data sets and large volumes of information. They look at many pieces to the puzzle of salmon management and ecosystem restoration. However in fisheries management, watersheds or ecosystems do not come with a picture clearly illustrating the functional ecological processes that lead to production of desirable fishes. Consequently, to interpret the relevance of those data sets, the picture (conceptual foundation) must be developed by scientists and managers from the best available scientific principles and assumptions. If the conceptual foundation underlying a restoration program is erroneous, it is equivalent to an attempt to complete a jigsaw puzzle using the wrong picture as a guide. Nevertheless, conceptual foundations should not be static, but should be revised continually as new theory emerges and new empirical information becomes available.

The power of the conceptual foundation to determine how information is interpreted, even to draw the wrong conclusion from otherwise sound data, is illustrated by the following example. At the beginning of the 20th century, biologists working with Pacific salmon were debating the "home stream theory." Some held that adult salmon had the ability to home back to the stream of their birth to spawn. Other biologists, including the eminent ichthyologist David Starr Jordan, rejected the home stream theory (Jordan 1904). In Jordan's conceptual foundation, the salmon's ecosystem did not extend much beyond the mouth of the natal river. He assumed that juvenile salmon migrated no more than 20 to 40 miles from the mouth of their natal stream. When the salmon reached maturity, they simply swam into the first river they came to, which, because they never migrated far from it in the first place, was almost always their home stream.

In 1896, juvenile salmon from the Clackamas hatchery (located on the Willamette River, the major tributary to the Columbia River flowing through Portland, Oregon) were fin-clipped for later identification and released into the river. Four years later, some of the tagged fish returned to the Columbia River. Instead of interpreting the recovery of tagged salmon in the Columbia River as evidence of homing, Jordan interpreted it as support for his assumption that the salmon did not migrate far from the mouth of their natal stream. Jordan's conceptual foundation contained at least one erroneous assumption, which caused him to misinterpret otherwise sound information.

The debate over the home stream theory was not an academic exercise. Whether or not salmon homed to their natal stream had important implications for salmon management, particularly the transfer of stocks between rivers through the hatchery program. By today's standards, Jordan derived his conceptual foundation from limited environmental data and from a rudimentary body of ecological theory. Nonetheless, his conceptual foundation was insufficient to allow new information to be correctly interpreted.

A robust conceptual foundation is derived from thorough analysis of the problem (i.e., breaking the problem into its components and their corollaries) and synthesis of available information (formalizing what is known).

A Brief History of Conceptual Foundations for Salmon Management

For millennia, humans have "managed" natural resources to supply the basic needs of society. Considerable spiritual, ethical, and scientific effort has been expended to define a sustainable relationship between humans and natural resources and to control the distribution and utilization of those resources.

For the native societies in the Columbia Basin that greeted the Lewis and Clark expedition in 1805, human well-being was linked to the abundance of salmon and steelhead. Complex social and ethical arrangements prescribed the interaction between humans and salmon (Gunther 1926; Waterman and Krober 1965; Martin 1978). Prime fishing areas such as Celilo Falls on the Columbia River (Figure 2.1) became cultural centers for native societies and the nexus of extensive trade and economic networks (Thomison 1987). For Native American societies, natural resource management had a spiritual basis that recognized the inexorable linkage between salmon, humans, and the world (Martin 1978; Highwater 1981; Lichatowich 1999). The natural world, including humans, was a single community of beings all with equal standing. Natural resources, such as beaver, cedar, salmon, clams, and salmonberry, were gifts that had to be treated with respect; otherwise, the giver would withhold them in the future (Mauss 1990). The well-being of human society was equated to the well-being of other elements of the natural system.

In the years following the arrival of Lewis and Clark, this native, salmon-based conceptual foundation was largely replaced by a new view of the relationship between humans and natural resources based on European culture. This new culture viewed natural resources from a much different perspective. Euro-Americans separated themselves from the natural world and looked on nature, not as a gift, but as something to be conquered and controlled. Nature was both an impediment to human purposes and an inexhaustible warehouse of raw material (Worster 1977; Callicott 1991). The natural world was to be "tamed" and brought under human control to maximize human survival and enjoyment. European development of the American West, in particular, was based on the goal of replacing what was viewed as a wasteful desert with a productive garden that would contribute to human economic betterment (Bottom 1997).

Early in the 20th century, Gifford Pinchot (1910) attempted to combine this worldview with the prevailing scientific approach to form a concept that became the basis for natural resource management; it continues to influence

Figure 2.1   Celilo Falls, located near The Dalles, Oregon, was an important historical fishing site for Native Americans. It was inundated after the completion of The Dalles Dam. (Photo from US Army Corp Digital Visual Library, website at http://images.usace.army.mil/photolib.html.)

our views of the natural world today (Worster 1977). Pinchot's Resource Conservation Ethic is encapsulated in his famous axiom that conservation of natural resources means achieving "the greatest good to the greatest number for the longest time" (Pinchot 1910, 48).

This view of natural resources was based on an agricultural perspective (Bottom 1997). Pinchot attempted to apply the knowledge of progressive agriculture to the management of natural resources, particularly in the context of public lands (Worster 1977). Science and technology under the guidance of bureaucracies staffed with experts would provide society with the tools to improve on nature, make its processes more efficient, and provide its abundance to society on a predictable basis (Hayes 1959; Worster 1977).

This view has been especially influential in the management of fishery resources. Technology in the form of hatcheries was equated to agriculture beginning with the first large-scale use of artificial propagation in the 1840s (Fry 1854; discussed more fully in Chapter 8 of this volume). Fish culture promised to replace the inefficiencies of nature with a fully controllable system that would maintain and enhance the supply of fish (Bottom 1997).

Organisms beneficial to humans, such as salmon, could be reared in huge numbers and released to feed in the virtually limitless pasture of the ocean. The high fecundity of salmon could be used to increase returns rather than being wasted in dealing with the rigors of the natural world (Bottom 1997). In a like manner, the environment could be controlled for the betterment of species of interest by removal of predators or undesirable species, introduction of other species, and construction of artificial habitats.

This perspective and the assumptions about nature from which it was derived led to a simple management model for fish and the lakes and rivers they inhabited. Hatcheries supplied fish to the fishermen, and their harvest was regulated so enough mature adults escaped to start the cycle over. This simple model rendered rivers and their ecological processes irrelevant. Oceans were viewed as inexhaustible pastures for hatchery-bred salmon. Fish were isolated from the river and protected from its ecological vagaries in the hatchery. Rivers became simple conduits to the sea for anadromous salmonids or stage props for fishermen angling for trout released from the hatchery at a catchable size. A natural extension of this management model was belief that hatcheries could "mitigate" for fish production lost due to habitat degradation (Bottom 1997; Lichatowich 1999; Williams et al. 2003).

By the middle of the 20th century, however, flaws in this conceptual foundation began to appear. During the late 1950s, hatcheries began using improved diets and better disease treatments. These improvements enabled the construction of numerous hatcheries throughout the region, most of which were intended to mitigate for dam construction and other activities that degraded the freshwater environment. Hatchery production soared and, coupled with favorable ocean conditions, resulted in greatly increased production of adult salmon. In the following decades, increasing numbers of fish were released from programs in the Columbia River and elsewhere with the expectation of nearly boundless harvest. However, in the late 1970s, as hatchery production released more than 250 million juvenile salmon, returns of most salmon to the Columbia River and elsewhere plunged (Figure 2.2). The dramatic decline of salmon abundance throughout the Pacific Northwest coincided with a dramatic change in ocean conditions (Beamish et al. 1999). While hatchery programs continue to play a major role in fisheries management in the Columbia River, no longer are they regarded as a panacea that permitted massive ecological change (the result of dam construction and irrigated agriculture), while maintaining fish abundance and harvest. Increasingly, the region is attempting to reconcile the role of hatcheries with an increased awareness of ecological complexity (Brannon et al. 1999; Mobrand Biometrics 2003).

This questioning of hatcheries and other icons of 20th-century fisheries management reflects significant change in the perspective guiding natural resource management, at least at the scientific level. The scientific community is moving away from the mechanistic, agricultural approach to the management

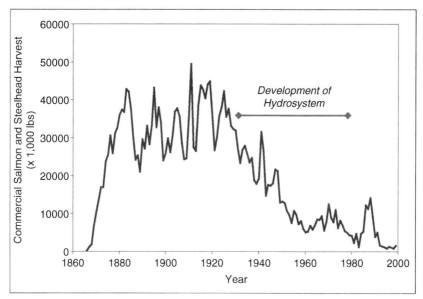

Figure 2.2 Commercial landings of salmon and steelhead in thousands of pounds in the Columbia River, 1866 to 2000 (modified from ISG 2000; Washington Department of Fish and Wildlife and Oregon Department of Fish and Wildlife 2001).

of natural resources to a perspective that recognizes the unique qualities of natural systems and the role of organisms and interrelationships within their ecosystems (e.g., Christensen et al. 1996). These views, collected under the term "Ecosystem Management," are being incorporated into state and federal natural resource management plans. If implemented, this shift in perspective could dramatically change how we manage natural resources.

However, for some natural resources, such as Pacific salmon, there are major impediments to implementing an ecosystem perspective. The anadromous life history of Pacific salmon encompasses an ecosystem so large and complex (rivers, estuary, ocean) that society has not been able to conceive of and bring into existence an institutional structure capable of effective management over the geographical area inhabited by the salmon (National Research Council [NRC] 1996). Management responsibility for the salmon's ecosystem is fragmented among several institutions and agencies. For many of those institutions, salmon management is not their primary mission. Institutions that control hydroelectric power production, timber harvest, grazing, irrigation, industrial, and urban development can have major impacts on salmon habitat, and yet the salmon's well-being is not their primary focus. An important consequence of this fragmented institutional structure is the barrier it raises to an effective ecosystem perspective in salmon management.

Evaluation of the Conceptual Foundation of Salmon Recovery

Conceptual foundations are often buried so deep in the culture of management agencies that they are rarely articulated, examined, or evaluated (Botkin
1990). Those unexamined principles, assumptions, and beliefs that make up
the conceptual foundations are now driving massive salmon restoration programs throughout the Pacific Northwest. The NPPC's (Council) Fish and
Wildlife Program is the largest recovery program in the Pacific Northwest,
and it is possibly the largest fishery restoration program in the world. An
evaluation of the conceptual foundation of the Council's salmon recovery
program provides a real example to illustrate the previous discussion in this
chapter and to compare with the alternative conceptual foundation we present in the next chapter. Before discussing the Council's conceptual foundation, a brief background on the program is in order.

Background on the Council's Fish and Wildlife Program

In setting up the NPPC in the early 1980s, Congress directed the Council to
base its fish and wildlife program on recommendations solicited from
throughout the region, paying particular attention to those provided by the
region's fishery managers and Indian Tribes. In creating the Council,
Congress sought to avoid creating a "super fish agency" but instead deferred
to the management authorities of the federal, Tribal, and state agencies
(Blumm 2002). Hence, by definition, the Council's program was to be a collection of views and opinions rather than a coherent program united by a
common conceptual foundation. As such, however, it was a reflection of the
prevailing paradigm that guided fishery management in the Pacific
Northwest. The Council's program was not to be developed by the Council,
but rather was to be assembled from recommendations submitted by any
party, with special deference given to recommendations from the federal,
Tribal, and state fishery managers. This method of program development
had several consequences for the final fish and wildlife program. First, it
ensured that the Council's program was inherently conservative in the sense
of toeing closely to the established norms of the region's fishery management. While the Council has provided the region with several innovative
ideas, such as protecting areas from hydropower development, adaptive management (Lee 1993), and ecosystem management (Marcot et al. 2002), in
practice the Council's program closely mirrors the prevailing natural resource
beliefs of the region.

Second, the Council's mandate does not naturally lend itself to development of a strategic and integrated program. The fish and wildlife program
has been a compilation of ideas submitted principally from the fish and

wildlife agencies. Each agency has a different legal mandate and its own (usually implicit) conceptual foundation based on tradition and constituency. Without strong guidance from an explicit, common scientific framework or conceptual foundation, the program is only loosely tied to specific goals. Typically, measures are proposed by various management agencies and interest groups, discussed in public forums, and adopted by the Council. The structure that exists within the program results more from *a posteriori* organization of measures than from an *a priori* concept and direction. While the Council has identified general goals and priorities, their level of generality is such that they provide little guidance or rationale for subsequent selection or prioritization of measures.

In the most recent permutation of its fish and wildlife program (NPPC 2000), the Council has attempted a radically different. The Council's program is now a relatively spare document that provides no specific restoration measures but, instead, outlines a conceptual foundation for fish recovery and provides a set of guiding principles for restoration measures. These broad, basin-scale guidelines are intended to be supplemented by specific measures developed through local processes in each of the 62 subbasins. While we applaud the Council's efforts to develop a conceptual foundation, at this point, it is not clear how successful the Council will be in melding this "top-down" framework with a "bottom-up" development of specific actions. The Council's current program remains a work in progress. Our review largely focuses on previous versions of the program as examples of the region's thinking about fish and wildlife recovery in the Columbia Basin over the last several decades of the 20th century.

Third, the conservative nature of the Council's program is further ensured by its cumbersome process of change. Any modification to the Council's program can occur only after a formal call for amendments and a lengthy review process that typically takes a year or more. While this has the benefit of preventing casual modification of the program, it also makes a very difficult vehicle to navigate through the changing topography of science and policy in the Columbia Basin.

Finally, and perhaps most telling, is that Congress severely limited the Council's authority to implement its program. The Council can affect state and tribal fishery management actions only by virtue of having some influence over the funding of projects by the Bonneville Power Administration. Federal agencies, although required to defer to the Council "to the extent practicable," in reality are driven by far more powerful legislation including the Endangered Species Act. The Council's main influence is through the public process and the political will of the body to create a regional vision. Over time, this will has diminished in favor of efforts by each state to secure its own interests.

While some of the institutional and procedural difficulties described above reflect the unique legislative basis for the Council and its fish and wildlife program, we suspect that similar problems can be found in various guises in other large-scale natural resource recovery programs. We view the Council's program as a metaphor for natural resource management in general, and its conceptual foundation reflects the principles, assumptions, and beliefs that underlie salmon recovery efforts throughout the region. Natural resource management programs are developed within a political structure that seeks to balance needed environmental protection with social, economic, and political realities. Public policy is developed through compromise, resulting in accommodation of competing views and diffusion of direction over time. To do this effectively requires a combination of political moxie and scientific innovation. Our expertise lies in the scientific realm, and we leave it to others to craft political structures that can remain fresh and innovative over time. However, we believe that any political structure will best succeed if it begins from an integrated approach based on an overall, scientifically credible conceptual foundation such as we propose in the next chapter. This provides a rational basis for actions and a standard for evaluation of measures based on general properties of the desired ecological outcome. It also provides an objective, explicit structure around which to shape the management program. In the case of the Columbia River, it would be naive to think that this would eliminate the traditional controversies that have divided the region's efforts for decades. However, this approach would place the Council's program on firmer scientific ground and provide a rational structure for the region's efforts.

Conceptual Framework of the Council's Fish and Wildlife Program

Most of this book outlines a conceptual foundation for salmon recovery based on contemporary scientific thought and the realities of the 21st century. It can be best understood by comparison with the prevailing conceptual foundation embodied in the Council's fish and wildlife program from 1982 to 1999[2]. The Council did not explicitly describe the principles, assumptions, and beliefs (conceptual foundation) that underlie its fish and wildlife program. Consequently, we used indirect methods to describe the conceptual foundation implied in the Council's program based on its collection of measures.

---

[2]As noted above, the Council's 2000 Fish and Wildlife Program proposes a spare, but scientifically credible conceptual foundation—the result of recommendations (and subsequent discussions) from the Council's two independent scientific advisory bodies throughout the last half of the 1990s. Many of the authors of this volume served or continue to serve on the Council's scientific advisory bodies.

Table 2.1 Summary of the general principles in the implied conceptual foundation for the NPPC's Fish and Wildlife Program. See Box 2.1 for explanation of level of proof and ISG (2000) for detailed descriptions of the principles and specific assumptions.

General Principles—Cause of the Salmon's Depletion	Level of Proof
1c. *Abundance of salmon and steelhead in the Columbia River Basin has, to a significant degree, declined due to, and is presently limited by, human actions.*	1

General Principles—Solutions to the Problem of Salmon Depletion	Level of Proof
1s. *The salmon-bearing ecosystem in the Pacific Northwest and Northeast Pacific Ocean has considerable excess carrying capacity.*	4
2s. *Ecosystem functions lost as a result of development of the Columbia River can be replaced by technological solutions to individual problems.*	4

While some of these are unique to the Columbia River, many are common to fisheries restoration in general. We identified three general principles (Table 2.1) and 29 specific assumptions (Table 2.2) implied by the measures included in the program. We used those principles and assumptions as a surrogate for the conceptual foundation. We then evaluated the scientific support for the principles and assumptions.

Thus, our review did not evaluate individual program measures, but instead focused on the biological rationale for measures or groups of related measures. For example, the region has devoted considerable effort to augment flow in the mainstem Columbia River, which reflects the belief that flow rates as modified by operation and development of the hydroelectric system have contributed to the declines in salmonid populations. Once articulated, such a statement is amenable to scientific analysis, whereas the individual measures may not be.

However, consideration of the scientific basis for individual assumptions may lead to focusing on the trees and missing the forest. It is quite possible for each individual measure or strategy to be based on sound scientific principles but for these measures collectively to be an inadequate response to the modification of the ecosystem that has occurred during this century. In fact, it is our conclusion that few really bad or wrong-headed ideas survive the test of public and scientific scrutiny. This is quite different, however, from saying that the sum of the parts will result in restoration of salmon in the Columbia Basin. Fitting all the correct puzzle pieces together does not necessarily yield a coherent picture. This is due to a predilection for politically and scientifically conservative actions and, as we contend is true of salmon restoration in general, a case where there is no unifying picture to shape the puzzle—no coherent conceptual foundation that links recovery actions into an overall recovery plan.

Table 2.2 Summary of the specific assumptions in the implied conceptual foundation for the NPPC's Fish and Wildlife Program. See Box 2.1 for explanation of level of proof and ISG (2000) for detailed descriptions of the principles and specific assumptions.

Specific Assumptions—Causes of the Salmon's Depletion	Level of Proof
1c. *Operation of the hydroelectric system is a major source of human-induced mortality, limiting numbers and diversity of salmonid populations.*	1
2c. *Operation and development of the hydroelectric system has altered the hydrologic profile of the river, which adversely affects survival of juvenile emigrants.*	1
3c. *There is a limited period of time within which yearling juvenile emigrants must reach the estuary to successfully move from the freshwater to the marine phase of the life cycle.*	2—3
4c. *Yearling Chinook emigrants utilize the mainstem Snake and Columbia Rivers primarily as an outmigration corridor linking tributary and marine areas.*	2
5c. *Survival of yearling juvenile emigrants is inversely related to the amount of time they spend in the impounded sections of the mainstem Snake and Columbia Rivers.*	3
6c. *The amount of time spent by yearling juvenile emigrants within the hydroelectric system is inversely related to the prevailing water velocity. Therefore, survival is positively related to the water velocity prevailing during the outmigration.*	3
7c. *Subyearling emigrants utilize the mainstem Snake and Columbia Rivers for both rearing and outmigration.*	1
8c. *Subyearling Chinook emigrants are less dependent on flow and water velocities as a physical aid to migration than yearling Chinook emigrants, but are affected by high summer water temperatures.*	1
9c. *Creation of reservoirs has enhanced native and exotic predator populations and increased the vulnerability of juvenile salmonids to predation.*	1
10c. *Impacts of alteration of the hydrologic cycle in the Columbia River on salmonid survival is not limited to the impounded section of the river, but extends to the conditions in the estuary and survival outside the impounded section.*	3
11c. *In addition to alteration of the hydrologic cycle and creation of reservoirs, the dams themselves form a second major impact of development and operation of the hydroelectric system.*	1
12c. *The primary source of mortality at dams occurs as juvenile fish pass through turbine generating units. This mortality occurs within the turbines and immediately downstream of the units.*	2—3
13c. *Operation and development of the hydroelectric system has been a major source of human-induced mortality to adult migrants, which has limited numbers and diversity of upriver salmonid populations.*	2
14c. *Present harvest rates are a significant factor limiting Chinook populations in the Columbia Basin.*	1
15c. *Adult return to spawning areas can be limited to some degree by illegal harvest in the Columbia and Snake Rivers.*	4

Table 2.2 *Continued*

Specific Assumptions—Causes of the Salmon's Depletion	Level of Proof
16c. *Natural populations are detrimentally affected by straying of returning hatchery fish.*	2
17c. *Overall survival of salmon and steelhead is decreased by exceeding the carrying capacity of the river, estuary, and/or ocean because of excessive releases of juvenile fish from production facilities.*	3
18c. *Absence of fish screens or inadequate screens on agricultural and municipal water intakes leads to increased mortality of juvenile salmon and steelhead.*	1
19c. *Productivity of naturally spawning populations is limited by habitat availability and habitat quality.*	1

Specific Assumptions—Solutions to the Problem of Salmon Depletion	Level of Proof
1s. *Water velocity can be enhanced either by augmenting flows from upstream reservoirs or by reducing the elevation of downstream reservoirs.*	1
2s. *Devices to collect juvenile fish before they pass into the turbines and deposit them downstream of the dam provide a benign means of passing the project.*	3
3s. *Spill provides the route of hydroelectric project passage with the lowest mortality to juvenile emigrants.*	3
4s. *Transportation of juvenile salmonids by barge can mitigate, in some fashion, for the biological impact of operation and development of the hydroelectric system for some species and life history types of juvenile salmonids in the mainstem Snake and Columbia Rivers, particularly in years of low runoff or other unusually bad conditions.*	3
5s. *Management of fisheries should be based on the amount of information available to managers regarding stock composition and abundance. Managers should be most restrictive on harvest when information on stock composition and abundance is the most uncertain so that errors do not occur at the cost of biological needs of the populations.*	2
6s. *Permanent loss of production capacity in the Columbia Basin as a result of operation and development of the hydroelectric system can be at least partially mitigated by improvements in habitat conditions in tributary areas.*	3
7s. *The watershed is the appropriate physical unit around which to organize efforts to improve conditions in the tributaries.*	1
8s. *Artificial production can be used to augment harvest without detrimental effects on naturally spawning populations.*	4
9s. *Artificially reared fish can be used to augment the production of natural fish populations (i.e., in supplementation projects) in a manner that minimizes genetic change or reductions of fitness in the population.*	3
10s. *Biological diversity can be stabilized or increased through habitat conservation.*	2

Box 2.1.	Levels of scientific support for implied assumptions in the Fish and Wildlife Program.

1. Thoroughly established, generally accepted, good peer-reviewed empirical evidence in its favor.
2. Strong weight of evidence in support, but not fully conclusive.
3. Theoretical support with some evidence from experiments or observations.
4. Speculative; little empirical support.
5. Misleading or demonstrably wrong; based on good evidence to the contrary.

For each principle or assumption, we assigned a qualitative rating that summarizes our assessment of the scientific support for the assumption based on the analysis presented in Chapters 4–9 (Box 2.1). The rating system is necessarily subjective and is intended to convey our judgment of the degree of scientific support available for each italicized statement, rather than representing a rigorous quantitative score. The general principles and assumptions, along with the level of scientific support, are shown in Tables 2.1 and 2.2. For a detailed explanation of each of the principles and assumptions see Independent Scientific Group (ISG) (2000).

In both tables, the principles and assumptions are divided into two categories: those that deal primarily with the causes of salmon depletion and those that deal primarily with solutions to the depletion problem. Within the general principles (Table 2.1), the cause of salmon depletion is known with a high degree of certainty. Human alteration and management of the watershed and its resources is the major cause of the declines in salmon abundance. On the other hand, principles in Table 2.1 that deal with solutions to the problem have little scientific support. The two principles dealing with solutions are closely related. The belief in excess capacity ignores the dynamics of the marine ecosystem and assumes that because current salmon production levels are below historical levels, the marine ecosystem has a large vacant capacity waiting to be filled by salmon from hatcheries. The artificial propagation of salmon in hatcheries has been a major technological solution to the problem of salmon depletion for over a century (Lichatowich 1999). At a very broad level, this analysis suggests that the cause of the problem has been identified, but the region has failed to come up with scientifically sound solutions.

The analysis of the specific assumptions (Table 2.2) is consistent with the general principles in that the causes have more scientific support than the proposed solutions. The average level of proof for the assumptions related to causes of the salmon's decline is 1.8, whereas the average score for the solution to the salmon problem is 2.5.

The Columbia Basin is subjected to all of the activities that alter aquatic habitat that are common to watersheds of the Pacific Northwest. However, it is the hydropower system—the large number of dams in the mainstem and tributaries—that is the obvious and pervasive alteration of the salmonid ecosystem in the Columbia Basin (Figures 2.3 and 2.4). An interesting pattern emerges from the specific assumptions about the causes of the salmon's decline that are directly or indirectly related to the hydropower system. From a general perspective, there is little scientific doubt that the hydroelectric

Figure 2.3 John Day Dam on the mainstem Columbia River. Photo from US Army Corp Digital Visual Library, website at http://images.usace.army.mil/photolib.html.

Figure 2.4 Little Goose Dam on the Lower Snake River. Photo from US Army Corp Digital Visual Library, website at http://images.usace.army.mil/photolib.html.

system is a major cause of the depletion of salmonids in the Columbia Basin (assumptions 1c, 2c, 9c, 11c, and 13c; average level of proof 1.2). The use of the river by juvenile salmonids, which is indirectly related to the hydropower system, is less certain (assumptions 3c, 4c, 7c, and 8c; average level of proof 1.7). However, when it comes to the specific ways that the hydropower system has reduced salmonid survival or productivity, the level of certainty drops further (assumptions 5c, 6c, and 12c; average level of proof 3).

The implied conceptual foundation for the Council's program has strong scientific support for its focus on the hydropower system as a major cause of the salmon's decline in the Columbia River, but the assumptions about the specific ways the hydropower system produces its effect on salmon are not as well supported. The assumptions about the specific ways that the hydropower systems affect salmonid survival and productivity lead directly to the kinds of solutions that are implemented. Assumptions related to those solutions, as might be expected, have the same level of scientific proof as the specific causes (assumptions 2s, 3s, 4s, and 6s; average level of proof 3).

Hatcheries are a major use of technology to overcome the effects of the hydropower system and the other factors degrading aquatic habitat in the basin (Figures 2.5 and 2.6). In terms of the budget, it is one of the most important restoration activities carried out in the Council's program. The other major activity is the use of technology to solve the problems of the hydropower system. These two activities, passage and hatcheries, consume roughly equal portions of the Council's program and together accounted for nearly 80% of the annual salmon recovery spending over the last two decades (General Accounting Office 1992, 1993, 2002). Assumptions about the use of hatcheries are unique among the specific assumptions in that the two assumptions about hatcheries are presented as a solution to the salmon's problem (assumptions 8s and 9s); however, in two other assumptions they are considered a part of the problem (assumptions 16c and 17c). This contradiction is a sign that the implied conceptual foundation may lack internal coherence, which might be expected given the way the program's measures are assembled.

The two areas that receive the largest share of funding in the Council's program—artificial propagation and technology to improve the passage of juveniles and adults past the mainstem dams (Figure 2.7)—are based on assumptions in the implied conceptual foundation that have low scientific support (assumptions 2s, 3s, 4s, 8s, and 9s; average level of proof 3.2). This is consistent with the general principle 3, which assumes technology can make up for habitat alteration in the basin caused by development. The level of proof for principle 3 is 4.

This analysis of the implied conceptual foundation underlying the Council's program suggests explanations for the lack of results after more than 20 years since the program was created. Most of the program's funding is focused on

Figure 2.5 Hatchery worker filling salmon egg incubation trays. Photo from US Army Corp Digital Visual Library, website at http://images.usace.army.mil/photolib.html.

technological solutions to the massive habitat alteration in the mainstem and tributaries of the basin. The level of proof in the general principle and specific assumptions related to the technological solutions is low–between 3 and 4. This suggests there is a high degree of uncertainty in those aspects of the program, which should have dictated a need for a strong monitoring and adaptive management program. Neither has materialized, in spite of many such recommendations supporting them from the Council's two independent scientific advisory groups, the Independent Scientific Advisory Board (ISAB) and Independent Scientific Review Panel (ISRP) (e.g., ISRP 1999).

Figure 2.6   Hatchery workers sorting coho salmon for within-hatchery matings at the Chiwawa Satellite Hatchery in central Washington. Photo by R. N. Williams.

Figure 2.7   Little Goose Dam face on the Lower Snake River showing passage routes and structures for outmigrating juvenile salmon and steelhead (the upper closed circular system) and returning adult salmon (the open fish ladder below the juvenile bypass system). Photo from US Army Corp Digital Visual Library, website at http://images.usace.army.mil/photolib.html.

The Council introduced adaptive management to its fish and wildlife program in 1987 in response to the deep divisions in the region that often revolved around technical questions of biology or hydrology. Adaptive management offered a way for the Council to take action in the face of significant scientific uncertainties (Lee and Lawrence 1986). However, its actual application to addressing scientific uncertainty appears quite limited (McConnaha and Paquet 1996). Adaptive management has been used to justify a variety of actions on the premise that something might be learned that could lead to improved management. However, such a passive approach to learning is at odds with the rigorous application of the scientific method that is central to the concept of adaptive management (Walters 1986; Hilborn and Winton 1993).

Volkman and McConnaha (1993) and McConnaha and Paquet (1996) noted that the Council's program is one of the first attempts to use adaptive management as part of an ecosystem scale restoration program. Previous applications focused on narrower, if often complex, problems such as harvest management (McAllister and Peterson 1992). Practical difficulties have resulted in only limited success in using adaptive management as part of the Council's fish and wildlife program; there appears to be no instance where adaptive management, in the sense of Holling (1978) and Walters (1986), has been used to address major uncertainties (Volkman and McConnaha 1993). The result after over 20 years of implementation is a program that has failed to achieve its goal and is still providing massive funding to recovery activities whose efficacies are still uncertain.

Literature Cited

Beamish, R. J., G. A. Noakes, G. A. McFarlane, L. Klyashtorin, V. V. Ivanov, and V. Kurashov. 1999. The regime concept and natural trends in the production of Pacific salmon. *Canadian Journal of Fisheries and Aquatic Sciences* 56:516–526.

Blumm, M. C. 2002. *Sacrificing the salmon. A legal and policy history of the decline of Columbia Basin salmon.* Book World Publications, Portland, Oregon.

Botkin, Daniel. 1990. *Discordant harmonies: A new ecology for the twenty first century.* Oxford University Press, New York.

Bottom, D. L. 1997. To till the water: A history of ideas in fisheries conservation. Pages 569–597 in R. J. Naiman and D. Stouder, eds. *Pacific salmon and their ecosystems: Status and future options.* Chapman and Hall, New York.

Brannon, E. L., K. P. Currens, D. Goodman, J. A. Lichatowich, W. E. McConnaha, B. E. Riddell, and R. N. Williams. 1999. Review of artificial production of anadromous and resident fish in the Columbia River Basin, Part I: A scientific basis for Columbia River Production programs. Northwest Power Planning Council, Portland, Oregon.

Christensen, N. L., A. Bartuska, J. H. Brown, S. Carpenter, C. D'Antonio, R. Francis, J. F. Franklin, J. A. MacMahon, R. F. Noss, D. J. Parsons, C. H. Peterson, M. G. Turner, and R. G. Woodmansee. 1996. The report of the Ecological Society of America Committee on the Scientific Basis for Ecosystem Management. *Ecological Applications* 6:665–691.

Cronon, W. 1995. Forward: With the best intentions. Pages vii–ix in N. Langston. *Forest dreams, forest nightmares: The paradox of old growth in the inland west.* University of Washington Press, Seattle, Washington.

Crutchfield, J. A., and G. Pontecorvo. 1969. The Pacific salmon fisheries: A study of irrational conservation. Resources for the future. Johns Hopkins Press, Baltimore, Maryland.

Fry, W. H. 1854. *Complete treatise on artificial fish-breeding.* D. Appleton and Company, New York.

Galbraith, J. K. 1959. *The affluent society.* New American Library, Times Mirror, New York.

General Accounting Office. 1992. Endangered species: Past actions taken to assist Columbia River salmon. General Accounting Office. Briefing Report to Congressional Requesters, GAO/RCED-92-173BR. Washington, DC.

General Accounting Office. 1993. Endangered species: Potential economic costs of further protection for Columbia River salmon. U. S. General Accounting Office. Report to Congressional Requesters, GAO/RCED-93-41. Washington, DC. 34 p.

General Accounting Office. 2002. Columbia River Basin salmon and steelhead: Federal agencies' recovery responsibilities, expenditures and actions. US General Accounting Office. Report to the U.S. Senate, GAO-02-612. Washington DC. 86 p.

Gunther, E. 1926. An analysis of the first salmon ceremony. *American Anthropology* 28:605–617.

Hayes, Samuel T. 1959. Conservation and the gospel of efficiency: The progressive conservation movement 1890–1920. Atheneum. New York.

Heilbroner, R., and W. Milberg. 1995. *The crisis of vision in modern economic thought.* Cambridge University Press, New York.

Highwater, J. 1981. *The primal mind: Vision and reality in Indian America.* Penguin Books, New York.

Hilborn, R., and J. Winton. 1993. Learning to enhance salmon production: Lesson from the Salmonid Enhancement Program. *Canadian Journal of Fisheries and Aquatic Sciences* 50:2043–2056.

Holling, C. S. 1978. *Adaptive environmental assessment and management.* John Wiley and Sons, New York.

Independent Scientific Group (ISG). 2000. Return to the river 2000: Restoration of salmonid fishes in the Columbia River ecosystem. Northwest Power Planning Council document 2000-12. Portland, Oregon.

Independent Scientific Review Panel (ISRP). 1999. Review of the Columbia River Basin Fish and Wildlife Program as directed by the 1996 amendment to the Power Act. Northwest Power Planning Council. Annual Report, ISRP 98-2. Portland, Oregon.

Jordan, D. S. 1904. The salmon of the Pacific. *Pacific Fisherman* 2:1.

Lee, K. N. 1993. *Compass and gyroscope: Integrating science and politics for the environment.* Island Press, Washington, DC.

Lee, K. N., and J. Lawrence. 1986. Adaptive management: Learning from the Columbia River Basin fish and wildlife program. *Environmental Law* 16:431–460.

Lichatowich, J. 1999. *Salmon without rivers: A history of the Pacific salmon crisis.* Island Press, Washington, DC.

Lichatowich, J. A., L. Mobrand, and T. Vogel. 1996. A history of frameworks used in the management of Columbia River Chinook salmon. Bonneville Power Administration. Portland. 117 p.

Marcot, B. G., W. E. McConnaha, P. H. Whitney, T. A. O'Neil, P. J. Paquet, L. E. Mobrand, G. R. Blair, L. C. Lestelle, K. M. Malone, and K. I. Jenkins. 2002. A multi-species framework approach to the Columbia River Basin. Northwest Power Planning Council, Portland, Oregon. http://www.edthome.org/framework/default.htm

Martin, C. 1978. *Keepers of the game: Indian-animal relationships and the fur trade.* University of California Press, Berkeley.

Mauss, M. 1990. *The gift: The form and reason for exchange in archaic societies.* W. W. Norton, New York.

McAllister, M. K., and R. M. Peterson. 1992. Experimental design in the management of fisheries: A review. *North American Journal of Fisheries Management* 12:1–18.

McConnaha, W. E., and P. J. Pacquet. 1996. Adaptive strategies for the management of ecosystems: The Columbia River experience. *American Fisheries Society Symposium* 16:410–421.

Mobrand Biometrics. 2003. Artificial Production Review and Evaluation. Northwest Power and Conservation Council, Portland, Oregon.

National Research Council (NRC). 1996. *Upstream: Salmon and society in the Pacific Northwest.* National Academy Press, Washington, DC.

Northwest Power Planning Council (NPPC). 2000. Columbia River Basin Fish and Wildlife Program. Council Document 2000-19, Northwest Power Planning Council, Portland, Oregon.

Pinchot, G. 1910. *The fight for conservation.* University of Washington Press, Seattle, Washington.

Thomison, P. 1987. When Celilo was Celilo: An analysis of salmon use during the past 11,000 years in the Columbia Plateau. Master's thesis, Oregon State University, Corvallis. 192 p.

Volkman, J. M., and W. E. McConnaha. 1993. Through a glass, darkly: Columbia River Salmon, the Endangered Species Act, and adaptive management. *Environmental Law* 23:1249–1272.

Walters, C. J. 1986. *Adaptive management of renewable resources.* MacMillan Publishing Company, New York.

Waterman, T., and A. Krober. 1965. The Kepel fish dam. *University of California publication in American archeology and ethnology* 35:49–80.

Williams, R. N., J. A. Lichatowich, P. R. Mundy, and M. Powell. 2003. Integrating artificial production with salmonid life history, genetic and ecosystem diversity: A landscape perspective. Issue Paper for Trout Unlimited, West Coast Conservation Office Portland, Oregon. 4 September 2003.

Worster, D. 1977. *Nature's economy: A history of ecological ideas.* Cambridge University Press, New York.

3

Developing a New Conceptual Foundation for Salmon Conservation

William J. Liss, Ph.D., Jack A. Stanford, Ph.D., James A. Lichatowich, M.S., Richard N. Williams, Ph.D., Charles C. Coutant, Ph.D., Phillip R. Mundy, Ph.D., Richard R. Whitney, Ph.D.

Introduction
The Traditional Management Perspective
A Dynamic Ecosystem Perspective
  Central Principles for Fish Conservation in the Columbia River Basin
    Conservation Principle 1
    Conservation Principle 2
    Conservation Principle 3
  The Normative Ecosystem
    Conservation Principle 1
  Habitat Variability and Complexity
    Conservation Principle 2
    Spatial and Temporal Habitat Variation: Processes and Patterns
  The Marine Environment
  Salmonid Life History and Population Diversity
    Conservation Principle 3
    Population Diversity
    Sustainability: Linkages to Habitat and Life History Diversity
    Salmonid Metapopulations
    Human Impacts on Regional Population Structure
  Strongholds: Remaining Conservation Reserves
Summary
  A Static View of Ecosystems
  A Dynamic View of Ecosystems
  Protection and Restoration of Natural Processes
  Sustaining Salmonid Populations
  The Next Steps
Literature Cited

"Conservation efforts must nurture the whole life history, not focus inordinate attention on elusive "bottlenecks" to production. I believe conservation efforts will fail if primary attention is not directed to providing the habitat opportunities that historically supported the stock in its natural state."

—*M. C. Healey. 1994. Variation in the life history characteristics of Chinook salmon and its relevance to conservation of the Sacramento winter run of Chinook salmon. Conservation Biology 8:876–877.*

Introduction

As the Columbia River was developed, fisheries management tended to emphasize improved fish production to enhance harvest for commercial and sport fishers and to offset production and habitat losses (Bottom 1996; Lichatowich 1999; Chapters 2 and 7). To accomplish this goal, managers attempted to reduce adult and juvenile mortalities through reliance on technological improvements to enhance passage through mainstem dams and reservoirs (Whitney et al. 1997), enhance artificial production efforts at the basin scale, liberalize harvest limits, and introduce non-native fish species and invertebrates such as *Mysis relicta* (Bottom 1997, Lichatowich 1999). With the multiple listings of salmon and steelhead stocks in the Columbia Basin in the early 1990s, it is easy to look back at these historical management decisions as naive or even misguided; however, it is important to realize that in large part the various actions represented accepted fisheries management practices at the time. Indeed, many of these approaches are still accepted practices within fisheries management, in spite of growing evidence of the shortcomings of this approach and the need for restoration strategies and tools that rely more on natural riverine processes and salmon life history diversity attributes than on technological substitutions.

The Traditional Management Perspective

Fisheries management was founded on the assumption that ecosystems tend toward static or steady-state conditions. The overriding goal of management in this view was to achieve the static state that resulted in optimal production (Cortner and Moote 1999). An underlying assumption of this view was that humans could exert a sufficient degree of control over natural systems to consistently achieve these optimal, static conditions (Holling and Meffe 1996). Management was reluctant to recognize the importance of natural environmental variability and complexity as essential features of healthy river systems and necessary for sustained fish production (e.g., Reeves et al. 1995; Stanford et al. 1996; Poff et al. 1997), and instead viewed environmental variability as an impediment to achieving optimal production (Cortner and

Moote 1999). Many management actions such as artificial production were intended to circumvent the impacts of natural environmental variability on fish survival (Bottom 1997, Lichatowich 1999). Protection of biodiversity and the natural processes that maintained it were viewed as constraints on production optimization, rather than as legitimate management goals themselves (Cortner and Moote 1999). For example, bounties were placed on bull trout, now a threatened species over much of its range, because managers feared these piscivores posed a threat to survival of juvenile salmon.

General public dissatisfaction with the state of fisheries resources in the Columbia River ensued as fish populations declined to the point that many were listed as threatened or endangered under the Endangered Species Act. Numerous court decisions handed down to rightfully protect Native American fishing rights added a new dimension to the management of weakening salmon stocks (Lichatowich 1999). The causes for these declines were complex, but involved human activities such as habitat degradation, poor hatchery practices, overharvest, and barriers such as mainstem and tributary dams (Figure 3.1) that not only hampered migration and reduced survival, but also altered river function including the seasonality of annual flow patterns (Figure 3.2; National Research Council 1996). Additionally, natural environmental changes, such as shifts in ocean conditions, exacerbated the effects of human alterations to freshwater systems. The need to respond to declines in fish populations to satisfy ecological, spiritual, cultural, and economic needs and to recover listed stocks improved understanding of complex natural ecosystems, and shifting public attitudes toward greater environmental protection and greater protection for biodiversity of native species (Cortner and Moote 1999) began to precipitate changes in the prevailing perspectives of fish conservation in the Columbia River Basin (e.g., Northwest Power Planning Council [NPPC] 2000).

In this chapter, we synthesize relatively recent knowledge pertinent to conservation of salmonid populations. The essential feature that clearly emerges from this synthesis is that ecosystems supporting salmonid species are dynamic rather than static systems, experiencing changes in state or structure that are driven by biological and physical processes operating at a variety of spatial scales. These natural processes create spatially and temporally diverse habitats with a high degree of connectivity among habitat patches. Habitat variation in space and time creates a template for development of diverse life histories and complexes of locally adapted populations that may be genetically and demographically linked by movement of individuals among populations. Life history and population diversity, both distinguishing features of salmonid species, are essential for sustaining productivity of salmonid species within a geographic region. Salmonid conservation should be directed at protection and restoration of both the physical processes that create diverse habitats and the biological processes that allow individuals, populations, and

ORNL 2005-01568/jcn

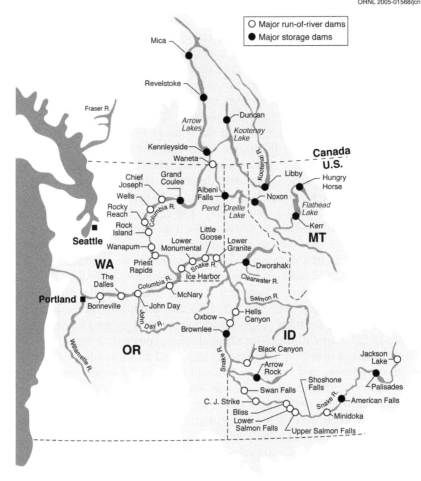

Figure 3.1 Map of the Columbia River Basin, showing major rivers, lakes, run-of-the-river dams, and storage dams.

population complexes to persist in those habitats. A dynamic view of ecosystems also suggests that restoration should not be directed at attempting to maintain ecosystems in a particular state defined by a specific (optimal) set of performance attributes or standards.

Other features relevant to the Columbia River ecosystem will be presented in chapters in this book dealing with genetic structure and diversity (Chapter 4), habitat (Chapter 5), juvenile migration (Chapter 6), hydrosystem mitigation (Chapter 7), artificial production (Chapter 8), harvest management (Chapter 9), and the estuarine and marine habitats (Chapter 10).

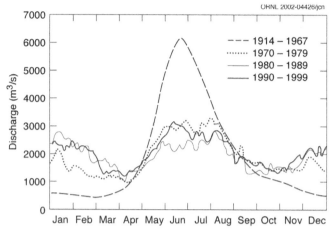

Figure 3.2 Changes in the annual hydrograph caused by regulation of the upper Columbia River watershed by dams, as exemplified by mean daily discharge at Birchbank, British Columbia, during the pre-regulation period (1914–1967), and for three post-regulation periods, including 1970–1979, 1980– 1989, and 1990–1998. From UCWSRI 2002.

A Dynamic Ecosystem Perspective

Central Principles for Fish Conservation in the Columbia River Basin

Here we summarize what we believe are three fundamental conservation principles that have emerged in fish conservation and restoration ecology. The concepts have only recently begun to be incorporated into fisheries management (but see Epilogue in Chapter 12; The Federal Approach to Salmon Recovery).

Conservation Principle 1

Restoration of Columbia River salmonids must address the entire ecosystem, which encompasses the continuum of freshwater, estuarine, and ocean habitats where salmonid fishes complete their life histories. This consideration includes human developments, as well as natural habitats.

Conservation Principle 2

Sustained salmonid productivity requires a network of complex and interconnected habitats, which are created, altered, and maintained by natural physical processes in freshwater, the estuary, and the ocean. These diverse and high-quality habitats, which have been extensively degraded by human activities, are crucial for salmonid spawning, rearing, migration, maintenance of food webs, and predator avoidance, and for maintenance of biodiversity. Ocean conditions, which are variable, are important in determining the overall patterns of productivity of salmon populations.

Conservation Principle 3

Genetic diversity, life history, and population diversity are ways salmonids respond to their complex and connected habitats. These factors are the basis of salmonid productivity and contribute to the ability of salmonids to cope with environmental variation that is typical of freshwater and marine environments.

The Normative Ecosystem

Conservation Principle 1

Restoration of Columbia River salmonids must address the entire natural and cultural ecosystem, which encompasses the continuum of freshwater, estuarine, and ocean habitats where salmonid fishes complete their life histories. This consideration includes human developments, as well as natural habitats.

The need for an ecosystem approach for restoring depleted fish populations is well recognized (e.g., Schlosser 1991; NRC 1996; Cortner and Moote 1999; Independent Scientific Group [ISG] 2000; NPPC 2000; Fausch et al. 2003). An ecosystem encompasses all the ecological and social processes that link organisms, including humans, with their environments. In the broadest sense, the ecosystem supporting Columbia River salmonids extends from headwater tributaries into the northeast Pacific Ocean and includes upland regions and riparian corridors, as well as surface and subsurface flow pathways and processes. The salmon-bearing ecosystem is characterized by processes that create and maintain a wide array of habitats in which fish grow and reproduce. These complex habitats with a high degree of spatial and temporal connectivity permit the development and expression of life history and population diversity (Healey and Prince 1995), which is an essential component of salmonid productive capacity (Reeves et al. 1995). Depleted populations of native salmonids cannot be expected to rebuild if any of the habitats required for successful completion of all life stages are seriously compromised by human activities.

A central question in salmonid conservation is how to restore fish populations in a landscape influenced and often dominated by human development (Figure 3.3). The common refrain that "we can't turn the clock back 150 years" is a truism (Poff and Ward 1990). On the other hand, we cannot afford to view salmon restoration as a lost cause. We need a view of an ecosystem

Figure 3.3 The Lower Columbia River Gorge and Bonneville Dam, the lowermost of the dams through which salmon and steelhead returning to the Columbia and Snake River systems must pass. Photo from US Army Corp Digital Visual Library, website at *http://images.usace.army. mil/photolib.html*.

as a dynamic mix of natural and cultural features that typify modern society, but that can still sustain all life stages of a diverse and productive suite of salmonid populations if the essential ecological conditions and processes necessary to maintain the populations exist within the ecosystem. We call this ecosystem, with its balanced mix of natural and cultural features, a "normative" ecosystem.

Normative refers to the norms of ecological functions and processes characteristic of salmon-bearing systems. These features, when balanced with society's needs and demands, would result in an ecosystem in which both natural and cultural elements exist in a balance that allows salmon to thrive and many of society's present uses of the river to continue, although not without modification. We emphasize that our description of the normative ecosystem is necessarily general and focuses on biological and physical processes and conditions characterizing the normative ecosystem. The normative ecosystem is not a static target or a single unique state of the river; rather, it is a continuum of conditions from slightly better than the current state of the river at one end of the continuum to relatively pristine at the other end (Figure 3.4).

The people of the Columbia Basin through their policy representatives will have to decide how much they are willing to improve ecological conditions for salmon based on their economic, cultural, and ecological values (Miller 1997).

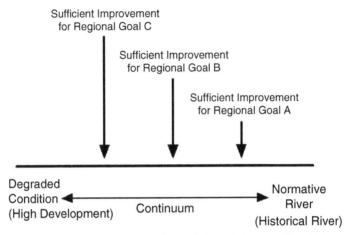

Sufficient Improvement
for Regional Goal C

Sufficient Improvement
for Regional Goal B

Sufficient Improvement
for Regional Goal A

Degraded
Condition ◄——————————————► Normative
(High Development)   Continuum   River
(Historical River)

Figure 3.4 Diagrammatic representation of the relationship between three different regional goals for salmon production and improvement in ecosystem conditions. The Normative River is defined in terms of the norms or standards that describe a set of conditions for a fully functional river ecosystem. The pre-development river, which is poorly understood or described, is not attainable due to irreversible ecological changes, such as large-scale habitat changes and species introductions.

The three hypothetical regional salmon production goals are as follows: Goal A relies on a nearly pristine refuge area, which is very close to the normative ideal, that is managed for salmon spawning; Goal B relies on a rehabilitated major subbasin, such as the lower Yakima, where appropriate flows and temperatures exist for juvenile salmon outmigration in the spring but not in the summer. The subbasin would continue to have irrigation dams and return water, yet still provide migration conditions sufficient to pass large numbers of salmonid smolts and adults; Goal C relies on improving fish bypass facilities at existing dam structures to incorporate fish behavior, thus making fish passage sufficiently normative to achieve passage and production goals.

Specific prescriptions, such as flow regimes, levels of stock and life history diversity, and so forth, will need to be developed to improve ecological conditions. We recognize that, because we are dealing with an ecosystem that has sustained extensive human development for over 150 years, numerous social and biophysical constraints exist for enhancing normative conditions (Blumm et al. 1998; Wood 1998). The challenge before the region is to reach consensus on the extent to which these constraints can be relaxed or removed to achieve fish and wildlife goals.[1] Nevertheless, we believe strongly that restoring and maintaining natural ecological processes and functions is the only way in which the Northwest Power and Conservation Council (NWPCC) Fish and

---

[1]Regional fish and wildlife goals are set forth in the Northwest Power and Conservation Council's Fish and Wildlife Program and in the various Biological Opinions and recovery plans (NOAA Fisheries) for ESA-listed salmonids in the Columbia River Basin.

Wildlife Program goals for recovery of salmonids and other fishes can be met fully. These processes create the habitat conditions necessary to maintain the productivity and diversity of salmonid resources (Reeves 1995; Stanford et al. 1996; Kaufmann et al. 1997; Poff et al. 1997; Beechie and Bolton 1999). Progress toward the restoration goal requires moving the system from the current, degraded condition to one that supports improved ecological conditions with regard to the most critical attributes for salmonids.

Habitat Variability and Complexity

Conservation Principle 2

Sustained salmonid productivity requires a network of complex and interconnected habitats, which are created, altered, and maintained by natural physical processes in freshwater, the estuary, and the ocean. These diverse and high-quality habitats, which have been extensively degraded by human activities, are crucial for salmonid spawning, rearing, migration, maintenance of food webs, and predator avoidance. Ocean conditions, which are variable, are important in determining the overall patterns of productivity of salmon populations.

Healey and Prince (1995) summarize a fundamental premise of the normative ecosystem concept—the linkage between habitat diversity and biodiversity:

"Maintaining a rich diversity of Pacific salmon genotypes and phenotypes depends on maintaining habitat diversity and on maintaining the opportunity for the species to take advantage of that diversity."

Diverse habitats are those that vary spatially, that is, from location to location, and temporally or with time. A complex and variable habitat provides the opportunity for development of diverse salmonid populations with life histories adapted to specific habitat conditions (Ebersole et al. 1996; Gresswell et al. 1997; Beechie and Bolton 1999). The magnitude and frequency of these habitat-forming events and processes vary with the spatial scale of the system (Frissell et al. 1986; Table 3.1). Large systems such as watersheds, entire streams, and geomorphic valley segments are altered by events and processes that occur relatively rarely but are of high magnitude, such as earthquakes and volcanism. In contrast, smaller scale systems such as riffles and pools are altered by events occurring more frequently but are of relatively small magnitude, such as floods and fires. Ocean conditions favorable for salmon growth and survival also vary on temporal cycles that are both long (decades or more; e.g., the Pacific Decadal Oscillation index) and short (El Niño events of one to a few years) in duration (Mantua et al. 1997).

Diverse habitats, both in Columbia River mainstem areas and tributaries, are essential for the persistence of individual species that require multiple habitats to complete their life cycles (e.g., Poff and Ward 1990; Schlosser 1990, 1991;

Table 3.1 Some events and processes controlling stream habitat on different spatiotemporal scales (after Frissell et al. 1996).

System level	Linear spatial scale (m)[a]	Evolutionary events[b]	Developmental processes[c]	Time scale of continuous potential persistence (years)[a]
Stream system	10^3	Tectonic uplift, subsidence; major volcanism; glaciation; climatic shifts	Planation; denudation; drainage network development	10^6–10^5
Segment system	10^2	Minor volcanism; earthquakes; very large landslides; alluvial or colluvial valley infilling	Migration of tributary junctions and bedrock nickpoints; channel floor downwearing	10^4–10^3
Reach system	10^1	Debris torrents; landslides; log input or washout; channel shifts, cutoffs; beaver damming; channelization, diversion, or damming by man	Aggradation/degradation associated with large sediment-storing structure; bank erosion; riparian vegetation succession	10^2–10^1
Pool/riffle system	10^0	Input, washout of wood, boulders; flood scour, deposition; thalweg shifts; numerous human and livestock activities	Small-scale lateral or elevational changes in bedforms; minor bedload resorting	10^1–10^0
Microhabitat system	10^{-1}	Annual sediment, organic matter transports; scour of stationary substrates; seasonal macrophyte growth and cropping	Seasonal depth, velocity, temperature changes; accumulation of fines; periphyton growth	10^0–10^{-1}

[a]Space and time scales indicated are appropriate for a second- or third-order stream.
[b]Evolutionary events change potential capacity, that is, extrinsic forces that create and destroy systems at that scale.
[c]Developmental forces are intrinsic, progressive changes following a system's genesis in an evolutionary event.

Poff et al. 1997; Fausch et al. 2002). They facilitate the expression of life history and population diversity and, in general, sustain biodiversity (e.g., Poff and Ward 1990; Schlosser 1990, 1991; Ebersole et al. 1996; Stanford et al. 1996; Fausch et al. 2002; Hilborn 2003); they also allow for local genetic adaptation to diverse habitat conditions (e.g., Beacham and Murray 1987; Taylor 1990; Halupka et al. 2003). Schlosser (1991) pointed out that heterogeneous habitats allow for separation of species and size classes, thus facilitating their coexistence, and they can provide refugia from severe environmental conditions (e.g., Ebersole 2001, 2003) and from predation. Thus, diverse and complex habitats are essential for sustaining salmonid productivity.

Habitats must be both suitable for survival and reproduction and accessible to organisms. Habitat connectivity provides the opportunity for species to reach habitats critical for completion of their life histories (Schlosser 1991) and is vitally important for (a) highly migratory species, such as salmon and lamprey; (b) species that move from one kind of freshwater habitat to another to spawn, such as the adfluvial and fluvial forms of bull trout and westslope cutthroat trout (Chapter 5); and (c) juveniles that move from the areas in which eggs were deposited and hatched to areas where rearing occurs. Connectivity also provides for routes by which individuals may recolonize habitats where populations have been extirpated or maintain the viability of populations that have reached dangerously low abundances (Cooper and Mangel 1999). Unfortunately, in many parts of the Columbia River Basin, habitats have been degraded and fragmented, and connectivity has been disrupted, often from construction of barriers to migration, such as mainstem and tributary dams, or through creation of lethal conditions, such as high temperatures that reduce survival of migrating adults or juveniles.

Spatial and Temporal Habitat Variation: Processes and Patterns

Like all large gravel-bed rivers, the Columbia River is a complex, dynamic gradient of habitat types from the headwaters to the estuary (see also Chapter 5). Salmonids, and all other riverine flora and fauna, are distributed rather predictably along that gradient according to the requirements of each stage in their life cycle (Vannote et al. 1980; Schlosser 1991). Each species or unique life history type will be present wherever there are enough resources to sustain growth and reproduction and thereby sustain the presence of the population in the river food web at that location (Hall et al. 1992). Some species can be maintained without much movement, and suites of organisms appear to occur in zones along the river continuum. Other species must move long distances in search of resources needed for each life stage, sometimes involving migrations into lakes (e.g., adfluvial bull trout and cutthroat trout), the ocean (e.g., Chinook salmon, coho salmon, chum salmon, and steelhead trout), or both (e.g., sockeye salmon).

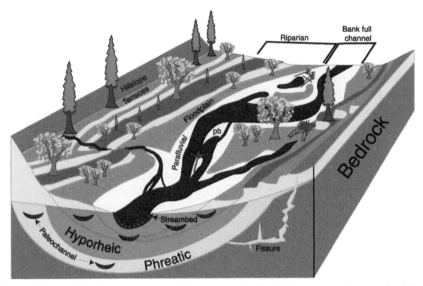

Figure 3.5   An idealized view of the 3-D structure of alluvial river ecosystems, emphasizing dynamic longitudinal, lateral, and vertical dimensions and the role of large wood debris eroded from the riparian zone. This landscape is produced by the legacy of cut and fill alluviation, which is linked to the natural-cultural setting of the catchment. Hence, the size of the flood plain and extent of hyporheic zone can be substantially larger than indicated here, extending to the valley walls. Biogeochemistry of surface and ground water interactions on such flood plains is mediated by complex interstitial flow paths or zones of preferential flow related to position of ancient river beds (paleochannels) within the alluvial bed sediments. From Stanford (1998).

Like all river ecosystems, the Columbia River has three important spatial dimensions (Figure 3.5; Ward 1989; Stanford 1998): (1) *Riverine*—a longitudinal continuum of runs, riffles, and pools from headwaters to mouth; (2) *Riparian*—a lateral array of habitats from the middle of the main channel through various side and flood channels and wetlands to flood plains and the uplands, including streamside vegetation and associated faunal assemblages; and (3) *Hyporheic*—a latticework of underground habitats associated with the flow of river water through the alluvium (bed sediments) of the channel and flood plains. These three interconnected habitat dimensions are constantly being reconfigured by physical (e.g., flooding, Table 3.1) and biological processes (e.g., salmon digging redds; beavers damming small streams and side channels on flood plains of larger rivers). Critical habitats for the various life stages of salmonids exist in all three dimensions. For example, the longitudinal continuum provides varied habitats for spawning and rearing, and a corridor for migration of adults and juveniles (Schlosser 1991). Leaves and wood debris eroded from the riparian zone into the channel energize the riverine food web, provide cover for fishes and refugia from floods, and cause localized cut-and-fill alluviation that provides additional habitat complexity.

Flow of river water through the hyporheic zone, that is, through interstitial pathways in gravel bars and floodplain alluvium and back to the surface, is an especially important habitat-forming process that may be overlooked with respect to salmonid ecology (Gibert et al. 1994). Salmonids select upwelling (water flowing upward through the gravel toward the gravel surface) and sometimes downwelling sites for spawning because their eggs are naturally aerated in those places (Geist and Dauble 1998).

Nutrients increase along interstitial flow pathways and stimulate production of food for insect larvae and juvenile salmon in upwelling zones. River temperatures are moderated by interstitial flow. Relative to surface temperatures, ground water from the hyporheic zone is cool in the summer and warm in the winter. Regional patterns of hyporheic flow appear to be critical to rivers in arid areas of the Columbia Plateau (e.g., Grande Ronde, John Day, Yakima), where late summer instream temperatures may be too high for salmonids (Li et al. 1995a, 1995b). The upwelling zones provide cool refugia for salmonids on hot summer days (e.g., Ebersole et al. 2001, 2003) and enhance winter growth by keeping the water warm and some habitats ice free. Upwelling ground water also mediates establishment of riparian plants.

The shapes of river channels are determined by bedrock geometry and geology and by the legacy of flooding. Big floods fill channels with inorganic and organic materials eroded from the stream banks and from areas upstream, thereby producing a continuum of pools, runs, riffles, gravel bars, avulsion channels, islands, and debris jams, and lateral floodplain terraces in many sizes and shapes (Poff et al. 1997). Much of the Columbia River and its tributaries within the Columbia Plateau are constrained by ancient basalts (lava rock). Thus, flood plains in the Columbia Basin are not expansive. In other areas of the basin, expansive flood plains are interspersed between canyon reaches. Channels with a greater sediment supply and frequent overbank flooding are constantly shifting, braiding, or meandering on the valley bottom from year to year as the channel fills with material in one place, causing the flow pathway to erode new channels into the flood plain.

Floodplain reaches and gravel-cobble bedded mainstem segments (e.g., alluvial reaches such as the Hanford Reach in central Washington, Figure 3.6) are especially important because habitat diversity and complexity is greatest in those locations. Alluvial reaches are arrayed along the stream between canyon segments like beads on a string (Stanford and Ward 1993) and appear to function as centers of biophysical organization (Regier et al. 1989). They are likely to be nodes of production and biological diversity that are structurally and functionally linked by the river corridor (Copp 1989; Zwick 1992; Stanford and Ward 1993; Ward and Stanford 1995a, 1995b). Worldwide, valley flood plains like the Hanford Reach of the Columbia River

Figure 3.6   Aerial view of Vernita Bar in the Hanford Reach. Vernita Bar is one of the major fall Chinook spawning areas within the Hanford Reach. Pacific Northwest National Laboratory file photo.

are characterized by nutrient-rich floodplain soils and diverse and productive backwater and mainstem fisheries (Welcomme 1979, 1995; Davies and Walker 1986; Lowe-McConnel 1987; Sparks et al. 1990; Junk and Piedade 1994). Not surprisingly, these areas are frequently centers of human activities within a watershed (Amoros et al. 1987; Petts et al. 1989; Wissmar et al. 1994).

The Marine Environment

One of the most important recent advances in salmon conservation has been improved understanding of changes in ocean conditions and how they affect anadromous fishes (Chapter 10). The Pacific Ocean and atmosphere do not move toward a steady state condition, but continually shift in response to changes in the global heat budget. These changes occur over decadal or longer time periods (e.g., the Pacific Decadal Oscillation index, Mantua et al. 1997) and subdecadal periods (e.g., El Niño-Southern Oscillation). Fluctuations in oceanic processes change the physical and biological environment and, thus, the composition of assemblages of marine biota and the survival of anadromous fishes.

Historically, in the development of management and restoration plans, as well as in the population models they used to set escapement and harvest levels, salmon managers acted as though ocean resources were unlimited and the survival of anadromous fishes in the ocean was relatively constant, a reflection of steady-state thinking. Salmon managers ignored the ocean because it is impossible to control the climatic patterns and physical factors that influence ocean productivity. Although we cannot control oceanic processes, it is possible to control and regulate our behavior and adjust management practices, including habitat protection in freshwater (Lawson 1993), in response to changes in the ocean. In that sense, the ocean is not beyond our capacity to act, but appropriate action will require better understanding of the linkages between freshwater and oceanic environments (Lawson 1993) and of the biophysical processes in the ocean that influence marine production of salmon.

Changes in the northeast Pacific Ocean that dramatically alter both freshwater and marine conditions for salmon (Miller et al. 1994; Hilborn et al. 2003) call into question management programs that emphasize constancy of the natural environment. Conservation programs designed under one climatic regime may not be appropriate under another. An ocean that is variable requires life history and genetic diversity in anadromous species to successfully respond to a wide variety of potential environmental conditions.

The performance of salmon in the estuary and ocean is not independent of management programs in freshwater. The belief that upswings in ocean productivity will continually rescue depleted stocks and so ensure persistence is misguided. For example, over the long term, variable ocean conditions, when combined with continuous habitat degradation, can eventually lead to the demise of fish stocks, even though there may be periods during which stocks appear to be recovering (Lawson 1993).

Salmonid Life History and Population Diversity

Conservation Principle 3

Life history diversity, genetic diversity, and metapopulation organization are ways salmonids have adapted to their complex and connected habitats. These factors are the basis of salmonid productivity and contribute to the ability of salmonids to cope with environmental variation that is typical of freshwater and marine environments.

Life history diversity is a fundamental component of salmonid biodiversity (e.g., Groot and Margolis 1991; Quigley and Arbelbide 1997). The rich mosaic of life histories so characteristic of salmonid fishes represents phenotypic responses to spatially and temporally variable habitat conditions that are at least in part genetically based (Holtby and Healey 1986; Beacham and Murray 1987; Taylor, 1991; Adkinson 1995; Healey and Prince 1995; Lichatowich et al. 1995; Quinn 1999; Hilborn et al. 2003). Some of this diversity, however, may result from genetic processes such as drift and founder effects (e.g., Adkinson 1995; Halupka et al. 2003). The tendency for salmonids, especially salmon, to return to their natal streams to spawn allows for local genetic adaptation and so contributes to geographic variation of life histories (Taylor 1990). In the Columbia River ecosystem, life history diversity should be substantial, owing to the ecosystem's large size, complex riverine physiography and geomorphology, highly variable flow regime, and complex oceanic circulation pattern, but this diversity has been seriously compromised by human actions.

W. F. Thompson (1959) visualized the salmon's habitat as "a chain of favorable environments connected within a definite season in time and place, in such a way as to provide maximum survival (p. 207)." We interpret Thompson's chain of interconnected habitats as temporal and spatial "pathways" through a diverse ecosystem (freshwater, estuarine, and marine). Salmonids following a particular chain of habitats—a particular pathway— exhibit a unique life history pattern. A life history pattern is the salmonid's response to problems of survival and reproduction in that chain of habitats. The complex, integrated set of phenotypic traits that comprise a fish's life history result from interaction of an individual's genotype and its environment (Taylor 1990; Healey and Prince 1995).

Like physical patterns and processes (Table 3.1), life history diversity can be manifest at various spatial scales (Healey and Prince 1995). A few examples will help illustrate this point. The well-known example of stream- and ocean-type salmon life histories represents diversity on a broad regional scale spanning the north Pacific from northern California to Kamchatka (Taylor 1990; Healey and Prince 1995; Chapter 4). The stream-type and ocean-type life histories may reflect variations in life history strategies that are thought to be related, in part, to the growth opportunity provided by rearing habitats and the distances juveniles must migrate to reach the ocean (Taylor 1990). Steam-type fish, those that migrate to the sea after spending a year rearing in freshwater, tend to inhabit areas that are more distant from the ocean and have relatively poor opportunities for growth, perhaps as a consequence of low stream temperatures. Stream-type life histories tend to be found mostly in northern areas such as Alaska and British Columbia and are common in tributaries of the upper Columbia River Basin. Ocean-type individuals that begin their migration to the sea in their first year of life tend to be found in streams closer to the ocean and where growth conditions may be more favorable. Ocean-type life histories occur mostly in streams that flow into the Pacific Ocean in more southern areas, such as the Oregon and Washington coasts. Historically, however, large populations of ocean-type salmon occurred in the upper reaches of the mainstem Columbia River (Dauble et al. 2003). Stream- and ocean-type fish also differ in other aspects of their life histories, such as oceanic distribution and timing of adult migration (Healey 1991).

Stream and ocean life histories are major life history themes, but variation in juvenile migration patterns occurs within each theme (Figures 3.7 and 3.8; Chapter 6). For example, stream-type juvenile Chinook salmon that migrate to sea in their second year exhibit variation in their migration pattern. Some stream-type Chinook salmon remain in headwater areas to rear, while others move into mainstem areas to rear in large pools over the winter (Healey

ORNL-DWG 96M-1233

Stream Flow and Juvenile Migration Patterns

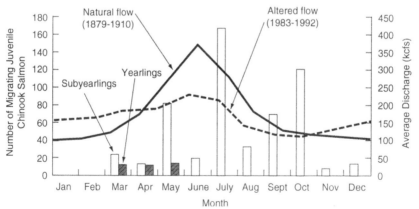

Figure 3.7 Data from Willis Rich (1939) showing counts of juvenile salmon by month for juvenile stream-type and ocean-type Chinook salmon observed near The Dalles, Oregon, prior to construction of any of the mainstem Columbia or Snake River dams.

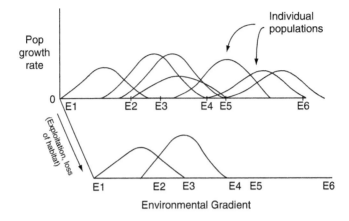

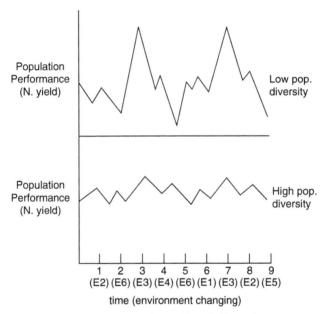

Figure 3.8 Hypothetical representation of spreading of risk. Each curve represents the growth rates of a population over a range of environmental conditions. The populations vary in the range of environmental conditions over which they can persist. As the environment fluctuates, the growth rates of some populations increase and those of other populations decline. The result is relative stabilization of abundance of the population group. When population diversity is reduced, the population group is capable of persisting only over a very narrow range of conditions. The resulting pattern of abundance is characterized by dramatic shifts, creating a high level of uncertainty, and increased probability of extinction.

1991). In the Columbia Basin, this pattern has been observed in the Yakima River (CTYIN, WDF, and WDW 1990), Grande Ronde River and Deschutes River (Ratliff 1981; Lindsey and McPhail 1986), and the Lemhi River (Kiefer and Lockhart 1995). Regulation of the annual hydrograph through development of the Columbia (and Snake) River hydroelectric system has also affected juvenile migration patterns. Under historical natural flow patterns (Figure 3.7), yearling stream-type (mostly spring) Chinook emigrated on the ascending limb of the spring runoff (March–May). Ocean-type (mostly fall) Chinook juveniles emigrated over a 10-month period (March–December), with the bulk of the juveniles emigrating on the descending hydrograph from July through October (Rich 1920, 1939).

Differences in life histories of species or races among watersheds or populations have been well documented (e.g., Ricker 1972; Beacham and Murray 1987; Groot and Margulas 1991; Healey and Prince 1995; Halupka et al. 2003). For example, Beacham and Murray (1987) present evidence that chum salmon from large rivers in British Columbia had higher ocean growth rates and were larger when they returned to spawn than fish of comparable age from smaller rivers. They hypothesized that small body size may reduce visibility to predators, while large body size may be advantageous for spawning in larger gravels characteristic of large rivers. Large fish also had morphological adaptations such as thicker, more muscular caudal peduncles that could aid swimming up larger, more turbulent rivers.

Life history variation within watersheds also has been recognized. Reimers (1973) documented five life history patterns of Chinook salmon in Sixes River, Oregon. The life history patterns were distinguished based on the length of time juveniles spent in various habitats, including tributaries, the river mainstem, and the estuary. The life histories also varied in their relative contribution to the adult spawning population.

Intrapopulational life history diversity can also occur in salmonid fishes. For example, both large and small coho adults exist within the Carnation Creek population in an apparently evolutionary stable strategy, seemingly contradicting the view that large size in Pacific salmon confers a selective advantage. Holtby and Healey's (1986) explanation for this size variation is related to heterogeneity of habitat conditions within the stream. They suggested that large females occupied areas of the stream with higher flows and large gravels that would be more susceptible to flood scour during egg incubation. Small females, on the other hand, tended to spawn in locations with smaller gravel and lower flows, sites apparently suitable for the smaller eggs and fry produced by small females. These sites also tended to be less susceptible to scour. Smaller females may have been forced into these spawning locations by competition from large fish.

Within-watershed life history variation is evident in resident salmonids as well. One example is the existence of both adfluvial, fluvial, and resident forms of bull trout and cutthroat trout within a single watershed (e.g., Rieman and McIntyre 1993; Quigley and Arbelbide 1997). Gresswell et al. (1997) documented variations in run timing among populations of Yellowstone cutthroat trout spawning in tributaries to Yellowstone Lake and related these variations to differences in timing of peak flows among the tributaries. Unfortunately, little is known about life history diversity of non-salmonid species such as Pacific lamprey, which are currently imperiled in the Columbia Basin.

Population Diversity

The changing pattern of habitat quality and availability across a landscape can be termed a "shifting mosaic" (Reeves et al. 1995). The implication of the shifting mosaic concept is that habitats will vary due to physiographic and hydrologic conditions and events. Some habitats will at times become unsuitable for survival and reproduction of organisms, while others become favorable. Salmonid populations occupying these habitats may fluctuate asynchronously, some increasing and others decreasing in response to changes in their habitats, and during extreme changes, when environmental constraints are strongest, individual populations may be extirpated (Reeves et al. 1995). However, as the effects of a natural disturbance at a particular location moderate over time, habitat quantity and quality will gradually be restored, and populations will increase, barring constraints imposed elsewhere is the system (e.g., the migration corridor). Habitats where local extinction has occurred will be recolonized by individuals from neighboring populations.

The strategy termed "spreading of risk" (den Boer 1968; Dunning et al. 1992) has been proposed as an explanation for stability and resilience of a population mosaic in a spatially and temporally variable environment. If the diverse populations within the mosaic fluctuate asynchronously, with some populations increasing and others decreasing, the mosaic as a whole, in effect, will be buffered against environmental change, thus stabilizing the dynamics of the group of populations in the mosaic and diminishing the risk of regional extinction (Figure 3.9). Diverse and asynchronously fluctuating populations thus contribute to the resilience of the population complex. A suitable analogy is the operation of an automobile in which the pistons of the car fire in an asynchronous manner, enabling the car to run smoothly.

Sustainability: Linkages to Habitat and Life History Diversity

Sockeye salmon in Bristol Bay, Alaska, have a long history of sustained productivity. Hilborn et al. (2003) examined the basis for this sustained productivity, and their work is worth discussing in some detail because it illustrates

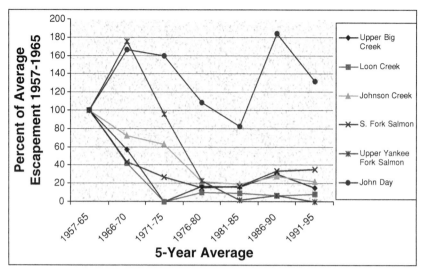

Figure 3.9 Escapement of spring and summer Chinook salmon into index streams of the Salmon River Basin and escapement of wild spring Chinook salmon into the John Day River illustrating synchronous declines of the populations. Upper Big Creek (wild spring Chinook, wilderness watershed); Loon Creek (wild summer Chinook, wilderness watershed); Johnson Creek (natural summer Chinook, managed watershed); South Fork Salmon River (hatchery-influenced summer Chinook, managed watershed); Upper Yankee Fork Salmon River (hatchery-influenced spring Chinook, managed watershed). Figure from Nemeth and Kiefer (1999).

the interplay of life history and population diversity in determining stability and sustainability of sockeye populations in a variable environment. None of the major systems in Bristol Bay has been affected by human actions other than harvest.

Sockeye salmon in Bristol Bay consist of a wide array of locally adapted populations with life histories adapting them to the diverse habitats in the rivers, streams, and lakes (Quinn 1999; Hilborn et al. 2003). Habitats in Bristol Bay in which sockeye spawn and their eggs incubate vary both spatially and temporally. Within the major lake systems composing Bristol Bay are spawning habitats that include streams, rivers, spring-fed ponds, groundwater-fed beaches, and along hillslopes and rocky beaches (Quinn 1999). These habitats differ substantially in depth, spawning substrate size, and flow patterns. Population structure within the major lake systems apparently is complex. For example, in Iliamna Lake, sockeye spawn on low-lying beaches of islands and on nearby mainland beaches and in spring-fed ponds. Circumstantial evidence garnered from transplants of mature adults and surveys of spawning dates suggests that fish using each island within the lake for spawning represent separate populations.

Sockeye populations spawning in these various habitats differ in life history characteristics that appear to adapt them to the particular natal habitat conditions. For example, sockeye spawning date appears to be keyed to the thermal regime of the river so that fry emergence is timed to coincide with the major period of food production in the spring. Salmon tend to spawn early (late June to mid-August) in smaller, colder streams where egg development is slower and time to emergence is lengthened, and later (late August–October) in rivers and lakes where water temperature may be warmer and time to emergence is shortened. Egg size also varies with habitat conditions, particularly with the type and size of the spawning substrate and oxygen supply. Salmon tend to produce larger eggs in areas such as rocky island beaches that provide the cold, well-oxygenated water needed for large eggs to successfully incubate. The large amount of yolk in the eggs helps sustain fry through the long period after emergence and possibly renders the eggs less vulnerable to predation by sculpin. Intermediate-sized eggs tend to be produced by female salmon spawning in rivers and streams, apparently conforming to the size of the incubation substrate found in these habitats. Small eggs are produced by female spawning in ponds and mainland beaches with finer substrate, where oxygen levels and water circulation through the substrate are reduced.

The body size of male and female salmon in a particular area appears to be related to the probability of predation by bears and by access to shallow spawning streams. Larger, deeper body size appears to be favored in salmon spawning in easily accessible areas with little or no predation, such as mainland islands and beaches. These larger fish tend to have a greater age a maturity. Salmon spawning in rivers and streams where accessibility may be a problem tend to be smaller in size. There is also variation among areas in freshwater residence of juveniles, smolt size, and predator avoidance as well as variation in time spent at sea.

The various spawning and incubation habitats are affected differentially by certain kinds of changes in environmental conditions. Temporal variation in freshwater habitats in Bristol Bay appears to be related to the same atmospheric factors that bring about interdecadal fluctuations in ocean conditions as indicated by the Pacific Decadal Oscillation (PDO), a climatic index that exhibits decadal scale oscillations and is associated with changes in physical conditions such as temperature and ocean productivity (Chapter 10). Hilborn et al. (2003) indicate that conditions in freshwater habitats, such as water flow, lake levels, ice cover, and temperature, were associated with major shifts in the PDO index. These changes can differentially affect salmon spawning in the various habitats within the major lake systems. For example, high flows in freshwater systems tend to be correlated with positive phases of the PDO. Temporal variation in hydrologic conditions among freshwater habitats within Bristol Bay can affect sockeye by, for example, limiting adult access to spawning areas in smaller streams during periods of low flows. In contrast, spawning areas at lake beaches are less likely to be affected by variations in flow, and therefore they remain readily accessible to spawning fish. High flows can also improve survival of downstream-migrating smolts because of reduced susceptibility to freshwater predators. The picture that emerges from Hilborn et al.'s (2003) work for Bristol Bay is that it is a system with spatially diverse habitats, whose favorability for spawning and rearing varies asynchronously due to variation in flow patterns and other environmental factors among streams, and with a diverse and complex population and life history structure is related, at least in part, to habitat complexity within the major lake basins.

Hilborn et al. (2003) organized biocomplexity of the Bristol Bay system at three scales: a coarse geographic scale encompassing major lake and river systems; a finer scale involving spawning streams, beaches, and ponds within the major systems; and life history variations among fish spawning in different areas. Catch (a reliable index of abundance for Bristol Bay stocks, Hilborn et al. 2003) from 1990 to the present and estimates of productivity (recruits per spawner) based on counts of returning fish from 1995 to the present were used to assess temporal population changes. The time periods used for the analysis spanned a major shift in the PDO index. Their analysis indicated that, at the scale of major lake and river systems, the two population performance indicators fluctuated out of phase, increasing for some lake and river systems in Bristol Bay and decreasing for others. Hilborn et al. (2003) concluded that asynchronous fluctuations in abundance among the major river and lake systems sustained productivity of the entire Bristol Bay stock complex. Analysis at the scale of major systems provides the strongest evidence for asynchronous variations in abundance.

Hilborn et al. (2003) also analyzed fish counts from three major types of spawning locations within the Iliamna Lake system: ponds and creeks, beaches, and rivers. Variations in spawner counts in these locations were associated with variations in the larger scale lake systems. Sockeye spawning at the different locations have different life histories, as indicated above, and Hilborn et al. (2003) suggest that they would respond differently to variations in freshwater habitat associated with changes in ocean-atmosphere system.

The remarkable habitat and life history complexity of Bristol Bay and its sockeye salmon stocks apparently has contributed significantly to the productivity and sustainability of the Bristol Bay stock complex. Diverse populations that tend to fluctuate out of phase with each other appear to buffer the complex of populations in the Bay against environmental change. This spreading of risk contributes to long-term stability and sustainability. Life history diversity, arising from adaptations of individual populations to their habitats, appears to be linked to the differential productivity of the stocks and also contributes to sustainability.

Salmonid Metapopulations

Recently, biologists have explored the application of the metapopulation theory to salmonid conservation (e.g., Li et al. 1995; Schlosser and Angermeier 1995; NRC 1996; Policansky and Magnuson 1998; Dunham and Rieman 1999; Rieman and Dunham 2000). The metapopulation concept is similar to the concept of spreading of risk. Metapopulations are spatially structured groups of local populations linked by dispersal (straying) of individuals (Hanski 1991; Hanski and Gilpin 1991). Some populations in the metapopulation may suffer extinction only to be reestablished through colonization from neighboring populations. Thus, metapopulation persistence is determined by the balance of local population extinction and reestablishment of extinct populations through recolonization from neighboring populations. The metapopulation concept is intuitively appealing because, like spreading of risk, it explains the persistence of a group of populations inhabiting a spatially and temporally variable environment.

Metapopulation structure is possible in salmon because they display high fidelity of homing to their natal streams, which allows them to establish locally adapted spawning populations (Taylor 1991), and they exhibit relatively low, but variable levels of straying (Quinn 1993), offering the opportunity for recolonization of habitats where local extinction has occurred (Cooper and Mangel 1998). High natal fidelity favors adaptation of specific breeding demes (i.e., local populations) to their environments via natural selection. In turn, this promotes population differentiation at the local level. However, because adjacent local populations are likely to occur in habitats that are similar (due simply to proximity), they may have very similar selection regimes. Therefore, differences or genetic divergence that accrues among them may be due largely to the effects of isolation and genetic drift. Straying (i.e., gene flow) between populations, even at very low levels, can tend to counteract the effects of isolation, thus retarding or even preventing genetic divergence among local populations.

Asynchronous population fluctuations enhance metapopulation persistence (Harrison and Quinn 1989; Hilborn et al. 2003; Isaak et al. 2003;). If, however, the local populations composing a metapopulation fluctuate synchronously, increasing and decreasing together so that their dynamics are temporally correlated (Harrison and Quinn 1989; Hanski 1991), theory suggests that the collective abundance of the population mosaic could fluctuate considerably (Figure 3.8) and the chance of metapopulation extinction would increase (Harrison and Quinn 1989). Populations can change synchronously, for example, if they occupy neighboring tributaries or watersheds, or share common rearing areas or migratory pathways, such as river mainstems, and so are subject to the same variations in environmental conditions. Synchrony could also be achieved if the populations experience mortality from a common fishery or if they are affected by large-scale environmental changes as could occur from some land use activities or climate change (Figure 3.9).

There have been few studies of synchrony in salmonid fishes (e.g., Peterman et al. 1998; Botsford and Paulsen 2000; Pyper et al. 2001). One such study was conducted by Isaak et al. (2003), who documented highly correlated fluctuations in spring/summer Chinook salmon redd counts from different streams within a wilderness area of the Middle Fork of the Salmon River in Idaho. Redd counts were more consistently correlated during the period when salmon abundance was low. Isaak et al. suggest that the correlations might be attributed to environmental factors that likely affected the habitats of multiple populations. These factors include reduced availability of nutrients resulting from decreased adult carcasses, decreased operation of density-dependent factors and greater control by density-independent factors as population abundances declined, mortality incurred as the salmon traversed the common migratory corridor of the Columbia and Snake Rivers that has been highly modified by construction of a series of mainstem dams, or increased mortality in the ocean that caused reduced abundances of spawners and thus indirectly set the stage for synchrony. Isaak et al. (2003) suggest that because synchronous fluctuations in local population abundance can increase the chance of metapopulation extinction, effort should be directed at desynchronizing population fluctuations by maintaining a diversity of habitats that support a broad range of life histories and ensuring that some of these habitats are sufficiently separated so as to be spared the common effects of large-scale catastrophic events. Obviously, a strategy to reduce the possibility of synchrony must be applied on a large geographic scale and will need to effectively address human actions that have widespread and contiguous impacts.

A number of general classes of metapopulation structure have been identified (Harrison 1991), and several of these classes are thought to be applicable to salmonids. Several analyses have suggested that regional salmonid population structure currently resembles a core–satellite metapopulation or its variant, a hybrid metapopulation (e.g., Rieman and McIntyre 1993; Li et al. 1995a; Schlosser and Angermeier 1995). Core populations occupy high-quality habitat and are generally large, productive populations that are less susceptible to extinction than smaller satellite populations (Hanski 1991; Harrison 1991, 1994). Core populations can serve as important sources of colonists that could both reestablish satellite populations in habitats where extinctions have occurred (Harrison 1991, 1994; Rieman and McIntyre 1993; Schlosser and Angermeier 1995) and sustain populations whose abundance has been severely depleted (Cooper and Mangel 1998). The concept of core populations is important because it suggests that certain large populations can buffer a population complex against environmental change and contribute to the resiliency of regional salmonid production.

A plausible metapopulation structure, generally speaking, is one that is composed of a core population and satellites that vary significantly in their probability of extinction. Salmonid populations can be subject to cyclic trends in environmental conditions, such as changes in ocean productivity that affect all the populations within the metapopulation. Under highly favorable environmental conditions, most of the local populations could persist, although some of the more vulnerable populations (sink populations) may need to be shored up by colonization from the core. As environmental conditions worsen, only the populations most susceptible to these changes would suffer extinction, still leaving a relatively intact metapopulation. But as conditions continue to deteriorate, more and more of the satellite populations may have reduced abundance or face extirpation. Under the most adverse conditions, only the core populations would persist. Then, as environmental conditions improve, vacant habitats could be colonized from the core, so long as habitat connectivity had been maintained, and the metapopulation could expand. Thus, as environmental conditions vary from favorable to adverse, metapopulation structure could expand and contract around the core population (s). In this case, the persistence of the whole metapopulation is virtually dependent on persistence of the core population.

In some cases, population structures that currently resemble metapopulations may have resulted from fragmentation of the habitat of a single continuous population resulting from human activities (McCullough 2000). The distinction is important because organisms that were members of a large continuous population may not have evolved the dispersal capabilities of organisms that were members of a metapopulation, where movement among disparate habitats is essential for recolonization of habitats where extinction has occurred (McCullough 2000). Thus, population fragments of a once continuous population may not truly function as a metapopulation.

Dauble et al. (2003) proposed that fall Chinook spawning in the Columbia and Snake Rivers resembled a core–satellite metapopulation. Their work was unique in that they were able to use historical records and geomorphic models to identify likely fall Chinook spawning areas. Dauble et al. (2003) suggested that groups of fish spawning in these areas represented local populations. Among these areas, they identified the ones where spawner abundance could have been unusually high and suggested that spawners in these areas represented core populations. Historically, the two principle fall Chinook spawning areas were below Swan Falls on the Snake River, an area now inaccessible to salmon due to the construction of the Hell's Canyon Dam complex, and an area adjacent to the confluence of the Columbia and Snake Rivers, an area largely inundated following construction of McNary Dam. Apparently, fall Chinook spawners also were abundant in the section of the mainstem Columbia presently inundated by the John Day Reservoir (Fulton 1968). As suggested by Stanford et al. (1996), the important mainstem fall Chinook spawning areas were located in relatively large alluvial flood plains with complex channel structures, including bars and islands that not only provided spawning habitat but also rearing areas for juveniles (Dauble et al. 2003).

Another area that appears to be of unusual importance as a center of salmon abundance in the Columbia Basin is the Salmon River, a tributary to the Snake River. Although the Salmon comprises only 6% of the land area of the Columbia River Basin, it provides more anadromous spawning area than any other subbasin. The Salmon River produces 39% of the spring Chinook salmon, 45% of the summer Chinook salmon, and 25% of the summer steelhead returning to the mouth of the Columbia River (NPPC 2000). Undoubtedly these areas of high abundance were important for persistence of populations within the Columbia Basin, but whether these areas functioned as core populations in the strict sense of metapopulations is uncertain.

Protection and restoration of salmon metapopulations has been recommended for conservation of salmon (NRC 1996; ISG 2000, McElhany et al. 2000;). Metapopulation theory is intuitively appealing, especially for salmon, and it fits nicely with a landscape view of a dynamic habitat. Rieman and Dunham (2000), however, reviewed empirical studies on salmonid metapopulation structure and concluded that little empirical support currently exists for metapopulation structure in salmonids.

Much research is necessary to determine the applicability of metapopulation theory to salmon conservation. The metapopulation structure of salmonids in the Columbia Basin, both historically and at present, is poorly understood. The role of straying in refounding populations, not to mention the empirical difficulties in assessing straying rates (McElhany et al. 2000), is largely unknown, and the function of asynchronous fluctuations (and the deleterious consequences of synchrony) have been determined for only a few systems, such as Bristol Bay. The existence of core populations, either in the past or at present, is somewhat speculative and based largely on habitat conditions (Stanford et al. 1996; Dauble et al. 2003) or on historical records of the abundance of spawners.

An important factor that confounds determination of metapopulation structure and processes has been the operation of hatcheries over the last century. Hatcheries, in many cases, have functioned as core populations (McElhany et al. 2000). Hatchery adults have been translocated to locations distant from the source facility; supplementation practices that involve stocking juveniles to enhance failing populations are extensively employed; and straying of hatchery adults into habitats where they interbreed with naturally spawning fish is a common occurrence. For example, in the upper Columbia steelhead ESU, over 50% of the spawners in the natural spawning grounds are estimated to be first-generation hatchery fish (Chapman et al. 1994, cited in McElhany et al. 2000). Many of the problems associated with hatchery practices are discussed in Chapter 4. In addition, land use actions have had a pervasive effect on habitat in many areas of the Columbia Basin, making it difficult to examine intact metapopulations outside of wilderness areas (e.g., Isaak et al. 2003) or other areas where habitat conditions have not been severely altered.

In spite of the uncertainty surrounding application of metapopulation theory to salmonid conservation, Reiman and Dunham (2000) conclude that metapopulation theory is valuable because it stimulates the quest for understanding the relationship between salmonid diversity and habitat complexity. It also provides a reasonable explanation for the persistence of populations in habitats that are altered by natural processes operating at a variety of scales. McElhany et al. (2000) suggest a precautionary approach for protection of population diversity that is relevant to metapopulations. They argue that because dynamic and diverse populations and life histories were a hallmark of the historical condition in the Columbia Basin and that this diversity was associated with productive populations, in lieu of perfect understanding of the role of diversity in sustaining salmonid productivity, the historical condition should serve as a guide or the "default" position for restoration.

Human Impacts on Regional Population Structure

Fragmentation and destruction of both mainstem and tributary habitats can disrupt regional population organization by reducing population and life history diversity, reducing population size and productivity, extirpating or fragmenting vital core populations, and isolating populations. In turn, this can significantly reduce the long-term persistence and stability of regional fish production (Rieman and McIntyre 1993; Harrison 1994; Li et al. 1995b; Schlosser and Angermeier 1995).

The extirpation and constriction of the distributions of Chinook salmon populations in the Columbia Basin provides a dramatic example of the consequences of large-scale habitat fragmentation and destruction. We offer a generalized example of the kinds of changes that could have occurred as a result of human development of the Columbia Basin over the past century. We suggest general distributional ranges of life history types to denote changes in distribution between the historical condition (Figure 3.10) and the present (Figure 3.11). Within the general range of each life history type, discrete patches of favorable spawning and rearing habitat undoubtedly occurred.

Chinook salmon in many tributaries to the Columbia Basin above Bonneville Dam prior to extensive human development likely consisted of a complex mosaic of spring, summer, and fall races of salmon distributed among mainstem and headwater spawning areas (Figure 3.10). Local

ORNL-DWG 96M-1226

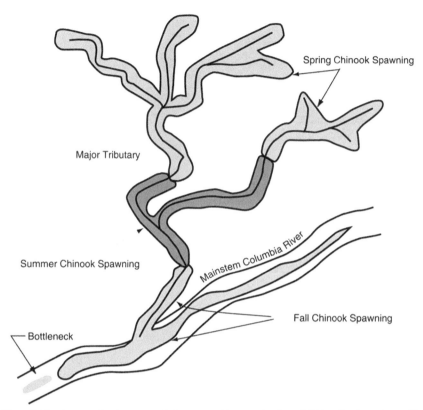

Figure 3.10   A generalized concept of the geographic organization of Chinook salmon in a mid-Columbia River subbasin prior to extensive human development. See text for explanation.

populations of fall Chinook whose juveniles migrated to the ocean as sub-yearlings spawned primarily in mainstem areas of the Columbia and Snake Rivers and lower mainstem segments of Columbia River tributaries (Fulton 1968; Howell et al. 1985; Mullan et al. 1992; Dauble et al. 2003). Spring and summer Chinook that also may have migrated as subyearlings reproduced in upper mainstem segments of major subbasins and lower reaches of tributaries to subbasin mainstems (Lichatowich and Mobrand 1995). Summer Chinook probably spawned lower in the subbasin mainstems than spring Chinook (French and Wahle 1959; Fulton 1968; Mullan et al. 1992; Lichatowich and Mobrand 1995). Populations of spring Chinook with yearling life histories reproduced in headwater streams of subbasin tributaries.

The decline of the ocean-type life history has been an important contributor to the overall decline in production of Chinook salmon within the

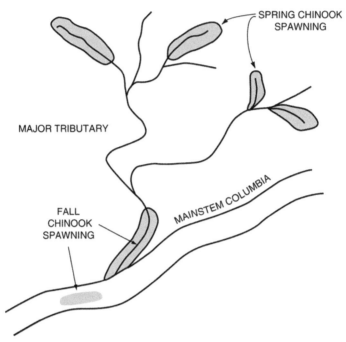

ORNL-DWG 96M-1225

Figure 3.11 Diagrammatic concept of the present geographically fragmented organization of Chinook salmon in the Columbia River Basin.

Columbia River Basin (Lichatowich and Mobrand 1995; Dauble et al. 2003). In sampling conducted in the lower Columbia River from 1914 to 1916, Rich (1920) observed a migration of ocean-type Chinook salmon throughout most months of the year. He attributed this to the sequential migration of juvenile Chinook salmon from tributaries progressively further upstream. Because the occurrence of the ocean-type life history pattern probably is related to areas in the watershed where stream temperatures afford enhanced growth opportunity (Taylor 1990), the ocean-type life history pattern likely would have originated from populations of fall Chinook salmon that spawned in the mainstems of the Snake and Columbia Rivers and in the lower reaches of some subbasins, and possibly from populations of summer and spring Chinook salmon that spawned in the warmer middle reaches of some of the subbasins (Figure 3.10; Lichatowich and Mobrand, 1995). In Rich's (1920) sample, ocean-type juveniles far outnumbered yearling stream-type migrants (Lichatowich and Mobrand 1995).

Inundation and isolation of alluvial habitats in the mainstem Columbia and Snake Rivers following construction of dams and degradation of mainstem habitats in major subbasins present a dramatic example of large-scale habitat degradation that has virtually eradicated many mainstem spawning fall Chinook populations (Figure 3.11). Dauble et al. (2003) estimated that 80% to 90% of the mainstem riverine habitat used for fall Chinook spawning has been eliminated. The major remaining mainstem spawning area for fall Chinook is the Hanford Reach (Becker 1985), an alluvial river segment in the mid-Columbia region. Remnant populations of fall Chinook also occur in the lower mainstems of some major subbasins, in the Snake River below Hell's Canyon Dam, and in the tailraces of Snake and Columbia river mainstem dams, but their abundance is much lower than in the past (Lavier 1976; Garcia et al. 1994).

In some subbasins, loss of habitat connectivity, due in part to high summer water temperatures that likely were lethal to migrating salmon, may have been a major contributor to the loss of the ocean-type life history (Lichatowich and Mobrand 1995). High temperatures in the lower sections of subbasins, resulting from a cumulative effect of watershed-wide habitat degradation, severed the connectivity of the chain of habitats linking the subbasins to the mainstem Columbia and Snake (Table 3.2), as happened in the Yakima River. In many subbasins, alteration of habitat in the migration corridor through excessive temperatures or other barriers could have eliminated life histories like the ocean-type that historically were dependent on migration in the summer and fall months (CTUIR and ODFW 1990 [Umatilla River]; CTYIN, WDF, and WDW 1990 [Yakima River]; ODFW, CTUIR, and NPT 1990 [Grande Ronde

Table 3.2   Habitat suitability for juvenile Chinook salmon in the lower reaches of Columbia River subbasins (after Lichatowich and Mobrand 1995).

Subbasin	Comments on Habitat	Source
Yakima	Lower river below Prosser (RM 47.1) frequently exceeds 75°F and occasionally reaches 80°F in July and August rendering the lower river uninhabitable by salmonids.	CTYIN et al. 1990
Tucannon	Water temperatures in lower river at or above lethal levels.	WDF et al. 1990
Umatilla	Lower 32 miles subject to irrigation depleted flows and temperatures exceeding upper lethal limits for salmonids.	CTUIR and ODFW 1990
John Day	Juvenile Chinook salmon generally not found in the river where temperatures reach 68°F. High stream temperature eliminates juvenile rearing habitat in the lower river.	Lindsay et al. 1981; ODFW et al. 1990
Deschutes	In the mainstem Deschutes River, summer temperatures are adequate for Chinook salmon. However, there are temperature problems in the lower reaches of the tributaries where spring Chinook salmon spawn. In addition, *Ceratomyxa shasta* limit the survival of juvenile Chinook salmon in the mainstem through the summer months.	Ratliff 1981; ODFW and CTWSR 1990

River]; ODFW, CTUIR, WDF, CTWSR 1990 [John Day River]; ODFW and
CTWSR 1990 [Deschutes River]); WDF et al. 1990 (Tucannon River)). Lethal
temperatures in the lower sections of the mainstem likely eliminated several
life history pathways in spring Chinook salmon in the Yakima River (Watson
1992). In contrast to Rich's (1920, 1939) findings early in this century, today,
stream-type juveniles are far more abundant than ocean-type migrants, sig-
naling a substantial loss of production of fall and summer Chinook (Figure
3.7). Even so, the distribution of stream-type juveniles often has been com-
pressed into small sections of favorable habitat in headwater areas of stream
whose lower reaches have been degraded. These populations may be quite isolated from other such populations (Figure 3.12).

Strongholds: Remaining Conservation Reserves

A fundamental premise of many aquatic conservation strategies is to identify
and protect remaining areas of high-quality habitat supporting abundant populations and a diverse number of native fish species (Bisson 1995; Li et al.

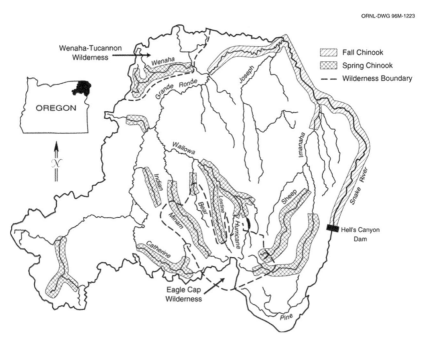

Figure 3.12 Spawning localities in northeastern Oregon of fall Chinook salmon (stippled) in
the undammed section of the Snake River below Hell's Canyon Dam, and for fall and spring
Chinook in two of its tributaries, the Grande Ronde and Imnaha. From Oregon Department of
Fish and Wildlife and Oregon Watershed Health (1994).

1995a; Quigley and Arbelbide 1997). These areas can serve as focal points for restoration (Williams 2001) and have been termed strongholds (Quigley and Arbelbide 1997). Unfortunately, the number of stronghold areas in the Columbia Basin is relatively small, and they tend to be found in high elevation areas that are relatively inaccessible or undesirable for human utilization (Li et al. 1995). Quigley and Arbelbide (1997) estimated that within the interior Columbia Basin, 0.01% of subwatersheds support three strong salmonid populations, 3% support two, and 20% support one. Most of the strongholds occur in areas with low road densities. It is unclear whether areas currently identified as strongholds supported populations that were historically core populations, remnants of larger more continuous populations, or relatively small populations that have persisted because of their isolation. Nevertheless, the supposition is that these areas essentially can function as core populations that are relatively resilient and perhaps could provide colonists to refound nearby habitats where populations have become extinct. However, many of these stronghold populations are relatively isolated from other strongholds because intervening habitats are unfavorable for movement of individuals between stronghold areas or because the distance between the strongholds is great. Reeves et al. (1995) cautioned that reserves such as these should not be viewed as being permanently suitable for sustaining populations because in a dynamic landscape even the quality of stronghold habitats can be diminished from both natural and anthropogenic disturbance. Thus, the persistence of stronghold populations is far from assured without some form of human intervention.

The requirements for strongholds that will be effective for sustaining salmonids over the long term are great. Ideally, stronghold habitats should possess characteristics of the intact ecosystems they represent, be large enough to accommodate the disturbance regime characteristic of the intact ecosystem, and be connected to neighboring suitable habitats (Bisson 1995; Li et al. 1995; Reeves et al. 1995; Quigley and Arbelbide 1997). They should also be large enough to reduce the chances of synchronized population behavior. It is probable that only a few strongholds meet these criteria.

The Hanford Reach is the largest mainstem stronghold for fall Chinook (Figures 3.6 and 3.13; Dauble et al., 2003). It supports 90% of the fall Chinook returning to the mid-Columbia (Figures 3.14 and 3.15). The Hanford Reach is the last undammed section of the mainstem Columbia in the United States, although flow through the Reach is regulated by dams upstream. Escapement to the Hanford Reach, where relatively high-quality spawning and rearing habitat is still available, has ranged from 15,000 to 36,000 fish from 1968 to 1983 and peaked at over 88,000 spawners in 1987 (Dauble and Watson, 1997). Redd counts have increased steadily since the late 1950s and exceeded 8,000 redds in the index areas in the late 1980s and again in 2002 and 2003.

The relatively high abundance of salmon spawning in the Hanford Reach could have been influenced by construction of Priest Rapids Dam at the

Figure 3.13 Hanford Reach fall Chinook salmon. Photo by G. McMichael, Pacific Northwest National Laboratory.

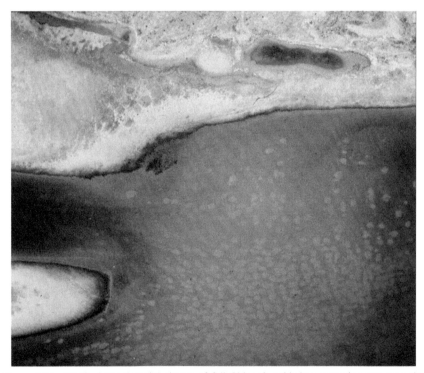

Figure 3.14 Hanford Reach aerial picture of fall Chinook redds in a spawning area near the 100-F nuclear reactor. Light colored circles and ovals are individual fall Chinook redds. Pacific Northwest National Laboratory file photo from 1994.

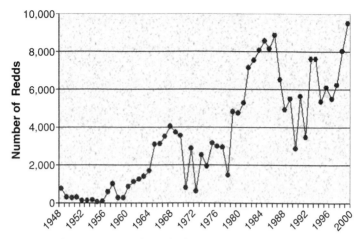

Figure 3.15 Spawning escapement of Hanford Reach fall Chinook in the Columbia Basin. Updated from Dauble and Watson (1997). Data and figure provided by Robert P. Mueller of Pacific Northwest National Laboratory, Richland, Washington.

upper end of the Reach in 1960, increases in smolt releases from Priest Rapids hatchery in the early 1980s (Dauble and Watson, 1997), and construction of mainstem Snake River dams which may have diverted salmon bound for the Snake into the Hanford Reach (Dauble et al. 2003). Nevertheless, fall Chinook in the Hanford Reach presently have characteristics typifying a core population. Observations of radio-tagged fall Chinook from the Hanford Reach reveal extensive movements throughout an area that includes the confluences of the Snake, Columbia, and Yakima Rivers (Figure 3.16). These patterns of movement of radio-tagged fish suggest that the Hanford population could serve as a source for colonization of adjacent tributaries if ecological conditions were improved or restored in them.

Summary

A Static View of Ecosystems

Like other areas of natural resource management, fisheries management has been dominated by an equilibrium view of ecosystems (Holling and Meffe 1996). In this view, human actions shift ecosystems away from the historical steady state toward a new degraded but nevertheless static condition. Restoration is viewed as a process directed at returning the system to a static condition more closely resembling the historical state (Holling and Meffe 1996). Monitoring assesses progress toward attaining these static conditions. The ability to exert a significant level of control over ecosystem processes is implicit in

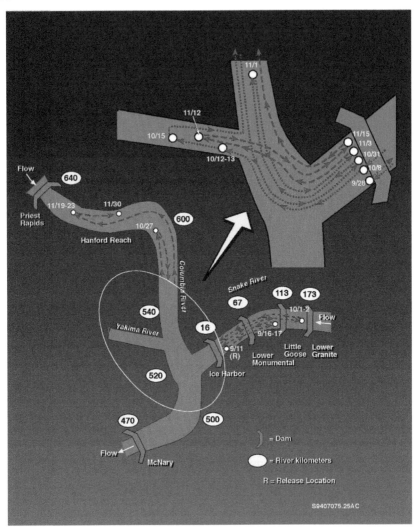

Figure 3.16 An example of the upstream movements of a female fall Chinook salmon monitored during radio telemetry study conducted in 1993 by the Washington Department of Fish and Wildlife and Pacific Northwest National Laboratory (Mendel and Milks 1997). After being tagged with a radio transmitter and released upstream of Ice Harbor Dam on September 11, 1993, biologists tracked this fish for 80 days as it moved between the Snake River, Yakima River, and the Hanford Reach. The fish (and tag) were finally recovered during spawner surveys on November 30, 1993, where it presumably spawned in the Hanford Reach in the vicinity of Vernita Bar. Graphic supplied by David R. Geist, Pacific Northwest National Laboratory, Richland, Washington.

the static view (Holling and Meffe 1996). A steady-state view of ecosystems is highly consistent with attempts to achieve certain human benefits from natural resources, and this is one of the important reasons that the view persists. Fire suppression and regulation of flood flows (Stanford et al. 1996) represent attempts to maintain ecosystems in a relatively static condition for protection of property, loss of valuable timber resources, flood control, and hydroelectric power production. Some hatchery practices also reflect the static view of ecosystems. The strategy of releasing increased numbers of smolts during periods of poor ocean conditions is an attempt to stabilize the size of the total run of returning adults. This approach often is unsuccessful, as the experience with Oregon coastal coho demonstrates, and has contributed to depletion of wild runs in mixed stock fisheries (Nickelson et al. 1986; Nickelson 2003).

To a considerable extent, habitat restoration actions have been predicated on the same static view of ecosystems where restoration actions are directed at creating and maintaining a specific habitat state that is intended to be stable in space and time (Beechie and Bolton 1999; Independent Scientific Advisory Board [ISAB] 2003). Fixed values of performance attributes or standards (e.g, a specific distribution and quantity of large wood, pool—riffle ratio, maximum stream temperature level) describe the static-state conditions that are to be achieved through restoration (ISAB 2003). Traditional approaches to habitat restoration, such as the nearly universal reliance on placement of instream structures (Figure 3.17), emphasize creating specific instream habitat conditions rather than focusing on the processes that create and change habitat state (Beechie and Bolton 1999). Often, however, floods and landslides can dislodge the structures, even in the best-designed instream restoration projects (Frissell and Nawa 1992; Kaufmann et al. 1997). The ISAB (2003) states that focusing on fixed performance standards redirects attention from protecting and restoring natural watershed processes and often does not take into account differences in habitat requirements among species and temporal changes in the habitat requirements of a given species. The static view is also reflected in conservation strategies that attempt to protect in perpetuity specific areas of currently favorable habitat or strongholds (Reeves et al. 1995). These strategies are absolutely essential because so few such areas remain; however, in the long term, even these areas could be subject to natural environmental changes that could lead to population declines or extirpation.

A Dynamic View of Ecosystems

The normative ecosystem concept is based on a dynamic view of aquatic ecosystems (e.g., Beechie and Bolton 1999; ISAB 2003; Poff et al. 1997; Poff and Ward 1990; Reice 1994; Reeves et al. 1995; Stanford et al. 1996)

Figure 3.17 Rocks forming a Rosgen "J" shape with a log and root wad cabled in below the rock structure in Asotin Creek, Idaho. Photo by R. N. Williams.

integrated within a landscape perspective (Schlosser 1991; Fausch et al. 2003) and cognizant of long-term as well as short-term changes in ecosystem state (Reeves et al. 1995; ISAB 2003). Dynamic ecosystems experience changes in state or structure that are driven by biological and physical processes operating within and external to the ecosystem (e.g., the disturbance regime). The focus of this dynamic view is on the processes producing change rather than solely on single states that are transitory either over short or long time periods (Bisson et al. 1997; Beechie and Bolton 1999; ISAB 2003). The dynamic view also applies to social dimensions of ecosystems, which will inevitably change due to processes such as population growth and the shift from resource- to technology-based local economies.

In the dynamic view, the focus is on natural processes, with ecosystem states simply being variable products of these processes (Reeves et al. 1995; Beechie and Bolton 1999; Bisson et al. 1997; Poff et al 1997; ISAB 2003). Thus, a particular value of a performance measure (e.g., a specific pool—riffle ratio or distribution and quantity of large wood) would receive less emphasis in the dynamic view than in the static view because it would be viewed as a transitory condition and would vary through time and from location to location.

This is not to say that measurements of specific habitat conditions have no utility. These measurements add to our understanding of the effects of landscape and instream processes on stream habitat, help define patterns of short- and long-term state changes resulting from the operation of the natural processes, assist in identifying life history requirements of individual species and fish community structure, and aid in understanding how human actions affect streams. However, no single state is "representative" of the complex and dynamic processes that give rise to that state.

Protection and Restoration of Natural Processes

In a dynamic ecosystem view, a cornerstone of salmonid restoration is protection and restoration of the natural processes that create and change habitat conditions (e.g., Reeves et al. 1995; Stanford et al. 1996; Poff et al. 1997; Beechie and Bolton 1999; ISAB 2003). These processes operate on a variety of spatial scales (Table 3.1) ranging from high-magnitude but infrequent events and processes that change whole watersheds and large stream segments, to more frequent but lower magnitude processes that alter smaller scale habitat elements such as riffles, pools, and large woody debris distribution and abundance. Geologic and climatic processes (Table 3.1) create the landscape mosaic that serves as the template for development of diverse locally adapted populations. Within major landscape units such as watersheds, smaller scale, higher frequency processes such as floods create and change habitat conditions necessary for completion of life histories (e.g., spawning, incubation, rearing, and migration).

Restoration of habitat-forming processes is crucial for reestablishing and sustaining diverse salmonid populations and life histories (Healey and Prince 1995; Ebersole et al. 1996; Reeves et al. 1996; Poff et al 1997; Beechie and Bolton 1999). Knowing how to restore natural processes is exceedingly challenging in a landscape that has been subject to over 100 years of human development and is highly fragmented, dominated by the inertia of established land and fish management practices, and in which public sentiment may not favor large-scale changes (e.g., reduced fire suppression and a more natural fire regime, re-regulation of rivers to achieve a more natural hydrograph (Stanford et al. 1996; Miller 1997; Poff et al. 1997).

A protocol for process-oriented restoration is predicated on establishing natural patterns and rates of habitat-forming processes either through evaluation of historical conditions or through assessments in unmanaged watersheds (Lichatowich et al. 1995; Beechie and Bolton 1999; ISAB 2003). Beechie and Bolton (1999) suggest that further steps involve assessment of how land use and other human activities have altered natural processes, identification of approaches necessary for protecting and restoring habitat-forming processes, estimation of the effectiveness of each of the approaches, and prioritization of the restoration actions.

While physical processes create diverse habitat condition, biological processes operating at several spatial scales enable salmonids to persist in a variable habitat (Ebersole et al. 1996; Poff and Ward 1997). Behavioral and physiological flexibility permit survival and reproduction under changing conditions (Poff and Ward 1990). Genetic diversity enables populations to adapt evolutionarily to long-term cyclic or directional environmental change. Individuals within populations could have somewhat different life histories (e.g., Holtby and Healey 1986), allowing them to survive and reproduce under somewhat different ranges of environmental conditions (McElhany et al. 2000). Thus, life history diversity within populations can enable populations to persist when faced with short-term environmental change (McElhany et al. 2000).

Sustaining Salmonid Populations

At a larger spatial scale, theory and some empirical evidence (e.g., Hilborn et al. 2003) indicate that the sustainability of population complexes of salmonid fishes or metapopulations depends on asynchronous fluctuations in abundance of populations and movement of individuals among population habitats. At any particular time, environmental conditions may favor some of the populations in the complex, fostering increases in their abundance, whereas other populations may not be well adapted to those same conditions, and their abundances may decrease, and they may not contribute significantly to productivity at that time. But at some time in the future in a dynamic landscape, habitats that currently may be favorable for population persistence could become unfavorable, and the populations in these habitats could decline while the productivity of those habitats that were of low quality and consequently the populations they support may increase, thus stabilizing abundance of the complex (Reeves et al. 1995; Hilborn et al. 2003). Thus, asynchronous population fluctuations contribute to resilience (Holling and Meffe 1996) of regional population groups.

Although it may occur at relatively low rates, for salmonids the process of straying can result in recolonization of suitable habitats where local population extinction has occurred, sustain weak salmon populations that otherwise would have suffered extinction (Cooper and Mangel 1999), and provide for genetic exchange. Connectivity among favorable habitats is essential not only for recolonization, but also to enable migration to and from spawning areas, movement of juveniles from areas where they hatched to rearing areas, and movement to refugia to escape unfavorable environmental conditions.

Core populations also can stabilize metapopulations. These populations occupy large, productive habitats and have relatively low probabilities of extinction. Core populations can serve as centers of abundance during periods of time when environmental conditions are unfavorable and the abundances of others populations within the complex are low. They also can stabilize core-satellite population complexes by serving as a relatively constant source of individuals that can refound and sustain populations. Protection of core-like populations and their habitats should be of the highest priority (ISG 2000; McElhany et al. 2000). For example, the Hanford stock of fall Chinook and the free-flowing river habitat where they spawn should receive the utmost protection possible (Williams 2001). Another critical implication of metapopulation thinking is that some habitats, although suitable, may be currently unoccupied, or the populations in these habitats may not be very abundant due to high mortality incurred, for example, during migration to the spawning area or from an intense ocean fishery. Protection should be extended to these habitats as well (McElhany et al. 2000).

Hilborn et al. (2003) warn, however, that concentrating conservation efforts on the most productive populations and ignoring the less productive ones may prove to be successful only in the short term, that is, while conditions favoring the currently productive populations last. For core-satellite metapopulations, over the long term, areas currently supporting core populations may become unfavorable while those supporting satellites or sinks may become favorable and thus, core populations and satellite or sink populations may exchange roles (Reeves et al. 1995; McElhany et al. 2000). What we recognize today as core populations may have, in fact, historically served as satellites, as could be the case for many strongholds, and so their function as a core may be overestimated. Few, however, will question the importance of protecting remaining productive populations even if they are not currently, or did not in the past, function as core populations.

The question of synchrony poses somewhat of a conundrum for salmonid restoration. Populations that are in close proximity are more likely to fluctuate synchronously because they are more likely to be affected by common environmental factors (e.g., land use, local catastrophic events) than populations that are far apart (Issak et al. 2003). However, because populations are in proximity, straying and thus the probability of refounding or rescuing populations may be more likely. On the other hand, the beneficial effects of asynchrony for populations that are distant could be offset by reduced levels of straying. Persistent metapopulations may require both populations that are close together and those that are far apart (McElhany et al. 2000; Isaak et al. 2003), coupled with special efforts to reduce human impacts that may result in synchronous population change, such as limiting the spatial extent on individual land use actions (see Hilborn et al. 2003).

The Next Steps

Significant progress has occurred over the last decade in recognizing the dynamic role of environmental variation and diversity within and among salmonid populations that produced the historically abundant and persistent salmon and steelhead runs of the Columbia River Basin. While incorporation of these concepts (diversity, variation, and natural processes) into fishery management programs has also occurred, it is doing so at a much slower pace—and of late has been threatened by federal policies that appear to be based more on political agendas (i.e., predetermined outcomes) than on the best scientific information (see Epilogue in Chapter 12). The salmon's best chance at long-term survival in the Columbia River Basin and the Pacific Northwest is for management of the fisheries, the watershed, and the hydrosystem to integrate the ecological and evolutionary attributes and needs of salmonids within the operational management of the developed watershed. Pacific Northwest salmon and steelhead evolved and adapted to the northwest river and ocean systems over many thousands of years. Our greatest challenge is to identify how to manage these large systems in a more natural manner that will positively benefit salmon and steelhead populations while maintaining important regional economic and cultural benefits (electrical production, flood control, transportation, etc.).

Literature Cited

Adkinson, M. D. 1995. Population differentiation in Pacific salmon: Local adaptation, genetic drift, or the environment. *Canadian Journal of Fisheries and Aquatic Sciences* 52:2762–2777.

Amoros, C., A. L. Roux, J. L. Reygrobellet, J. P. Bravard, and G. Pautou. 1987. A method for applied ecological studies of fluvial hydrosystems. *Regulated Rivers* 1:17–36.

Beacham, T. D., and C. B. Murray. 1987. Adaptive variation in body size, age, morphology, egg size, and developmental biology in chum salmon (*Oncorhychus keta*) in British Columbia. *Canadian Journal of Fisheries and Aquatic Sciences* 44:244–261.

Beechie, T. and S. Bolton. 1999. An approach to restoring salmonid habitat-forming processes in Pacific Northwest watersheds. *Fisheries* 4: 6–15.

Bisson, P. A. 1995. Ecosystem and habitat conservation: More than just a problem of geography. *Evolution and the aquatic ecosystem: Defining unique units in population conservation* 17:329–333.

Bisson, P. A., G. H. Reeves, R. E. Bilby, and R. J. Naiman. 1997. Watershed management and Pacific salmon: Desired future conditions. Pages 447–474 in D. J. Stouder, P. A. Bisson, and R. J. Naiman, eds. *Pacific salmon and their ecosystems: Status and future options.* Chapman and Hall, New York.

Blumm, M. C., L. J. Lucas, D. B. Miller, D. J. Rohlf, and G. H. Spain. 1998. Saving Snake River water and salmon simultaneously: The biological, economic, and legal case for breaching the Lower Snake River dams, lowering John Day Reservoir, and restoring natural river flows. *Environmental Law* 28:101–153.

Bottom, D. L. 1997. To till the water: A history of ideas in fisheries conservation. Pages 569–597 in R. J. Naiman and D. Stouder, eds. *Pacific salmon and their ecosystems: Status and future options.* Chapman and Hall, New York.

Chapman, D., C. Pevan, T. Hillman, A. Giorgi, and F. Utter. 1994. Status of summer steelhead in the mid-Columbia River. Don Chapman Consultants Inc., Boise, Idaho, 235 p.

Confederated Tribes of the Umatilla Indian Reservation (CTUIR), and Oregon Department of Fish and Wildlife (ODFW). 1990. Umatilla River subbasin salmon and steelhead production plan. Northwest Power Planning Council, Portland, Oregon.

Confederated Tribes Yakima Indian Nation (CTYIN), Washington Department of Fisheries (WDF), and Washington Department of Wildlife (WDW). 1990. Yakima River subbasin salmon and steelhead production plan. Northwest Power Planning Council, Portland, Oregon.

Cooper, A. B. and M. Mangel. 1998. The dangers of ignoring metapopulation structure for the conservation of salmon. *Fishery Bulletin* 97:213–226.

Copp, G. H. 1989. The habitat diversity and fish reproductive function of floodplain ecosystems. *Environmental Biology and Fisheries* 26:1–27.

Cortner, H. J. and M. A. Moote. 1999. *The politics of ecosystem management.* Island Press, Washington DC.

Dauble, D. D., T. P. Hanrahan, D. R. Geist, and M. J. Parsley. 2003. Impacts of the Columbia River hydroelectric system on main-stem habitats of fall Chinook salmon. *North American Journal of Fisheries Management* 23:641–659.

Dauble, D. D., and D. G. Watson. 1997. Status of fall Chinook populations in the Mid-Columbia River, 1948–1992. *North American Journal of Fisheries Management* 17:283–300.

Den Boer, P. J. 1968. Spreading of risk and stabilization of animal numbers. *Acta Biotheoretica* 18:165–194.

Dunning, J. B., B. J. Danielson, and H. R. Pulliam. 1992. Ecological processes that affect populations in complex landscapes. *Oikos* 65:169–175.

Ebersole, J. L., W. J. Liss, and C. F. Frissell. 1997. Restoration of stream habitat in the western United States: Restoration as reexpression of habitat capacity. *Environmental Management* 21:1–14.

Ebersole, J. L., W. J. Liss, and C. A. Frissell. 2003a. Coldwater patches in warm streams: Physicochemical characteristics and the influence of shading. *Journal of the American Water resources Association* 39:355–368.

Ebersole, J. L., W. J. Liss, and C. A. Frissell. 2003b. Thermal heterogeneity, stream channel morphology, and salmonid abundance in northeastern Oregon streams. *Canadian Journal of Fisheries and Aquatic Sciences* 60:1266–1280.

Fausch, K. D., C. E. Torgersen, C. V. Baxter, and H. W. Li. 1997. Landscapes to riverscapes: Bridging the gap between research and conservation of stream fishes. *Bioscience*: 52:1–15.

French, R. R., and R. J. Wahle. 1959. Biology of Chinook and blueback salmon and steelhead in the Wenatchee River system. US Fish and Wildlife Service, Special Scientific Report, Fisheries No. 304.

Frissell, C. A., W. J. Liss, C. E. Warren, and M. D. Hurley. 1986. A hierarchical framework for stream habitat classification: Viewing streams in a watershed context. *Environmental Management* 10:199–214.

Frissell, C. A. and R. K. Nawa. 1992. Incidence and causes of physical failure of artificial habitat structures in streams of western Oregon and Washington. *North American Journal of Fisheries Management* 12:182–197.

Fulton, L. A. 1968. Spawning areas and abundance of Chinook salmon (*Oncorhynchus tshawytscha*) in the Columbia River Basin—Past and present. US Fish and Wildlife Service, Special Scientific Report, Fisheries No. 571.

Garcia, A. P., W. P. Connor, and R. H. Taylor. 1994. Fall Chinook salmon spawning ground surveys in the Snake River. Annual Report for 1993. Pages 1–21 in D. W. Rondorf and K. F. Tiffan, eds. *Identification of the spawning, rearing, and migratory requirements of fall Chinook salmon in the Columbia River Basin.* Bonneville Power Administration, Portland, Oregon.

Geist, D. R., and D. D. Dauble. 1998. Redd site selection and spawning habitat use by fall Chinook salmon: The importance of geomorphic features in large rivers. *Environmental Management* 22:655–669.

Gibert, J., J. A. Stanford, M.-J. Dole-Oliver, and J. V. Ward. 1994. Basic attributes of groundwater ecosystems and prospects for research. Pages 7–40 in J. Gibert, D. L. Danielopol, and J. A. Stanford, eds. *Groundwater ecology*. Academic Press, San Diego, California.

Gresswell, R. E., W. J. Liss, and G. L. Larson. 1994. Life history organization of Yellowstone cutthroat trout (*Oncorhychus clarki bouvieri*) in Yellowstone Lake. *Canadian Journal of Fisheries and Aquatic Sciences* 51:298–309.

Groot, C., and L. Margolis, eds. 1991. *Pacific salmon life histories*. University of British Columbia, Vancouver, Canada.

Hall, C. A. S., J. A. Stanford, and F. R. Hauer. 1992. The distribution and abundance of organisms as a consequence of energy balances along multiple environmental gradients. *Oikos* 65:377–390.

Halupa et al. 2003.

Hanski, I. 1991. Single-species metapopulation dynamics: Concepts, models, and observations. *Biological Journal of the Linnean Society* 42:17–38.

Hanski, I. A., and M. E. Gilpin. 1997. *Metapopulation biology: Ecology, genetics, and evolution*. Academic Press, New York.

Harrison, S. 1991. Local extinction in a metapopulation context: An empirical evaluation. *Biological Journal of the Linnean Society* 42:73–88.

Harrison, S. 1994. Metapopulations and conservation. Pages 111–128 in P. J. Edwards, R. M. May, and N. R. Webb, eds. *Large-scale Ecology and Conservation Biology*. Blackwell Scientific Publications, Oxford, UK.

Harrison, S., and J. F. Quinn. 1989. Correlated environments and the persistence of metapopulations. *Oikos* 56:293–298.

Healey, M. C. 1991. Life history of Chinook salmon. Pages 313–393 in C. Groot and L. Margolis, eds. *Pacific salmon life histories*. University of British Columbia Press, Vancouver, BC, Canada.

Healey, M. C., and A. Prince. 1995. Scales of variation in life history tactics of Pacific salmon and the conservation of phenotype and genotype. *Evolution and the aquatic ecosystem: Defining unique units in population conservation* 17:176–184.

Hilborn, R., T. P. Quinn, D. E. Schindler, and D. E. Rogers. 2003. Biocomplexity and fisheries sustainability. *Proceedings of the National Academy of Sciences* 100:6564–6568.

Holling, C. S. and G. K. Meffe. 1996. Command and control and the pathology of natural resource management. *Conservation Biology* 10:328–337.

Holtby, L. B. and M. C. Healey. 1986. Selection for adult size in female coho salmon (*Oncorhychus kisutch*). *Canadian Journal of Fisheries and Aquatic Sciences* 43:1946–1959.

Howell, P., K. Jones, D. Scarnecchi, L. LaVoy, W. Kendra, and D. Ortmann. 1985. Stock assessment of Columbia River anadromous salmonids. Volume I: Chinook, coho, chum and sockeye salmon stock summaries. Bonneville Power Administration and Oregon Dept. of Fish and Wildlife. Final Report.

Independent Scientific Group (ISG). 1999. Scientific issues in the restoration of salmonid fishes in the Columbia River. *Fisheries* 24:10–19.

Independent Scientific Advisory Board (ISAB). 2003. A review of strategies for recovering tributary habitat. Northwest Power and Conservation Council ISAB 2003-2.

Isaak, D. J., R. F. Thurow, B. E. Reiman, and J. B. Dunham. 2003. Temporal variation in synchrony among Chinook salmon (*Oncorhynchus tshawytscha*) redd counts from a wilderness area in central Idaho. *Canadian Journal of Fisheries and Aquatic Sciences* 60:840–848.

Kauffman, J. B., R. L. Beschta, N. Otting, and D. Lytjen. 1997. An ecological perspective of riparian and stream restoration in the western United States. *Fisheries* 22:12–24.

Kiefer, R. B., and J. N. Lockhart. 1995. Intensive evaluation and monitoring of Chinook salmon and steelhead trout production, Crooked River and Upper Salmon River sites. Bonneville Power Administration. DOE/BP-21182-5. Portland, Oregon.

Lavier, D. C. 1976. Major dams on Columbia River and tributaries. Washington Department of Game. Investigatve Reports of Columbia River Fisheries Project, Olympia, Washington.

Lawson, P. W. 1993. Cycles in ocean productivity, trends in habitat quality, and the restoration of salmon runs in Oregon. *Fisheries* 18:6–10.

Li, H. W., K. Currens, D. Bottom, S. Clarke, J. Dambacher, C. Frissell, P. Harris, R. M. Hughes, D. McCullough, A. McGie, K. Moore, R. Nawa, and S. Thiele. 1995a. Safe havens: Refuges and evolutionarily significant units. *Evolution and the aquatic ecosystem: Defining unique units in population conservation* 17:371–380.

Li, H. W., G. A. Lamberti, T. N. Pearsons, C. K. Tait, and J. L. Li. 1995b. Cumulative impact of riparian disturbances in small streams of the John Day Basin, Oregon. *Transactions of the American Fisheries Society* 123:627–640.

Lichatowich, J. 1999. *Salmon without rivers: A history of the Pacific salmon crisis.* Island Press, Washington, DC.

Lichatowich, J. A., and L. E. Mobrand. 1995. Analysis of Chinook salmon in the Columbia River from an ecosystem perspective. Mobrand Biometrics. Research Report.

Lichatowich, J., L. Mobrand, L. Lestelle, and T. Vogel. 1995. An approach to the diagnosis and treatment of depleted Pacific salmon populations in Pacific Northwest watersheds. *Fisheries* 20:10–18.

Lindsey, C. C., and J. D. McPhail. 1986. Zoogeography of fishes of the Yukon and Mackenzie basins. Pages 639–674 in C. H. Hocutt and E. O. Wiley, eds. *Zoogeography of North American Freshwater Fishes.* John Wiley and Sons, New York.

Lowe-McConnel, R. H. 1987. *Ecological Studies in Tropical Fish Communities.* Cambridge University Press, Cambridge, UK.

Mantua, N. J., S. R. Hare, Y. Zhang, J. M. Wallace, and R. C. Francis. 1997. A Pacific inter-decadal climate oscillation with impacts on salmon production. *Bulletin of the American Meteorological Society* 78:1069–1079.

McCullough, D. R. 1996. Introduction. Pages 1–10 in D. R. McCullough, ed. *Metapopulations and Wildlife Conservation.* Island Press, Washington.

McElhany, P., M. H. Ruckelshaus, M. J. Ford, T. C. Wainwright, and E. P. Bjorkstedt. 2000. Viable salmonid populations and the recovery of evolutionarily significant units. US Department of Commerce, NOAA Tech. Memo. NMFS-NWFSC-42. 156 p.

Miller, A. J., D. R. Cayan, T. P. Barnett, N. E. Graham, and J. M. Oberhuber. 1994. The 1976-77 climate shift of the Pacific Ocean. *Oceanography* 7:21–26.

Miller, D. B. 1997. Of dams and salmon in the Columbia/Snake Basin: Did you ever have to make up your mind? *Rivers: Studies in the Science, Environmental Policy, and Law of Instream Flow* 6:69–79.

Mullan, J. W., A. Rockhold, and C. R. Chrisman. 1992. Life histories and precocity of Chinook salmon in the Mid-Columbia River. *Progressive Fish-Culturist* 54:25–28.

National Research Council. 1996. Upstream: Salmon and society in the Pacific Northwest. Report on the Committee on Protection and Management of Pacific Northwest Anadromous Salmonids for the National Research Council of the National Academy of Sciences. National Academy Press, Washington DC.

Nemeth, D. J., and R. B. Kiefer. 1999. Snake River spring and summer Chinook salmon—The choice for recovery. *Fisheries* 24:16–23.

Nickelson, T. 2003. The influence of hatchery coho salmon (*Oncorhynchus kisutch*) on the productivity of wild coho salmon populations in Oregon coastal basins. *Canadian Journal of Fisheries and Aquatic Sciences* 60:1050–1056.

Nickelson, T. E., M. F. Solazzi, and S. L. Johnson. 1986. Use of hatchery coho salmon (*Oncorhynchus kisutch*) presmolts to rebuild wild populations in Oregon coastal streams. *Canadian Journal of Fisheries and Aquatic Sciences* 43:2443–2449.

Northwest Power Planning Council (NPPC). 2000. Columbia River Basin Fish and Wildlife Program: A Multi-species Approach for Decision Making. Northwest Power Planning Council. 2000-19. Portland, Oregon.

Oregon Department of Fish and Wildlife, Confederated Tribes of the Umatilla Indian Reservation, Nez Perce Tribe, Washington Department of Fisheries, and Washington Department of Wildlife. 1990. Grande Ronde River subbasin, salmon and steelhead production plan. Northwest Power Planning Council. Portland, Oregon.

Oregon Department of Fish and Wildlife (ODFW), and Confederated Tribes of the Warm Springs Reservation (CTWSR). 1990. Deschutes River subbasin: Salmon and steelhead production plan. Northwest Power Planning Council. Portland, Oregon.

Oregon Department of Fish and Wildlife (ODFW), Confederated Tribes of the Umatilla Indian Reservation (CTUIR), and Confederated Tribes of the Warm Springs Reservation (CTWSR). 1990. John Day River Subbasin: Salmon and steelhead production plan. Northwest Power Planning Council. Portland, Oregon.

Petts, G. E., H. Moller, and A. L. Roux, eds. 1989. *Historical change of large alluvial rivers: Western Europe.* John Wiley & Sons, Chichester, UK.

Poff, N. L., J. D. Allan, M. B. Bain, J. R. Karr, K. L. Prestegaard, B.D. Richter, R. E. Sparks, and J. C. Stromber. 1997. The natural flow regime. *Bioscience* 47:769–784.

Poff, N. L. and J. V. Ward. 1990. Physical habitat template of lotic systems: Recovery in the context of historical pattern of spatial heterogeneity. *Environmental Management* 14:629–645.

Policansky, D., and J. J. Magnuson. 1998. Genetics, metapopulations, and ecosystem management of fisheries. *Ecological Applications* 8(1) Supplemental: S119–S123.

Quigley, T. M. and S. J. Arbelbide. 1997. An assessment of ecosystem components in the Interior Columbia basin and portions of the Klamath and Great Basins. General Technical Report PNW-GTR-405. Portland, Oregon. US Department of Agriculture, Forest Service, Pacific Northwest Research Station.

Quinn, T. P. 1993. A review of homing and straying of wild and hatchery-produced salmon. *Fisheries Research* 18:29–44.

Quinn, T. P. 1999. Revisiting the stock concept in Pacific salmon: Insights from Alaska and New Zealand. *Northwest Science* 73:312–324.

Ratliff, D. E. 1981. *Ceratomyxa shasta*: Epizootiology in Chinook salmon of central Oregon. *Transactions of the American Fisheries Society* 110:507–513.

Reeves, G. H., L. E. Benda, K. M. Burnett, P. A. Bisson, and J. R. Sedell. A disturbance-based approach to maintaining and restoring freshwater habitats of evolutionarily significant units of anadromous salmon in the Pacific Northwest. Pages 334–339 in J. L. Nielsen, ed. *Evolution and the Aquatic Ecosystem.* American Fisheries Society Symposium 17, Bethesda, Maryland.

Regier, H. A., R. L. Welcomme, R. J. Stedman, and H. F. Henderson. 1989. Rehabilitation of degraded river ecosystems. Pages 86–89 in D. P. Dodge, ed. *Proceedings of the International Large River Symposium.* Canadian Special Publications in Fisheries and Aquatic Sciences.

Reimers, P. E. 1973. The length of residence of juvenile fall Chinook salmon in Sixes River, Oregon. *Research Reports of the Fish Commission of Oregon* 4:2. Portland, Oregon.

Reice, S. R. Nonequilibrium determinants of biological community structure. *American Scientist* 82:424–435.

Rich, W. H. 1920. Early history and seaward migration of Chinook salmon in the Columbia and Sacramento Rivers. *Bulletin of the US Bureau of Fisheries, Washington DC* 37.

Rich, W. H. 1939. Fishery problems raised by the development of water resources. Pages 176–181. *Dams and the Problems of Migratory Fishes.* Fish Commission of the State of Oregon, Stanford University.

Ricker, W. E. 1972. Hereditary and environmental factors affecting certain salmonid populations. Pages 19–160 in R. C. Simon and P. A. Larkin, eds. *The stock concept in Pacific salmon.* University of British Columbia, Vancouver, Canada.

Rieman, B. E., and J. B. Dunham. 2000. Metapopulation and salmonids: A synthesis of life history patterns and empirical observations. *Ecology of Freshwater Fish* 9:51–64.

Rieman, B. E., and J. D. McIntyre. 1993. Demographic and habitat requirements for conservation of bull trout. US Forest Service, Intermountain Research Station. General Technical Report, INT-308. Ogden, Utah. 38 p.

Schlosser, I. J. 1991. Stream fish ecology: A landscape perspective. *BioScience* 41:704–712.

Schlosser, I. J., and P. L. Angermeier. 1995. Spatial variation in demographic processes of lotic fishes: Conceptual models, empirical evidence, and implications for conservation. *American Fisheries Society Symposium* 17:392–401.

Sparks, R. E., P. B. Bayley, S. L. Kohler, and L. L. Osborne. 1990. Disturbance and recovery of large floodplain rivers. *Environmental Management* 14:699–709.

Stanford, J. A. 1998. Rivers in the landscape: Introduction to the special issue on riparian and groundwater ecology. *Freshwater Biology* 40:402–406.

Stanford, J. A., and J. V. Ward. 1993. An ecosystem perspective of alluvial rivers: Connectivity and the hyporheic corridor. *Journal of the North American Benthological Society* 12:48–60.

Stanford, J. A., J. V. Ward, W. J. Liss, C. A. Frissell, R. N. Williams, J. A. Lichatowich, and C. C. Coutant. 1996. A general protocol for restoration of regulated rivers. *Regulated Rivers* 12:391–413.

Taylor, E. B. 1990. Environmental correlates of life-history variation in juvenile Chinook salmon, *Oncorhynchus tshawytscha* (Walbaum). *Journal of Fish Biology* 37:1–17.

Taylor, E. B. 1991. A review of local adaptation in Salmonidae, with particular reference to Pacific and Atlantic salmon. *Aquaculture* 98:185–207.

Thompson, W. F. 1959. An approach to population dynamics of the Pacific red salmon. *Transactions of the American Fisheries Society* 88:206–209.

UCWSRI (Upper Columbia White Sturgeon Recovery Initiative). 2002. Upper Columbia White Sturgeon Recovery Plan. Nelson, BC: British Columbia Ministry of Water, Land, and Air Protection.

Vannote, R. L., G. W. Minshall, K. W. Cummins, J. R. Sedell, and C. E. Cushing. 1980. The river continuum concept. *Canadian Journal of Fisheries and Aquatic Sciences* 37:130–137.

Ward, J. V. 1989. The four-dimensional nature of lotic ecosystems. *Journal of the North American Benthological Society* 8:2–8.

Ward, J. V., and J. A. Stanford. 1995a. Ecological connectivity in alluvial river ecosystems and its disruption by flow regulation. *Regulated Rivers: Research and Management* 11:105–119.

Ward, J. V., and J. A. Stanford. 1995b. The serial discontinuity concept: Extending the model to floodplain rivers. *Regulated Rivers: Research and Management* 10:159–168.

Washington Department of Fisheries (WDF), Confederated Tribes of the Umatilla Indian Reservation (CTUIR), Nez Perce Tribe (NPT) of Idaho, and Washington Department of Wildlife (WDW). 1990. Tucannon River subbasin salmon and steelhead production plan. Northwest Power Planning Council. Portland, Oregon.

Watson, B. 1992. Using "Patient/Template Analysis" in the design of projects to increase natural production of anadromous salmonidss. *Salmon management in the 21st century: Recovering stocks in decline. Proceedings of the Coho / Chinook Workshop.* Northeast Pacific Chinook and Coho Workshop, Boise, Idaho.

Welcomme, R. L. 1979. Fisheries Ecology of Floodplain Rivers. Longman, London.

Welcomme, R. L. 1995. Relationships between fisheries and the integrity of river systems. *Regulated Rivers* 11:121–136.

Whitney, R. R., L. D. Calvin, J. Michael W. Erho, and C. C. Coutant. 1997. Downstream passage for salmon at hydroelectric projects in the Columbia River Basin: Development, installation, and evaluation. Northwest Power Planning Council. Technical Report, 97–15. Portland, Oregon. 101 p.

Williams, R. N. 2001. Refugia-based conservation strategies: Providing safe havens in managed river systems. Pages 59–63 in Oregon Trout, ed. *Oregon Salmon: Essays on the state of the fish at the turn of the Millennium*. Oregon Trout, Portland, Oregon.

Wissmar, R. C., J. E. Smith, B. A. McIntosh, H. W. Li, G. H. Reeves, and J. R. Sedell. 1994. A history of resource use and disturbance in riverine basins of eastern Oregon and Washington (early 1800s–1990s). *Northwest Science* 68:1–35.

Wood, M. C. 1998. Reclaiming the natural rivers: The Endangered Species Act as applied to endangered river ecosystems. *Arizona Law Review* 40:197–286.

Zwick, P. 1992. Stream habitat fragmentation—A threat to biodiversity. *Biodiversity and Conservation* 1:80–97.

Return
to the
River

4

Diversity, Structure, and Status of Salmon Populations

Richard N. Williams, Ph.D., James A. Lichatowich, M.S., Madison A. Powell, Ph.D.

Origin, Diversity, and Structure of Pacific Salmon and Steelhead Populations
   Origins
   The Stock Concept
      Stock Concept in Fisheries
      Stock Concept in Pacific Salmon
      Salmonid Life Histories and Habitat
      Stock Conservation
   Population Structure of Anadromous Salmonid Populations
      Salmonid Metapopulations
      Human Impacts
      Regional Stochasticity
      Geographic Organization of Chinook Salmon
   Genetic Structure of Anadromous Salmonid Populations
      General Patterns
      Genetic Structure of Columbia Basin Chinook Salmon
      Genetic Structure of Columbia Basin Steelhead
      Conclusions
   Status of Pacific Salmon Species
      Chinook Salmon
      Coho Salmon
      Chum Salmon
      Sockeye Salmon
      Pink Salmon
   Status of Columbia River Trout and Char
      Overview
      Rainbow and Steelhead Trout
      Cutthroat Trout
      Coastal Cutthroat Trout
      Westslope Cutthroat Trout
      Bull Charr

Status of Indigenous Species other than Salmonids
  Sturgeon
  Pacific Lamprey
  Conclusions and Implications
Literature Cited

> *"There is grandeur in this view of life, with its several powers, having been originally breathed into a few forms or into one; . . . from so simple a beginning endless forms most beautiful and most wonderful have been, and are being, evolved."*
> —Charles Darwin. 1859. *The Origin of Species.* Penguin Books, Baltimore.
> *"Nothing in Biology makes sense except in the light of evolution."*
> —Theodosius Dobzhansky. 1973. Nothing in biology makes sense except in the light of evolution. *American Biology Teacher* 35:125–129.
> *"Evolution was not a change in gene frequencies, but the twin processes of adaptive change and the origin of diversity."*
> —Ernst Mayr. 1991. *One Long Argument: Charles Darwin and the Genesis of Modern Evolutionary Thought.* Harvard University Press, Cambridge, Massachusetts.

Origin, Diversity, and Structure of Pacific Salmon and Steelhead Populations

Salmonids are well recognized for exhibiting diverse life history strategies, ecological adaptations, and genetic characteristics (Groot and Margolis 1991; Taylor 1991; Behnke 2002). These attributes are thought to be linked to productivity, long-term fitness, and persistence. This chapter is organized into two major sections. In the first section, we describe the origin, diversity, genetic structure, and population structure of Pacific salmon and steelhead trout in the Columbia River Basin, emphasizing genetic and life history diversity observed within and among populations and species. In the second half of the chapter, we review the current status of salmonids, sturgeon, and lamprey native to the Columbia River.

Origins

The evolutionary tale of salmonids is still most often told through fossilized bones that have been left behind. However, details of their evolutionary history are being teased out of the immense tangle of some 3 billion nucleotides that make up their genome with ever-increasing frequency as our development of molecular genetic techniques marches on. We now know early, ray-finned fishes gave rise to the most prolific group among jawed fishes—the teleosts (Strickberger 1989). Today, teleosts comprise some 35,000 species, more than any other vertebrate taxon. If one were to sample 20 fish species randomly, 19 of those sampled would be teleosts. Within that diverse group resides the family Salmonidae. *Eosalmo driftwoodensis* is the earliest known fossil salmonid dating back to the Eocene epoch. We can infer an unbroken lineage between *Eosalmo* and its contemporaries, some 50 million years ago, to the salmonids present in the Columbia River today. Figure 4.1 shows the generalized evolutionary relationships between salmonids in the genus *Oncorhynchus*.

The Stock Concept

From the time of Plato until the 19th century, western scientists viewed species as fixed types, based on an idealized set of characters that described each species (i.e., the Essentialists' view) (Mayr 1970). Individuals that varied from this fixed type or ideal were viewed as errors attributable to developmental processes. Thus, biological diversity within a species had little positive meaning.

The transition from the concept of species as a fixed type, to species comprising many populations, each containing individuals that vary slightly from each other, was a major advancement in biology. It was this shift that gave Charles Darwin the point of view he needed to see the struggle for existence taking place between individuals and not species. Population thinking paved the way for Darwin's work on natural selection and the revolution of biological sciences that followed (Mayr 1982).

Diversity is an inherent attribute of salmonids in naturally functioning ecosystems (Groot and Margolis 1991; Taylor 1991; Behnke 1992; Soverel 1996). Salmonid diversity is expressed as population, life history, and genetic

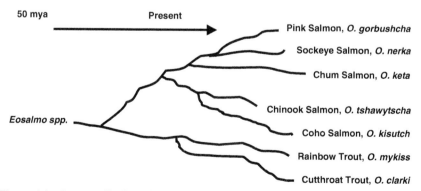

Pacific Salmon Radiation, Genus *Onchorhnynchus*

50 mya   Present

Pink Salmon, *O. gorbushcha*

Sockeye Salmon, *O. nerka*

Chum Salmon, *O. keta*

Chinook Salmon, *O. tshawytscha*

Eosalmo spp.

Coho Salmon, *O. kisutch*

Rainbow Trout, *O. mykiss*

Cutthroat Trout, *O. clarki*

Figure 4.1 A generalized evolutionary relationship between salmonids in the genus *Oncorhynchus* (Steelquist 1992; Behnke 2002).

diversity and results in part from the ability of salmonid fishes to adapt to a wide array of habitat conditions (Taylor 1991; Healey 1994). Since habitats vary in space (i.e., from location to location) and through time at each location, diversity is likely not constant, but changes as conditions in the environment change. Diversity probably contributes to the resilience and stability of regional groups of salmonid populations.

A population, or stock, can be defined as a self-sustaining breeding group within a species that is relatively reproductively isolated from other breeding groups (Ricker 1972). In population biology, this is also called a deme (Hedrick 2000). However, the term "population" has been used to define other kinds of aggregations of plants and animals. For example, fishery managers often define a stock as all the fishes of a species in a management area whose boundaries are set for administrative or regulatory purposes. Unfortunately these management stocks do not always coincide nor necessarily encompass breeding structure of "biological" populations within that stock. Administrative and biological definitions of stock often come into conflict in salmon management. The implications of that conflict are discussed later in this section. The generally accepted definition of a salmon stock comes from Ricker (1972):

> "fish spawning in a particular lake or stream (or portion of it) at a particular season, which fish to a substantial degree do not interbreed with any group spawning in a different place, or in the same place at a different season."

Stock Concept in Fisheries

Population thinking was recognized earlier and has undergone greater development in fisheries than in any other field of biology (Sinclair 1988). All species of fish do not have the same level of complexity in their population structure. A comparison of population richness among marine fish species from the north Atlantic (Table 4.1) placed Atlantic salmon at one end of the range (a large number of populations) and the European eel at the other end (a single population) (Sinclair and Iles 1989). Pacific salmon would fall on the left-hand side of Table 4.1 at a level similar to the Atlantic salmon (Ricker 1972). This can be attributed for the most part to an aspect of salmonid behavior, known as phylopatry or homing to a natal stream.

Stock Concept in Pacific Salmon

Not long after Pacific salmon came under commercial harvest, careful observers on the West Coast recognized that salmon from different rivers varied in important life history or morphometric characteristics. R. D. Hume (1893), who operated salmon canneries in California and Oregon and was an early proponent of the artificial propagation of salmon, observed:

Table 4.1 The Continuum of population richness in anadromous and marine species in the northern Atlantic (Sinclair and Iles 1989).

(Decreasing population richness $\rightarrow \rightarrow \rightarrow \rightarrow \rightarrow \rightarrow$)

Atlantic salmon
    Atlantic shad
        Atlantic herring
            Atlantic cod
                rainbow smelt
                    haddock
                        Atlantic mackerel
                            Atlantic menhaden
                                European eel

"The fact that in rivers which enter the sea within a few miles of each other, as well as the different tributaries of the same river, the fish (salmon) will have local characteristics which enable those who are familiar with the various streams to distinguish to which river or tributary they belong.

I firmly believe that like conditions must be had in order to bring about like results, and that to transplant salmon successfully they must be placed in rivers where the natural conditions are similar to that from which they have been taken."

After reviewing results of tagging experiments that supported the hypothesis that Pacific salmon homed to their natal stream, Rich (1938) concluded that the species of Pacific salmon were divided into local populations:

"In the conservation of any natural biological resource it may, I believe, be considered self-evident that the population must be the unit to be treated. By population I mean an effectively isolated, self perpetuating group of organisms of the same species. Given a species that is broken up into a number of such isolated groups or populations, it is obvious that the conservation of the species as a whole resolves into the conservation of every one of the component groups; that the success of efforts to conserve the species will depend, not only upon the results attained with any one population, but upon the fraction of the total number of individuals in the species contained within the populations affected by the conservation measures."

At least some fish culturists recognized the implications of stock structure as early as 1939 and realized the transfer of salmon between rivers was not a desirable management activity (Oregon Fish Commission 1933). Although conclusive proof was lacking, biologists working in the Columbia Basin began to recognize that salmon species were composed of populations adapted to their local habitat (Craig 1935). Management had to take each population into consideration if it was going to be successful.

"Knowing further that each race is self-propagating, it becomes perfectly apparent that all parts of the salmon run in the Columbia River must be given adequate protection if the run as a whole is to be maintained. The protection of only one or two portions of the run will not be sufficient, inasmuch as certain races will be left entirely unprotected." (Oregon Fish Commission 1931)

Elements of a conceptual foundation that recognized the importance of stocks and local adaptation emerged in the 1930s (Rich 1938). However, progress in this direction was truncated by the development of the hydroelectric potential of the basin and the plan devised to mitigate for that development, the Lower Columbia River Fishery Development Program (LCRFDP). Although the LCRFDP had six phases, the overall approach was to shift salmon and steelhead production to the lower river below the proposed McNary Dam. The desired level of production would be achieved by a combination of enhanced lower river stocks and transfer of upper river stocks to the lower river (Laythe 1948). The belief that such a transfer could be successful seems to contradict the understanding biologists had at the time regarding the importance and management implications of the stock structure of Pacific salmon. It should be noted that the Fraser River restoration program initiated a few years before the LCRFDP did emphasize the importance of individual stocks.

"Management of the Fraser River sockeye population by individual genetic races was developed and perfected by the commission (International Pacific Salmon Fisheries Commission). This management philosophy was an important component of the rehabilitation of the runs in combination with the contribution of fishways and commercial fishing closures." (Roos 1991)

The importance of stock structure in salmon management received renewed emphasis in the late 1960s and early 1970s (Calaprice 1969; Paulik 1969; Ricker 1972). The Endangered Species Act (ESA) focused attention on the stock structure of Pacific salmon in the Columbia River in the late 1980s up to the present. Recently, stocks of Pacific salmon have been inventoried and their status described (Howell et al. 1985; Nehlsen et al 1991; Washington Department of Fisheries et al. 1993; Busby et al. 1996; Huntington et al. 1996; Washington Department of Fish and Wildlife and Oregon Department of Fish and Wildlife 1994, 2001)]

Scientists often refer to local adaptation in salmon populations, although the evidence for it is circumstantial (Taylor 1991). Phylopatry may provide an adaptive evolutionary advantage to salmonids through a higher average survivorship of juveniles (through access to parental resources), access to other close relatives as mates and the resultant maintenance of locally adapted gene complexes (Shields 1987). Evolutionary theory states that differentiation or diversity, which may ultimately lead to speciation, arises in isolation by accumulation of selectively advantageous traits adapting in concert with a changing environmental template (Mayr 1966). Thus, local adaptation and population diversity may well serve as a focal point for continued speciation potential (Soule 1988). The term "local adaptation" can be misleading if, in the case of salmonids, adaptation is interpreted to occur only within a specific local environment, such as a spawning area in a tributary stream, rather than to all the habitats in which salmon complete their life cycle. Although salmon exist in populations that typically home to their natal stream and spawn in relative isolation from other salmon populations, they are adapted to the habitats (river, estuary, and ocean) where individuals in a population complete their life cycles, as well as to the variability that occurs in these habitats over both short- and long-term time scales. Such variation encompasses annual and decadal variations in climate, Pacific Decadal Oscillation (PDO) cycles, El Niño-Southern Oscillations (ENSOs), and geologic events (see a more detailed discussion of these marine phenomena in Chapter 10). Riddell (1993) emphasized the importance of not overlooking locally adapted salmonid populations in the long-term persistence of metapopulations. When climates change or oscillate, peripheral populations may become significantly adaptive and able to colonize or recolonize habitats. These populations may also contribute diversity back into a "core" of diversity during times of contraction (Scudder 1989). The main implication of local adaptation, and thus population differentiation, resides in management decisions regarding the scale at which populations are protected from anthropogenic hazards (Riddle 1993; National Research Council [NRC] 1996).

Salmonid Life Histories and Habitat

Adaptation to the locally varied habitat may be expressed through variation in life history traits, although not all variation in traits among populations is adaptive. A trait exhibited by a local population is adaptive if it has a genetic basis and if it enhances survival or reproductive success (Taylor 1991). Life histories are composed of demographic traits such as age at maturity, mortality schedules, size, and growth (Stearns 1995). These traits are affected by behavior both learned and fixed, which are collectively referred to as behavioral ecology and often dictate many aspects of the expression of life history variability (Gould 1982). Salmonid life history traits also include (a) the age and size that juveniles migrate within the river system (resident, riverine [fluvial]), into lakes (resident, adfluvial), or to the sea (anadromous); (b) growth and maturity during riverine and laucustran migrations; (c) spawning habitat preferences; (d) emigration patterns; and (e) age and timing of spawning migration. The expression of many of these traits varies in response to environmental variation as well. For successful completion of the life history, quality habitat must exist for each life stage, or mortality ultimately will exceed productivity and that life history type will be extinguished. In tributaries flowing through the shrub and shrub-steppe region of the Columbia

River Basin, the loss of summer-migrating, underyearling Chinook salmon due to habitat degradation may have been a major cause of decline in spring and summer Chinook salmon (Lichatowich and Mobrand 1995).

Salmon habitat simply may be thought of as seasonally important places where salmon carry out their life histories (Thompson 1959). The presence of these places is important, but so is the ability to move between them at appropriate times. Complex habitats with a high degree of spatial and temporal connectivity permit the development and expression of life history diversity, which is an essential component of salmonid productive capacity. In a life history context, salmon restoration implies reestablishment of life history diversity and all the complexity and stochasticity within a given set of environmental variables.

Stock Conservation

While the conservation of local populations or stocks of Pacific salmon and the preservation of their genetic resources is an important goal (Riggs 1990; Altukhov and Salmenkova 1991; Kapuscinski et al. 1991), achieving that goal is neither simple nor easy. Merely verifying that a local stock has different traits (size, time of spawning, time of juvenile migration, etc.) compared with other nearby stocks is not sufficient, since these traits could merely be a reflection of phenotypic variation within a "reaction norm"[1] and not indicative of genetic variation (Schlichting and Pigliucci 1998). Documenting that the observed differences between populations is adaptive requires the trait's genetic basis be documented. Variation in the trait must be related to measurable, significant differences in survival or reproductive success among individuals in a common environment, and the mechanism that maintains the trait in the population must be demonstrated (Taylor 1991). These are not easy criteria to meet.

One might assume, since the extinction of a stock could represent a substantial loss of genetic diversity, managers would give evidence of local adaptation, even circumstantial evidence, the benefit of the doubt when setting stock boundaries. This would be consistent with a precautionary, or risk-averse, management philosophy. However, the size of a stock's boundary can have critical impacts on management programs. Narrowly defined boundaries complicate or prohibit harvest management in marine and lower river areas where stocks are mixed, and they restrict the use of hatchery fish in outplanting programs.

---

[1]Populations are composed of multiple genotypes; each genotype can be expressed in a variety of different phenotypes via interactions between the genotype and environmental variation. A reaction norm measures the breadth of response (i.e., the number and variety of phenotypes) by a genotype to environmental variation.

The need to conserve biodiversity between and within locally adapted stocks of salmon and the conflict between that goal and traditional management programs has created two strongly held positions characterized by the terms "lumpers" and "splitters." Lumpers tend to see a few large stocks, whereas splitters tend to see a large number of small stocks. Driving this debate is the underlying question: How much weight should we give to management strategies, as opposed to biological criteria, when setting stock boundaries? Biologists that manage salmon harvest and hatchery programs often define stocks as aggregates of populations (Thompson 1965; Wright 1965). Traditional harvest and hatchery practices based on that approach have contributed to a homogenization of the genetic differences between stocks (Calaprice 1969; Nelson and Soule 1987; Reisenbichler and Phelps 1989), reduced the productivity, and threatened the existence of populations in smaller, less productive streams (Ricker 1958; Thompson 1965; Wright 1993).

To a large degree, the debate over the size of stock boundaries is driven by the search for the "ideal" stock designation. Managers are looking for stock boundaries that lead to the conservation of biodiversity and at the same time fit conveniently into existing harvest and hatchery management strategies. Waples (1991, 1995) defined criteria for listing salmonid populations under the U.S. ESA based upon a concept of an evolutionary significant unit (ESU). ESUs are generally broadly defined and may represent metapopulations, races, and even multiple races. Because such broad definitions are employed in listing decisions, some constituent populations within the ESU may still be imperiled and ultimately drop out (Knudsen 2002).

There is no ideal stock designation. Even the definition used by Ricker (1972) leads to different interpretations because there is so little hard information on reproductive isolation or genotypic or phenotypic descriptions of spawning aggregations of salmon, particularly in the smaller streams. In addition, the species of Pacific salmon are organized in a hierarchical structure (this chapter, below). The biological units in the hierarchy (species, metapopulation, population, or stock, subpopulation, individual) and their associated geographical units (region, subregion, river, tributary, and redd) persist for different time intervals. The objective for most management actions should be to select the most inclusive population/geographic unit for which a management action will not cause the loss of genetic diversity contained in less inclusive groups (Mundy et al. 1995). However, sustainability of economically and socially important fisheries along with the salmonid populations that support them will likely require an emphasis toward functioning watershed ecosystems and allowances for natural variation and diversity, the most inclusive of groupings (Frissell et al. 1997; Knudsen 2002; MacDonald et al. 2002).

The debate over the home stream theory has been settled for several decades, but the stock concept still stimulates debate. The biological implications of the stock concept to fisheries management are profound. Disregarding the smaller populations or managing them collectively, as we often do in our mixed stock salmon fisheries, can lead to disintegration of the stock system (Altukhov 1981). It is important to consider the fate of small subunits of a stock during management of routine harvest, hatcheries, river flows, and habitat protection. It is also critical that they be considered during years of crisis (Thompson 1965; Paulik 1969). For example, during periods of sustained drought, focusing management entirely on the larger stocks or stock aggregations will quickly drive the smaller subpopulations to extinction. However, the small populations that inhabit marginal habitats within the range of a metapopulation may be an important source of genetic diversity of the species (Mayr 1970; Scudder 1989). W. F. Thompson (1965) described the problem 40 years ago:

> "We regulate our fisheries. But we concentrate them on the best races and one by one these shrink or vanish and we do not even follow their fate because we have not learned to recognize their independent component groups or to separate them one from the other. We continue our unequal demands, knowing only that our total catches diminish, as one by one small populations disappear unnoticed from the greater mixtures which we fish."

Population Structure of Anadromous Salmonid Populations

Salmonid Metapopulations

The metapopulation concept is a relatively new approach to various issues in population biology (Hanski and Gilpin 1997) and has important management and conservation implications for Columbia River and Pacific Northwest salmonids (see also Chapter 3 for metapopulation discussion). Two key assumptions drive the metapopulation approach. The first is that populations are spatially structured into assemblages of local breeding populations (i.e., the metapopulation), while the second is that migration among the local populations has some effect on local dynamics, including the possibility of population reestablishment following extinction (Hanski 1991; Hanski and Gilpin 1991).

Metapopulation structure is likely in salmonids (Rieman and McIntyre 1993; Mundy et al 1995; Schlosser and Angermeier 1995; NRC 1996; Brannon et al 2002) because adult salmonids display high fidelity of homing to their natal streams (Helle 1981) while at the same time exhibiting relatively low, but variable, levels of straying or dispersal (Quinn and Fresh 1984). Nevertheless, it has proven difficult to demonstrate in studies of natural salmonid populations (Rieman and Dunham 2000). Studies show only weak genetic evidence of metapopulation structure, suggesting that while the metapopulation processes of extinction, dispersal, and recolonization may be important in the long-term maintenance and structuring of salmonid populations, they are likely to operate over evolutionary scales of time rather than over shorter ecological time frames.

Metapopulation structure is thought to be hierarchical in nature and, in Columbia River salmonids, may exist at various scales ranging from linkages among populations within a tributary of a watershed to aggregations at the subbasin or even ecological province level (Northwest Power Planning Council [NPPC] 1997, 1998, 2000). Linkages among populations are likely to be the strongest at the smallest geographic scale, because dispersal among local populations is more likely than dispersal between geographically distant populations (MacArthur and Wilson 1967; Diamond 1984; Rosenzweig 2002). The persistence of a metapopulation is determined by the balance of local population extinction and the reestablishment of extinct populations through recolonization. Dispersal from neighboring local populations functions in recolonization of habitats where local extinction has occurred.

Human Impacts

Extinctions of Chinook salmon populations have increased over the last 100 years (Williams et al. 1992; NRC 1996; Lichatowich 1999; Thurow et al 2000) and altered the organization of regional systems of populations in the Columbia Basin (Figures 3.4, 3.5, and 3.8 and associated discussion). Many remaining stocks occur at low levels and are vulnerable to stochastic extinction (Nehlsen et al. 1991; Huntington et al. 1996). Fragmentation and degradation of habitat can disrupt regional production and linkages among populations through extirpation of vital core populations and isolation of remaining populations (Rieman and McIntyre 1993; Harrison 1994; Schlosser and Angermeier 1995). As a result, regional production declines, and the vulnerability of individual populations to extinction increases (Rieman and McIntyre 1993; Harrison 1994; Li et al. 1995; Schlosser and Angermeier 1995).

While most fall Chinook populations spawning in the mainstem reaches of the Columbia and Snake Rivers have been driven extinct, one remaining viable mainstem population is the fall Chinook population spawning in the Hanford Reach (Becker 1985; Geist 1995). This population, averaging 40,000 to 50,000 returning adult fish since the mid-1960s, is the largest naturally spawning population of Chinook salmon above Bonneville Dam (see Figure 3.6). It has been stable during years when populations in other parts of the basin have undergone severe decline (Oregon Department of Fish and Wildlife and Washington Department of Fish and Wildlife 1995).

Regional Stochasticity

The probability of metapopulation extinction is enhanced if the dynamics of local populations and their individual probabilities of extinction become temporally correlated or synchronized (Harrison and Quinn 1989; Hanski 1991). Regional stochasticity refers to the correlated or synchronized dynamics of local populations resulting from the operation of common environmental factors (Hanski 1991). An important likely consequence of human development in watersheds is increased synchrony in the dynamics of naturally and artificially produced salmon.

Adjacent local populations are more likely to respond synchronously to environmental factors, whereas local populations that are more geographically distant are more likely to experience asynchronous dynamics (Figure 3.10; Harrison and Quinn 1989; Hanski 1991; Rieman and McIntyre 1993).

Redd counts in index areas in the Salmon River and its tributaries in central Idaho suggest that some local spring and summer Chinook populations have been experiencing synchronous decline since the late 1960s mid-1970s (Figure 3.9; Nemeth and Kiefer 1999). Redd counts for these populations appear to be acting synchronously, as compared to redd counts of Chinook salmon in the John Day River, a downstream tributary system to the middle Columbia River. Since the habitat where the Salmon River stocks spawn is of relatively high quality, and much of it within or adjacent to wilderness areas, the synchronizing influence is likely downstream from the spawning areas, either in lower mainstems of the Salmon River, in the mainstem Snake River or Columbia River, or in the ocean.

Human impacts may have shifted metapopulation structure from core-satellite to non-equilibrium metapopulations. In non-equilibrium metapopulations, extinction rates are consistently greater than recolonization rates and the metapopulations are undergoing regional decline (Harrison 1991). Once stabilizing core populations have been driven extinct, recolonization and reestablishment of extinct local populations is limited or does not occur, and only isolated satellite populations remain. Isolated populations have little chance of being refounded after a local extinction compared to a population that is close to other populations. As populations become isolated, local extinctions become permanent and the entire metapopulation moves incrementally toward extinction (Rieman and McIntyre 1993).

Geographic Organization of Chinook Salmon

The complex of spatially distributed local spawning populations associated with major subbasins and contiguous areas of the mainstem Columbia or Snake River probably formed metapopulations. Fall Chinook spawning in mainstem reaches of the Columbia and Snake and the lower reaches of major subbasins could have formed one type of metapopulation, while summer and spring Chinook spawning in the upper mainstems of major subbasins and spring Chinook spawning in headwater areas could have formed another type.

Both genetic and life history evidence distinguish spring Chinook from fall Chinook in the Columbia and Snake Basins. Additionally, genetic and tagging data show that Columbia River Chinook are well differentiated from Snake River Chinook, suggesting significant long-term reproductive isolation between the two groups (Utter et al 1989, 1995; Matthews and Waples 1991; Waples et al. 1991). In the Snake River, fall Chinook are differentiated from the spring and summer races with respect to life history characteristics, such as annual timing of adult migration, geographic distribution of spawning habitat, and genetic attributes (Matthews and Waples 1991; Waples et al. 1991). French and Wahle (1959) observed summer and spring Chinook on the spawning grounds of the Wenatchee and Methow Rivers, whereas Mullan et al. (1992b) reported mixing of summer and fall fish on the spawning areas of mid-Columbia River tributaries. In contrast to the Snake River, there has been a tendency to group Columbia River summer Chinook and fall Chinook because they both migrate downstream as subyearlings.

Most recent genetic evidence, which examined matrilineal lines (mitochondrial DNA) of both fall and spring Chinook in the Snake River and Columbia mainstem, suggests long-term separation of the two races (Brannon et al. 2002). The study also revealed a significant level of divergence among fall Chinook matrilines apart from the modest divergence in spring Chinook populations, suggesting fall Chinook or their progenitors have been around longer and spring/summer Chinook as a race may be a result of adaptations to exploit differing ecological conditions present in lower order streams and rivers (Brannon et al. 2002).

Historically, Chinook population sizes and probabilities of extinction most likely varied along a continuum determined in part by habitat size and quality. At one end of the continuum were the large core-type populations spawning in high-quality mainstem habitats. Other local populations likely had characteristics similar to satellite populations. Local Chinook populations most prone to extinction, and probably most variable in abundance, may have been those inhabiting smaller streams in arid terrain. In periods of drought, salmon populations inhabiting these streams may have had difficulty in persisting. Chinook populations intermediate in size and sensitivity to extinction may have occupied streams in regions with higher precipitation and streams draining mountainous terrain whose headwaters are in high-elevation areas. In these streams, both flows and temperatures may be more suitable for juvenile rearing.

Genetic Structure of Anadromous Salmonid Populations

"Sustainable increases in salmon and steelhead productivity in the Columbia River Basin can only be achieved if the genetic resources required for all forms of production, present and future, are maintained in perpetuity."

—Riggs, Larry A. 1990. Principles for Genetic Conservation and Production Quality. Northwest Power Planning Council, Portland, Oregon.

Anadromous salmonids occur widely throughout the northern hemisphere in river systems north of approximately 40°N latitude. Native species in the genus *Salmo* occur across the northern arc of the Atlantic Basin, while species in *Oncorhynchus* occur throughout the northern arc of the Pacific Basin. Separation of trouts from Atlantic salmonids has in the past been problematic, but morphological evidence and now genetic evidence support the inclusion of Pacific trouts (rainbow and cutthroat trout) into the genus *Oncorhynchus* rather than *Salmo* (Smith and Stearly 1989; Behnke 1992, 2002). Species in *Salvelinus* occur in both Atlantic and Pacific Basin river systems. As a group, the salmonid species exhibit a remarkable range of diversity in life history characteristics, ecological attributes, and molecular genetic variability (Groot and Margolis 1991; Taylor 1991; Quinn and Unwin 1993). Although the exact mechanisms and relationships are poorly understood, genetic diversity is recognized as a major contributor to productivity, fitness, and adaptability (Allendorf and Leary 1986; Quattro and Vrijenhoek 1989; Liskauskas and Ferguson 1991; Beatty 1992). Therefore, it is important to understand how genetic variation in salmonids is structured within each species and among its populations, in order to preserve existing genetic diversity and to ensure the persistence of evolutionarily derived aggregates of populations (Allendorf and Phelps 1981a; 1981b; Allendorf and Leary 1988; Allendorf and Waples 1996). The importance of local adaptation, and the microgeographic scale under which it may occur (e.g., Philipp and Clausen 1995), is only now receiving increasing attention, in spite of its recognition by early fisheries managers (Rich 1939; Schuck 1943). Increasing recognition is also occurring that a biologically and economically feasible way to increase salmonid production is to utilize the natural productive capacity of existing native stocks that are adapted to their local environments, rather than attempting to rely on hatchery-reared fish that may not be wholly adapted to specific local environments for production boosts.

Significant population genetic research, most of it relying on allelic variation at protein coding loci (i.e., allozymes), has occurred on salmonids in the last 20 years. These studies have described general patterns of genetic variation that are common to both anadromous and resident forms of salmonids. More recent direct analyses of mitochondrial and nuclear DNA, while frequently providing additional resolution beyond that provided by allozyme analysis, have largely revealed the same general principles of genetic structure within and among populations.

General Patterns

Due to the commercial value of Pacific salmonids and problems related to harvest, culture, and conservation (Utter 1991), considerable effort has been

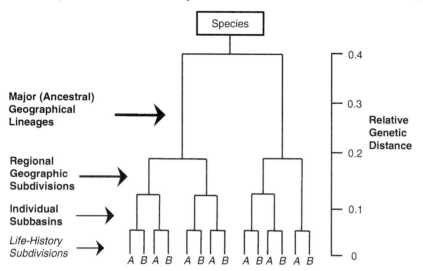

Figure 4.2 Conceptual diagram of the hierarchical genetic structuring in salmonids from species down to the local population level (modified from Utter et al. 1989).

directed into large-scale genetic studies. The initial purpose of these studies was to identify genetic differences among geographic populations within different species, such that samples from a mixed stock fishery, as typically occurs in ocean catches, could be examined for contributions by each of the geographic populations (Fournier et al. 1984; Milner et al. 1985; Utter et al. 1987; Shaklee et al. 1990; Utter et al. 1987; Utter and Ryman 1993). An extensive multi-agency program has resulted in the creation of very large data sets that can be used to assess genetic structure within some species over much or all of their natural distributions. For example, geneticists from a number of federal, state, and provincial agencies, as well as universities, have compiled a data set for chum salmon (*O. keta*) that examines 50 to 75 gene loci from over 150 populations throughout the Pacific Rim (Washington, British Columbia, Alaska, Russia, and Japan) (Beacham et al. 1985; Kondzela et al. 1994; Wilmot et al. 1994; Winans et al. 1994; Phelps et al. 1994, 1995).

Such studies of genetic variation commonly reveal strong patterns of geographic structuring in salmonid species (Allendorf and Utter 1974, 1979; Loudenslager and Gall 1980; Stoneking et al. 1981; Utter et al. 1989; Bartley and Gall 1990; Bernatchez and Dodson 1994; Gall et al. 1992; Kondzela et al. 1994; Phelps et al. 1994; Shaklee and Varnavskaya 1994; Varnavskaya et al. 1994a; Wood et al. 1994). Geographically adjacent populations are typically less distinct from one another than from geographically distant populations based on suites of molecular genetic characters. Thus, genetic structuring among most salmonid species is hierarchical in nature, with the first level of differentiation occurring as geographical aggregates of populations (Figure 4.2).

Phylogenetic or evolutionary analysis of such data often reveals that the primary geographic groupings correspond to major evolutionary or ancestral lineages within each species (Utter et al. 1989, 1995; Busack and Shaklee 1995; Williams et al. 1997). These lineages reflect clear evolutionary divergence from other lineages within the species as a result of genetic differences that have accumulated over evolutionary time between the lineages. For example, pink salmon (*O. gorbuscha*) can be separated into two major evolutionary lineages, based on even-year and odd-year occurrence (see species summaries below). The two lineages exhibit large genetic differences that are an expected consequence of the rigid 2-year life history of pink salmon, leading to the nearly complete reproductive isolation of the even- and odd-year broodlines.

Below the level of the major evolutionary lines, salmonid species exhibit further genetic structuring that is also typically geographic in nature (Figure 4.2). The major evolutionary lines shown in Figure 4.2 are equivalent to the major ancestral units of Utter et al. (1995) and the major ancestral lineages (MALs) of Busack and Shaklee (1995). Such regional differentiation has been observed in chum salmon in Washington (Kondzela et al. 1994), British Columbia (Beacham et al. 1985), Alaska (Phelps et al. 1994; Wilmot et al. 1994), and the western Pacific Basin (Russia and Japan) (Winans et al. 1994), where populations clustered on the basis of major islands, major river systems, and along major contiguous coastlines.

Typically, the next level of genetic structuring observed in salmonids is that of the individual watershed or subbasin, within which populations are usually closely related to one another (Figure 4.2; Utter et al. 1989). Nevertheless, populations within an individual subbasin may exhibit diverse life history strategies that include differences in run timing, age and size at maturity, and so on. Presently, we do not know the lower limit of genetic structuring within salmon populations (Utter et al. 1993); however, recent work by Gharrett and colleagues (Gharrett and Smoker 1991, 1993) on pink salmon in a small creek near Juneau, Alaska, has revealed heretofore unexpected levels of genetic substructuring within a single salmon population. Although salmonids are known for their ecological and behavioral plasticity, results such as these suggest a very strong role for local adaptation (with fitness implications) for many populations.

Genetic Structure of Columbia Basin Chinook Salmon

Genetic structure of individual Columbia Basin salmonid species are presented later in this chapter; however, it is instructive to briefly review the genetic structure of Chinook salmon in the Columbia Basin for two reasons. In general, they demonstrate the hierarchical patterns discussed above; however, they also reveal a second specific pattern that is an exception

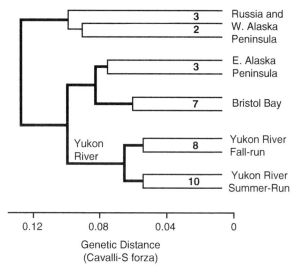

Figure 4.3 An example of geographic structure in the genetic architecture of Chinook salmon from North America and Asia surveyed by protein electrophoresis (Wilmot et al. 1994). Numbers on the minor branches equal the number of populations surveyed.

to the general discussion above that seems to occur only in large river systems.

Across their geographic distributions, Chinook salmon form a genetically complex network of populations that are structured primarily on the basis of geography (true phylogeographic assemblages) into large regional groups (see Figure 4.3; Wilmot et al. 1994), which correspond to the large regional groups identified for many other Pacific salmon species (Utter et al. 1989).

Within the large regional groups, Chinook also show substantial geographic substructuring, largely on the basis of subbasins or individual watersheds. Time of adult return to the river was not a major factor in establishing relationships of stocks among areas. Instead, populations with different run timings from the same stream were more similar genetically to one another than to populations with similar run timing from different areas (Utter et al., 1989). Thus, one major conclusion from these observations is that run-timing differences between stocks within subbasins have evolved via life history diversification from a single founding stock, regardless of the run timing of the founder stock. Therefore, the seasonal races of Chinook salmon that occur in many subbasins have evolved many times as independent events. Evidence from introductions of Pacific salmon into exotic locations further supports this idea. For example, Kwain and Thomas (1984) observed the development of spring-spawning Chinook salmon in the Great Lakes from introductions of fall-spawning Chinook, while Quinn and Unwin (1993)

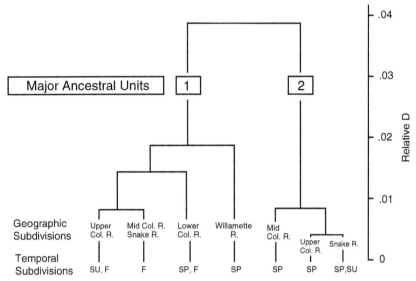

Figure 4.4 Geographic and temporal associations in the genetic structure of Chinook salmon populations from the upper Columbia and Snake Rivers (Matthews and Waples 1991; Utter et al. 1991; Waples et al. 1991).

described five different life history strategies in Chinook salmon introduced to New Zealand from a single founding source.

In contrast to the general pattern described above, Chinook populations in the upper Columbia and Snake Rivers (Matthews and Waples 1991; Waples et al. 1991; Utter et al. 1995) exhibit substructuring on the basis of run timing first, then geography (Figure 4.4). In these instances, populations with similar run timing were more similar to each other than they were to geographically proximate populations with different run timing. In the Columbia Basin, these differences result in four distinct evolutionary groupings: (1) Snake River spring and summer Chinook, (2) Snake River fall Chinook, (3) Columbia River spring Chinook, and (4) Columbia River summer and fall Chinook. Work form Brannon et al. (2002) also demonstrates major divisions between fall and spring summer races based upon matrilineal lines and that these races have been separated for a long period of time based upon their genetic divergence (Figure 4.5).

Run-timing associations among populations, which are not apparent throughout most of the Chinook salmon's range, occur in the Columbia and Yukon Rivers. These are among the largest of the river systems draining into the North Pacific, and their large size and nature may have provided Chinook salmon core habitats that were stable over long periods of time, allowing local adaptation and divergence of populations based on run timing. Because

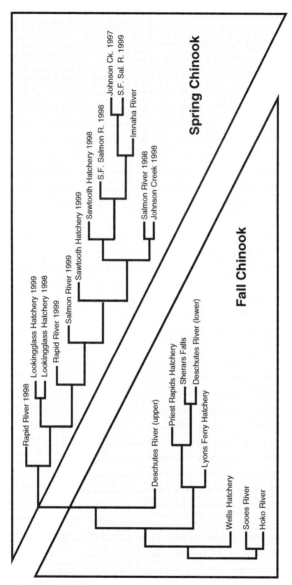

Figure 4.5 Dendrogram of Chinook salmon populations constructed using the Fitch and Margoliash (1967) least squares method and a nucleotide substitution matrix (d values) generated using Nei and Tajima (1981). Mitochondrial data from two out-of-basin populations, Sooes (n=28) and Hoko (n=23) Rivers were used as "outgroups."

strong heritability is associated with run timing (Helle 1981), divergence between run types seems likely over evolutionary time. However, the lack of association in run timing among populations throughout most of the Chinook salmon's range suggests the turnover time for populations may be short enough to counteract the processes of local adaptation, isolation, and divergence. Stochastic events (e.g., volcanic eruptions) in the relatively unsettled environment of the Pacific Northwest no doubt contribute to dispersal of founders during favorable conditions in one location and poor conditions in another. In 1998, a major landslide on the Stewart River, British Columbia, resulted in poor water conditions locally. As a consequence, a large number of sockeye salmon returning to the Stewart River simply dispersed among other drainages. One such sockeye was observed (origin verified by genetic testing) in the Clearwater River in Idaho, and several others were observed returning to the National Marine Fisheries Service (NMFS) research facility in Manchester, Washington (Powell unpublished data). This process can be looked at much like ecological succession, where there is an endpoint toward which things move if the system is left undisturbed long enough. However, frequent disturbance events continually reset the system back to or back toward its starting point, which for salmonids is the initial colonization of a watershed. Very large river systems like the Columbia and Yukon, although highly variable temporally, may have allowed long persistence of some populations and the emergence of run-timing associations among populations.

With respect to metapopulation theory, Chinook salmon populations in the Columbia and Snake Basins overall may be viewed in this context. Brannon et al. (2002) presented combined life history and genetic evidence for Chinook salmon population structure that appear to support this idea. Based upon calculations of genetic distance among populations (Nei 1987), one can arrange Chinook salmon populations in the Columbia and Snake Basins into a hierarchical set of related populations and their geographic distributions (Figure 4.6). In this instance, first-order metapopulations group within their geographic areas with a level of divergence < 0.8%, second-order populations range from 0.8% to 1.8%, and third-order populations have genetic distances higher than 1.8%. Ecological variation, life history variation, adaptation, and dispersal all act in concert to drive population differentiation, the results of which can be measured in one way as the genetic patterning observed in Figure 4.6. In general, this analysis is consistent with the ESUs identified by National Oceanic and Atmospheric Administration (NOAA) Fisheries; however, the first-order metapopulation groupings depicted in Figure 4.6 show finer-scale subdivisions within ESUs that may be useful for management purposes. For example, spring Chinook in the upper Columbia River are listed as a single ESU, but comprise four first-order metapopulations (Figure 4.6).

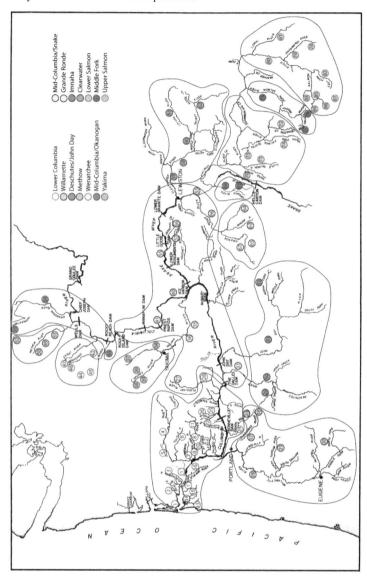

Figure 4.6 Chinook salmon population structure in the Columbia River Basin shown in clusters as 14 first-order metapopulations. Membership in the clusters are color coded and outlined in red to represent metapopulation boundaries (Brannon et al. 2001).

Genetic Structure of Columbia Basin Steelhead

The evolutionary history of steelhead populations follows much the same pattern as Chinook salmon. They are thought to have survived only within Pacific refugia prior to the last glaciation about 12,000 years ago (Behnke 1992; 2002). However, during this period, rainbow trout populations are thought to have segregated into their respective inland and coastal forms (Allendorf and Utter 1979; Hershberger 1992). Both forms exhibit anadromous and resident life histories. It can be argued for iteroparous fish that the freshwater resident form observed in *O. mykiss* is a tactic of "bet-hedging" against the possibility of extinction in the stochastic environments they inhabit (Miller and Brannon 1982). Thus, structure among steelhead populations may be more influenced by survival of resident contributors to the effective population size than in other salmonids. This, however, depends upon the level of gene flow between resident and anadromous forms, which has been demonstrated to be highly variable (McCusker et al. 2000b, Zimmerman et al. 2000).

Steelhead population structure also seems to be influenced by habitat preference, in part by avoiding competitive interaction in overlapping niches with salmon (Brannon et al. 2002). Unfortunately, competitive interaction between *O. mykiss* and other salmonids has likely been exacerbated due to large-scale, historical stocking of *O. mykiss* throughout the Columbia and Snake Basins (Lee et al. 1997). Evidence of non-indigenous rainbow trout introgressing into local populations is abundant (Campton and Johnston 1985; Currens and Shreck 1993; Phelps et al. 1994; Currens 1997). This too influences population structure and serves to artificially homogenize *O. mykiss* distributions within the interior basins.

Based upon allozyme frequency data from the extensive coast-wide database for steelhead, the distribution of *O. mykiss* populations throughout the Columbia and Snake Basins still show moderate amounts of genetic divergence, as represented in Figure 4.7 from Waples et al. (1993) and Brannon et al. (2002).

Amazingly, if steelhead populations in the Snake and Columbia Basins are portrayed in the ecological context of a metapopulation (Lande and Barrowclough 1987), they form groupings that are very similar to Chinook salmon in respect to their genetic partitioning versus geographic distributions (Figure 4.8). The metapopulation structure apparent among steelhead show slightly different levels of genetic distance than Chinook metapopulations do in Figure 4.6. First-order metapopulations exhibit genetic distances of < 1.0%, second-order groups show distances between 1.0% and 1.5%, and third-order structure is apparently above 1.5% genetic distance. As with Chinook, this analysis is generally consistent with the ESUs identified by NOAA Fisheries, but the first-order metapopulation groupings show finer-scale

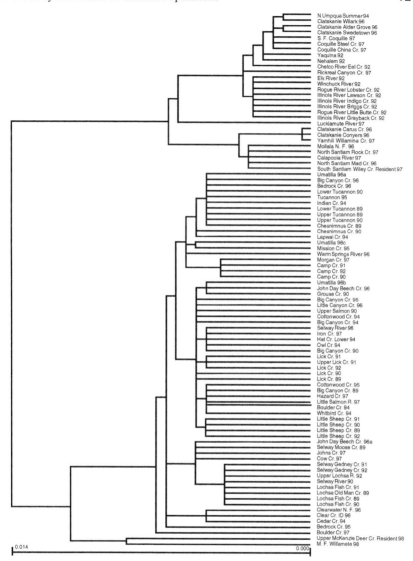

Figure 4.7 Dendrogram of *O. mykiss* populations constructed for spawning sites and years using hierarchical analysis of Nei (1987). Data sources from Waples et al. (1993) and NMFS Coastwide Database (Brannon et al. 2002).

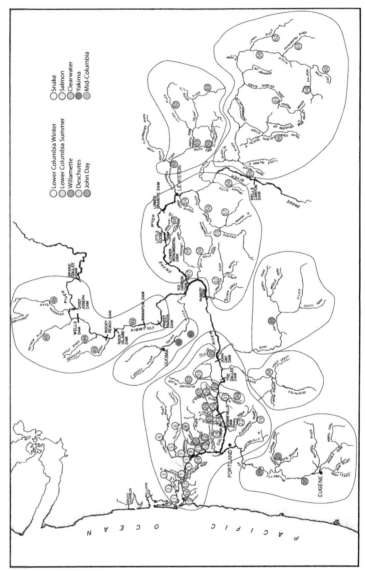

Figure 4.8 Steelhead population structure in the Columbia River Basin shown in clusters as 10 first-order metapopulations. Membership in the clusters are color coded and outlined in red to represent metapopulation boundaries (Brannon et al. 2001).

subdivisions within an ESU that may be useful for management purposes. For example, Snake River Basin steelhead are listed as a single ESU, but comprise first-order metapopulations in the Clearwater, South Fork Salmon, Middle Fork Salmon, and upper Main Salmon River subbasins (Figure 4.8).

Conclusions

Though steelhead life history patterns and the forces that influence their population structure are manifestly different than those of Chinook salmon, remarkably they show what appears to be the same archetype with respect to distribution within a metapopulation framework (hatchery influence notwithstanding). Thus, we may be able to apply metapopulation theory to other salmonid populations and species within the Columbia and Snake Basins. This would provide predictive power to conservation measures aimed at restoring extirpated populations, reestablishing or preserving connectivity between isolated demes, and give a temporal (based on gene flow) and spatial (based on geographic partitioning) reference from which to do so.

Status of Pacific Salmon Species

Pacific salmon, as well as resident salmonids, have disappeared from much of their historic range, and in many locations once-abundant populations have been extirpated or are severely depressed. For overviews of the decline and current status of Columbia River Basin stocks, see Nehlsen et al. (1991), Huntington et al. (1996), the NRC report (1996) and Thurow et al. (2000). Detailed status reviews also are available for mid-Columbia and Snake River salmon and steelhead stocks (Craig and Hacker 1940; Chapman et al. 1994a, 1994b, 1995; Busby et al. 1996). At the present time, only Lewis River (Washington) and Hanford Reach (Washington) fall Chinook, Wenatchee River (Washington) sockeye, and five summer steelhead stocks in the John Day River (Oregon) can be classified as healthy (Huntington et al. 1996). Indigenous resident salmonids are now restricted to between 10% and 30% of their original range (Trotter 1987; Behnke 1992; Quigley et al. 1996). We also review the status of white and green sturgeon and Pacific lamprey in this chapter. These species are important food web corollaries of anadromous salmonids in freshwaters, and recovery actions for the endangered Kootenay River sturgeon seem to be at odds with actions for endangered salmon (Marotz et al. 1996).

Chinook Salmon (*Oncorhynchus tshawytscha*)

Background

Chinook salmon (Figure 4.9) are distributed in Asia from Hokkaido, Japan, north to the Anadyr River, Russia, and on the Pacific Coast of North America from central California to Kotzebue Sound in Alaska (Healey 1991). North of the Columbia River, the post glacial radiation of Chinook salmon came from two principal refugia: The two thirds of the Columbia River that remained ice free and Beringia, an ice free area in the lower Yukon River and adjacent coastal areas of the Bering Sea (Lindsay et al. 1986; McPhail and Lindsey 1986).

Figure 4.9   Chinook salmon. Illustration from actual specimen, by J. R. Tomelleri.

Chinook salmon radiated south from Beringia to about 56°N and Chinook salmon from the Columbia River recolonized deglaciated streams north to 56°N.

Chinook salmon may enter rivers of the northwest in any month of the year (Healey 1991). In the Columbia River, the spawning migration is divided into three distinct races: spring, summer and fall. At the present time, the largest run enters the river in the fall. Historically, the spring and summer runs were much larger than they are today, but they were depleted by over-harvesting and habitat degradation (Chapman et al. 1991, 1994a, 1994b; Lichatowich and Mobrand 1995; NRC 1995; Oregon Department of Fish and Wildlife and Washington Department of Fish and Wildlife 1995; Taylor 1999; Lichatowich 1999).

Gilbert (1912) divided the juvenile life histories of Chinook salmon into ocean and stream types. The ocean type migrates to sea in the first year, often within three months after emergence. The stream type migrates to sea in the spring after a year or more in freshwater (Healey 1991). The ocean type is the dominant life history in streams south of the Columbia River. Both ocean and stream types occur from the Columbia River north to 56°N, with the ocean type predominantly in the coastal areas and the stream type in inland areas. North of 56°N, the stream-type life history is dominant (Taylor 1990). After an analysis of the distribution of stream- and ocean-type life histories, Taylor (1990) concluded that variability in life history is in part a response to growth opportunity (environmental conditions) and selection for size at migration.

Evolutionary History and Genetic Structure of Chinook Salmon

Neave (1958) argued that the Pacific salmon diverged into seven species entirely within the Pleistocene, 0.5 to 1 million years ago. However, analysis of mitochondrial DNA suggests the ancestral line that produced Chinook salmon is 2 to 3 million years old (Thomas and Beckenbach 1989).

Genetic data exist for Chinook salmon populations ranging from California (Utter et al. 1989; Gall et al. 1992), Oregon, Washington, British Columbia (Utter et al. 1989, 1995; Matthews and Waples 1991; Waples et al. 1991), to Alaska (Gharrett et al. 1987). Chinook salmon form a genetically complex network of populations that are structured primarily on the basis of geography into large regional groups (Utter et al. 1995), that correspond to the large regional groups identified for many other Pacific salmon species. Within the large regional groups, Chinook also show substantial geographic substructuring, largely on the basis of subbasins or individual watersheds. Although Chinook salmon show hierarchical levels of geographically based genetic structuring throughout most of their range, Chinook populations in the upper Columbia and Snake Rivers (Matthews and Waples 1991; Waples et al. 1991; Utter et al. 1995) exhibit substructuring first on the basis of run timing, then geographically (Figure 4.3). In these instances, populations with similar run timing were more similar to each other than they were to geographically proximate populations with different run timing.

Research on the genetic structure of Chinook salmon in the Columbia Basin has focused on the higher levels in the hierarchy of genetic organization, which have been defined variously as major ancestral lineages (Utter et al. 1995), genetic diversity units (Busack and Shaklee 1995), ESUs, (Waples 1991), and stocks. These efforts, largely driven by questions of genetic stock identity, have provided information critical to our understanding of genetic structure within species and for the identification of genetic conservation units, such as ESUs. However, little effort has been expended on the genetic infrastructure within populations or stocks.

The genetic infrastructure of a stock allows the population to adapt to fluctuating environments and to survive long-term environmental change (Gharrett and Smoker 1993). One visible indication of variation within a population and an indication of infrastructure is the existence of life history diversity. Studies of Chinook salmon have shown considerable variation in life history patterns (Reimers 1973; Schluchter and Lichatowich 1977; Carl and Healey 1984). However, only one of these studies examined both life history and genetic diversity in the same population, and that study did demonstrate a relationship between juvenile migration patterns and genetic diversity (Carl and Healey 1984). In the Columbia Basin, Lichatowich (1995) hypothesized that the observed loss of life history diversity in spring and summer Chinook salmon was due to depletion of the runs.

Historic and Present Distribution of Chinook Salmon

The predevelopment abundance of Chinook salmon in the Columbia Basin was estimated at 4.7 to 9.2 million fish (NPPC 1986). In 1994, 400,000

Chinook salmon of both hatchery and wild origin entered the river (Oregon Department of Fish and Wildlife and Washington Department of Fish and Wildlife 1995).

Chinook salmon generally spawn in the mainstem and larger tributaries in the Columbia Basin (see Figures 3.5, 3.6, and 3.8). Therefore, the construction of mainstem dams has had a major impact on their spawning distribution and production. The spring/summer runs of Chinook salmon migrated to and spawned throughout the Columbia and Snake Rivers, generally in the upper reaches of the watershed and its tributaries (Figure 4.10). Summer Chinook spawned in the mainstem below the outlet of Windermere Lake in British Columbia, 1,200 miles from the sea (Fulton 1968). In the Snake River, spring Chinook migrated to Rock Creek, a tributary below Shoshone and Auger Falls, 900 miles from the sea (Fulton 1968). Historically, the Salmon River (a Snake Basin tributary) alone produced 39% to 45% of the spring/summer Chinook salmon (Figure 4.11) in the Columbia Basin (NMFS 1995). Spring and summer Chinook salmon are totally blocked in their upstream migration in the Columbia River at Chief Joseph Dam and in the Snake River by Hells Canyon Dam. Fulton (1968) described the historical spring/summer Chinook salmon spawning areas, which were eliminated by development in the basin:

> "Major areas of the John Day and Umatilla rivers, parts of the Clearwater and Powder rivers, all of the Payette, Owyhee, Boise, and Bruneau, major portions of the Walla Walla, Yakima and Okanogan rivers, important tributaries above Chief Joseph Dam including the San Poil, Spokane, Kettle, Pend Oreille, and Kootenay rivers."

The fall run of Chinook salmon spawned in the lower tributaries and in the lower and middle mainstem of the Columbia River and in the Snake River up to Augur Falls (Fulton 1968). Some of the most valuable spawning areas were in the mainstems of the Columbia River, nearly all of which were inundated by construction of dams. The Hanford Reach and the Snake River below the Hells Canyon complex of dams are the only remaining free-flowing reaches in the Columbia Basin; however, the only significant remaining mainstem spawning area for fall Chinook salmon is the Hanford Reach. Recent surveys in the mainstem Snake River show small but increasing fall Chinook production from site-specific spawning areas (Figure 4.12) that include the tailraces below each mainstem dam (Garcia et al. 1994; Dauble and Watson 1997; Dauble et al. 2003). Irrigation and habitat degradation eliminated spawning areas in many of the lower reaches of tributaries such as the John Day, Umatilla, and Walla Walla Rivers. From 1957 to 1960, the largest group of fall Chinook (41,000 fish) spawned in the Snake River and the second largest (34,000 fish) spawned in the mainstem Columbia River in the area now inundated by the John Day Dam.

Figure 4.10 Spring Chinook spawning habitat in the upper Imnaha River, a tributary to the Snake River in northeastern Oregon. Photo by R. N. Williams.

Figure 4.11 Summer Chinook spawning in Johnson Creek, a tributary to the Salmon River in central Idaho. Photo by R. N. Williams.

Figure 4.12 Fall Chinook spawning habitat on the main Snake River above Lewiston, Idaho. Photo by W. Smoker.

Life History Diversity in Chinook Salmon

The geographic organization of Chinook salmon in the Columbia Basin prior to extensive human development likely consisted of a complex mosaic of spring, summer, and fall races of salmon distributed among mainstem and headwater spawning areas (see Figures 3.5 and 3.9). Local populations of fall Chinook salmon whose juveniles migrated as subyearlings spawned in mainstem areas of the Columbia and Snake Rivers and lower mainstem segments of Columbia River tributaries (Fulton 1968; Howell et al. 1985; Mullan et al. 1992a). Spring and summer Chinook that migrated as subyearlings reproduced in upper mainstem segments of major subbasins and lower reaches of tributaries to subbasin mainstems (Lichatowich and Mobrand 1995). Summer Chinook probably spawned lower in the subbasin mainstems than did spring Chinook (French and Wahle 1959; Fulton 1968; Lichatowich and Mobrand 1995). Populations of spring Chinook with yearling life histories reproduced in headwater streams of subbasin tributaries.

The complex of spatially distributed local spawning populations associated with major subbasins and contiguous areas of the mainstem Columbia or Snake River may have formed metapopulations composed of local populations connected, at least to some degree, by dispersal. One type of metapopulation was composed of fall Chinook spawning in mainstem reaches of the Columbia and Snake Rivers and the lower reaches of major subbasins, while summer and spring Chinook spawning in the upper mainstems of major subbasins and spring Chinook spawning in headwater areas comprised another type of metapopulation.

Present metapopulations organization, which is fragmented as compared to probable historic organization, may result in reduced resilience; however, in theory at least, metapopulations have the ability to recover from catastrophic decline. Habitat fragmentation has increased isolation of populations and probably reduced dispersal rates, due to both increased distances between populations and the degraded quality of connecting habitats. Most mainstem spawning populations, which may have served as stable sources of colonists, are virtually extinct, and viable naturally spawning populations are confined to relatively isolated headwater areas. Thus, dispersal among populations may be restricted, making "rescue" of severely depleted populations and recolonization of habitats where extinction has occurred much less likely. Moreover, confining populations to headwater areas may increase their susceptibility to habitat alterations from land use such as grazing and logging (see Chapter 5) unless the populations inhabit areas protected from adverse land use.

Both genetic and life history evidence suggests that spring Chinook are distinguished from fall Chinook in the Columbia River Basin. Fall Chinook are differentiated from the spring races with respect to life history characteristics, such as annual timing of adult migration, geographic distribution of spawning habitat, and genetic attributes (Waples et al. 1991). Summer Chinook in the upper Columbia River appear to be more closely related to fall Chinook than to spring Chinook, whereas in the Snake River, summer Chinook are more closely related to spring Chinook (Utter et al. 1995). French and Wahle (1959) observed summer and spring Chinook on the spawning grounds of the Wenatchee and Methow Rivers, whereas Mullan et al. (1992a) reported mixing of summer and fall fish on the spawning areas of mid-Columbia River tributaries. There may be a tendency to group summer Chinook and fall Chinook because they both migrate downstream as subyearlings. Conventional wisdom suggests that all spring Chinook exhibit yearling juvenile migration even though there is evidence to the contrary.

Redd (salmon nests) counts in index areas of the Imnaha and Grande Ronde Rivers and their tributaries suggest that spring Chinook populations have been experiencing synchronous decline since the late 1960s–mid-1970s (Figure 3.9). In 1994 and 1995, no redds were located in the index areas in Bear, Hurricane, Indian, and the North and South Forks of Catherine Creek. No redds were recorded in Sheep Creek from 1993 to 1995. Since the habitat where these stocks spawn is of relatively high quality, and considering that the Wenaha River is nearly entirely within a wilderness area, the synchronizing influence may be downstream from the spawning areas, either in lower mainstems of the Grande Ronde and Imnaha, in the mainstem Snake or Columbia River, or in the ocean.

Harvest Summary of Chinook Salmon

Intensive fisheries did not begin until cannery technology reached the Columbia River in 1866 (Craig and Hacker 1940). Chinook salmon, and especially the spring or summer run fish, brought the highest price and made the highest quality canned product, so the early fisheries targeted those runs (Craig and Hacker 1940). After 1866, the catch of Chinook salmon increased rapidly and peaked in 1883 at 19,413 metric tons (Beiningen 1976). The harvest of Chinook salmon can be divided into four phases (Figure 4.13):

A. Initial development of the fishery (1866–1888);
B. A period of sustained harvest with an average annual catch of about 25 million pounds (1889–1922);
C. Resource decline with an average annual harvest of 15 million pounds (1923–1958);
D. Maintenance at a depressed level of production of about 5 million pounds (1958 to the present).

Recent declines may indicate the system is slipping to a new, lower level of productivity.

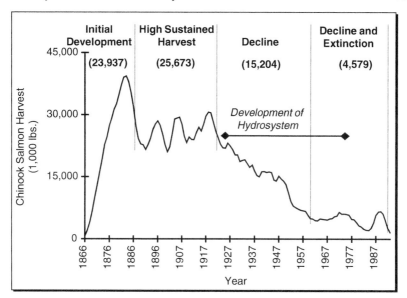

Figure 4.13 Five-year running average of Chinook salmon harvest in thousands of pounds in the Columbia River, 1866 to 1992, with the average harvest for four periods of development.

Between 1889 and 1920, the harvest of Chinook salmon was relatively stable; however, catch data alone mask a major qualitative shift in the fishery (Figure 4.13). During that period, the spring and summer races of Chinook salmon were declining, and harvest was maintained by a shift from the spring/summer fish to fall Chinook salmon. In 1892, fall Chinook made up 5% of the harvest and by 1912, it had risen to 25%. In 1920, fall Chinook salmon made up 50% of the catch. The harvest of all Chinook salmon underwent a rapid decline after 1923; however, the decline in the spring and summer races started as early as 1911 (Craig and Hacker 1940). One of the factors contributing to this decline was the development of the offshore troll fishery which started in 1910 and expanded in the 1920s. Decline in abundance reached the point that two in-river fisheries were closed: 1965 was the last summer Chinook season and 1977 was the last spring Chinook season (Oregon Department of Fish and Wildlife and Washington Department of Fish and Wildlife 1993)[2]. In 1994, the Young's Bay fishery accounted for 81% of the commercial landings below Bonneville Dam (Oregon Department of Fish and Wildlife and Washington Department of Fish and Wildlife 1995).

---

[2]However, recent increases in spring Chinook returns (2000–2003), largely due to large-scale hatchery releases coupled with good smolt outmigration and good ocean conditions, have allowed limited spring Chinook harvests by tribal and sport fishers in the Salmon River in central Idaho. Whether such increases are sustainable remains to be seen.

Propagation Efforts for Chinook Salmon

Chinook salmon were the first fish to be artificially propagated in the Columbia Basin. In 1877, a private company, the Oregon and Washington Propagation Company, constructed the first hatchery on the Clackamas River. The hatchery program grew rapidly and remained an important management activity, even though there was little evidence that artificial propagation was in fact enhancing Chinook salmon in the basin. After 1960, with the introduction of better feeds and hatchery practices, artificially propagated Chinook salmon began making significant contributions to the fisheries. The hatchery program for Chinook salmon has grown from releasing 61 million juveniles in 1960 to 160 million in 1988. For more detailed discussion of artificial propagation see Chapter 8.

Status under the Endangered Species Act

Throughout the range of Chinook salmon south of the Canadian border, NMFS recognizes 17 ESUs, 8 of which lie in the Columbia River drainage. ESA status of Chinook salmon within the Columbia Basin is shown in Table 4.2.

Coho Salmon (*Oncorhynchus kisutch*)

Background

The spawning distribution of coho salmon (Figure 4.14) in the western Pacific extends from as far south as Chongjin on the east coast of North Korea, north to the Anadyr River. In the eastern Pacific, coho salmon are distributed from the San Lorenzo River on Monterey Bay to Point Hope in Alaska (Sandercock 1991). Coho salmon generally enter the rivers to spawn in late summer or fall, although spawning migrations in other seasons have been noted. More than one seasonal spawning migration into a single river is rare (Sandercock 1991).

Table 4.2   ESA status for Chinook salmon in the Columbia and Snake Rivers.

River system	Run timing	ESA status
Upper Columbia River	spring	Endangered
	summer-fall	Not Warranted
Mid-Columbia	spring	Not Warranted
	summer-fall	Not Warranted
Lower Columbia	fall	Threatened
Upper Willamette	spring	Threatened
Snake River	summer-spring	Threatened
	fall	Threatened

Figure 4.14   Coho salmon. Illustration from actual specimen, by J. R. Tomelleri.

Evolutionary History and Genetic Structure of Coho Salmon

Coho salmon spawn in small tributary and headwater streams more frequently than other salmon species (Aro and Shepard 1967). Coho exhibit low levels of genetic variation as compared to the other Pacific salmon species (Utter et al. 1973; Olin 1984; Wehrhahn and Powell 1987) but still show large regional geographic differentiation. Analysis of mitochondrial DNA suggest that three phyletic lines of salmonids diverged more than 2 million years ago, and in one of those lines a subsequent divergence 1 to 1.5 million years ago led to rainbow, coho, and Chinook salmon (Thomas and Beckenbach 1989). Weitkamp et al. (1995) identified six potential coho salmon ESUs in California, Oregon, and Washington: central California coast, southern Oregon/northern California coasts, Oregon coast, lower Columbia/southwestern Washington coast, Olympic Peninsula, and Puget Sound/Strait of Georgia. The lower Columbia/southwestern Washington coast ESU contains the remaining stocks of coho salmon in the Columbia Basin. Unfortunately, most of the native Columbia River coho stocks were extinct before an analysis of their genetic structure could be completed.

Historic and Present Distribution of Coho Salmon

The predevelopment run size of coho salmon was estimated at 903,000 to 1,780,000 fish (NPPC 1986). In 1994, the minimum number of coho salmon entering the Columbia River was 178,900 fish (Oregon Department of Fish and Wildlife and Washington Department of Fish and Wildlife 1995), nearly all of which were of hatchery origin.

The principal spawning areas for coho salmon were in the tributaries to the lower river; however, Fulton (1970) also identified coho spawning in tributaries above Bonneville Dam, including Hood, John Day, Grande Ronde, Spokane, Entiat, Wenatchee, and Methow Rivers. All coho stocks above

Bonneville Dam, with the exception of the Hood River stock, were classified as extinct by Nehlsen et al. (1991).

At present, production of coho salmon is almost entirely from artificial propagation. The NMFS could not identify any remaining natural populations of coho salmon in the lower Columbia River that warranted protection under the ESA (Johnson 1991). The possible exception is the late run of coho salmon into the Clackamas River. Whether the Clackamas stock is the last remaining wild stock in the Columbia River or a stock similar to the other hatchery stocks in the lower river is uncertain (Weitkamp et al. 1995). Remnant wild populations may also exist in the Hood River and Klickitat River. Habitat degradation and overharvesting contributed to the depletion and extinction of the wild coho salmon stocks in the Columbia River. The massive hatchery program, which included interstock transfers, was an additional factor in the decline of coho salmon (Flagg et al. 1995).

Harvest Summary of Coho Salmon

Coho salmon were not as abundant as Chinook salmon in the Columbia River. Coho salmon were considered inferior by the cannery operators, so they were not harvested in the early years of the intensive fishery in the Columbia River (DeLoach 1939; Craig and Hacker 1940). The first coho salmon were commercially harvested in 1892 (Figure 4.15) in conjunction

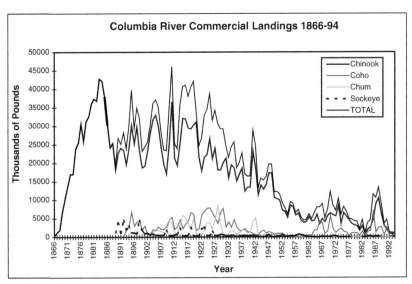

Figure 4.15   Columbia River commercial salmon fishery landings, 1866–1994 (Beiningen 1976; ODFW and WDF 1993).

with a shift in harvest to fall-running fish when the prime spring run of Chinook salmon became depleted (Lichatowich and Mobrand 1995). The fishery for coho salmon intensified after 1920 when Chinook salmon went into rapid decline; however, by the mid-1930s, coho salmon were also in a steep decline that persisted for 30 years (Figure 4.15). The decline was real, but part of the apparent decline was due to a shift to offshore fishing by the growing troll fleet. After 1930, harvest in the Oregon Production Index (OPI) is a better indication of the pattern of abundance of Columbia River coho salmon. The OPI includes in-river and ocean catch of coho salmon from southwestern Washington to northern California (Figure4.16).

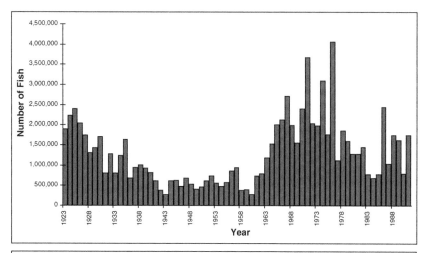

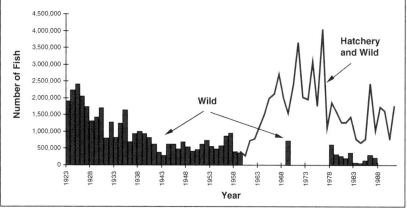

Figure 4.16 Annual ocean harvest of coho salmon in numbers of fish for the Oregon Production Index area, 1923–1991. Plot A is for total catch, while Plot B shows wild and hatchery produced components of total catch (Borgerson 1992; Oregon Department of Fish and Wildlife 1982; Pacific Fishery Management Council 1992).

In the mid-1960s, improved hatchery practices and favorable ocean conditions combined to produce an apparent recovery of coho salmon production, which persisted until 1976 (Figures 4.15 and 4.16). The recovery was primarily due to increased survival of hatchery-reared fish. The wild component of the OPI harvest remained depressed throughout the 1960s and 1970s. By 1991, habitat degradation and fisheries on mixed stocks of wild and hatchery coho salmon led to the conclusion that no viable wild stocks of coho salmon existed in the lower Columbia River (Johnson 1991).

Propagation Efforts for Coho Salmon

The first plant of artificially propagated coho salmon in the Columbia Basin took place in 1896. Coho salmon have been propagated continuously since 1900 (Cobb 1930). In the early 1990s, there were 16 hatcheries operating in the lower Columbia River (Johnson 1991), which released 29 to 54 million juvenile coho salmon annually from 1984 to 1992. The origin of the coho salmon brood stocks in Oregon's lower Columbia River hatcheries is uncertain. Johnson et al. (1991) described the brood stocks as mixtures of fish from a variety of sources, including coastal populations, Washington stocks, and native stocks. These mixed stocks have been extensively outplanted throughout the basin (Flagg et al. 1995). For additional information on the hatchery program see Chapter 8.

Status under the Endangered Species Act

Throughout the range of coho salmon south of the Canadian border, NMFS recognizes six ESUs, only one of which lies in the Columbia River drainage. Oregon coastal Coho populations form a threatened ESU, while Coho in the lower Columbia/Southwest Washington ESU have been considered candidates for listing since the mid 1990s. In June 2005, they were reclassified as a threatened ESU.

Chum Salmon (*Oncorhynchus keta*)

Evolutionary History and Genetic Structure of Chum Salmon

Chum salmon (Figure 4.17) populations exhibited the kind of geographic and regional differentiation described in Figure 4.2 (Beacham et al. 1985; Kondzela et al. 1994; Phelps et al. 1994, 1995; Wilmot et al. 1994; Winans et al. 1994), where populations clustered on the basis of major islands, major river systems, and along major contiguous coastlines (see Figure 4.3 showing the genetic distances among Chinook salmon populations from northern Alaska and Russia). In some instances, however, chum salmon populations in

Figure 4.17   Chum salmon. Illustration from actual specimen, by J. R. Tomelleri.

regional aggregates sorted by run timing rather than by subbasin. In other words, as is the case for Chinook salmon outside the Columbia River Basin, chum salmon populations in several regions, including the Yukon River (see Figure 4.2; summarized from Wilmot et al. 1994), Hood Canal, and Puget Sound (Phelps et al. 1995), were more similar to distant populations with similar run timing than they were to adjacent populations (within the same subbasin) with different run timing.

Historic and Present Distributions of Chum Salmon

The three remaining spawning areas for chum salmon are in Washington state in tributaries to the lower river below Bonneville Dam (Oregon Department of Fish and Wildlife and Washington Department of Fish and Wildlife 2001) in Hamilton Creek, Hardy Creek, and the Grays River.

Harvest Summary of Chum Salmon

Chum salmon were not as abundant as Chinook salmon in the Columbia River and were considered inferior by the cannery operators, so they were not harvested in the early years of the intensive fishery in the Columbia River (DeLoach 1939; Craig and Hacker 1940). Chum salmon entered the fishery in 1894 (Figure 4.13) in conjunction with a shift in harvest to fall running fish as the prime spring run Chinook became depleted (Lichatowich and Mobrand 1995). From the early 1900s through the 1950s, the harvest of chum salmon was more variable but generally followed the trend in harvest of coho salmon (Figures 4.13 and 4.15). Since chum salmon were the lowest grade of canned salmon in the Columbia River, some of the variability in harvest was due to a fluctuating demand for cheap fish (Craig and Hacker 1940).

Propagation Efforts for Chum Salmon

Chum salmon were not propagated extensively in hatcheries and their abundance did not increase in the 1960s as observed for coho and Chinook salmon, which were propagated (Figures 4.13 and 4.15). The collapse of Columbia River chum salmon in the 1940s and 1950s paralleled the decline of coastal chum salmon populations in Oregon, Washington, and British Columbia, suggesting that an overall decline was occurring due to a regional climatological or oceanic factor (Oakley 1996).

Status under the Endangered Species Act

Throughout the range of chum salmon south of the Canadian border, NMFS recognizes four ESUs, with the Columbia River ESU listed as threatened.

Sockeye Salmon (*Oncorhynchus nerka*)

Background

Sockeye salmon (Figure 4.18) are distinguished from other Pacific salmon species by their use of lakes for the freshwater rearing of juveniles. Sockeye are widely distributed in western North America and eastern Asia (Burgner 1991); however, sockeye have been extirpated from most of the localities formerly occupied in the contiguous United States (California, Oregon, Washington, and Idaho).

Substantial information exists on reproductive biology, age structure, growth, and productivity of Columbia River sockeye. Columbia River sockeye salmon spawn in tributaries and outlets of Lakes Wenatchee and Osoyoos in August and September (Mullan 1986; Hatch et al. 1995) and

Figure 4.18   Sockeye salmon. Illustration from actual specimen, by J. R. Tomelleri.

hatch and swim into rearing lakes in the late winter and spring of the following year. Depending on growth, sockeye juveniles will spend one to three winters in the rearing lake and one to three winters in the ocean. Slower-growing sockeye take longer to pass through each life history stanza than faster-growing sockeye. The typical Columbia River sockeye spends one winter in freshwater and two winters in the ocean to return as an adult in its fourth year of life.

Lake Osoyoos (Okanogan River) sockeye are unique among sockeye populations in occasionally having 3-year-old adults as the dominant age class (one winter in freshwater and one in the ocean). Size at age, growth, productivity, and historical zoogeography are reviewed by Fryer (1995). Columbia River sockeye from Lake Osoyoos tend to be large as smolts (greater than 10 cm) and small as adults (less than 45 cm and less than 2 kg), whereas Wenatchee sockeye tend to be smaller than Osoyoos sockeye as smolts and larger and older as adults. Differences between the attributes of Wenatchee and Osoyoos sockeye are ascribed to the physical and biological differences in the characteristics of the rearing lakes (Fryer 1995).

Evolutionary History and Genetic Structure of Sockeye Salmon

Sockeye salmon occur in rivers associated with nursery lakes or in groundwater-dominated streams widely along the Pacific coast north of the Columbia River and within a limited distribution in Russia along the Kamchatka Peninsula and the northern coast of the Bering Sea (Varnavskaya et al. 1994a). Like coho salmon, sockeye exhibit a low level of genetic variation as compared to pink, chum, or Chinook salmon (Varnavskaya et al. 1994a, 1994b; Wood et al. 1994). This may be the result of inbreeding related to the greater extent of reproductive isolation between spawning populations, a consequence of well-developed homing behavior in sockeye as demonstrated by tagging experiments and gene flow calculations (Quinn et al. 1987; Altukhov and Salmenkova 1991).

Nevertheless, the genetic architecture of sockeye salmon shows large-scale geographic differentiation, with groups from Kamchatka, western Alaska, southeastern Alaska, northern British Columbia, southern British Columbia, and Washington being well differentiated (Varnavskaya et al. 1994b; Wood et al. 1994). Large genetic differences occur between sockeye from some of the different regions, reflecting major ancestral or evolutionary lineages, which appear to have been influenced by recent historical glaciation events. Present distributions and genetic relationships among sockeye populations appear to be related to historical expansion and recolonization from a few ice-free refugia (Wood et al. 1994). Within each of these larger regions, sockeye salmon populations show additional geographical substructuring; however, populations within regions are well differentiated from one another, reflecting the relative reproductive isolation of individual sockeye populations from one another. Genetic studies of sockeye and kokanee populations in British Columbia show evidence of parallel life history evolution (Taylor et al. 1996), as well as differentiation among reproductive ecotypes of kokanee (Taylor et al. 1997).

Historic and Present Distribution of Sockeye Salmon

At least 27 lakes originally supported populations of Columbia River sockeye in Oregon, Washington, and Idaho (Fryer 1995). Loss of access to spawning areas due to construction of small agricultural storage and diversion dams has reduced the number of lakes open to sockeye, resulting in a reduction of 96% in juvenile rearing habitat between settlement during the 1840s and the present (Rich 1941; Mullan 1986; NPPC 1986; Fryer 1995). Sockeye now occur in the Columbia River Basin in three localities: Lake Wenatchee, Washington; Lake Osoyoos, Washington and British Columbia; and Redfish Lake, Idaho. However, the Idaho population is a federally listed endangered species as of December, 1991.

Age, growth, and stock identification studies and spawning ground surveys are conducted by the Columbia River Inter-Tribal Fish Commission under the auspices of the Pacific Salmon Treaty.

Harvest Summary of Sockeye Salmon

Historical annual abundances of 2 to 3 million adult sockeye salmon (Chapman 1986; NPPC 1994) supported annual commercial landings that twice exceeded 4.5 million pounds during the 1890s. As measured by commercial catches, adult returns of Columbia River sockeye declined sharply after 1900 (Figure 4.13). Present levels of returns are in the tens of thousands; in 1995, spawning escapements to Osoyoos and Wenatchee were less than 10,000 adults each. The number of adults deemed by the management entities to be sufficient for fully seeding sockeye spawning grounds (i.e., the escapement goal) for the Columbia River Basin is presently 75,000 (Oregon Department of Fish and Wildlife and Washington Department of Fish and Wildlife 1995). No commercial fishing has occurred since 1988, and the annual commercial season has often been canceled during the past 25 years. A sport fishery occurs on Lake Wenatchee when abundances permit, yielding a catch of 7,000 sockeye as recently as 1993 (Oregon Department of Fish and Wildlife and Washington Department of Fish and Wildlife 1995). Subsistence and ceremonial harvests by treaty Indian Tribes occur above Bonneville Dam, with harvests being in the area of 5,000 adults per year, prior to the limiting of sockeye salmon as a federally endangered species in Idaho.

Propagation Efforts for Sockeye Salmon

Current propagation efforts for sockeye occur at Lake Wenatchee, Redfish Lake, and Lake Osoyoos. Sockeye have proven difficult to culture with standard hatchery methods due to their susceptibility to disease. The alternative that has been developed is to move the fry soon after hatching into net pens in the lake where they are reared for a time and then released into the lake to overwinter before spring outmigration.

Propagation efforts at Lake Wenatchee, funded by the Chelan County P.U.D. as part of a FERC agreement, have added about 15% to the sockeye outmigration from Lake Wenatchee.

The program for the recovery of the endangered Redfish Lake, Idaho sockeye is administered by the NMFS in cooperation with the Bonneville Power Administration, the State of Idaho, and other concerned fisheries agencies, including Indian tribes. The program involves and relies heavily on artificial production using all returning Redfish Lake sockeye, along with genetic input from the resident beach spawning kokanee, which are part of the same ESU.

Considerable propagation efforts have been directed at kokanee, the resident form of sockeye salmon. Stocks have been widely transferred throughout the basin, and kokanee populations in most large lakes or reservoirs are genetic mixtures of multiple stocks (M. Powell and R. Williams, unpublished data). There is interest in the basin in attempting to reestablish anadromous sockeye runs from residualized kokanee populations; however, the probability of that occurring may decrease as a consequence of the mixed genetic heritage of most kokanee populations.

Status under the Endangered Species Act

Throughout the range of sockeye salmon south of the Canadian border, NMFS recognizes seven ESUs, three of which lie in the Columbia River drainage. Only the Snake River sockeye ESU is listed (endangered), while the Okanogan River and Lake Wenatchee ESUs were not deemed warranted for listing.

Pink Salmon (*Oncorhynchus gorbuscha*)

Evolutionary History and Genetic Structure

Pink salmon (Figure 4.19) can be separated into two major evolutionary lineages, based on even-year and odd-year occurrences that exhibit large genetic differences (Gharrett et al. 1988). This is an expected consequence of their rigid 2-year life history and results in the nearly complete reproductive isolation of the even- and odd-year broodlines. Within the even- and odd-year

Figure 4.19   Pink salmon. Illustration from actual specimen, by J. R. Tomelleri.

broodlines, pink salmon populations show typical hierarchical geographic differentiation as described above (Beacham et al. 1988; Gharrett et al. 1988; Shaklee et al. 1995; Varnavskaya and Beacham 1992; Shaklee and Varnavskaya 1994).

In spite of the near reproductive isolation of the two broodlines throughout their native distribution, Kwain and Chappel (1978) reported the development of even-year pink salmon runs from a single release of odd-year breeding pink salmon into the Great Lakes.

Historic and Present Distributions of Pink Salmon

Pink salmon occur irregularly along the Oregon and Washington coasts, including the Columbia River, but spawning distributions occur from Puget Sound and the Olympic Peninsula north to Norton Sound in Alaska (Groot and Margolis 1991; Hard et al. 1996).

Status under the Endangered Species Act

Throughout the range of pink salmon south of the Canadian border, NMFS recognizes only two ESUs, one for even-year and another for odd-year broodlines. Neither is deemed as warranted for listing.

Status of Columbia River Trout and Char

Overview

Background

Rainbow and cutthroat trout exist in both anadromous and resident forms. Anadromous rainbow trout, known as steelhead (Figure 4.20), have histori-

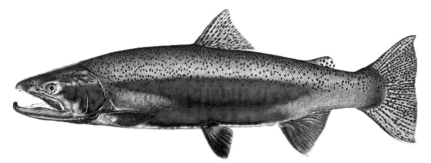

Figure 4.20 Steelhead trout, adult male from the South Fork of the Salmon River, Idaho. Illustration from actual specimen, by J. R. Tomelleri.

cally been managed more in concert with the five anadromous salmon species described above than with resident rainbow and cutthroat trout. This distinction continues today: NMFS supervises ESA concerns for steelhead and anadromous coastal cutthroat trout, while the U.S. Fish and Wildlife Service (USFWS) provides ESA oversight for all other rainbow and cutthroat trout taxa, as they are not anadromous. Because of the importance of steelhead trout within the Columbia River Basin fisheries resource, they will be discussed further separately from resident (non-anadromous) rainbow trout.

Distributions and abundance of the many species and forms of rainbow and cutthroat trout have declined in the last 150 years to fractions of their historic ranges (10%–30% depending upon species) (Trotter 1989; Behnke 1992, 2002; Quigley et al. 1996). Reasons for declines are similar for all taxa. The declines are reasonably well documented, and concerns over the status of various species, subspecies, or distinct local populations have prompted a series of petitions for review or listing under the ESA. These include reviews of the status of steelhead populations coast-wide and sea-run coastal cutthroat trout from the Umpqua River for anadromous forms, as well as reviews of the status of interior rainbow trout (i.e., redband trout) and bull trout (species wide). At present, bull trout, Umpqua River coastal cutthroat trout, five ESUs of steelhead trout, and the Kootenai River sturgeon are listed as threatened or endangered under the ESA.

Rainbow trout and cutthroat trout have suffered primarily from habitat degradation and competition with introduced non-native salmonids, usually hatchery rainbow trout (Behnke 1992). In addition to habitat degradation, steelhead distributions and abundance have been impacted by hydroelectric construction, which eliminated access to large spawning areas above Grand Coulee Dam and the Hells Canyon complex of dams, as well as inducing passage mortalities on both adult and juvenile migrants. Bull charr or bull trout have suffered primarily from habitat degradation, but also from past fisheries management practices and from the introduction of non-native brook trout (actually a charr also) (Leary et al. 1993; Rieman and McIntyre 1993; Dambacker and Jones 1997; Goetz 1997; Watson and Hillman 1997; Jakober et al. 1998; Dunham et al. 2003).

Introduction of non-native salmonids impacts native salmonids in two major ways. First, introduced salmonids may serve as ecological competitors with native salmonids and reduce their abundance through competition for food or specific microhabitats (Fausch 1988). Second, non-native salmonids are frequently able to hybridize with native salmonids. This results in the introduction of non-native genes into the native population, which can reduce the reproductive fitness of the progeny. The degree to which the native population is affected depends, among other things, on the degree of outbreeding depression (i.e., reduction in fitness) that occurs after hybridization. For brook trout and bull trout hybrids, genetic and abundance data from Leary et al. (1993) suggests that brook × bull hybrids are strongly selected against. Brook trout appear to be replacing bull trout in several index streams in Montana and Idaho, probably due to earlier sexual maturation by brook trout and aggressive breeding behavior by brook trout males.

Hybridization and genetic introgression have also been documented many times for native rainbow and cutthroat trout populations (Campton and Johnston 1985; Campton and Utter 1985b; Currens et al. 1990; Carmichael et al. 1993; Williams et al. 1996; Campbell et al. 2002); however, this work has rarely been extended into an examination of fitness consequences of introgression. Nevertheless, introductions of non-native salmonids are generally recognized as one of the major factors in the decline of native salmonids in the Interior West (see indigenous species lists in Tables 4.3 and 5.1). Most states have taken steps to inventory native trout populations and protect those that

Table 4.3   Indigenous species-level taxa of trout and char with coastal or interior distributions.

Coastal species (anadromous and resident forms)

   Coastal Cutthroat Trout (*O. clarki clarki*)
   Coastal Rainbow Trout (*O. mykiss irideus*)
   Dolly Varden Char (*Salvelinus malma*)

Interior species (non-anadromous forms only)

   Yellowstone Cutthroat Trout (*O. c. bouvieri*)
   Westslope Cutthroat Trout (*O. c. lewisi*)
   Interior Rainbow Trout ("redband" trout) (*O. mykiss gairdneri*)
   Bull Trout or Bull Char (*S. confluentus*)

Interior species (also exhibiting anadromy)

   Interior Rainbow Trout ("redband" steelhead trout) (*O. mykiss gairdneri*)

---

are identified as remnant native stocks free of introgression from non-native salmonids.

All of these taxa exhibit a range of life history strategies, which include both migratory and resident (i.e., non-migratory) forms. All of the coastal salmonids as well as interior rainbow trout exhibit anadromy. Other interior salmonids exhibit resident and migratory life history strategies; the latter may include adfluvial and fluvial forms.

Evolutionary History and Genetic Structure

Genetic structure has been examined in some detail in cutthroat trout (Loudenslager and Gall 1980; Campton and Johnston 1985; Martin et al. 1985; Leary et al. 1987; Allendorf and Leary 1988; Behnke 1992) and bull trout (Leary et al. 1993; Kanda et al. 1997; Williams et al. 1997), but less so in rainbow trout (Wishard et al. 1984; Campton and Johnston 1985; Currens et al. 1990; Williams et al. 1996). Nevertheless, all three species show geographic patterns of genetic variation and divergence into major evolutionary lineages. Cutthroat trout and bull trout show additional geographic substructuring within major evolutionary lines; however, such patterns are less clear in rainbow trout, probably due to the more recent evolutionary derivation of many of the inland rainbow forms.

Rainbow and Steelhead Trout (*Oncorhynchus mykiss*)

Background

The rainbow trout group, which includes the rainbow trout (Figure 4.21) and allied forms, such as the Mexican golden (*O. chrysogaster*), Gila (*O. gilae gilae*), Apache (*O. g. apache*), California golden (*O. m. aquabonita*), and the redband trout, occurs throughout coastal rivers from northern Mexico to the Kuskokwim River in Alaska. In the Pacific Northwest, rainbow trout can be divided (by the Cascade Crest) into coastal and inland forms. Behnke (1992) has suggested subspecies-level distinction for the two groups and refers to the coastal form as *O. mykiss irideus* and the interior form as *O. m. gairdneri*. Inland rainbow trout (e.g., redbands) occur throughout the Columbia River Basin to barrier falls on the Snake, Spokane, Kootenay, and Clark Fork Rivers. Both coastal and interior rainbow trout exhibit anadromous and resident forms. Anadromous rainbow trout, regardless or coastal or interior origin, are referred to as steelhead.

Evolutionary History and Genetic Structure of Rainbow Trout

Rainbow (and steelhead), rainbow-like, and cutthroat trout evolved from a common ancestor that diverged from Pacific salmon approximately 5 million

Figure 4.21   Coastal rainbow trout. Illustration from actual specimen, by J. R. Tomelleri.

years ago (Behnke 1988, 1992, 2002). The rainbow and cutthroat lines diverged from one another about 2 million years ago. Substantial evolutionary divergence has occurred in each species; however, considerable controversy exists among systematists concerning delineation of species and subspecies forms in rainbow trout. Taxa relationships in the rainbow group are less clear than within the cutthroat species, probably due to the more recent evolutionary derivation of many of the inland rainbow forms (McCusker et al. 2000a, 2000b).

Historical and Present Distributions of Rainbow Trout

Native populations of rainbow trout, including coastal rainbow trout, have been reduced from their historic distributions (Currens et al. 1990; Behnke 1992; Quigley et al. 1996). Coastal and interior forms of rainbow trout have been dramatically affected by habitat degradation and by widespread introductions of hatchery-reared rainbow trout. In many larger river systems in the Interior West, such as the Kootenay and its tributary creeks, hatchery rainbow trout have survived in many instances and interbred with native interior rainbow trout populations (Sage and Leary 1995; Williams and Jaworski 1995). In contrast, hatchery rainbow trout stocked into small desert streams in southern Idaho and northern Nevada have had almost no genetic effect on native rainbow trout populations (Williams et al. 1996). Survival of hatchery rainbow trout is probably extremely low in the harsh environmental conditions of these cold desert stream systems.

Propagation Efforts of Rainbow Trout

Rainbow trout have been extensively propagated (Behnke 1992, 2002). The majority of hatchery rainbow trout strains appear to have been developed

Figure 4.22 Steelhead trout, adult male from the North Umpqua River, Oregon. Illustration from actual specimen, by J. R. Tomelleri.

from coastal rainbow trout, including both resident and anadromous forms, from the northern California area (Needham and Behnke 1962; Busack et al. 1979; Crawford 1979; Busack and Gall 1980). Hatchery-reared rainbow trout have been widely planted throughout the western United States and are thought to be one of the major factors, along with habitat degradation, in the decline of interior rainbow (i.e., redband) and cutthroat trout populations.

Status under the Endangered Species Act

Throughout the range of steelhead trout (Figure 4.22) south of the Canadian border, NMFS recognizes 15 ESUs, 5 of which occur in the Columbia River Basin. The ESUs and their statuses are listed in Table 4.4. Redband trout were petitioned for listing under the ESA in 1995; however, the listing was judged as not warranted at that time, in part due to lack of information. Future petitions for listing are expected to occur.

Table 4.4 ESA status for steelhead trout in the Columbia and Snake Rivers.

ESU	ESA status
Upper Columbia River	Endangered
Middle Columbia River	Threatened
Lower Columbia River	Threatened
Upper Willamette River	Threatened
Snake River Basin	Threatened

Cutthroat Trout (*Oncorhynchus clarki*)

Background

The cutthroat trout is a polytypic species that occurs over a wide geographic range of coastal and interior waters in the western United States and Canada. Sixteen subspecies have been recognized in the recent literature (Loudenslager and Gall 1980; Leary et al. 1987; Behnke 1992, 2002). Eight of these have large geographic distributions, while another eight are either undescribed subspecies, native to a very small geographic area, or both. Four subspecies occur within the Columbia River drainage. Three of these (coastal cutthroat, *O. c. clarki*; westslope cutthroat, *O. c. lewisi*; and Yellowstone cutthroat, *O. c. bouvieri*) have large geographic distributions, while the Snake River Finespot, *O. c.* spp., has a restricted distribution in the upper Snake River and its tributaries in eastern Idaho and western Wyoming. Yellowstone (Figure 4.23) and Snake River Finespot cutthroat trout occur only above Shoshone Falls (near Twin Falls, Idaho), and therefore, rarely figure into resident fish concerns in the Columbia River drainage. However, water abstractions from reservoirs upstream of Shoshone Falls (e.g., Palisades Reservoir) can affect populations of these subspecies (Thurow et al. 1988).

Evolutionary History and Genetic Structure of Cutthroat Trout

Cutthroat and rainbow trout diverged from one another about 2 million years ago (Behnke 1992, 2002). Substantial evolutionary divergence has occurred in cutthroat trout, resulting in great diversity in morphology, phenotypic traits, behavior, genetic attributes, and ecological adaptations (Leary et al. 1987; Trotter 1987; Allendorf and Leary 1988; Behnke 1992).

Figure 4.23   Yellowstone cutthroat trout. Illustration from actual specimen, by J. R. Tomelleri.

Cutthroat trout invaded the Columbia River before rainbow trout and diverged into four major evolutionary lines between 0.5 and 1 million years ago. The evolutionary lines are represented by the present subspecies of coastal, westslope, Yellowstone, and Lahontan (*O. c. henshawi*) cutthroat trout. The Columbia River drainage, including the Snake River above Shoshone Falls, includes populations of the first three of these subspecies, and therefore contains a substantial portion of the genetic diversity and evolutionary heritage of the cutthroat trout species. No other major river system in the western United States or Canada contained such taxonomic diversity among western trout. Our discussion is restricted to coastal and westslope cutthroat only, as they occur in portions of the Columbia River basin accessible to anadromous salmonids.

Subsequent evolution of the four major lines of cutthroat trout into the approximately 16 subspecies recognized today occurred quite recently, that is, within the last 100,000 years or less. Genetic divergence among the more recently evolved subspecies is low to non-existent (Leary et al. 1987; Shiozawa and Evans 1995), reflecting their recent evolutionary separation. Patterns of genetic structure within subspecies are not uniform. Some subspecies appear to have little divergence among populations (e.g., Yellowstone, Snake River Finespot, Lahontan, Humboldt), whereas others appear to have undergone local adaptation resulting in greater divergence among populations (e.g., westslope and coastal) (Loudenslager and Gall 1980; Leary et al. 1987; Shiozawa and Evans 1995). Obviously, strategies to conserve genetic diversity would differ for these two groups of subspecies. Where little divergence occurs among populations, preservation of a small number of populations is likely to conserve a large portion of the genetic diversity that exists within that subspecies. In contrast, where substantial divergence occurs among populations within a subspecies, conservation efforts are going to have to be directed at the local population level in order to conserve genetic diversity.

Coastal and westslope cutthroat trout appear to contain substantial amounts of genetic variation that is highly structured as compared to most other inland subspecies of cutthroat trout. Genetic studies of coastal cutthroat trout (Campton and Utter 1985a) revealed genetic differences among groups of populations from different geographic locations, suggesting a lack of gene flow among populations over geographic scales and the likelihood of substantial local adaptation for populations. Genetic variation among westslope cutthroat populations (Leary et al. 1987; Allendorf and Leary 1988) showed significant differences among populations, but did not reveal any particular geographic structuring to the variation. Nevertheless, for both subspecies, genetic structuring is apparent among local populations. Thus, conservation efforts for both subspecies must be directed at least at the local watershed scale, if not at the population level.

Propagation Efforts for Cutthroat Trout

Most of the interior subspecies of cutthroat trout were propagated at one time or another (Behnke 1992); however, little recognition was given to the uniqueness of each subspecies, so that stocks from different subspecies were frequently mixed or transplanted. For example, because of the ease of collection of spawning adults from tributary streams of Yellowstone Lake, the Yellowstone cutthroat trout has had more propagation effort and has been more widely distributed via stocking than other cutthroat trout subspecies (Gresswell 1979, 1988). In spite of these early, large-scale hatchery and stock transfer programs, genetic assays of present-day cutthroat trout populations reveal little incidence of genetic introgression (M. Powell and R. Williams unpublished data; Shiozawa and Evans 1995; Campbell et al. 2002). Thus, it appears that most stock transfers of cutthroat trout outside their native distribution did not result in hybridization with the indigenous trout (Williams 1991; Williams and Jaworski 1995).

Coastal Cutthroat Trout (*Oncorhynchus clarki clarki*)

Background

Coastal cutthroat trout (Figure 4.24) occur from Prince William Sound in Alaska south to the Eel River in California with their distribution corresponding closely to the Pacific coastal rainforest belt (Trotter et al. 1993). Typically, coastal cutthroat do not occur east of the Cascade Range in Washington and Oregon. Throughout its range, both anadromous and nonmigratory resident forms exist. Anadromous forms show little differentiation across the range, whereas isolated resident forms exhibit considerable divergence in morphological characters. Like many of the other cutthroat trout subspecies, coastal cutthroat exhibit a diversity of life history strategies, even among resident forms (Trotter 1989; Behnke 1992; Johnson et al. 1994). Trotter et al. (1993) identify, at minimum, three life history strategies among

Figure 4.24   Coastal cutthroat trout. Illustration from actual specimen, by J. R. Tomelleri.

resident populations (resident, fluvial, and adfluvial), in addition to the anadromous form.

Historic and Present Distribution of Coastal Cutthroat Trout

Like all subspecies of cutthroat trout, coastal cutthroat trout distributions and abundance have declined dramatically since historic times. The subspecies probably suffers more from decreases in abundance than decreases in distribution. Nehlsen et al. (1991) considered almost all native populations of sea-run cutthroat in the western United States to be at some risk of extinction due largely to pervasive continuing declines in stock size.

Causes of decline are typical for cutthroat trout in general: habitat degradation due to logging, urban development, or mainstem passage; competition or hybridization from non-native and hatchery trout; and overharvesting by anglers (Trotter 1987, 1989; Nehlsen et al. 1991). Coastal cutthroat throughout its range and westslope cutthroat in the Columbia drainage co-evolved with rainbow trout. Although low levels of gene flow probably occur between the two species (Leary et al. 1987), hybridization with non-native rainbow trout has probably had little effect on coastal cutthroat. In contrast, hybridization with non-native rainbow trout is one of the major factors in the decline of other interior cutthroat trout subspecies, which historically had allopatric distributions from rainbow trout.

Status under the Endangered Species Act

Throughout the range of coastal cutthroat trout south of the Canadian border, NMFS recognizes seven ESUs, two of which occur in the Columbia River Basin. The southwestern Washington/Columbia River ESU is proposed for threatened status, while the upper Willamette ESU is deemed not warranted for listing. The Umpqua River coastal cutthroat ESU was listed as endangered in 1996, but in 1999 was delisted, as genetic evidence suggested it was part of the larger Oregon coastal cutthroat trout ESU.

Westslope Cutthroat Trout (*Oncorhynchus clarki lewisi*)

Background

Westslope cutthroat trout (Figure 4.25) are native to the upper Missouri and Columbia drainages. West of the Continental Divide, the natural distribution includes the following rivers: upper Kootenay, Clark Fork, Spokane, Coeur d'Alene, St. Joe, Clearwater, and Salmon. Isolated disjunct populations of westslope also occur in the John Day River and various tributaries of the middle Columbia, including the Lake Chelan drainage, and numerous

Figure 4.25   Westslope cutthroat trout. Illustration from actual specimen, by J. R. Tomelleri.

tributaries of the Methow River. These disjunct populations may be remnants from the late-Pleistocene flooding of Lake Missoula.

Historical and Present Distribution of Westslope Cutthroat Trout

Westslope cutthroat trout have undergone dramatic range reductions. Liknes
and Graham (1988) estimated that genetically pure westslope cutthroat trout
in Montana currently occur in 2.5% of their historical range. The Salmon,
Clearwater, St. Joe, and upper Flathead Rivers all appear to be strongholds
for westslope cutthroat trout. In Idaho, their occurrence is strongly correlated
with federal land status; that is, most strong populations of westslope cutthroat occur in designated or proposed wilderness areas.

Causes of decline are typical for inland cutthroat trout in general: habitat
degradation, competition or hybridization from non-native and hatchery trout,
and overharvesting by anglers (Nehlsen et al. 1991; Trotter et al. 1993; Behnke
2002). Fisheries agencies have realized the greater vulnerability of cutthroat
trout to angling harvest than rainbow or brown trout (*Salmo trutta*), and frequently, westslope cutthroat populations are now protected by special regulations, specifically catch-and-release. These special regulations have helped
maintain westslope cutthroat trout populations in the St. Joe River, Kelly
Creek in the Clearwater River, and the Middle Fork of the Salmon River. All
three locations are well known in the angling world and function as fishing des-
tinations because of the cutthroat trout and the special regulations.

Bull Charr (*Salvelinus confluentus*)

Background

Bull trout (Figure 4.26), more correctly know as bull charr, are one of five currently recognized species in the genus *Salvelinus* in North America. It has been

Figure 4.26 Bull trout. Illustration from actual specimen, by J. R. Tomelleri.

recognized as a "species of special concern" by the American Fisheries Society (Williams et al. 1989) and by many state agencies. Concern for the bull charr's status prompted petitions for review or listing under the ESA in October 1992 and January 1993. Review by the USFWS resulted in a decision that listing was warranted, but precluded. A more recent review by the USFWS resulted in the listing of bull charr populations in the Columbia and Klamath rivers as threatened.

Evolutionary History and Genetic Structure of Bull Charr

The genus *Salvelinus* includes a number of species complexes that have confounded systematists for some time. As many as 45 different scientific names have been applied to North American char (Bond 1992); however, most current systematists recognize only five species. The bull charr was formally described by Cavender (1978) after he examined them and Dolly Varden (*S. malma*) specimens from throughout their respective ranges and identified species-level diagnostic morphological characters. Cavender (1978) suggested that bull charr originated in the Columbia River and has extended and constricted its range according to climate changes. Its recent historic distribution extends from the McCloud River in northern California through inland western North America to the upper Yukon and MacKenzie drainages in Canada (Bond 1992).

Genetic studies of bull charr populations throughout the Columbia and Klamath River drainages (Leary et al. 1993; Williams et al. 1997) show evidence of macrogeographic genetic structure. Both allozyme and mitochondrial DNA analyses differentiated bull charr in the Klamath drainage from bull charr in the Columbia drainage at a level typical of the major subspecies in cutthroat trout. Within the Columbia River drainage, bull charr from the lower Columbia (Deschutes and Lewis Rivers) formed an evolutionarily distinct group from bull charr populations in the remainder of the Columbia River above the John Day River (Williams et al. 1997). Bull charr populations in the Columbia River system above the Deschutes River shared a common mitochondrial DNA pattern that is suggestive of a single founding population (Williams et al. 1997). Allozyme data for the same populations (Leary et al. 1993), in spite of showing little overall genetic variation, revealed significant differences among upper Columbia River bull charr populations. Taken together, the mtDNA and allozyme data show that populations were once linked genetically, but have been separated long enough to accrue population-specific allozyme profiles. Thus, historic linkages among bull charr populations in the upper Columbia River have been broken. The genetic data do not provide insight into whether fragmentation of the historic metapopulation structure is a result of natural processes (gradual warming and drying of climate in the Intermountain West) or human-induced changes in habitat quality.

Riemen and McIntyre (1993) advocate a conservation approach for bull charr protection and restoration that focuses on identifying core areas that contain linked bull charr populations in high-quality habitat. The Flathead River system in northwestern Montana above Flathead Lake may represent one such potential core area. Genetic studies of bull charr within the Flathead subbasin (Kanda et al. 1997) suggest intact metapopulation structure within most of the major drainages, but little gene flow among populations from different drainages.

Historical and Present Distribution of Bull Charr

The current distribution of bull charr in the Pacific Northwest and Intermountain West is fragmented. Populations occur coldwater primarily in pristine or nearly pristine headwater regions of the Columbia and Klamath drainages (Rieman and McIntyre 1993). Many populations have undergone significant declines in recent years (Howell and Buchanan 1992; Thomas 1992). Because bull charr populations are now restricted to headwater regions and much of the historic metapopulation structure is now fragmented, vulnerability to extinction has increased for individual populations (Rieman and McIntyre 1993).

Propagation Efforts for Bull Charr

Bull charr have been little used in propagation efforts; however, the USFWS initiated some exploratory propagation efforts at the Creston National Fish Hatchery in Montana in the mid-1990s.

Status under the Endangered Species Act

Throughout the range of bull charr south of the Canadian border, the USFWS recognizes three ESUs, two of which occur in the Columbia River Basin. All three ESUs (Klamath River, Columbia River, and Jarbidge River) are listed as threatened.

Status of Indigenous Species other than Salmonids

Sturgeon (Green Sturgeon [*Acipenser medirostris*], White Sturgeon [*Acipenser transmontanus*])

Background

Sturgeon (Figure 4.27) are an ancient anadromous fish, which were formerly widely distributed on all continents in the northern hemisphere. Two species of sturgeon occur in the Columbia River Basin. During the 20th century, extensive disruption of freshwater and estuarine habitats coupled with heavy exploitation severely reduced populations of sturgeon throughout their range.

Historic and Present Distribution of Sturgeon

Green sturgeon are found in the lower 40 miles of the Columbia River, in its estuary, and in the adjacent marine waters. The green sturgeon has not been reported in the Columbia River above Bonneville Dam, River Mile 145, and it is thought to be concentrated in the lower 40 miles of the main river (Oregon Department of Fish and Wildlife and Washington Department of Fish and Wildlife 1995). Green sturgeon reach lengths of up to 7 feet, and females are sexually mature at 5 to 6 feet in length. Information on the spawning period, spawning behavior, and other details of the reproductive

Figure 4.27 White sturgeon. Illustration from actual specimen, by J. R. Tomelleri.

biology of green sturgeon in the Columbia River is lacking (Oregon Department of Fish and Wildlife and Washington Department of Fish and Wildlife 1995).

White sturgeon were once widely distributed among the watersheds of the Columbia River Basin, and they still enjoy a higher abundance and wider geographic distribution than the green sturgeon. White sturgeon below Bonneville Dam exhibit the anadromy characteristic of the species; however, sturgeon in the reservoirs above Bonneville Dam may be capable of completing their reproductive cycle within a single reservoir (Parsley et al. 1993; Parsley and Beckman 1994). Sexual maturity is found in males of 4 feet and longer and in females 6 feet and longer. Females have fecundity proportional to length, with 100,000 to 300,000 eggs per female. Spawning does not occur annually, but at 2- to 4-year intervals. Fecundity may be proportional to the length of time between spawnings. Spawning requires fast flowing waters over rocky substrate at temperatures of 48° to 62°F in May and June.

White sturgeon in the lower Columbia River 3 feet long or less grow at the rate of about 3 inches per year. Sturgeon beyond 3 feet in length grow at 3 inches per year until sexual maturation, when annual growth slows substantially. Sturgeon are about 8 inches long at 1 year of life and attain the length of 6 feet at 23 years of age. The time span between the lengths of 3.5 to 5.5 feet in length is about 10 years.

Dams constrain the movements of white sturgeon, creating isolated populations in the reservoirs of the Columbia River power system (Beamesderfer and Nigro 1993; Parsley et al. 1993; Parsley and Beckman 1994). Productivity of the isolated populations is lower than in the unimpounded river system due to impacts of hydroelectric system operation on the reproductive activities. Low flows in May and June inhibit spawning and subsequent recruitment. Appropriate rearing habitats for juvenile and adult sturgeon are provided within the reservoirs. However, severe population reductions occurred during the early 1980s in the John Day and The Dalles Reservoirs as a result of fishing.

Harvest of Sturgeon

Commercial white sturgeon fisheries began in the 1880s reaching a peak of 6 million pounds in 1892, with catches declining sharply by 1899. During this time, the average individual in the harvest was 7 feet and 150 pounds. With protection of the broodstock afforded by maximum size limits on harvests imposed in 1950, recovery of the populations became possible. Sturgeon stocks appeared to rebound in the 1970s approximately 20 years after the maximum size limit on harvests was imposed. Contemporary fisheries harvest the same number of sturgeon harvested during the 1890s; however, the average size is much lower, so the annual harvest is about 1 million pounds.

Population levels in the John Day and The Dalles pools have declined sharply, probably in response to levels of exploitation. In the upper Columbia River and in the Snake River, sturgeon population abundance varies from one impounded section to another, with populations in some sections perhaps approximating historic numbers.

Research and Propagation of Sturgeon

Ongoing research programs are conducted by the Columbia River Inter-Tribal Fish Commission, and the states of Washington, Oregon, Idaho, and the Nez Perce Tribe of Idaho. Research is focused on understanding the harvest, population dynamics, and reproductive biology of white sturgeon, following recommendations made by Beamesderfer and Nigro (1993). Some hatchery production of sturgeon has occurred in Oregon and Idaho.

Status under the Endangered Species Act

The Kootenai River sturgeon is listed by the USFWS as an endangered ESU.

Pacific Lamprey (*Lampetra tridentata*)

Background, Distribution, and Status

The Pacific lamprey is a jawless anadromous fish, which is widely distributed in western North America and eastern Asia. It is one of three species of lamprey in the Columbia River Basin along with the anadromous river lamprey (*L. ayresii*) and the resident brook lamprey (*L. richardsoni*). Numerous factors, including loss of freshwater habitat and construction of hydroelectric dams, have contributed to its near extirpation in the Snake River portion of the Columbia River Basin, and to the reduction in numbers of adults seen at the counting windows on the hydroelectric dams (Close et al. 1995)

During its marine residency, adult lamprey are obligate parasites on adult bony fishes, including salmon (Scott and Crossman 1973). Because of this, management agencies have either ignored them or attempted to eradicate them. In any event, specific data on the age, growth, and productivity of Pacific lamprey in the Columbia River Basin is limited (Kan 1975). In general, adults spawn in small tributaries at an age of about 7 years. The young rear in tributaries in the form of early juveniles called ammocoetes, and in the main river as late juveniles, neither of which are parasitic life history stages. As adults in the marine environment, lamprey attach themselves to hosts where they subsist on bodily fluids extracted through a hole bored in the host's side. Lamprey may return to spawn at around age 7.

Lamprey have had difficulty adapting to the hydroelectric dams. Because lamprey utilize much of the same freshwater spawning habitat as do spring Chinook salmon, it may be inferred that lamprey have been reproductively disadvantaged to the same extent, as have the Chinook, due to logging, grazing, agriculture, mining, and other natural resource extraction activities. However, upstream passage at mainstem hydroelectric projects has been identified as an additional life history bottleneck for adult lamprey as compared to spring Chinook salmon.

The role of lamprey in the ecosystem as a prey item, and as a force in the biogeochemical cycle, merits consideration. Their role in bringing nutrients into the predominantly oligotrophic Snake River Basin may have contributed directly to salmon production in that region.

Research and Propagation of Pacific Lamprey

Native Americans prize the lamprey as a ceremonial food item, and annual subsistence and ceremonial harvests on the order of several thousand "eels" are taken by the tribes. The Council has called for a lamprey research program, and several institutions have developed background information and recommended an approach to monitoring and management (Close et al. 1995).

Conclusions and Implications

Different species and populations of salmonids in the Columbia River and elsewhere exhibit remarkable phenotypic, life history, ecological, behavioral, and genetic diversity. The diversity described in this chapter, which is a hallmark of salmonids in general, arose from differential or local adaptation to the varied and variable environments within the complex landscapes of the Columbia Basin. The diversity has resulted from the plasticity, adaptability, productivity, and long-term persistence of salmonids in the fluctuating geological and environmental landscapes of the Pacific Northwest. Such diversity, which buffers salmonid populations against both short- and long-term scales of environmental variation, has become even more important today as human activities have increased the rate and amplitude of environmental fluctuations over those salmon experienced historically. We believe diversity (phenotypic, life history, genetic, ecological, etc.) within and among salmon populations is critical to the long-term persistence of salmon in the Columbia River ecosystem. We also believe salmon populations in the Columbia River today can form the base for rebuilding salmon abundance and diversity.

Literature Cited

Allendorf, F. W., and R. F. Leary. 1986. Heterozygosity and fitness in natural populations of animals. Pages 57–76 in M. Soulé, ed. *Conservation biology: The science of scarcity and diversity.* Sinauer Associates, Sunderland, Massachusetts.

Allendorf, F. W., and R. F. Leary. 1988. Conservation and distribution of genetic variation in a polytypic species, the cutthroat trout. *Conservation Biology* 2:170–184.

Allendorf, F. W., and S. R. Phelps. 1981a. Isozymes and the preservation of genetic variation in salmonid fishes. Pages 37–52 in N. Ryman, ed. *Fish gene pools.* Ecological Bulletins, Stockholm, Sweden.

Allendorf, F. W., and S. R. Phelps. 1981b. Use of allelic frequencies to describe population structure. *Canadian Journal of Fisheries and Aquatic Sciences* 38:1507–1514.

Allendorf, F. W., and F. M. Utter. 1974. Biochemical systematics of the genus *Salmo. Animal Blood Groups Biochemistry and Genetics* 5:1–33.

Allendorf, F. W., and F. M. Utter. 1979. Population genetics. *Fish Physiology* 8:407–454.

Allendorf, F. W., and R. S. Waples. 1996. Conservation and genetics of salmonid fishes. Pages 238–280 in J. C. Avise, ed. *Conservation genetics: Case histories from nature.* Chapman and Hall, New York.

Altukhov, Y. P. 1981. The stock concept from the viewpoint of population genetics. *Canadian Journal of Fisheries and Aquatic Sciences* 38:1523–1528.

Altukhov, Y. P., and E. A. Salmenkova. 1991. The genetic structure of salmon populations. *Aquaculture* 98:11–40.

Aro, K. V., and M. P. Shepard. 1967. Pacific salmon in Canada. *International North Pacific Fishery Commission Bulletin* 23:225–327.

Bartley, D. M., and G. A. E. Gall. 1990. Genetic structure and gene flow in Chinook salmon populations of California. *Transactions of the American Fisheries Society* 119:55–71.

Beacham, T. D., R. E. Withler, and A. P. Gould. 1985. Biochemical genetic stock identification of chum salmon (*Oncorhynchus keta*) in southern British Columbia. *Canadian Journal of Fisheries and Aquatic Sciences* 42:437–448.

Beacham, T. D., R. E. Withler, C. B. Murray, and L. W. Barner. 1988. Variation in body size, morphology, egg size, and biochemical genetics of pink salmon in British Columbia. *Transactions of the American Fisheries Society* 117:109–126.

Beamesderfer, R. C., and A. A. Nigro. 1993. Status and habitat requirements of the white sturgeon populations in the Columbia River downstream from McNary Dam. Volume I. Bonneville Power Administration. Final Report, Project No. 86–50, Contract Number DE-AAI79-86BP63584. Portland, Oregon.

Beatty, R. E. 1992. Changes in size and age at maturity of Columbia River upriver bright fall Chinook salmon (*Oncorhynchus tshawytscha*): Implications for stock fitness, commercial value, and management. Master's thesis, Oregon State University, Corvallis. 270 p.

Becker, C. D. 1985. Anadromous salmonids of the Hanford Reach, Columbia River: 1984 status. Pacific Northwest Laboratories. PNL-5371. Richland, Washington.

Behnke, R. J. 1988. Phylogeny and classification of cutthroat trout. *American Fisheries Society Symposium* 4:1–7.

Behnke, R. J. 1992. *Native trout of western North America.* American Fisheries Society, Bethesda, Maryland.

Behnke, R. J. 2002. *Trout and salmon of North America.* Free Press, New York.

Beiningen, K. T. 1976. Fish Runs, Report E. *Investigative Reports of Columbia River Fisheries Project.* Pacific Northwest Regional Commission, Portland, Oregon.

Bernatchez, L., and J. J. Dodson. 1994. Phylogenetic relationships among Palearctic and Nearctic whitefish (*Coregonus sp.*) populations as revealed by mitochondrial DNA variation. *Canadian Journal of Fisheries and Aquatic Sciences* 51:240–251.

Bond, C. E. 1992. Notes on the nomenclature and distribution of the bull trout and the effects of human activity on the species. Pages 1–4 in P. J. Howell and D. B. Buchanan, eds. *Gearhart Mountain Bull Trout Workshop, August, 1992*. Oregon Chapter of the American Fisheries Society, Gearhart Mountain, Oregon.

Borgerson, L. 1992. Memo to Jim Lichatowich.

Brannon, E, M. Powell, T. Quinn, and A. Talbot. 2002. Population structure of Columbia River Basin Chinook salmon and steelhead trout. Report to Bonneville Power Administration. Portland, Oregon, 178 p.

Brown, J. H., and A. Kodric-Brown. 1977. Turnover rates in insular biogeography: Effects of immigration on extinction. *Ecology* 58:445–449.

Burgner, R. L. 1991. Life history of sockeye salmon (*Oncorhynchus nerka*). Pages 3–117 in C. Groot and L. Margolis, eds. *Pacific salmon life histories*. UBC Press, Vancouver, British Columbia.

Busack, C., and J. B. Shaklee. 1995. Genetic diversity units and major ancestral lineages of salmonid fishes in Washington. Washington State Department of Fish and Wildlife. Technical Report, RAD-95–02. Olympia, Washington.

Busack, C. A., and G. A. E. Gall. 1980. Ancestry of artificially propagated California rainbow trout strains. *California Fish and Game* 66:17–24.

Busack, C. A., R. Halliburton, and G. A. E. Gall. 1979. Electrophoretic variation and differentiation in four strains of domesticated rainbow trout (*Salmo gairdneri*). *Canadian Journal of Genetics and Cytology* 21:81–94.

Busby, P. J., T. C. Wainwright, G. J. Bryant, L. J. Lierheimer, and R. S. Waples. 1996. Status review of west coast steelhead from Washington, Idaho, Oregon, and California. National Marine Fisheries Service, Northwest Fisheries Science Center. NOAA Technical Memorandum, NMFS-NWFSC-10. Seattle, Washington.

Calaprice, J. R. 1969. Production and genetic factors in managed salmonid populations. Pages 377–388 in T. G. Northcote, ed. *Symposium on salmon and trout in stream*. Institute of Fisheries, University of British Columbia, Vancouver.

Campbell, M. R., J. Dillon, and M. S. Powell. 2002. Hybridization and introgression in a managed, native population of Yellowstone cutthroat trout: genetic detection and management implications. *Transactions of the America Fisheries Society* 131:364–375.

Campton, D. E., and J. M. Johnston. 1985. Electrophoretic evidence for a genetic admixture of native and nonnative rainbow trout in the Yakima River, Washington. *Transactions of the American Fisheries Society* 114:782–793.

Campton, D. E., and F. M. Utter. 1985a. Genetic structure of anadromous cutthroat trout (*Salmo clarki clarki*) populations in the Puget Sound area: Evidence for restricted gene flow. *Canadian Journal of Fisheries and Aquatic Sciences* 42:110–119.

Carl, L. M., and M. C. Healey. 1984. Differences in enzyme frequency and body morphology among three juvenile life history types of Chinook salmon (*Oncorhynchus tshawytscha*) in the Nanaimo River, British Columbia. *Canadian Journal of Fisheries and Aquatic Sciences* 41:1070–1077.

Carmichael, G. J., J. N. Hanson, M. E. Schnidt, and D. C. Morizot. 1993. Introgression among apache, cutthroat, and rainbow trout in Arizona. *Transactions of the American Fisheries Society* 122:121–130.

Cavender, T. M. 1978. Taxonomy and distribution of the bull trout, *Salvelinus confluentus* (Suckley), from the American Northwest. *California Fish and Game* 64:139–174.

Chapman, D., A. Giorgi, M. Hill, A. Maule, S. McCutcheon, D. Park, W. Platts, K. Pratt, J. Seeb, L. Seeb, and F. Utter. 1991. Status of Snake River Chinook salmon. Don Chapman Consultants, Boise, Idaho. 520 p.

Chapman, D., A. Giorgi, T. Hillman, D. Deppert, M. Erho, S. Hays, C. Peven, B. Suzumoto, and R. Klinge. 1994a. Status of summer/fall Chinook salmon in the Mid-Columbia region. Don Chapman Consultants, Boise, Idaho. 411 p.

Chapman, D., C. Peven, A. Giorgi, T. Hillman, and F. Utter. 1995. Status of spring Chinook salmon in the mid-Columbia region. Don Chapman Consultants, Boise, Idaho. 477 p.

Chapman, D., C. Peven, T. Hillman, A. Giorgi, and F. Utter. 1994b. Status of summer steelhead in the Mid-Columbia river. Don Chapman Consultants, Boise, Idaho. 235 p.

Chapman, D. W. 1986. Salmon and steelhead abundance in the Columbia River in the nineteenth century. *Transactions of the American Fisheries Society* 115:662–670.

Close, D. A., M. Fitzpatrick, H. Li, B. Parker, D. Hatch, and G. James. 1995. Status report of the Pacific lamprey (*Lampetra tridentata*) in the Columbia River basin. US Department of Energy, Bonneville Power Administration, Division of Fish and Wildlife. Contract No. 95BI-39067. Portland, Oregon.

Cobb, J. N. 1930. Pacific salmon fisheries. Bureau of Fisheries. Document No. 1092. Washington, DC.

Craig, J. A. 1935. The effects of power and irrigation projects on the migratory fish of the Columbia River. *Northwest Science* 9:19–22.

Craig, J. A., and R. L. Hacker. 1940. The history and development of the fisheries of the Columbia River. *U.S. Bureau of Fisheries Bulletin* 32:133–216.

Crawford, B. A. 1979. The origin and history of trout broodstocks of the Washington Department of Game. Washington Department of Game. Fisheries Research Report. Olympia, Washington.

Currens, K. P., C. B. Schreck, and H. W. Li. 1990. Allozyme and morphological divergence of rainbow trout (*Oncorhynchus mykiss*) above and below waterfalls in the Deschutes River, Oregon. *Copeia* 1990:730–746.

Dambacker, J. M., and K. Jones. 1997. Stream habitat of juvenile bull trout populations in Oregon and benchmarks for habitat quality. Pages 353–360 in W. C. Mackay, M. K. Brewin, and M. Monita, eds. *Friends of the bull trout conference proceedings*. Bull Trout Task Force (Alberta), c/o Trout Unlimited, Calgary.

Dauble, D. D., T. P. Hanrahan, D. R. Geist, and M. J. Parsley. 2003. Impacts of the Columbia River hydroelectric system on main-stem habitats of fall Chinook salmon. *North American Journal of Fisheries Management* 23:641–659.

Dauble, D. D., and D. G. Watson. 1997. Status of fall Chinook populations in the Mid-Columbia River, 1948–1992. *North American Journal of Fisheries Management* 17:283–300.

DeLoach, D. B. 1939. The salmon canning industry. *Oregon State Monographs* Economic Studies No. 1.

Diamond, J. M. 1984. "Normal" extinctions of isolated populations. Pages 191–246 in M. N. Nitecki, ed. *Extinctions*. Chicago Press, Chicago, Illinois.

Dunham, J., B. Rieman, and G. Chandler. 2003. Influences of temperature and environmental variables on the distribution of bull trout within streams at the southern margin of its range. *North American Journal of Fisheries Management* 23:894–904.

Fausch, K. D. 1988. Tests of competition between native and introduced salmonids in streams: What have we learned? *Canadian Journal of Fisheries and Aquatic Sciences* 45.

Fitch, W. M., and M. Margoliash. 1967. Construction of phylogenetic trees. *Science* 155:279–284.

Flagg, T. A., F. W. Waknitz, D. J. Maynard, G. B. Milner, and C. V. W. Mahkhen. 1995. The effect of hatcheries on native coho salmon populations in the lower Columbia River. *Uses and Effects of Cultured Fishes in Aquatic Ecosystems Symposium* 15:366–375.

Fournier, D. A., T. D. Beacham, B. E. Riddell, and C. A. Busack. 1984. Estimating stock composition in mixed stock fisheries using morphometric, meristic, and electrophoretic characteristics. *Canadian Journal of Fisheries and Aquatic Sciences* 41:400–408.

French, R. R., and R. J. Wahle. 1959. Biology of Chinook and blueback salmon and steelhead in the Wenatchee River system. US Fish and Wildlife Service. Special Scientific Report, Fisheries No. 304.

Frissell, C. A., W. J. Liss, C. E. Warren, and M. D. Hurley. 1986. A hierarchical framework for stream habitat classification: Viewing streams in a watershed context. *Environmental Management* 10:199–214.

Fryer, J. K. 1995. Columbia River Sockeye Salmon. Ph.D. dissertation, University of Washington, Seattle.

Fulton, L. A. 1968. Spawning areas and abundance of Chinook salmon (*Oncorhynchus tshawytscha*) in the Columbia River Basin—past and present. USDI, Fish and Wildlife Service. Special Scientific Report, Fisheries No. 571.

Fulton, L. A. 1970. Spawning areas and abundance of steelhead trout and coho, sockeye, and chum salmon in the Columbia River Basin—past and present. USDC, NOAA, NMFS. Special Scientific Report, Fisheries No. 618.

Gall, G. A. E., D. Bartley, B. Bentley, J. Brodziak, R. Gomulkiewicz, and M. Mangel. 1992. Geographic variation in population genetic structure of Chinook salmon from California and Oregon. *Fishery Bulletin* 90:77–100.

Garcia, A. P., W. P. Connor, and R. H. Taylor. 1994. Fall Chinook salmon spawning ground surveys in the Snake River. Annual Report for 1993. Pages 1–21 in D. W. Rondorf and K. F. Tiffan, eds. *Identification of the Spawning, Rearing, and Migratory Requirements of Fall Chinook Salmon in the Columbia River Basin*. Bonneville Power Administration, Portland, Oregon.

Geist, D. R. 1995. The Hanford Reach: What do we stand to lose? *Illahee* 11:130–141.

Gharrett, A. J., S. M. Shirley, and G. R. Tromble. 1987. Genetic relationships among populations of Alaskan Chinook salmon (*Oncorhynchus tshawytscha*). *Canadian Journal of Fisheries and Aquatic Sciences* 44:765–774.

Gharrett, A. J., and W. W. Smoker. 1991. Two generations of hybrids between even- and odd-year pink salmon (*Oncorhynchus gorbuscha*): A test for outbreeding depression? *Canadian Journal of Fisheries and Aquatic Sciences* 48:1744–1749.

Gharrett, A. J., and W. W. Smoker. 1993. Genetic components in life history traits contribute to population structure. Pages 197–202 in J. G. Cloud and G. H. Thorgaard, eds. *Genetic conservation of salmonid fishes*. Plenum Press, New York.

Gharrett, A. J., C. Smoot, A. J. McGregor, and P. B. Holmes. 1988. Genetic relationships of even-year northwestern Alaskan pink salmon. *Transactions of the American Fisheries Society* 117:536–545.

Gilbert, C. H. 1912. Age at maturity of the Pacific coast salmon of the genus *Oncorhynchus*. US Bureau of Fisheries Bulletin, Washington, DC.

Goetz, F. A. 1997. Habitat use of juvenile bull trout in Cascade mountain streams of Oregon and Washington. Pages 339–352 in W. C. Mackay, M. K. Brewin, and M. Monita, eds. *Friends of the bull trout conference proceedings*. Bull Trout Task Force (Alberta), c/o Trout Unlimited, Calgary.

Gotelli, N. J. 1991. Metapopulation models: The rescue effect, the propagule rain, and the core-satellite hypothesis. *The American Naturalist* 138:768–776.

Gresswell, R. E. 1979. Yellowstone Lake—A lesson in fishery management. Pages 143–147 in W. King, ed. *Wild Trout II*. Federation of Fly Fishermen, Yellowstone National Park.

Gresswell, R. E., ed. 1988. *Status and management of interior stocks of cutthroat trout*. American Fisheries Society, Bethesda, Maryland.

Groot, C., and L. Margolis, eds. 1991. *Pacific salmon life histories*. University of British Columbia, Vancouver.

Hanski, I. 1991. Single-species metapopulation dynamics: concepts, models, and observations. *Biological Journal of the Linnean Society* 42:17–38.

Hanski, I., and M. Gilpin. 1991. Metapopualtion dynamics: brief history and conceptual domain. *Biological Journal of the Linnean Society* 42:3–16.

Hanski, I. A., and M. E. Gilpin. 1997. *Metapopulation biology: Ecology, genetics, and evolution.* Academic Press, New York.

Hard, J. J., R. G. Kope, W. S. Grant, F. W. Waknitz, L. T. Parker, and R. S. Waples. 1996. Status review of pink salmon from Washington, Oregon, and California. US Department of Commerce. NOAA Tech Memo, NMFS-NWFSC-25. Seattle, Washington. 131 p.

Harrison, S. 1991. Local extinction in a metapopulation context: an empirical evaluation. *Biological Journal of the Linnean Society* 42:73–88.

Harrison, S. 1994. Metapopulations and conservation. Pages 111–128 in P. J. Edwards, R. M. May, and N. R. Webb, eds. *Large-scale ecology and conservation biology.* Blackwell Scientific Publications, Oxford, UK.

Harrison, S., and J. F. Quinn. 1989. Correlated environments and the persistence of metapopulations. *Oikos* 56:293–298.

Hatch, D. R., D. R. Pederson, J. K. Fryer, and M. Schwartzberg. 1995. Wenatchee River Salmon Escapement Estimates using Video Technology in 1994. Columbia River Inter-Tribal Fish Commission. Technical Report 95–3. Portland, Oregon.

Healey, M. C. 1991. Life history of Chinook salmon. Pages 313–393 in C. Groot and L. Margolis, eds. *Pacific salmon life histories.* University of British Columbia Press, Vancouver, BC.

Healey, M. C. 1994. Variation in the life history characteristics of Chinook salmon and its relevance to conservation of the Sacramento winter run of Chinook salmon. *Conservation Biology* 8:876–877.

Helle, J. H. 1981. Significance of the stock concept in artificial propagation of salmonids in Alaska. *Canadian Journal of Fisheries and Aquatic Sciences* 38:1665–1671.

Howell, P., K. Jones, D. Scarnecchi, L. LaVoy, W. Kendra, and D. Ortmann. 1985. Stock assessment of Columbia River anadromous salmonids. Volume I: Chinook, coho, chum and sockeye salmon stock summaries. Bonneville Power Administration and Oregon Department of Fish and Wildlife. Final Report.

Howell, P. J., and D. B. Buchanan, eds. 1992. *Proceedings of the Gearhart Mountain Bull Trout Workshop, August, 1992, Gearhart Mountain, Oregon.* Oregon Chapter of the American Fisheries Society, Corvallis, Oregon.

Hume, R. D. 1893. Salmon of the Pacific Coast. Schmidt Label & Lithographic Co., San Francisco.

Huntington, C., W. Nehlsen, and J. Bowers. 1996. A survey of healthy native stocks of anadromous salmonids in the Pacific Northwest and California. *Fisheries* 21:6–14.

Jakober, M. J., T. E. McMahon, R. F. Thurow, and C. G. Clancy. 1998. Role of stream ice on fall and winter movements and habitat use by bull trout and cutthroat trout in Montana headwater streams. *Transactions of the American Fisheries Society* 127:223–235.

Johnson, O. W. 1991. Status Review for Lower Columbia River Coho Salmon. National Marine Fisheries Service, Northwest Fisheries Center. Seattle, Washington.

Johnson, O. W., R. S. Waples, T. C. Wainright, K. G. Neely, F. W. Waknitz, and L. T. Parker. 1994. Status Review for Oregon's Umpqua River Sea-Run Cutthroat Trout. Seattle National Marine Fisheries Service Northwest Fisheries Science Center. NOAA Technical Memorandum, NMFS-NWFSC-15. Seattle, Washington.

Kan, T. T. 1975. Systematics, variation, distribution, and biology of lampreys of the genus Lampetra in Oregon. Ph.D. dissertation, Oregon State University, Corvallis.

Kanda, N., R. F. Leary, and F. W. Allendorf. 1997. Population genetic structure of bull trout in the Upper Flathead River drainage. Pages 299–308 in W. C. Mackay, M. K. Brewin, and M. Monita, eds. *Friends of the bull trout conference proceedings.* Bull Trout Task Force (Alberta), c/o Trout Unlimited Canada, Calgary.

Kapuscinski, A. R., C. R. Steward, M. L. Goodman, C. C. Krueger, J. H. Williamson, E. Bowles, and R. Carmichael. 1991. Genetic conservation guidelines for salmon and steelhead supplementation. Draft. 55 p.

Kondzela, C. M., C. M. Guthrie, S. L. Hawkins, C. D. Russell, J. H. Helle, and A. J. Gharrett. 1994. Genetic relationships among chum salmon populations in southeast Alaska and northern British Columbia. *Canadian Journal of Fisheries and Aquatic Sciences* 51:50–64.

Kwain, W., and J. A. Chappel. 1978. First evidence for even-year spawning pink salmon, Oncorhynchus gorbuscha, in Lake Superior. *Journal of the Fisheries Research Board of Canada* 35:1373–1376.

Kwain, W., and E. Thomas. 1984. First evidence of spring spawning by Chinook salmon in Lake Superior. *North American Journal of Fisheries Management* 4:227–228.

Laythe, L. L. 1948. The fishery development program in the lower Columbia River. Pages 42–55. *American Fisheries Society*. American Fisheries Society, Atlantic City, New Jersey.

Leary, R. F., F. W. Allendorf, and S. H. Forbes. 1993. Conservation genetics of bull trout in the Columbia and Klamath River drainages. *Conservation Biology* 7:856–865.

Leary, R. F., F. W. Allendorf, S. R. Phelps, and K. L. Knudsen. 1987. Genetic divergence and identification of seven cutthroat trout subspecies and rainbow trout. *Transactions of the American Fisheries Society* 116:580–587.

Li, H. W., K. Currens, D. Bottom, S. Clarke, J. Dambacher, C. Frissell, P. Harris, R. M. Hughes, D. McCullough, A. McGie, K. Moore, R. Nawa, and S. Thiele. 1995. Safe havens: Refuges and evolutionarily significant units. *Evolution and the aquatic ecosystem: defining unique units in population conservation* 17:371–380.

Lichatowich, J. A. 1999. Salmon without rivers: A history of the Pacific salmon crisis. Island Press, Washington, DC.

Lichatowich, J. A., and L. E. Mobrand. 1995. Analysis of Chinook salmon in the Columbia River from an ecosystem perspective. Mobrand Biometrics. Research Report.

Liknes, G. A., and P. J. Graham. 1988. Westslope cutthroat trout in Montana: Life history, status, and management. *American Fisheries Society Symposium* 4:53–60.

Lindsay, R. B., W. J. Knox, M. W. Flesher, B. J. Smith, E. A. Olson, and L. S. Lutz. 1986. Study of wild spring Chinook salmon in the John Day River system. Bonneville Power Administration, Portland, Oregon.

Lindsey, C. C., and J. D. McPhail. 1986. Zoogeography of fishes of the Yukon and Mackenzie basins. Pages 639–674 in C. H. Hocutt and E. O. Wiley, eds. *Zoogeography of North American freshwater fishes*. John Wiley and Sons, New York.

Liskauskas, A. P., and M. M. Ferguson. 1991. Genetic variation and fitness: A test in a naturalized population of brook trout (*Salvelinus fontinalis*). *Canadian Journal of Fisheries and Aquatic Sciences* 48:2152–2162.

Loudenslager, E. J., and G. A. E. Gall. 1980. Geographic patterns of protein variation and subspeciation in cutthroat trout, *Salmo clarki*. *Systematic Zoology* 29:27–42.

Marotz, B. L., C. Althen, B. Lonon, and D. Gustafson. 1996. Model development to establish Integrated Operational Rule Curves for Hungry Horse and Libby Reservoirs—Montana. Bonneville Power Administration. Final Report, DOE/BP-92452-1. Portland, Oregon. 114 p.

Martin, R. C., L. A. Mehrhoff, J. E. Chaney, and S. Sather-Blair. 1985. Status review of wildlife mitigation at 14 of 27 major hydroelectric projects in Idaho. Bonneville Power Administration. Final report, BPA 83–478, DE-AI79–84BP12149. Portland, Oregon.

Matthews, G. M., and R. S. Waples. 1991. Status review for Snake River spring and summer Chinook salmon. National Marine Fisheries Service, Northwest Fisheries Science Center. NOAA Technical Memorandum, NMFS F/NWC-200. Seattle, Washington. 75 p.

Mayr, E. 1970. *Populations, species and evolution.* Belknap Press, Cambridge, Massachusetts.

Mayr, E. 1982. *Toward a new philosophy of biology.* Harvard University Press, Cambridge, Massachusetts.

McCusker, M. R., E. Parkinson, and E. B. Taylor. 2000a. Mitochondrial DNA variation in rainbow trout (*Oncorhynchus mykiss*) across its native range: Testing biogeographical hypotheses and their relevance to conservation. *Molecular Ecology* 9:2089–2108.

McCusker, M. R., E. A. Parkinson, and E. B. Taylor. 2000b. Phylogenetic conservation units for rainbow trout in British Columbia. Province of British Columbia. Fisheries Management Report, 112. Vancouver, British Columbia. 35 p.

McPhail, J. D., and C. C. Lindsey. 1986. Zoogeography of the freshwater fishes of Cascadia (the Columbia system and rivers north to the Stikine). Pages 615–637 in C. H. Hocutt and E. O. Wiley, eds. *Zoogeography of North American freshwater fishes*. John Wiley and Sons, New York.

Milner, G. B., D. J. Teel, F. M. Utter, and G. A. Winans. 1985. A genetic method of stock identification in mixed populations of Pacific salmon, *Oncorhynchus spp. Marine Fisheries Review* 47:1–8.

Mullan, J. W. 1986. Determinants of sockeye salmon abundance in the Columbia River, 1880's–1982: A review and synthesis. US Fish and Wildlife Service. Biological Report, 86(12). 136 p.

Mullan, J. W., A. Rockhold, and C. R. Chrisman. 1992a. Life histories and precocity of Chinook salmon in the Mid-Columbia River. *Progressive Fish-Culturist* 54:25–28.

Mullan, J. W., K. R. Williams, G. Rhodus, T. W. Hillman, and J. D. McIntyre. 1992b. Production and habitat of salmonids in mid-Columbia River tributary streams. US Fish and Wildlife Service. Monograph I. Washington, DC.

Mundy, P. R., T. W. H. Backman, and J. M. Berkson. 1995. Selection of conservation units for Pacific salmon: lessons from the Columbia River. *Evolution and the aquatic ecosystem: defining unique units in population conservation* 17:28–40.

Naiman, R. J., ed. 1992. *Watershed management: Balancing sustainability and environmental change*. Springer-Verlag, New York.

National Marine Fisheries Service (NMFS). 1995. Draft Biological Opinion on reinitiation of consultation on 1994–1998 operation of the federal Columbia River Power System and juvenile transportation program in 1994–1998. US Army Corps of Engineers, Bonneville Power Administration, Bureau of Reclamation, and National Marine Fisheries Service. Portland, Oregon. 146 p.

National Research Council (NRC). 1995. *Science and the Endangered Species Act*. National Academy Press, Washington.

NRC. 1996. Upstream: Salmon and society in the Pacific Northwest. Report on the Committee on Protection and Management of Pacific Northwest Anadromous Salmonids for the National Research Council of the National Academy of Sciences. National Academy Press, Washington DC.

Neave, F. 1958. The origin and speciation of *Oncorhynchus. Proc. Trans. R. Soc. Can. Ser. 3* 52:25–39.

Needham, P. R., and R. J. Behnke. 1962. The origin of hatchery rainbow trout. *Progressive Fish Culturist* 24:156–158.

Nehlsen, W., J. E. Williams, and J. A. Lichatowich. 1991. Pacific salmon at the crossroads: Stocks at risk from California, Oregon, Idaho, and Washington. *Fisheries* 16:4–21.

Nei, M. 1987. *Molecular evolutionary genetics*. Columbia University Press, New York.

Nei, M., and F. Tajima. 1981. DNA polymorphism detectable by restriction endonucleases. *Genetics* 97:145–163.

Nelson, K., and M. Soule. 1987. Genetical conservation of exploited fishes. Pages 345–368 in N. Ryman and F. Utter, eds. *Population Genetics and Fishery Management*. University of Washington Press, Seattle, Washington.

Nemeth, D. J., and R. B. Kiefer. 1999. Snake River spring and summer Chinook salmon—the choice for recovery. *Fisheries* 24:16–23.

Northwest Power Planning Council (NPPC). 1986. Council Staff Compilation of Information on Salmon and Steelhead Losses in the Columbia River Basin. Northwest Power Planning Council. Portland, Oregon.

NPPC. 1994. Columbia River Basin Fish and Wildlife Program. Northwest Power Planning Council. Portland, Oregon.

NPPC. 1997. An integrated framework for fish and wildlife management in the Columbia River. Northwest Power Planning Council. NPPC 97–2. Portland, Oregon.

NPPC. 1998. A proposed scientific foundation for the restoration of fish and wildlife in the Columbia River. Northwest Power Planning Council. NPPC 98–16. Portland, Oregon.

NPPC. 2000. Columbia River Basin Fish and Wildlife Program. Northwest Power Planning Council. Portland, Oregon.

Oakley, A. L. 1996. A summary of information concerning chum salmon in Tillamook Bay. *Oregon Fish Commission Research Briefs* 12:5–21.

Olin, P. G. 1984. Genetic variability in hatchery and wld populations of coho salmon (*Oncorhynchus kisutch*) in Oregon. Master's thesis, University of California at Davis. 73 p.

Oregon Department of Fish and Wildlife. 1982. Comprehensive plan for production and management of Oregon's anadromous salmon and trout. Part II: Coho salmon plan. Portland, Oregon.

Oregon Department of Fish and Wildlife, and Washington Department of Fish and Wildlife. 1993. Status report: Columbia River fish runs and fisheries, 1938–92. Olympia, Washington.

Oregon Department of Fish and Wildlife, and Washington Department of Fish and Wildlife. 1995. Status report: Columbia River fish runs and fisheries, 1938–94. Portland, Oregon.

Oregon Department of Fish and Wildlife, and Washington Department of Fish and Wildlife. 2001. Status report: Columbia River fish runs and fisheries, 1938-2000. Portland, Oregon.

Oregon Fish Commission. 1931. Biennial Report of the Fish Commission of the State of Oregon to the Governor and the Thirty-Sixth Legislative Assembly. Salem, Oregon.

Oregon Fish Commission. 1933. Biennial Report of the Fish Commission of the State of Oregon to the Governor and the Thirty-Seventh Legislative Assembly, 1933. Salem, Oregon.

Pacific Fishery Management Council. 1992. Review of the 1991 Ocean Salmon Fisheries. Pacific Fishery Management Council. Portland, Oregon.

Parsley, M. J., and L. G. Beckman. 1994. White sturgeon spawning and rearing habitat in the lower Columbia River. *North American Journal of Fisheries Management* 14:812–827.

Parsley, M. J., L. G. Beckman, and G. T. McCabe. 1993. Spawning and rearing habitat use by white sturgeon in the Columbia River downstream from McNary Dam. *Transactions of the American Fisheries Society* 122:217–227.

Paulik, G. J. 1969. Computer simulation models for fisheries research, management, and teaching. *Transactions of the American Fisheries Society* 98:551–559.

Phelps, S., J. Uehara, D. Hendrick, J. Hymer, A. Blakley, and R. Brix. 1995. Genetic diversity units and major ancestral lineages for chum salmon in Washington. Pages C-1–C-55 in C. Busack and J. B. Shaklee, eds. *Genetic diversity units and major ancestral lineages of salmonid fishes in Washington.* Washington Department of Fish and Wildlife, Olympia, Washington.

Phelps, S. R., L. L. LeClair, S. Young, and H. L. Blankenship. 1994. Genetic diversity patterns of chum salmon in the Pacific Northwest. *Canadian Journal of Fisheries and Aquatic Sciences* 51:65–83.

Philipp, D. P., and J. E. Clausen. 1995. Fitness and performance differences between two stocks of largemouth bass from different river drainages within Illinois. *Uses and effects of cultured fishes in aquatic ecosystems* 15:236–243.

Quattro, J. M., and R. C. Vrijenhoek. 1989. Fitness difference among remnant populations of the endangered Sonoran Topminnow. *Science* 245:976–978.

Quigley, T. M., R. W. Haynes, and R. T. Graham. 1996. Integrated scientific assessment for ecosystem management in the Interior Columbia Basin and portions of the Klamath and

Great Basins. US Department of Agriculture, Forest Service, Pacific Northwest Research Station. General Technical Report, PNW-GTR-382. Portland, Oregon.

Quinn, T. P., and K. Fresh. 1984. Homing and straying in Chinook salmon (*Oncorhynchus tshawytscha*) from Cowlitz River Hatchery, Washington. *Canadian Journal of Fisheries and Aquatic Sciences* 41:1078–1082.

Quinn, T. P., and M. J. Unwin. 1993. Variation in life history patterns among New Zealand Chinook salmon (*Oncorhynchus tshawytscha*) populations. *Canadian Journal of Fisheries and Aquatic Sciences* 50:1414–1421.

Quinn, W. H., V. T. Neal, and S. E. A. de Mayolo. 1987. El Niño occurrences over the past four and a half centuries. *Journal Geophysical Research* 92:14,449–14,461.

Reeves, G. H., L. E. Benda, K. M. Burnett, P. A. Bisson, and J. R. Sedell. 1995. A disturbance-based ecosystem approach to maintaining and restoring freshwater habitats of evolutionarily significant units of anadromous salmonids in the Pacific Northwest. *Evolution and the aquatic ecosystem: Defining unique units in population conservation* 17:334–349.

Reimers, P. E. 1973. The length of residence of juvenile fall Chinook salmon in Sixes River, Oregon. *Research Reports of the Fish Commission of Oregon* 4:3–43.

Reisenbichler, R., and S. R. Phelps. 1989. Genetic variation in steelhead (*Salmo gairdnerii*) from the North Coast of Washington. *Canadian Journal of Fisheries and Aquatic Sciences* 46:66–73.

Rich, W. H. 1938. Local populations and migration in relation to the conservation of Pacific salmon in the western states and Alaska. Fish Commission of the State of Oregon. Contribution No. 1. 6 p.

Rich, W. H. 1939. Fishery problems raised by the development of water resources. Pages 176–181 in *Dams and the problems of migratory fishes*. Fish Commission of the State of Oregon, Stanford University.

Rich, W. H. 1941. The present state of the Columbia River salmon resources. Pages 425–430 in *Sixth Pacific Science Congress*. Fish Commission of the State of Oregon, Berkeley.

Ricker, W. E. 1958. Handbook of computations for biological statistics of fish populations. Fisheries Research Board of Canada, Ottawa. 300 p.

Ricker, W. E. 1972. Hereditary and environmental factors affecting certain salmonid populations. Pages 19–160 in R. C. Simon and P. A. Larkin, eds. *The stock concept in Pacific salmon*. University of British Columbia, Vancouver.

Rieman, B. E., and J. B. Dunham. 2000. Metapopulation and salmonids: A synthesis of life history patterns and empirical observations. *Ecology of Freshwater Fish* 9:51–64.

Rieman, B. E., and J. D. McIntyre. 1993. Demographic and habitat requirements for conservation of bull trout. US Forest Service, Intermountain Research Station. General Technical Report, INT-308. Ogden, Utah. 38 p.

Rieman, B. E., and J. D. McIntyre. 1995. Occurrence of bull trout in naturally fragmented habitat patches of varied size. *Transactions of the American Fisheries Society* 124:285–296.

Rieman, B. E., and J. D. McIntyre. 1996. Spatial and temporal variability in bull trout redd counts. *North American Journal of Fisheries Management* 16:132–141.

Riggs, L. A. 1990. Principles for genetic conservation and production quality. Northwest Power Planning Council. Contract C90–005. Portland, Oregon.

Roos, J. F. 1991. Restoring Fraser River salmon. Pacific Salmon Commission, Vancouver, British Columbia.

Sage, G. K., and R. F. Leary. 1995. Electrophoretic analysis of fifteen rainbow trout populations from the Kootenai River in Northern Idaho. US Fish and Wildlife Service, Spokane, Washington. 8 p.

Sandercock, F. K. 1991. Life history of coho salmon (*Oncorhynchus kisutch*). Pages 397–445 in C. Groot and L. Margolis, eds. *Pacific salmon life histories*. UBC Press, Vancouver.

Schlosser, I. J., and P. L. Angermeier. 1995. Spatial variation in demographic processes of lotic fishes: conceptual models, empirical evidence, and implications for conservation. *American Fisheries Society Symposium* 17:392–401.

Schluchter, M., and J. A. Lichatowich. 1977. Juvenile life histories of Rogue River spring Chinook salmon *Oncorhynchus tshawytscha* (Walbaum), as determined from scale analsysis. Oregon Department of Fish and Wildlife. Information Report Series, Fisheries No. 77–5. Corvallis, Oregon.

Schoener, T. W. 1983. Field experiments on interspecific competition. *American Naturalist* 122:240–285.

Schoener, T. W. 1991. Extinction and the nature of the metapopulation: a case system. *Acta Oecologica* 12:53–76.

Schuck, H. A. 1943. Survival, population density, growth and movement of the wild brown trout in Crystal Creek. *Transactions of the American Fisheries Society* 73:209–230.

Scott, W. B., and E. J. Crossman. 1973. Freshwater Fishes of Canada. Fisheries Research Board of Canada, Ottawa.

Scudder, G. G. E. 1989. The adaptive significance of marginal populations: a general perspective. Pages 180–185 in C. D. Levings, L. B. Holtby, and M. A. Henderson, eds. *Proceedings of the national workshop on effects of habitat alteration on salmonid stocks.* Canadian Special Publication of Fisheries and Aquatic Sciences, Bethesda, Maryland.

Shaklee, J. B., J. Ames, and D. Hendrick. 1995. Genetic diversity units and major ancestral lineages for pink salmon in Washington. Pages B-1–B-36 in C. Busack and J. B. Shaklee, eds. *Genetic diversity units and major ancestral lineages of salmonid fishes in Washington.* Washington Department of Fish and Wildlife, Olympia, Washington.

Shaklee, J. B., S. R. Phelps, and J. Salini. 1990. Analysis of fish stock structure and mixed-stock fisheries by the electrophoretic characterization of allelic enzymes. Pages 173–196 in D. H. Whitmore, ed. *Electrophoretic and isoelectric focusing techniques in fisheries management.* CRC Press, Boca Raton, Florida.

Shaklee, J. B., and N. V. Varnavskaya. 1994. Electrophoretic characterization of odd-year pink salmon (*Oncorhynchus gorbuscha*) populations from the Pacific coast of Russia, and compmparison with selected North American populations. *Canadian Journal of Fisheries and Aquatic Sciences* 51:158–171.

Shiozawa, D. K., and R. P. Evans. 1995. The use of DNA to identify geographical isolation in trout stocks. Pages 125–131 in R. H. Hamre, ed. *Wild Trout V: Proceedings of a Symposium.* Trout Unlimited, Vienna, Virginia.

Sinclair, M. 1988. *Marine populations: An essay on population regulation and speciation.* University of Washington Press, Seattle, Washington.

Sinclair, M., and T. D. Iles. 1989. Population regulation and speciation in the oceans. *J. Cons. Int. Explor. Mer.* 45:165–175.

Soverel, P. W. 1996. Kamchatka steelhead project: Short report—1996. Wild Salmonid Center. Annual Report. Seattle, WA.

Stacey, P. B., and M. Taper. 1992. Environmental variation and the persistence of small populations. *Ecological Applications* 2:18–29.

Stanford, J. A., J. V. Ward, W. J. Liss, C. A. Frissell, R. N. Williams, J. A. Lichatowich, and C. C. Coutant. 1996. A general protocol for restoration of regulated rivers. *Regulated Rivers* 12:391–413.

Stearns, S. C. 1995. *The evolution of life histories.* Oxford University Press, New York.

Stoneking, M., D. J. Wagner, and A. C. Hildebrand. 1981. Genetic evidence suggesting subspecific differences between northern and southern populations of brook trout (*Salvelinus fontinalis*). *Copiea* 1981:810–819.

Taylor, E. B. 1990. Environmental correlates of life-history variation in juvenile Chinook salmon, *Oncorhynchus tshawytscha* (Walbaum). *Journal of Fish Biology* 37:1–17.

Taylor, E. B. 1991. A review of local adaptation in Salmonidae, with particular reference to Pacific and Atlantic salmon. *Aquaculture* 98:185–207.

Taylor, E. B., C. J. Foote, and C. C. Wood. 1996. Molecular evidence for parallel life-history evolution within a Pacific salmon (sockeye salmon and kokanee, *Oncorhynchus nerka*). *Evolution* 50:401–416.

Taylor, E. B., S. Harvey, S. Pollard, and J. Volpe. 1997. Postglacial genetic differentiation of reproductive ecotypes of kokanee *Oncorhynchus nerka* in Okanagan Lake, British Columbia. *Molecular Ecology* 1997:503–517.

Taylor J. E., III. 1999. Making salmon: an environmental history of the Northwest fisheries crisis. University of Washington Press, Seattle, Washington.

Thomas, G. 1992. Status report: Bull trout in Montana. Montana Department of Fish, Wildlife and Parks. Helena, Montana. 68 p.

Thomas, W. K., and A. T. Beckenbach. 1989. Variation in salmonid mitochondrial DNA: Evolutionary constraints and mechanisms of substitution. *Journal of Molecular Evolution* 29:233–245.

Thompson, W. F. 1959. An approach to population dynamics of the Pacific red salmon. *Transactions of the American Fisheries Society* 88:206–209.

Thompson, W. F. 1965. Fishing treaties and salmon of the North Pacific. *Science* 150:1786–1789.

Thurow, R. F., C. E. Corsi, and V. K. Moore. 1988. Status, ecology and management of Yellowstone cutthroat trout in the Upper Snake River drainage, Idaho. Pages 25–36 in R. E. Gresswell, ed. *Status and management of interior stocks of cutthroat trout, Symposium 4.* American Fisheries Society, Bethesda, Maryland.

Thurow, R.F., D. C. Lee, and B. E. Rieman. 2000. Status and distribution of Chinook salmon and steelhead din the interior Columbia River basin and portions of the Klamath River basin. Pages 133–160 in E. Knudsen, C. Steward, D. MacDonald, J. Williams, and D. Reiser, eds. *Sustainable fisheries management: Pacific salmon.* CRC Press, Boca Raton, Florida.

Trotter, P. C. 1987. *Cutthroat: Native trout of the west.* Colorado Associated University Press, Boulder, Colorado.

Trotter, P. C. 1989. Coastal cutthroat trout: A life history compendium. *Transactions of the American Fisheries Society* 118:463–473.

Trotter, P. C., P. A. Bisson, and B. Fransen. 1993. Status and plight of searun cutthroat trout. Pages 203–212 in J. G. Cloud and G. H. Thorgaard, eds. *Genetic Conservation of Salmonid Fishes.* Plenum Press, New York.

Utter, F. M. 1991. Biochemical genetics and fishery management: An historical perspective. *Journal of Fish Biology* 39 (Supplement A):1–20.

Utter, F. M., P. Aebersold, and G. Winans. 1987. Interpreting genetic variation detected by electrophoresis. Pages 21–45 in N. Ryman and F. Utter, eds. *Population genetics and fishery management.* University of Washington Press, Seattle, Washington.

Utter, F. M., F. W. Allendorf, and H. O. Hodgins. 1973. Genetic variability and relationships in Pacific salmon and related trout based on protein variations. *Systematic Zoology* 22:257–270.

Utter, F. M., D. W. Chapman, and A. R. Marshall. 1995. Genetic population structure and history of Chinook salmon of the upper Columbia River. *Evolution and the aquatic ecosystem: Defining unique units in population conservation* 17:149–168.

Utter, F., G. Milner, G. Stahl, and D. Teel. 1989. Genetic population structure of Chinook salmon, *Oncorhynchus tshawytscha,* in the Pacific Northwest. *Fishery Bulletin* 87:239–264.

Utter, F. M., and N. Ryman. 1993. Genetic markers and mixed stock fisheries. *Fisheries* 18:11–21.

Utter, F. M., J. E. Seeb, and L. W. Seeb. 1991. Complementary uses of ecological and biochemical genetic data in identifying and conserving salmon populations. *Fisheries Research* 18:59–76.

Utter, F. M., J. E. Seeb, and L. W. Seeb. 1993. Complementary uses of ecological and biochemical genetic data in identifying and conserving salmon populations. *Fisheries Research* 18:59–76.

Varnavskaya, N. V., and T. D. Beacham. 1992. Biochemical genetic variation in odd-year pink salmon (*Oncorhynchus gorbuscha*). *Canadian Journal of Zoology* 70:2115–2120.

Varnavskaya, N. V., C. C. Wood, and R. J. Everett. 1994a. Genetic variation in sockeye salmon (*Oncorhynchus nerka*) populations in Asia and North America. *Canadian Journal of Fisheries and Aquatic Sciences* 51:132–146.

Varnavskaya, N. V., C. C. Wood, R. J. Everett, R. L. Wilmot, V. S. Varnasky, V. V. Midanaya, and T. P. Quinn. 1994b. Genetic differentiation of subpopulations of sockeye salmon (*Oncorhynchus nerka*) within lakes of Alaska, British Columbia, and Kamchatka, Russia. *Canadian Journal of Fisheries and Aquatic Sciences* 51:147–157.

Waples, R. S. 1991. Definition of "species" under the endangered species act: Application to pacific salmon. National Marine Fisheries Service. Seattle, Washington. 29 p.

Waples, R. S., O. W. Johnson, P. B. Aebersold, C. K. Shiflett, D. M. VanDoornik, D. J. Teel, and A. E. Cook. 1993. A genetic monitoring and evaluation program for supplemented populations of salmon and steelhead in the Snake River Basin. Bonneville Power Administration. Annual Report, DE-A179–89BP0091.

Waples, R. S., J. R. P. Jones, B. R. Beckman, and G. A. Swan. 1991. Status review for Snake River fall Chinook salmon. National Marine Fisheries Service, Northwest Fisheries Science Center. NOAA Technical Memorandum, NMFS F/NWC-201. Seattle, Washington. 71 p.

Washington Department of Fish and Wildlife, and Oregon Department of Fish and Wildlife. 1994. Status report: Columbia River fish runs and fisheries, 1938–1993. Washington Department of Fish and Wildlife. Olympia, Washington. 271 p.

Washington Department of Fish and Wildlife, and Oregon Department of Fish and Wildlife. 2001. Status report: Columbia River fish runs and fisheries, 1938–2000. Washington Department of Fish and Wildlife and Oregon Department of Fish and Wildlife. Olympia, Washington, and Clackamas, Oregon.

Washington Department of Fisheries, Washington Department of Wildlife, and Western Washington Treaty Indian Tribes. 1993. 1992 Washington State salmon and steelhead stock inventory. Olympia, Washington.

Watson, G., and T. W. Hillman. 1997. Factors affecting the distribution and abundance of bull trout: an investigation at hierarchical scales. *North American Journal of Fisheries Management* 17:237–252.

Wehrhahn, C. F., and R. Powell. 1987. Electrophoretic variation, regional differences, and gene flow in the coho salmon (*Oncorhynchus kisutch*) of southern British Columbia. *Canadian Journal of Fisheries and Aquatic Sciences* 44:822–831.

Weitkamp, L. A., T. C. Wainwright, G. J. Bryant, G. B. Milner, D. J. Teel, R. G. Kope, and R. S. Waples. 1995. Status review of coho salmon from Washington, Oregon, and California. US Department of Commerce. NOAA Technical Memorandum, NMFS-NWFSC-24. Seattle, Washington. 285 p.

Williams, J. E., J. E. Johnson, D. A. Hendrickson, S. Contreras-Balderas, J. D. Williams, M. Navarro-Mendoza, D. E. McAllister, and J. E. Deacon. 1989. Fishes of North America endangered, threatened, or of special concern: 1989. *Fisheries* 14:2–21.

Williams, J. E., J. A. Lichatowich, and W. Nehlsen. 1992. Declining salmon and steelhead populations: New endangered species concerns for the west. *Endangered Species Update* 9:1–8.

Williams, R. N. 1991. Genetic analysis and taxonomic status of cutthroat trout from Willow Creek and Whitehorse Creek in southeastern Oregon. Department of Biology, Boise State University. BSU Evolutionary Genetics Lab Report 91–3. Boise, Idaho. 15 p.

Williams, R. N., R. P. Evans, and D. K. Shiozawa. 1997. Mitochondrial DNA diversity patterns of bull trout in the Upper Columbia River Basin. Pages 283–297 in W. C. Mackay, M. K. Brewin, and M. Monita, eds. *Friends of the bull trout conference proceedings*. Bull Trout Task Force (Alberta), c/o Trout Unlimited Canada, Calgary.

Williams, R. N., and M. Jaworski. 1995. Genetic analysis of two rainbow trout populations from the Kootenai River in Northern Idaho. Clear Creek Genetics. Lab Report 95–2. Meridian, Idaho. 22 p.

Williams, R. N., D. K. Shiozawa, J. E. Carter, and R. F. Leary. 1996. Genetic detection of putative hybridization between native and introduced rainbow trout populations of the upper Snake River. *Transactions of the American Fisheries Society* 125:387–401.

Wilmot, R. L., R. J. Everett, W. J. Spearman, R. Baccus, N. V. Vanaskaya, and S. V. Putivkin. 1994. Genetic stock structure of western Alaska chum salmon and a comparison with Russian Far East stocks. *Canadian Journal of Fisheries and Aquatic Sciences* 51:84–94.

Winans, G. A., P. B. Aebersold, S. Urawa, and N. V. Varnavskaya. 1994. Determining the continent of origin of chum salmon (*Oncorhynchus keta*) using genetic stock identification techniques: Status of allozyme baseline in Asia. *Canadian Journal of Fisheries and Aquatic Sciences* 51:95–113.

Wishard, L. N., J. E. Seeb, F. M. Utter, and D. Stefan. 1984. A genetic investigation of suspected redband trout populations. *Copiea* 1984:120–132.

Wood, C. C., B. E. Riddell, D. T. Rutherford, and R. W. Withler. 1994. Biochemical genetic survey of sockeye salmon (*Oncorhynchus nerka*) in Canada. *Canadian Journal of Fisheries and Aquatic Sciences* 51:114–131.

Wright, S. 1965. The interpretation of population structure by F—Statistics with special regard to systems of mating. *Evolution* 19:395–420.

Wright, S. 1993. Fishery management of wild Pacific salmon stocks to prevent extinctions. *Fisheries* 18:3–4.

5

The Status of Freshwater Habitats

Jack A. Stanford, Ph.D., Christopher A. Frissell, Ph.D., Charles C. Coutant, Ph.D.

Physiography of the Columbia River

Characteristics of High-Quality Riverine Habitat
  Habitat for Reproduction
  Juvenile Rearing and Movements
  Habitat for Adult Migration
  Influence of Wood on Development of High-Quality Habitat
  Temperature Pattern: The Primary Determinant of Water Quality
  Groundwater Upwelling: A Key Attribute of High-Quality Habitat

Causes and Consequences of Habitat Degradation
  Conclusions from Recent Reviews and Synoptic Studies

Similar Effects from Different Causes: Human Activities and Consequences
  Beaver Trapping
  Logging
  Grazing
  Mining
  Irrigation and Cropland Agriculture
  Urban and Other Sources of Excessive Nutrients and Toxic Pollutants
  Stream Regulation: Effects of Dams, Reservoirs, and Diversions
    Dams
    Reservoirs
    Diversions

Coherence of Habitat Loss and Structural Changes in Columbia River Fisheries

Changed Food Production and Feeding during Migration
  The Riverine Food Web
  Hydrosystem Alterations of Food Webs
    Loss of Riverine Insect Production
    Loss of Riparian Flooding
    Altered Timing of Production and Consumption
    Invasion of Reservoirs by Estuarine Invertebrate Species
    Nutritional Status of Juvenile Migrants

Habitat Quantity and Quality: A Summary and Synthesis
    The Effects of River Regulation
        Habitat Diversity Is Substantially Reduced as a Consequence of Regulation
        Native Biodiversity Decreases and Non-native Species Proliferate as a Consequence of
            Regulation
        Normative Conditions Are Reexpressed Predictably as Distance Downstream from the Dam
            Increases and in Relation to Influences of Tributaries
        Reestablishing Normative Ecological Conditions
Habitat Conclusions and Recommendations
    Conclusions
    Critical Uncertainties
    Recommendations
Literature Cited

"Maintaining a rich diversity of Pacific salmon genotypes and phenotypes depends on maintaining habitat diversity and on maintaining the opportunity for the species to take advantage of that diversity." (Healey and Prince 1995)

In this chapter we examine the loss of freshwater habitat suitable for salmon in the Columbia Basin. We do this in the context that salmon life history diversity and productivity are driven mainly by environmental variation. This theme requires an ecosystem view, which we equate with a catchment- or watershed-level analysis and synthesis. We provide details for what is obvious: that human stressors such as dams, agriculture and forestry, and invasions of non-native biota have vastly reduced freshwater habitat quantity and quality throughout the system, and that abundance, productivity, and life history variation of the native fishes have declined simultaneously. We start with a description of the system and the habitats that are germane to the fishes. We then examine the stressors and consider possible solutions. We conclude that it is time to seriously consider restoration of habitat-forming (normative) flows in entire subbasins, at least from the mid-Columbia downstream. This assumes that ocean and estuaries mainstem bottlenecks will not compromise productivity in restored tributaries, an assumption that is supported by the continuing viability of fall Chinook salmon in the Hanford Reach.

Physiography of the Columbia River

The Columbia River is one of the larger rivers of the world and also one of the most developed, with ten major dams on the main river within the United States and five in Canada (Table 5.1; Figure 5.1). All of the major dams in the United States and three in Canada have hydroelectric generation; the other dams are primarily water storage reservoirs for regulating flow for downstream power production and flood control. All the big natural lakes in the Columbia River system have dams on the outlets: Flathead, Pend Orielle, Coeur d'Alene,

Table 5.1 Discharge statistics for the Columbia River and selected tributaries.

River: Basin Name and Location	Average Discharge (cms)	Discharge Extremes (cms) Maximum Minimum		Drainage Area above Station (sq. km.)	Average Discharge per Sq. Km. above Station (cms)	Period of Record
Columbia: Columbia at Beaver Army Terminal, Quincy, OR.	7,730	24,466 (1996)	1,801 (2001)	724,025 (including Canada)	0.01	1968–2001
Clearwater: Clearwater near Spalding, ID.	433	4,701 (1948)	14 (1937)	31,080	0.01	1910–2000
Cowlitz: Cowlitz at Castle Rock, WA.	261	3,794 (1933)	30 (1935)	8,870	0.03	1926–2000
Flathead: Flathead at Perma, MT.	340	1,512 (1997)	76 (1984)	24,350	0.01	1983–2000
Grande Ronde: Grande Ronde at Troy, OR.	88	1,195 (1996)	9.7 (1977)	10,360	0.008	1944–2000
John Day: John Day at McDonald Ferry, OR.	60	1211 (1965)	0.003 (1977)	20,980	0.003	1904–2000
Methow: Methow near Pateros, WA.	45	770 (1972)	4.2 (1974)	4,831	0.009	1959–1972
Owyhee: Owyhee below Owyhee Dam.	12	617 (1952)	.03 (1952)	28,617	0.0004	1932–2001
Snake: Snake below Ice Harbor Dam, WA.	1,565	8,637 (1974)	76 (1979)	281,000	0.006	1912–1999
Spokane: Spokane at Long Lake, WA.	225	1,339 (1948)	2.5 (1994)	15,590	0.01	1939–2000
Willamette: Willamette at Portland, OR.	917	11,893 (1996)	119 (1978)	29,728	0.03	1972–2000
Yakima: Yakima at Kiona, WA.	102	1,795 (1906)	3.1 (1906)	15,940	0.006	1905–2000

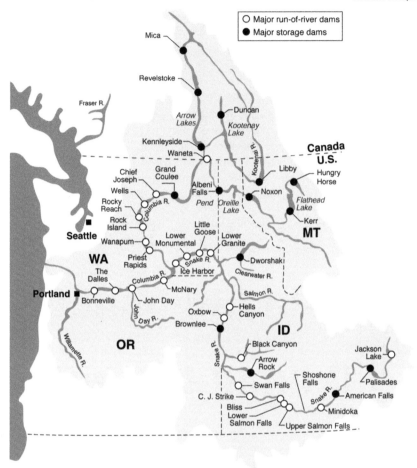

Figure 5.1 Map of the Columbia and Snake River mainstems, showing the major dams and tributaries.

Priest, Kootenay, and Okanagan. There are numerous smaller dams throughout the river basin that are not shown on the map in Figure 5.1, but which block spawning migrations.

The catchment basin encompasses many different environments and climates encompassed by the wet coastal, Cascade and Rocky Mountain ranges and the semi-arid Columbia Plateau. The latter lies in the rain shadow of the Cascades. The extreme environmental diversity of the basin is underscored by the fact that the Columbia Basin includes parts of 18 of the 43 physiographic provinces identified for the western United States (Omernik 1987). Parts of 13 terrestrial (Ricketts et al. 1999) and 3 freshwater (Abell et al. 2000) ecoregions are included in the Columbia Basin. Runoff comes from snowpack in the headwaters and seasonal rainfall in the lower elevations and coastal areas.

The Columbia River system is composed of steep gradient headwater streams that coalesce to form the major tributary rivers of the basin. The tributary rivers flow alternately through mountain valleys, where large alluvial flood plains occur and through deep canyon reaches (Figure 5.2). In the alluvial reaches, the slope of the valley bottom is 2% or less, allowing the river to spread out with a braiding or meandering channel system with a wide variety of aquatic habitats, many entirely off-channel connected by ground water flux. These complex alluvial flood plains occur within the river continuum from headwaters to mainstem confluence like beads on a string. They are important with respect to salmonid ecology because they provide critical habitats (described below) that are much less available within the constrained channels of many of the canyon reaches. On the Columbia Plateau, the lower reaches of tributary rivers, like the Deschutes and John Day, are partially or completely constrained by ancient lava flows and flood plains there are less well developed. The same is true for much of the mainstem Columbia and lower Snake Rivers. Indeed, the only segments of the mainstem with extensive flood plains occur in the free-flowing Hanford Reach downstream of Priest Rapids Dam and historically the long segment from the Umatilla River confluence near McNary Dam downstream to near John Day Dam, a reach that is impounded today. All of the Columbia Gorge, from about John Day Dam to Portland, is constrained and now impounded. Most of the channel of the Columbia River on its coastal plain below the gorge is constrained by revetments built since the 1920s, and lateral movement of floodwaters occur only on very high flow years.

Prior to extensive regulation by dams (Figure 5.1), the river was a gravel bed system from headwaters to mouth, although sand-sized substratum became progressively more common in a downstream direction. Pre-regulation photos of the mainstem river in the Columbia Gorge show large sand dunes along the river. Gravel and cobble were deposited extensively on the intermontane flood plains during high flows. The constrained canyon reaches slowed the flow of floods allowing sand, gravel, and cobble, along with tree boles and root wads eroded upstream, to be deposited in the alluvial reaches upstream of the constriction. This process of cut-and-fill alluviation created a wide variety of instream and floodplain features of many sizes and shapes. Gravel bars and associated features also occurred to a lesser degree in the constrained reaches, except at rapids created by exposed bedrock.

The bed load of the mainstem river is now largely retained in reservoirs. Only the finest sediments associated with spring runoff and other flooding reach the estuary, owing to retention behind the many dams that have been constructed in the basin (Table 5.2, Figure 5.1). The mainstem retains only

Figure 5.2 North Fork of the Flathead River in Montana, looking downstream, in April 2003, showing alluvial (foreground) and canyon (in distance, between mountain ranges) reaches. Cottonwoods and other vegetation on the flood plains have not yet leafed out. There is a potential dam site in the canyon at the top of the picture where the river goes through the two mountain ranges, which would impound the valley and all of the aquatic and riparian floodplain habitat. Such habitat loss has been typical of much of the Columbia River system. Photo by J. A. Stanford.

Table 5.2 Year in operation, length of reservoir, year in service of juvenile salmon collection facilities, and year in service for hydroelectric dams of the Columbia Basin in the United States.

Dam	Year of Initial Service	Length of Reservoir (miles)
Columbia River		
Rock Island (RM 453.4)	1933	21
Bonneville (RM 145.5)	1938	46
Grand Coulee (RM 596.6)	1941	151
McNary (RM 292)	1953	61
Collection Facilities	*1979*	
Chief Joseph (RM 545.1)	1955	52
The Dalles (RM 191.5)	1957	24
Priest Rapids (RM 397.1)	1959	18
Rocky Reach (RM 473.7)	1961	42
Wanapum (RM 415.8)	1963	38
Wells (RM 515.1)	1967	29
John Day (215.6)	1968	76
Snake River		
Brownlee (SRM 285)	1958	57
Oxbow (SRM 273)	1961	12
Ice Harbor (SRM 9.7)	1961	32
Hells Canyon (SRM 247)	1967	22
Lower Monumental (SRM 41.6)	1969	29
Collection Facilities	*1992*	
Little Goose (SRM 70.3)	1970	37
Collection Facilities	*1975*	
Lower Granite (SRM 107.5)	1975	39
Collection Facilities	*1976*	

Barge and Truck Transportation: 1976, truck transport began; 1977, barge (2) use began; 1981, barges (3) and trucks (5) expanded; 1982, barges (4) expanded, trucks (5); 1990, new barges (2) added; now at full capacity: 6 barges (296,000 pounds of fish), 5 trucks, 3 mini-tankers. Collection facilities have been used since 1979 to capture juvenile salmon for transport in tanks on barges or trucks to a release point below Bonneville Dam in attempt to reduce migration mortality. Detectors for fish implanted with passive integrated tags (PIT) at upstream sites are located at Lower Granite, Little Goose, Lower Monumental, McNary, John Day, and Bonneville Dams. These tags allow mortality of juveniles to be quantified routinely. Source: Corps of Engineers 1984; Athearn 1994.

one free-flowing segment in the United States, the Hanford Reach (Figure 5.3), and in Canada, the reach from Columbia Lake to the reservoir behind Mica Dam. Many of the tributaries also are regulated, either by high storage dams used for hydropower production and flood control or by low head diversion dams for irrigation withdrawals. Bypass devices for migrating fishes have been built only on the mainstem and tributaries below Chief Joseph Dam on the Columbia River and downstream of Hells Canyon Dam on the Snake River. No passage facilities exist for the large dams in the Canadian portion of the basin, apparently because migration of anadromous fishes was

Figure 5.3   The White Bluffs area of the Hanford Reach of the mainstem Columbia River, the last mainstem spawning area for a relatively stable population of fall Chinook salmon. Photo courtesy of D. Geist.

blocked by dams in the United States. This ignores the fact that many so-called resident fishes may need to migrate within the freshwater system to complete life histories naturally. Further information on the Columbia River Basin is available in Stanford et al. (2005).

Characteristics of High-Quality Riverine Habitat

Salmonid fishes of all species require cold, clean water for survival and growth, and clean, stable, and permeable gravel substrate, usually in running-water environments, for reproduction. The specific habitat requirements of various species are discussed in detail elsewhere (Salo and Cundy 1987; Groot and Margolis 1991; Meehan 1991; Rhodes et al. 1994). In this chapter, we summarize the major habitat requirements of various life stages.

Habitat for Reproduction

Incubation of salmonid eggs and fry occurs within the interstitial spaces of alluvial gravels in the beds of cool, clean streams and rivers. Native species of salmon, trout, and whitefish in the Columbia Basin are all lotic (running water) spawners in alluvial reaches of rivers and streams, except for sockeye salmon and kokanee (landlocked sockeye), which historically spawned on shallow, groundwater-effluent shoals or beaches on isolated shorelines of deep, cold lakes (e.g., Redfish, Idaho; Wenatchee, Washington; Ketcheless, Washington; Okanagan, British Columbia; and Kootenay, British Columbia) or similar environments in tributary streams of the lakes (Evermann 1895). A few sockeye populations and some kokanee populations have residualized (adopted a landlocked life cycle) from stocking in lakes where they are not naturally established (e.g., Pend Oreille, Idaho) (Rieman and Bowler 1980). Introduced salmonids, such as lake trout and brook trout in particular, use running-water habitats for reproduction very differently than salmonids native to the Columbia Basin, by being much less migratory throughout their life cycles. This natural propensity is favored by impoundments and other migration barriers, which allows non-native salmonids to occupy much of the reproductive "space" of the basin. For example, lake trout have taken over Flathead Lake, Montana, and with brook trout and rainbow trout, the non-native species are more abundant than native cutthroat trout and bull trout throughout the Flathead River system upstream of the lake, much of which is protected wilderness (Stanford and Ellis, 2002).

The season of spawning, egg incubation, and fry emergence varies among species of salmonids, with many rivers and streams historically supporting both fall-spawning (winter incubating) and spring-spawning (spring and early summer incubating) populations. The relative success of fall-versus spring-spawning strategies can vary, depending on climate and hydrologic regime, catchment stability and sedimentation, water temperature patterns, the relative influence and availability of groundwater efflux zones, and controls exerted by seasonal flow conditions and physical barriers on the ability of adult fish to gain access to spawning sites. In general, spring-spawning species (e.g., steelhead, rainbow and redband trout, and cutthroat trout) concentrate their reproductive activities in smaller, headwater streams and in spring snowmelt-fed streams (Figure 5.4). Reproduction of fall-spawning species (chum, Chinook, and coho salmon and bull trout) occurs most frequently in alluvial reaches of larger streams and rivers (Figure 5.5) where groundwater efflux strongly buffers local interstitial and surface water conditions (e.g., Baxter and Hauer 2000). Long-term patterns of local variation in the seasonality of flow and sediment transport, the availability of clean, dynamic gravel-cobble bars, patterns in groundwater-surface water exchange, and seasonally dynamic thermal patterns exert high-magnitude, density-independent effects on the survival and recruitment of salmonids that strongly constrain the abundance of all later life stages, including harvestable adults. Salmonid fish populations display a diversity of local adaptation of life histories and behaviors in concordance with this local environmental complexity (Chapter 4). These components of the freshwater environment are highly vulnerable to alteration by most kinds of human activities and natural events (with sedimentation of spawning gravels being the foremost threat).

Figure 5.4 Spring Chinook spawning habitat in the upper Imnaha River in eastern Oregon. Photo by R. N. Williams.

Juvenile Rearing and Movements

Upon emergence from the gravel, young salmonids are mobile, which increases their individual flexibility to cope with environmental variation by seeking suitable habitat conditions. Mobility is limited, however, particularly

Figure 5.5 Spawning summer Chinook and habitat in Johnson Creek, a tributary to the south fork of the Salmon River in central Idaho. Photo by R. N. Williams.

for fry, and they can easily be washed downstream even if the river is at base flow. Thus, suitable habitat and food resources must be available in proximity to spawning areas for successful first-year survival. Moreover, movement may come with high metabolic cost and high risk of mortality, such as through exposure to predators, unless movements are tied closely to patches of predictable, high-quality habitat. Thus, regulated stream flows that fluctuate rapidly can be devastating because they represent a continuously variable disturbance. Natural floods also redistribute juveniles, but that effect is a rare event or pulse type of disturbance in the context of natural life stage timing and the fish are thus adapted to the effects (if not too severe and too frequent). In any case, juveniles necessarily seek low-velocity habitats within a complex river that afford cover, a steady supply of small food particles, and refuge from larger predatory fishes, birds, and mammals. Examples of such habitats include quiet-water areas, backwaters, small spring-fed channels along stream margins, floodplain ponds and sloughs and alcoves within structural complexes created by woody debris, bank structures, and riparian vegetation or aquatic plants (Figure 5.6; Bisson et al. 1997). These critical habitats ("critical" in the sense that if they are not present or are compromised by stream regulation or pollution, the juvenile life stage may suffer high mortality) are most abundant and structurally diverse on aggraded,

Figure 5.6 A wall-based channel located on the valley floor of the Hoh River Valley on the Olympic Peninsula, Washington. Wall-based channels are spring-fed streams created though a combination of groundwater recharge and spring seeps located at the base of valley walls. The small streams have a relatively stable flow and temperature and provide critical winter refugia to juvenile coho. Photo by John McMillan, Wild Salmon Center.

floodplain reaches where they are created and maintained by cut-and-fill alluviation. Thus, middle reaches of the Columbia are particularly critical habitat because they contain the greatest degree of flood plain development in the watershed, as discussed earlier.

Another important component when considering salmon spawning is that the availability of critical habitats must not only be evaluated in space, but also in time. An example of the importance of temporal patterns can be seen in salmon, which were never very common in the Columbia River, and chum salmon, which were common, move quickly to the ocean after hatching, and sockeye salmon move into lakes as small fry. However, even these earlier migrating species that seemingly use the river simply as an outmigration corridor from spawning grounds also require shallow resting habitats with cover from predators, as provided by the complex features of naturally functioning flood plains (Groot and Margolis 1991).

Natal (very young fry) salmonids typically feed on aquatic and terrestrial invertebrates and small vertebrates, and they can grow rather rapidly in high-quality alluvial habitats with good riparian margins. As water temperature increases beyond about 15°C, metabolic costs escalate rapidly and available food resources support progressively lower densities of juvenile salmonids (Li et al. 1995b). Summer temperatures in most Columbia River tributaries, particularly the floodplain reaches that have been extensively altered by human activities, typically far exceed this value and, in many cases, exceed the lethal thermal maximum for salmonids (juveniles near 24°C; adults near 21°C) (Rhodes et al. 1994). Suspended sediments impair the ability of salmonids to see and capture prey, and accelerated deposition or transport of sediments on streambeds can deplete populations of stream invertebrates that are most important for salmonid growth.

Juveniles of some salmonid populations and species are known to successfully move long distances (many tens or even hundreds of kilometers) from their natal habitats, and some, such as pink, chum, some sockeye and fall or ocean-type Chinook salmon, virtually never are resident; they move downstream after emergence, progressively stopping to feed and grow in lower velocity habitats created by eddies in constrained (canyon) segments and, in particular, the complex habitats of flood plains. Migratory behaviors may be genetically determined, in part, if fidelity of movement is consistent from year to year, but these movements are closely tied to availability of preferred habitats. For example, some populations of juveniles in larger rivers tend to move downstream in late fall to seek wintering habitat in low-velocity backwaters and spring-fed channels of large flood plains or deep pools, whereas others tend to move upstream where wintering habitat is available in aggraded headwater areas (e.g., in beaver ponds).

In any case, each local breeding population likely evolved site- and season-specific patterns of early life history behavior that allow juveniles to efficiently locate and exploit the locally available patchwork of habitat. For example, sockeye fry resulting from spawning in lake inlets move downstream to rear in the lake, while those spawned in the outlet usually move upstream to rear in the same lake (Brannon and Salo 1982). Lack of such a locally adapted genetic heritage is one likely reason that hatchery-origin fish, including all forms of cultured and translocated fish and their offspring, typically exhibit lower fry to adult survival rates than indigenous fish of the same species, age, and size (Ricker 1972, Riddell and Legget 1981). Although total survival of some hatchery fish in very rare cases may be higher than wild fish (Shreamer 1995) if the hatchery releases are in coastal streams near the ocean. This increased survival may be due simply to the mass numbers of cultured fish that can be released. Moreover, destruction or alteration of available habitat mosaics created by natural biophysical processes (e.g., as a consequence of cumulative effects of flow regulation and fine sediment, thermal and chemical pollution, and upland and riparian land misuse) almost always impairs the survival of indigenous fish by compromising their inherited ability to anticipate and "track" high-quality habitats (Stanford et al. 1996).

Long-distance migration of juveniles or sub-adults downstream to lakes, rivers, or the ocean for maturation may intuitively seem maladaptive. However, migration can allow increased opportunities for access to concentrated food resources that allow rapid growth, permit escape from localized concentrations of predators or marginal habitats, and mediate longer life span and larger body size, which confers selective advantage. While some species retain a diversity of rearing strategies that allow them to persist in headwater populations even when opportunities for downstream migration are good (e.g., steelhead, kokanee salmon), other species are completely dependent on long-distance migration for maturation and survival (other salmon, and to a lesser extent, bull trout). On the other hand, some populations do not move much, staying in the same stream reaches or lakes throughout their life cycle (e.g., bull trout in Upper Kintla Lake, Montana) (Hauer et al. 1980). However, the great historical abundance of migration-dependent species in the Columbia River indicates this system (until recently) provided habitat favorable for a wide array of anadromous and river-migrant salmonid life histories for many centuries. Owing to the complex physiography of the Columbia River, opportunities for adaptation to particular rivers and even river segments was historically high and, as noted in Chapters 3 and 4, metapopulations were composed of suites of interactive, but locally adapted, stocks.

Habitat for Adult Migration

After growth and maturation, salmonid adults generally return to natal spawning areas for reproduction. The timing of adult entry and movement in rivers and tributary streams, and even the size, shape, and strength of adult fish represent adaptations to the specific physical and biological challenges presented by the upstream pathway to a specific spawning area. For example, waterfalls and similar physical barriers may be passable only at a specific range of flows that typically occur during one month of the year, and then only by fish that have particular physical capabilities for jumping or "scooting" (with burst swimming) over the barrier. The entire sequence of migration behavior must be properly timed to meet such windows of opportunity. For fall-spawning fish, prevailing warm water conditions in late summer often present strong thermal barriers to movement, and suitable habitats for resting may be few and far between (Berman and Quinn 1991). Therefore, again at the adult life stage, population-specific behavioral patterns, closely attuned to the mosaic of habitats that are available for migrant adults, may be critical for survival and successful reproduction.

Faithful homing to natal spawning areas is typical, but straying does occur. Adult fish also exhibit a remarkable ability to locate and select high-quality habitat patches for spawning (i.e., areas of suitably sized gravel and cobble with high rates of interstitial flow to modulate temperatures and oxygenate the nest or redd). They will actively stray from natal habitats to spawn elsewhere if suitable habitat is not available in the natal area due to natural or human causes, such as landslides, dams, or pollution that may negatively impact reproduction.

However, little evidence exists for the generally accepted notion that straying occurs when spawners are very abundant on apparently limited spawning areas. More often, later spawning occurs on top of earlier redds. In some locations, this superposition is detrimental to survival of previously spawned eggs. Elsewhere, the repeat spawning drives the early eggs deeper into the sediments away from predation, and it repeatedly cleans the fines that the river is constantly depositing in the low-velocity environments of the redd pocket. That the characteristically intense fighting for the best spawning locations, as is typical of salmonids, somehow elicits long-distance straying is unfounded. Expansion to nearby marginal habitat for spawning under intense competition does occur, however. Salmonids apparently have an innate propensity to stray from natal populations in modest but significant rates, owing to the legacy of dynamic environmental conditions related to glaciation that has characterized their range around the Pacific Rim for at least the last 100,000 years. This propensity allows gene flow from one population to another, which strengthens species composed of many populations. Moreover, Pacific salmonid species substantially radiated during the Pleistocene (Behnke et al. 2002). They certainly can adapt rapidly to new environments after artificial introductions, supporting the idea of a natural propensity for straying. For example, while many kokanee salmon stocked in Flathead Lake in 1910 and thereafter returned to the shoreline hatchery facility where they originated, many others strayed to suitable natural spawning grounds in widely segregated but very focused areas of groundwater upwelling on the lakeshore and upstream areas within three to four decades (Stanford and Ellis 2002). Chinook salmon in the Rio Santa Cruz in Patagonia (Argentina) in South America were recently discovered spawning in a single headwater reach apparently migrating over 1,000 km from the South Atlantic Ocean where they mature (and compete with native penguins for demersal forage fish!). This population apparently was derived from a single stocking of juveniles in the lower river around 1930 (M. Pasquel, Conicet Universities de Patagonia, personal communication August 28, 2003).

In any case, once they enter spawning areas, large adult migrant fish often are highly visible and quite vulnerable to terrestrial (including human) and avian predators. The availability of deep resting pools, riparian forest canopy, undercut banks, and large woody debris accumulations in the proximity of spawning habitats can be critical for survival and successful reproduction of migratory salmonids, particularly those that venture far upstream and that are required to spend long periods holding in small river and stream environments. The complex spatial and temporal interactions associated with diverse channel systems, natural obstructions, and dead and living vegetation provides protective and productive habitat for spawners (Figure 5.7). This underscores the fact that both adult and juvenile salmonids are dependant on the river-riparian interface (ecotone) that dominates the landscape of naturally functioning flood plains (Stanford 1998).

Figure 5.7   Wild summer steelhead spawning in the South Fork of the Hoh River, on the Olympic Peninsula, Washington. Photo by John McMillan, Wild Salmon Center.

Influence of Wood on Development of High-Quality Habitat

Productive salmonid rivers worldwide tend to be clogged with living and dead wood (Figures 5.8 and 5.9), a fact that is vastly underemphasized in management and restoration. Indeed, much of the historical habitat complexity of streams throughout the Columbia Basin was associated with accumulations of large wood. Virtually all Columbia Basin streams, including rivers of the high desert, traversed riparian forest mosaics that usually included stands of large-diameter, older trees (Wissmar et al. 1994; Gregory et al. 2001). These riparian forests (including downed trees in the channel) were often the most accessible source for high-quality logs during settlement and later proliferation of timber markets. Large wood was eliminated early on, and its renewal has subsequently been suppressed by continued logging and grazing in riparian areas (McIntosh et al. 1994; Wissmar et al. 1994). Clearance of rivers to facilitate log drives and other forms of navigation also contributed to loss of natural debris jams in many rivers (Sedell and Froggatt 1984; Sedell et al. 1991).

Large, downed boles and wood debris aggregates in the channel and on floodplain surfaces are absolutely integral to the development of habitat in Columbia Basin streams, particularly in the alluvial reaches where substratum

Figure 5.8   Large woody debris piled onto a mid-river island by spring floods, Bulkley River, Canada. Photo by R. N. Williams.

Figure 5.9   The Hoh River is a large glacially influenced river on the west side of Washington's Olympic Peninsula with an extensive and complex flood plain with a well-developed riparian component. The river is home to an abundant and diverse salmonid assemblage. Photo by John McMillan, Wild Salmon Center.

size is smaller and interstitial cover more limited than in the boulder-dominated channels of high gradient streams. In concert with the bank stability and flow resistance conferred by living riparian vegetation, coarse woody debris acts to deflect flows, creating low-velocity flow refugia, scouring deep pools, locally trapping sediments and fine organic material that contributes to aquatic food webs, and providing a diverse and stable habitat mosaic used heavily by many kinds of organisms, including salmonid fishes (Sedell and Froggatt 1984; Naiman 1992). Wood accumulations may play a direct role in forcing surface flows into alluvial aquifers and promoting efflux of hyporheic flow and shallow groundwater back into surface waters (Ebersole 1994). At a larger scale, debris jams cause temporary obstructions to the river course that, during peak flows, promote local channel switching and floodplain inundation, primary processes that create and rejuvenate the diverse mosaic of main channel, backwater, slough, springbrook, and hyporheic habitats common to natural alluvial rivers (Sedell and Froggatt 1984; Stanford and Ward 1993; Hauer et al. 2003). Such channel movement and floodplain inundation also sustains diversity in floodplain vegetative communities.

Big wood in and along the margins of the channel likely was a very important feature of the mainstem river as well. Lewis and Clark noted numerous large jams of huge tree boles in eddies and side channels as they passed through the rain forest zone of the lower river below the Columbia Gorge (Moulton 1988). Today little or no large wood recruitment is possible, because the system is so altered by dams and water diversions. Moreover, river and fisheries managers have actively removed dead wood for decades, along with beaver dams, which are natural landscape features because they erroneously thought they hindered fish movement.

Temperature Pattern: The Primary Determinant of Water Quality

Salmonid physiology, like that of most aquatic organisms, is controlled by environmental temperature. Growth increases logistically as temperatures warm up to the thermal optimum of the genome and then declines rapidly with further warming (Brett 1971). Many other physiological functions show similar temperature relationships (Figure 5.10). Niche segregation in aquatic habitats is governed by physiological adaptations to the natural annual thermal temperature pattern (Ward and Stanford 1982). For example, altitudinal species replacements that are coherent with increasing amplitude in the annual thermal pattern from headwaters downstream repeatedly have been demonstrated for stream insects (Ward 1986; Hauer et al. 2000). Thus, riverine food webs and productivity of salmon habitat largely are controlled by temperature patterns. A particular habitat that seemingly has ample food and cover may

ORNL 98-128294/mhr

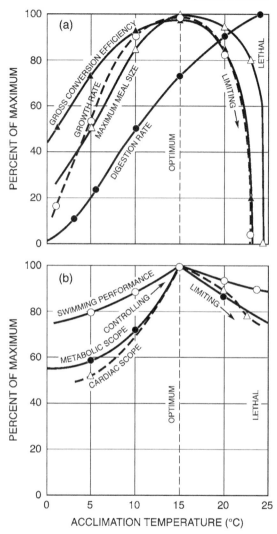

Figure 5.10 Physiological performance of juvenile sockeye salmon as a function of holding (acclimation) temperature. These data indicate optimum temperatures and distinguish between the rate-controlling role of temperatures below the optimum temperature and the decline of physiological performance (performance-limiting role) as temperatures exceed the optimum. After Brett (1971).

not be a preferred or even usable habitat if the temperature pattern is not consistent with the evolved tolerances of a particular fish population. On the contrary, fish may be present or even abundant when food and cover are sub-par, if the temperature pattern is tolerable. Hence, natural temperature patterns are crucial to river conservation in general and to sustainability of salmonid productivity in particular (Stanford et al. 1996; Coutant 1999).

Temperatures in the Columbia Basin generally are lowest in January and February and highest in August and September (Ebel et al. 1989). Thermal patterns in tributaries throughout the basin differ widely with location, elevation, and input from rainfall, snowmelt, glaciers, and aquifers (Coutant 1999). In general, cold runoff from mountainous tributaries gradually warms as the water progresses downstream. The principal flow of mainstem rivers is warmest near the Columbia River outlet, where temperatures peak near 21°C (70°F). Clearly, there are exceptionally higher temperatures in dry, low-flow years. Development of tributaries such as the Yakima, Okanagan, and Umatilla Rivers for agriculture and urbanization has resulted in their outlets to the mainstem reaching summer temperatures about 4°C above levels expected otherwise. Historically, average temperatures at the mouth of the Snake River during August and September have always been a few degrees higher than those in the mainstem Columbia (Roebeck et al. 1954; Jaske and Synoground 1970). High summer temperatures may be substantially moderated by cold water discharges from alluvial aquifers during dry base flow periods. The reverse may occur in winter due to the moderating effect of flux of river water through the ground (Ward 1985). Thus, the general pattern that existed historically and therefore influenced population structure of salmonids in the Columbia, ranged from low amplitude cold headwaters (0–18°C) to higher amplitude warm waters (0–22°C) in the dry interior areas of the basin. In a few tributaries, such as the John Day and Owyhee Rivers during very low water periods in late summer, temperatures may have approached lethal levels, but fish likely had thermal refugia in localized upwelling areas of cold groundwater (Torgersen et al. 1999).

Salmonid populations or stocks have evolved or been selected through both natural selection and hatchery practice to tolerate the divergent environmental conditions and habitats of the Columbia and other Pacific Rim rivers. Chinook salmon occur from Alaska to the Central Valley of California, and other salmonids similarly are widely distributed and therefore subjected to variable temperature patterns. Moreover, temperature requirements differ by life stage. For Chinook salmon, the following responses have been documented (Groot et al. 1995):

- *adult migration and spawning:*
  optimum 50°F (10°C), with a range of about 46.4°F–
  55.4°F (8°C–13°C);
  stressful > 60°F (15.6°C); lethal >70°F (21°C);
- *egg incubation:*
  optimum < 50°F (< 10°C), with a range of about 46.4°F–
  53.6°F (8°C–12°C);

stressful > 56°F (13.3°C); lethal > 60°F (15.6°C); and,
*   *juvenile rearing:*
    optimum 59°F (15°C) with a range of about 53.6°F–
    62.6°F (12°C–17°C);
    stressful > 65°F (18.3°C); lethal 77°F (25°C).

In these data, "optimum" generally means several degrees above and below the stated value. "Stressful" is performance markedly below optimum, and "lethal" is for standard 1-week exposures (higher temperatures may be tolerated for short-duration exposures). A detailed compendium of temperature requirements of Chinook salmon was prepared by McCullough (1999). Other salmon species are not markedly different. Because growth is temperature dependent, within the relatively wide thermal tolerances of salmon species, particular populations may be associated with particular temperature patterns (Beacham and Murray 1990). In the Columbia Basin, fall Chinook typically occupy the lower reaches or mainstem river where annual temperature amplitude is great, whereas spring Chinook are adapted to the colder headwater areas. Summer Chinook are intermediate. At least this was historically the case prior to substantial alternation of natural population structure by dams and other anthropogenic effects. In fact, these observed distributions may be nothing more than growth responses to variable temperature patterns that, coupled with fidelity for spawning sites, have produced a cline of life history types. Bull trout and lake trout are especially adapted to cold water and are rarely observed in water above 10°C. The conclusion is that, while tolerances in stress tests vary and considerable life history variation exists, Columbia River and other salmonids segregate in relation to temperature pattern just as do other aquatic organisms. Disruption of temperature pattern produces abnormal conditions that affect the fish directly through influences on metabolism and indirectly through food web changes. There is increasing knowledge about overall thermal habitat requirements and selection by salmonids and the relationships to temperatures of these habitats (e.g., Templeton and Coutant 1971; Torgerson et al. 1999). This understanding has not, however, been adequately accommodated in water management decisions.

Groundwater Upwelling: A Key Attribute of High-Quality Habitat

Large woody debris accumulations and other structures confer to natural alluvial rivers a high degree of morphological complexity that results: contributes to in highly connected subsurface and surface flow paths (Figure 5.11). Deep pools, low-velocity backwaters and springbrooks isolated from main channel flows are common zones of upwelling and concentration of groundwater in ways that create diverse thermal refugia for fishes and other

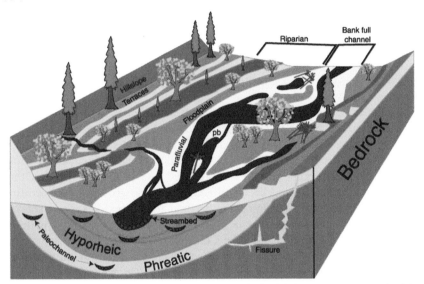

Figure 5.11   An idealized view of the 3-D structure of alluvial river ecosystems, emphasizing dynamic longitudinal, lateral, and vertical dimensions and the role of large wood debris eroded from the riparian zone. This landscape is produced by the legacy of cut-and-fill alluviation, which is linked to the natural-cultural setting of the catchment. Hence, the size of the flood plain and extent of hyporheic zone can be substantially larger than indicated here, extending to the valley walls, as in Figure 5.2. Biogeochemistry of surface water and groundwater interactions on such flood plains is mediated by complex interstitial flow paths or zones of preferential flow related to position of ancient river beds (paleochannels) within the alluvial bed sediments. From Stanford (1998).

organisms (Sedell et al. 1990, Stanford and Ward 1993; Ebersole 1994). These habitats are cold relative to warm surface waters in summer, warmer than surface waters in winter, and can sometimes be nutrient-rich and highly bio-productive (Stanford and Ward 1993).

In winter, groundwater-influenced stream habitats (Figure 5.6; upwelling zones in main channels and backwaters; back bar and wall-based channels; low-terrace springbrooks), especially on alluvial flood plains, often remain free of anchor and surface ice, buffering them from the stresses of winter freezing and thawing processes that can be highly disruptive of biota, including wintering fishes. Groundwater-influenced habitats are well known to provide important spawning habitats for fall-spawning salmon and bull trout. Historical data are clear on where the fish spawned (Fulton 1968, 1970), and it is apparent that these were areas of groundwater upwelling. Groundwater-rich pools, beaver ponds, and springbrooks also appear to provide critical winter habitat for juvenile and adult salmonids, which may move long distances to congregate in these areas (Cunjak and Power 1986; Chisolm et al. 1987; Cunjak and Randall 1993). In summer, cool water refugia, maintained by groundwater upwelling, are used heavily by adult spring Chinook (Berman and Quinn 1991), resident trout (Li et al. 1995a), and virtually all other salmonids that inhabit warmer river reaches. In large portions of the Columbia Basin at lower elevations and in desert areas, it is likely that native salmonids would not persist, except for the availability of cool refugia at groundwater upwelling sites.

Upwelling areas on alluvial reaches are hot spots of bioproduction because the plant-available nutrients accumulate in groundwater flow pathways (Figure 5.12; Vervier and Naiman 1992; Stanford et al. 1994). These nutrients greatly stimulate primary production and increase protein content of emergent hydrophytes and riparian woody vegetation (Harner and Stanford 2003). Hence, riverine habitats influenced by groundwater provide a more consistent and abundant food supply for all life stages of salmonid fishes and

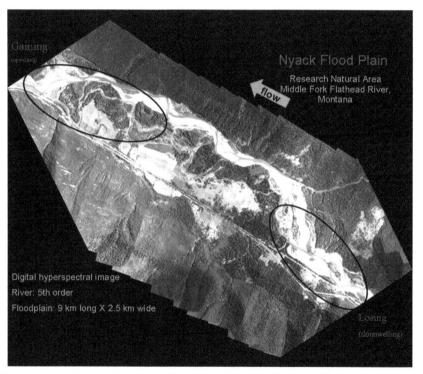

Figure 5.12 An image of the Nyack Flood Plain of the Middle Fork of the Flathead River in northwest Montana created from multispectral data obtained from a sensor in an aircraft. The image is spatially explicit and shows the channel structure including spring channels and wetlands created by upwelling of hyporheic water (see Figure 5.10) in low contour areas created by flood scour. Circles indicate areas of regional downwelling of river water into the alluvial aquifer at the upstream end and an area of regional upwelling from the aquifer to the surface at the downstream end. From Harner and Stanford (2003).

other food web components. This is in contrast to steep gradient canyon segments, where surface and groundwater interactions are limited due to bedrock controls on channel geomorphology. Salmonid habitats there are limited to main channel features such as rapids, runs, pools, and eddy complexes associated with rock and woody debris and there is less riparian development (Stanford et al. 2005).

Complex interactions between groundwater and surface water are key attributes of high-quality riverine habitat for salmonid fishes in the Columbia River Basin. Since these habitats are created by the interrelationship between flow and bedload movement (cut-and-fill alluviation) in relation to the slope of the main river channel, these habitats are not distributed uniformly. They are most well developed on aggraded flood plains. Floodplain segments also are human foci within the river continuum, because these tend to be the most productive nodes for agriculture and water diversion and impoundment in the watershed landscapes of the Columbia Basin (Stanford et al. 2005). As a consequence, maintenance of critical salmonid habitat has been problematic owing to conflict of uses. Moreover, owing to the early development of floodplain reaches, salmonid habitat in these areas was compromised to varying extents many years ago, and measures to protect rivers have tended to focus on much less productive canyon and high mountain segments that were not only of less importance to humans, but also were less important habitats for salmonids in general.

We realize that protection was accorded to various river reaches, not in relation to use by salmonids, but because these segments had wilderness or other values. However, the restoration of habitat complexity in the alluvial reaches should be emphasized in recovery efforts.

Causes and Consequences of Habitat Degradation

Conclusions from Recent Reviews and Synoptic Studies

Vast quantities of information on habitat conditions in Columbia Basin watersheds and streams have been collected by numerous agencies. Unfortunately, the quality and scientific value of much of this information is questionable or very limited. Much is inaccessible and perhaps permanently lost, and very little of the remainder that is potentially useful has been comprehensively analyzed, summarized, or otherwise made available in reports or publications. Fortunately, over the recent past (especially in the mid-1990s), several important reviews and a few comprehensive research studies of salmonid habitat status, trends, and threats in the Pacific Northwest and Columbia Basin have been published. Hence, comprehensive review herein would be redundant; readers are urged to consult the citations given below for details. Our objective in this section is to underscore the general conclusions from these reviews: (1) Considerable degradation of salmonid habitat has occurred in the Columbia River Basin; (2) habitat conservation and restoration has not been a priority for management; and (3) where habitat restoration has been attempted, the results generally have been unsuccessful or counterproductive (General Accounting Office 1992).

Regional reviews of salmonid population status (Nehlsen et al. 1991; Frissell 1993; Rieman and McIntyre 1993; Wilderness Society 1993; National Research Council [NRC] 1995; Stouder et al. 1997; NMFS 2000) strongly implicate habitat degradation as a major contributing cause of population decline. Frissell (1993) and the Wilderness Society (1993) point out that regional patterns in decline of salmonids and other fishes in areas not subject to the impacts of dams and major diversions indicate the pervasive importance of general, catchment-wide habitat degradation as a threat to fish populations. This is not to say that construction and operation of the many dams and reservoirs in the Columbia Basin are not important factors in run declines (they are); but, clearly other kinds of human land use and associated freshwater habitat degradation can and do endanger salmonid populations.

General discussions of some mechanisms of habitat change in response to human activities and its effects on riverine ecosystems can be found in Elmore (1992), Naiman (1992), Stanford and Ward (1992), and briefly in earlier sections of this chapter. Bisson et al. (1993, 1997) Forest Ecosystem Management Assessment Team ([FEMAT] 1993), Meehan (1991), and Salo and Cundy (1987) comprehensively reviewed some of the multifarious pathways by which human land management activities are known to degrade habitat and affect salmonid populations. These processes are no different in the Columbia River Basin than elsewhere in the world.

Lichatowich and Mobrand (1995), Lichatowich et al. (1995), and Wissmar et al. (1994) discuss many early references to habitat degradation and its consequences for salmon runs in the Columbia Basin. Many of the historical sources date from prior to the turn of the 20th century, and it is clear that degradation of freshwater habitat (e.g., by beaver activity and early mining activity) was well underway in the basin soon after its colonization by Europeans. Available data indicate that coho salmon in the Columbia River Basin were in serious decline by the 1930s, long before construction of mainstem dams began, in part reflecting the effects of extensive human-caused changes in low-elevation habitats (Pacific Rivers Council et al. 1993).

Theurer et al. (1985) developed a modeling procedure to relate riparian vegetation to thermal regimes of streams, and tied this to relationships between temperature and salmon abundance developed in previous Columbia River research. They applied the models to assess fish habitat in the Tucannon River drainage (Figure 5.13), estimating losses of salmon production caused by riparian land uses in the drainage, and predicting potential gains in salmon populations that might result from protection and

Figure 5.13 Spring Chinook spawning and rearing habitat in the Upper Tucannon River, Washington. Photo by T. Iverson.

restoration of historical levels of riparian forest cover. Though necessarily based on limited data, this work is notably one of the first credible attempts to understand natural Columbia River salmon production and habitat status in the context of large-scale human alteration of ecosystem pattern and process. The results strongly suggest that salmon recovery in the alluvial reaches of the Tucannon, and likely in other mid-Columbia rivers (our conclusion), is directly tied to substantial improvements in habitat condition. Li et al. (1995b) reported the results of extensive observational and experimental studies demonstrating impacts on the aquatic system of human disturbances in riparian zones in the John Day Basin, including thermal alteration shown to be highly adverse to salmonid fishes.

Wissmar et al. (1994) and McIntosh et al. (1995a, 1995b, 1995c, 1995d, 1995e) demonstrate the pervasive, adverse impacts on fish habitat that human activities have caused in the Columbia Basin. They document damaging changes in channel morphology and stability and progressive and persistent loss of riparian vegetation, pools, large wood, and other biologically important habitat elements in streams whose catchments experienced extensive logging, grazing, mining, and other human extractive development. By contrast, streams in catchments dominated by relatively undisrupted wilderness or roadless areas exhibited little long-term change, or even showed improvement in fish habitat conditions over the study period (the past several decades).

Obviously, natural disturbance processes (floods, droughts, diseases) occur episodically in roadless and natural areas, but, in general, natural disturbances appear to have much less adverse effect on native fishes than do human disturbances. For example, catchments affected by large-scale wildfire since the 1940s, as long as they were not also affected by extensive human activities, appeared to maintain high-quality habitat or improving trends in habitat conditions (McIntosh et al. 1995a, 1995b, 1995c, 1995d, 1995e).

Based on regional assessment of biological and federal-land resources in the Columbia Basin, Henjum et al. (1994) strongly advocated the protection of roadless and late-successional forest lands to provide watershed-level refugia for fishes and other aquatic species. Rhodes et al. (1994) presented an extensive synthesis of previously unexamined or fragmentary data from various agency sources to demonstrate the extreme importance of roadless and little-impacted catchments as *de facto* strongholds for declining Snake River salmon. In an analysis of a large field data set on habitat condition and fish populations in the Clearwater River Basin in the Upper Columbia, Huntington (1995) showed that roadless catchments, even those that had intensely burned earlier this century, provided higher quality habitat to more diverse and abundant native fish populations than did nearby, heavily "managed" catchments. Huntington's analysis also indicated that non-native species such as the brook trout (*Salvelinus fontinalis*), which can displace native trout and interbreed with bull trout, flourish in catchments where habitats have been more extensively impacted by humans. Henjum et al. (1994) and Li et al. (1995b) pointed out that even though remaining relatively undisturbed headwater areas may afford marginal habitats relative to the historical distribution of fish species in Oregon, protection of these areas appears critical for near-term persistence and long-term restoration of native fishes, including salmon. In a multispecies biodiversity assessment of the Swan River Basin in Montana, Frissell et al. (1995) found that tributaries draining roadless areas, especially those that have not been extensively stocked with non-native fishes, appeared to be disproportionately important for native trout and other aquatic and wetland-dependent species.

Several independent scientific reviews were conducted in the 1990s of habitat improvement projects, funded by the Bonneville Power Administration, in the Columbia Basin (Beschta et al. 1991; Frissell and Nawa 1992; Kauffman et al. 1993; Henjum et al. 1994; NRC 1995; Naiman and Bilby 1998). The reviews were highly critical of habitat management that has (unsuccessfully) emphasized the installation of costly artificial structures in lieu of full protection and restoration of natural vegetation and ecological processes that create and maintain fish habitat.

Doppelt et al. (1993) offers a lucid critique of misplaced priorities in past policy and habitat management programs, and provides a road map for a more comprehensive and ecologically credible approach to restoration of salmon ecosystems. Their recommendations and Wissmar et al. (1994) suggest that managers focus on identifying existing high-quality watersheds and downstream "nodal" habitats or "hotspots" that are disproportionately important for protecting existing native species populations and protecting them from proposed or recent human disturbances (e.g., through removal of existing logging roads or removal of livestock from riparian areas). The second priority in terms of urgency of action, but equally necessary for long-term success, is restoring adjacent and selected downstream habitat patches that can increase basin-wide biological connectivity and allow expansion, life history diversification, and demographic and genetic reconnection of existing population fragments.

The loss of salmon from freshwater habitats has created a negative feedback loop for restoration of these habitats. It is becoming increasingly clear that historically the nutrients from decaying carcasses of adult salmon brought nutrients from the ocean to the often nutrient-deficient upper tributaries (Larkin and Slaney 1997; Bilby et al. 1998). The decaying carcasses fed both aquatic life (e.g., insect larvae and their predators, young salmon) and terrestrial animals, and the nutrients they released were used by aquatic and riparian plants. Much of the productivity of tributary ecosystems has been attributed to nutrients imported by adult salmon. With declining salmon runs, nutrient supplies have diminished, and whole ecosystems are more depauperate.

Similar Effects from Different Causes: Human Activities and Consequences

A theme of this chapter is that many kinds of human activities tend to result in similar changes in aquatic ecosystems, although the magnitude, persistence, interactions, and biological outcome of the effects can vary widely, according to local conditions and history. While site-specific prediction of impacts can be difficult and uncertain (e.g., influences of a particular forestry prescription; outcome of fish stocking), catchment-scale trends and spatial patterns in freshwater habitat condition in the mainstem Columbia River and its tributaries are generally well documented, predictable, and stereotypical (Salo and Cundy 1987; Rhodes 1995; Rhodes et al. 1994). In this section, we provide a very general sketch of the typical effects of various human activities in the Columbia Basin.

Beaver Trapping

Perhaps the earliest exploitative land use in the Columbia Basin was large-scale trapping of beaver, which began in the mid-1800s (NRC 1995). Beaver dams were historically very extensive in nearly all alluvial and low-gradient segments of Columbia River tributaries and were common in branches and backwaters of the larger tributaries and Columbia itself. Alluvial flood plains were sites of heavy beaver activity, causing streams to meander and braid, thereby maximizing the mosaic structure of salmonid habitats.

Beaver dams (Figure 5.14a) and their foraging activities (Figure 5.14b) created storage sites that buffered flows of water and downstream transport of organic matter, nutrients, and sediment. Beaver ponds were important rearing and wintering areas for many species of salmonids, and promoted channel switching and geomorphic complexity that encourages extensive exchange of surface and subsurface waters in alluvial aquifers (Naiman and Fetherston 1993). Another underappreciated function of the beaver may be its unique role as an upstream vector of vegetative propagules of willow and other important riparian species, allowing their recolonization following debris flows, severe drought, and other catastrophes that can wipe out riparian plant communities in tributary catchments.

The activity of trapping permanently reduced or extirpated most beaver populations, with resulting widespread loss of structural elements, floodplain processes, and vegetative diversity that had developed as a result of centuries of ongoing beaver activity. Throughout the Columbia Basin, beaver-mediated creation of salmonid habitat is nowhere near its zenith in the river system that Europeans discovered, even though beaver have been included in state wildlife management programs for at least the latter half of this century.

Logging

Early settlement in the Columbia River Basin was concentrated in alluvial bottomlands along lower elevation tributary rivers and streams, where arable soils and water were plentiful and transportation was most feasible. Logging in flood plains and bottomlands accompanied the earliest settlement for purposes of land clearance, access to and through stream channels for transport, and for construction materials. A sawmill was operating in Vancouver, Washington, as early as 1827 (NRC 1995). As regional and national markets and transport systems developed, timber grew rapidly to become a major commercial component of the Pacific Northwest's economy. Cutting of timber remains a widespread industrial activity in the Columbia Basin today, although most large, valuable stands of old-growth forest are long gone. As shown by Henjum et al. (1994), timber cutting in the Columbia Basin has, in many areas, been disproportionately concentrated in low-elevation valleys and riparian areas, where high-value species and older trees were historically most abundant.

Logging of trees from riparian areas directly eliminates the source of large woody debris that is so central to many ecosystem processes and the

Figure 5.14 Beavers modify the riparian landscape and affect salmonids in several ways: (a) through building dams and blocking small tributaries, and (b) through cutting riparian trees and shrubs for food and building materials. Photos by R. N. Williams.

maintenance of habitat complexity and productivity in streams and rivers. Moreover, it directly reduces shade and alters near-surface microclimatic conditions that protect streams from climate-driven warming in summer and freezing in winter (Salo and Cundy 1987; Maser et al. 1988; FEMAT 1993; Naiman and Fetherston 1993). Removal of standing live trees or downed wood can jeopardize the long-term stability of channel banks, floodplain, and toeslope surfaces. In areas where tree regeneration is dependent on seed sources or specific ecological conditions, logging has resulted in the permanent loss of ecologically valuable tree species, such as western red cedar and ponderosa pine. In addition, operation of machinery necessary to cut and remove trees can directly damage soils, vegetation, and channel features, altering ecological processes in these sensitive areas.

Despite speculation to the contrary, no study has demonstrated that "safe" or "beneficial" levels or methods of logging in riparian and floodplain areas exist from the standpoint of maintaining the many natural ecological functions of forests. This is exacerbated by the massive, regional scale at which previous logging has caused long-term impoverishment or impairment of ecological components and processes in the Columbia Basin. Therefore, recent scientific assessments have recommended no removal of trees from these key areas (FEMAT 1993, Henjum et al. 1994; Rhodes et al. 1994).

Although today the logging of riparian and floodplain forests continues in many areas, most timber harvest volume is produced by more extensive and frequent cutting of smaller, lower value trees over larger, upland areas, which requires extensive road networks and results in major alteration of forest cover conditions across large catchments. These landscape alterations have different, but equally threatening effects on catchment processes and freshwater habitat.

Although humans build roads for many purposes, the vast majority of roads in the Columbia Basin have been (and continue to be) constructed for purposes of timber logging transport and access for silvicultural management of commercial forest lands. In the spectrum of natural disturbance processes, road networks have no known natural analogue. Road networks are direct sources of accelerated sediment production and efficient delivery to the stream network (Meehan 1991; Rhodes et al. 1994). Roads also permanently intercept and redirect surface and subsurface flow of water, altering hydrologic and thermal regimes in streams (Meehan 1991; Rhodes et al. 1994). Roads can serve as vectors for forest pathogens and increase the spatial extent of a wide range of human activities, such as legal fishing, poaching, and deliberate or unintentional introduction of non-native species that threaten native biodiversity (Frissell and Adams 1995; Frissell et al. 1995; Noss and Murphy 1995).

Logging often results in the removal of forest cover in patterns and at rates far exceeding the scope of natural events that have historically dominated forest landscapes in the Columbia Basin. Moreover, unlike fire, disease, windthrow, and other natural forest disturbances, logging causes the large-scale removal of largest size fractions of woody debris from forests (Maser et al. 1988). The mechanical means used for cutting and removing large trees can create unnatural soil disturbance and compaction that accelerates surface erosion and alters hydrologic relations. Opening the forest canopy, especially if it occurs across a significant portion of a catchment, alters microclimate, snow accumulation and melt and other aspects of precipitation, and can change the routing and slope storage of water, often resulting in downstream changes in stream flow and channel stability that are detrimental to fishes. Such changes typically include increased flashiness of discharge, increased peak flows, and accompanying increases in sediment load due to erosion of channel margins and heads. On steep and unstable terrain, changes in subsurface flows and soil moisture, perhaps together with reduced root strength, can increase the frequency and alter the style of landslide and gully erosion (Salo and Cundy 1987; Meehan 1991; Naiman and Fetherston 1993; Rhodes et al. 1994). Increased transpiration and reduced moisture-capturing and -retaining efficiency of second-growth forests following extensive logging can result in long-term depletion of summer and fall low stream flows, even as winter and spring peak flows increase (Hicks et al. 1991; Rhodes et al. 1994).

Despite the vast spatial extent of past and present logging activities in the Columbia Basin, few of these ecosystem changes are satisfactorily explained or accounted for in existing models of cumulative watershed effects employed by land management agencies to assess environmental impacts, and these processes are rarely monitored on a site-specific or watershed basis. Indeed, the most pervasive impact on fish and fish production, especially the large bodied adfluvial and anadromous species, is increased access for human predators by the logging road network. For example, Baxter et al. (1999) showed that adfluvial bull trout were largely missing from roaded tributaries of the Swan River in Montana, whereas unroaded areas retained populations.

Grazing

Grazing by domestic livestock can make changes to riparian areas and stream channels that are detrimental to salmonids. Kauffman and Krueger (1984) and Rhodes et al. (1994) provide valuable reviews of this subject. While grazing by domestic species began very early in some areas with cultivation of horse herds by Indian tribes (e.g., the Grande Ronde Basin), large numbers of sheep and cattle arrived with European settlers during the late 1800s. Even though peak numbers of livestock probably occurred prior to the turn of the century, grazing impacts on aquatic systems since then have continued relatively unabated. More than a century of continuous grazing in many areas has caused progressive deterioration of range and riparian conditions throughout the Columbia Basin (Rhodes et al. 1994; Wissmar et al. 1994; Lichatowich and Mobrand 1995).

Livestock impacts to streams occur through three major vectors: (1) direct trampling of channels, banks, and soils; (2) removal and alteration of vegetation, particularly in riparian areas; and (3) direct introduction or overland flow of fecal wastes and urine into surface waters. The direct effects of large, grazing animals include trampling and sloughing of stream banks, loss of overhanging banks, accelerated bank erosion, compaction of soils, and increased sediment input to adjacent and downstream reaches. Grazing and trampling of vegetation in riparian areas and flood plains generally reduces vegetative cover and vigor, suppresses or eliminates some vegetation species (especially palatable but ecologically critical woody species, such as willows), and reduces canopy cover over the channel. The result is typically widened and open channels, with lower, warmer, more turbid surface flows in summer, more extensive and damaging ice conditions in winter, and flashier, more turbid flows in winter and spring runoff periods (Figure 5.15). Fine sediment concentrations increase and channel stability decreases (Meehan 1991; Rhodes et al. 1994; Li et al. 1995b). Eutrophic enrichment from livestock wastes can cause depletion of oxygen required by fishes and their principal food organisms. These changes are adverse to salmonids at virtually all life stages.

Figure 5.15 Cattle grazing in the stream bed in McIntyre Creek in eastern Oregon. Grazing impacts have eroded the stream bank, reducing the water retention capacity of the local landscape, causing the stream to flow subsurface. Photo by R. N. Williams.

Mining

The many effects of mining are discussed in general terms in Nelson et al. (Meehan 1991). Although extensive mining has occurred in many areas of the Columbia River Basin, the history and effects of these activities have not been comprehensively compiled and described in any single source. Mining effects, although difficult to sort out from those of many other simultaneous and subsequent disturbances, were no doubt extensive in some major tributaries (e.g., John Day River, Salmon River, Coeur d'Alene River, Upper Clark Fork River) by the late 1800s. It is likely that the historic impacts of mining on salmon and native trout (Figure 5.16) have been given short shrift in recent reviews (NRC 1995) because of the relative paucity of information, and perhaps because mining today is less widespread an activity than logging, grazing, and irrigated and cropland agriculture. However, the old disturbances and their effects remain. Some disturbed dredge mining sites are being reclaimed, such as in the John Day subbasin (Figure 5.17) where localized dredge mining occurred within historic spring Chinook spawning areas; however, reclamation is expensive and requires monitoring to assess whether reclamation actions led to increased use by spawning adults.

Mining activities of various kinds inflict intense soil disturbance and erosion (Figure 5.18). In addition to very large sediment inputs to downstream

Figure 5.16 The legacy of dredge mining on the Yankee Fork, a tributary system of the Main Salmon River, in central Idaho. Photo by R. N. Williams.

Figure 5.17 Dredge mine reclamation site on Granite Creek on the North Fork of the John Day in spring Chinook spawning habitat. Photo by R. N. Williams.

Figure 5.18 Small-scale instream mining on a tributary system in the Clearwater River, Idaho. Photo by R. N. Williams.

reaches, placer mining causes direct, wholesale destruction of natural channels, flood plains, valley floor soils and vegetation, and alluvial aquifers.
Natural recovery is inhibited, perhaps permanently. Areas of the Upper
Grande Ronde Basin (McIntosh et al. 1994) and elsewhere subjected to
placer mining, for example, have not recovered to any semblance of natural
structure and function in more than a century. Leaching of toxic materials
from mining wastes and milling sites can permanently contaminate and
impair the productivity of stream and riverine ecosystems many kilometers
downstream, as is most evident in the Upper Clark Fork and Coeur d'Alene
River Basins in the headwaters of the Columbia, where native salmonids
have been virtually eliminated from the affected waters for a century or
longer. It appears likely that other mining districts may suffer more subtle,
not yet documented depressions in biological productivity from mining
waste toxicity.

Irrigation and Cropland Agriculture

Cropland agriculture affects vast areas of the Columbia River Basin (Figure
5.19) although this activity is perhaps most concentrated on arid basalt
plateaus and Palouse prairie country where surface waters are scarce.
No comprehensive review of the effects of cropland agriculture on fish habitat in the Columbia Basin exists, as far as we know, but the NRC found that
dewatering associated with agriculture irrigation was the primary cause of
the vastly reduced riparian condition throughout the United States (NRC
1999). Farming can significantly alter hydrology and increase erosion and
sedimentation processes many-fold over natural rates. Where farming
impinges on wetlands, flood plains, and riparian areas, it directly destroys
riparian vegetation and channel structure. The principal effects of cropland
agriculture on fish in the Columbia Basin no doubt stem from flow diversion
and withdrawal for irrigation (NRC 1995). Some irrigation also occurs to
support grazing of pasture. Irrigated agriculture began with early settlement
in the mid-1800s, but rapidly accelerated with the assistance of large, govern-
ment-subsidized projects starting in the early 1900s and continuing to the
present.
    Although the effects of irrigation on freshwater habitat and fish popula-
tions are a widespread problem globally, few good review papers are available
that address the scope this problem. There has not been a comprehensive ecological assessment of the consequences of irrigation for fish in the Columbia
Basin.
    Dams and diversions to provide water for irrigation can block movements
of migratory fishes and divert fish from natural habitats into ditches or
onto fields, killing them (Figure 5.20). Diversions dewater natural habitats,
reducing habitat available in streams and sometimes rendering it entirely hos-

Figure 5.19 Irrigated fields in the middle Snake River plain, south central Idaho. Photo by R. N. Williams.

tile (e.g., through warming) or serving as a barrier to fish passage (e.g., loss of surface flow through riffle crests). Water in storage ponds typically warms much more than free-flowing water in streams and natural aquifers. Water that is returned to streams from irrigated fields is typically warm and often laden with very high concentrations of sediments, nutrients, and pesticides. Vaccaro (1988) developed a simulation model of the effects of irrigation diversions on surface water temperature in the Yakima River Basin. Vaccaro projected that removing the effects of irrigation diversion could cool summertime temperatures in critical salmon habitats by up to 2°C; this effect was most pronounced at lower elevations where larger, alluvial reaches of the

Figure 5.20   Irrigation diversion dam on Bear Creek, a tributary in the Grande Ronde subbasin in northeastern Oregon. Photo by R. N. Williams.

river and its tributaries once supported abundant salmon production (Lichatowich and Mobrand 1995).

Urban and Other Sources of Excessive Nutrients and Toxic Pollutants

Concentrations of dissolved solids and pollutant loads generally increase from headwaters to oceanic confluence in most of the nation's large rivers, including the Snake and Columbia Rivers (Smith et al. 1987), as a consequence of the cumulative loads of pollutants from all land use activities. A primary source is treated sewage effluents and storm drainage from the urban areas along the river corridors. Oxygen depletions and other indicators of severe organic and nutrient pollution from point sources near and within urban centers were once common in the lower mainstem reaches of the Columbia River and many of the larger tributaries (Stober and Nakatani 1992). Owing to the U.S. Federal Clean Water Act, treatment of sewage and effluents from pulp mills and other industrial sources has been substantially improved in the last two decades. Creation of many reservoirs within the continuum also contributed to the decline in pollution because pollutants are processed by food webs and retained in reservoir sediments. However, continuing concern exists for loading of plant growth nutrients in the large on-channel lakes and reservoirs. Loads are being legally allocated to sources through actions to limit the total maximum daily load to lakes, for example, Flathead Lake in Montana (Flathead Basin Commission 1994). The U.S. Environmental Protection Agency, in cooperation with state water quality agencies, has an extensive program to determine total maximum daily loadings of pollutants. Metals and organic carcinogens are present in fish tissues throughout the basin (Stober and Nakatani 1992), even in headwater systems like Flathead Lake (where there are detectable PCB concentrations in fish tissues) (Flathead Basin Commission 1994), underscoring the need for continued vigilance. There is also concern that if pollutant levels are lowered, contaminants sequestered in the sediments could leach back into the water column, frustrating restoration efforts. On the Columbia River it has been shown that fluoride effluent from an aluminum plant upstream from John Day Dam caused delays of as much as 4 days in passage of Chinook past John Day Dam.

Nonetheless, water pollutants, other than from fine sediments, increased temperature and metals from mining districts as discussed elsewhere in this chapter, generally are not considered a major factor in salmonid declines nor particularly problematic for recovery. We are not sure that the available data have been examined well enough to agree with this consensus. Indeed, data on pollution loads, particularly from diffuse (non-point) sources and interactions between maintenance of salmonid critical habitats for all life stages, has not been examined extensively in the Columbia River system, at least not in the context of salmonid restoration.

Stream Regulation: Effects of Dams, Reservoirs, and Diversions

Flow regulation for purposes of hydropower production and flood control has been the primary issue for salmonid conservation and restoration in the Columbia Basin since construction of the first small tributary dams for power generation early this century. The effects of dams in the basin were magnified by construction of mainstem projects since 1938 that directly affect virtually all migratory fishes in the middle and upper Columbia Basin.

Dams

Mortality of salmonid fishes caused by dam passage (e.g., through turbines and bypass facilities) has dominated discussion and actions for salmonid recovery. Many millions of dollars have been spent on facilities and research to increase bypass efficiency in the absence of accurate mortality estimation (see following chapters). Recent studies on the Snake River suggest far lower mortality associated with reservoir transit and dam bypass by wild fall Chinook salmon than previously thought. These issues are discussed in detail elsewhere in this book (Chapters 6 and 7). We note that recovery efforts also have focused heavily on decreasing the transit time for smolts in the highly regulated mainstem, either by use of storage releases to move smolts out of the system or by barging, even though such actions clearly are selective of specific life history types. We conclude that greater attention to habitat-related effects of stream regulation is needed.

The ecology of regulated streams has been summarized in several volumes (Ward and Stanford 1979; Lillehammer and Saltveit 1984a, 1984b; Craig and Kemper 1987; Petts and Wood 1988; Petts et al. 1989; Calow and Petts 1992; Hauer 1993). Principles from a very diverse and detailed literature (Stanford et al. 1996) directly apply to the Columbia River. In this section, we point out that dams not only directly block fish movement and migrations but also have many important consequences for salmonid habitat and populations, including (1) destruction of riverine habitat upstream of dams and its conversion to novel, reservoir habitats; and (2) the creation of highly artificial flow, thermal, and sediment regimes downstream of dams.

Reservoirs

Reservoirs represent massive loss of the once highly productive riverine habitat that occurred above most dam sites. Optimal dam sites are often located at narrow bedrock constrictions below wide, aggraded valleys, which allow large storage ratios for a given dam size. As described above, these aggraded, alluvial reaches correspond to flood plains or to highly productive riverine habitats for fishes and other native biota, where lateral habitat complexity is high, interaction between groundwater and surface waters is great, and natural riparian vegetation is extensive, heterogeneous, and productive (Ward 1973; Ward and Stanford 1995). Not only was key habitat lost to inundation, but also flow regulation has vastly changed riverine habitats downstream. Operation of the dams limits peak flows and increases base flow causing channels to degrade and disconnect from flood plains and channel substratum to armor with large rocks and cobbles. Even in the relatively constricted mainstem Columbia River, alluvial features prevailed in the form of complex island, point and eddy reattachment bars composed of sand, gravel, and cobble. Back bar channels and sloughs were common features of the mainstem channels and flood plains. All of the mainstem habitat open to anadromous fishes above Bonneville Dam are now lacustrine except for the Hanford Reach. However, bars in the Hanford Reach are composed of very large cobble, the fines having been sluiced out, and back bar channels and sloughs are largely filled in with riparian vegetation owing to years of rapidly fluctuating base flow and lack of peak flows.

Based on early inventories of salmonid habitat (Fulton 1968, 1970), it is clear that native salmonids exploited these lost alluvial habitats heavily as spawning, nursery, refuge, and resting areas. Migratory salmon that originated all over the upper Columbia passed through these river segments as juveniles and adults, and these fish almost certainly took advantage of such riverine habitats to varying degrees. It is largely unknown to what extent reservoirs replace the ecological functions of these lost riverine habitats (although see below), but the status and trend of many fish populations suggests to a large degree they do not (Lichatowich and Mobrand 1995).

Reservoir storage and dam operations dramatically alter flow regimes of rivers downstream of the projects as well. Typically, natural seasonal flow peaks are reduced and delayed or eliminated, and low flow periods are continuously or intermittently augmented by controlled releases. These changes in hydrology, coupled with the effects of limnetic processes (e.g., seasonal, vertical stratification of temperature and biotic productivity) that affect reservoir water before its release, substantially alter thermal and nutrient regimes, which are typically highly predictable in natural, free-flowing large rivers of the Columbia Basin (Stanford and Hauer 1992; Brusven et al. 1995). As a consequence, high-quality rearing habitat associated with shallow low-velocity floodplain features becomes progressively disconnected from the channel. Owing to lack of scour associated with flooding, these key habitats fill with sediments and dense vegetation. In many cases, the vegetation is non-native, and regulated streams are active corridors for spread of noxious weeds and woody plants. For example, 70% of the connected wetlands of the five main floodplain reaches in the Yakima Basin have been disconnected from the main channels (Snyder et al. 2003). Indeed, an axiom of the ecology of regulated streams is that artificial regulation of flow, temperature, and nutrients favors a select few species, often non-native ones (see below), over the majority of native species whose life histories are evolutionarily adapted to the naturally prevailing (and seasonally fluctuating) thermal and hydrologic template. Accompanying these flow alterations are changes in turbidity and sediment transport caused by storage of sediments behind the dam or lack of scour to move fine sediments influent below the dams, which also can stress native fishes and their natural prey base by altering riverine habitat dynamics and reducing habitat diversity. Moreover, short-term baseflow fluctuations associated with hydropower peaking operations produce a large zone along each side of the river where aquatic biota cannot live. This so-called varial zone, which includes all of the shallow, low-velocity habitats within the river channel, occurs on all regulated river segments in the Columbia Basin and substantially compromises instream food webs and productivity. Juvenile salmonids cannot feed and rest in fluctuating flows and are washed downstream, whether they want to or not (Chapter 6). Moreover, shallow-water food supplies for juveniles is limited or non-existent.

The most pervasive impact of these interactive changes on river structure and function concerns changes in diel and annual temperature patterns, which is the primary determinant of fish distribution (Hall et al. 1992).

Indeed, lethal high temperatures occur in many places throughout the basin, based on generally accepted thermal criteria (National Academy of Sciences/National Academy of Engineering 1973) and measured temperatures (cf. Karr 1992; Lilga 1998).

Effects of the Columbia River dams on temperature patterns have been investigated at several scales. Studies in the 1960s (Jaske and Goebel 1967) showed that the construction of river-run reservoirs on the mainstem of the Columbia River caused no significant changes in the average annual water temperature. However, storage and release of water from Lake Roosevelt had delayed the timing of peak summer temperatures below Grand Coulee Dam since 1941. This delay was about 30 days at Rock Island Dam and was reflected as far downstream as Bonneville Dam near the river's outlet. Temperature extremes in the mainstem were moderated by the reservoir complex, so that the river below Grand Coulee is now somewhat cooler in summer and warmer in winter. This trend is particularly evident in tailwaters of major storage reservoirs such as Brownlee, Hells Canyon, and Dworshak, where high storage-to-flow ratios hold cold bottom water in the reservoir for release through deep outlets until well into summer. Mainstem reservoirs in the Snake and Columbia Rivers have created shallow, slowly moving reaches of shorelines where solar heating has raised temperatures of salmon-rearing habitat (Figure 5.21), especially for underyearling fall Chinook above tolerable levels, negating this as usable habitat for much of the summer (Curet 1993; Key et al. 1995).

Karr et al. (1998) have shown the advantages of using cold hypolimnetic releases from Dworshak Reservoir on the Clearwater River for cooling the lower Snake River in summer (Figure 5.22). The stratified flows (Figure 5.23) can be followed to the lowermost dams, and the temperatures are lowered into a range more suitable for salmonids.

Water temperatures in lower tributaries were generally low enough prior to European settlement to allow summer outmigrations of subyearling smolts. This life history strategy is greatly diminished today when we have intolerably high late spring and summer temperatures. There is an increased tendency for Snake River fall Chinook salmon to remain upstream of the lower river's reservoirs over the fall and winter rather than migrate in summer as is usual for this race (Connor et al. 2002). This tendency appears to be the result of well-documented poor survival of migrants in summer (Connor et al. 2003a, 2003b) and apparently higher survival of those that delay migration until the following spring. Analyses of returning adults indicate that nearly half are spring migrants. Thus, the fall Chinook salmon that had migrated out of the Snake River as underyearlings are increasingly developing a new "reservoir" life history. Juveniles remain in the cooler refuge near the mouth of the Clearwater River in summer and resume migration as yearlings in spring.

Watson et al. (1992) describes characteristics of the lower Yakima River that coincided with a subyearling smolt life history. Much of the lower main-

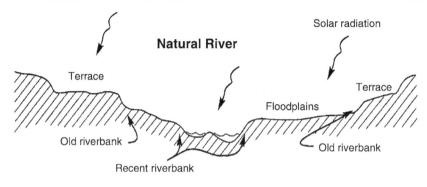

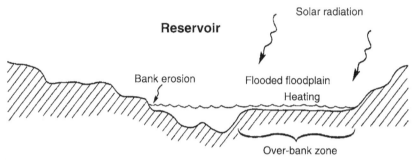

Figure 5.21 Creation of shallow overbank and erosional zones in alluvial valleys by damming that provide large areas to collect solar radiation and cause temperature increases in river water. Figure from Coutant (1999).

stem Yakima River consisted of intricately braided channels flowing through dense riparian forests. The shading, combined with lack of warm irrigation water now prevalent probably resulted in water temperatures considerably lower than today. Haggett (1928, cited in Watson et al. 1992) reported that heavy outmigrations of underyearling smolts began in June, peaked in mid-July, and continued through September. Similar timing is still found in the Rogue River (Schluchter and Lichatowich 1977). Today, any smolts leaving the Yakima must do so nearly 2 months earlier. The original braiding and complex of side channels probably retained cooler water from springtime flows in river gravels so that overall lower river temperature was cooled (Stanford et al. 2002).

A similar problem occurs in the lower reaches of other tributaries. Largely because of water withdrawals for irrigation and removal of riparian vegetation, water temperatures in summer are higher than those known to be lethal or debilitating to salmonids. Streams known to be affected include the Umatilla, Grande Ronde, and Okanagan. These high temperatures have prevented juvenile fish from migrating or redistributing downstream or to tributary branches. Adult fish have been prevented from ascending to suitable

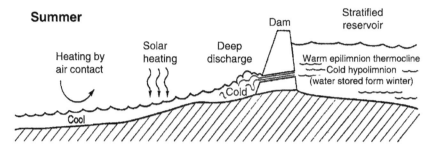

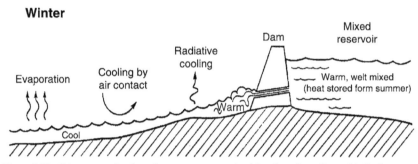

Figure 5.22 Diagrammatic view of a storage reservoir with a deep (hypolimnetic) discharge and its influences on downstream temperatures in summer (top) and winter (bottom). Figure from Coutant (1999).

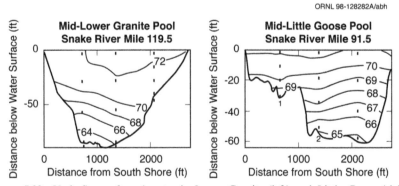

Figure 5.23 Underflows of cool water in Lower Granite (left) and Little Goose (right) Reservoirs on the Snake River from addition of 63°F (17.2°C) Clearwater River water from Dworshak Reservoir with 68°F (20°C) water from the Snake River. Figure from Coutant (1999).

spawning areas. Unsuitable temperatures have served to fragment the habitat of tributary basins (see metapopulation discussion in Chapter 4).

Maximum temperatures in the mainstem Snake River, where salmon survival is most tenuous, are generally lower in summer than before the series of storage and mainstem reservoirs was installed. But the peak temperatures have been shifted to later in the year due to the latency of reservoir storage. The natural pattern has similarly changed in the mainstem Columbia River (Figure 5.24). Temperatures at this time are higher than adults can survive for long periods during late summer and fall migrations (Figure 5.25).

Stanford et al. (1996) proposed a protocol for restoring temperature, flow, wood deposition, food web, and other lost ecosystem functions to regulated rivers. They proposed that channel-floodplain connectivity and revitalization of instream habitat structures can be accomplished by re-regulation of flows and temperatures (e.g., by selective release structures) to more normative regimes, assuming that pollution is not also a problem. Scouring flows are possible in most regulated reaches in at least average to wet years. Reduction of base flow fluctuation to mimic natural conditions can be accomplished by base loading the turbines; to reduce revenue lost from loss of peaking capability, base flows may be higher than historically occurred, but they cannot be fluctuated if a productive food web is to develop in the varial zone. In the Columbia River system, revenue lost by base loading some dams perhaps could be offset by peaking other dams that do not have riverine segments downstream (e.g., the mid-Columbia dams could be operated as re-regulation

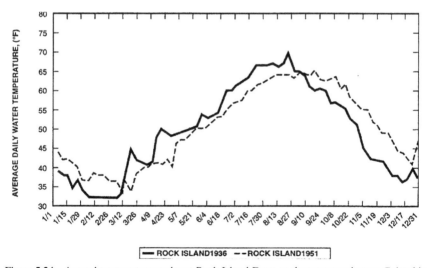

Figure 5.24 Annual temperature cycles at Rock Island Dam on the upper mainstem Columbia River before (1936) and after (1951) Grand Coulee Dam was in operation. Figure from Coutant (1999).

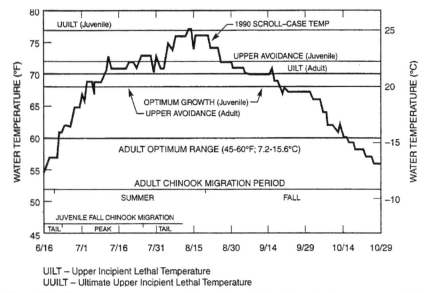

Figure 5.25 Temperature requirements of juvenile and adult Chinook salmon, compared with scroll-case temperatures of the mainstem Snake River at Lower Granite Dam in 1990. Normal adult and juvenile migration periods are shown near the bottom of the figure. Scroll-case temperatures represent the main channel flow, which is generally cooler in summer than shallow-water habitats normally occupied by juvenile salmon. From Coutant (1999).

systems for Grand Coulee peaking operations). Moreover, peaking flows provide turbulent waves that may assist movement of juvenile salmonids through reservoir-dominated reaches (see Chapter 6). Obviously, re-regulating the Columbia River system in order to lessen the effects of daily flow fluctuations requires careful analysis. This was attempted in the System Operations Review of the Bureau of Reclamation; however, the analysis itself and none of the alternatives embraced the principles of the ecology of regulated streams in the manner discussed here.

Diversions

All of these principles apply equally well to the many regulated tributaries of the mainstem Columbia River. Establishment of more normal flow and temperature conditions is possible in many tributaries by re-regulation of discharge schedules from the storage reservoirs in the headwaters. Rivers such as the Yakima, Walla Walla, and Umatilla are heavily impacted by irrigation withdrawals and high temperatures during periods of very low base flows, which are far lower now than occurred historically. Indeed, some segments are dry part of the time (Figure 5.26). Salmon and other aquatic biota cannot exist in these key tributaries in any sustainable numbers until base flows are elevated to

Figure 5.26   The Methow River near Arrowleaf in central Washington in late October, showing the lack of surface flows in both (a) the main channel and (b) an adjacent side channel. This area is important spawning and rearing habitat for spring Chinook earlier in the year. Photo by R. N. Williams.

a stage that allows productive food webs to persist in channel and shallow floodplain habitats. Higher base flows will also allow effective interstitial flow through gravel bars and flood plains, which likely will substantially cool surface waters in upwelling zones during critical, late-summer hot periods.

No mitigation effort can be manifested at the ecosystem level, however, without recognition of the watershed as the appropriate scale for river resource management. It is well established in the scientific literature that the land and aquatic area comprising watershed or catchment basin exerts strong physical and biological controls on the development of stream and lake ecosystems (Schuman and Lichty 1965; Hynes 1975; Frissell et al. 1986; Sheldon 1988; Moyle and Sato 1991; Stanford et al. 1996). In the past, attempts to protect and restore aquatic habitat and populations have often met with failure because they disregarded the overriding role of catchment conditions and processes in shaping aquatic ecosystems (Plattsand Nelson 1985; Frissell and Nawa 1992; Doppelt et al. 1993; Rhodes et al. 1994; Sear 1994).

Coherence of Habitat Loss and Structural Changes in Columbia River Fisheries

The Columbia River ecosystem is home to a wide variety of native resident and anadromous fishes. The ichthyofauna of the river system includes 65 native species, including three lamprey species. All of the eastern Pacific anadromous salmon species (pink, chum, sockeye, Chinook, coho), plus cutthroat and steelhead, historically were common in the Columbia River. At least 53 non-native fishes have become well established (Ward and Ward 2003; Wydoski and Whitney 2003). In general, native fishes have declined in range and abundance, while non-natives have proliferated. These changes are coherent with habitat loss and alteration associated with the development of the Columbia system (Netboy 1980; White 1995).

Through homing and natural selection, each native salmonid population is closely adapted to the particular array of habitats that is available to it, as we have described above. Non-native fishes have been widely introduced in the Columbia Basin, but it is notable that introduced fishes tend to be most successful in streams and rivers where natural habitat has been altered and native fishes depleted. Large-scale human disruption of historical habitat mosaics can create novel ecological niches that native fishes have not evolved to fill, providing a toehold for the invasion and eventual proliferation of introduced species (Balz and Moyle 1993). For example, Huntington (1995) reported that non-native brook trout (Figure 5.27) were more abundant in streams draining extensively logged areas of the Clearwater National Forest in Idaho than in streams whose catchments were predominantly roadless and unlogged. The presence of the brook

Figure 5.27 Non-native brook trout from a coldwater spring lake in the headwaters of the Snake River, Idaho. Photo by R. N. Williams.

trout may have been due to easier access afforded by roading, which likely facilitated planting of brook trout as well as increased fishing pressure on the native species. The logging itself may have been only indirectly involved in non-native invasions, as related to habitat modification associated with increased water and sediment yield.

Direct human alteration of riverine ecosystems in the Columbia Basin has massively promoted the proliferation of non-native fish species. Li et al. (1987) documented fish assemblage structure in major reaches of the mainstem Columbia and found that all reservoirs were strongly dominated by non-native species, such as smallmouth bass, walleye, yellow perch, and channel catfish. Most of these species are voracious predators on other fishes, and many are known to consume young salmon. These species also tend to prefer different thermal conditions than do native salmonids, so that they may be favored by the many human activities that alter thermal regimes.

By contrast, while the free-flowing Hanford Reach of the Columbia includes small numbers of virtually all the same species, its overall fish numbers and biomass remain dominated by salmonids and other native fish species (Li et al. 1987). This is strong evidence that maintaining (or restoring) a semblance of historical seasonal flow regime and habitat diversity can benefit native fishes and select against introduced species that prey on or otherwise adversely interact with salmon and other native species.

Non-native fishes are proliferating throughout the Columbia River system, especially in the reservoirs and intensely regulated river segments (Stanford et al., 2005). They are establishing metapopulations with cores and satellites as discussed for native salmonids in Chapter 3. This process could complicate proposals for restoration, if flow augmentation, drawdowns, and other schemes that strongly affect reservoir levels result in displacement or emigration of large numbers of non-native fishes from mainstem habitats into tributary streams. The result could be temporarily, if not permanently, increased interaction between wild salmonids and non-native fishes in tributary environments that have so far remained mostly free of dominance by non-native fishes. Continued degradation of habitats in tributary streams and possible climate changes also promote the possibility of wider invasion and establishment of non-native, warmwater and coolwater fishes in the basin.

Changed Food Production and Feeding during Migration

Juvenile salmonids use the mainstems of the Columbia and Snake Rivers as both migration corridors and habitats for feeding. How well we understand feeding (and resulting growth) may be as important as how well we understand migration. The feeding function is especially important for underyearling fall Chinook salmon, which grow as they slowly migrate downstream (see Chapter 6). Yearling salmon and steelhead also feed during migration, as documented in Chapter 6, although their transit through the mainstem is more rapid. The Columbia River Basin mainstem, however, has changed greatly in recent years and appears to have lost a major portion of its normal, riverine-carrying capacity for feeding juvenile salmonids, particularly outmigrants. Clearly, the food-producing and feeding habitats of the mainstem Snake and Columbia Rivers differ considerably today from those that shaped the evolution of anadromous salmonids. Some alternative foods more typical of slower water have replaced the normal riverine food chain, with unresolved questions of the adequacy of that replacement for feeding migrating salmonids. With a feeding habitat greatly changed and probably much depleted, release of large numbers of hatchery fish into it may exacerbate an already tenuous situation for wild stocks.

The Riverine Food Web

Juvenile salmonids in a riverine environment feed primarily on drifting
aquatic insects and terrestrial insects that fall into the water. For smaller
salmonids, midges (chironomids) are the predominant source; as fish grow,
they eat more of the larger aquatic insects, such as caddisflies and mayflies.
Chironomids and other aquatic insects were highlighted by the earliest studies of Chinook salmon in the Columbia River (Chapman and Quistorff
1938). Coho salmon fry in British Columbia were shown to eat pupae, adults,
and pupal exuviae (shed skins) of chironomids as they drift downstream
(Mundie 1971). Becker (1973a, 1973b) and Rondorf et al. (1990) established
that newly emerged adult midges composed more than half of the diet of
underyearling Chinook salmon in the Hanford Reach of the Columbia River.
Dauble et al. (1980) found midge larvae and pupae accounted for 78% by
number and 59% by volume of the total ingested items in the Hanford Reach.
Caddisfly adults became more important as food items there in June and July,
as did shallow-water cladocerans (*Daphnia*). Loftus and Lenon (1977) found
chironomids were the most important food for Chinook salmon in an Alaska
river and that heavy feeding occurred during downstream migration. In the
lower Columbia River below Bonneville Dam, Craddock et al. (1976) found
insects, both adult and larvae, to be the dominant food in spring and fall,
although zooplankton from upstream reservoirs was important in summer.
In New Zealand, Sagar and Glova (1987) found introduced Chinook salmon
eating drifting chironomid larvae and pupae, and mayflies in spring and more
terrestrial insects in summer. Some other studies in small streams have shown
young Chinook salmon to eat mostly drifting terrestrial insects (Johnson
1981; Sagar and Eldon 1983). Rondorf et al. (1990) found caddisflies to be
the main food item for subyearling Chinook salmon (64% by weight) in the
lower Hanford Reach from May to August. There is less information for
yearling salmon, but Schreck et al. (1995) found a wide range of aquatic and
terrestrial insects in mostly full stomachs of yearling Chinook salmon in the
free-flowing Willamette River, with Diptera (including chironomids) being
either the principal or an abundant component. Rondorf et al. (1985) considered migrating smolts to be actively feeding to offset the depletion of
energy reserves during seaward migration. Kolok and Rondorf (1987)
reported on food components of juvenile spring Chinook salmon in John
Day Reservoir. Thus, the general food and feeding relationships of young
salmonids in rivers seem well established, although more information could
be useful for yearlings.

Riverine environments tend to produce aquatic insects adapted to flowing
waters, while terrestrial insects fall to the water from the riparian zone. The
pre-dam mainstem Columbia and Snake Rivers were classic gravel-bed
rivers, dominated by gravel and cobble (rounded rock) substratum variously

constituted as bars, low islands, runs, and pools with back channels and sloughs. These are the habitats that produce large numbers of aquatic insects. Alluvial gravel reaches alternated with more canyon-like reaches where bedrock was exposed. Riparian vegetation was typically restricted to a narrow shoreline zone in the upper arid region that constitutes much of the migration corridor (Buss and Wing 1966; Fickeisen et al. 1980a, 1980b; Lewke and Buss 1977; Rickard et al. 1982). Different floral communities colonized shifting sands at the river's edge, alluvial fans at the mouths of tributary canyons, cobble and gravel slopes, outcroppings of basalt and granite, and disturbed areas caused by annual erosion, rock slides, grazing, and flooding that resulted in seral plant stages. In the entire mainstem, these features remain only in the Hanford Reach of the mid-Columbia and transition zones of the lower reaches of the Clearwater River and the Snake River below Hells Canyon Dam to the upper reaches of the Lower Granite pool.

The salmonid life cycles were intimately linked to an annual flooding cycle of the mainstem. Although it is widely understood that juvenile migrants use the spring freshet for downstream migration, it is less well recognized that feeding is also aided by flooding. Fall Chinook salmon fry emerging from gravels in spring typically began their feeding and rearing phase in shorelines and sloughs as mainstem water levels rose across cobble bars and into riparian vegetation with the melt of winter snowpacks in the tributaries. The most active rearing period for Chinook underyearlings in the mainstem often occurred in late spring and early summer when waters were highest and the most riparian vegetation was flooded. The underyearlings moved gradually downstream through the summer, rearing as they went. Yearlings moved downstream relatively quickly during this same spring freshet period, but there is evidence that they, too, paused periodically in back eddies to feed (Schreck et al. 1995).

Submerged riparian vegetation was probably important for young salmon as a substrate for production of invertebrate food, although this has been shown only anecdotally for the mainstem Columbia and Snake Rivers (C. C. Coutant, personal observations). There is ample evidence from other scientific studies that submerged plant material may be related generally to prey abundance and fish growth. Submerged wood is clearly an important habitat in other aquatic systems for growing invertebrates, especially aquatic insects such as chironomids (Nilsen and Larimore 1973; Stites and Benke 1989). More abundant submerged surfaces generally translate to more invertebrates, as with submerged stream macrophytes (Gregg and Rose 1985).

Flooding provides not only the submerged surface areas for aquatic insects, but it also provides the colonizers. Larval chironomids of all sizes are a common component of stream drift (Mundie 1971), especially during periods of flooding. Although larvae are not commonly eaten by young salmon, drift of chironomid larvae seems to serve largely to colonize the submerged gravel and plant surfaces, where the larvae feed on periphyton and attached organic silt and grow rapidly (Oliver 1971). Drifting chironomid larvae loosened from the streambed or as newly hatched instars quickly colonize previously exposed cobbles and submerged vegetation when waters rise. They develop within a few weeks to the pupae and emerging adults that are the preferred food for young salmonids (many chironomid species have short generation times and very high annual productivity). Timing of flood events was probably important for feeding salmon because (1) the flooding needed to be maintained for several weeks to allow for colonization and growth of chironomids, and (2) flooding needed to occur in spring, because chironomids have their normal peak production in the spring at the time of peak abundance of downstream-migrating juvenile salmon (especially underyearling fall Chinook salmon).

The flooded riparian vegetation also provides terrestrial insects (e.g., ants and spiders) used as salmon food. Because young salmon are at the edges of rivers (underyearlings) and in back eddies (yearlings), they are away from most of the drifting benthic (lithic) invertebrates and in a zone where aquatic and terrestrial drift derived from overhanging brush and flooded riparian vegetation would be most valuable to them. The importance of flood pulses in riverine ecosystems in general is becoming more recognized and is described by Power et al. (1988), Welcomme (1988, 1995), Junk et al. (1989), Bayley (1995), and Gutreiter et al. (1999).

Historically, positive flow-survival relationships for salmon in the Columbia and Snake Rivers may have related, at least in part, to the amount of riparian vegetation flooded during high-flow years. More flooding, when it occurs in a high-volume peak that lasts several weeks, as in the mid-Columbia in 1965, would make a large amount of riparian substrates available for aquatic insect colonization and production of abundant food. Better recruitment of white sturgeon in the Columbia, Sacramento, and Fraser Rivers has been related to the abundance of flooded riparian habitat at time of spawning and early life stages (Coutant 2004). This hypothesis is being tested for salmon in a suite of pristine Pacific Rim rivers that vary naturally in floodplain extent and productivity; preliminarily, it appears that those systems with extensive areas of off-channel floodplain habitat are substantially more productive (Stanford et al. in press; *http://www.umt.edu/flbs/Research/ SaRON.htm*).

Hydrosystem Alterations of Food Webs

The result of impoundment and flow regulation of the mainstem Columbia and Snake Rivers is a mainstem ecosystem that does not appear capable of producing nearly as much high-quality food for juvenile salmon as did the free-flowing and annually flooding river. The success of fall Chinook salmon in the still-riverine Hanford Reach compared to the endangered status of this race in the fully dammed lower Snake River is perhaps partly a result of the differences in food production in the rearing-migration corridor. Research has identified physical and biological causes for the decline and change in food availability.

Loss of Riverine Insect Production

Hydroelectric development has transformed riverine reaches into reservoirs with slow currents, silt bottoms, and fairly stable water elevations. River-like conditions exist in dam tailwaters and persist for a few miles into each reservoir, but most riverine habitat has been lost. With the loss of flowing-water habitat has gone the hard-substrate community of chironomids, caddisflies, mayflies, and other insects that fed juvenile salmon (the riverine food chain). In its place have come midges characteristic of soft substrates and aquatic worms, with planktonic zooplankton becoming a major replacement food (Bennett et al. 1993). Slowly moving shoreline waters of reservoirs warm rapidly in summer, forcing juvenile salmon to move out of their normal shoreline habitat and to the cooler channel (Curet 1993). Fish that relied on shoreline-oriented food production are now obliged to feed on reservoir zooplankton (Muir and Emmett 1988; Rondorf et al. 1985).

Loss of Riparian Flooding

Impoundment and flow regulation by upstream reservoirs have reduced historical flood pulses that previously had inundated vegetated shoreline areas and produced abundant food for salmon. Shorelines once fringed with vegetation are now lined with rock riprap (US Army Corps of Engineers 1976), which produces little insect life suitable as salmonid food (Janecek and Moog 1994). Other shorelines are eroding banks. Even where riparian vegetation has developed as reservoir shorelines age, the stability of reservoir surface elevations during salmon outmigration prevents significant flooding and food production.

Current knowledge specific to the Columbia and Snake Rivers falls short of quantifying the benefit of flooded riparian vegetation in the ecology of juvenile salmon (although research is still possible at Hanford) and its loss through most of the mainstem. Such knowledge would, however, be useful for the contemporary problem of rehabilitating the carrying capacity of salmon-rearing habitats. The reasonable, but locally undemonstrated, importance of riparian habitat for invertebrate (especially chironomid) food production could be a working hypothesis for studies of carrying capacity in the

Columbia River Basin mainstem rivers and lower tributaries. Useful comparisons could be made between the Hanford Reach and various reservoir reaches to quantify, as best we can today, the losses through impoundment. If the hypothesized benefits are substantiated and high, then proposals for flow augmentation and reservoir drawdowns could logically take into consideration a restoration of more natural shoreline vegetation and its seasonal flooding.

Altered Timing of Production and Consumption

Hydropower development has apparently altered the match in timing between food production and demand. Whereas much attention has been given to timing of juvenile outmigrations to match food-production cycles in the estuary and ocean, the "window of opportunity" (Walters et al. 1978), little attention has been paid to correlations between timing of fish abundance, flooding of riparian habitats, and alternative food-production cycles in the mainstem. Evidence suggests that the reservoir zooplankton on which salmon now feed develops primarily later in summer. This topic needs more research and analysis.

Invasion of Reservoirs by Estuarine Invertebrate Species

Mainstem reservoirs have been colonized by invertebrate species usually associated with estuaries (Figure 5.28; Haertel and Osterberg 1967; Jones et al. 1990), and one of these species, *Corophium salmonis*, has become a prominent part of the salmonid food chain in lower Columbia River reservoirs. In the lower mainstem, *Corophium* has become the predominant food for downstream migrants (Muir and Emmett 1988). *Corophium* now occurs in the headwaters of Lower Granite Reservoir, where it was the most prevalent invertebrate species in both numbers and biomass between August 1993 and September 1995 (Nightengale and Bennett 1996). It does not appear to be eaten by juvenile salmonids there, however. Another estuarine species, *Neomysis mercedis*, has also become abundant in John Day pool and other mainstem reservoirs (Haskell and Stanford in press) and likely has reduced zooplankton forage that otherwise would be available to migrating salmon. *Mysis relicta,* introduced to freshwater lakes such as Flathead Lake, Montana, has caused much ecological havoc, including detrimental competition with kokanee for planktonic food (Lasenby et al. 1986; Spencer et al. 1991). An "estuarinization" of the mainstem Columbia and Snake Rivers has apparently taken place, which may be related to the poor strength of wild salmonid populations (Haskell and Stanford in press).

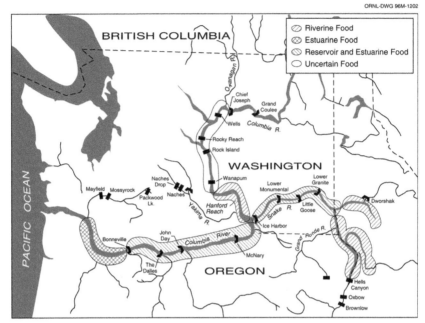

Figure 5.28   The types of main river feeding areas available for emigrating juvenile salmonids in the portion of the Columbia River Basin accessible to anadromous fishes. Note the estuarine component of the food web in the lower Snake River, as well as in the middle and lower Columbia River.

Nutritional Status of Juvenile Migrants

There are indications that the nutritional status of outmigrants is poor. Curet (1993) found that subyearling fall Chinook salmon in Lower Granite and Little Goose Reservoirs were feeding at only 27% of their maximum ration during April—July. This was only 7% greater than the estimated maintenance ration that would provide no growth, and it indicated to Curet that there were food limitations in the habitat. The Smolt Monitoring Program observed outmigrant fall Chinook salmon in poor physical condition during drought years. This question of nutritional status deserves more research attention than it has been given. Migrants that exist at just above the starvation level can hardly be expected to have good long-term survival. Outmigrating salmon have been shown to now lose between 52% and 71% of their lipids and 10% to 21% of their protein content by the time they reach the lowermost dam, with an overall decline in condition factor (James Congleton, University of Idaho, unpublished). The results correlate well with laboratory studies of starvation.

Whether the newly established *Corophium* is an adequate food substitute for the reduced native riverine food items is an important question. Kolok and

Rondorf (1987) showed that the consumption of *Corophium* was associated with reduced caloric densities in stomach contents. That is, the amount of usable energy per volume of food material was less than that of other foods being consumed. Muir and Emmett (1988) noted, however, that the low caloric density might be compensated by ease of availability, meaning less energy had to be expended to capture the prey. De La Noue and Choubert (1985) compared the food values of chironomid larvae, *Daphnia*, and a freshwater gammarid amphipod (similar to *Corophium*) for rainbow trout and found the amphipod to rate poorly. Total and essential amino acids were lowest in the amphipod. Both daphnia and chironomids met the amino acid requirements for salmonids, but the amphipod was deficient in arginine and lysine. The digestibility was also poorest, with the highest percentage of consumed material being passed as feces. The authors rated the amphipod as "much inferior" to either daphnia or chironomids (which were generally equally good) as an aquaculture food. Thus, the replacement of riverine chironomids by estuarine *Corophium* may well be having a detrimental effect on the nutrition of downstream-migrating juvenile salmonids. This inference from published studies needs to be tested by studies of feeding and nutritional status of Columbia River Basin fish. Nutritional status can be evaluated by examination of whole body energy content (cal/g). Whole body lipid and fatty acid content are variables that could be very useful (Brett and Muller-Navarra 1997).

Habitat Quantity and Quality: A Summary and Synthesis

The Effects of River Regulation

At least three fundamental principles emerge from the large literature on the ecology of regulated rivers (Poff et al. 1997; Stanford et al. 1996). These principles are particularly germane to derivation of restoration strategies for Columbia River salmonids.

Habitat Diversity Is Substantially Reduced as a Consequence of Regulation

The dams of the Columbia River have inundated many of the piedmont and mountain valley flood plains, thereby severing the river continuum. Mass transport dynamics that create instream and floodplain habitats for riverine biota in remaining free-flowing reaches have been drastically altered. Flood peaks have been eliminated, discharges are more variable, and temperature seasonality has been altered.

As a consequence of reservoir storage of peak flows for flood control, navigation, irrigation, and hydropower production, base flows have increased substantially and, in many places, fluctuate so erratically that aquatic biota cannot survive in shallow, near-shore habitats. Persistent shallow- or slack-water habitats are especially important for survival of early life history stages of fishes that cannot survive in the strong currents of the channel thalweg. Storage of bedload in the reservoir and constant clear-water flushing downstream artificially has depleted gravel and finer sediments in the tailwaters, causing armoring of the bed with large cobble and boulder substratum. Channel constrictions and habitat simplification is nearly universal, except in headwater areas (e.g., Figure 5.29). Vegetation has clogged backwaters owing to loss of scouring flood flow. Riparian communities have been altered by deforestation and agricultural activities, which interact with effects of regulation to reduce habitat heterogeneity.

The general conclusion is that regulation has created a discontinuum of environmental conditions and severed the connectivity of channel, groundwater, floodplain, and upland components of the catchment ecosystem. Habitats for riverine biota have become spatially homogeneous, limited to the permanently wetted portion of the channel thalweg that is dominated by conditions dictated by operations of upstream storage reservoirs. Indeed, serial construction of low-head dams has converted virtually the entire mainstems of the lower Snake and Columbia Rivers into shallow reservoir habitat that is neither truly laucustrine, nor riverine.

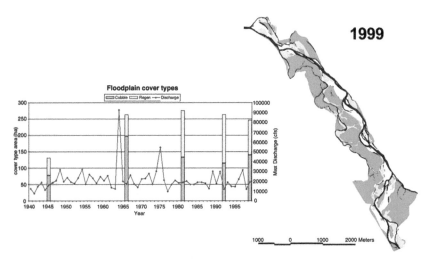

Figure 5.29 Expansion and contraction of the parafluvial zone (see Figure 5.11) of the Nyack Flood Plain, Montana, in relation to the long-term hydrograph. Extreme flooding mediates intense scour followed by vegetation succession on the scoured surfaces in succeeding years. This process creates the habitat mosaic that fosters salmon production in alluvial rivers. From D. Whited, Flathead Lake Biological Station, unpublished data.

Native Biodiversity Decreases and Non-native Species Proliferate as a Consequence of Regulation

Native biodiversity has decreased substantially in the last 120 years (Behnke 2002; Huntington et al. 1996). Most salmon populations spawning in the mainstem Columbia and Snake Rivers have been extirpated (Nehlsen et al. 1991). In the headwaters of tributaries, salmon populations have become increasingly isolated by flow regulation, diversion, and habitat degradation, especially in the lower reaches. Moreover, for anadromous species, mortality resulting from passage through dams and reservoirs in the mainstem may not affect all species and life histories equally, selecting against certain life history types, thereby reducing biodiversity, increasing habitat fragmentation, and increasing the vulnerability of populations to extinction.

Altered temperature patterns and continual export of very fine organic matter and dissolved nutrients, coupled with simplification of the channel, stabilization of bottom substratum, and loss of floodplain inundation, has promoted environmental conditions to which native species are maladapted. This has created opportunities for non-native plants and animals to establish robust populations. In some cases, one or a few native species are more abundant than they were before regulation (Poe et al. 1991). Non-native invertebrates, fishes, and plants are consistently more abundant in regulated river reaches compared to unregulated reaches (Li et al. 1987). Reasons for non-native proliferation vary, but in general non-native species are often better competitors in the homogeneous habitats of regulated river reaches. A wide array of non-natives have been introduced into the Columbia River system.

Normative Conditions Are Reexpressed Predictably as Distance Downstream from the Dam Increases and in Relation to Influences of Tributaries

The Serial Discontinuity Concept (SDC) (Ward and Stanford 1983, 1995) predicts that the conditions described above that are attributable to flow regulation will ameliorate in river reaches downstream of storage reservoirs, as a natural consequence of the biophysical energetics of rivers. The distance downstream of the dam needed to reset normative conditions is related to the limnological attributes (depth, volume, water retention time, trophic state) of the reservoir, the mechanics of water release (surface, bottom, or depth selective), the mode of dam operations, and the influence of tributaries entering downstream from the dam. If the tributaries are large and unregulated, they may substantially accelerate the reset (Stanford and Hauer 1992). In any case, given enough distance, conditions at some point downstream from the dam will closely approximate original conditions.

Reset toward natural conditions has been demonstrated in Columbia River tributaries downstream of storage reservoirs, for example, Flathead River (Hauer and Stanford 1982; Stanford et al. 1988), Kootenai River (Perry et al. 1986), and Clearwater River (Munn and Brusven 1991). For the lower Snake and Columbia Rivers, however, little reset of riverine conditions can be expected, because almost no river environment remains due to nearly continuous impoundment. The Hanford Reach, which contains a nearly 50-mile, undammed free-flowing stretch between Priest Rapids and McNary Dams, is the single exception in the mainstem.

The negative influences of logging, grazing, dams, irrigation withdrawals, urbanization, exotic species introductions, and other human activities have been documented for all of the Columbia River tributaries (NRC 1996; Stanford, et al 2005). Many, if not all of the larger tributaries are degraded by streamside uses that fail to recognize the importance of riparian vegetation and local upwelling areas (e.g., springbrooks, ponds, and wetlands) on flood plains as essential ecological features. Some of these habitat problems can be normalized by re-regulation of flows, as discussed above, and detailed in Stanford et al. (1996) and Poff et al. (1997). In addition, new incentives for streamside stewardship that conserve and enhance connectivity and productivity of floodplain habitats need to be fostered for tributaries as well as mainstem reaches (Poff et al. 2003). Special incentives for protection of riparian zones are needed in reaches where historically intact streamside conditions presently occur as a consequence of limited human influences. Owing to the dramatic escalation of intermountain valley development, in the next decade we could lose remaining productive and connected salmonid habitats (e.g., Stanley Basin on the Salmon River, North and Middle Forks of the Flathead River). For example, cool water refugia exist in the Grand Ronde (Li et al. 1995a) that are critical for salmonids. Such refugia should be accorded special protection before they are purposefully or inadvertently degraded. The Quartz Lake watershed in Glacier National Park is the only example we are aware of in the entire Columbia Basin where an entirely native food web, including the full compliment of native headwater salmonids, remains intact. This and other native fish refuges should be completely protected as native salmonid fishery reserves. Plans for protection of remaining quality habitat and stabilization and normalization of degraded habitats are needed for every tributary in the Columbia Basin. However, it may be prudent to focus actions on those tributaries that have the greatest likelihood of playing a key role in salmonid recovery (e.g., those that are in proximity to currently functional habitats that are producing salmonids, such as the Yakima and its potential connectivity to the Hanford Reach of the mainstem Columbia River).

Reestablishing Normative Ecological Conditions

Subbasin planning for fisheries enhancement in the Columbia River has emphasized hatchery operations and habitat restoration (see *www.nwcouncil. org* for subbasin planning documents) and a great deal of money has been spent. Overall, there is little to show for this expenditure in terms of improving fisheries or fish habitat. Indeed, 7 fishes (12 stocks or populations), 6 snails and 2 aquatic plants have been listed as endangered or threatened under the Federal Endangered Species Act and we know of no habitat restoration that has resulted in quantified increases in fisheries that were targeted *a priori*. Habitat loss is only part of the overall problem as discussed in other chapters of this volume; however, population viability is inexorably tied to availability of enough habitat of sufficient quality to allow reproduction to exceed mortality. For much of the Columbia Basin, such conditions simply do not exist, especially for anadromous salmonids, as reported time and again (Stanford, et al. 2005). Dams and dewatered or dry reaches prevent spawner access to relatively good habitat higher in the subbasins. Smolts encounter reservoirs or dewatered landscapes in place of functional flood plains with productive rearing habitat. An inventory of remaining fish habitat in the contemporary floodplain ecology context emphasized herein has not been done for the entire Columbia Basin, and it is badly needed. Perhaps the Hanford Reach is the only place left upstream of the Columbia Gorge with enough viable habitat for wild salmon and steelhead, even though some accessible tributaries such as the John Day, Grand Ronde, Salmon, and Clearwater (except the North Fork) are free flowing. Even the Hanford Reach is subjected to fluctuating flows that routinely strand juveniles. Habitat restoration in the basin is highly localized, and inane activities such as fencing of cattle and cottonwood and willow plantings on dewatered or substantially regulated tributaries are done under the rubric of habitat enhancement, usually without success (a pervasive national problem; see Bernhardt et al., 2005). We believe that habitat salvation in the Columbia lies first in restoration of normative flow patterns that allow dynamic habitat formation and maintenance and second in allowing wild fish to naturally stray in from the remaining remnant populations without threat of harvest. Let the river and the fish do the work. However, for anadromous species this must be done by "bottom-up restoration"; that is, the most downstream problems must be solved before investing higher up in the stream corridor so that wild fish can access newly restored areas. Such whole-river restoration is possible only in subbasins of the mid-Columbia, such as the Yakima (Stanford et al. 2002), and that assumes that the continuing viability of the Hanford Chinook means that the mainstem river downstream is not severely compromised by impoundment.

Habitat Conclusions and Recommendations

Conclusions

1. Habitat required for salmonid migration, spawning, incubation, and juvenile rearing has been severely degraded in the Columbia Basin by the cumulative effects of flow regulation by dams and diversions, sedimentation and contamination from forestry and agricultural activities, and massive introduction of non-native biota (fish, invertebrates, and riparian plants).

2. Owing to the diverse climates and food web assemblages of the different ecoregions that make up the Columbia River catchment, native salmonids displayed great diversity of life history types (stocks or populations) specifically adapted to the wide array of natural habitats. Diversity has been substantially depleted by habitat loss, fragmentation, and degradation.

3. Habitat fragmentation and loss is extensive throughout the Columbia River Basin, except in those few areas where human activities are limited, particularly in roadless and wilderness areas in the upper portions of some subbasins. Incremental loss of incubation, rearing, and spawning sites has reduced or eliminated production of salmonid stocks and disrupted natural metapopulation structure and dynamics.

4. Most alluvial floodplain reaches and associated habitats historically supporting large, productive spawning populations and providing essential, high-quality rearing habitats for maturing and migrating juveniles have been destroyed by reservoir inundation, substantially degraded by altered flows associated with hydropower operations, or disconnected from the salmon ecosystem by dams that block migratory pathways.

5. Habitat restoration using artificial structures (e.g., weirs, logs cabled into streams, coffer dams, rock gardens) has failed to mitigate the adverse effects of temperature alteration, sedimentation, and simplification of habitat structure and processes caused by upland and riparian land use activities.

6. Presence of non-native fishes is a strong indicator of habitat degradation and is problematic for any restoration effort. Non-native fishes are far less abundant and reproductive in free-flowing segments where native habitats remain in relatively good shape (e.g., Hanford Reach). Native salmonids (and other aquatic vertebrates) remain healthy in less than 5% of the headwater streams of the Columbia River tributaries in Idaho, Montana, Oregon, and Washington, owing to genetic introgression and displacement by non-native species, which has been mediated by a long history of stocking of cultured brook, rainbow, brown trout, and lake trout in headwater lakes and streams. Adfluvial populations of bull trout and cutthroat trout have been vastly compromised by food web changes

in the big valley bottom lakes (e.g., Pend Oreille, Flathead) as a result of misguided stocking of non-native mysid shrimp and the interactions of these shrimp with non-native fishes.

7. In the Hanford Reach and other alluvial river segments, highly productive, flooded riparian zones provided much of the riverine food production (in the form of rapidly colonizing and growing chironomids) for migrants in spring (both underyearlings and yearlings); these critical food web components do not exist in mainstem reservoirs and are substantially reduced or eliminated in riverine segments that are regulated by dams.

8. Typical lake food items (zooplankton) provide an inadequate food source, and fish in the Snake River reservoirs are energetically deficient.

9. Food abundance in the mainstem during rearing and migration, which is higher with flooding, may affect salmon survival.

10. Estuarine invertebrates have colonized the lower Columbia River reservoirs and may provide a food source, but more likely they are competing with migrating juveniles for limited zooplankton forage.

11. Availability of riverine habitat for producing food and the longitudinal continuity of riverine and estuarine food webs are major differences between successful Hanford stocks (riverine, continuity) and unsuccessful Snake River stocks (reservoirs, discontinuous).

12. Habitat restoration allowing new populations and sustained new productivity requires a catchment approach that starts with restoration of normative flows; for anadromous species, this means normative flow restoration in selected subbasins, such as the Yakima, from headwaters to mainstem confluence.

Critical Uncertainties

1. The exact magnitude and timing of restored flows and temperature regimes need to be empirically determined for specific free-flowing segments and requires a broadly multidisciplinary approach. However, no uncertainty exists with respect to the need to reestablish flow and temperature seasonality and to stabilize base flow and temperature fluctuations.

2. Although "best management practices" (BMPs) may reduce impacts to habitat compared to unregulated land use, uncertainty about effectiveness of present BMPs must be resolved by scientific evaluation at both site-specific and watershed scales; some results will not be known for decades after implementation. In the face of uncertainty about the sufficiency of current land use practices, designation and protection of a well-distributed network of reserve areas and habitat patches from new land-disturbing activities is necessary to establish experimental natural

baselines and to establish a biological hedge against possible failure of BMPs in treated areas.

3. Habitat restoration may be ineffective at restoring native species where introduced non-native species are well established. Available science suggests that non-natives will be most vulnerable, and many can be effectively suppressed, where habitats are maintained by natural range of flow and temperature variation. However, abrupt changes in reservoir management could temporarily drive existing populations of some non-native fishes into tributary habitats, increasing the risk of their colonization of tributaries. On the other hand, reservoir changes also will likely create new mainstem habitat refugia for native fishes. The risk of dispersal and establishment of non-native fishes will be lowest where tributaries retain relatively natural stream flows, thermal regimes, habitat diversity, and intact native fish assemblages.

4. Restoration of normative habitat using the whole-river model for tributaries still accessible to anadromous fish assumes that ocean and mainstem bottlenecks are within a range that will allow sufficient successful migrations of spawners and juveniles to allow sustained productivity in the restored subbasins.

5. The mainstem Columbia River may have too many hydropower and irrigation storage reservoirs to ever allow sufficient habitat restoration to allow native salmonid diversity and productivity to substantially recover. However, the surprising resilience and salmon productivity of the Hanford Reach suggests that restoration of critical salmonid habitat is possible without impractical alteration of dam operations.

Recommendations

1. Free-flowing reaches downstream of hydroelectric dams should be re-regulated to reestablish normative flow and temperature regimes and thereby allow the river to naturally restore instream and floodplain habitats and food webs.

2. Restoration of substantial mainstem habitat likely can be accomplished by drawdown of selected reservoirs to expose and restore alluvial reaches (e.g., upper ends of John Day and McNary pools). These options should be quantitatively examined.

3. Habitat restoration should be framed in the context of measured trends in water quality because functional salmonid habitats are characterized by high-quality (pure, cool, clear) water, and few people will argue with the actions to sustain attributes of high-water quality.

4. New timber harvest prescriptions (e.g., selective cutting, attempted fire simulations, salvage logging, road retirement), sustainable agriculture

practices, and other land use practices for upland and riparian areas, commonly referred to as best management practices (BMPs), need to be empirically tested and demonstrated as effective in credible short- and long-term studies before they can be considered sufficient for conserving and enhancing water quality and salmonid habitats.

5. If the restoration goal of the Fish and Wildlife Program and other efforts includes conservation and enhancement of remaining native and naturally reproducing salmonids, all stocking of non-native biota should be stopped in habitats used by or hydrologically connected to habitats required by all life stages of native salmonids (resident and anadromous). Carefully evaluated mechanisms to reduce or eliminate the reproductive capacity or dispersal of non-native species in native salmonid habitats should be implemented if riverine controls (e.g., by restoration of flushing flows) prove ineffective in controlling non-native species.

6. A well-distributed network of reserve watersheds and riverine habitat patches, based on the current distribution of strong subpopulations of native salmonids, should be designated and protected from new land-disturbing activities in order to establish experimental natural baselines for evaluation of effectiveness of management practices and to establish a biological hedge against possible failure of BMPs to conserve and enhance aquatic habitat in treated areas.

7. A study plan should be developed for evaluating the importance of food production to the success of juvenile rearing and outmigration in the Columbia River Basin. This plan should include the following activities:

   a. Test, through field studies, the nutritional state of migrating Snake River salmonids identified by Curet in relation to that of mid-Columbia stocks, to estimate the importance of food availability to salmon survival.

   b. Estimate, through field studies of insect colonization and growth during flooding and spatial analyses of flood plains, the quantity of salmonid food potentially produced by flooded riparian lands in the lower Columbia-Snake Basin and lost by river regulation, and relate quantitatively to the food requirements of migrating juvenile salmon.

   c. Determine, through field studies, the current extent of colonization of reservoirs by estuarine species and their role in reservoir food webs.

   d. Establish, through laboratory feeding experiments, the suitability of estuarine organisms as food for downstream migrants relative to riverine food organisms.

   e. Estimate, through field studies and laboratory feeding experiments, the importance of longitudinal continuity of food for relative survival of mid-Columbia (Hanford) and Snake River migrants.

   f. Estimate, through field studies, the value of macrophytes for producing food for mid-Columbia salmonids.

g. Evaluate the nutritional status of juvenile salmonids during transportation from upper river dams to below Bonneville Dam.
8. Provide an integrated assessment of the role of food and feeding on the nutrition of downstream migrants, leading to conclusions regarding action options for restoration of riverine food chains (e.g., induced flooding, riparian habitat restoration) and promotion of estuarine food chains (e.g., species stocking).

Literature Cited

Abell, R. A., D. M. Olson, E. Dinerstein, P. T. Hurley, J. T. Diggs, W. Eichbaum, S. Walters, W. Wettengel, T. Allnutt, C. J. Loucks, P. Hedao, and W. W. Fund. 2000. *Freshwater ecoregions of North America: A conservation assessment.* Island Press, Washington, DC.

Balz, D. M., and P. B. Moyle. 1993. Invasion resistance to introduced fishes by a native assemblage of California stream fishes. *Ecological Applications* 3:246–255.

Baxter, C. V., C. A. Frissell, and F. R. Hauer. 1999. Geomorphology, logging roads and the distribution of bull trout (*Salvelinus confluentus*) spawning in a forested river basin: Implications for management and conservation. *Transactions of the American Fisheries Society* 128:854–867.

Baxter, C. V., and F. R. Hauer. 2000. Geomorphology, hyporheic exchange, and selection of spawning habitat by bull trout (*Salvelinus confluentus*). *Canadian Journal of Fisheries and Aquatic Sciences* 57:1470–1481.

Bayley, P. B. 1995. Understanding large river-floodplain ecosystems. *BioScience* 45:153–158.

Beacham, T. D., and C. B. Murray. 1990. Temperature, egg size, and development of embryos and alevins of five species of Pacific salmon: A comparative analysis. *Transactions of the American Fisheries Society* 119:927–945.

Becker, C. D. 1973a. Development of Simulium (Psilozia) vittatum Zett. (Diptera: Simuliidae) from larvae to adults at thermal increments from 17.0 to 27.0 C. *American Midland Naturalist* 89:246–251.

Becker, C. D. 1973b. Food and growth parameters of juvenile Chinook salmon, *Oncorhynchus tshawytscha*, in central Columbia River. *Fishery Bulletin* 71:387–400.

Behnke, R. J. 2002. *Trout and salmon of North America.* Free Press, New York.

Bennett, D. H., T. J. Dresser, Jr., T. S. Curet, K. B. Lepla, and M. A. Madsen. 1993. Lower Granite Reservoir in-water disposal test: Results of the fishery, benthic and habitat monitoring program-year 3 (1990). US Army Corps of Engineers, Walla Walla District, Walla Walla, Washington.

Berman, C. H., and T. P. Quinn. 1991. Behavioral thermoregulation and homing by spring Chinook salmon, *Oncorhynchus tshawytscha* (Walbaum), in the Yakima River. *Journal of Fish Biology* 39:301–312.

Bernhardt, E. S., M. A. Palmer and 22 others. 2005. Synthesizing U. S. river restoration efforts. Science 308:636–637.

Beschta, R. L., W. S. Platts, and B. Kauffman. 1991. Field review of fish habitat improvement projects in the Grande Ronde and John Day River Basins of eastern Oregon. COE/BP-21493-1, Bonneville Power Administration, Portland, Oregon.

Bilby, R. E., B. R. Fransen, P. A. Bisson, and J. K. Walter. 1998. Response of juvenile coho salmon (*Oncorhynchus kisutch*) and steelhead (*Oncorhynchus mykiss*) to the addition of salmon carcasses in two streams in southwestern Washington, U.S.A. *Canadian Journal of Fish Aquatics And Science* 55:1909–1918.

Bisson, P. A., T. P. Quinn, G. H. Reeves, and S. V. Gregory. 1993. Best management practices, cumulative effects, and long-term trends in fish abundance in Pacific Northwest river systems.

Pages 189–232 in R. J. Naiman, ed. *Watershed management: Balancing sustainability and environmental change.* Springer-Verlag, New York.

Bisson, P. A., G. H. Reeves, R. E. Bilby, and R. J. Naiman. 1997. Watershed management and Pacific salmon: Desired future conditions. Pages 447–474 in D. J. Stouder, P. A. Bisson, and R. J. Naiman, eds. *Pacific salmon and their ecosystems: Status and future options.* Chapman and Hall, New York.

Brannon, E. L., and E. O. Salo. eds. 1982. Proceedings of the Salmon and Trout Migratory Behavior Symposium. University of Washington School of Fisheries, Seattle.

Brett, J. R. 1971. Energetic responses of salmon to temperature: A study of some thermal relations in the physiology and freshwater ecology of sockeye salmon (*Oncorhynchus nerka*). *American Zoologist* 11:99–113.

Brett, M. E., and D. C. Muller-Navarra. 1997. The role of highly unsaturated fatty acids in aquatic foodweb processes. *Freshwater Biology* 38:483–499.

Brusven, M. A., D. J. Walker, K. M. Painter, and R. C. Biggam. 1995. Ecological-economic assessment of a sediment-producing stream behind Lower Granite Dam on the Lower Snake River, USA. *Regulated Rivers* 10:373–388.

Buss, I. O., and L. D. Wing. 1966. Pre-impoundment observations of wintering mallards and nesting Canada geese on the Snake River, southeast Washington. *Research Studies of the University of Washington* 34:1–36.

Calow, P., and G. E. Petts, eds. 1992. *The rivers handbook, Volume 1: Hydrological and ecological principles.* Blackwell Scientific, Oxford, UK.

Chapman, W. M., and E. Quistorff. 1938. The food of certain fishes of north central Columbia River drainage, in particular, young Chinook salmon and steelhead trout. *Washington Department of Fisheries Biological Report* 37A:1–14.

Chisolm, I. M., W. A. Hubert, and T. A. Wesche. 1987. Winter stream conditions and use of habitat by brook trout in high-elevation Wyoming streams. *Transactions of the American Fisheries Society* 116:176–184.

Connor, W. P., H. L. Burge, and R. Waitt. 2002. Juvenile life history of wild fall Chinook salmon in the Snake and Clearwater rivers. *North American Journal of Fisheries Management* 22:703–712.

Connor, W. P., H. L. Burge, J. R. Yearsley, and T. C. Bjornn. 2003a. Influence of flow and temperature on survival of wild subyearling fall Chinook salmon in the Snake River. *North American Journal of Fisheries Management* 23:362–375.

Connor, W. P., R. K. Steinhorst, and H. L. Burge. 2003b. Migrational behavior and seaward movement of wild subyearling fall Chinook salmon in the Snake River. *North American Journal of Fisheries Management* 23:414–430.

Coutant, C. C. 1999. Perspectives on temperature in the Pacific Northwest's fresh waters. ORNL/TM-1999/44, Oak Ridge National Laboratory, Oak Ridge, Tennessee.

Coutant, C. C. 2004. A riparian habitat hypothesis for successful reproduction of white sturgeon. *Review of Fisheries Science* 12:23–73.

Craddock, D. R., T. H. Blahm, and W. D. Parente. 1976. Occurrence and utilization of zooplankton by juvenile Chinook salmon in Lower Granite and Little Goose reservoirs, Washington. Master's thesis, University of Idaho, Moscow.

Craig, J. F., and J. B. Kemper, eds. 1987. *Regulated streams: Advances in ecology.* Plenum Press, New York.

Cunjak, R. A., and G. Power. 1986. Winter habitat utilization by stream resident brook trout and brown trout. *Canadian Journal of Fisheries and Aquatic Sciences* 43:1970–1981.

Cunjak, R. A., and R. G. Randall. 1993. In-stream movements of young Atlantic salmon (*Salmo salar*) during winter and early spring. *Canadian Journal of Fisheries and Aquatic Sciences* Special Publication 118:43–51.

Curet, T. S. 1993. Habitat use, food habits, and the influence of predation on subyearling Chinook salmon in Lower Granite and Little Goose Reservoirs, Washington. Master's thesis, University of Idaho, Moscow.

Dauble, D. D., R. H. Gray, and T. L. Page. 1980. Importance of insects and zooplankton in the diet of 0-age Chinook salmon (*Oncorhynchus tshawytscha*) in the central Columbia River. *Northwest Science* 54(4):253–258.

De La Noue, J., and G. Choubert. 1985. Apparent digestibility of invertebrate biomass by rainbow trout. *Aquaculture* 50:103–112.

Doppelt, B., M. Scurlock, C. Frissell, and J. Karr. 1993. *Entering the watershed: A new approach to save America's river ecosystems.* Island Press, Covelo, California.

Ebel, W. J., C. D. Becker, J. W. Mullan, and H. L. Raymond. 1989. The Columbia River: Toward a holistic understanding. Pages 205–219 in D. P. Dodge, ed. Proceedings of the International Large River Symposium (LARS). Special Publication of the *Canadian Journal of Fisheries and Aquatic Sciences.*

Ebersole, J. L. 1994. Stream habitat classification and restoration in the Blue Mountains of northeast Oregon. Master's thesis, Oregon State University, Corvallis.

Elmore, W. 1992. Riparian responses to grazing practices. Pages 442–457 in R. J. Naiman, ed. *Watershed management: Balancing sustainability and environmental change.* Springer-Verlag, New York.

Evermann, B. W. 1895. A preliminary report upon salmon investigations in Idaho in 1894. *Bulletin of US Fisheries Commission* 15:253–284.

Forest Ecosystem Management Assessment Team (FEMAT). 1993. Forest ecosystem management: An ecological, economic, and social assessment. Interagency Supplemental Environmental Impact Statement Team, Portland, Oregon.

Fickeisen, D. H., D. D. Dauble, D. A. Neitzel, W. H. Rickard, R. L. Skaggs, and J. L. Warren. 1980a. Aquatic and riparian resource study of the Hanford Reach, Columbia River, Washington. Battelle Pacific Northwest Laboratories, Richland, Washington.

Fickeisen, D. H., R. E. Fitzner, R. H. Sauer, and J. L. Warren. 1980b. Wildlife usage, threatened and endangered species and habitat studies of the Hanford Reach, Columbia River, Washington. Battelle Pacific Northwest Laboratories, Richland, Washington.

Flathead Basin Commission. 1994. 1993-1994 Biennial Report. Flathead Basin Commission, Kalispell, Montana.

Frissell, C. A. 1993. A new strategy for watershed restoration and recovery of Pacific salmon in the Pacific Northwest. Pacific Rivers Council, Eugene, Oregon.

Frissell, C. A., and S. B. Adams. 1995. Factors affecting distribution and co-occurrence of eastern brook trout and other fishes in the northern Rocky Mountains. Progress Report, USDA Forest Service, Intermountain Research Station, Missoula, Montana.

Frissell, C. A., J. Doskocil, J. T. Gangemi, and J. A. Stanford. 1995. Identifying priority areas for protection and restoration of aquatic biodiversity: A case study in the Swan River Basin, Montana, USA. Open File Report 136-95, Flathead Lake Biological Station, University of Montana, Polson.

Frissell, C. A., W. J. Liss, C. E. Warren, and M. D. Hurley. 1986. A hierarchical framework for stream habitat classification: Viewing streams in a watershed context. *Environmental Management* 10:199–214.

Frissell, C. A., and R. K. Nawa. 1992. Incidence and causes of physical failure of artificial habitat structures in streams of western Oregon and Washington. *North American Journal of Fisheries Management* 12:182–197.

Fulton, L. A. 1968. Spawning areas and abundance of Chinook salmon (*Oncorhynchus tshawytscha*) in the Columbia River Basin–Past and present. USDI, Fish and Wildlife Service, Special Scientific Report–Fisheries No. 571.

Fulton, L. A. 1970. Spawning areas and abundance of steelhead trout and coho, sockeye, and chum salmon in the Columbia River Basin–Past and present. USDC, NOAA, NMFS, Special Scientific Report–Fisheries No. 618.

General Accounting Office. 1992. Endangered species: Past actions taken to assist Columbia River salmon. Briefing Report to Congressional Requesters GAO/RCED-92-173BR, General Accounting Office, Washington, DC.

Gregg, W. W., and F. L. Rose. 1985. Influence of aquatic macrophytes on invertebrate community structure, guild structure, and microdistribution in streams. *Hydrobiologia* 128:45–56.

Gregory, S. V., L. R. Ashkenas, and P. Minear. 2001. Application of analysis of historical channel change in the restoration of large rivers. *Verhandlungen Internationale Vereinigung fur Theoretische und Angewandte Limnologie* 27:4077–4086.

Groot, C., and L. Margolis, eds. 1991. *Pacific salmon life histories.* University of British Columbia Press, Vancouver.

Groot, C., L. Margolis, and W. C. Clarke, eds. 1995. *Physiological ecology of Pacific salmon.* University of British Columbia Press, Vancouver.

Gutreuter, S., A. D. Bartels, K. Irons, and M. B. Sandheinrich. 1999. Evaluation of the flood-pulse concept based on statistical models of growth of selected fishes of the upper Mississippi River system. *Canadian Journal of Fisheries Aquatics and Science* 56:2282–2291.

Haertel, L. S., and C. L. Osterberg. 1967. Ecology of zooplankton, benthos, and fishes in the Columbia River estuary. *Ecology* 48:459–472.

Hall, C. A. S., J. A. Stanford, and F. R. Hauer. 1992. The distribution and abundance of organisms as a consequence of energy balances along multiple environmental gradients. *Oikos* 65:377–390.

Harner, M.J. and J. A. Stanford. 2003. Difference in cottonwood growth between a losing and gaining reach of an alluvial floodplain. Ecology 84:1453–1458.

Haskell, C. A. and J. A. Stanford. In press. Ecology and upstream invasion of an estuarian mysid shrimp in the Columbia River (USA). *River Research and Applications.*

Hauer, F. R. 1993. Artificial streams for the study of macroinvertebrate growth and bioenergetics. *Journal of the North American Benthological Society* 12:333–337.

Hauer, F. R., C. N. Dahm, G. A. Lamberti, and J. A. Stanford. 2003. Landscapes and ecological variability of rivers in North America: Factors affecting restoration strategies. Pages 81–105 in P. A. Bisson, ed. *Strategies for restoring river ecosystems: Sources of variability and uncertainty in natural and managed systems.* American Fisheries Society, Bethesda, Maryland.

Hauer, F. R., and J. A. Stanford. 1982. Ecological responses of hydropsychid caddisflies to stream regulation. *Canadian Journal of Fisheries and Aquatic Sciences* 39:1235–1242.

Hauer, F. R., J. A. Stanford, J. J. Giersch, and W. H. Lowe. 2000. Distribution and abundance patterns of macroinvertebrates in a mountain stream: An analysis along multiple environmental gradients. *Verhandlungen Internationale Vereinigung fur Theoretische und Angewandte Limnologie* 27:1485–1488.

Hauer, F. R., E. G. Zimmerman, and J. A. Stanford. 1980. Preliminary investigations of distributional relationship of aquatic insects and genetic variation of a fish population in the Kintla Drainage, Glacier National Park. Pages 71–84 in *Proceedings of the Second Symposium on Research in National Parks.* American Institute of Biological Sciences, Washington, D.C.

Healy, M. C., and A. Prince. 1995. Scales of variation in life history tactics of Pacific salmon and the conservation of phenotype and genotype. *Evolution and the aquatic ecosystem: Defining unique units in population conservation.* 17:176–184.

Henjum, M. G., J. R. Karr, D. L. Bottom, J. C. Bednarz, S. G. Wright, S. A. Beckwitt, and E. Beckwitt. 1994. *Interim protection for late-successional forests, fisheries, and watersheds: National forests east of the Cascade Crest, Oregon and Washington.* The Wildlife Society, Bethesda, Maryland.

Hicks, B. J., R. L. Beschta, and R. D. Harr. 1991. Long-term changes in streamflow following logging in western Oregon and associated fisheries implications. *Water Resources Bulletin* 27:217–226.

242   Stanford *et al.*

Huntington, C. W. 1995. Fish habitat and salmonid abundance within managed and unroaded landscapes on the Clearwater National Forest, Idaho. USFS Eastside Ecosystem Management Project, Walla Walla, Washington.

Huntington, C., W. Nehlsen, and J. Bowers. 1996. A survey of healthy native stocks of anadromous salmonids in the Pacific Northwest and California. *Fisheries* 21:6–14.

Hynes, H. B. N. 1975. The stream and its valley. Verh. Internat. Verein. Limnol. 19:1–15.

Independent Scientific Advisory Board (ISAB). 2003. A review of strategies for tributary habitat recovery. ISAB Report 2003-2. Northwest Power and Conservation Council, NOAA-Fisheries, and Columbia Indian Tribes, Portland, Oregon. (www.nwcouncil.org)

Janecek, B. F. U., and O. Moog. 1994. Origin and composition of the benthic invertebrate riprap fauna of impounded rivers. *Verhandlungen Internationale Vereinigung fur Theoretische und Angewandte Limnologie* 25:1624–1630.

Jaske, R. T., and J. B. Goebel. 1967. Effects of dam construction on temperatures of the Columbia River. *Journal of the American Water Works Association* 59:935–942.

Jaske, R. T., and M. O. Synoground. 1970. Effect of Hanford plant operations on temperature of the Columbia River, 1964 to present. BNWL-1345, Pacific Northwest Laboratory, Richland, Washington.

Johnson, J. H. 1981. Comparative food selection by coexisting subyearling coho salmon, Chinook salmon, and rainbow trout in a tributary of Lake Ontario. *NY Fish and Game Journal* 28:150–161.

Jones, K., C. A. Simenstad, D. L. Higley, and D. L. Bottom. 1990. Community structure, distribution, and standing stock of benthos, epibenthos, and plankton in the Columbia River estuary. *Progressive Oceanography* 25:211–241.

Junk, W. J., P. B. Bayley, and R. E. Sparks. 1989. The flood pulse concept in river-floodplain systems. *Canadian Special Publication of Fisheries and Aquatic Sciences* 106:110–127.

Karr, J. R. 1992. Ecological integrity: Protecting earth's life support systems. Pages 223–238 in R. Costanza, B. G. Norton, and B. D. Haskell, eds. *Ecosystem health. New goals for environmental management.* Island Press, Washington, DC.

Karr, M. H., J. K. Fryer, and P. R. Mundy. 1998. Snake River Water Temperature Control Project. Phase II. Methods for managing and monitoring water temperatures in relation to salmon in the lower Snake River. Columbia River Inter-Tribal Fish Commission, Portland, Oregon.

Kauffman, J. B., R. L. Beschta, and W. S. Platts. 1993. Fish habitat improvement projects in the Fifteenmile Creek and Trout Creek basins of central Oregon: Field review and management recommendations. BPA Project BPA Project No. 86-079; 84-062, Bonneville Power Administration, Division of Fish and Wildlife, Portland, Oregon.

Kauffman, J. B., and W. C. Krueger. 1984. Livestock impacts on riparian ecosystems and streamside management implications–A review. *Journal of Range Management* 37:430–438.

Key, L. O., R. Garland, and E. E. Kerfoot. 1995. Nearshore habitat use by subyearling Chinook salmon in the Columbia and Snake Rivers. Pages 74–107 in D. W. Rondorf and K. F. Tiffan, eds. *Identification of the spawning, rearing, and migratory requirements of fall Chinook salmon in the Columbia River Basin.* Bonneville Power Administration, Portland, Oregon.

Kolok, A. S., and D. W. Rondorf. 1987. Effects of differential gastric evacuation and multispecies prey items on estimates of daily energy intake in juvenile Chinook salmon. *Environmental Biology of Fishes* 19:131–137.

Larkin, G. A., and P. A. Slaney. 1997. Implications of trends in marine-derived nutrient influx to south coastal British Columbia salmonid production. *Fisheries* 22(11):16–24.

Lasenby, D. C., T. G. Northcote, and M. Furst. 1986. Theory, practice and effects of *Mysis relicta* introductions to North American and Scandinavian Lakes. *Canadian Journal of Fisheries and Aquatic Sciences* 43:1277–1284.

AU: coexi
ing? or coe
iting?

Lewke, R. E., and I. O. Buss. 1977. Impacts of impoundment to vertebrate animals and their habitats in the Snake River canyon, Washington. *Northwest Science* 51:219–270.

Li, H. W., K. Currens, D. Bottom, S. Clarke, J. Dambacher, C. Frissell, P. Harris, R. M. Hughes, D. McCullough, A. McGie, K. Moore, R. Nawa, and S. Thiele. 1995a. Safe havens: Refuges and evolutionarily significant units. *Evolution and the aquatic ecosystem: Defining unique units in population conservation* 17:371–380.

Li, H. W., G. A. Lamberti, T. N. Pearsons, C. K. Tait, and J. L. Li. 1995b. Cumulative impact of riparian disturbances in small streams of the John Day Basin, Oregon. *Transactions of the American Fisheries Society* 123:627–640.

Li, H. W., C. B. Schreck, C. E. Bond, and E. Rexstad. 1987. Factors influencing changes in fish assemblages of Pacific Northwest streams. Pages 193–202 in W. J. Matthews and D. C. Heins, eds. *Community and evolutionary ecology of North American stream fishes.* University of Oklahoma Press, Norman.

Lichatowich, J. A., and L. E. Mobrand. 1995. Analysis of Chinook salmon in the Columbia River from an ecosystem perspective. Research Report, Mobrand Biometrics.

Lichatowich, J., L. Mobrand, L. Lestelle, and T. Vogel. 1995. An approach to the diagnosis and treatment of depleted Pacific salmon populations in Pacific Northwest watersheds. *Fisheries* 20:10–18.

Lilga, M. C. 1998. Effects of flow variation on stream temperatures in the lower Yakima River. Master's thesis, Washington State University, .

Lillehammer, A., and S. J. Saltveit. 1984a. The effect of the regulation on the aquatic macroinvertebrate fauna of the River Suldalslagen, western Norway. Pages 540–000 in A. Lillehammer and S. J. Saltveit, eds. *Regulated rivers.* Oslo University Press, Oslo, Norway.

Lillehammer, A., and S. J. Saltveit, eds. 1984b. *Regulated rivers.* Oslo University Press, Oslo, Norway.

Loftus, W. F., and H. L. Lenon. 1977. Food habits of the salmon smolts, *Oncorhynchus tshawytscha* and *O. keta*, from the Salcha River, Alaska. *Transactions of the American Fisheries Society* :235–240.

Maser, C., R. F. Terrant, J. M. Trappe, and J. F. Franklin. 1988. From the forest to the sea: A story of fallen trees. USDA Forest Service, Pacific Northwest Research Station, Portland, Oregon.

McCullough, D. A. 1999. A review and synthesis of effects of alterations to the water temperature regime on freshwater life stages of salmonids, with special reference to Chinook salmon. Columbia River Inter-Tribal Fish Commission, Portland, Oregon.

McIntosh, B. A., S. E. Clarke, and J. R. Sedell. 1995a. Summary report for Bureau of Fisheries stream habitat surveys: Clearwater, Salmon, Weiser, and Payette River basins, 1934–1942. DOE/BP-02246-2, Bonneville Power Administration, Portland, Oregon.

McIntosh, B. A., S. E. Clarke, and J. R. Sedell. 1995b. Summary report for Bureau of Fisheries stream habitat surveys: Cowlitz River Basin, 1934-1942. DOE/BP-02246-4, Bonneville Power Administration, Portland, Oregon.

McIntosh, B. A., S. E. Clarke, and J. R. Sedell. 1995c. Summary report for Bureau of Fisheries stream habitat surveys: Umatilla, Tucannon, Asotin, and Grande Ronde River basins, 1934-1942. DOE/BP-02246-1, Bonneville Power Administration, Portland, Oregon.

McIntosh, B. A., S. E. Clarke, and J. R. Sedell. 1995d. Summary report for Bureau of Fisheries stream habitat surveys: Willamette River basin, 1934-1942. DOE/BP-02246-3, Bonneville Power Administration, Portland, Oregon.

McIntosh, B. A., S. E. Clarke, and J. R. Sedell. 1995e. Summary report for Bureau of Fisheries stream habitat surveys: Yakima River basin, 1934-1942. DOE/BP-02246-5, Bonneville Power Administration, Portland, Oregon.

McIntosh, B. A., J. R. Sedell, J. E. Smith, R. C. Wissmar, S. E. Clarke, G. H. Reeves, and L. A. Brown. 1994. Historical changes in fish habitat for select river basins of eastern Oregon and Washington. *Northwest Science*, Special Edition 68:36–53.

Meehan, W. R., ed. 1991. *Influences of forest and rangeland management of salmonid fishes and their habitats.* American Fisheries Society, Bethesda, Maryland.

Moulton, G. E., ed. 1988. *The journals of the Lewis and Clark expedition July 28-November 1, 1805.* University of Nebraska Press, Lincoln.

Moyle, P. B., and G. M. Sato. 1991. On the design of preserves to protect native fishes. Pages 155–169 in W. L. Minckley and J. E. Deacon, eds. *Battle against extinction: Native fish management in the American West.* University of Arizona Press, Tucson.

Muir, W. D., and R. L. Emmett. 1988. Food habits of migrating salmonid smolts passing Bonneville Dam in the Columbia River, 1984. *Regulated Rivers* 2:1–10.

Mundie, J. H. 1971. Sampling benthos and substrate materials, down to 50 microns in size, in shallow streams. *Journal of the Fisheries Research Board of Canada* 28:849–860.

Munn, M., and M. A. Brusven. 1991. Benthic macroinvertebrate communities in nonregulated and regulated waters of the Clearwater River, Idaho. *Regulated Rivers* 6:1–11.

Naiman, R. J., ed. 1992. *Watershed management: Balancing sustainability and environmental change.* Springer-Verlag, New York.

Naiman, R. J., and R. E. Bilby, eds. 1998. *River ecology and management: Lessons from the Pacific coastal ecoregion.* Springer-Verlag, New York.

Naiman, R. J., and K. Fetherston. 1993. Restoration of watersheds and naturally-spawning salmon populations in the Pacific Northwest. Testimony before the US House of Representatives Subcommittee on Environment and Natural Resources, Committee on Merchant Marine and Fisheries, March 9, 1993.

National Academy of Sciences/National Academy of Engineering. 1973. Water Quality Criteria. EPA-R3-73-033, Environmental Protection Agency, Washington, DC.

National Marine Fisheries Service (NMFS). 2000. Conservation of Columbia Basin fish: All-H paper. US Department of Commerce, National Oceanic and Atmospheric Administration, in consultation with the Federal Caucus, Portland, Oregon. 3 vol.

National Research Council (NRC). 1991. *Nutrient requirements of coldwater fishes.* National Academy of Sciences Press, Washington, DC.

NRC. 1995. *Science and the Endangered Species Act.* National Academy Press, Washington, D.C.

NRC. 1999. *New strategies for America's watersheds.* National Academy Press, Washington, D.C.

Nehlsen, W., J. E. Williams, and J. A. Lichatowich. 1991. Pacific salmon at the crossroads: Stocks at risk from California, Oregon, Idaho, and Washington. *Fisheries* 16:4–21.

Netboy, A. 1980. *The Columbia River salmon and steelhead trout: Their fight for survival.* University of Washington Press, Seattle.

Nightengale, T., and D. Bennett. 1996. Hard substrate benthic invertebrates in Lower Granite Reservoir. Presentation to Idaho Chapter American Fisheries Society, Coeur d'Alene, Idaho, March 2, 1996.

Nilsen, H. C., and R. W. Larimore. 1973. Establishment of invertebrate communities on log substrates in the Kaskaskia River, Illinois. *Ecology* 54:366–374.

Noss, R. F., and D. D. Murphy. 1995. Endangered species left homeless in Sweet Home. *Conservation Biology* 9:229–231.

Oliver, D. R. 1971. Life history of the Chironomidae. *Annual Revue of Entomology* 16:211–230.

Omernik, J. M. 1987. Ecoregions of the conterminous United States. *Annals of the Association of American Geographers* 77:118–125.

Pacific Rivers Council et al. 1993. Petition to the National Marine Fisheries Service for a rule to list, for designation of critical habitat, and for a status review of coho salmon (*Oncorhynchus kisutch*) throughout its range in Washington, Oregon, and California under the Endangered Species Act. Pacific Rivers Council, Eugene, Oregon.

Perry, S. A., W. B. Perry, and J. A. Stanford. 1986. Effects of stream regulation on density, growth and emergence of two mayflies (Ephemeroptera: Ephemerellidae) and a caddisfly

(Trichoptera: Hydropsychidae) in two Rocky Mountain rivers (USA). *Canadian Journal of Zoology* 64:656–666.

Petts, G. E., H. Moller, and A. L. Roux, eds. 1989. *Historical change of large alluvial rivers: Western Europe.* John Wiley & Sons, Chichester, UK.

Petts, G. E., and R. Wood, eds. 1988. *River regulation in the United Kingdom.* John Wiley & Sons, .

Platts, W. S., and R. L. Nelson. 1985. Stream habitat and fisheries response to livestock grazing and instream improvement structures, Big Creek, Utah. *Journal of Soil and Water Conservation* 40:374–379.

Poe, T. P., H. C. Hansel, S. Vigg, D. E. Palmer, and L. A. Prendergast. 1991. Feeding of predaceous fishes on out-migrating salmonids in John Day Reservoir, Columbia River. *Transactions of the American Fisheries Society* 120:405–420.

Poff, N. L., J. D. Allan, M. B. Bain, J. R. Karr, K. L. Prestegaard, B. D. Richter, R. E. Sparks, and J. C. Stromberg. 1997. The natural flow regime: A paradigm for river conservation and restoration. *BioScience* 47:769–784.

Poff, N. L., J. D. Allan, M. A. Palmer, D. D. Hart, B. D. Richter, A. H. Arthington, J. L. Meyer, K. H. Rogers, and J. A. Stanford. 2003. River flows and water wars: Emerging science for environmental decision making. *Frontiers in Ecology and the Environment* 1:298–306.

Power, M. E., R. J. Stout, C. E. Cushing, P. P. Harper, F. R. Hauer, W. J. Matthews, P. B. Moyle, B. Statzner, and I. R. Wais De Badgen. 1988. Biotic and abiotic controls in river and stream communities. *Journal of the North American Benthological Society* 7:456–479.

Rhodes, J. J. 1995. A comparison and evaluation of existing land management plans affecting spawning and rearing habitats of Snake River Basin salmon species listed under the Endangered Species Act. Report prepared for the National Marine Fisheries Service, Columbia Inter-Tribal Fish Commission, Portland, Oregon.

Rhodes, J. J., D. A. McCullough, and J. F. A. Espinosa. 1994. A coarse screening process for evaluation of the effects of land management activities on salmon spawning and rearing habitat in ESA consultations. Technical Report 94-1, Columbia River Inter-Tribal Fish Commission, Portland, Oregon.

Rickard, W. H., W. C. Hanson, and R. E. Fitzner. 1982. The non-fisheries biological resources of the Hanford Reach of the Columbia River. *Northwest Science* 56:62–76.

Ricker, W. E. 1972. Hereditary and environmental factors affecting certain salmonid populations. Pages 19–160 in R. C. Simon and P. A. Larkin, eds. *The stock concept in Pacific salmon.* University of British Columbia, Vancouver.

Ricketts, T. H., Dinnerstein, E., Olson, D. M., Loucks, C. L., Eichbaum, W., DellaSala, D., Kavanagh, K., Hedao, P., Hurley, P. T., Carney, K. M., Abell, R., Walters, S., and Fund, W. W. 1999. *Terrestrial ecoregions of North America: A conservation assessment.* Island Press, Washington, DC.

Riddell, B. E., and W. C. Legget. 1981. Evidence for an adaptive basis variation in body morphology and time of downstream migration juvenile Atlantic salmon (*Salmo salar*). *Canadian Journal of Fisheries and Aquatic Sciences* 38:308–320.

Rieman, B. E., and B. Bowler. 1980. Kokanee trophic ecology and limnology in Pend Oreille Lake. Forest, Wildlife and Range Exp. Station, University of Idaho, Moscow.

Rieman, B. E., and J. D. McIntyre. 1993. Demographic and habitat requirements for conservation of bull trout. General Technical Report INT-308, US Forest Service, Intermountain Research Station, Ogden, Utah.

Roebeck, G. G., C. Henderson, and R. C. Palange. 1954. Water quality studies on the Columbia River. Special Report, US Department of Health, Education and Welfare, Public Health Service, Washington, DC.

Rondorf, D. W., M. S. Dutchuk, A. S. Kolok, and M. L. Gross. 1985. Bioenergetics of juvenile salmon during the spring outmigration. DF-A179-82BP35346, US Fish and Wildlife Service, Cook, Washington.

Rondorf, D. W., G. A. Gray, and R. B. Fairley. 1990. Feeding ecology of subyearling Chinook in riverine and reservoir habitats of the Columbia River. *Transactions of the American Fisheries Society* 119:16–24.

Sagar, P. M., and G. A. Eldon. 1983. Food and feeding of small fish in the Rakaia River, New Zealand. *New Zealand Journal of Marine and Freshwater Research* 17:213–226.

Sagar, P. M., and G. J. Glova. 1987. Prey preferences of a riverine population of juvenile Chinook salmon, *Oncorhynchus tshawytscha*. *Journal of Fish Biology* 31:661–673.

Salo, E. O., and T. W. Cundy. 1987. Streamside management: Forestry and fishery interactions. University of Washington, Seattle.

Schluchter, M., and J. A. Lichatowich. 1977. Juvenile life histories of Rogue River spring Chinook salmon *Oncorhynchus tshawytscha* (Walbaum), as determined from scale analsysis. Information Report Series Fisheries No. 77-5, Oregon Department of Fish and Wildlife, Corvallis, Oregon.

Schreck, C. B., J. C. Snelling, R. E. Ewing, C. S. Bradford, L. E. David, and C. H. Slater. 1995. Migratory characteristics of juvenile spring Chinook salmon in the Willamette River. DOE/BP-92818-5, Bonneville Power Administration, Portland, Oregon.

Schumm, S. A., and R. W. Lichty. 1965. Time, space, and causality in geomorphology. *American Journal of Science* 263:110–119.

Sear, D. A. 1994. River restoration and geomorphology. *Aquatic Conservation: Marine & Freshwater Ecosystems* 4:169–177.

Sedell, J. R., and J. L. Froggatt. 1984. Importance of streamside forests to large rivers: The isolation of the Willamette River, Oregon, U.S.A., from its floodplain by snagging and streamside forest removal. *Verhandlungen Internationale Vereinigung fur Theoretische und Angewandte Limnologie* 22:1828–1834.

Sedell, J. R., F. N. Leone, and W. S. Duvall. 1991. Water transportation and storage of logs. *American Fisheries Society* Special Publication 19:325–367.

Sedell, J. R., G. H. Reeves, F. R. Hauer, J. A. Stanford, and C. P. Hawkins. 1990. Role of refugia in recovery from disturbances: Modern fragmented and disconnected river systems. *Environmental Management* 14:711–724.

Sheldon, A. L. 1988. Conservation of stream fishes: patterns of diversity, rarity and risk. *Conservation Biology* 2:149–156.

Smith, R. A., R. B. Alexander, and M. G. Wolman. 1987. Water quality trends in the nation's river. *Science* 235:1607–1615.

Snyder, E. B., C. P. Arango, D. J. Eitemiller, J. A. Stanford, and M. L. Uebelacker. 2003. Floodplain hydrologic connectivity and fisheries restoration in the Yakima River, U.S.A. *Verhandlungen Internationale Vereinigung fur Theoretische und Angewandte Limnologie* 28:1653–1657.

Spencer, C. N., B. R. McClelland, and J. A. Stanford. 1991. Shrimp stocking, salmon collapse, and eagle displacement: Cascading interactions in the food web of a large aquatic ecosystem. *BioScience* 41:14–21.

Stanford, J. A. 1998. Rivers in the landscape: introduction to the special issue on riparian and groundwater ecology. *Freshwater Biology* 40:402–406.

Stanford, J. A., and B. K. Ellis. 2002. Natural and cultural influences on ecosystem processes in the Flathead River Basin (Montana, British Columbia). Pages 269–284 in J. S. Baron, ed. *Rocky mountain futures: An ecological perspective.* Island Press, Washington DC.

Stanford, J. A., and F. R. Hauer. 1992. Mitigating the impacts of stream and lake regulation in the Flathead River catchment, Montana, USA: An ecosystem perspective. *Aquatic Conservation: Marine and Freshwater Ecosystems* 2:35–63.

Stanford, J. A., S. V. Gregory, F. R. Hauer and E. B. Snyder. 2005. Columbia River basin. pp. 591–654. In: Benke, A. V. and B. Cushing. Rivers of North America. Elsvier, New York.

Stanford, J. A., F. R. Hauer and M. S. Lorang. In press. The shifting habitat mosaic of river ecosystems. *Verhandlungen Internationale Vereinigung fur Theoretische und Angewandte Limnologie.*

Stanford, J. A., F. R. Hauer, and J. V. Ward. 1988. Serial discontinuity in a large river system. *Verhandlungen Internationale Vereinigung fur Theoretische und Angewandte Limnologie* 23:1114–1118.

Stanford, J. A., E. B. Snyder, M. N. Lorang, D. C. Whited, P. L. Matson, and J. L. Chaffin. 2002. The Reaches Project: Ecological and geomorphic studies supporting normative flows in the Yakima River Basin, Washington. Open File Report 170-02, Prepared for Yakima Office, Bureau of Reclamation, US Department of the Interior, Yakima, Washington by Flathead Lake Biological Station, University of Montana, Polson.

Stanford, J. A., and J. V. Ward. 1992. Management of aquatic resources in large catchments: Recognizing interaction between ecosystem connectivity and environmental disturbance. Pages 91–124 in R. J. Naiman, ed. *Watershed management: Balancing sustainability with environmental change.* Springer-Verlag, New York.

Stanford, J. A., and J. V. Ward. 1993. An ecosystem perspective of alluvial rivers: Connectivity and the hyporheic corridor. *Journal of the North American Benthological Society* 12:48–60.

Stanford, J. A., J. V. Ward, and B. K. Ellis. 1994. Ecology of the alluvial aquifers of the Flathead River, Montana. Pages 367–390 in J. Gibert, D. L. Danielopol, J. A. Stanford, ed. *Groundwater Ecology.* Academic Press, San Diego, California.

Stanford, J. A., J. V. Ward, W. J. Liss, C. A. Frissell, R. N. Williams, J. A. Lichatowich, and C. C. Coutant. 1996. A general protocol for restoration of regulated rivers. *Regulated Rivers* 12:391–413.

Stites, D. L., and A. C. Benke. 1989. Rapid growth rates of chironomids in three habitats of a subtropical blackwater river and their implications for P:B ratios. *Limnological Oceanography* 34:1278–1289.

Stober, Q. J., and R. E. Nakatani. 1992. Water quality and biota of the Columbia River system. Pages 51–84 in C. D. Becker and D. A. Neitzel, eds. *Water quality in North American river systems.* Battelle Press, Columbus, Ohio.

Stouder, D. J., P. A. Bisson, and R. J. Naiman, eds. 1997. *Pacific salmon and their ecosystems: Status and future options.* Chapman and Hall, New York.

Templeton, W. L., and C. C. Coutant. 1971. Studies on the biological effects of thermal discharges from nuclear reactors to the Columbia River at Hanford. Pages 591–614 in *Environmental aspects of nuclear power stations.* International Atomic Energy Agency Symposium, Vienna, Austria.

Theurer, F. D., I. Lines, and T. Nelson. 1985. Interaction between riparian vegetation, water temperature and salmonid habitat in the Tucannon River. *Water Resources Bulletin* :53–64.

Torgersen, C. E., D. M. Price, H. W. Li, and B. A. McIntosh. 1999. Multiscale thermal refugia and stream habitat associations of Chinook salmon in northeastern Oregon. *Ecological Applications* 9:301–319.

US Army Corps of Engineers. 1976. Inventory of riparian habitats and associated wildlife along Columbia and Snake rivers. North Pacific Division, Portland, Oregon. North Pacific Division, U.S. Army Corps of Engineers, Portland, Oregon.

Vaccaro, J. J. 1988. Simulation of stream temperatures in the Yakima River Basin, Washington, April-October 1991. Water Resource Investigations Report 85-4232, US Geological Survey.

Vervier, P., and R. J. Naiman. 1992. Spatial and temporal fluctuations of dissolved organic carbon in subsurface flow of the Stillaguamish River (Washington, USA). *Archiv fur Hydrobiologie* 123:401–412.

Walters, C. J., R. M. Hilborn, R. M. Peterman, and M. J. Staley. 1978. Model for examining early ocean limitation of Pacific salmon production. *Journal of the Fisheries Research Board of Canada* 35:1303–1315.

Ward, J. V. 1973. Molybdenum concentrations in tissues of rainbow trout (*Salmo gairdneri*) and kokanee salmon (*Oncorhynchus nerka*). *Journal of the Fisheries Research Board of Canada* 30:841–842.

Ward, J. V. 1985. Thermal characteristics of running waters. *Hydrobiologia* 125:31–46.

Ward, J. V. 1986. Altitudinal zonation in a Rocky Mountain stream. *Arch. Hydrobiol* Suppl. 74:133–199.

Ward, J. V., and J. A. Stanford, eds. 1979. *The ecology of regulated streams*. Plenum Press, New York.

Ward, J. V., and. J. A. Stanford. 1982. Thermal responses in the evolutionary ecology of aquatic insects. *Ann. Rev. Entomol* 27:97–117.

Ward, J. V., and J. A. Stanford. 1983. The intermediate disturbance hypothesis. An explanation for biotic diversity patterns in lotic ecosystems. Pages 347–356 in T. D. Fontaine and S. M. Bartell, eds. *Dynamics of lotic ecosystems*. Ann Arbor Scientific Publishers, Ann Arbor, Michigan.

Ward, J. V., and J. A. Stanford. 1995. The serial discontinuity concept: Extending the model to floodplain rivers. *Regulated Rivers* 10:159–168.

Ward, N. E. and Ward, D. L. (2004). Resident fish in the Columbia River Basin: Restorations, enhancement, and mitigation for losses associated with hydroelectric development and operations. *Fisheries* 29:10–18.

Watson, S., E. McCauley, and J. A. Downing. 1992. Sigmoid relationships between phosphorus, algal biomass and algal community structure. *Canadian Journal of Fisheries and Aquatic Sciences* 49:2605–2610.

Welcomme, R. L. 1988. Concluding remarks: I. On the nature of large tropical rivers, floodplains, and future research directions. *Journal of the North American Benthological Society* 7:525–526.

Welcomme, R. L. 1995. Relationships between fisheries and the integrity of river systems. *Regulated Rivers* 11:121–136.

White, R. 1995. *The organic machine: The remaking of the Columbia River*. Hill and Wang, New York.

Wilderness Society. 1993. *The living landscape: Vol. 2. Pacific salmon and federal lands*. The Wilderness Society, Bole Center for Forest Ecosystem Management.

Wissmar, R. C., J. E. Smith, B. A. McIntosh, H. W. Li, G. H. Reeves, and J. R. Sedell. 1994. A history of resource use and disturbance in riverine basins of eastern Oregon and Washington (early 1800s-1990s). *Northwest Science* 68:1–35.

Wydoski, R. S., and R. R. Whitney. 2003. *Inland fishes of Washington*. 2nd ed. American Fisheries Society, Bethesda, Maryland.

Return to the River

6

Hydroelectric System Development:
Effects on Juvenile and Adult Migration

Charles C. Coutant, Ph.D., Richard R. Whitney, Ph.D.

Development of the Hydroelectric System
  Effects of Dams on Anadromous Fishes
  Dams and Other Obstacles to Migrations of Salmon
  Other Effects of Dams on Spawning of Anadromous Fishes
  Rearing of Juvenile Salmon
Downstream Migration of Juvenile Salmon and Steelhead
  Surface Orientation
  Daily Migration Cycles
  Use of Flow Dynamics in Migration
  Downstream Migration: Active versus Passive
  Subyearling Chinook Migrants
    Subyearling Migration in Rivers
    Subyearlings in the Snake River
    Subyearling Migrations in Reservoirs
    Subyearling Migrations in the Freshwater Estuary
    Experimental Research on Subyearlings
    Management Implications for Hanford and Snake River Subyearling Chinook
    Management Risks for Subyearlings: Passive versus Active Migration Modes
  Yearling Chinook Migrants
    Evidence for Flushing of Yearlings
    Flow Structure as an Aid to Migration for Yearling Chinook
    Yearling Chinook in Reservoirs
    Yearling Chinook in the Estuary
  Population Contrasts for Yearling Chinook: Snake and Willamette Rivers
  Other Salmon and Steelhead Trout
    Coho Salmon
    Sockeye Salmon
    Steelhead Trout
Effects of Flow on Rate of Migration
  Snake River
    Chinook and Steelhead

Mid-Columbia Reach
   Yearling Chinook
   Subyearling Chinook
   Steelhead and Sockeye
Discussion of Travel Time Studies
   Principles of Water/Fish Movement
Alternative Flow Hypotheses: Salmonid Smolt Migration and Survival
   Velocity Hypothesis
   Fluctuating Flow Hypotheses
   Fish Orientation Hypothesis
   Dam Obstacle Hypothesis
Directions for Future Research
Effects of Hydrosystem Development on Life History Diversity
Literature Cited

"Dam construction presents a serious threat to the continued expansion–and indeed the very existence–of the commercial and recreational value of the Fraser River fisheries resource. . . . Although the fish-dam problem has existed for centuries in many countries, no practical solutions have yet been found that afford complete protection for anadromous fish in rivers obstructed and altered by large dams."
—F. J. Andrew and G. H. Geen. (Andrew and Green, 1960, p.2).

Development of the Hydroelectric System

Development of the hydropower system in the Columbia River Basin began in the late 19th century on the tributaries. The first dam on the mainstem Columbia River was Rock Island Dam, completed in 1933. From 1933 to 1975, full development proceeded (Figure 6.1), to the point that Grand Coulee Dam blocked the Columbia River mainstem first in 1941, and Hells Canyon Dam blocked the Snake River in 1967. In 1955, Chief Joseph Dam was constructed downstream of Grand Coulee Dam. There are now thirteen hydroelectric dams on the Columbia River and Snake River mainstem that are passable by salmon: five on the mid-Columbia reach (i.e., the mainstem from Chief Joseph Dam to the confluence of the Snake River), four on the Snake River, and four on the lower Columbia River mainstem. Two Canadian mainstem dams, as well as three on tributaries there, two projects in Montana, and one in Idaho, realized as a result of a 1964 treaty, provide the primary capability for storage of water within the Columbia Basin (Bonneville Power Administration 1980). Hydroelectric power generation, flood control, and irrigation were the benefits expected from the full development of the potential of the Columbia Basin (Logie 1993). The Hanford Reach, the one remaining undammed portion of the river, was debatable as a potential dam site, due to the potential for flooding of underground storage

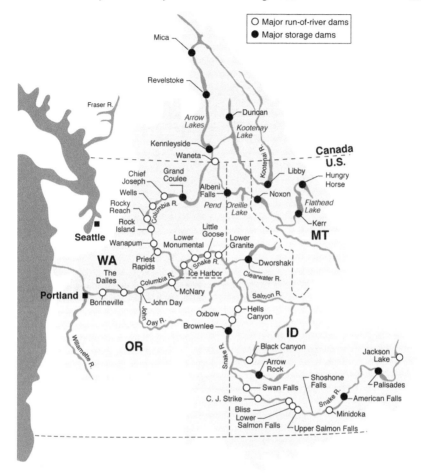

Figure 6.1 Map of the Columbia River Basin, showing major dams and lakes.

facilities for atomic wastes at the Hanford Reservation. The impounded Columbia River looks very different from the unimpounded Hanford Reach (Figure 6.2).

Seventy dams located on tributaries in the basin are also part of the coordinated hydroelectric system. Some of these dams, such as those on the Cowlitz River, are not passable by salmon, and others, such as those above Hells Canyon Dam, lie above impassable dams. Another 128 dams on tributaries, while not part of the coordinated hydroelectric system, present passage and water quality problems for anadromous and resident fish species.

Total storage capacity of the reservoirs in the system amounts to 55.3 million acre feet (68.2 billion m³), which is about 25% of the basin's average total

(a)

(b)

Figure 6.2 Photographs of (a) a typical mainstem Columbia River dam and its reservoir, and (b) a portion of the unimpounded Hanford Reach, which is similar to the historical river. Photos from (a) US Army Corp Digital Visual Library, website at *http://images.usace.army.mill photolib.html*, and (b) courtesy of D. Geist.

annual runoff (Logie 1993). This capacity is used to store a portion of the spring freshet for the benefit of later power production, and drawdown in late winter and early spring for the benefit of downstream flood control and other purposes (Logie 1993). As a comparison, storage capacity in the Colorado River is about four times the average annual runoff in that system. In addition to the seasonal shifts in flow made possible by the storage capacity provided by dams in the basin, the storage capacity has made possible production of power on demand, a practice known as "load following." Load following can lead to rapid and large changes in river flow as power plants are managed to take advantage of short-term markets for power. For example, many plants routinely reduce flows at night to the minimum necessary to keep turbines running that are required for plant operations. As a result, river flow can change by a magnitude of four times or more in a matter of an hour (Independent Scientific Advisory Board [ISAB] 1998).

Effects of Dams on Anadromous Fishes

As the nearest large river to the north, the Fraser River stands as an example where experience with salmon is useful for comparison with experience in the Columbia. In 1960 at the behest of the International Pacific Salmon Fisheries Commission, Andrew and Geen (1960) undertook an analysis of the probable effects of hydroelectric development in the Fraser River, British Columbia, on salmon production in the Fraser system. The proposed development would have involved construction of 18 dams on the mainstem and 44 dams on tributaries. They concluded that dam construction presented a serious threat to the continued existence of the Fraser River fishery and noted that no practical solutions had yet been found that provide protection for anadromous fish in rivers obstructed and altered by large dams (see quote at beginning of this chapter). Largely because of Andrew and Geen's conclusions, the Fraser River mainstem remains undammed to this day. Although their study was completed 37 years ago, their conclusion that no practical solution to the fish-dam problem has yet been found still applies to a certain extent, as borne out by experience in the Columbia River, which is summarized below.

Dams and Other Obstacles to Migrations of Salmon

Construction and operation of dams for hydroelectric power production has produced a change in flow conditions encountered by adult salmon migrating upstream. Adult salmon have always had to counteract effects of a variety of flows. High flows that result from natural events can present problems in passage for adult salmon. For example, it is well established that sockeye adults are unable to pass upstream through the Tumwater canyon on the Wenatchee River until spring flows decline to below about 4,000 cfs (French and Wahle 1968; Allen and Meekin 1980; Mullan et al. 1986; Chapman et al. 1995). Burgner (1991) reported that sockeye adults tend to move upstream in slower water and eddies along the stream banks. Such habitat is not available in Tumwater Canyon at flows above 4,000 cfs.

Dams have created some unique changes. Formerly, flow was spread over the entire cross section of the river, with the volume being concentrated in a channel cut by scouring action of the river. At present, flow near dams is usually concentrated through the powerhouse and into the tailrace. During times when flow exceeds powerhouse capacity, spillways are operated that, in most cases, but not all, split the flow away from the powerhouse. At some point downstream, water from the two sources meets to continue downstream. Passage facilities (fish ladders) for adult salmon are normally located at each end of the powerhouse and at the bank opposite the powerhouse. Usually, access to the passage is available at openings across the powerhouse. While this arrangement attempts to simulate natural conditions, it is necessary to adjust operations of individual turbines and spill bays at times to compensate for adverse flow conditions that may be produced at adult passage entrances under some flows (Bjornn and Peery 1992; Dauble and Mueller 1993; Mendel et al. 1994). The Fish Passage Center (FPC) annually produces a plan for adult passage for each project on the mainstem Columbia and Snake Rivers. It includes measures specific to each project, which are designed to optimize conditions for passage of adults (FPC 1994).

In some tributaries, irrigation removals have created problems for passage of adults. For example, low flows at the mouth of the Yakima River below Rosa Dam during the summer months lead to high water temperatures that are a barrier to passage of adult salmon. The same occurs for sockeye (and perhaps Chinook) in the Okanogan River, and elsewhere. This is discussed further in the section on habitat. Another example among many, is the Umatilla River, where flow in the lower river is insufficient to provide passage for salmon.

The typical mainstem dam on the Columbia River presents challenges to the migrations of both adult and juvenile anadromous fishes. The mainstem dams are for the most part around 100 feet high, although Rock Island Dam is about 50 feet high. In contrast, Grand Coulee and Hells Canyon Dams are over 700 feet high and impassable to fish. Adult salmon and steelhead moving upstream from right to left in Figure 6.3 may pass the project by way of fish ladders or by way of the navigation channel in those dams equipped with one (the mid-Columbia dams are not so equipped). Careful adjustments of flow and other characteristics of the ladders are required to keep them functioning properly. For example, changes in elevation of the forebay or tailrace

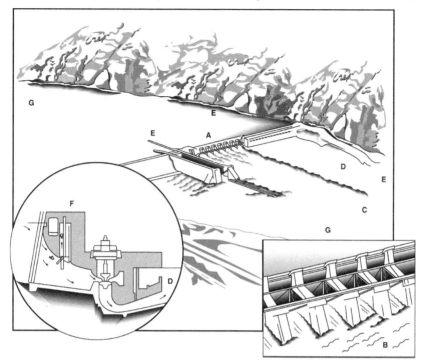

Figure 6.3 Diagram of a typical hydroelectric dam in the Columbia River Basin. The figure shows the spillway (A and inset B), the powerhouse to the right of the spillway, powerhouse cross section (area F in the circular inset), and the navigation lock (E) to the left of the spillway (not present in mid-Columbia dams). In the powerhouse cross section, fish are shown moving up into a bypass inside the powerhouse, while the water continues on through the turbine. The diagram also shows the spillway tailrace (c) and the powerhouse tailrace (D), the adult fish ladder exit and entrance (E on the right), and navigability (G).

require corresponding adjustments in the ladder. Levels of spill at some projects may produce conditions in the tailrace that create irregular patterns of flow that can confuse adults attempting to find ladder entrances. Many of these adjustments are now computerized.

Juvenile emigrants, moving downstream in the direction from left to right in Figure 6.3, may pass the project by one of four basic routes: the powerhouse, the spillway, the navigation channel, or the fish ladders. The four lower Columbia River and four Snake River dams are equipped with turbine intake screens that divert juvenile salmon away from the turbines into bypass systems. Juvenile salmon migrating downstream past those projects with fish passage facilities for juveniles may use several routes. They either pass through the turbines, spillways, turbine intake bypass systems, navigation locks, or ice and trash sluiceways, which have been modified for fish passage.

A few juvenile salmon may pass by way of fish ladders designed for adult passage, but these are not designed, located, or operated in ways that attract juveniles.

Other Effects of Dams on Spawning of Anadromous Fishes

Maturing salmon typically ascend freshwater streams where they deposit their eggs in gravel of suitable size. In the case of sockeye, some populations also may spawn in near-shore areas of lakes, where spring water provides the interstitial water movement in the gravel required for successful incubation of eggs. Survival of eggs and alevins depends upon movement of clean water of suitable temperature and oxygen content through the permeable gravels in which the redds are constructed. Information on specific requirements for flow, permeability of gravel, and effects of temperature on development may be found in Groot and Margolis (1991), Meehan (1991), Rhodes et al. (1994), and Salo and Cundy (1987).

Some stocks of Chinook are adapted to spawn in the mainstem of the Snake and Columbia Rivers and the lower ends of tributary streams. Other stocks of Chinook and the principal stocks of other species of salmon have life histories adapted to spawning and rearing farther up in the tributaries. Observations of Chinook spawning in the mainstem indicate that velocity of flow may be more important than water depth in determining location of their spawning (Chambers 1955). Chinook have been observed spawning in water as deep as 30 or 35 feet (W. M. Chapman 1943; Chambers 1955; Meekin 1967; D. Chapman et al. 1983; Garcia et al. 1994). While the areas in the reservoirs are no longer suitable due to low water velocities, Chinook have been observed spawning in the tailraces immediately below most Columbia and Snake River mainstem dams (Horner and Bjornn 1979; Dauble and Watson 1990; Garcia et al. 1994).

Chambers (1955) observed Chinook that spawned in water velocities ranging from 2.75 to 3.75 ft/s, with very few above 3 ft/s. Bovee et al. (1978) developed probability-of-use curves for Chinook spawning in various water velocities. His curves indicate spawning is most likely to occur at velocities between 0.67 ft/s and 4 ft/s. At Vernita Bar, velocities on the spawning grounds varied with flow, but were probably chosen by the fish at low flows when velocities were in the range of 1 to 2 ft/s (Chapman et al. 1983). Groves (1993) determined that the suitable velocities for spawning of fall Chinook salmon in the Hells Canyon Reach of the Snake River were 1.3 to 6.6 ft/s. In the normal Columbia Basin stream, substrate size is affected by velocity.

The Hanford site, where the fall Chinook salmon population is presently most successful, has the most intact habitat and ecological processes of any mainstem site in the Columbia Basin. It is characterized by broad gravel spawning bars located about 5 to 12 km (3.1–7.5 miles) and 40 km (25 miles) downstream of Priest Rapids Dam, which are occupied from October to March by thousands of salmon constructing their redds (Dauble and Watson 1990). Annual spawning surveys were conducted by D. Watson beginning in the 1940s. Bauersfeld (1978) and Chapman et al. (1983) have characterized the effects of gravel size and flow regimes for the most densely occupied spawning area at Vernita Bar. These redds generally lie upriver of a 48-km (30-mile) zone of islands, side channels, backwaters, and sloughs that extends to the city of Richland (especially the White Bluffs, F-Area, and Hanford townsite areas). These diverse habitats provide a mixture of critical rearing, resting, and feeding areas for juvenile Chinook salmon and are, without a doubt, a key component in the success of the Hanford Reach fall Chinook.

The mainstem inundated areas were formerly used for spawning by Chinook and perhaps by other salmonids. The reservoirs created by the dams reduced the area where suitable water velocities and substrate are to be found. At present, the main salmon spawning in the mainstem Snake and Columbia Rivers above Bonneville Dam is the ocean-type fall Chinook salmon (Healey 1991), although there were other stocks that spawned on mainstem gravel bars in the past (Fulton 1968; Lichatowich and Mobrand 1995). The basin's healthiest population is in the mid-Columbia in the undammed Hanford Reach (Dauble and Watson 1990). The remaining mainstem-spawning populations are now confined to small numbers of fall Chinook salmon that spawn in the tailraces of each of the dams on the mainstem Columbia and Snake Rivers and the main channels of some lower Columbia River tributaries. (Horner and Bjornn 1979; Dauble et al. 1989; M. Erho, personal communication; Garcia et al. 1994). Other, troubled populations spawn in what remains of the undammed Snake and Clearwater Rivers (Garcia et al. 1994) between Lower Granite Reservoir and migration-blocking storage dams (Hells Canyon Dam on the Snake River and Dworshak Dam on the Clearwater River). Before Brownlee and Hells Canyon Dams were constructed beginning in the late 1950s, fall Chinook salmon spawned in the mainstem Snake River well above the dam sites (Krcma and Raleigh 1970). By the time Brownlee Dam was built, salmon had already been blocked from the upper reaches of the Snake River by hydroelectric and irrigation dams that date back as early as 1901 (Swan Falls).

In the unimpounded Snake River above Lewiston, Idaho, Chinook salmon spawn in scattered redds at rapids between river kilometers 238.6 (head of Lower Granite Reservoir) and 396.6 (Hells Canyon Dam). They also spawn in the lower Clearwater River. There is more suitable spawning area than there is spawning activity (Garcia et al. 1994).

In the Snake River, there are several tributaries with productive spring Chinook salmon populations, although populations are in decline and the stock is listed as endangered. One of the most far-removed tributaries from the ocean is the upper Salmon River in Idaho, which is still a major natural salmon production area (Kiefer and Lockhart 1995). Before construction of Brownlee and Hells Canyon Dams beginning in the late 1950s, spring Chinook salmon spawned in Eagle Creek and the Weiser River, both upstream tributaries to the Snake River (Krcma and Raleigh 1970). In the mid-Columbia River, there are naturally reproducing populations of spring Chinook and summer Chinook that are not abundant. A large number of yearling Chinook juveniles come from upper Columbia River hatcheries (Dauble et al. 1989). Spring Chinook salmon also occur in the Willamette and Yakima Rivers.

In the 1970s, researchers became aware of the importance of load following during Chinook spawning, which led to exposure of redds and incubating eggs in the Hanford Reach (Watson et al. 1969; Bauersfeld 1978; Chapman et al. 1983). Because of the great importance of this spawning area, studies were conducted which led to identification of flow control measures that could improve spawning success. The result was a long-term (1988–2005) Vernita Bar Settlement Agreement among the fishery agencies and the power, flood control, and irrigation interests to stabilize flows. The agreement was approved by the Federal Energy Regulatory Commission (FERC) in December 1988. In 1999, following a recommendation of the ISAB for further study to identify measures that would protect emerged juveniles from stranding as a result of load variations, an agreement was reached to further stabilize hourly fluctuations in flow during the time when emerged fry are still in the area, preparing to migrate downstream. This has led to a reduction in juvenile mortalities caused by strandings (Douglas Ancona, Grant County Public utility District [PUD]., personal communication.) Chapter 7 includes more details on this effort.

In the mainstem Columbia and Snake Rivers, the basin is fully developed for hydropower production to the extent that, other than in the Hanford Reach, the reservoir of each downstream project impinges upon the tailrace of the one upstream. However, even so, extreme reductions of flow have led to dewatering of redds in the immediate tailwater area, such as at Chief Joseph Dam in 1967 (Meekin 1967), and elsewhere in unusual circumstances, such as in reaction to load rejection[1] at a project (mid-Columbia P.U.D. biologists, personal communications).

In the tributaries, rapid, large fluctuations in flow that are associated with load following have been shown to produce adverse effects on resident fishes

---

[1]"Load rejection" is the term used for powerhouse shutdown resulting from an unforeseen problem in transmission. The lack of generation leads to containment of river flow above the dam until spill gates can be opened. As a result, water elevation in the tailrace will be lowered. FERC requires that each project have an emergency plan prepared to deal with such situations.

downstream. For example, in the Kootenai River, the Proposed Recovery Plan for Endangered Kootenai River Sturgeon notes that there has been no successful spawning of this fish since Libby Dam was put into operation as part of the hydropower system. The indication is that fluctuations in flow have adversely affected reproductive success of the sturgeon. The plan calls for stable flows during the spawning and incubation period of this fish (U.S. Fish and Wildlife Service 1995). Adverse effects on other resident fishes in the Kootenai River have been documented (Perry and Perry 1991; ISAB 1997). In the Hungry Horse/Flathead Lake Basin, adverse effects on biota in the Flathead River system and downstream have been documented, as a result of which the FERC has issued an order calling for stable flows out of Kerr Dam (Stanford and Hauer 1992; ISAB 1997).

Rearing of Juvenile Salmon

Rearing of sockeye juveniles takes place in lakes, of which three remain in the Columbia Basin, representing about 5% of the lake area formerly available to them (Mullan 1986). Rearing of chum salmon is short-term, as the juveniles move downstream immediately after emergence from the gravel. Coho and steelhead typically spawn and rear in the tributaries. Chinook rear both in the tributaries and mainstem. There are two major life-history types of Chinook in the basin, generally distinguished by the relative lengths of freshwater rearing (Gilbert 1912; Groot and Margolis 1991). Ocean-type Chinook are usually mainstem or coastal river spawners with short migration distances to the sea, whereas stream-type stocks spawn in the tributaries and thus have longer migration routes (Taylor 1990). Ocean-type Chinook exhibit a short freshwater residence for rearing (feeding and growing), usually leaving the river ecosystem within 6 months of emergence from the spawning gravel. Stream-type Chinook, on the other hand, reside in the stream for one year or longer before emigrating rapidly to the ocean. Steelhead and coho may be thought of as exhibiting the "stream type" life history, as they typically spend a year to several years in the stream.

The amount of space available for rearing of juveniles has an effect on the rate of survival of fry, and this, of course, is a function of flow. In examining counts of adult Chinook at Bonneville Dam during a time prior to full development of the hydropower system (1935–1945), Silliman (1950) found that 27% of the variability in numbers of returning adults could be explained by volume of flow during April and May of the year of their outmigration as juveniles. Considering the large number of factors that can affect the number of returning adults, Silliman concluded that flow was a significant factor in production of Chinook. Because there was little storage capacity in the basin at that time, flows in April and May must have been strongly correlated with flows during the rearing phase of Chinook juveniles, as well as during their outmigration during the time period considered by Silliman. We have discussed the necessary features of desirable salmon habitat already. The point to be made here is that volume of flow affects the amount of available habitat and its quality. The historical record generally shows better salmon production in wet years (Anderson et al. 1996). Droughts have been particularly devastating for survival of juvenile salmonids and for returns of adults in subsequent years in this and other river basins (e.g. California). A number of studies have demonstrated that production of coho smolts, and in some cases of Chinook smolts, is related to average flow in the nursery stream and varies from year to year with flow in the particular tributary (Jager et al. 1997). The effects of flow in the nursery areas are in any case difficult to separate from effects of temperature because flow and water temperature are themselves correlated. A more complete discussion of temperature effects is provided in another chapter of this book.

The mainstem of the Columbia and Snake Rivers contains both ocean- and stream-type migrating Chinook. Presently, ocean-type fish are represented by fall Chinook salmon (and less abundant summer Chinook in the mid-Columbia) that spawn in the mainstem and lower reaches of tributaries and rear (feed and grow) in the mainstem as they move slowly in spring and summer toward the sea. Stream-type Chinook undergo a year or more of rearing in tributary headwaters and move rapidly through the mainstem in spring (Figure 6.4). Stream type Chinook include spring Chinook salmon (and summer Chinook in the Snake River drainage). Coho salmon, which often rear for 2 years in tributaries, and steelhead/rainbow trout, which most often rear for 1 to 2 years but can rear up to 7 years before leaving tributaries to migrate to sea, move out rapidly in the spring, along with the spring Chinook (Peven et al. 1994).

Anadromous salmonids have in common the need to migrate downstream through the mainstem. Thus, the mainstem typically hosts stream-type (yearling) Chinook, steelhead, and coho during the spring when river flows are normally the highest of the year. Some subyearling (ocean-type) Chinook may be seen at that time as well, but their peak in abundance comes later in the summer, in July or August in the mid-Columbia reach and the lower Snake River and later in the lower Columbia River.

Downstream Migration of Juvenile Salmon and Steelhead

Central to evaluation of any flow management strategy to aid migrating juvenile salmonids is an understanding of how juveniles migrate downstream. As we seek ways to modify the hydrosystem toward a more normative condition discussed earlier, we must have a sound scientific basis for what is normal in

Spiraling Migration of Yearling & Subyearling Salmon

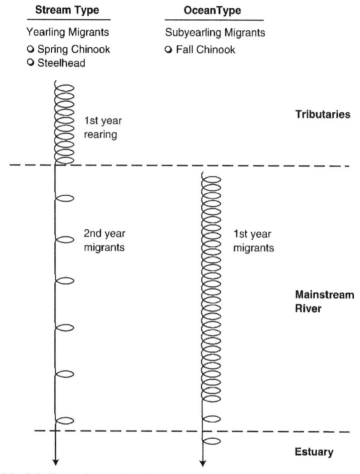

Figure 6.4   Spiraling or zig-zag migrations of stream-type and ocean-type salmonids showing occupancy of tributaries and mainstem, and relative amount of time spent holding and moving (spiraling lengths) in these habitats. Each spiral loop indicates a period of holding and feeding.

the lives of the fish we wish to preserve. What is their normal behavior? What habitats in the migratory corridor do they normally use and need? How has this normal behavior been altered by the hydropower system and related water uses? With much of the original landscape now missing, can we deduce fish needs by comparing environments in which stocks are doing fairly well with those where they are doing poorly?

Because so much of the debate over survival of downstream migrants has revolved around issues of travel time, the ways fish interact with water flow to accomplish their downstream movement is important. Are they simply passive particles being flushed downstream? If so, then water travel time may be preeminent. Or is there, as seems likely, a more complex behavior? Are fish selecting particular portions of the migratory corridor (large-scale spatial complexity)? Are they orienting to particular hydraulic features of the moving water (fine-scale spatial complexity)? Are they moving at different times of day (temporal complexity)? What differences in these behaviors constitute inter- and intraspecific diversity that we may want to preserve? How do our existing and proposed hydropower mitigation schemes, such as intake screening, flow augmentation, reservoir drawdown, dam breaching, and surface bypasses (among others), fit with the normal fish uses of their environment? This section briefly discusses our rather primitive understanding of fish migration behavior during downstream migration for the purpose of evaluating effects and designing mitigation alternatives.

Surface Orientation

Most studies of juvenile salmon migration in rivers and reservoirs have shown a surface orientation during movement. Smoltification is accompanied by a transition to more pelagic behavior and surface orientation (Schreck and Li 1984). Netting of fish in the unimpounded mainstem Snake and Columbia Rivers showed a predominantly surface orientation (Mains and Smith 1963; Dauble et al. 1989), as did studies in Snake River reservoirs (L. S. Smith 1982).

Early studies of passage at dams showed accumulation of fish at the surface in dam forebays and a preference for surface outlets (Andrew and Geen 1960; J. R. Smith et al. 1968; Coutant and Whitney 2000). The development of fish bypasses at Columbia basin dams was influenced greatly by observations that fish drawn into deep turbine entrances sought to return to the surface through gatewells (Long 1968b; Marquett et al. 1970; Bentley and Raymond 1976). The natural surface orientation of juvenile salmonids, especially at dam forebays, is presumed to be a principal reason why a surface flow bypass at Wells Dam on the mid-Columbia River has been so successful at passing fish (G. E. Johnson et al. 1992).

Numerous studies have shown that juvenile salmon do not readily find their way downstream past dams with deep outlets (Whitney et al. 1997; Coutant and Whitney 2000). This was the primary factor in failure of fish passage facilities at Brownlee Dam on the Snake River, as well as at a number of other dams in the basin (see the additional discussion in Chapter 7 of juvenile fish passage facilities and mitigation activities at the dams).

Daily Migration Cycles

There is an abundant literature demonstrating alternating movement and holding periods by migrating juveniles within a daily cycle. Northcote (1984), in summarizing research on the mechanisms of fish migration in rivers, noted that most downstream movement is not constant, but nocturnal, except during periods of high turbidity. Jonsson (1991) reviewed the effects of water flow, temperature, and light on fish migration in rivers and noted that many authors have found downstream migrations to occur mainly during darkness. When migration is not completed in a single night, as it might be in coastal rivers, the migrants occupy holding areas during daylight (McDonald 1960; Hartman et al. 1967; Solomon 1978; Hansen and Jonsson 1985). This is when most feeding occurs. These observations have often been confirmed experimentally (see references in Jonsson 1991).

Daily cycles are evident in the Columbia River Basin. Mains and Smith (1963) identified diel periodicity in studies of the undammed Snake and Columbia Rivers in the 1950s (Figure 6.5). There was a notable diurnal periodicity when juvenile salmonid passage was examined at John Day Dam in 1986 (Johnsen et al. 1987). Most juveniles (Chinook yearlings, Chinook subyearlings, steelhead, coho, and sockeye) were caught between sunset and sunrise (Johnsen et al. 1987; Figure 6.6). Although perhaps an artifact of dam passage, the similarity to movement in the undammed reaches studied by

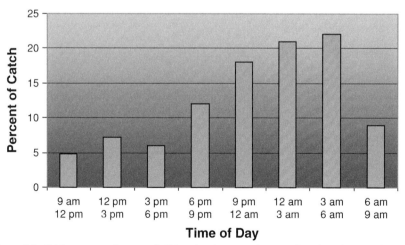

Figure 6.5 Diel patterns of seasonal Chinook salmon catch per unit volume in experimental fyke nets placed in the unimpounded Columbia River at Byer's Landing (near Richland, Washington) in 1955 by 3-hour periods. From Mains and Smith (1963).

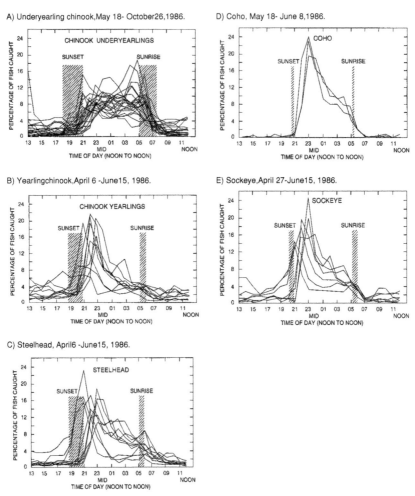

Figure 6.6 Composites of weekly diel patterns of the passage of juvenile salmonids through John Day Dam in 1986, as measured by gatewell counts (Johnson and Wright 1987): (A) sub-yearling Chinook salmon, (B) yearling Chinook salmon, (C) steelhead trout, (D) coho salmon, and (E) sockeye salmon.

Mains and Smith (1963) suggests this is an innate behavior. Laboratory flume studies with fall Chinook subyearlings show day–night differences in tendency to be displaced downstream in changing water velocities (Nelson et al. 1994). This was also seen in New Zealand subyearling Chinook salmon (Irvine 1986), indicating an innate basis for nighttime movement.

Use of Flow Dynamics in Migration

There is increasing evidence that juvenile salmon make use of certain features of flow hydrodynamics in their migration (Coutant 1998, 2001). For example, accelerating flows appear to foster fish movement. Wild and hatchery yearling Chinook salmon and steelhead at the Salmon River and Snake River traps and steelhead at the Clearwater trap show increases in sample counts during and shortly after flow increases (visual inspection of graphs) (FPC 1994; Buettner and Brimmer 1995).

The fluid dynamics literature for rivers suggests many features that may be used by migrating salmonids to assist their migration. These features include surges or stage waves, turbulent bursts, and vortices. Presently though, the advanced development of hydrodynamic theories and practices has not been matched by parallel studies in fish behavior. The somewhat confusing literature on juvenile salmonid responses to flow (rheotaxis) might be clarified if the focus of attention were to be directed to the fluid dynamic structure of flows as orienting mechanisms. The suspicion that swimming fish will exploit vortices to decrease muscle activity and conserve energy (Coutant and Whitney 2000) has recently been demonstrated in laboratory studies of rainbow trout by Liao et al. (2003). The effectiveness of flow baffles for guiding fish at certain spill sites (e.g., Wells Dam) are likely the result of inducing features of fluid flow that are naturally important for fish migration. Future studies of these factors might suggest ways that flow fields in bypass structures and dam forebays might be modified in ways that assist guidance of migrants, as suggested by Coutant (1998, 2001).

Downstream Migration: Active versus Passive

Downstream migration of juvenile salmonids is more complex than their simply being washed downstream by river flows (sometimes referred to as "flushing"). Once migration is initiated, downstream migration is more aptly characterized as a discontinuous movement rather than as the continual linear progression characteristic of a water particle (Figure 6.7).

Physiological and behavioral changes in most anadromous juvenile salmonids cue their increased tendency to move downstream. Larger juveniles approach a time when they are ready to move from the system. There is a large but rather inconclusive literature concerning the environmental and biological cues that stimulate migration (Groot and Margolis 1991). Several studies have shown a general relationship between increased size of juvenile salmonids and selection of greater water depth and/or current velocity (Dauble et al. 1989), although these studies have generally been made in small streams rather than mainstems of large rivers. Fish in deeper, swifter water of

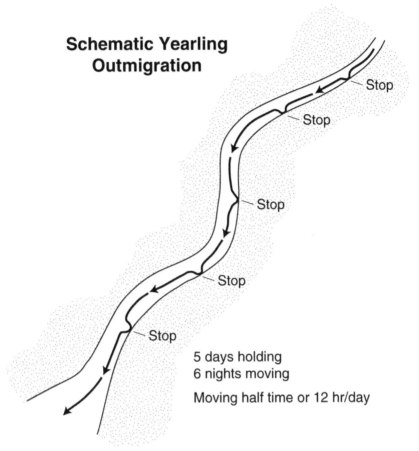

Figure 6.7 A conceptual view of juvenile salmonid downstream migration, which involves periods of movement in the mid-channel followed by stops, which are periods of resting and feeding along shorelines and in backeddies.

tributary streams would thus be more readily transported downstream passively.

When young salmon reach a certain size (or receive other cues, such as length of day), they also transform physically (silvery color, deciduous scales, and change in body shape), physiologically, and behaviorally from the parr stage to the smolt stage that is better adapted to make the transition to saline water, a process referred to as "smoltification" (Hoar 1976). On the other hand, there is evidence that the process of moving downstream can itself lead to development of the characteristics associated with smoltification (Beeman et al. 1990).

Along with the characteristics mentioned above, these transformations include changed swimming behavior and proficiency, lower swimming stamina, and increased buoyancy that also make the fish more likely to be passively transported by currents (Saunders 1965; Folmar and Dickhoff 1980; L. S. Smith 1982). In general, the smoltification process is timed to be completed as fish are near the freshwater to salt water transition. Too long a migration delay after the process begins is believed to cause the fish to miss the "biological window" of optimal physiological condition for the transition (Walters et al. 1978). Nonetheless, the smoltification process is usually identifiable among yearlings after the time they leave their tributary rearing areas.

The concept of migration as mostly passive, taking advantage of downstream displacements by water currents, is initially attractive for interpreting migrations of fish in the Columbia River Basin. Hoar (1954) favored the idea of passive migration of sockeye and coho salmon, which he reasoned were carried by currents when their heightened activity at migration time brought them to zones of water movement. L. S. Smith (1982), using experimental observations of coho salmon, supported the idea of fish orienting mostly head-upstream during emigration while drifting seaward. More recent laboratory flume experiments by Nelson et al. (1994) confirmed swimming behavior by Chinook salmon subyearlings at about one body length per second (bl/s) heading into the current during downstream displacement. This behavior, in experimental fish taken from migrating populations in McNary pool and McNary and John Day Dams throughout the main 4-month migration period, would allow fairly passive displacement. Passive migration has been the predominant view for Atlantic salmon that migrate from Scotland (Thorpe and Morgan 1978; Thorpe et al. 1981) and Maine (McCleave 1978). Thorpe (1982) reasoned that there should be little biological advantage in a migrant expending scarce energy resources by actively swimming. High water discharge in rivers correlates with downstream movement of juveniles in a variety of fish species (Jonsson 1991).

Passive displacement may account for downstream movement, but this seems insufficient for explaining the full migratory behavior of juvenile salmonids. Active downstream movement of sockeye salmon fry after hatching was observed and even attributed to a compass orientation mechanism rather than to simply following currents (Groot 1965; Brannon et al. 1981). Complex behavioral changes both stimulate and maintain migration behavior (Hoar 1976). Many migration studies have involved Atlantic salmon, in which response to currents is complex, and includes a mix of passive and oriented movement (Arnold 1974). Atlantic salmon studies showed that active swimming is used for a considerable portion of the distance traveled even though it may be a small proportion of the time (Fangstam et al. 1993). Most studies just cited identified at most 6 to 9 hours of the 24 during which juveniles moved with the current at a speed more or less consistent with current velocity, often at night. There is an active process of transition between day-time feeding and nighttime movement. L. S. Smith (1982) acknowledged active swimming for only about a third of the time as a possibility in Columbia River salmon smolts. Adams et al. (1995) found that yearling steel-head moved about 50% faster than yearling Chinook salmon through Lower Granite Reservoir under the same flow rates, indicating migration mecha-nisms other than passive drift. The experimental studies by Liao et al. (2003) suggest how the fish can use hydraulic vortices to aid migration rate and con-serve energy.

Subyearling Chinook Migrants

The actual or probable historical distribution of fall Chinook subyearlings in space and time during migration can be reconstructed from several sources. Early accounts (Rich 1920), including quantitative observations at unim-pounded Hanford and Snake River sites (Mains and Smith 1963; Dauble et al. 1989), shoreline seining surveys in unimpounded reaches (Becker 1973; Dauble et al. 1980; Key et al. 1994, 1995), and studies from the estuary below Bonneville Dam (Dawley et al. 1986), provide useful information on unim-pounded conditions. Information on the spatial and temporal distribution in the impounded Snake River is available from Curet (1993), Key et al. (1994, 1995), and J. R. Smith (1974); in the impounded Columbia River at McNary Reservoir from Key et al. (1994, 1995); and in the John Day Reservoir from Giorgi et al. (1990).

Subyearling Migration in Rivers

Before dams, subyearling Chinook salmon used the lower river throughout the summer (Figure 6.8) for a combination of rearing and seaward migration (Rich 1920). Even after dams were built, subyearling Chinook salmon migrated through the reservoirs at relatively slow migration rates through the summer and into autumn (Raymond et al. 1975; Miller and Sims 1984; Johnsen et al; Giorgi et al. 1994). There has been concern over the demon-stration that the time of seaward migration has been lengthened by the effects of lower water velocities in reservoirs than found in unimpounded river con-ditions (Raymond 1968, 1979; Park 1969). The lengthened migration times coincide with general population declines of Snake River fish. Temporal pat-terns of counts of fish passing dams has provided most of this information; there has been little investigation of what behavioral changes may have occurred to the fish in the reservoirs during the delay. The importance of this delay for survival is unclear. Giorgi et al. (1990) have attempted to consoli-date some of this information for John Day Reservoir. More recent syntheses

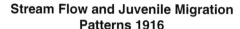

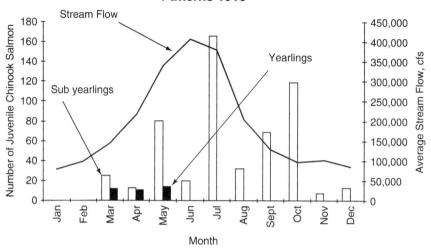

Figure 6.8   Stream flow and juvenile migration patterns from 1916 near The Dalles, Oregon. Note the relationship of the yearling and subyearling emigrations to the natural hydrograph and the 10-month duration of the subyearling emigration period. From Rich (1920).

by S. G. Smith et al. (2002) have helped clarify the situation (as discussed below).

The use of shoreline habitats by juveniles is well demonstrated in the Hanford Reach and is probably a key component in the success of that stock (Figures 6.9 and 6.10; see also discussion of shoreline habitats in Chapter 5). Subyearling Chinook salmon fry drift downstream throughout the river cross section from March to May after they emerge from redds (Dauble et al. 1989; Key et al. 1994, 1995) and move to shoreline areas where they begin to rear. Young Chinook parr occupy large expanses of shoreline areas of reduced current velocity (Dauble et al. 1989; Key et al. 1995) where they feed primarily on emerging chironomids and terrestrial insects (Becker and Coutant 1970; Becker 1973; Dauble et al. 1980). Shoreline or bank aggregations of early Chinook salmon juveniles have been observed in other systems (e.g., Big Qualicum River, British Columbia), with deeper water used as fish grow (Lister and Genoe 1970). Production of aquatic chironomids and terrestrial insects dropping into the water is probably facilitated in the Columbia River Basin by rising waters of the freshet, which inundate large areas of gently sloping cobble bars, sandy shores, and vegetated riparian zones of sloughs and high-water channels (see Figures 6.9 and 6.10, and discussion in Chapter 5). Because laboratory studies have shown that Chinook salmon feeding rates were highest in moderate turbidities and low in clear water

Proximity of Holding and Moving Habitats

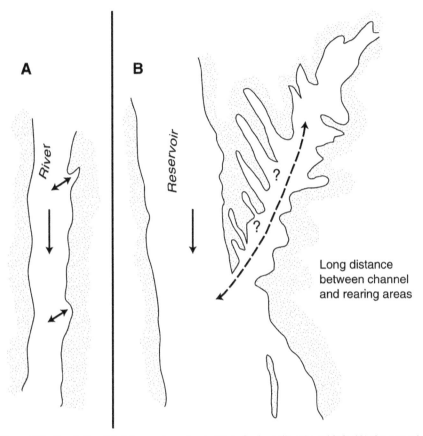

Figure 6.9 Proximity of holding and moving habitats for juvenile salmonids in (A) the natural river, where coves and backeddies are near the main channel, and (B) a reservoir, where flooded tributaries or old river channels create long distances between shoreline feeding locations and the main channel.

(Gregory and Northcote 1993), the turbidity of freshets was probably also important for rearing.

There is a daily cycle of movement. The chronology of subyearing Chinook movement through the nearly 90-km Hanford Reach can be deduced from catches in fyke nets suspended at different depths across the river cross section and shoreline seining and electrofishing (Dauble et al. 1989). Fish move downstream gradually in a diurnal cycle, feeding in shallows in the daytime and moving downstream in deeper, swifter water at night. Peak fyke-net catches in the channel occur at 2200 to 2400 h with fish distributed through-

Figure 6.10   Researchers seining for juvenile fall Chinook salmon in the shoreline habitat of the Columbia River's unimpounded Hanford Reach. Photo courtesy of D. Geist.

out the water column, particularly during the later phases of rearing and migration (Dauble et al. 1989). Fish collections identified an activity pattern that included migration, feeding, and resting periods. Much of the pattern seems to be daily, although an individual fish could spend more than one day in a shoreline area. This rearing-migration pattern moves the fish down-stream at night and to the shoreline during the day, where ample food exists for sustained growth. Because the Hanford Reach is undammed and flows are regulated for the benefit of the fish, the present pattern of juvenile fall Chinook salmon distribution may approximate the historical condition. Chapman et al. (1994a) referred to the phenomenon as "zigzagging." The behavior pattern was described in detail by Hillman and Chapman (1989). We referred to the pattern as "spiraling" in the prepublication draft of *Return to the River*, consistent with concepts of material spiraling in the riverine ecology literature (see Figures 6.4 and 6.7).

Hatchery-released fall Chinook salmon smolts may be less oriented to shorelines than are wild fish. They were less abundant in near-shore areas than were wild fish in studies at Hanford (Dauble et al. 1989). These artifi-cially reared fish may be less inclined to alternate between feeding and migrating, at least in the initial weeks following release from the Priest Rapids hatchery just upstream of the Hanford Reach. This behavioral difference may be significant in determining relative survival during emigration.

Subyearlings in the Snake River

Snake River fall Chinook salmon emerge from the gravel later than at Hanford, with peaks occurring in late April to late May (Connor et al. 1995a, 1995b). They rear in near-shore areas from mid-March through mid-July in the Snake River and in the Clearwater River, depending on emergence dates, with a mid-May to mid-June peak. Based on high percentages of recaptures of tagged fish, they appear to concentrate in particular shoreline areas and stay there for some time (Connor et al. 1995a). As water warms and flows begin to decline, rearing fish move downstream. Since 1991, flow augmentation from Hells Canyon Dam and above has been used to assist these fish in moving past Lower Granite Dam during the summer. Migration past Lower Granite Dam of juvenile salmon tagged with radio frequency identification tages (known as passive integrating transponders [or PIT tags]) that allow identification of individual fish has sometimes been protracted (into early September) but sometimes truncated by late July. These studies, which are continuing, have not yet sought daily patterns in movement.

A daily pattern of downstream migration of subyearlings was documented in the Snake River before it was impounded. Mains and Smith (1963) observed a pattern that was similar to that at the Hanford Reach. During their study, most migration occurred at night, although there seemed to be some subyearlings moving downstream in the main channel at all times of day. Mains and Smith did not examine diurnal patterns of horizontal distribution, but noted high overall catches near shore, where shoreline proximity, not velocity, was stated as the main factor. Daily patterns were also evident in catches of fall Chinook subyearlings emigrating downstream in the Snake River as it entered Brownlee Reservoir in the 1960s, before this population was extirpated (Krcma and Raleigh 1970). This stock migrated mostly from sunrise to 1000 hrs and from 1500 hrs to 1900 hrs. The fact that this timing contrasts with mostly nighttime migration elsewhere suggests that there once might have been stock differences in diurnal timing.

Subyearling Migrations in Reservoirs

Early studies at dams showed that more subyearling Chinook salmon moved through the dams themselves at night than in the day. In research using special bypasses at Bonneville Dam, Gauley et al. (1958) found significantly more subyearling Chinook moving from 1800 hrs to 0600 hrs in four out of five seasons (1946, 1949, 1950, and 1953). Diel movement of migrating subyearling Chinook salmon was shown by Long (1968a) in samples from the turbine intakes at The Dalles Dam in 1960, where the passage at night was 60% to 70% of the daily total. The clear diurnal pattern for subyearlings was evident at John Day Dam in 1986 in all weeks from mid-May to the end of October, although there were always some fish moving during the day (L. Johnson and Wright 1987).

Studies in Snake River impoundments show similar behavioral patterns for subyearlings under reservoir conditions. Snake River fall Chinook were captured in impounded waters upstream from Lower Monumental Dam during emigration. Migrating fish were sampled by gill nets set in relatively shallow (48 ft deep) and deep (96 ft) areas of the reservoir (but there was no sampling along shore). Most Chinook (92%) were taken at night in the upper 12 feet of the central, deep portion of the reservoir (80% of these in the upper 6 ft). Few were collected in the reservoir during the day in either the deep or shallow reservoir station, suggesting that the Chinook salmon were elsewhere, most likely near the unsampled shoreline. These data seem to indicate migration with a daily pattern of high abundance in upper pelagic waters of the reservoir at night (for active migration) and resting or feeding in the shoreline area not sampled in the daytime. This pattern would be consistent with observations at Hanford.

A shoreline distribution of subyearling juvenile Chinook in the impounded Snake River in daytime was confirmed over several years of shoreline seining (agency reports 1986–1993 by D. H. Bennett, Idaho State University) and through 3 years of shoreline seining and open water trawling of Lower Granite and Little Goose Reservoirs by Curet (1993). Slow-velocity, sandy shores were preferred, and artificial shorelines of rock rip-rap were strongly avoided. Curet observed that fish became more oriented to channel waters during the day once shoreline temperatures exceeded 18° to 20°C. Thus, diurnal warming of near-shore shallows could cause some change in inshore–offshore movements in reservoirs during the later spring and summer migration times. Curet (1993) linked these high shoreline temperatures to reduced feeding and higher than normal metabolic demands. It appeared to Curet that subyearling Chinook did not just pass quickly through, but used the shoreline and open water areas of the reservoirs for rearing before migrating farther downriver.

The Snake River and Hanford fish both share the same emigration path in reservoirs of the Columbia River below the confluence of the Snake River. From a study of juveniles feeding in the McNary pool (including a riverine section below Hanford, an intermediate section below the Snake River confluence, and the dam forebay), Beeman et al. (1990) concluded that the river and reservoirs are not used as a conduit for rapid migration, but that there is summer rearing and gradual downstream movement in the reservoir system in much the same way as these juveniles historically used the free-flowing Columbia River. Similarly, Giorgi et al. (1986) found that subyearling Chinook salmon in John Day Reservoir in the early 1980s did not exhibit consistent downstream movement indicative of continual, directed seaward migration. A majority of fish captured by purse seine, marked, and released at transects throughout the reservoir were recaptured at or upstream from the site of release. They were not consistently displaced passively downstream via the current. Although Giorgi et al. (1986) felt that their observed upstream movement was not consistent with the tail-first drift model of migration, there could be more consistency than was appreciated. A scenario can be visualized in which nighttime "drift" in the pelagic zone, alternating with shoreline feeding in the day, actually moves the fish upstream as it weakly swims against a non-existent (or very slow) current. With no orientation other than suspended objects nearby, the fish may be behaving quite normally. Flume experiments by Nelson et al. (1994) showed daytime swimming behavior could exceed the test water velocity (especially in August), thus displacing fish upstream.

Key et al. (1994, 1995) found that the shoreline orientation of subyearling juvenile Chinook salmon in the daytime and low numbers there at night occurred also in a slough of McNary Reservoir, just downstream of the Snake-Columbia confluence. At this point in time and space, the fish had transformed to the smolt stage. Key et al. concluded that the shoreline orientation was more related to fish behavior than to either fish size or environmental conditions (temperatures were not sufficiently high to force fish away from shallows). Their analysis of fish distribution led them to hypothesize that subyearlings in the reservoir situation now move to the bottom in intermediate depths (rather than to the channel), where they become torpid during the night. This hypothesis has not been tested by field sampling at night.

Subyearling Migrations in the Freshwater Estuary

A pattern of spatial distribution of fall Chinook salmon subyearlings somewhat similar to that at Hanford was seen in the tidal freshwater Columbia River estuary below the most downstream dam, where conditions more nearly approximate the pre-dam condition (Dawley et al. 1986; Ledgerwood et al. 1990). Here, the subyearlings from both upriver sources and lower river tributaries were most abundant from May through September, when beach seines were the most effective gear for capturing juveniles (indicating shoreline orientation). Dawley et al. (1986) caught most fish (90%) during daylight hours with peaks during early morning and at dusk. Subyearlings caught in pelagic (open-water) habitats were larger than those collected in intertidal areas. They were in the top 3 m of the water column, and had fewer food items in their stomachs, suggesting active emigration (Dawley et al. 1986). These larger fish tended to be from upriver sources, which suggested they had completed their rearing. Generally, feeding was most intense in the shallow, intertidal areas (McCabe et al. 1986). Subyearlings in shore areas tended to move gradually downstream as they fed in the daytime (McCabe et al. 1986). Ledgerwood et al. (1990) also found a clear daily pattern of abundance of subyearlings in beach seine catches, with a peak about 1.5 hours after sunrise followed by steady catches during daylight and a minor peak 1.5 hours before sunset. Night catches along the shoreline were low. Purse seine catches in the river channel peaked just before sunrise and decreased throughout the day. In general, low night catches in the channel suggested that there was no pronounced nighttime movement.

Migration timing in the upper estuary and the sizes of migrants indicate a migration pattern that is not characterized by constant flushing by high flows. The annual pattern of movement of subyearlings seen by Dawley et al. (1986), in which a few fish moved through the area as early as June and many moved in August, showed that these fish were not migrating with high early-summer flows.

Marked hatchery releases in the upper estuary summarized by Dawley et al. (1986) showed no relationship between rate of downstream migration and river flows, despite an earlier migration of subyearlings from upriver in high-water years than in low-water years. There was, however, an increased rate of movement with increasing fish size. The evidence supports fish remaining in the river until reaching a length of 7 to 8 cm before entering the estuary. The trend toward later timing of migrants in the estuary (Dawley et al. 1986) might be partially explained by a slower growth rate in the river (because of less abundant preferred food and higher than optimum temperatures), rather than changes in river velocity.

For each of these estuary studies, daytime shoreline feeding and night (or twilight) migration would seem to fit the distribution most accurately (perhaps with less nighttime movement in the estuary than in upriver sites), and is consistent with longer estuarine residence shown by Reimers (1973) and slower estuarine than riverine movement shown by Dawley et al. (1986). River flow and velocity seem to be little involved.

Experimental Research on Subyearlings

Experimental results on subyearling swimming behavior by Nelson et al. (1994) were more complex than could be explained by continual, passive, or directed movement. Orientation with the head into the current was the most common observation. As water velocities increased, the number of fish exhibiting this orientation increased. At slower velocities in the 5 to 50 cm/s range studied, fish swam upstream at rates comparable to the experimental water velocity, thus maintaining their position in the flume. As velocities were increased, a threshold velocity of 25 to 40 cm/s was passed at which fish reduced their swimming to speeds of 0.5 to 1.5 body lengths per second (bl/s) and they were displaced downstream. This displacement was not "passive," as even during times of displacement experimental fish were never displaced downstream as far as they would have been by drifting with the current. During all trials, fish rarely drifted without locomotor control. These experimental results are consistent with a holding behavior in low flows (typical of the shoreline feeding part of a spiral) and controlled downstream displacement at high flows (consistent with the downstream movement part of a daily cycle). The experiments also showed that fish tended to swim slower at night, which is the normal time of downstream displacement. This change in threshold for displacement could provide the necessary twice-daily transitions for a spiral or zig-zag migration. Nelson et al. (1994) cite convincing literature to support a behavioral explanation for these observations rather than one based on fatigue (fish would not have become physiologically fatigued by the velocities and length of time exposed in their tests, based on published studies of salmon fatigue).

There were also hints of other relevant behaviors not yet fully explored in the tests by Nelson et al. (1994). There was one day of directed downstream swimming in late May during the normal peak emigration and a selection of highest velocities in the flume for downstream displacement during dates of most active emigration. The authors propose an increased "disposition to emigrate" during this time that would coincide with a change to lower threshold water velocities for a fish to reduce its swimming speed to the minimum orientation velocity of about 1 bl/s. Perhaps the migratory zig-zag or spiral for subyearlings has a seasonal change in periodicity, with a behavioral basis for a longer spiraling length at the times (related to day lengths?) of normal peak river flows.

Management Implications for Hanford and Snake River Subyearling Chinook

The difference in the success of fall Chinook salmon in the Snake River versus the Columbia River at Hanford provides useful contrasts that may be related to rearing and migration habitats. The Hanford stock flourishes (Dauble and Watson 1990), whereas the Snake River stock is listed as endangered and continues to decline (National Marine Fisheries Service [NMFS] 1995). Understanding differences in the habitats and behaviors that promote survivorship of these two stocks may be useful for stemming the decline of Snake River salmon. These stocks share habitat from the confluence of the Columbia and Snake Rivers to the ocean, but differ in their upstream habitats. They may also differ in locations of their ocean residence, which could affect overall population success (A. Giorgi, personal communication).

Beyond differences in the amount and quality of spawning habitat available to the returning adults, the relative success of the two stocks of fall Chinook may be related to the quality and diversity of mainstem habitats available to their juveniles. Hanford Reach fall Chinook have access to shorelines with abundant insect food in the riparian vegetation and flooded cobble beaches, and benefit from stabilization of flows at critical periods. However, Snake River fall Chinook, soon after entering Lower Granite Reservoir, move to reservoir shorelines characterized by eroding soil banks or rock rip-rap, both of which are poor habitats for producing abundant insect prey (Janecek and Moog 1994). By late May or early June, shoreline waters in the Snake River reservoirs are often too warm for young salmonids, and juvenile feeding must occur in pelagic waters where their preferred food is scarce. In these reaches, pelagic Cladocera, not shoreline chironomids, are the dominant food item for subyearlings, even though chironomids provide the greatest caloric value (Rondorf et al. 1990). Subyearlings shift their diet to smaller, less preferred Daphnia species in embayments of Lake Wallula (behind McNary Dam) due to the prey's higher densities and ease of capture in the pelagic environment. Curet (1993) demonstrated that juvenile fall Chinook in Lower Granite and Little Goose Reservoirs were not obtaining sufficient food to account for much more than basal metabolism (7% greater than estimated maintenance ration), which could be one of the factors contributing to their lack of population success.

The shoreline-feeding portion of the migration behavior may be most critical for long-term survival in the early stages of rearing and migration of subyearling Chinook salmon. It is at this time when the Snake River and Hanford stocks differ most. It could be argued that superior growth and energetic reserves of Hanford fish acquired in the high-quality riverine habitat of the undammed reach with stabilized flows within and just below the spawning areas are enough to carry them through the poorer food resources of downstream reservoirs, whereas the Snake River subyearlings are impoverished nearly from the start by barren shorelines of Lower Granite and Little Goose Reservoirs. Even though subyearlings are well fed and have grown rapidly in the reach below Hells Canyon Dam (Rondorf, personal communication), they may not endure the poor migration habitats of the Snake River reservoirs. This hypothesis is controversial and is presently under investigation. Studies by Muir et al. (1996) suggest that the condition of smolts emigrating out of the Lower Snake River is good, whereas recent work by James Congleton and students of the University of Idaho show physiological stress and poor condition factors for Snake River smolts, particularly for those migrating in the summer months (presentation at the US Army Corps of Engineers' Anadromous Fish Evaluation Program, 1999 Annual Research Review, Walla Walla, November 17, 1999).

As the migration behavior of subyearling Chinook is better understood in relation to smoltification, parts of McNary Reservoir may be found to be critically important to survival of the Snake River stocks. From the mouth of the Snake River to nearly the Walla Walla River (a distance of about 14.5 km), the Snake River side of the Columbia River (i.e., the east shore) is a series of sloughs and wetlands unlike the opposite shore (Asherin and Claar 1976). These wetlands are probably the combined result of an ancient Snake River channel (Burbank Slough) and sediments from the present Snake River confluence that have been distributed in two major sets of bars down the Columbia River. Key et al. (1995) conducted diurnal sampling of subyearling Chinook salmon in Villard Slough in this complex, and much of the remainder of sampling appears to have been carried out in this reach. Smolts from the Snake River appear to be drawn into these long slough areas to feed during the day, but are apparently unable to return to the channel at night to resume downstream drift. One can speculate that this trapping on the Snake River side (but not on the side occupied by flows from the upper Columbia River), in combination with the advanced state of smolt development of Snake River emigrants, could be responsible for a disproportionate loss of Snake River fall Chinook at this point compared with the Hanford stock coming down the Columbia channel (along the west shore) at the same time.

The Snake River Canyon reach, which is physically dominated by the canyon itself, may never have had the ecological complexity, habitat diversity, and food web productivity that existed in the downstream lower-gradient alluvial reaches, such as the Hanford Reach. Consequently, Snake River fall Chinook may have evolved mechanisms to partially compensate for naturally poor feeding habitat during emigration through the lower Snake River mainstem. Taylor (1990), in his review of 160 Chinook salmon populations ranging from California to Alaska, Kamchatka, and New Zealand, indicated that increased migration distance selects for larger size at seaward migration, due to increased metabolic demands of migration. Recent research has, indeed, found the Snake River subyearlings in the unimpounded reach between Lower Granite Reservoir and Hells Canyon Dam to be larger than Hanford fish at comparable dates despite emerging later from the gravel and having more distance yet to travel (Key et al. 1994).

The extent to which the dissimilarity between stocks in their emergence timing and early size could be due to temperature differences has not been determined (Hells Canyon Dam discharges are warmer in winter and cooler in spring and summer than temperatures at Hanford). But despite this apparent growth rate and size advantage, the Snake River stock now does poorly.

Management Risks for Subyearlings: Passive versus Active Migration Modes

There may be risks for subyearling salmon associated with management actions based on a constant flushing (passive drift) model. Because subyearlings spend a large amount of time feeding in shoreline habitats, management alternatives for the mainstem that focus on increasing water velocities in the main channel through reservoir drawdowns or flow augmentation need careful evaluation. In the early 1990s, lowering of reservoir elevations in the spring freshet season was one of the principal methods proposed for attaining high water velocities thought to be conducive to constant flushing in the mainstem

Snake and Columbia Rivers (NPPC 1994a, 1994b; NMFS 1995). The logic behind the seasonal drawdown proposal is that a smaller volume of water in a reservoir would translate to a more rapid movement of a unit volume of water through it, including contained fish. However, seasonal reservoir drawdowns to attain the presumed benefits of spring flows for constant flushing behavior in yearling emigrant salmon are likely to negatively impact the shoreline habitat needed by subyearling salmon. Because the critical habitat for subyearling survival most likely is flooded shorelines, complex backchannels, and other vegetated habitats that are productive of invertebrate food, temporary seasonal drawdowns that extend into late spring and summer could be counterproductive and actively decrease food availability for the later emigrating juveniles. As an experimental drawdown of Lower Granite Reservoir in 1992 showed, drawdowns created long expanses of muddy shorelines that would have little or no food available for subyearling salmon during the shoreward portion of their daily migratory spiral. Moderate flooding of a stable, vegetated riparian shoreline is more compatible with the fall Chinook salmon's migration behavior and ecology. In contrast to seasonal drawdowns, permanent drawdown would allow riparian vegetation to develop. Seasonal flooding of this habitat would enhance the river's productivity during emigrations, presuming flows were stabilized to some degree.

High levels of flow, when not coupled with flooding productive shoreline areas, would appear to reduce food availability for juvenile fall Chinook in the present reservoir system. Rondorf et al. (1990) observed a reduction of the present main food item, pelagic cladocerans, in mid-reservoir and dam forebay stations during June, which coincided with peak seasonal flows. High flows apparently flushed away these planktonic food items, which were the main replacements for the insects (midges and caddisflies) eaten in the riverine section below Hanford.

Yearling Chinook Migrants

Most spring and summer Chinook salmon from the Snake River drainage are of the stream type, migrating to sea rapidly after one year in freshwater. However, Curet (1993) notes personal observations by Idaho Department of Fish and Game personnel that some subyearlings in the Snake River are of spring Chinook origin. Mattson (1962) observed three distinct migrations in Willamette River spring Chinook in the 1940s: in their first spring and summer as subyearlings, in fall as a migration of subyearlings at time of heavy rains, and in spring as a movement of yearlings. Spring Chinook from the mid-Columbia tributaries migrate as yearlings, as do the spring Chinook reared in mid-Columbia hatcheries. There are suggestions that some now-extirpated stocks of spring Chinook had primarily subyearling emigrations

(J. Lichatowich, personal communication). Summer Chinook salmon in the mid-Columbia above Hanford are allied with the fall runs rather than with the spring runs, as in the Snake River system. Whereas subyearling Chinook salmon exhibit a slow downstream migration that we have seen is composed of downstream movement interspersed with shoreline feeding on a daily cycle, the yearlings are commonly thought to have a very different migratory pattern, consisting of a rapid emigration of fish from the river during the spring freshet, which is consistent with flushing behavior.

Evidence for Flushing of Yearlings

Yearlings are normally in the process of smoltification as they migrate downstream. This process of physiological change begins 20 to 30 days after river migration begins (Beeman et al. 1990). Decreased swimming performance (and greater ease of passive movement by currents) during smoltification seems to be a part of their emigration strategy (L. S. Smith 1982).

Wild/natural spring Chinook from Idaho move rapidly downstream with spring flow in the unimpounded tributaries. In all years studied (1988–1992) by Kiefer and Lockhart (1995), wild spring Chinook salmon smolts from the upper Salmon River were stimulated to migrate in spring by increases in discharge (often storm events), and their peak of arrival at Lower Granite Dam coincided with peaks in flow there. Such results suggest a flushing mechanism, whether passive or active. Similar results were obtained for spring salmon smolts tagged in the Middle Fork Salmon River (Matthews et al. 1992). There was also a downstream movement of parr in autumn, stimulated by rapid declines in temperature (Kiefer and Lockhart 1995). Higher percentages of parr emigrated from higher elevations (harsher climate). Natural migration in Snake River tributaries must be somewhat slower than water flow; otherwise, smolts stimulated to emigrate at the first increase in discharge would not arrive at the first mainstem dam on the Snake River at peak flow (Kiefer and Lockhart 1995).

Similarly, rapid emigration of wild yearling smolts was observed between an outmigrant trap on the Salmon River and either a Snake River trap at Lewiston or Lower Granite Dam in 1993 (Buettner and Brimmer 1995). A two-fold increase in discharge increased migration rate to Lower Granite Dam by 5.2 times. Hatchery and wild Chinook were shown to be capable of traveling between the Salmon River and Snake River traps (164 km) in 24 to 30 hours.

Telemetry studies by Schreck et al. (1995) showed clear periods of flushing and directed downstream swimming. A majority of fish at these times moved at rates faster than measured water velocities, particularly in 2 years when the radiotelemetry was conducted during prominent high-water freshets. When flows were low or declining, fish usually moved more slowly than the water.

Many fish moved uniformly as a group, although the lead fish and the order of the others changed numerous times, suggesting differing lengths of time spent in resting and feeding. Some fish migrated considerably more slowly than the majority, remaining in the upper river for considerable lengths of time following tagging and release.

Migration rates varied with water velocities (Schreck et al. 1995). This occurred along the Willamette River as fish generally moved more rapidly in the upstream zones of more rapid water flow. They also moved more rapidly during times of high flow than during times of lower flow in any one year. During non-freshet spring periods (3 of 5 years studied), fish moved more slowly than the water over 24-hour periods. High and rising flows, however, appeared to stimulate an emigration of fish from the river in a manner consistent with flushing behavior. At freshet times, fish appear to have long spiraling lengths and thus exit from the system quickly.

Flow Structure as an Aid to Migration for Yearling Chinook

Accelerating flows and hydrodynamic features such as waves, surges, turbulent bursts (Figure 6.11), or vortices (Figure 6.12) appear likely to assist the migration rate of yearling migrants (Coutant 2001). Yearling Chinook salmon on the unimpounded Snake River (Mains and Smith 1963) and the Willamette River (Schreck et al. 1995) have been observed to move on the increasing arm of the freshet. Similarly, Hesthagen and Garnas (1986) showed that significantly more Norwegian Atlantic salmon migrated when the discharge was increasing (with a drop in temperature) than under the opposite conditions. Laboratory studies on rainbow trout have shown that these fish can use vortices to decrease muscle activity and swim more efficiently (Liao et al. 2003), which probably occurs in Chinook salmon, also.

Our analysis of data from the FPC (1994; Berggren and Filardo 1993) and Buettner and Brimmer (1995) suggest that fish movement increases in the Snake River system with accelerating flow (Figure 6.13). Wild and hatchery yearling Chinook salmon and steelhead at the Salmon River and Snake River traps and steelhead at the Clearwater trap show increases in sample counts during and shortly after flow increases. The effect seems to be present still at Lower Granite Dam, but not downstream at Snake River dams (the wild yearling Chinook index was not included in the 1993 report for Columbia River dams). Wild steelhead seem to show the effect in FPC data from McNary, John Day, and Bonneville Dams.

Achord et al. (1995b) noted a historical pattern of migration on rising water flow in Snake River Chinook yearlings, with the pattern still evident in PIT-tag detections at Lower Granite Dam of spring Chinook tagged the previous summer as parr. Lower dams did not show the historical pattern; migration coincided with peak flows. For summer Chinook yearlings, the

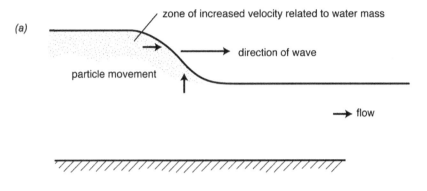

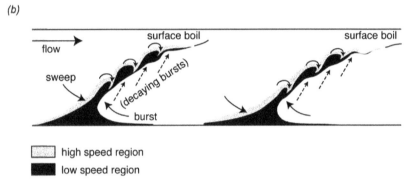

Figure 6.11   Side views of hydraulic features of rivers that probably aid downstream smolt migration. (a) Stage waves and surges have higher velocity at their leading edges relative to the water mass; (b) turbulent bursts have a high-speed ejection of water away from a rough riverbed that creates "boils" at the surface. Both features provide zones of increased velocity that smolts may select to increase migration speed with minimal swimming. Not to scale. Turbulent bursts after Leeder (1983).

main passage of tagged fish was during rising flows at all three dams. The evidence for a flushing mechanism of migration (discussed above) generally includes observations of migration on rising flows, especially freshets.

   With increasing evidence that yearling Chinook salmon move downstream on rising flows (see references to migration with freshets cited above and by Northcote 1984), and similar observations for steelhead, it is tempting to suggest that they may be adapted to catching the stage wave (flood surge) as well as the water mass. Rapid increases in flow or other disturbances in a channel generate a moving surge or stage wave downstream that is recognized in the field of fluid dynamics (Albertson and Simons 1964). Such surges or waves

(a)

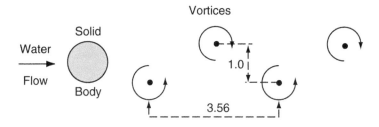

(b)

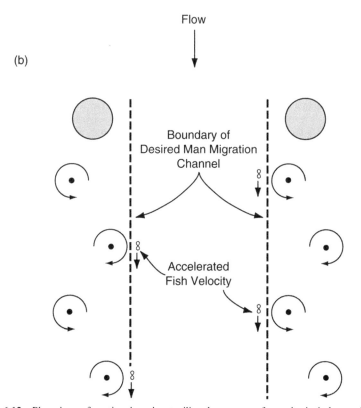

Figure 6.12 Plan views of vortices in a river trailing downstream from physical obstructions. (a) A double row of vortices shed from a single obstruction (von Karman train). The distance between sequential vortices is generally about 3.5 tomes the lateral spread. (b) A set of vortices from paired obstructions in a channel, with a zone of accelerated flow between obstructions. Downstream migrating fish are suspected to catch the accelerated flows to speed migration.

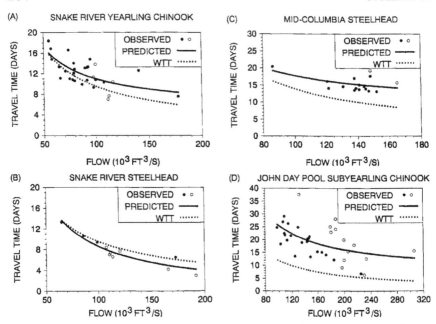

Figure 6.13 Observed fish travel times with summary regression lines and an estimate of water travel time (WTT) over a range of river flow rates for (a) Snake River yearling Chinook salmon, (b) Snake River steelhead trout, (c) middle Columbia River steelhead trout, and (d) lower Columbia River (John Day pool) subyearling Chinook salmon. Open circles 1981–1983; solid circles 1984–1991. From Berggren and Filardo (1993).

move ahead of the main water mass and at rates faster than water particle movement (which also accelerates as stage increases). Koski (1974) found that the velocity of the wave in the Snake River in Hells Canyon was 12.9 fps at 7,700 cfs and 11.4 fps at 5,000 cfs, whereas the average velocity of the water mass was 2.3 fps at 7,700 cfs and 1.7 fps at 5,000 cfs.

Smolts adapted to migrating on moving surges would get both a directional cue and an assist that could move them, too, at rates faster than water particle movement. Telemetry studies in the Willamette River (Schreck et al. 1995) showed spring Chinook yearlings accelerating to faster than water velocities in swift, shallow reaches (where small waves would be expected to merge into larger surges). Conversely, as depth increases, the wave height decreases and waves have less tendency to pile up as surges or bores. Reservoirs would thus inhibit the continuation of surges begun in riverine sections. Waves in slowly moving reservoir waters could easily propagate upstream as well as downstream. A fish could lose both the directional cue and assistance in moving downstream. Some specific hypotheses for this phenomenon are discussed later in the chapter.

The role of hydrodynamic features other than thalweg velocity in fish emigration needs to be further explored for a proven link to such features as stage waves, turbulent bursts, and vortices. A greater understanding of hydrodynamic features may offer opportunities for water management that could be more effective in moving fish with less water than would current applications, such as flow augmentation. Some ways to induce and use turbulent flow to enhance fish guidance have been proposed (Coutant 2001).

Yearling Chinook in Reservoirs

Our concepts of how yearling Chinook salmon migrate through reservoirs is undergoing significant change. Several studies in the 1990s left a somewhat conflicting picture of smolt movement in reservoirs. A particular problem has been reconciliation of studies of bulk water and fish movement with more detailed studies of individual fish, using PIT tags and radiotelemetry. Some hypotheses to sort out these data are given later in the chapter.

The FPC, for example, summarized water transit time as a function of discharge (Figure 6.14) and fish median travel times over 6 years for yearlings passing through Snake River and mid-Columbia River reservoirs (Figure 6.15) that show fairly clearly that in the Snake River and lower Columbia mainstem, fish move faster at higher flows during the migration period, especially evident at lower flow ranges (FPC 1994). They have continued to add years to this analysis (see www.fpc.org). These studies use average flows and median travel times for large groups of fish. L. S. Smith's (1982) postulation that smolts swim weakly upstream, and thereby move downstream tail-first at a velocity less than that of water movement, has been used to explain the difference between water particle travel time and smolt travel time (Berggren and Filardo 1993).

Other studies have emphasized individual fish. For example, Buettner and Brimmer (1995) chronicled travel time and migration rate of PIT-tagged wild Chinook salmon through Lower Granite Reservoir. They calculated that a two-fold increase in discharge increased migration rate by 4.1 times. This change occurred when flows were increasing from about 60 to 160 kcfs. However, as flows decelerated later in the season, travel rate slowed markedly in a pattern that did not conform to the flow-migration rate relationship seen during accelerating flows. Thus, it was thought that a pattern of migrating largely on accelerating flows might persist in reservoirs as well as rivers. Radiotelemetry of individual Chinook salmon smolts has shown a diel periodicity of movement. For fish tagged and released upstream of John Day Dam, both arrival at the dam and passage through it occurred on a diel cycle, with peaks near dusk (Giorgi et al. 1986). A pattern of alternating movement and rest appears to be well established for reservoirs close to these dams.

Early studies at dams identified a clear diurnal periodicity in passage of yearlings. Gauley et al. (1958) found significantly more yearlings migrating through a Bonneville Dam bypass in 4 out of 5 years in the 1940s and 1950s during nighttime hours than during daytime hours. Long (1968a) found about 94 % of yearling Chinook salmon passed The Dalles Dam in nighttime hours in 1960. Yearling Chinook salmon passed John Day Dam mostly at night, with prominent peak movement between sunset and midnight in all weeks between early April and mid-June 1986 (Johnsen et al. 1987). Radiotelemetry of individual Chinook salmon smolts has shown a diel periodicity of movement. For fish tagged and released upstream of John Day Dam, both arrival at the dam and passage through it occurred on a diel cycle, with peaks near dusk (Giorgi et al. 1986). A pattern of alternating movement and rest appears to be well established for reservoirs close to these dams.

The otherwise consistent diel pattern was not borne out in studies of PIT-tagged spring and summer Chinook yearlings at two Snake River dams and McNary Dam in 1992 or 1993. Achord et al. (1995b) found diel patterns in the fish bypass systems to be weak, inconsistent between dams, and often the reverse of the normal pattern; that is, peaks often occurred in the daytime. The anomaly, although not well understood, could signal a breakdown or a variation of the usual diel migration in these reservoirs.

Radio-tagged smolts released at John Day Dam traversed the The Dalles pool at speeds of about 2.0 m/h and usually did not stop in the reservoir before arriving at The Dalles Dam forebay (Snelling and Schreck 1994). After passing the dam volitionally (through the ice-and-trash sluiceway or through spillways), nearly one third held in downstream areas.

The progressive increase in smoltification of Chinook salmon yearlings with time in the migration season appears to correlate with depth of travel and, thus, changes in fish guidance efficiency at dams (Giorgi et al. 1988). More thoroughly smolted fish were caught in the tops of fyke net screens over turbine intakes, whereas less thoroughly smolted ones were caught nearer the bottom in three of four test dates. Decreases in swimming performance observed during smoltification of coho salmon (Glova and McInerney 1977; Flagg and Smith 1982) also are consistent with the results of these collections. These tests suggest an increased tendency of more developed fish to flush, at least during the movement period. The results also are consistent with studies of Atlantic salmon, which increase their buoyancy by filling the swim bladder in an apparent effort to aid the transition from bottom dwelling to pelagic existence during migration (Giorgi et al. 1988).

Degree of smoltification clearly affected travel times of yearling Chinook through Lower Granite pool and the correlation of travel times to changes in flow (Beeman et al. 1990; Giorgi 1993). Whereas fish with low levels of ATPase (beginning of smoltification) traveled the reservoir length slowly and showed a marked increase in travel times at lower flows, the more smolted fish with high ATPase levels had a nearly uniformly rapid rate of movement over all flows. Slowing was seen only at the lowest flows. Cramer and Martin

(1978, as reported in Giorgi 1993) observed that larger Rogue River Chinook salmon migrated fastest. Viewing migration as a spiraling or zig-zag event suggests that the less smolted fish could stop to rest more often or for longer durations than the more smolted fish, which may move more continuously (rather than just at a faster speed). These alternatives could be tested with radiotelemetry.

Reconciliation of numerous disparate study results of yearling Chinook in reservoirs seems to depend on a better understanding of reservoir hydraulics in relation to hydropower operations. The hypotheses that follow later in the chapter may help resolve the differences.

Yearling Chinook in the Estuary

Studies of migration in the upper (tidal fresh water) estuary are generally consistent with the riverine studies. A diel pattern of movement in the upper estuary seems to be prevalent, although somewhat different from that in the mainstem river. In the upper Columbia River estuary at Jones Beach (Rm 46), Dawley et al. (1986) and Ledgerwood et al. (1990) found that the majority of yearling Chinook salmon migrated mid-river (few were caught in beach seines; more were caught in pelagic purse seines). Their migration rates were about the same in the estuary as in the river. Peak catch was mid- to late morning. After a period of low catches between dusk and midnight, there were larger catches (but still fairly low) during the rest of the night. The authors conclude that because mid-river-oriented yearling fish do not appear in shoreline areas during darkness, when migration rates are low, they probably hold near the bottom in deep areas of low current velocity. The yearlings were feeding, as evidenced by stomach contents. From release to recapture, groups of yearlings analyzed by Dawley et al. (1986) did not show movement rates that were well correlated with river flows (in data that spanned very high to very low flow years). Despite differences in timing between the river and estuary, there is evidence of a periodic, rather than a constant, flushing character to the migration.

Population Contrasts for Yearling Chinook: Snake and Willamette Rivers

As with subyearling Chinook, it is useful to look for well-studied populations that differ in their success and compare their migratory behavior and habitats. A contrast as clear as between Hanford Reach and upper Snake River subyearling fall Chinook salmon populations is not available for yearlings. It seems reasonable, though, to compare the successful Willamette River spring Chinook salmon (a population that does not pass mainstem dams) with the endangered Snake River spring/summer Chinook that pass eight dams on the

Snake and Columbia Rivers. Some comparable study techniques (telemetry) have been used, although data are sparse.

Alternating periods of downstream movement and resting, followed by periods of resting and feeding was an evident behavior in the Willamette River spring Chinook tracked in their downstream migration through most of the undammed river from Dexter Dam upstream of Eugene to Willamette Falls near Portland (Schreck et al. 1995). Fish fed well, predominantly on immature insects characteristic of drift. In contrast, yearlings from the Snake or upper Columbia swam the length of The Dalles pool without stopping (Schreck et al. 1995). Migration was interrupted at the dam forebay, but fish maintained an active searching behavior, rather than a holding (resting/feeding) one. Only one route of passage at the dam allowed fish to find and use holding areas near islands. Examination of the Snake River reservoirs shows few, if any, habitats that would qualify as normal holding areas, based on the limited data on habitat suitability from the Willamette River and The Dalles tailwater. Although lack of a flow appropriate to support constant flushing behavior in the Snake River has been viewed as the critical missing habitat factor for its unsuccessful salmon populations, it may be that the lack of both high, accelerating velocities and suitable habitats for resting and feeding are equally important. Further data collection and analysis of the situation with these two populations may lead to results useful for management in the Snake River.

Other Salmon and Steelhead Trout

Coho Salmon

Coho salmon migrations have been little studied in the Columbia and Snake Rivers. Coho were declared extinct in the Snake River Basin in the late 1980s. They were also absent from the mid-Columbia reach for decades until efforts at reintroduction by the Yakama Tribe, with funding from the Bonneville Power Administration, have brought some returns of adults in 2003 and 2004. Most fish in recent years have originated from hatchery stocks in the lower and mid-Columbia River and migrate as yearlings. Due to poor returns, mid-Columbia hatchery rearing of coho was terminated in the early 1990s.

In the Columbia River estuary at Jones Beach, Dawley et al. (1986) and Ledgerwood et al. (1990) found coho salmon in both beach seine and channel purse seine catches. There were erratic changes in numbers of beach seine catches through the day and generally low catches at night. Most fish were caught in beach seines between 0830 and 1430 h, with peak catches in mid-day. Channel samples showed little day–night differences except for a sharp peak just after sunrise. The data suggest schools of fish moving in both areas, but near shore in the daytime. Marked releases of coho showed travel in the estuary at rates about 40% faster than in the river, suggesting some use of tidal currents to aid migration. Movement rate was not correlated with river flow.

As with other species of salmon, coho showed a diurnal passage pattern at dams. Studies at John Day Dam in 1986 revealed almost all coho moving at night with peak passage shortly before midnight (Johnsen et al. 1987). Considerable passage occurred through the night until shortly after sunrise. There is much uncertainty regarding this species, but its minor status in the Columbia River mainstem and rather complete hatchery dependence makes study and management less important than for other species.

Sockeye Salmon

Juvenile sockeye salmon emigrate as one-year-olds from the upper Columbia River, principally from Lakes Osoyoos and Wenatchee in the Okanagan and Wenatchee Rivers respectively (Fryer 1995). One other stock, the Snake River stock from lakes in the Stanley Basin of Idaho, now restricted to Redfish Lake, are on the endangered species list. Their abundance is extremely low, so the juveniles are rarely seen at downstream sampling locations. Historically, sockeye salmon existed in all moraine lakes in the Stanley Basin of Idaho (Salmon River drainage) (Evermann 1895), in lakes in the Yakima River Basin, and in the numerous large lakes in the upper Columbia River Basin, as well as the three named above. The lake area in the Columbia Basin now open to sockeye in the Columbia Basin is approximately 5% of the area formerly available to them (Mullan 1986).

Netting in the Hanford Reach found most emigrating juvenile mid-Columbia sockeye salmon at night (2200 to 0400 h) in the deepest part of the channel, along with yearling Chinook (Dauble et al. 1989). Where these fish were located in daylight hours was unexplained.

Sockeye smolts at John Day Dam migrated with a distinct diurnal cycle in studies in 1986 by Johnsen et al. (1987). There were daily peaks shortly after sunrise. Passage rates during much of the night were similar to daytime rates early in the migration (late April–early May) but much higher in all weeks thereafter until mid-June. Earlier dam passage studies (Gauley et al. 1958; Long 1968b) did not tally sockeye. Giorgi (1993) observed that the current low level of the Snake River stock, despite some PIT tagging of Redfish Lake juveniles, meant that it was unlikely that there would be sufficient data to investigate effects of flow on migration times and survival for many years afterward.

Much of what we know about sockeye salmon migration has come from extensive research on the species in British Columbia. Sockeye smolt migration in British Columbia has been shown to peak at dusk and dawn (Groot

1965; Hartman et al. 1967). Speed of migration in British Columbia sockeye smolts changed with time of day, and the net displacement of fish increased as the season progressed (W. E. Johnson and Groot 1963). Downstream migrating fish tend to rise to the surface (Groot 1965; McCart 1967). Smolts entering a river from a lake swim actively with the currents (Groot 1982). Groot (1982) considered sockeye salmon migration to be a number of "hops" during which fish rise to the surface during peak times of activity and return to greater depths during periods of lower activity.

Steelhead Trout

Steelhead populations have been crossbred and transferred extensively throughout the streams of both Oregon and Washington (Royal 1972; Reisenbichler and Phelps 1989; Reisenbichler et al. 1992). They spawn widely throughout the Columbia River basin tributaries. Thus, the ability to distinguish stock-specific migratory behaviors has been compromised. Therefore, generalized species' responses are the most germane. The steelhead has the reputation of being a fast migrator and a species that would be aided by flows appropriate to support constant flushing behavior (Berggren and Filardo 1993).

Yearling or age 2 steelhead migrate downstream in the mid-Columbia River from spawning tributaries and upstream plantings from hatcheries (Dauble et al. 1989). As in the case of spring Chinook and sockeye salmon, steelhead were found at night (2400 to 0400 h) in the deep part of the Hanford main channel (Dauble et al. 1989). Some were electroshocked in shoreline areas but not enough to establish a diurnal pattern. Diurnal variation in appearance in the deep main channel suggests that there may be a cyclic pattern of migration.

Massey (1967) observed diurnal periodicity in steelhead emigration at Willamette Falls, Oregon, based on sampling of industrial shoreline water intakes. Peak movement was noon to 3 pm, with a minimum from midnight to 3 am. The majority of these fish moved downstream near the center of the river. Andrews (1958) noted that wild steelhead smolts in the Alsea River, Oregon, moved both day and night, but the most rapid movement was just after sunset and just before sunrise.

Northcote (1962) observed the downstream movement of rainbow trout in streams with infra-red light and concluded that the majority were heading downstream, many were at or near the water surface, and they swam at a speed greater than the surrounding water. This agrees with travel time data for Snake River steelhead presented by Berggren and Filardo (1993) that showed movement faster than water travel time. Rainbow/steelhead thus appear to be adapted to the flush, and to improve upon it by active swimming, at least for part of the day. As suggested above for yearling Chinook salmon, the downstream migrants may be adapted to catching the stage wave as well as the moving water mass (Pacific Northwest River Basins Commission 1974).

In the upper Salmon River, which is a major production area for natural summer steelhead, smolts behaved similarly to spring Chinook (Kiefer and Lockhart 1995). They began to emigrate in spring with the first rising flows and arrived at Lower Granite Dam with the peak flows. There was also an autumn downstream displacement of age 2 fish from higher elevations that seemed stimulated by falling temperatures.

In PIT-tag studies by Buettner and Brimmer (1995), wild steelhead moved rapidly downstream in the upper Snake River system and increased their migration rate about proportionately to changes in flow. A two-fold increase in discharge increased migration rate by two times between the Clearwater trap and Lower Granite Dam and 2.1 times between the Salmon River trap and the dam. Both river and reservoir passage were included in these estimates.

Migrating steelhead smolts feed on their way to the ocean. Royal (1972) found most migrating steelhead in the Alsea River, Oregon, both wild and hatchery, had food in their stomachs. Aquatic insects were the main food items.

As with Chinook salmon smolts, radiotelemetry of steelhead smolts has identified holding behavior as well as rapid downstream migration. Ward et al. (1994) observed holding behavior in some steelhead smolts even though most migrated through the 15.3-km Portland harbor in 1 to 2 days. Snelling and Schreck (1994) found that smolts released upstream and downstream of The Dalles Dam searched out a place to hold in the riverine sections just downstream. The holding areas were eddies near islands, the same places used by yearling Chinook. These sites contrasted with the migration corridor in the deep channel. The authors related holding to stress, but it may reflect a normal pattern of migration.

In the estuary, Dawley et al. (1986) observed that steelhead traveled 50% faster than they did in the river. This observation is especially interesting in light of riverine migrations by steelhead being more rapid than water travel (Berggren and Filardo 1993). These fish may use tidal flows to their advantage, as has been seen in other species.

In Lower Granite Reservoir, Buettner and Brimmer (1995) found the rate of migration of wild steelhead also to be flow dependent. Statistical analysis of 5 years of data showed that a two-fold increase in flow increased migration rate by 2.5 times. Such data have been interpreted as support for a constant flushing mode of migration. As with Chinook salmon yearlings, however, detailed analysis of the data for 1993 shows a slowing of migration on deceleration of flows that does not conform to the flow-rate relationship during accelerating flows.

In the impounded Snake River, J. R. Smith found most steelhead migrating in the upper 36 feet (1974). About three quarters of those caught were taken at night (between dusk and dawn). There was no indication of where these fish were in the daytime.

Yearling steelhead were identified in early studies at dams as having a diurnal pattern of migration, with most passing at night. Studies at a Bonneville Dam bypass by Gauley et al. (1958) showed this pattern in four out of five seasons in the 1940s and 1950s. Long's (1998a) studies of turbine passage at The Dalles Dam showed 80% to 90% of yearling steelhead passed in the night. The steelhead pattern of passage at John Day Dam from early April to mid-June 1986 showed most fish traveling at night with prominent peak migration times shortly before midnight (Johnsen et al. 1987). These consistent patterns strongly suggest a spiraling migration behavior in which habitat other than main channel flow is also important.

Effects of Flow on Rate of Migration

Snake River

Chinook and Steelhead

Within the Snake River and as far as McNary Dam or John Day Dam on the mainstem, the downstream migration for both yearling Chinook salmon (Figure 6.13a) and steelhead (Figure 6.13b) is faster at high flows than at low flows (Raymond 1968; Sims and Ossiander 1981; Berggren and Filardo 1993; McConnaha 1993; Connor et al. 1994; Maule et al. 1994; Achord et al. 1995a, 1995b; Bvettner and Brimmer 1995; S. G. Smith et al. 1997a, 1997b). Impoundment of the Columbia and Snake Rivers has decreased the migration speed of yearling Chinook salmon and steelhead (Giorgi et al. 2002). According to these authors, Ebel and Raymond (1976) estimated that following dam construction on the lower Snake River yearling Chinook salmon and steelhead took almost twice the time (about 65 days) to reach The Dalles Dam from the Salmon River in Idaho, as had been required prior to dam construction.

Buettner and Brimmer (1995) found that during the early part of the season when flows were increasing, travel time of radio-tagged spring (yearling) Chinook in Lower Granite Reservoir was reduced by high flows, but that later when flows were declining, travel time slowed. Achord et al. (Achord et al. 1995a, 1995b) also concluded that the principal portion of the outmigration of spring Chinook smolts in Lower Granite Reservoir occurred during the early phase when flows were increasing. On the other hand, using data from recovery of PIT tags and a larger body of data and measuring travel time through a longer reach (from above Lower Granite Dam to McNary and John Day Dams), S. G. Smith et al. (1997a) speculated that faster travel time was associated with changes in fish physiology as smoltification progressed.

Connor et al. (1995a) found travel time for hatchery origin subyearling Chinook through Lower Granite Reservoir was reduced with larger size at release, higher volume of flow, and higher water temperature. Giorgi (1993) found no relationship of flow with travel time of subyearling Chinook in any of the 3 years they conducted their study in John Day Reservoir, but found release date and/or temperature did affect travel time.

In tributaries, incremental increases in flow also have been found to stimulate movement of steelhead (Maule et al. 1994) and of spring Chinook in the Salmon River, Idaho (Matthews et al. 1992; Kiefer and Lockhart 1995). Schreck et al. (1995) found that spring Chinook in the Willamette River moved more rapidly during times of high flow than in times of low flow in a given year.

Analyses have been made more difficult by the fact that the level of smolti-fication of the fish can also affect their rate of migration, and this often varies together with flow (Beeman et al. 1990; Giorgi et al. 1990; Berggren and Filardo 1993).

Mid-Columbia Reach

Yearling Chinook

Unlike the situation in the Snake River, there is no effect of flow on travel time in the mid-Columbia reach (Chapman et al. 1995). Chinook from the Winthrop, Entiat, and Leavenworth hatcheries and from the Rock Island sampler showed no effect of flow on travel time to McNary Dam. The FPC (1994) and Maule et el. (1994) found a weak effect of flow on travel time. However, in the latter two studies, the authors found that degree of smoltifi-cation of the fish was more important than flow in determining travel time. Giorgi et al. (1997), using a large body of data from recoveries at McNary Dam of Chinook, steelhead, and sockeye that were PIT tagged at Rock Island Dam, found that neither ocean-type (subyearlings) nor stream-type Chinook (yearlings) showed any measurable response to flow in the mid-Columbia.

Subyearling Chinook

Subyearling Chinook (ocean type) from the mid-Columbia reach move downstream more slowly than yearling Chinook (stream type) or steelhead (Figure 6.13). For example, yearling hatchery Chinook took an average of about 4.4 days to pass through John Day Reservoir, compared to 14 days for subyearling Chinook at the same flow, 250 kcfs (Chapman et al. 1994a). John Day, having the largest reservoir in the lower river, probably is the place where travel time is the longest. In the mid-Columbia reach, subyearling (sum-mer/fall) Chinook, on average traveled an estimated 4.4 to 10.0 mi./d in the years from 1984 through 1992 (Chapman et al. 1994a). While there was no relationship between travel time and flow, release date, or size at release, there was a significant effect of water temperature on rate of travel.

Steelhead and Sockeye

Travel time of sockeye and steelhead was reduced as flow increased in the mid-Columbia reach (Giorgi et al. 1997). For mid-Columbia steelhead (Figure 6.13c), travel time from the mouth of the Methow River to McNary Dam, a distance of 232 miles, was reduced 2.3 days (from 20 to 17.7 days in transit) by an increase in flow from a base of 80 kcfs to 100 kcfs (Chapman et al. 1994b). The reduction was estimated to be less at higher base flows (reduced 1 day by an increase in flow from 140 to 160 kcfs) (Berggren and Filardo 1993; Chapman et al. 1994b).

At flows of 80 kcfs in the mid-Columbia reach, the predicted travel time of sockeye from the tailrace at Priest Rapids to McNary Dam was about 10 days, and an increase to 100 kcfs reduced predicted travel time by about 3.5 days over the 161-mile reach (Chapman et al. 1990, 1995). At higher flows there was less effect on travel time, according to the equation that was developed, amounting to only half a day reduction with an increase in flow from 160 to 180 kcfs.

In the Snake River, Chapman et al. (1990) concluded that the flow augmentation, which is aimed at hatchery Chinook salmon, is early for the sockeye outmigration.

Discussion of Travel Time Studies

Principles of Water/Fish Movement

There has been considerable effort devoted to the collection of data on migration rates of downstream-migrating salmonids and the statistical relationships to environmental variables (Buettner and Brimmer 1995). There has been less effort expended in conceptual thinking about migration speed, including consideration of the fundamental principles of animal and water movement, and relationships of these principles to the observed migratory timing. Even less attention has been given to whether and how different migration rates affect salmon survival (i.e., relationships between timing of movement and the innate behavioral patterns and ecological needs of the species and life stage). The exception to survival linkage has been attempted connections between initiation and rate of movement and the physiological processes of smoltification (Wedemeyer et al. 1980).

There is notable disagreement over what the empirical evidence about the rate of migration timing and river discharge tells us. McNeil (1992) found no positive relationship between flow and passage time. However, the preponderance of evidence clearly supports the links between flow and migration rate. Within the Snake River and as far as McNary or John Day Dams on the mainstem, the downstream migration for both steelhead and yearling Chinook salmon is faster at high flows than at low flows (Raymond 1968; Sims and Ossiander 1981; Berggren and Filardo 1993; McConnaha 1993; Connor et al. 1994; Maule et al. 1994; Achord et al. 1995a, 1995b; Bvettner and Brimmer 1995; S. G. Smith et al. 1997a, 1997b). This view is reflected in salmon restoration plans (NPPC 1994c; NMFS 1995, 2000). Some of the disagreement relates to the time periods selected for statistical analyses by McNeil (1992), in which inclusion of dates outside the actual migration period can severely affect the results.

The level of smoltification of the fish affects the rate of outmigration, making it difficult to separate the effect of this factor from the effect of flow (Berggren and Filardo 1993; Muir et al. 1995; Giorgi et al. 1997). In a multivariate analysis, Berggren and Filardo (1993) found only one variable, release date, explained the variation in measured travel time of yearling Chinook in the mid-Columbia reach to McNary Dam. Obviously, degree of smoltification is a function of release date. Fish released later from hatcheries are more likely to be further along in development than earlier releases, resulting in lower travel times later in the season. Muir et al. (1995) demonstrated that the degree of smoltification affects the rate of migration of juvenile salmon. They found that smolt development responded to photoperiod and temperature so that as day length and temperature increased later in the season, the fish moved more rapidly.

It is evident that other factors besides flow or rate of movement of water masses affect travel time of smolts (Figure 6.13) because, with the exception of steelhead, smolts' rate of movement is slower than the concurrent average water travel time (Beeman et al. 1990; Berggren and Filardo 1993; Buettner and Brimmer 1993). There is evidence that juvenile salmon make use of certain features of flow hydrodynamics in their migration. For example, accelerating flow seems to foster fish movement. Wild and hatchery Chinook salmon and steelhead captured at the Salmon River and Snake River traps and steelhead captured at the Clearwater trap show increases in sample counts during and shortly after flow increases (FPC 1994; Buettner and Brimmer 1995).

Flow has attributes of volume (amount of discharge) and velocity (which is related to the shape of the channel). Average velocity in the Snake River between Lower Granite Dam and Ice Harbor Dam, calculated from the average cross section, increases linearly with flow from 0.25 fps at a flow of 20 kcfs to about 1 fps at 100 kcfs (Chapman et al. 1994a). In the mid-Columbia reach, average volumes of flow are usually much higher, but corresponding velocities between Wells Dam and Priest Rapids Dam are estimated to be similar to those in the Snake River at flows up to the level of normal highs in the Snake River. In the mid-Columbia reach, velocities are associated with flows in the upper reservoirs. On the other hand, in the reach from Wells Dam to Rock Island Dam (dams that have relatively little storage capacity), velocities are higher (1–3.2 fps) over the range of flows from 80 to 240 kcfs, than in the Rock Island to Priest Rapids stretch of river (0.6–2.1 fps), where there is somewhat more storage capacity. While these average flow calculations can be misleading because velocity will not be uniform across the reservoir at any given flow, the point is that the more water impounded behind a dam, the larger the volume of flow required to reach velocities formerly reached in the unimpounded reaches.

Migration timing depends upon the fish's orientation and behavior in the water as well as whether downstream migrating salmonids flush or stop and go. There has been much debate over whether downstream migrations, in general, are active or passive (see literature reviewed by Jonsson 1991; discussion above). Downstream swimming in the direction of water flow would generate quite rapid downstream movement, with travel times shorter than those for water during periods of active migration. This behavior, as observed in rainbow trout by Northcote (1962), especially when it might be coupled with accelerating flows as in a flood surge, could be very effective in moving fish rapidly. Orientation upstream at a stabilizing swimming velocity, as suggested by L. S. Smith (1982) and Williams and Matthews (1994), would generate a downstream drift at rates less than water movement. Totally passive migration is also possible, in which undirected (or no) fish movements result in net displacement at the rate of the water mass. Coupled with a possible spiraling migratory behavior having alternating times of displacement and resting or feeding, these orientation alternatives could give considerably different migration rates over distances of kilometers. Should these orientations differ temporally, such as in a daily cycle or between early and late migrants in a cohort or whether or not a stage wave is passing, the resulting travel times could be expected to differ in ways that would confound conventional statistical approaches.

Alternative Flow Hypotheses: Salmonid Smolt Migration and Survival

It has long been assumed that river flow rate is a good predictor of fish travel time and their survival, with a near linear relationship (e.g., Berggren and Filardo 1993). However, the adequacy of flow (river discharge) as a predictor of travel time and survival for outmigrant salmon and steelhead smolts for purposes of flow management decisions has been questioned by many water managers. An overriding importance of river discharge for determining migration rates and smolt survival is professed by the FPC, which has produced graphs showing the relation of salmon/steelhead smolt travel time and survival to water travel time, which is a function of river flow (see *www.fpc.org*) (Figures 6.14, 6.15, 6.16). Similar attempts to show linkages among survival, water travel times, and flow have been made by the NMFS (S. G. Smith et al. 2002). When data are aggregated over sufficient years to

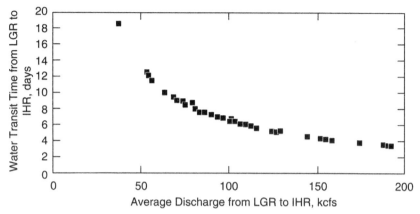

Figure 6.14 Water transit time from Lower Granite Dam to Ice Harbor Dam as a function of the daily average river discharge (flow). Provided by the Fish Passage Center, Portland, Oregon.

attain a wide range of average river flows (including the especially low flow year of 2001), statistically significant linear relationships can be demonstrated using standard regression techniques. However, as more years and more survival estimates have been added over time, the purported linear relationship takes on a clearer "broken stick" appearance (first suggested by

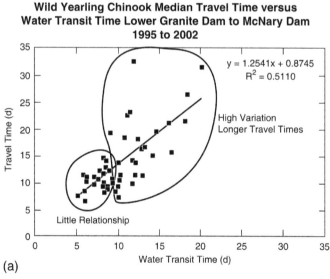

Figure 6.15 Fish transit time as a function of water transit time from Lower Granite Dam to Ice Harbor Dam 1995 to 2002, for (a) wild yearling Chinook salmon,

(Continued)

**Hatchery Yearling Chinook Median Travel Time versus
Water Transit Time Lower Granite Dam to McNary Dam
1995 to 2002**

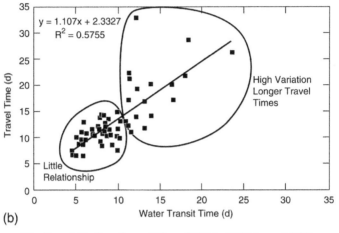

(b)

**Steelhead Median Travel Time 1996 to 2002 from LGR to
MCN versus Water Transit Time through Same Reach**

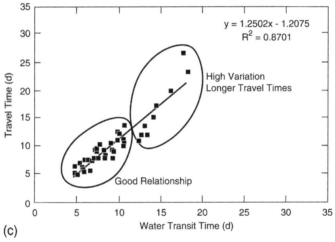

(c)

Figure 6.15 *Continued* (b) hatchery-origin yearling Chinook salmon, and (c) steelhead. See Figure 6.14 to relate water transit time to river flow. Data from the Fish Passage Center, with annotations by the authors of this chapter.

Chapman et al. 1990). In this model (a spline regression), the portion of the survival data at lower river flows (longer water travel times) shows a relationship with flow, whereas the portion at higher flows (shorter water travel times) shows little relationship (FPC data discussed above; NMFS unpublished analyses by Steven Smith) (Figure 6.17). The break point has been determined mathematically by S. Smith. For yearling Chinook in the Snake

River, based on PIT tag detections over the years 1995 to 2001, the break point lies in the vicinity of 96.4 kcfs; for fall Chinook, it lies in the range 40 to 50 kcfs (less well defined), and for steelhead at 101 kcfs. That is, survival is highly dependant on flow (higher survival at higher flows) over a range of low daily average flows but is independent of flow above this range.

Under the broken-stick model, water in spring might be managed for successful yearling smolt migration without much regard for water flow rates

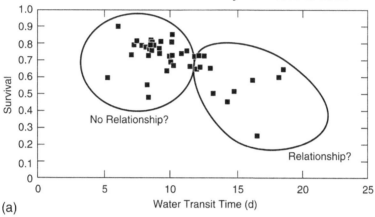

(a)

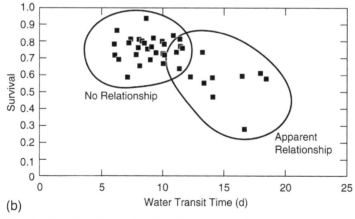

(b)

Figure 6.16 Survival of smolts as a function of water transit time from Lower Granite Dam to Ice Harbor Dam 1995 to 2002, for (a) wild yearling Chinook salmon, (b) hatchery-origin yearling Chinook salmon, and (c) steelhead. See Figure 6.14 to relate water transit time to river flow. Data from the Fish Passage Center, with annotations by the authors of this chapter.

(Continued)

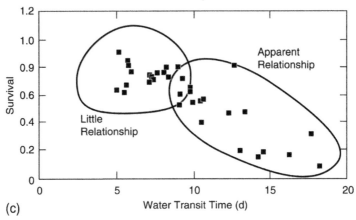

Figure 6.16 *Continued*

when flows exceed a threshold of about 100 kcfs, yet be intimately tied to flow for fish migrating at lower flow rates. For the summer underyearling migrations, the threshold would be near 40 to 50 kcfs.

Management of the hydrosystem would be benefited by knowing what environmental or fish behavioral changes occur in the vicinity of the break point that affect fish migration rates and survival. With such knowledge, we may be able to find causative factors other than just the quantity of water that could be managed to benefit fish. The mechanisms by which fish survival

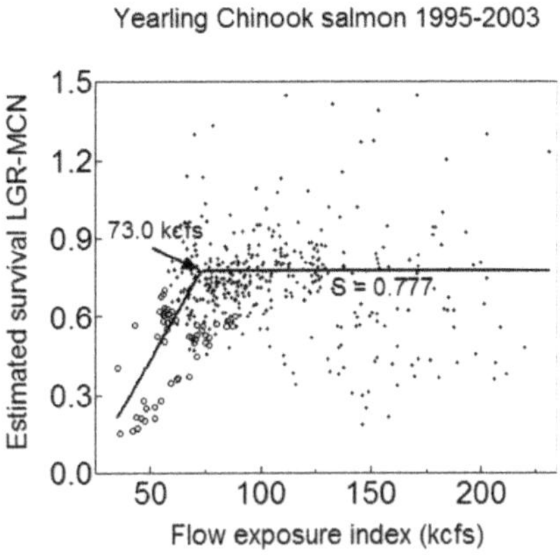

Figure 6.17 "Broken stick" piecewise linear regression model for relating survival of individual PIT-tagged yearling Chinook salmon smolts to river flow, 1995 to 2003, from Lower Granite Dam to McNary Dam. Data from 2001 are plotted with Circles; crosses designate data from all other years. Break point, or threshold, is not precisely estimated. From Unpublished analyses by S. Smith, National Marine Fisheries Service, Seattle, Washington.

differs according to river flow are not clear. Therefore, we discuss several alternative hypotheses and the relevant evidence.

Improvement of in-river survival of juvenile salmonids depends upon the specific needs of fish that go beyond simply moving them downstream as though they were passive objects, as discussed previously. River flow occurs in the context of a total river environment, one that historically provided a number of diverse habitats for use by different salmon species and stocks as they moved downstream as juveniles and upstream as adults. Nonetheless, a more detailed look at the details of fish migration should be informative.

Our focus at this point in the text is on mainstem flows as they might affect juvenile migration through both riverine and reservoir reaches and in the vicinity of dams. Hydropower production has changed riverine habitats, but the basic needs of the fish remain the same. There are several relevant details of the life history requirements of juvenile salmonid fishes. To begin, it must be recognized that (1) there are different life history patterns that interact with mainstem flows in different ways, (2) emigration is not a passive riding of currents straight to the sea, (3) high-quality mainstem habitat for the resting and feeding stages is necessary (see also discussions of habitat in Chapter 5 and earlier in this chapter), (4) juvenile salmonids are generally surface oriented when moving downstream, and (5) they probably use the complex unsteady and turbulent flow of river environments as migration guides and assists, rather than relying upon either mass water movement or their swimming abilities alone. Simplification of a relationship between flow and survival that centers on average water velocities and travel times for juveniles in the hydropower system is probably inappropriate for the full range of life history types of salmon, nor does it provide a holistic view of measures needed for recovery and reestablishment of salmonid populations.

The following section discusses several alternative hypotheses or premises about the possible effects of "flow" on travel time and survival of juvenile salmonids, and the relevant evidence. These hypotheses were proposed by the ISAB (2003).

Velocity Hypothesis

The prevailing hypothesis has pointed to different river velocities at different flow rates as the primary influence on smolt travel time and survival, as discussed above (Berggren and Filardo 1993; FPC 2002; Giorgi et al. 2002). Analysts have presumed that the river-reservoir system flows at a faster velocity when there is higher volume of discharge. The premise under which migration management by flow augmentation operates is that faster smolt migration means less time available for negative effects of predation, disease, high temperature, high dissolved gas, or other damaging factors, and thus higher survival at higher flows (Giorgi et al. 2002). In support of the velocity assumption, the FPC has recently calculated probable average water travel times through the lower Snake River as a function of flow (average daily river discharge) and the estimated volume of the river-reservoir system during times when migrating salmon and steelhead are present (Figure 6.14). The volume differs somewhat from one migration period to another, depending on reservoir elevations during actual migrations. In other words, water travel times are calculated from gross replacement of the available water volume in the river-reservoir reach by the amount of inflow. Similarly, the FPC has assumed that fish migrate essentially passively and that their travel times will be similar to the water travel times, or at least directly related to them. This assumption is tested by plots of fish travel times in relation to water travel times, which generally show a statistically significant relationship, with fish traveling faster at short water travel times (= higher water velocities) (Figure 6.15). Survival shows an inverse relationship (Figure 6.16). Thus, management for higher flows through flow augmentation or other means is expected to lead to higher smolt survival during the outmigration.

We believe there is more to the relationship than is revealed by this analysis. The estimates of water velocity can be faulted on several grounds. First, there are few actual measured velocities in the reservoirs of the lower Snake River at different flow rates (but see Venditti et al. 2000). We substantiated this by inquiries among regional researchers. Second, the gross water replacement formula for estimating velocities does not consider details of channel hydraulics or reservoir circulation, as has been pointed out by Dreher et al. (2000). Such details will influence how river discharge translates to water velocities actually present that might be detected by fish (and resulting in fish responses). Third, the analyses have depended upon average flows over some time period as the variable that affects travel time and/or survival of the fish.

Flows (and therefore channel velocities and water travel times) are not constant through daily cycles. This is true particularly when base flows are below hydraulic capacity of the powerhouses, which leads to rapidly fluctuating flows from the hydropower facilities. At the lowest flows, we determined that pulses of high and low powerhouse discharge in the lower Snake River induce an oscillation (seiche) in the reservoirs that can induce reverse flow (ISAB 2003). This is discussed further below.

The broken-stick pattern describing the relationship between flow and survival suggests that there is more involved than simple average velocity relationships across the range of possible flows. This pattern suggests some difference in mechanisms at different flows that gross velocity alone is not resolving.

Fluctuating Flow Hypotheses

As we have pointed out, one change in hydrosystem operations that occurs across a range of seasonal daily average flow rates is the amount of within-

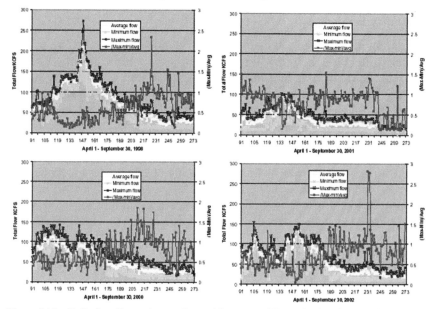

Figure 6.18 Daily flow fluctuations at Lower Monumental Dam, Snake River, April 1 through September 30, 1998, 2000, 2001, and 2002. In high-flow periods, the fluctuations are a small percentage of the flows (maximum, minimum, or average), whereas at low flows, the fluctuations are often greater than the average flow. From ISAB (2003).

day fluctuation caused by variable hydropower generation, especially in the Snake River. Particularly, at flows less than hydraulic capacity of the lower Snake River powerhouses (about 110 kcfs)[2], each day has variable flow on an hourly basis (Figure 6.18). The ranges between minimum and maximum flows do not differ greatly among different average flows, but the fluctuation range is a higher proportion of the daily average flow when the daily average flow is low (Figure 6.18). This means that the fractional within-day variation in flow (and probably river velocities and fish travel times) in a river or reservoir is greater at lower average flows than it is at higher average flows. For 1998, for example, the daily variation in flow from Lower Monumental Dam was a small percentage of the daily average flow (10%–20%) during May and June, which contrasts with the fluctuations in daily average flow in July and August, the period of outmigration of fall Chinook, when variation was a high percentage, often exceeding 100%.

The mechanism by which this flow fluctuation might affect smolt survival is open to speculation, although circumstantial evidence points to some likely mechanisms we can discuss. Furthermore, the relationships of fluctuating flows to the break points in flow-survival curves deserve some discussion in terms of mechanisms. We examined several alternatives.

(A) Fish Stranding

In the Hanford Reach of the mainstem Columbia River, stranding of under-yearling Chinook salmon by fluctuating flows has been identified as an important factor causing mortalities (Wagner et al. 1999; Tiffan et al. 2002). Research has identified that fry of fall Chinook, in particular, use shoreline areas for feeding and rearing during the high flows of daytime, but are often left stranded in isolated pools or on large flats and gravel bars when the river flow rapidly decreases.

Because of these mortalities, the Vernita Bar Agreement that regulated Priest Rapids Dam discharges during spawning of adults and incubation of eggs was modified in the year 2000 to provide more stable flows during the time when underyearlings are most abundant in the Hanford Reach. A similar problem, but to a smaller degree, might occur in the lower Snake River and elsewhere. It is likely to be a smaller problem in the lower Snake River, which is nearly exclusively a simple channel having steep sides and few flat zones for stranding. The reach is predominantly reservoirs, with elevations managed to avoid wide fluctuations. Fluctuations in discharge may still affect survival in reaches dominated by reservoirs, but probably not by stranding to the degree observed in the Hanford Reach.

(B) Unstable Reservoir Hydraulics Confusing to Fish

We hypothesized that rapid changes in river discharge at an upstream dam (both increases and decreases) would cause hydraulic instability in the down-stream reservoir. This instability would potentially affect local hydraulic patterns in the reservoirs and thus migration rates of fish in these unstable flows. The consequence could be increased susceptibility to many survival-reducing factors (e.g., predation). Venditti et al. (2000) observed wandering and upstream swimming by radio-tagged fall (ocean-type) Chinook salmon in Little Goose pool in July and August of 1995, 1996, and 1997, as well as large differences in migration rates. Plumb et al. (2003) observed similar behavior in the Lower Granite pool during the period from 1996 to 2001.

We initially tested the hypothesis that flow fluctuations destabilized the reservoir hydraulics by obtaining mid-January 2003 hourly data (from the U.S. Corps of Engineers website [http://www.nwd-wc.usace.army.mil/nww/rreports.htm]) on discharges, forebay elevations, and tailwater elevations for the four lower Snake River dams at low flows. Mid-January typically has the lowest Snake River flows and the highest likelihood of showing hydraulic effects of fluctuating dam operations. Operations during those low flows may be representative of operations during similar low flows in the period from July to August. We discovered seiches (periodic oscillations of the water surface that involve water movements upstream and downstream; i.e., the "sloshing bathtub") in the lowermost three Snake River reservoirs (Little Goose, Lower Monumental, and Ice Harbor) during weekdays but not in Lower Granite Reservoir (Figure 6.19). The observed seiches, reported by the ISAB (2003) for the lower Snake River reservoirs, are consistent with the oscillation being caused by pulsing outflows at the three upstream dams (Lower Granite, Little Goose, and Lower

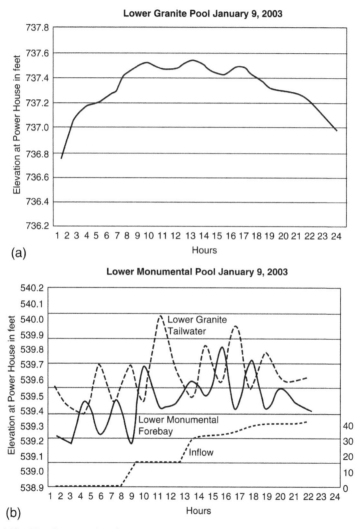

(a)

(b)

Figure 6.19 Hourly water elevations on January 9, 2003, in (a) the forebay of Lower Granite Dam, where there is no dam immediately upstream with fluctuating flow releases and no seiche is evident, and (b) Lower Monumental Dam forebay and the tailwater of Little Goose Dam immediately upstream, where flows vary between 0 and 40,000 cfs daily, and a seiche with a period of 2 to 3 hours is evident. From ISAB (2003).

Monumental, respectively). Lower Granite Reservoir has an undammed portion of river above the head of the reservoir and receives side flow from tributaries, both of which are factors that would dampen the effects of oscillations. Dam outflows ranged from near zero at night to between 10 and 30 kcfs in the daytime. Further support for the oscillations being forced by pulsed discharges came from observation that the oscillations dampened during a weekend of stable flows (about 13 kcfs), but were reinitiated when flows again fluctuated markedly in the following week. Oscillations in Little Goose and Lower Monumental pools had periods of somewhat over 2 hours, whereas Ice Harbor pool had a complex oscillation with many apparent harmonics or suboscillations (befitting the Ice Harbor pool with its more complex morphometry). The oscillations of Little Goose and Lower Monumental pools had single or odd-numbered nodes (centers of oscillation), because the upswings occurred in reverse sequence (mirror image) at forebays and tailwaters. Figure 6.20 illustrates surface elevation changes and seiche-induced flows in a theoretical lake basin (Lemmin and Mortimer 1986). The tailwater oscillations had higher amplitudes, consistent with shallower water and narrower channels than in the forebays.

We calculated that during hours of zero discharge from Lower Granite Dam, the necessary water displacement for the observed elevation changes

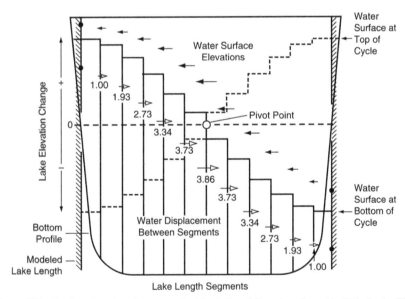

Figure 6.20 Surface elevation changes and seiche-induced flows in a theoretical lake basin. The example basin is tub-shaped, and idealized to a rectangle (hatched vertical lines). Water elevations oscillate around a single node at the middle at the average water elevation (dashed horizontal line). Water elevation oscillates between the solid histograms peaking at the left and the dotted ones peaking at the right, with relative water displacement indicated by the arrows. From Lemmin and Mortimer (1986).

would cause flows alternately upstream and downstream of 10 kcfs with a periodicity of about 1.4 hours. When the dam discharged its peak of 20 kcfs for the day, the downstream flow at the node would be accelerated to about 30 kcfs and decelerated to about 10 kcfs downstream every 1.4 hours.

These fluctuating dam outflows and their potential for producing seiches, flow reversals, and complex mixing at low flows could be disruptive for downstream migrants and lead to the wandering and upstream swimming of radio-tagged smolts that were observed in Little Goose pool in the years 1995 to 1997 by Venditti et al. (2000) and in Lower Granite pool from 1996 to 2001 by Plumb et al. (2003). There is a large literature on seiches in lakes and coastal waters, much of it from European limnological research of the late 1800s and early 1900s (e.g., see Hutchinson 1957).

Subsequent to making these analyses and discovering the winter seiches, we obtained hourly records of flows and elevations of tailwaters and forebays for the years 1995 to 2002 from DART (courtesy of Chris Van Holmes, University of Washington DART). It appears that flows do not get as low as zero during the months of fish migrations, but that considerable pulsing occurs at times when downstream migrants are present. During 2001, the year of the telemetry study by Plumb et al. (2000), there were periodic episodes of flow reduction every day but two during their study. Most episodes, 46 of 60, occurred in the hours from 0100 to 0400, during which the minimum flow reached about 65% of flow in the preceding and following time intervals, and lasted an average of about 7 hours (some as long as 15 and 18 hours). No seiches were apparent in the records of water elevations during the weeks of fish migrations, although a thorough examination of the data set should be conducted.

The documented, large variations in flow through a daily cycle undoubtedly create complex hydraulics that probably are disruptive to the downstream migration of fish. The flows at which daily cycling predominates may be one cause of the "broken stick" relationship between daily average flow and travel time or survival. This alternative hypothesis for decreased survival at low flows should be investigated further.

(C) Fluctuating Forebay Flows

The fluctuation of dam discharges within a day by power operations affects the dam's forebay as well as its tailwater. As water in a Snake River reservoir nears an operating powerhouse, its velocity increases progressively within a zone extending about 300 ft upstream of the dam (Adams et al. 1997). The velocities attained are proportional to the powerhouse discharge rate. Thus, downstream migrants are exposed to hourly changes in forebay velocities when dam discharges fluctuate.

Fish behavior in the near forebay has been shown to strongly affect the time taken to pass a dam (Plumb et al. 2003). This behavior is related to the daily average river flow rate, with major changes documented to occur for spring Chinook salmon and steelhead near 100 kcfs. It may not be coincidence that this flow is approximately the calculated break point in flow-survival curves (Smith et al. 2002). Therefore, it is likely that short-term (i.e. hourly), within-day variations in forebay velocities could cause variations in behavior in the forebays and affect dam passage times. The variable velocity environment could foster some of the behavioral anomalies seen by Plumb et al. in 2001, a drought year (e.g., upstream movement). These authors noted that powerhouse operations included daily shifts of load from turbines at one end of the powerhouse to the other end. These shifts could also create conditions confusing to migrating fish, which might increase their vulnerability to predators that are known to be present in the forebays of most dams.

Although the field observations of velocities and fish behavior were conducted at times when flows were at seasonal highs and yearling Chinook salmon and steelhead were migrating, the relative intensity of flow fluctuations is greater at lower flows in summer when underyearling fall Chinook are migrating (survival of underyearlings during migration is generally poorer than for yearlings). The flows experienced in spring 2001 were somewhat representative of flows typically seen later in the year when juvenile fall Chinook are migrating. Thus, the strong behavioral responses seen for yearlings are probably magnified for underyearlings.

Fish Orientation Hypothesis

Fish behavior in different river conditions of velocity and turbulence has been proposed as a guidance mechanism for migration (Coutant 1998, 2001). This mechanism may help explain the broken-stick relationship of fish survival to water travel time and flow. Behavioral responses to detailed flow hydraulics may provide opportunities for water management to benefit fish that include factors other than increasing daily average river flow rates through augmentation. For years, it has been believed that downstream-migrating salmon and steelhead make use of river currents to assist their downstream movement (Thorpe 1982). As shown in telemetry studies by Venditti et al. (2000) and Plumb et al. (2003), smolts sometimes lose their downstream orientation in reservoir forebays where flow rates are diminished and show milling behavior and upstream forays extending several miles. It would appear that the smolts have lost their cues for "downstream" and are searching to relocate the downstream flow (Venditti et al. 2000).

Salmon smolts exhibit several modes of migration, as discussed previously. Often salmon smolts swim facing upstream near the surface of the main river channel or thalweg at a very slow velocity relative to the surrounding water (see references cited in Coutant and Whitney 2000).

Despite a small component of forward swimming, their net movement is downstream. Most smolts will, at times, leave this migration mode and move to shallower water, presumably to feed. (Underyearling Chinook salmon, especially, demonstrate a daily cycle of daytime feeding along shorelines and nighttime movement to river channels where they are displaced downstream.) Some, especially steelhead, will also actively swim downstream at times, thus moving at a rate faster than the average water travel time (Berggren and Filardo 1993; Muir et al. 1995; Peak and McKinley, 1998). Nonetheless, downstream displacement appears to be primarily a combination of riverine flow and the intentional positioning of a smolt to maintain its controlled orientation in it.

The telemetry studies of wandering smolts in dam forebays and other locations with low river velocities suggest a disappearance of the smolts' attachment to river flows at some combination of river conditions, or an inability to orient with it. One possibility is that the water velocity (or turbulent cues to that velocity) may simply fall below the velocity that smolts recognize. Riverine turbulence has been suggested as an important behavioral cue (Coutant 1998, 2001). It is hypothesized that smolts detect small pressure changes along their bodies (by the lateral line sensory system), which result from turbulence that is characteristic of a flowing river. So long as this turbulence is sensed, the fish recognizes that it is in a moving flow and it maintains its orientation in the water, resulting in net downstream movement. When the appropriate level of turbulence is not sensed (a level yet to be determined experimentally or by field observation, which we can call the "critical turbulence"), then the fish leaves its flow-attachment mode and actively swims to relocate the zone of flow. As with other hypotheses, slowed migration presumably results in increased exposure to predators, disease, high temperatures, high dissolved gas, or other damaging factors that reduce survival.

We hypothesize that a major factor determining smolt travel time through reservoirs such as those in the Snake River is the distance downstream that the critical turbulence level extends into the reservoir. Throughout a wide range of rather high flows, the critical turbulence level is met or exceeded all the way to the dam (Figure 6.21a). In this case, fish travel time varies with the water travel time through the reservoir, and survival is high because the fish maintain rapid downstream displacement through the whole reservoir. There is no relationship of survival to flow at these high river discharge rates because speedy migration is maintained to and through the dam regardless of flow rate above a certain level. This conjecture conforms to the flat (no relationship) portion of the broken-stick model. On the other hand, at low flows the critical turbulence level is not maintained all the way to the dam (Figure 6.21b). In this situation, fish lose their orientation cue before reaching the dam and are left to wander through the downstream portion of the reservoir,

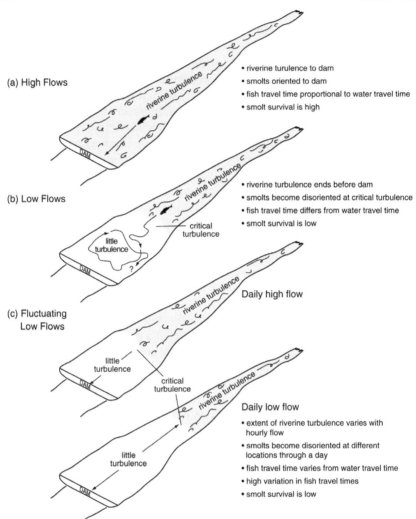

Figure 6.21 Hypothesized effect of turbulence on smolt migration in reservoirs. (a) At high flows, smolts are guided by turbulence cues all the way to the dam. (b) At low flows, turbulence cues are lost and smolts lose direction and wander, causing slowed migrations and additional predation. (c) In daily fluctuating low flows, travel time is variable among smolts due to the hourly differences in extent of turbulent cues. From ISAB (2003).

slowing overall migration rate and exposing fish to factors that likely reduce survival. Also, more stored energy would be used by the smolts as they must swim actively rather than be displaced passively, thus possibly reducing survival directly or indirectly. Fish travel time and survival are both related to flow at these lower flow rates because at higher flow rates, the critical turbulence level will extend closer to the dam and the zone of fish wandering is smaller. The lower the flow, the more the wandering. This conforms to the lower portion of the broken-stick model for yearling Chinook and steelhead, wherein both fish travel time and survival are related to water travel time and flow; this would help explain the relationship for subyearling (fall) Chinook.

Fluctuating flows are important in this hypothesis. Relationships between fish travel time and water travel time, as graphed by the FPC (Figure 6.15) and others, are variable among individually PIT-tagged smolts, especially at longer water travel times (lower flows). This variability is hypothesized here to result from hourly fluctuating flows from dams throughout a day at lower daily average flows (Figure 6.21c). Different smolts migrating at different times of day can be exposed to different rates of water movement and different locations along the length of reservoir where the critical turbulence level is no longer met. This would yield very different fish travel times for the same daily average flow and daily average water travel time, as has been demonstrated in the FPC analyses.

River modifications other than raising daily average flows have been suggested to assist smolts in their migration through reservoirs (Coutant 1998, 2001). These modifications generally consist of passive or active ways to increase turbulence and extend the distance that the critical turbulence level occurs in reservoirs like those on the Snake River. Structural features have been suggested to focus flows and generate riverine turbulence. Active approaches (pumps, propellers) may be useful in certain situations such as approaches to fish bypasses. These approaches for enhancing migration rates through reservoirs deserve attention. We recognize that their application over the large reaches involved might pose significant engineering problems. Further hydrologic studies would be needed to identify places where they might be useful or necessary. Fish behavioral responses would need to be identified as well. When the alternative to such studies is flow augmentation from limited upstream sources of water (and competition with irrigation and other uses with high economic value), the additional attention to such river management approaches may be economically justified.

Dam Obstacle Hypothesis

Migration rates of yearling steelhead and Chinook salmon in spring declined to near zero within 400 ft of Lower Granite Dam when daily average flows were less than about 100 kcfs (Plumb et al. 2003). This result emerged from radiotelemetry studies in 1996 to 2001 at annual average flows that ranged from above the average for recent years to a near record low. The 100 kcfs value is essentially the same as the apparent flow break point seen in flow-survival curves for these species (e.g., FPC data and S.G. Smith et al. 2002). This observation suggests that the dam structure itself or some hydraulic features at the dam (e.g., lack of attraction flow, daily discharge cycle, lack of spill, noise) act as a more prominent barrier to migration at flows below 100 kcfs. These data suggest that it would be valuable to examine dam structures or operations in more detail at flows below 100 kcfs.

In summary, there is increasing evidence, especially from radiotelemetry studies of the behavior of individual migrants, that details of the interactions of smolts with hydraulics in rivers and reservoirs are important for understanding the relationships between river discharge, smolt travel times, and smolt survival. A perpetual accumulation of data on gross aspects such as average flow rates and water displacement times, as seems to be the ongoing strategy, will not lead to better management of fish migration and survival. A concerted effort to study these details is required. The cost of such an effort is miniscule compared to current expenses for flow augmentation and spill.

Directions for Future Research

Our review in the preceding chapter shows that both passive and active migrations occur. As Jonsson (1991) noted, fish must actively initiate emigration. Clearly, fish that are holding during a diel cycle, either at the bottom or in shoreline backwaters, must actively swim to get themselves oriented into the main current for what might later be passive movement. Both avoiding obstacles during downstream movement (e.g., being swept into backeddies) and ending the movement phase of spiraling would require an active component. All of these complicate a simple interpretation of migration rates between widely separated points.

It may be useful in the future to compare the different implications for rivers and reservoirs with the results of field studies of fish passage to see which implications (and thus behaviors) are supported by the evidence. It has already led us to consideration of vortices, stage waves, or surges in affecting migration. However, the effectiveness of hydraulic patterns remains to be demonstrated in impoundments. Another area of investigation might be to test different fish behaviors with river management alternatives in hydrodynamic models of river and reservoirs to develop computer simulations of fish passage timing. The simulations under different combinations of behaviors and water flow regimes can be compared to the field data. Additional scenarios can be examined, more than is possible with the actual historical record of flows, migration times, and other factors. For example, the effects on passage rates of different lengths of time spent in displacement and stationary resting/feeding can be examined for a range of flows even though there are few field studies of diurnal behavior. The objective of such analyses would be to indicate the possible habitat requirements of each species/stock and their projected gain (or loss) from velocity increases from managed reservoir drawdown or augmented flows.

We would especially like to see specific tests of the hypotheses we presented for explaining the apparent "broken stick" model for smolt travel times and survival as functions of river discharge (flow). Telemetry studies of fish traveling in reservoirs, detailed analyses of individual fish with PIT tags, and aggregate studies of flows and fish movements over various time periods all have unique information to tell us. It appears to be particularly important to move away from analyses that use average flows in an attempt to explain fish responses. Hourly records of discharge are available for the entire hydropower system. With this detailed information, we can more fruitfully plan mitigation actions to foster both successful fish migration and their survival.

Effects of Hydrosystem Development on Life History Diversity

The Columbia-Snake River Basin, at the time Europeans arrived, was characterized by an assemblage of Pacific salmon species and stocks with highly divergent life history strategies (see Chapters 3 and 4). This diversity developed as the Wisconsin glaciation retreated and the exposed landscape was recolonized by stream-type salmon from northern refugia and ocean-type fish from southern refugia (Lindsey and McPhail 1986; McPhail and Lindsey 1986). Differentiation probably occurred within stocks as they adapted to the peculiarities of specific tributary systems and the migration corridors to and from them. It is believed that migration distance and growth opportunity in the vicinity of spawning (a combination of water temperature and day length) were major factors in this differentiation (Taylor 1990).

Overall stock diversity was probably reflected in a diversity of migration behaviors related to constant flushing or spiraling, as well. It follows logically from the diversity of tributary habitats and flows that salmon as a group would diversify to make full use of different migratory corridors, as Rich (1920) observed. The differences in diurnal migratory behavior of the now-extirpated upper Snake River stocks of spring Chinook salmon studied by Krcma and Raleigh (1970) and other stocks is just one example. The primal river had spring freshets of varying magnitudes and durations that afforded quick passage, open channels for quick flush, backwaters for lingering, eddies and deep pools for resting, riparian habitats that afforded stragglers with abundant food and shelter, and so forth. Each habitat niche was probably occupied by a species or stock (often overlapping). Because each salmon species in the Columbia-Snake system has a multiyear life cycle and attainment of maturity can vary across several ages, each population was buffered from unfavorable conditions in any one or few years, as the riverine environment varied from year to year. Good years for one species' or stock's migration strategy (habitat use) may have been bad for another one's strategy. Because the relative benefit could switch from year to year, the diversity of stocks would persist.

Any strategy that manages river flows consistently is likely to favor fish stocks with one migratory behavior or habitat use to the detriment of others. Some stocks might, therefore, be pushed to extinction or very low levels while others are protected and fostered. For example, consistently high flows in the Snake River in May coupled with reservoir drawdown may create a fast-flushing, bare channel highly suited for moving yearling spring Chinook downstream rapidly (begging the question of any daytime resting require-ments), but at the same time be inconsistent with the requirements of under-yearling fall Chinook salmon for slow-water areas with riparian vegetation for their characteristically slow downstream movement. It can be hypothe-sized that one factor contributing to the present sad state of the Snake River fall run fish might be because of the poor riparian habitat of the present Snake River (in contrast to the riparian vegetation-rich Hanford reach).

The most favorable flow strategy for a diverse assemblage of salmonids would be one that varies, favoring some stocks at one time and other stocks another time. In the normative river concept, this variability should mimic natural variability, although replacing a climate-driven variability with a planned one (assuming the reservoirs are not permanently drawn down to natural riverbed). Although not easy, we can envision flow management in which reservoirs are drawn down temporarily in different ways in successive years: for example, one year in three for maximal support of constant flush-ing behavior, and another in which floods are created to overtop riparian zones to create maximal shoreline habitat. The third year could be main-tained stable. These flow strategies could be coupled with non-flow measures for salmon such as replacement of shoreline rock rip-rap with vegetation. The occasional exceptionally dry year (that restricts planned flooding) or wet year (that floods no matter what the plan) would add a certain primal vari-ability. This cycle might eventually be matched with predictions of good or poor ocean conditions for salmon under the El Niño-Southern Oscillation (ENSO) and Pacific Decadal Oscillations (PDO) that now are recognized as causing some of the variability in salmon survival (Mote et al. 2003).

Literature Cited

Achord, S., D. J. Kamikawa, B. P. Sandford, and G. M. Matthews. 1995a. Monitoring the migra-tions of wild Snake River spring and summer Chinook salmon smolts, 1993. Bonneville Power Administration. DOE/BP-18800-2. Portland, Oregon.

Achord, S., G. M. Matthews, D. M. Marsh, B. P. Sandford, and D. J. Kamikawa. 1995b. Monitoring the migrations of wild Snake River spring and summer Chinook salmon smolts, 1992. Bonneville Power Administration. DOE/BP-18800-1. Portland, Oregon.

Adams, N., S. Evans, J. Kelly, M. Banach, and M. Tuell. 1995. Behavior of radio-tagged juvenile Chinook salmon and steelhead in Lower Granite Reservoir, Washington, during spring 1994. Pages 30–75 in D. Rondorf and M. Banach, eds. *Migrational characteristics of juvenile Chinook salmon and steelhead in Lower Granite Reservoir and tributaries.* US Army Corps of Engineers, Walla Walla, Washington.

Adams, N. S., D. W. Rondorf, E. E. Kofoot, J. J. Banach, and M. A. Tuell. 1997. Migrational characteristics of juvenile spring Chinook salmon and steelhead in the forebay of Lower Granite Dam relative to the 1996 surface bypass collector tests. 1996 annual report to the US Army Corps of Engineers. Contract E-86930151. Walla Walla, Washington.

Albertson, M. L., and D. B. Simons. 1964. Fluid mechanics. Pages 7-1–7-49 in V. T. Chow, ed. *Handbook of applied hydrology.* McGraw-Hill Book, New York.

Allen, R. L., and T. K. Meekin. 1980. Columbia River sockeye salmon study, 1971–1874. Washington Department of Fisheries. Progress Report, 120. Olympia, Washington. 75 p.

Anderson, D. A., G. Christofferson, R. Beamesderfer, B. Woodard, M. Rowe, and J. Hansen. 1996. StreamNet: Report on the Status of Salmon and Steelhead in the Columbia River Basin–1995. Bonneville Power Administration. DOE/BP-65130-1. Portland, Oregon. 76 p.

Andrew, F. J., and G. H. Geen. 1960. Sockeye and Pink Salmon Production in Relation to Proposed Dams in the Fraser River System. International Pacific Salmon Fisheries Commission. Bulletin XI. 259 p.

Andrews, R. E. 1958. Factors influencing the seaward migration of smolt steelhead trout in the Alsea River, Oregon. Master's thesis, Oregon State University, Corvallis.

Arnold, G. P. 1974. Rheotropism in fishes. *Biological Review* 49:515–576.

Asherin, D. A., and J. J. Claar. 1976. Inventory of Riparian Habitats and Associated Wildlife along the Columbia and Snake Rivers. Volume 3B, Snake River-McNary Reservoir. US Army Corps of Engineers, Walla Walla, Washington.

Bauersfeld, K. 1978. The effect of daily flow fluctuations on spawning fall Chinook in the Columbia River. Washington Deptartment of Fisheries. Technical Report, No. 38. Olympia, Washington. 32 p.

Becker, C. D., and C. C. Coutant. 1970. Experimental drifts of juvenile Chinook salmon through effluent at Hanford in 1968. Battelle-Northwest. USAEC R&D Report, BNWL-1499. Richland, Washington.

Becker, C. D. 1973. Columbia River thermal effects study: Reactor effluent problems. *Journal of the Water Pollution Control Federation* 45:850–869.

Beeman, J. W., D. W. Rondorf, J. C. Faler, M. E. Free, and P. V. Haner. 1990. Assessment of smolt condition for travel time analysis. Bonneville Power Administration. Annual Report. Contract DOE/BP-35245-3. Portland, Oregon. 103 p.

Bentley, W. W., and H. L. Raymond. 1976. Delayed migrations of yearling Chinook salmon since completion of Lower Monomental and Little Goose Dams on the Snake River. *Transactions of the American Fisheries Society* 105:422–424.

Berggren, T. J., and M. J. Filardo. 1993. An analysis of variables influencing the migration of juvenile salmonids in the Columbia River basin. *North American Journal of Fisheries Management* 13:48–63.

Bjornn, T. C., and C. A. Peery. 1992. A review of literature related to movements of adult salmon and steelhead past dams and through reservoirs in the lower Snake River. University of Idaho, US Fish and Wildlife Service Cooperative Fish and Wildlife Research Unit, Moscow. Technical Report, 92-1. US Army Corps of Engineers, Walla Walla, Washington. 80 p.

Bonneville Power Administration. 1980. The role of the Bonneville Power Administration in the Pacific Northwest power supply system including its participation in a hydro-thermal power program. US Department of Energy. EIS-0066. Portland, Oregon.

Bovee, K. D., J. A. Gore and A. J. Silverman. 1978. Field testing and adaptation of a methodology to measure in-stream values in the Tongue River, northern Great Plains (NGP) region. US Environmental Protection Agency. EPA Report, EPA-908/4-78-004A.

Brannon, E. L., T. P. Quinn, G. L. Luccetti, and B. D. Ross. 1981. Compass orientation of sockeye salmon fry from a complex river system. *Canadian Journal of Zoology* 59:1548–1553.

Buettner, E. W., and A. F. Brimmer. 1993. Smolt monitoring at the head of Lower Granite Reservoir and Lower Granite Dam: Annual Report 1992. Bonneville Power Administration. DOE/BP-11631-9. Boise, Idaho. 64 p.

Buettner, E. W., and A. F. Brimmer. 1995. Smolt monitoring at the head of Lower Granite Reservoir and Lower Granite Dam. Annual Report for 1993 Operations. Bonneville Power Administration. Portland, Oregon.

Burgner, R. L. 1991. Life history of sockeye salmon (*Oncorhynchus nerka*). Pages 3–117 in C. Groot and L. Margolis, eds. *Pacific salmon life histories*. University of British Columbia Press, Vancouver.

Chambers, J. S. 1955. Research relating to study of spawning grounds in natural areas. Pages 88–94 in Washington Department of Fisheries. Report to US Army Corps of Engineers, 1955. Olympia, Washington.

Chapman, D., A. Giorgi, T. Hillman, D. Deppert, M. Erho, S. Hays, C. Peven, B. Suzumoto, and R. Klinge. 1994a. Status of summer/fall Chinook salmon in the mid-Columbia region. Don Chapman Consultants, Boise, Idaho. 411 p.

Chapman, D., C. Peven, T. Hillman, A. Giorgi, and F. Utter. 1994b. Status of summer steelhead in the mid-Columbia river. Don Chapman Consultants, Boise, Idaho. 235 p.

Chapman, D., C. Peven, A. Giorgi, T. Hillman, and F. Utter. 1995. Status of spring Chinook salmon in the mid-Columbia region. Don Chapman Consultants, Boise, Idaho. 477 p.

Chapman, D. W., D. E. Weitkamp, T. L. Welsh, and T. H. Schadt. 1983. Effects of minimum flow regimes on fall Chinook spawning at Vernita Bar 1978-82. Report to Grant County PUD, Ephrata, Washington. Don Chapman Consultants., Boise, Idaho. 123 p.

Chapman, D. W., W. S. Platts, D. Park, and M. Hill. 1990. Status of Snake River sockeye salmon. Don Chapman Consultants, Boise, Idaho. 90 p.

Chapman, W. M. 1943. The spawning of Chinook salmon in the main Columbia River. *Copeia* 1943:158–170.

Connor, W. P., H. L. Burge, D. Steele, C. Eaton, and R. Bowen. 1995a. Rearing and emigration of naturally produced Snake River fall Chinook salmon juveniles. Pages 41–73 in D. W. Rondorf and K. F. Tiffan, eds. *Identification of the spawning, rearing, and migratory requirements of fall chinook salmon in the Columbia River Basin*. Bonneville Power Administration, Portland, Oregon.

Connor, W. P., A. P. Garcia, A. H. Connor, R. H. Taylor, C. Eaton, D. Steele, R. Bowen, and R. D. Nelle. 1995b. Fall Chinook salmon spawning habitat availability in the free-flowing reach of the Snake River. Pages 22–40 in D. W. Rondorf and K. F. Tiffan, eds. *Identification of the spawning, rearing, and migratory requirements of fall chinook salmon in the Columbia River Basin*. Bonneville Power Administration, Portland, Oregon.

Coutant, C. C. 1998. Turbulent attraction flows for juvenile salmonid passage at dams. Oak Ridge National Laboratory. ORNL/TM-13608. Oak Ridge, Tennessee. 28 p.

Coutant, C. C., and R. R. Whitney. 2000. Fish behavior in relation to passage through hydropower turbines: A review. *Transactions of the American Fisheries Society* 129:351–380.

Coutant, C. C. 2001. Turbulent attraction flows for guiding juvenile salmonids at dams. *American Fisheries Society Symposium* 26:57–78.

Cramer, S., and J. Martin. 1978. Rogue River Basin evaluation program. Oregon Department of Fisheries and Wildlife. Progress Report to US Army Corps of Engineers. DAC-57-75-C-0109. Portland, Oregon.

Curet, T. S. 1993. Habitat use, food habits, and the influence of predation on subyearling Chinook salmon in Lower Granite and Little Goose Reservoirs, Washington. Master's thesis, University of Idaho, Moscow.

Dauble, D. D., R. H. Gray, and T. L. Page. 1980. Importance of insects and zooplankton in the diet of 0-age Chinook salmon (*Oncorhynchus tshawytscha*) in the central Columbia River. *Northwest Science* 54:253–258.

Dauble, D. D., T. L. Page, and J. R. W. Hanf. 1989. Spatial distribution of juvenile salmonids in the Hanford Reach, Columbia River. *Fishery Bulletin* 87:775–790.

Dauble, D. D., and D. G. Watson. 1990. Spawning and abundance of fall Chinook salmon (*Oncorhynchus tshawytscha*) in the Hanford Reach of the Columbia River, 1948-1988. U S Department of Energy, Pacific Northwest Laboratory. PNL-7289, UC-600. Richland, Washington.

Dauble, D. D., and R. P. Mueller. 1993. Factors affecting the survival of upstream migrant adult salmonids in the Columbia River Basin. Bonneville Power Administration. Recovery Issues for Threatened and Endangered Snake River Salmon, 9 of 11. Portland, Oregon. 72 p.

Dawley, E. M. et al. 1986. Migrational characteristics, biological observations, and relative survival of juvenile salmonids entering the Columbia River estuary, 1966-1983. Bonneville Power Administration. DOE/BP-39652-1. Portland, Oregon.

Dreher, K. J., C. R. Petrich, K. W. Neely, E. C. Bowes, and A. Byrne. 2000. Review of survival, flow, temperature, and migration data for hatchery raised, subyearling fall Chinook salmon above Lower Granite Dam, 1995-1998. Idaho Department of Water Resources, Boise, Idaho.

Ebel, W. and H. Raymond. 1976. Effects of atmosphere gas saturation on salmon and steelhead trout of the Snake and Columbia rivers. *Marine Fisheries Review* 38(7):1–14.

Evermann, B. W. 1895. A preliminary report upon salmon investigations in Idaho in 1894. *U.S. Fisheries Commission Bulletin* 15:253–284.

Fangstam, H., I. Berglund, M. Sjoberg, and H. Lundqvist. 1993. Effects of size and early sexual maturity on downstream migration during smolting in Baltic salmon (*Salmo salar*). *Journal of Fish Biology* 43:517–529.

Fish Passage Center (FPC). 1994. Fish Passage Center Annual Report 1993. Bonneville Power Administration. DOE/BP-38906-3. Portland, Oregon.

FPC. 2002, October 14. Preliminary update on juvenile migration characteristics. Memorandum from Michelle DeHart to Members' Management Ad Hoc Group. Portland, Oregon.

Flagg, T. A., and L. S. Smith. 1982. Changes in swimming behavior and stamina during smolting in coho salmon. Pages 191–195 in E. L. Brannon and E. O. Salo, eds. *Salmon and trout migratory behavior symposium*. University of Washington School of Fisheries, Seattle.

Folmar, L. C., and W. W. Dickhoff. 1980. The parr-smolt transformation (smoltification) and seawater adaptation in salmonids. A review of selected literature. *Aquaculture* 21:1–37.

French, R., and R. J. Wahle. 1968. Study of loss and delay of salmon passing above Rock Island Dam, Columbia River, 1954-56. *Fishery Bulletin* 65:339–368.

Fryer, J. K. 1995. Columbia River sockeye salmon. Ph. D. dissertation, University of Washington, Seattle.

Fulton, L. A. 1968. Spawning areas and abundance of Chinook salmon (*Oncorhynchus tshawytscha*) in the Columbia River Basin–Past and present. US Department of the Interior, Fish and Wildlife Service, Special Scientific Report–Fisheries No. 571.

Garcia, A. P., W. P. Connor, and R. H. Taylor. 1994. Fall Chinook salmon spawning ground surveys in the Snake River. Annual Report for 1993. Pages 1–21 in D. W. Rondorf and K. F. Tiffan, eds. *Identification of the spawning, rearing, and migratory requirements of fall Chinook salmon in the Columbia River Basin*. Bonneville Power Administration, Portland, Oregon.

Gauley, J. E., R. E. Anas, and L. C. Schlotterbeck. 1958. Downstream Movement of Salmonids at Bonneville Dam. US Fish and Wildlife Service, Special Scientific Report–Fisheries 236.

Gilbert, C. H. 1912. Age at maturity of the Pacific coast salmon of the genus *Oncorhynchus*. US Bureau of Fisheries Bulletin, Washington, DC.

Giorgi, A. E., L. C. Stuehrenberg, D. R. Miller, and C. W. Sims. 1986. Smolt Passage Behavior and Flow-Net Relationship in the Forebay of John Day Dam. Bonneville Power Administration. DOE/BP-39644-1. Portland, Oregon.

Giorgi, A. E., G. A. Swan, W. S. Zaugg, T. Coley, and T. Y. Berila. 1988. Susceptibility of Chinook salmon smolts to bypass systems at hydroelectric dams. *North American Journal of Fisheries Management* 8:25–29.

Giorgi, A. E., D. R. Miller, and B. P. Sanford. 1990. Migratory behavior and adult contribution of summer outmigrating subyearling Chinook salmon in John Day Reservoir, 1981–1983. Bonneville Power Administration DOE/BP-39645-3. Portland, Oregon. 68 p.

Giorgi, A. E. 1993. Flow augmentation and reservoir drawdown: Strategies for recovery of threatened and endangered stocks of Salmon in the Snake River Basin. Bonneville Power Administration. DOE/BP-99654-2. Portland, Oregon.

Giorgi, A. E., D. R. Miller, and B. P. Sanford. 1994. Migratory characteristics of juvenile ocean-type Chinook salmon, *Oncorhynchus tshawytscha*, in John Day Reservoir on the Columbia River. *Fishery Bulletin* 92:872–879.

Giorgi, A. E., T. W. Hillman, J. R. Stevenson, S. G. Hays, and C. M. Pevan. 1997. Factors that influence the downstream migration rates of juvenile salmon and steelhead through the hydroelectric system in the mid-Columbia River Basin. *North American Journal of Fisheries Management* 17:268–282.

Giorgi, A. E., M. Miller, and J. Stevenson. 2002. Mainstem Passage Strategies in the Columbia River System: Transportation, Spill and Flow Augmentation. Doc. 2002-3. Northwest Power Planning Council, Portland, Oregon. 97 pp.

Glova, G. J., and J. E. McInerney. 1977. Critical swimming speeds of coho salmon (*Onchorhynchus kisutch*) fry to smolt stages in relation to salinity and temperatures. *Journal of the Fisheries Research Board of Canada* 34:151–154.

Gregory, R. S., and T. G. Northcote. 1993. Surface, planktonic, and benthic foraging by juvenile Chinook salmon (*Oncorhynchus tshawytscha*) in turbid laboratory conditions. *Canadian Journal of Fisheries and Aquatic Sciences* 50:233–240.

Groot, C. 1965. On the orientation of young sockeye salmon (*Oncorhynchus nerka*) during their seaward migration out of lakes. *Behaviour* 14(Suppl.):1–198.

Groot, C. 1982. Modifications on a theme–A perspective on migratory behavior of Pacific salmon. Pages 1–21 in E. L. Brannon and E. O. Salo, eds. *Salmon and trout migratory behavior symposium.* University of Washington School of Fisheries, Seattle.

Groot, C., and L. Margolis, eds. 1991. *Pacific salmon life histories.* University of British Columbia Press, Vancouver.

Groves, P. A. 1993. Habitat available for, and used by, fall Chinook salmon within the Hells Canyon Reach of the Snake River. Annual Progress Report 1992. Environmental Affairs Deartment, Idaho Power Company, Boise, Idaho. 127 p.

Hansen, L. P., and B. Jonsson. 1985. Downstream migration of hatchery-reared smolts of Atlantic salmon (*Salmo salar*) in the River Imsa, Norway. *Aquaculture* 45:237–248.

Hartman, W. L., W. R. Heard, and B. Drucker. 1967. Migratory behaviour of sockeye salmon fry and smolts. *Journal of the Fisheries Research Board of Canada* 24:2069–2099.

Healey, M. C. 1991. Life history of Chinook salmon. Pages 313–393 in C. Groot and L. Margolis, eds. *Pacific salmon life histories.* University of British Columbia Press, Vancouver.

Hesthagen, T., and E. Garnas. 1986. Migration of Atlantic salmon smolts in River Orkla of central Norway in relation to management of a hydroelectric station. *North American Journal of Fisheries Management* 6:376–382.

Hillman, T., and D. Chapman. 1989. Abundance, growth and movement of juvenile Chinook salmon and steelhead. Pages 1–41 in Summer and winter ecology of juvenile Chinook salmon and steelhead trout in the Wenatchee River, Washington. Report to Chelan County Public Utility District, Boise, Idaho. Don Chapman Consultants, Boise, Idaho.

Hoar, W. S. 1954. The behavior of juvenile Pacific salmon, with particular reference to the sockeye (*Oncorhynchus nerka*). *Journal of the Fisheries Research Board of Canada* 11:69–96.

Hoar, W. S. 1976. Smolt transformation: Evolution behavior and physiology. *Journal of the Fisheries Research Board of Canada* 33:1233–1252.

Horner, N., and T. C. Bjornn. 1979. Status of Upper Columbia River Fall Chinook Salmon. US Fish and Wildlife Service, Boise, Idaho. 45 p.

Hutchinson, G. E. 1957. *A treatise on limnology. Volume I. Geography, physics, and chemistry.* John Wiley & Sons, New York.

Independent Scientific Advisory Board (ISAB). 1997. Ecological impacts of the Biological Opinion for endangered Snake River salmon on resident fishes in the Hungry Horse and Libby systems in Montana, Idaho, and British Columbia. Northwest Power Planning Council. ISAB Report, 97-3. Portland, Oregon.

ISAB. 1998. Recommendation for stable flows in the Hanford Reach during the time when juvenile fall Chinook are present each spring. Northwest Power Planning Council and the National Marine Fisheries Service. ISAB Report, 98-5. Portland, Oregon.

ISAB. 2003. Review of flow augmentation: Update and clarification. Northwest Power and Conservation Council and National Marine Fisheries Service. ISAB Report, 2003-1. Portland, Oregon.

Irvine, J. R. 1986. Effects of varying discharge on the downstream movement of salmon fry, *Oncorhynchus tshawytscha* Walbaum. *Journal of Fish Biology* 28:17–28.

Jager, H. I., H. E. Cardwell, M. J. Sale, M. S. Bevelhimer, C. C. Coutant, and W. V. Winkle. 1997. Modelling the linkages between flow management and salmon recruitment in rivers. *Ecological Modeling* 103:171–191.

Janecek, B. F. U., and O. Moog. 1994. Origin and composition of the benthic invertebrate riprap fauna of impounded rivers. *Verhandlungen Internationale Vereinigung fur Theoretische und Angewandte Limnologie* 25:1624–1630.

Johnsen, R. C., L. A. Wood, and W. W. Smith. 1987. Monitoring of Downstream Salmon and Steelhead at Federal Hydroelectric Facilities–1986. Bonneville Power Administration. DOE/BP-20733-2. Portland.

Johnson, G. E., C. M. Sullivan, and J. M. W. Erho. 1992. Hydroacoustic studies for developing a smolt bypass system at Wells Dam. Pages 221–237 in *Fisheries Research* 14.

Johnson, L., and R. Wright. 1987. Hydroacoustic evaluation of the spill program for fish passage at John Day Dam in 1987. Associated Fish Biologists, Inc. Report to the US Army Corps of Engineers. Contract No. CADW57-87-C-0077. Portland, Oregon. 77 p.

Johnson, W. E., and C. Groot. 1963. Observations on the migration of young sockeye salmon, *Oncorhynchus nerka*, through a large complex lake system. *Journal of the Fisheries Research Board of Canada* 20:919–938.

Jonsson, N. 1991. Influence of water flow, water temperature and light on fish migration in rivers. *Nordic Journal of Freshwater Research* 66:20–35.

Key, L. O., J. A. Jackson, C. R. Sprague, and E. E. Kerfoot. 1994. Nearshore habitat use by subyearling Chinook salmon in the Columbia and Snake Rivers. Pages 120–150 in D. W. Rondorf and W. H. Miller, eds. *Identification of the spawning, rearing, and migratory requirements of fall Chinook salmon in the Columbia River Basin.* Bonneville Power Administration, Portland, Oregon.

Key, L. O., R. Garland, and E. E. Kerfoot. 1995. Nearshore habitat use by subyearling Chinook salmon in the Columbia and Snake Rivers. Pages 74–107 in D. W. Rondorf and K. F. Tiffan, eds. *Identification of the spawning, rearing, and migratory requirements of fall Chinook salmon in the Columbia River Basin.* Bonneville Power Administration, Portland, Oregon.

Kiefer, R. B., and J. N. Lockhart. 1995. Intensive evaluation and monitoring of Chinook salmon and steelhead trout production, Crooked River and Upper Salmon River sites. Bonneville Power Administration. DOE/BP-21182-5. Portland, Oregon.

Koski, C. H. 1974. Time of travel. Pages 33–43 in *Anatomy of a river.* Pacific Northwest River Basins Commission, Vancouver, Washington.

Krcma, R. F., and R. F. Raleigh. 1970. Migration of juvenile salmon and trout into Brownlee Reservoir, 1962-65. *Fishery Bulletin* 68:203–217.

Ledgerwood, R. D., F. P. Thrower, and E. M. Dawley. 1990. Diel sampling of migratory juvenile salmonids in the Columbia River estuary. *U.S. Fishery Bulletin* 89:69–78.

Lemmin, U., and C. H. Mortimer. 1986. Tests of an extension to internal seiches of Defant's procedure for determination of surface seiche characteristics in real lakes. *Limnology and Oceanography* 31:1207–1231.

Liao, J. C., D. N. Beal, G. V. Lauder, and M. S. Triantafyllou. 2003. Fish exploiting vortices decrease muscle activity. *Science* 302:1566–1569.

Lichatowich, J. A., and L. E. Mobrand. 1995. Analysis of Chinook salmon in the Columbia River from an ecosystem perspective. Mobrand Biometrics. Research Report.

Lindsey, C. C., and J. D. McPhail. 1986. Zoogeography of fishes of the Yukon and Mackenzie basins. Pages 639–674 in C. H. Hocutt and E. O. Wiley, eds. *Zoogeography of North American Freshwater Fishes*. John Wiley and Sons, New York.

Lister, D. B., and H. S. Genoe. 1970. Stream habitat utilization by cohabiting underyearlings of Chinook (*Oncorhynchus tshawytscha*) and coho (*O. kisutch*) salmon in the Big Qualicum River, British Columbia. *Journal of the Fisheries Research Board of Canada* 27:1215–1224.

Logie, P. 1993. Power System Coordination. A Guide to the Pacific Northwest Coordination Agreement. Columbia River System Operation Review. Joint Project of US Department of Energy, Bonneville Power Administration; US Department of the Army, Corps of Engineers, North Pacific Division; and US Department of the Interior, Bureau of Reclamation, Pacific Northwest Region. 46 p.

Long, C. W. 1968a. Diel movement and vertical distribution of juvenile anadromous fish in turbine intakes. *U.S. Fishery Bulletin* 66:599–609.

Long, C. W. 1968b. Research on Fingerling Mortality in Kaplan Turbines. NMFS/NOAA Northwest Fisheries Science Center, Seattle, Washington.

Mains, E. M., and J. M. Smith. 1963. The distribution, size, time and current preferences of seaward migrant Chinook salmon in the Columbia and Snake Rivers. *Washington Department of Fisheries Research Paper* 2:5–43.

Marquett, W., F. J. Ossiander, R. Duncan, and C. W. Long. 1970. Research on the Gatewell-Sluice Method of Bypassing Downstream Migrant Fish around Low Head Dams. NMFS/NOAA Northwest Fisheries Science Center. Report to U.S. Army Corps of Engineers. 32 p.

Massey, J. B. 1967. The downstream migration of juvenile anadromous fish at Willamette Falls, Oregon. Oregon State Game Commission, Fishery Division. Columbia River Fishery Development Program Progress Report, Portland, Oregon. 17 p.

Matthews, G. M., S. Achord, J. R. Harmon, O. W. Johnson, D. M. Marsh, B. P. Sandford, N. N. Paasch, K. W. McIntyre, and K. L. Thomas. 1992. Evaluation of transportation of juvenile salmonids and related research on the Columbia and Snake Rivers, 1990. US Army Corps of Engineers. Contract No. DACW68-84-H0034. Portland, Oregon.

Mattson, C. R. 1962. Early Life History of Willamette River Spring Chinook Salmon. Fish Commission of Oregon. Portland. 50 p.

Maule, A. G., J. W. Beeman, R. M. Schrock, and P. V. Harner. 1994. Assessment of smolt condition for travel time analysis: Annual Report 1991-1992. Bonneville Power Administration. DOE/BP-35245-5. Portland, Oregon. 192 p.

McCabe, G. T., Jr., R. L. Emmett, W. D. Muir, and T. H. Blahm. 1986. Utilization of the Columbia River estuary by subyearling Chinook salmon. *Northwest Science* 60:113–124.

McCart, P. 1967. Behaviour and ecology of sockeye salmon fry in the Babine River. *Journal of the Fisheries Research Board of Canada* 24:375–428.

McCleave, J. D. 1978. Rhythmic aspects of estuarine migration of hatchery-reared Atlantic salmon (*Salmo salar*) smolts. *Journal of Fish Biology* 12:559–570.

McConnaha, W. E. 1993. History of flow survival studies in Columbia Basin. Northwest Power Planning Council. Memorandum to Don Bevan, Chairman, NMFS Salmon Recovery Team, Portland, Oregon. 7 p.

McDonald, J. 1960. The behavior of Pacific salmon fry during their downstream migration to freshwater and saltwater nursery areas. *Journal of the Fisheries Research Board of Canada* 17:655–676.

McNeil, W. J. 1992. Timing of passage of juvenile salmon at Columbia River dams. Report for Direct Service Industries, Portland, Oregon. 60 p.

McPhail, J. D., and C. C. Lindsey. 1986. Zoogeography of the freshwater fishes of Cascadia (the Columbia system and rivers north to the Stikine). Pages 615–637 in C. H. Hocutt and E. O. Wiley, eds. *Zoogeography of North American Freshwater Fishes.* John Wiley and Sons, New York.

Meehan, W. R., ed. 1991. *Influences of forest and rangeland management of salmonid fishes and their habitats.* American Fisheries Society, Bethesda, Maryland.

Meekin, T. K. 1967. Observations of exposed fall Chinook redds below Chief Joseph Dam during periods of low flow. Washington Department of Fisheries. Report to Douglas County P.U.D., No. 1. East Wenatchee, Washington. 25 p.

Mendel, G., D. Milks, M. Clizer, and R. Bugert. 1994. Upstream passage and spawning of fall Chinook salmon in the Snake River. Washington Department of Fisheries. Report to Bonneville Power Administration. Contract Number DE-BI 79-92 BP60415. Olympia, Washington.

Miller, D. R., and C. W. Sims. 1984. Effects of flow on the migratory behavior and survival of juvenile fall and summer Chinook salmon in John Day Reservoir. National Marine Fisheries Service. Annual Report to Bonneville Power Administration. Contract DE-A179-83BP39645. Portland, Oregon.

Mote, P. W., and 10 co-authors. 2003. Preparing for climate change: The water, salmon and forests of the Pacific Northwest. *Climatic Change* 61:45–88.

Muir, W. D., S. G. Smith, R. N. Iwamoto, D. J. Kamikawa, K. W. McIntyre, E. E. Hockersmith, B. P. Sandford, P. A. Ocker, T. E. Ruehle, and J. G. Williams. 1995. Survival estimates for the passage of juvenile salmonids through Snake River Dams and Reservoirs, 1994. Bonneville Power Administration. DOE/BP-10891-2, Contract No. DE-A179-93BP10891, COE Walla Walla District Delivery Order E86940119. Portland, Oregon. 187 p.

Muir, W. D. et al. 1996. Survival estimates for the passage of yearling Chinook salmon and steelhead through Snake River dams and reservoirs, 1995. Northwest Fisheries Science Center, National Marine Fisheries Service, Seattle, Washington.

Mullan, J. W. 1986. Determinants of sockeye salmon abundance in the Columbia River, 1880's–1982: A review and synthesis. US Fish and Wildlife Service. Biological Report, 86(12). 136 p.

Mullan, J. W., M. B. Dell, S. G. Hays, and J. A. McGee. 1986. Some factors affecting fish production in the Mid-Columbia River 1934-1983. US Fish and Wildlife Service. FRI/FAO-86-15. 69 p.

National Marine Fisheries Service (NMFS). 1995. Biological Opinion on reinitiation of consultation on 1994-1998 operation of the federal Columbia River Power System and juvenile transportation program in 1994-1998. US Army Corps of Engineers, Bonneville Power Administration, Bureau of Reclamation, and National Marine Fisheries Service. Portland, Oregon. 146 p.

NMFS. 2000. Endangered Species Act, Section 7 Consultation. Biological Opinion. US Army Corps of Engineers, Bonneville Power Administration, Bureau of Reclamation, and National Marine Fisheries Service. Portland, Oregon.

Nelson, W. R., L. K. Freidenburg, and D. W. Rondorf. 1994. Swimming performance of subyearling Chinook salmon. Pages 39–62 in D. W. Rondorf and W. H. Miller, eds. *Identification of the spawning, rearing, and migratory requirements of fall Chinook salmon in the Columbia River Basin.* Bonneville Power Administration, Portland, Oregon.

Northcote, T. G. 1962. Migratory behaviour of juvenile rainbow trout, *Salmo gairdnerii*, in outlet and inlet streams of Loon Lake, British Columbia. *Journal of the Fisheries Research Board of Canada* 19:201–270.

Northcote, T. G. 1984. Mechanisms of fish migration in rivers. Pages 317–355 in J. D. McCleave, G. P. Arnold, J. J. Dodson, and W. H. Neill, eds. *Mechanisms of Migration in Fishes.* Plenum, New York.

Northwest Power Planning Council (NPPC). 1994a. Columbia River Basin Fish and Wildlife Program. Northwest Power Planning Council, Portland, Oregon.

NPPC. 1994b. Discussion Paper on Mainstem Passage Hypotheses. Northwest Power Planning Council, 94-11. Portland, Oregon. 71 p.

NPPC. 1994c. Harvest Management. Northwest Power Planning Council. Staff Briefing Paper, 94-44. Portland, Oregon.

Pacific Northwest River Basins Commission. 1974. Anatomy of a River. Pacific Northwest River Basins Commission, Vancouver, Washington.

Park, D. L. 1969. Seasonal changes in downstream migration of age-group 0 Chinook salmon in the upper Columbia River. *Transactions of the American Fisheries Society* 98:315– 317.

Peak, S., and R. S. McKinley. 1998. A re-evaluation of swimming performance in juvenile salmonids relative to downstream migration. *Canadian Journal of Fish and Aquatic Sciences* 55: 682–687.

Perry, S. A., and W. B. Perry. 1991. Organic carbon dynamics in two regulated rivers in north-western Montana, USA. *Hydrobiologia* 218:193–203.

Peven, C. M., R. R. Whitney, and K. R. Williams. 1994. Age and length of steelhead smolts from the mid-Columbia basin, Washington. *North American Journal of Fisheries Management* 14:77–86.

Plumb, J. M., M. S. Novick, A. C. Braatz, J. N. Lucchesi, J. M. Sprando, N. S. Adams, and D. W. Rondorf. 2003. Behavior and migratory delay of radio-tagged juvenile spring Chinook salmon and steelhead through Lower Granite Dam and reservoir during a drought year. USGS Columbia River Research Laboratory, Cook, Washington. Report for US Army Corps of Engineers, Contract W68SBV00104592, Walla Walla, Washington.

Raymond, H. L. 1968. Migration rates of yearling Chinook salmon in relation to flows and impoundments in the Columbia and Snake Rivers. *Transactions of the American Fisheries Society* 97:356–359.

Raymond, H. L., C. W. Sims, R. C. Johnsen, and W. W. Bently. 1975. Effects of power peaking operations on juvenile salmon and steelhead trout migrations, 1974. Northwest Fisheries Science Center, National Marine Fisheries Service. Report to US Army Corps of Engineers, Seattle, Washington.

Raymond, H. L. 1979. Effects of dams and impoundments on migrations of juvenile Chinook salmon and steelhead from the Snake River, 1966 to 1975. *Transactions of the American Fisheries Society* 108:505–529.

Reimers, P. E. 1973. The length of residence of juvenile fall Chinook salmon in Sixes River, Oregon. *Research Reports of the Fish Commission of Oregon* 4:3–43.

Reisenbichler, R., and S. R. Phelps. 1989. Genetic variation in steelhead (*Salmo gairdnerii*) from the North Coast of Washington. *Canadian Journal of Fisheries and Aquatic Sciences* 46:66–73.

Reisenbichler, R. R., J. D. McIntyre, M. F. Solazzi, and S. W. Landino. 1992. Genetic variation in steelhead of Oregon and Northern California. *Transactions of the American Fisheries Society* 121:158–169.

Rhodes, J. J., D. A. McCullough, and J. F. A. Espinosa. 1994. A coarse screening process for evaluation of the effects of land management activities on salmon spawning and rearing habitat in ESA consultations. Columbia River Inter-Tribal Fish Commission. Technical Report, 94-1. Portland, Oregon.

Rich, W. H. 1920. Early history and seaward migration of Chinook salmon in the Columbia and Sacramento Rivers. *Bulletin of the US Bureau of Fisheries* 37.

Rondorf, D. W., G. A. Gray, and R. B. Fairley. 1990. Feeding ecology of subyearling Chinook in riverine and resevoir habitats of the Columbia River. *Transactions of the American Fisheries Society* 119:16–24.

Royal, L. A. 1972. An examination of the anadromous trout program of the Washington State Game Department. Washington State Department of Game. Olympia, Washington. 176 p.

Salo, E. O., and T. W. Cundy. 1987. Streamside management: Forestry and fishery interactions. University of Washington, Seattle.

Saunders, R. L. 1965. Adjustment of buoyancy in young Atlantic salmon and brook trout by changes in swim bladder volume. *Journal of the Fisheries Research Board of Canada* 22:335–352.

Schreck, C. B. and. H. W. Li. 1984. Columbia River salmonid outmigration: McNary Dam passage and enhanced smolt quality. Bonneville Power Administration, Project 82-16. Portland, Oregon. 117 p.

Schreck, C. B., J. C. Snelling, R. E. Ewing, C. S. Bradford, L. E. David, and C. H. Slater. 1995. Migratory characteristics of juvenile spring Chinook salmon in the Willamette River. Bonneville Power Administration. DOE/BP-92818-5. Portland, Oregon.

Silliman, R. P. 1950. Fluctuations in abundance of Columbia River Chinook salmon (*Oncorhynchus tschawytscha*) 1935-45. *Fishery Bulletin* 51:364–383.

Sims, C. W., and F. J. Ossiander. 1981. Migrations of juvenile Chinook slamon and steelhead trout in the Snake River, from 1973 to 1979: A research summary. NMFS/NOAA Northwest and Alaska Fisheries Center. Final Report to US Army Corps of Engineers. Contract No. DACW69-78-C-0038. Portland, Oregon. 31 p.

Smith, J. R., J. R. Pugh, and G. E. Monan. 1968. Horizontal and vertical distribution of juvenile salmonids in Upper Mayfield Reservoir, Washington. US Fish and Wildlife Service Special Scientific Report–Fisheries No. 566.

Smith, J. R. 1974. Distribution of seaward-migrating Chinook salmon and steelhead trout in the Snake River above Lower Monumental Dam. *Marine Fisheries Review* 36:42–45.

Smith, L. S. 1982. Decreased swimming performance as a necessary component of the smolt migration in salmon in the Columbia River. *Aquaculture* 28:153–161.

Smith, S. G., W. D. Muir, E. E. Hockersmith, S. Achord, M. Eppard, T. E. Ruehle, and J. G. Williams. 1997a. Survival estimates for the passage of juvenile salmon through Snake River dams and reservoirs, 1996. Northwest Fisheries Science Center, NMFSNOAA. Report to Bonneville Power Administration, DE-AI79-93BP10891. Seattle, Washington. 192 p.

Smith, S. G., W. D. Muir, E. E. Hockersmith, M. B. Eppard, and W. P. Connor. 1997b. Passage survival of hatchery subyearling fall Chinook salmon to Lower Granite, Little Goose, and Lower Monumental dams. NMFS/NOAA and USFWS. Northwest Fisheries Science Center and Idaho Fisheries Resource Office. Annual Report to US Army Corps of Engineers and Bonneville Power Administration, Contract No. 93AI10891. Seattle, Washington. 65 p.

Smith, S. G., , W. D. Muir, J. G. Williams, and J. R. Skalski. 2002. Factors associated with travel time and survival of migrant young Chinook salmon and steelhead in the lower Snake River. *North American Journal of Fisheries Management* 22:385–405.

Snelling, J. C., and C. B. Schreck. 1994. Movement, Distribution, and Behavior of Juvenile Salmonids Passing through Columbia and Snake River Dams. Bonneville Power Administration. 82-003. Portland, Oregon.

Solomon, D. J. 1978. Some observations on salmon smolt migration in a chalkstream. *Journal of Fish Biology* 12:571–574.

Stanford, J. A., and F. R. Hauer. 1992. Mitigating the impacts of stream and lake regulation in the Flathead River catchment, Montana, USA: An ecosystem perspective. *Aquatic Conservation: Marine and Freshwater Ecosystems* 2:35–63.

Taylor, E. B. 1990. Environmental correlates of life-history variation in juvenile Chinook salmon, *Oncorhynchus tshawytscha* (Walbaum). *Journal of Fish Biology* 37:1–17.

Thorpe, J. E., and R. I. G. Morgan. 1978. Periodicity in Atlantic salmon *Salmo salar L.* smolt migration. *Journal of Fish Biology* 12:541–548.

Thorpe, J. E., L. G. Ross, G. Struthers, and W. Watts. 1981. Tracking Atlantic salmon smolts, *Salmo salar L.*, through Loch Voil, Scotland. *Journal of Fish Biology* 19:519–537.

Thorpe, J. E. 1982. Downstream movements of juvenile salmonids: a forward speculative view. Pages 387-395 in J. D. McCleave, G. P. Arnold, J. J. Dodson, and W. H. Neill, eds. *Mechanisms of migration in fishes*. Plenum, New York.

Tiffan, K. F., R. D. Garland, and D. W. Rondorf. 2002. Quantifying flow-dependent changes in subyearling fall Chinook salmon rearing habitat using two-dimensional spatially explicit modeling. *North American Journal of Fisheries Management* 22:713–726.

US Fish and Wildlife Service. 1995. Fish and Wildlife Coordination Act Report on the Columbia River system operation review. US Fish and Wildlife Service. Portland, Oregon. 45 p.

Venditti, D. A., D. W. Rondorf, and J. M. Kraut. 2000. Migratory behavior and forebay delay of radio-tagged juvenile fall Chinook salmon in a lower Snake River impoundment. *North American Journal of Fisheries Management* 20:41–52.

Wagner, P. G., J. Nugent, W. Price, R. Tudor, and P. Hoforth. 1999. 1997-1999 evaluation of juvenile fall Chinook stranding on the Hanford Reach. Annual Report to Bonneville Power Administration, Contract 97B130417. Portland, Oregon.

Walters, C. J., R. M. Hilborn, R. M. Peterman, and M. J. Staley. 1978. Model for examining early ocean limitation of Pacific salmon production. *Journal of the Fisheries Research Board of Canada* 35:1303–1315.

Ward, D. L., A. A. Nigro, R. A. Farr, and C. J. Knutsen. 1994. Influence of waterway development on migrational characteristics of juvenile salmonids in the lower Willamette River, Oregon. *North American Journal of Fisheries Management* 14:362–371.

Watson, D. G., C. E. Cushing, C. C. Coutant, and W. L. Templeton. 1969. Effect of Hanford reactor shutdown on Columbia River biota. Pages 291–299 in D. J. Nelson and F. C. Evans, editors. USAEC CONF-670503, National Technical Information Center, Springfield, Virginia. *Proceedings of the Second National Symposium on Radioecology, May 15-17, 1967, Ann Arbor, Michigan.*

Wedemeyer, G. A., R. L. Saunders, and W. C. Clarke. 1980. Environmental factors affecting smoltification and early marine survival of anadromous salmonids. *Marine Fishing Review* 42:1–14.

Whitney, R. R., L. D. Calvin, M. W. Erho Jr., and C. C. Coutant. 1997. Downstream passage for salmon at hydroelectric projects in the Columbia River Basin: Development, installation, and evaluation. Northwest Power Planning Council. Technical Report, 97-15. Portland, Oregon. 101 p.

Williams, J. G., and G. M. Mathews. 1994. A review of flow/survival relationships for juvenile salmonids in the Columbia River Basin. *Fishery Bulletin* 93:732–740.

Return *to the* River

7

Mitigation of Salmon Losses Due to Hydroelectric Development

Richard R. Whitney, Ph.D., Charles C. Coutant, Ph.D., Phillip R. Mundy, Ph.D.

Introduction
  Mitigation Approach
  Legal Authorities for Mitigation
  Background
Dams as Obstacles for Anadromous Fish
  Adult Upstream Passage
  Downstream Passage of Juveniles
  Mortalities at the Dams
  Bypass Provisions for Juvenile Salmon and Steelhead at the Dams
  Diverting Juvenile Salmonids at the Turbine Intakes
    Turbine Intake Screens
    Application of Turbine Intake Screens for Bypass of Juveniles
    Fish Behavior and Fish Passage Efficiency
  Diverting Juvenile Salmonids into Spill and Surface Bypass Systems and Devices
    Provision of Spill as a Means of Smolt Bypass
    Spill Effectiveness and Spill Efficiency
    Ice and Trash Sluiceways—Effectiveness in Passing Juvenile Salmonids
    Application of Surface Collection at Other Projects
    Attracting Juvenile Salmon Directly into Spill for Passage
    Survival of Juvenile Salmonids Passing the Dam in Spill
    Supersaturation of Gas due to Spill
  Mortality of Juvenile Salmonids in Reservoirs
    Predation
  Proposed, but Largely Untried, Methods for Juvenile Bypass
  Transportation of Juvenile Salmonids
  Managing Passage of Juvenile Salmonids through the Hydrosystem
Dams as Regulators of Flow
  Effects of Flow Regulation on Adult Anadromous Fishes
  Effects of Flow Regulation on Juvenile Salmonids
    Mortalities of Juvenile Salmonids in River Reaches
    Water Budget and Flow Augmentation

   Mitigation of Effects of Load Following on Survival of Juvenile Salmonids
  Effects of the Hydroelectric System on Water Temperature
   Mitigation of Effects on Temperature
 Summary of Flow Effects
  Mitigation Efforts and Their Effectiveness
  Passage Goals and Attempts to Achieve Them
Summary
Literature Cited

> "Our experience with these adult returns has made us believe we can work wonders with this transportation system. We believe that . . . combining the traveling screens, . . . expanding the transportation effort on schedule, and adding spillway deflectors at the dams to reduce nitrogen concentration, we can restore adult steelhead trout to their former levels within two or three years. After the Snake River Mitigation Plan is approved by Congress, it seems possible that we can establish adult runs of both steelhead trout and salmon in far greater numbers than ever existed before."
>
> —Wes Ebel, 1977, presenting the National Marine Fisheries Service's perspective on major fish passage problems and their solutions in Columbia River Salmon and Steelhead, American Fisheries Society.
>
> "Never have so many capable people labored so hard for so long to produce so little."
>
> —Al Wright, independent consultant, referring to Columbia River restoration efforts at a conference titled "Columbia River Management—Time to Rewire the System," held December 15, 1997, in Seattle, Washington.

Introduction

The quotations at the heading of this chapter represent two extremes of opinion concerning the effectiveness of measures being taken to restore populations of salmon and steelhead in the Columbia Basin. The truth lies somewhere in between. There has been some degree of success; however, it is also clear that development and operation of the hydroelectric system have resulted in lasting adverse effects on salmon and steelhead abundance, productivity, and diversity. Rigorous quantification of the adverse effects has not been possible because of the numerous contributing and interacting causes of the declines in salmon and steelhead abundance, such as overfishing, general habitat degradation due to poor land management practices, pollution, and other causes (National Research Council [NRC] 1996; Chapters 5 and 9).

From 1933 to 1975, full development of the hydroelectric system proceeded (Figure 7.1) to the point that Grand Coulee Dam first blocked the Columbia River mainstem in 1941, and Hells Canyon Dam blocked the Snake River in 1967. Later (1955), Chief Joseph Dam was constructed 52 miles downstream of Grand Coulee Dam and blocked the river there. Of the original 1,200 miles of river available to salmon and steelhead in the mainstem Columbia River, only 545 miles remain accessible below Chief Joseph Dam, and only 55 of those miles, in the Hanford Reach, remain undammed.

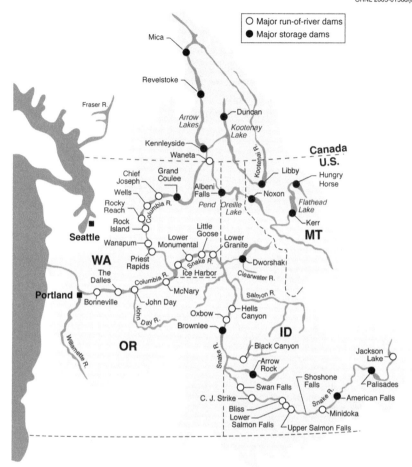

ORNL 2005-01568/jcn

Figure 7.1   Map of the Columbia River Basin showing mainstem run-of-the-river and storage dams. Map shows headwater areas in Canada, Montana, and Wyoming.

In the Snake River Basin, salmon formerly had access to 500 miles of river up to Shoshone Falls; however, that was halved with the construction of Hells Canyon Dam in 1967 at river mile 247. About 100 miles of river immediately below Hells Canyon Dam remain as fluvial habitat down to Lower Granite Dam but with flow regulated by the series of hydropower dams upstream. Of the remaining habitat accessible to salmon, 80% consists of reservoirs and 20% remains riverine habitat although regulated by upstream dams. In terms of drainage area, including the mainstem Columbia and

328   Whitney *et al.*

Snake River tributaries, about 45% of the original total in the Columbia Basin remains accessible to anadromous salmonids (Figure 7.2) (NWPPC 1982). Development of the 13 Columbia and Snake River mainstem dams has had profound impacts not only on Columbia Basin salmon and steelhead stocks but also on the region's economy and society. The mainstem hydroelectric projects provided the region with abundant electricity (Figures 7.3 and 7.4) and irrigation water (Figure 7.5), yet inundated or blocked vast amounts of salmon and steelhead spawning and rearing habitat (Figure 7.6),

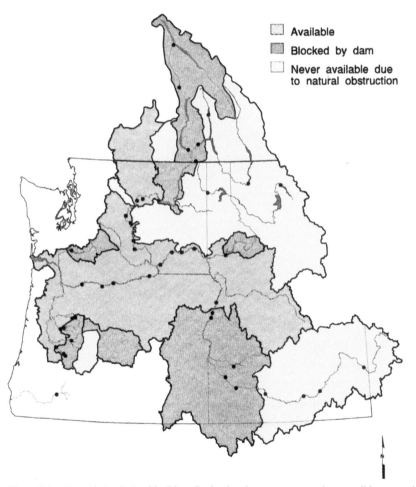

Figure 7.2 Map of the Columbia River Basin showing areas presently accessible to anadromous salmonids, areas now blocked, but previously accessible, and areas never accessible. Courtesy of Bonneville Power Administration.

Figure 7.3　Bank of electricity-generating turbines in Bonneville Dam. Photo from U.S. Army Corps of Engineers Digital Visual Library, website at *http://images.usace.army.mil/photolib.html*.

Figure 7.4　Power lines coming from the powerhouse at Rocky Reach Dam in the mid-Columbia in central Washington. Photo by R. N. Williams.

Figure 7.5   Irrigation diversion dam on Taneum Creek in the Yakima Basin. Note the rotating drum screens that divert outmigrating juvenile salmon and steelhead away from the irrigation canal and into a bypass flow that returns them to the river to continue their downstream migration. Photo by R. N. Williams.

and inundated historic tribal fishing grounds (Figure 7.7) and important tribal cultural or archaeological sites (Figure 7.8). Socioeconomic benefits of the electric power generated were offset to a degree by losses to the commercial, recreational, and tribal fisheries.

Mitigation Approach

In this chapter, we describe the adverse effects of that development and operation on anadromous fishes and discuss mitigation measures that have been undertaken to restore the fish. The effort to restore populations of anadromous salmon and steelhead in the Columbia Basin is perhaps one of the most ambitious environmental undertakings in modern times and is a response to development of the basin for hydroelectric power, irrigation, and flood control. These developments took place in the early 1900s, culminating in the construction of the four federal power dams on the lower Snake River in the mid-1970s. Today, flow in the river is highly regulated for these purposes.

Figure 7.6 Fall Chinook spawning and rearing habitat in the lower Snake River inundated by Lower Granite Dam. Photo from U.S. Army Corps of Engineers Digital Visual Library, website at *http://images.usace.army.mil/photolib.html*.

Figure 7.7 Historic Native American tribal fishing site at Celilo Falls, near The Dalles, Oregon, inundated by The Dalles Dam. Photo from U.S. Army Corps of Engineers Digital Visual Library, website at *http://images.usace.army.mil/photolib.html*.

Figure 7.8 Native American petroglyph found near The Dalles, Oregon. Photo from U.S. Army Corps of Engineers Digital Visual Library, website at *http://images.usace.army.mil/photolib.html.*

Early efforts to mitigate the effects of hydrosystem development on salmon and steelhead abundance focused primarily on engineered or mechanical fixes, rather than solutions centered on salmon behavior and ecology. For example, solutions for passage of juvenile salmon have focused upon providing hatcheries, installing turbine intake screens to divert juvenile salmon away from the turbines, providing barges for transportation of diverted salmon to the river below Bonneville Dam to avoid as many dams as possible, and the like. In contrast, adult passage solutions were a mix of engineering and biological approaches, as it was necessary to study salmon behavior in order to identify appropriate flows and velocities that would help migrating adults find and use fish ladders at the various hydroelectric projects.

The primary mitigation response to development of the Columbia and Snake River hydroelectric system has been investments in passage technology to decrease adult and juvenile mortalities associated with upstream and downstream migrations and development of a widespread system of hatcheries designed to produce fish to offset the losses in production that occurred through blockage, inundation, and degradation of habitat. Reports to Congress from the General Accountability Office (GAO) note that nearly equal amounts have been spent on passage and hatchery technologies that total to approximately 75% of the more than $3 billion spent on Columbia Basin salmon recovery efforts through 1991 (GAO 1992, 2002).

The engineering approach remains the primary mitigation approach today, although recent mitigation efforts are increasingly incorporating fish behavior and normative ecological conditions into the proposed solutions. For example, surface collector devices that do not depend upon flow into the turbines to attract juveniles into a bypass channel are being examined as alternative passage routes for juvenile salmon. While this approach relies on substantial engineering and planning, it also takes advantage of the normal behavior of juvenile salmon to orient themselves within the surface waters during their downstream migrations and therefore, provides passage conditions that are more normal (normative) for outmigrating juvenile salmon than passage through the turbine intakes and the bypass systems.

Legal Authorities for Mitigation

The U.S. Congress has served as the legal authority for measures aimed at mitigating for losses caused to fish and wildlife by development and operation of the federal portion of the hydroelectric system, through provisions in the legislation authorizing its construction and operation. Construction and operation of hatcheries were the measure of choice in the past (Lichatowich 1999; Chapter 8). Hatcheries intended for mitigation of losses of salmon and steelhead have been financed, constructed, and operated with Federal funds as part of the Grand Coulee Fish Maintenance Program, the Mitchell Act, the Lower Snake River Compensation Plan, and others.

The Federal Energy Regulatory Commission (FERC) has the responsibility of licensing non-federal power projects in the Columbia Basin. On the mainstem, there are five projects owned and operated by three public utility districts (PUDs), licensed by FERC and established according to State of Washington statutes. As provided in their FERC licenses, each of the three PUDs constructed and financed the operation of hatcheries as part of their mitigation responsibilities. The present chapter focuses on mitigation measures other than hatcheries, which are discussed in detail in Chapter 8 of this volume.

Congress adopted the Pacific Northwest Electric Power Planning and Conservation Act in 1980 (the Power Act), which created a coordinating body, the Northwest Electric Power Conservation Council, subsequently renamed the Northwest Planning and Conservation Council (NPCC) (note that references to council publications in this chapter will use whichever acronym was accurate at the time of publication). The Act authorized construction and operation of an electricity intertie grid that allows flexibility in exportation and importation of power in the Pacific Northwest, through power lines controlled by the Bonneville Power Administration (BPA). In addition, the Act called for the Council to develop a Fish and Wildlife Program to address the losses of fish and wildlife resulting from construction and operation of the hydroelectric system.[1] Federal funds to accomplish mitigation of losses were specified to be supplied by BPA. The Council adopted its first Fish and Wildlife Program in 1982 and has revised it periodically since then, most recently in 2000. Currently, BPA is spending $139 million annually on fish and wildlife projects arising from the Council's recommendations. There are more than 200 projects in the Council's Fish and Wildlife Program. Expenditures by the PUDs under FERC authority are in addition to the BPA amount.

The Act established the Council as a policymaking body, but gave it no regulatory authority. The Council, for example, has no authority over regulation of the many diverse fisheries that take place on stocks of salmon and steelhead originating in the Columbia Basin. That task requires multijurisdictional coordination that is still lacking a regional focal point that has regulatory authority (e.g., Salmon and Steelhead Advisory Commission [SSAC], 1984).

Prior to the Power Act, coordination of efforts to overcome adverse effects of development and operation of the hydroelectric system took place through the Columbia River Fisheries Council and various other mechanisms, including constituents lobbying the U.S. Congress.

A series of listings under the Endangered Species Act (ESA) in the early 1990s has complicated mitigation responsibilities in the Columbia River Basin. Listings included Chinook salmon and steelhead in the Snake River, upriver spring Chinook salmon and steelhead in the upper Columbia Reach, and bull trout and Kootenai River sturgeon. Recovery planning for anadromous species falls to the National Oceanic and Atmospheric Administration (NOAA) Fisheries, while the U.S. Fish and Wildlife Service has responsibility for bull trout and Kootenai River sturgeon.

Since the early 1990s, NOAA Fisheries and the U.S. Fish and Wildlife Service have served as lead agencies in development of Federal Columbia River Power System (FCRPS) plans for the operation of the federal portion of the Columbia River Power System, to benefit ESA-listed populations. As required by the ESA, the FCRPS plan is developed by consultation among the federal action agencies in the Columbia Basin affected by management of fish and wildlife (NOAA Fisheries and Fish and Wildlife Service), flood control (U.S. Army Corps of Engineers), irrigation (Bureau of Reclamation), and hydropower (BPA).

The FCRPS plans for recovery of listed fishes are designated as Biological Opinions (BiOp). In these, the measures proposed for federal action expected

---

[1]The Power Act established the principle that hydropower users are responsible for protecting and restoring fish and wildlife resources that are adversely affected by the federal Columbia River power system (NRC 1996). Lee (1993) provides a perspective on development of the Act and its effects.

to lead to recovery of the fish run to hundreds of pages dealing with system configuration, river flows, reservoir management, passage improvements, spill, juvenile fish transportation, predator management, and more. The BiOps have undergone several revisions and are still in flux as a result of law suits and rulings by the Federal District Court.

In May, 2003, the Court ruled that the 2000 BiOp was "invalid", because it did not satisfy requirements of the Endangered Species Act. The Court then instituted a process of "remand", which allowed NOAA Fisheries time to revise it.[2] NOAA Fisheries submitted a revised BiOp in 2004, which the Court rejected in June 2005 as not being in conformity with the requirements of the ESA (National Wildlife Federation et al. v National Marine Fisheries Service et al. CV 01-640-RE (Lead Case) CV 05-23-RE (Consolidated Cases). Opinion and Order, Hon. J. A. Redden, United States District Court, District of Oregon. June 10, 2005). Some details are provided later in the text in this Chapter. Notably, the Court ordered that the FCRPS provide large amounts of spill at the Federal dams in 2005 to improve survival of juvenile fall Chinook salmon during their downstream migration.

Thus, the Federal District Court has been brought into assert, where necessary, its authority to supervise management of mitigation measures to assure recovery of listed species, though NOAA Fisheries must adopt measures that will satisfy the law as interpreted by the Court. The Council retains its authority to develop plans for unlisted species of fish and wildlife.

Background

Development of the Columbia and Snake River mainstem hydroelectric system dramatically altered the river environment in which salmon and steelhead had existed for thousands of years (Figure 7.9). Juveniles migrating downstream to the ocean and adults returning to upriver natal spawning grounds encountered new environmental conditions and numerous sources of mortality beyond those previously experienced by salmonids in the Columbia Basin. In many instances, only by integrating the behavior and life history attributes of salmon and steelhead into the operation of the hydrosystem and its individual projects have we been able to reduce the mortalities experienced by salmonids moving through the hydrosystem (Figure 7.10).

Various mitigation measures have been adopted. These measures may be viewed as mainstem habitat improvements, in the sense that the dams and their reservoirs are unnatural obstacles to migration of salmon. The

---

[2]NOAA Fisheries maintains a website with progress reports applying to the remand at www.salmonrecovery.gov/remand

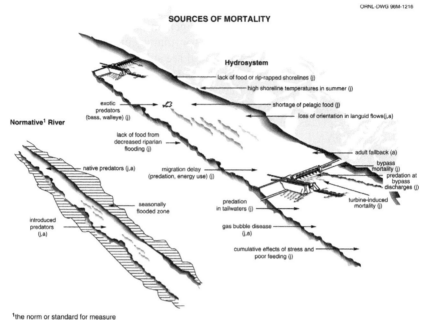

ORNL-DWG 96M-1216

Figure 7.9   Sources of mortality for juvenile and adult migrating salmon and steelhead in the predevelopment natural river (lower left in diagram) and in the developed river (upper right), where the changed river conditions and the hydroelectric project introduce numerous additional sources of mortality.

measures are intended to (1) expedite upstream movement of adults, (2) divert juveniles away from turbine intakes at the powerhouses and provide an alternate route of passage, (3) speed the downstream movement of juveniles and improve temperatures and other water quality characteristics to improve their survival, and (4) stabilize flows for the benefit of spawning salmon, incubation of their eggs, and protection from desiccation of newly hatched and migrating fry.

As more was learned about effects of the hydroelectric system, studies were conducted to measure rates of mortality in reservoirs, in forebays, at the dams, and in the tailraces. Identification of the sources of mortality has helped develop priorities for actions aimed at mitigating for losses. Mitigation for losses of juveniles thus involves actions throughout the river, including modifications of system operations, as well as individual project configurations.

Determining the extent of responsibility for mitigation has continued as an open question, given that numerous factors have together contributed to losses of fish and wildlife in the basin. Two mid-Columbia PUDs, Chelan

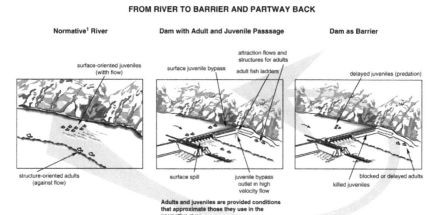

ORNL-DWG 96M-1211

FROM RIVER TO BARRIER AND PARTWAY BACK

Figure 7.10 Upstream and downstream migration conditions for adult and juvenile salmon, respectively, in the predevelopment river (far left panel in diagram), the river with a newly developed hydroproject (far right panel), and in the developed river, where fish behavior and life history attributes are integrated into the operations of the hydroelectric project (middle panel).

County PUD No. 1[3] and Douglas County PUD No. 1, approached this problem by negotiation of Habitat Conservation Plans (HCPs), which establish survival goals for anadromous fish in passage through their projects (Rock Island Dam, Rocky Reach Dam, and Wells Dam), relative to agreed upon most likely "pre-dam" survival rates. The difference, where it cannot be mitigated in place is to be made up by hatchery production or natural production in the tributaries. With this approach, the hydnopower operators are free to choose alternatives that best accomodate their interests.

The NPCC also has underway a process for subbasin planning that is intended to identify limiting factors and potential for mitigation activities appropriate for BPA funding in the tributaries. In this process, boundaries of the 58 subbasins are defined by drainages of the individual tributaries to the Columbia River. The entire Columbia Basin is included in the scope of this effort, which is expected to culminate in adoption of the individual plans as part of the Fish and Wildlife Program of the Council through its amendment

---

[3]The HCPs are a mechanism provided for as a process in NOAA Fisheries application of the ESA. In effect, they are permits for the "taking" of endangered species. Numerous parties, in addition to the PUDs and NOAA Fisheries are signatories, including Washington Department of Fish and Wildlife, the Confederated Tribes of the Colville Reservation, and others. They join in requesting that FERC adopt the HCPs as part of the operating licenses for the dams.

process in late 2004 or early 2005. As of May 2005 a total of 48 subbasin management plans have been adopted by the council. For the most part, these call for actions in the tributaries to mitigate for losses in the mainstem.

In the basin as a whole, each project has its own particular set of mitigation measures; we discuss the combinations in the text that follows. Decision makers have given priority to ameliorating the effects of the hydroelectric system through mainstem measures affecting individual projects. When these are judged not to be sufficient to make up for the total losses, a policy has evolved that can shift efforts to the tributaries where certain measures, particularly habitat improvement, may be expected to lead to increased fish production to make up the difference.

Dams as Obstacles for Anadromous Fish

Flow in the Columbia and Snake Rivers is tightly regulated by operations for power production, resulting in effects on fish. In addition, the dams themselves are obstacles to migration. The typical mainstem dam on the Columbia River presents challenges to the migrations of both adult and juvenile anadromous fishes. The mainstem dams (Figure 7.1) are for the most part around 100 feet high (Figure 7.11). In contrast, other dams in the region, particularly those with large storage reservoirs, are much higher and create impassable barriers to fish. Examples include Chief Joseph Dam at nearly 200 feet high, Hells Canyon Dam at over 200 feet high, and Dworshak Dam at 717 feet height (Figure 7.12). Impassable dams have blocked anadromous fishes from about 31% of their former habitat in the Columbia Basin in terms of stream miles, and 55% in terms of area (NWPPC, 2003).

Adult Upstream Passage

Passage at the dams is utilized by fall Chinook salmon and American shad spawning in the mainstem and by summer and spring Chinook, steelhead, sockeye, and coho salmon spawning in the tributaries. Pacific lamprey recently have been discovered to experience difficulty passing the ladders (Figures 7.13 and 7.14), and the U.S. Army Corps of Engineers is studying the problem. Some adult sturgeon are too large for the ladder pools and occasionally are discovered stranded within a ladder.

When Congress authorized the dams constructed and operated by federal agencies (U.S. Army Corps of Engineers, BPA, and U.S. Bureau of Reclamation), they required adult fish ladders at all except Grand Coulee Dam and Chief Joseph Dam, which were thought to be too high for effective ladders. Passage for adult salmon was provided at the time of construction at the five dams in the mid-Columbia Reach licensed by FERC. On the Snake

Figure 7.11 Lower Monumental Dam on the Lower Snake River. This is a typical run-of-the-river dam where elevation differences between the forebay and tailrace are less than 100 feet. Photo from U.S. Army Corps of Engineers Digital Visual Library, website at *http://images. usace.army.mil/photolib.html*.

Figure 7.12 Dworshak Dam on the North Fork of the Clearwater River in north central Idaho. This is a typical high dam and storage reservoir where elevation differences between the forebay and tailrace are greater than 300 feet. Photo from U.S. Army Corps of Engineers Digital Visual Library, website at *http://images.usace.army.mil/photolib.html*.

Figure 7.13 Adult fish ladder at Ice Harbor Dam on the lower Snake River. Photo from U.S. Army Corps of Engineers Digital Visual Library, website at *http://images.usace.army. mil/photolib.html.*

River, FERC did not require fish ladders at Hells Canyon Dam, a privately owned facility, on the basis that it was thought to be too high for the technology available. Thus, eight of the Corps of Engineers projects downstream of Chief Joseph Dam on the mainstem Columbia and on the lower Snake Rivers, as well as the five PUD projects in the mid-Columbia reach, include provisions for passage of adult anadromous fishes. Early studies focused upon improvements of upstream passage for adults, leading to development of optimum standards for design and operation. Mighetto and Ebel (1995)

Figure 7.14   Adult fish ladder at Bonneville Dam on the lower Columbia River. Photo by R. N. Williams.

describe early research that was undertaken to improve passage for adult salmon and steelhead at the federal dams. Similar efforts took place later at the PUD dams and elsewhere. Situations affecting the ability of adult salmon to find and ascend fish passage facilities are site specific, due to differences in configuration of powerhouses, spillways, and river conditions.

Adult migrating salmon and steelhead may pass a project by way of fish ladders or by way of the navigation channel in those dams equipped with navigation facilities (Figure 7.15), although few fish choose the navigation channels as passage routes. Their operation is intermittent, and they lack a strong, regular flow of "attraction water" that will draw fish to them. The five mid-Columbia dams from Priest Rapids Dam upstream are not equipped with navigation channels.

Improvements in the technology of adult fish passage continue to this day. Tracking of radio-tagged adult salmon and steelhead has made it possible to study movements of fish in detail, resulting in recommendations for improvement in configurations of fish ladders. These studies are well summarized by Ferguson et al. 2003. With this technology, it has been possible to identify previously undetected problems adult salmon encounter in finding the entrances to some ladders and in proceeding upward, versus "falling back" At some facilities few fish proceeded directly upward. In other cases, fish "fell back" over spill ways or other routes after ascending the ladders Delay in migration is one focus of these studies along with rates of survival in transit.

Figure 7.15   Navigational locks at Bonneville Dam. Photo from U.S. Army Corps of Engineers Digital Visual Library, website at *http://images.usace.army.mil/photolib.html.*

These situations continue to be addressed by modifications of the ladder configurations. The 2000 BiOp prescribes performance standards for adult survival during migration. Recent installation of adult Passive Integrated Transponder (PIT) tag detectors in some ladders should make it possible to obtain better estimates of adult survival during upstream migration.

Fish ladders throughout the Columbia Basin are now operated according to criteria developed jointly by the project operators, NOAA Fisheries, and the state and tribal fisheries managers. Annual detailed operating plans are developed that specify such things as water temperatures to be maintained, water depths and head to be maintained on the entrance gates, powerhouse collection channels, floating orifices, ladder flow, counting windows, and ladder exits (Ferguson et al. 2003). Examples of such ladders and associated structures are illustrated in Mighetto and Ebel (1995).

Downstream Passage of Juveniles

Measures adopted to improve conditions for juvenile salmonids as they emigrate to the sea have emphasized two strategies: (1) diverting juveniles away from turbine intakes at the powerhouses into an alternate route of passage and (2) speeding their downstream movement and improving temperatures and other water quality characteristics by means of special flow provisions.

Mortalities at the Dams

Juvenile salmonids may pass a hydroelectric project by a variety of routes (Figure 7.16). Juvenile emigrants, moving downstream in the direction from left to right in Figure 7.16, may pass the project by one of four primary routes: the powerhouse, the spillway, the navigation locks, or the fish ladders. Secondary routes are available at projects equipped with ice and trash sluiceways and/or turbine intake screens that lead into bypass flumes, a process we describe later. When Bonneville Dam was constructed, the need for provision of bypass facilities specifically designed for juvenile salmonids was anticipated, but there was little or no information available on what the desirable characteristics of such a bypass might be (however, see reviews by Mighetto and Ebel 1995; Whitney et al. 1997; Ferguson et al. 2003).

ORNL-DWG 96M-1175

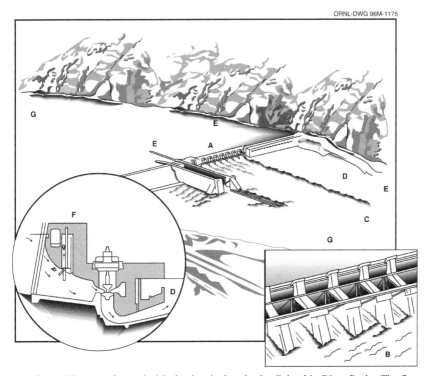

Figure 7.16 Diagram of a typical hydroelectric dam in the Columbia River Basin. The figure shows the spillway (A and inset B), the powerhouse to the right of the spillway, powerhouse cross section (area F in the circular inset), and the navigation lock (E) to the left of the spillway (not present in mid-Columbia dams). In the powerhouse cross section, fish are shown moving up into a bypass inside the powerhouse, while the water continues on through the turbine. The diagram also shows the powerhouse tailrace (D), the adult fish ladder exit and entrance (E on the right), and navigability (G), which is accomplished by a system of locks on one side of the dam, as in a canal.

When Rock Island and Bonneville Dams were built in the 1930s, there was a difference of opinion as to whether there might be a need to provide special passage routes for juvenile salmon at the dams. One school of thought held that the turbines were so large and the blades spaced so far apart that it was unlikely that small fish would be injured by passage through them (Petersen 1995, p. 110). Nevertheless, at Bonneville Dam, four surface outlets were provided for juvenile salmon at the time of its construction. However, they proved to be ineffective and were used primarily to sample fish as they passed the dam (Mighetto and Ebel 1995).

An elaborate technology has developed around the effort to divert juvenile salmonids away from turbine intakes (Mighetto and Ebel 1995; Whitney et al. 1997; Ferguson et al. 2003). The motivation for this effort derives from a study conducted in the late 1940s by Harlan Holmes, a biologist with the federal agency then known as the U.S. Fish and Wildlife Service. Holmes marked some juvenile salmon by clipping fins, and released two groups, one in the tailrace below Bonneville Dam and the other in a place where they would have to pass through the turbines. He then examined the adults in later years as they returned to the hatchery at Bonneville. He found that passage through turbines led to substantial mortalities of juvenile salmonids—about 15% (Mighetto and Ebel 1995).[4] That work and subsequent verification by others (see summaries by Whitney et al. 1997 and Ferguson et al. 2003) led to a focus upon diverting juvenile salmonids away from the turbine intakes. Bickford and Skalski (2000) found that for 102 replicate releases of salmonid juveniles migrating through Snake and Columbia River dams, the average survival through Kaplan turbines was 87.3%.[5] It is apparent that mortality is project specific.

A number of the studies, which produced low estimates in the range of 3.9% to 7% mortality, were conducted using a balloon tag (Heisey et al. 1992) or radio tags designed so that fish could be recovered quickly in the tailrace after passage through the turbine or spillway or other passage routes (Skalski et al. 1998). In this way, an estimate was obtained of direct mortality in turbine passage, while the higher estimates of loss were derived from studies designed with paired releases of marked fish, where the first group was released into the turbine intakes and the second, "control" group was released in the tailrace, and recoveries of both groups took place at a facility downstream. The latter studies included an additional element of mortality in and beyond the tailrace that was associated with delayed effects of turbine passage. This phenomenon

---

[4]Curiously, the study remains unpublished to this day, although it is referred to by Bell et al. (1967) and Ferguson et al. (2003). Ferguson et al. state that Holmes' manuscript is available on request from files of the U.S. Fish and Wildlife Service, Vancouver, Washington.

[5]All of the powerhouses in the Snake and Columbia Rivers, with one exception, are equipped with modified Kaplan-type turbines. The exception is Powerhouse 2 at Rock Island Dam, which is equipped with horizontally oriented bulb turbines.

occurs if fish that have passed through the turbines are disoriented for a distance downstream and are thus more vulnerable to predation than fish released directly into the tailrace as controls. Several recent studies have used two methods simultaneously (radio tags, balloon tags and/or PIT tags) at the same project to better isolate sources of mortality (Ferguson et al. 2003).

In addition to losses of juvenile salmon in direct turbine passage, losses have been identified in intake and discharge structures, the tailrace, or reservoir. Losses due to predation as an incidental effect of turbine passage and other losses not directly assignable to turbine effects have also been found (Long et al. 1975). Mortality (both direct and indirect) of juvenile salmon and steelhead in passage through turbines is variable among projects, as it is dependent upon operating conditions, inherent efficiencies of the turbines, and other factors. Turbine efficiency is thought to have a direct effect on passage survival of fish (Ferguson et al. 2003). Operation of turbines within 1% of their efficiency ratings is called for in the NPCC's Fish and Wildlife Program and in the 2000 BiOp. Hydroelectric system operators are investigating "fish friendly" turbines that would include such features as minimum gap runners, higher efficiency ratings, and other modifications (Brookshire and McKinnon 1995). As turbines age and need to be replaced (Figure 7.17), it makes sense where possible to replace older designs with newer, fish friendly designs.

Bypass Provisions for Juvenile Salmon and Steelhead at the Dams

This section discusses development and operation of bypass systems and the provision of spill, measures employed to divert fish away from the turbine intakes. Transportation of juveniles by barge and by truck is associated with the turbine intake bypass systems, since it depends upon fish collected by the intake screen systems. We describe and evaluate specific mitigation measures undertaken at the dams to reduce adverse effects on survival of juvenile salmon and steelhead as they migrate to the sea and encounter the dams and effects of their operations.

We use several terms, described here, to evaluate the effectiveness of various passage routes and bypass systems for juvenile salmon migrating downstream. Fish passage efficiency (FPE) is the percentage of total fish approaching a project that pass by routes other than the turbines. Fish guidance efficiency (FGE) is the percentage of fish approaching a turbine intake that is successfully diverted by screens into a bypass system. Similarly, spill effectiveness and effectiveness of the ice and trash sluiceways are evaluated in terms of the percentages of total fish that are diverted into spill or the sluiceways, respectively. FPE includes all of the fish except those that continue their passage through the turbines. Thus, FPE includes FGE as well as spill effectiveness and effectiveness of any ice and trash sluiceway. In each

Figure 7.17 Installation of a turbine fan into Grand Coulee Dam. Photo from U.S. Army Corps of Engineers Digital Visual Library, website at *http://images.usace.army.mil/photolib.html.*

route, there may be associated injury or mortality of fish. An additional standard for fish survival is therefore applied to those fish that are successfully passed away from the turbines.

The Council's Fish and Wildlife Program (NPCC 1996) sets a standard of 80% FPE at each project and 95% survival[6]. Fish passage efficiency is the percentage of juvenile fish that approach a dam and are diverted away from the turbine intakes into other passage routes. Providing alternative passage

---

[6]The FCRPS BiOp of 1995 specified the same objectives, but these were modified in the 2000 BiOp by specification of spill levels after it was learned that spill amounts required to accomplish those standards led to unacceptable supersaturation of gsses.

routes has been the focus of the efforts to accomplish this mitigation goal. The primary measures adopted to accomplish this have been screens placed in the upper portion of the turbine intakes to divert the fish into bypass flumes, surface collectors, and provision of spill.

Diverting Juvenile Salmonids at the Turbine Intakes

Turbine Intake Screens

The development of powerhouse bypass systems has a history dating from the 1960s (Whitney et al. 1997). Mighetto and Ebel (1995) summarized the decades of work by NOAA Fisheries and the U.S. Army Corps of Engineers to develop a satisfactory turbine intake screen. Screen design and testing has moved from the initial development of submerged traveling screens, to fixed bar screens, to extended length screens. It continues to evolve in order to achieve higher FGEs. NOAA Fisheries has developed a set of criteria for designs of mechanical screen bypass systems, including guidelines for locating bypass outfalls (National Marine Fisheries Service [NMFS] 1995b). Ferguson et al. (2003) provide further sources of information on systems design.

The basic principle of operation is that fish, normally oriented toward the surface, are drawn downward by flow into the powerhouse at the face of the dam toward the upper portion of the turbine intake (Figure 7.16). The screens (Figure 7.18a) are lowered through the gatewell into position in the upper portion of the intake to divert fish upward into the gatewell instead (Figure 7.16). The first prototype was tested in the field at Ice Harbor Dam in 1969 (Long et al. 1975). While the frame itself was fixed, once it was in place, the screen rotated along a track with rollers at the upper and lower ends in order to continually backflush debris off its face, thus the name *submerged traveling screen* (STS) (Figure 7.18b). Screens were evaluated in terms of their FGE, defined as the percentage of fish entering the intake that were diverted by the screens upward into the gatewells. Initial tests involved placement of fyke nets below the screens to capture fish that were not guided (Figure 7.19). Later tests have depended upon hydroacoustic assessment, radio tracking, or PIT tag estimates of the numbers of fish not guided.

Experience with the traveling screens had shown them to be costly to build and maintain. Tests at Bonneville Dam in the 1970s used a fixed screen concept that would be less complex and less costly. Results were promising, leading to testing of a full-scale device at McNary Dam in 1978 (Krcma et al. 1978). Cleaning could be accomplished by periodically raising the angle of the screen to create a backflush through the mesh. Results were favorable (Ruehle et al. 1978; Krcma et al. 1980). Subsequent tests of a fixed bar screen

a) McNary Dam screen

b) Inspecting traveling screen at Rocky Dam

Figure 7.18 Turbine screens: (a) installation of a turbine screen at McNary Dam, and (b) Jim Mullan checking a traveling screen at Rocky Reach Dam. Photos: (a) from U.S. Army Corps of Engineers Digital Visual Library, website at *http://images.usace.army.mil/photolib.html* and (b) by R. Whitney.

design (Figure 7.20) in prototype at Priest Rapids Dam and Wanapum Dam confirmed the favorable results of the tests at Bonneville Dam and McNary Dam (Mid-Columbia Coordinating Committee 1989).

Development of extended length screens with improved FGE favored the fixed bar screen design over the STSs, which were more subject to mechanical problems. As a result, extended length bar screens are now the technology

Figure 7.19 Fyke net array for estimating juvenile turbine bypass mortality at Priest Rapids Dam. Photo by R. Whitney.

of choice at the U.S. Army Corps of Engineers projects equipped with screens. With extended screens, significant increases in FGE were measured: 66% for yearling Chinook compared to 57% with the standard STS, and 83% for steelhead compared to 77% with a standard STS (Swan et al. 1990).

Encouraging results at Lower Granite Dam led to the design of two types of prototype extended length screens—a bar screen and an STS—that were tested at McNary Dam from 1991 to 1994. Tests of full extended screens

Figure 7.20   Bar screen on the deck of Priest Rapids Dam with impinged (i.e., dead) juvenile salmonids. Photo by R. Whitney.

were also initiated at The Dalles and Little Goose Dams in 1993 (Gessel et al. 1995). Side-by-side comparisons showed that the extended length bar screens yielded a higher FGE than did STSs at both McNary Dam and The Dalles Dam (McComas et al. 1994; Ferguson et al. 2003).

Effectiveness of Turbine Intake Screens for Bypass of Juvenile Salmon

The effectiveness of turbine intake screens to bypass juvenile salmon has been difficult to evaluate. The measure of effectiveness of turbine intake screens is their FGE, the percentage of fish approaching the turbine intakes that is diverted into a bypass of some kind. Values of FGE measured in tests of prototype devices at various projects are shown in Whitney et al. (1997) and Ferguson et al. (2003) and are variable from one test to another. A primary reason for the variability is that the tests typically are conducted to evaluate changes in configurations and not for the purpose of developing a statistically rigorous estimate for use in evaluating total project (or system) guidance efficiency. In addition to differing with respect to the design and configuration of the apparatus, the estimates differ from project to project, the fish species and stocks (stream-type vs. ocean-type Chinook), their degree of smoltification, time of day (particularly day vs. night), and progress of the season (Hays and Truscott 1986; Swan and Norman 1987; Giorgi and Stuehrenberg 1988; Peven and Keesee 1992; Peven et al. 1995; Swan et al. 1985, 1986, 1987).

FGEs shown in Table 7.1 (from Ferguson et al. 2003) range from 38% to 95% for yearling Chinook, 16% to 62% for subyearling Chinook, and 40%

Table 7.1 Fish guidance efficiency (FGE) at Columbia and Snake River dams for 1999 configuration[d] (from Ferguson et al. 2003).

Site (Screen type)	Species	PATH FGE (%)[a]	NMFS FGE (%)[b]
Lower Granite Dam (ESBS)	Yearling Chinook	78	78
	Subyearling Chinook	-	53
	Steelhead	-	81
Little Goose Dam (ESBS)	Yearling Chinook	82	82
	Subyearling Chinook	-	45
	Steelhead	-	81
Lower Monumental Dam (STS)	Yearling Chinook	61	61
	Subyearling Chinook	-	49
	Steelhead	-	82
Ice Harbor Dam (STS)	Yearling Chinook	71	71
	Subyearling Chinook	-	46
	Steelhead		93
McNary Dam (ESBS)	Yearling Chinook	95	95
	Subyearling Chinook	-	62
	Steelhead	-	89
John Day Dam (STS)	Yearling Chinook	67	64
	Subyearling Chinook	-	34
	Steelhead	-	85
The Dalles Dam (None)[c]	Yearling Chinook	46	46
	Subyearling Chinook	-	46
	Steelhead	-	40
Bonneville Dam			
Powerhouse 1 (STS)	Yearling Chinook	41	38
	Subyearling Chinook	-	16
	Steelhead	-	41
Powerhouse 2 (STS)	Yearling Chinook	43	44
	Subyearling Chinook	-	18
	Steelhead	-	48

[a]Based on report to PATH from Anderson et al. (1998).
[b]Based on NMFS sensitivity run #1 (assumes $FGE_{ESBS} > FGE_{STS}$ for wild yearling Chinook salmon).
[c]FGE values for The Dalles are based on passage through the ice and trash sluiceway.
[d]These estimates likely have range, but that range changes with a number of factors and is not easily estimated. ESBS, extended length submerged bar screen.

to 93% for steelhead. Most poorly guided of all the salmonids are sub-yearling Chinook and sockeye salmon, with FGEs near half those of the other salmon. FGE estimates have improved with time, due to improvement of the screens. Up to the time of the report by Whitney et al. (1997), estimated FGE for yearling Chinook in prototypes ranged from 26% (Bonneville Powerhouse 1) to 88%, with most (6 of 11) in the range of 65% to 80%. For subyearling Chinook, FGE ranged from 20% to a maximum of 67%. For steelhead, FGE ranged from 76% to 93%, with most (6 of 8) above 80%.

Since 1995, with the ESA listing of Snake River stocks, hydroacoustics, radiotelemetry, and PIT tag methods have been explored for measurement of FGE at the Snake River and lower Columbia River dams, rather than fyke nets previously used, because the fyke nets kill fish and some are listed as threatened or endangered species. Furthermore, questions were raised about the possible effects of the nets themselves, that they might affect the flow pattern at the screen and thus the measured FGE (Magne et al. 1989; Stansell et al. 1990, 1991; Thorne and Kuehl 1989, 1990).

Extended length screens generally have shown higher FGEs than have the standard screens, by variable amounts depending upon fish species and stock (ocean-type vs. stream-type Chinook) and features of the specific project. For example, at The Dalles Dam, extended length screens produced estimates of FGE for yearling Chinook of 69%, compared to 44% to 56% for the standard screens tested, and 83% for steelhead, compared to 71% to 80% for standard screens (Krcma 1985; Absolon et al. 1995; see also Whitney et al. 1997). At McNary Dam, FGE for subyearling Chinook was measured as 67% with the extended screen and 34% to 46% with the standard screen (Krcma et al. 1982; McComas et al. 1994, see also Whitney et al. 1997).

Conduit to the Tailrace

Once diverted by the screens into the gatewells, the juvenile salmon and steelhead are led through orifices in the gatewells into a conduit to the tailrace below the dam. At Bonneville Dam's first powerhouse, orifices were cut from the gatewells to the ice and trash sluiceway to provide an exit for fish. A vertical barrier screen was installed in the gatewell to create an upward flow to encourage movement of the fish toward the orifices near the surface. A dewatering system was provided at the end of the ice and trash sluiceway, where water was pumped back into the forebay in order to reduce the volume of water that entered a 20-inch conduit leading to the tailrace. At McNary Dam and other projects of the U.S. Army Corps of Engineers, a separate bypass flume was constructed within the ice and trash sluiceway. Evaluations of effectiveness of the systems led to improvements in designs (Krcma et al. 1984, 1985, 1986; Krcma 1985; Swan et al. 1987, 1990, 1992; Swan and Norman 1987).

Mortality and Descaling of Salmonids in Bypass Systems

While the primary criterion in evaluating the effectiveness of mechanical bypass systems is their FGE, concerns were raised that injuries to fish could occur within the bypass itself. The bypass system can be a source of mortality for juveniles at each stage in their passage, beginning with their encounter with shear forces associated with the flow pattern created by the screen, encounter with the screen itself, or with trash or structures within the gatewells, within the conduit, within the sampling system that is present at most facilities, or at the outfall and below where predators may concentrate. Impingement on the screens and injury of diverted fish are problems that have had to be addressed by manipulations of screen openings, angle of deployment of the screen, velocity at the screen, and other factors.

A percentage of the approaching fish may strike the screen in passing and lose some scales, while others, particularly small subyearling Chinook, may become impinged on the screen. Impingement rates of yearling Chinook are negligible in properly tuned systems but may be "high" in prototypes (Peven 1993). For example, at Lower Granite Dam, impingement that had ranged from 0.04% to 3% was reduced to less than 1% by design changes to the extended length screen tested in 1990 (Wik and Barila 1990). Survival rate in bypass systems were 97% to 98% (U.S. Army Corps of Engineers 1992).

Standards defining descaling have been developed, and a threshold level of a percentage of missing scales has been set to meet the criteria of the definition of "descaled" and to what degree (Koski et al. 1986). Implications of descaling are not clear, since no direct relationship with survival has yet been established. Nevertheless, descaling has been given serious attention as a possible mortality factor. For example, system improvements at Lower Granite Dam decreased descaling rates between 1982 and 1988 from 15.5% to 2.4% for Chinook and from 16.8% to 1.4% for steelhead (Koski et al. 1986). Information on individual species observed at the bypasses at the Snake River dams and McNary Dam is provided in Ferguson et al. (2003).

Mortality rates are less than descaling rates. In the years from 1998 to 2002, mortality rates of yearling Chinook, steelhead, and coho were generally in the range of 0.2% to 0.4% at the Snake River dams whose systems are monitored[7] and McNary Dam (systems that are used for collecting salmonids for transportation), while mortality rates for subyearling Chinook were from 1% to 3% except for at Little Goose Dam where the rate was 6% (Ferguson et al. 2003). Figures reported for John Day Dam are similar (Ferguson et al. 2003).

---

[7]The sampling facility at Ice Harbor Dam is only operated for short periods during parts of each outmigration each year. Data collected are useful primarily for detecting unusual problems in-season.

To arrive at a general conclusion as to the degree to which turbine intake screens have improved survival in passing the dams, it is necessary to consider the FGE for the particular species at the particular dam and then calculate the survival probabilities separately for the group that is successfully diverted and for the group that is not. As one example, at Lower Granite Dam, 78% might be diverted by screens from the turbine intakes, leaving 22% that are not diverted. The 78% might experience, for example, 2% mortality as a result of bypass effects (including direct and indirect effects), while the group not diverted might experience, for example, 12% mortality in turbine passage. Thus, the end result might be about 4% of the total number passing the dam that are lost during passage. In the absence of screens, the end result might have been a 12% loss due to turbine passage. The reader might carry on this exercise for fish that show lower and higher FGEs, but the illustration serves our purpose here. Fish originating above Lower Granite Dam have to pass eight dams before they reach the estuary. Without screens and without spill only 40% would survive.

Stress Measured During Passage

Because studies have demonstrated increases in stress indicators as fish are handled, or crowded, as in bypass systems, it was prudent to consider whether stress is a significant factor in the mortality of fish that use the bypass systems. Several indicators of stress levels, such as plasma cortisol and glucose concentrations, have been shown to alter as juvenile Chinook passed successive points in the bypass system (Maule et al. 1988, as summarized in Ferguson et al. 2003). Nevertheless, reviews by Whitney et al. (1997) and Ferguson et al. (2003) concluded that stress is not a significant contributing factor to mortality of fish that are bypassed. It was postulated that cumulative effects of stressors might lead to delayed mortality (i.e., effects felt outside of the hydroelectric system where they might not be measured); however, recent analysis provides a full accounting of mortality and leaves little or no room for such delayed effects (Williams et al. 2003).

Bypass Outfall

An additional source of mortality to guided fish is the portion of the conduit leading from the dewatering screens at the sampler to the tailrace. Marked fish released out of the north shore outfall at McNary Dam were recovered at half the rate of other release groups, suggesting that predation in the vicinity of the outfall was responsible for added mortality (Sims and Johnson 1977).

Studies conducted on tagged juveniles at Bonneville Dam's Powerhouse 2 in the early 1990s showed that recovery rates of subyearling Chinook that had transited the bypass were significantly lower than for fish that had passed through the turbines, suggesting higher mortality of juvenile salmon in the bypass than in the turbines (Ledgerwood et al. 1991). The location of the outfall, in a place where predators could congregate, was identified as the most likely source of the high mortality that had been measured. Predation at the outfall by northern pikeminnows was thought to be the principal source of mortality, rather than predation within the system itself.

A solution was proposed to relocate the outfall to a place with higher water velocities where the predators would be less able to maintain themselves in the current. In 1999, the two outfalls were combined into a flume that carries the fish downstream a distance of 1.7 miles, where it empties into midchannel where high velocities of flow had been observed (Figure 7.21). Laboratory studies had shown that northern pikeminnow were unlikely to be able to maintain their positions in that current.

Application of Turbine Intake Screens for Bypass of Juveniles

Studies by NOAA Fisheries and others developed a set of criteria for successful bypass systems. The criteria establish maximum velocities, advise open rather than closed conduit in order to avoid pressurization, set appropriate

Figure 7.21 Bonneville Dam juvenile bypass outfall. The juvenile bypass system at Bonneville Dam carries juveniles over 2 miles downstream of the dam and releases them into the main current where they are less susceptible to predation by northern pikeminnow. The bypass system has both high and low exit outfalls that can be selected depending upon river flows and river elevations. Photo by R. N. Williams.

angles for curves and changes in elevation, set standards for dewatering, and identify other factors in the design (Bates 1992). NMFS has adopted a policy statement that provides for development and evaluation of new technology under controlled conditions (Office of Technology Assessment 1995, Appendix B).

Successful tests of prototype screens led either to their installation or to schedules for installation at projects in the lower Snake and lower Columbia Rivers. Currently, all of the U.S. Army Corps of Engineers projects except The Dalles Dam are equipped with turbine intake screens. STSs are installed at Lower Monumental, Ice Harbor, John Day and Bonneville Dams, and extended length bar screens are installed at Lower Granite, Little Goose, and McNary Dams. None of the mid-Columbia PUD projects is equipped with intake screens, but Wells Dam and Rocky Reach Dam are equipped with surface collectors.

In contrast to screening, Rock Island, Wanapum, and Priest Rapids Dams use spill amounts to attempt to reach the 80% fish passage goal set by a FERC order. At Wanapum Dam and Priest Rapids Dam, the required spill amounts are quite large. We discuss spill as a bypass measure below. Prototype intake screens were tested with some success, and schedules were set for installation at Priest Rapids and Wanapum Dams, but these are being delayed while alternative passage systems are being explored. At Rocky Reach Dam, prototype tests of intake screens annually from 1985 to 1994 did not produce satisfactory results (Peven et al. 1995), and testing of a surface collector proceeded to full installation of a device in 2003. More information on this is provided below, under the heading of surface collection.

Finally, although it is not associated with a dam, the steam electric facility at Hanford, Washington (Hanford Generating Plant) should be mentioned here, as it has a cooling water intake with six bays, each equipped with a traveling screen designed to protect juvenile fish (Stone and Webster Engineering Corp. 1982). Average survival of Chinook yearlings encountering the screen was found to be 97.9% (Page et al. 1976).

Fish Behavior and Fish Passage Efficiency

The preponderance of evidence demonstrates that juvenile salmon migrating downstream are oriented to the upper portion of the water column (Chapter 6). Coutant and Whitney (2000) summarized the effects of fish behavior on turbine passage and the ability to divert fish from the intakes. Giorgi and Stevenson (1995) reviewed much of the evidence at U.S. Army Corps of Engineers projects, and Johnson (1995) reviewed the evidence from salmon literature world-wide.

When they encounter a dam, juvenile salmon prefer surface outlets, when they are available, and are reluctant to sound into deeper water. As seen in a cross section of the powerhouse (see inset circle, Figure 7.16), to follow the flow of the water onto the upstream face of the powerhouse, the juveniles are forced to dive (note arrows below point F in the inset, Figure 7.16) into the turbine intakes. As a result of their surface orientation, juvenile salmon accumulate in gatewells of unscreened turbine intakes, as first noted in the early 1960s (Mighetto and Ebel 1995). Eicher (1988) reviewed studies of passage efficiency at deep intakes. The studies of Regenthal and Rees (1957, cited in Eicher 1988) were particularly informative. They showed 55% of Chinook would exit the reservoir when the only route was 118 feet deep or less, 48% when it was at 146 feet, and 8% when it was 160 feet (as summarized in Eicher 1988). Eicher (1988) concluded that *"it has been accepted that fish sound to great depths as a last resort, and if an alternative, such as an artificial outlet, is available, they will use it preferentially and can be collected in that way."* (p. 1–15) In contrast to juvenile salmonid passage routes associated with turbine intake systems, routes that are more surface oriented include ice and trash sluiceways, spillways, and surface collectors. We discuss these in the following sections.

Diverting Juvenile Salmonids into Spill and Surface Bypass Systems and Devices

Surface bypass systems operate by skimming water from the surface layer of the reservoir at the face of the dam and diverting fish into water that is spilled either through a flume or through the spillway directly to the river below. They take advantage of the general surface orientation of juvenile salmonids migrating downstream, in contrast to the turbine intake screens, which operate by diverting upward fish that have sounded to enter the intakes back upward into gatewells where they are led to orifices that connect to bypass flumes. The objective of surface bypass systems is to attract or intercept the fish before they sound toward the turbine intakes. If flows are sufficient to be detected by the fish, they will be led into a body of water in which they continue as it is released through the dam. The routes followed may be by ice and trash sluiceways or chutes, or through spill that ends up below the dam. In some situations, where release into existing spillways or chutes is not feasible, it may be necessary to provide a flume for passage to the river below the dam.

Surface spill (Figures 7.22 and 7.23) as a means to efficiently bypass migrating juvenile salmon arose partly out of observations made at Wells Dam in the mid-Columbia River. The success of juvenile spill efficiencies at Wells Dam stimulated studies of the possibility of applications elsewhere, as recommended by the Snake River Salmon Recovery Team (Bevan et al. 1993). Unfortunately, the technology used at Wells Dam is not directly transferable

Figure 7.22   Spill at Lower Monumental Dam in the lower Snake River. Lower Monumental Dam is second of the Snake River dams and is the sixth dam up from the Columbia River mouth. Photo from U.S. Army Corps of Engineers Digital Visual Library, website at *http://images.usace.army.mil/photolib.html*.

Figure 7.23   Spill at Bonneville Dam in the lower Columbia River. Bonneville Dam is the lowermost dam on the Columbia River. Photo from U.S. Army Corps of Engineers Digital Visual Library, website at *http://images.usace.army.mil/photolib.html*.

to any other mainstem or Snake River project in the basin, due to the unique hydrocombine configuration of Wells Dam (Figure 7.24), where the spillway is located directly above the turbine intakes. Examination of surface spill potential at Columbia and Snake River projects has focused on modifications of existing ice and trash sluiceways, spillway entrances to enhance surface openings, or on the development of surface bypass collection facilities.

Provision of Spill as a Means of Smolt Bypass

The following is an overview of spill provided through spillways, as it has been implemented for passage of juvenile salmon at projects in the mainstem Columbia and Snake Rivers[8]. Depending upon the hydraulic capacity of the individual projects and the river flow in the particular year, spill will normally occur during the spring freshet when the largest numbers of juvenile salmon are moving downstream. Added storage capacity in the Columbia River hydroelectric system made possible a reduction in the spring runoff, resulting in less spill and forcing more juvenile salmon through the turbines where rates of mortality are much higher.

The first formal application of spill as a bypass measure for juvenile salmon in the Columbia Basin came in the spring of 1980 as a result of a settlement agreement among the parties to the mid-Columbia Proceeding, which came as a result of a suit filed by the state of Washington in the Federal Energy Regulatory Commission (FERC). The agreement provided, among other things, for spill of 10% of the river flow at each of the five mid-Columbia projects during the period in the spring when the middle 80% of the migrating juvenile salmon were determined to be present.

Spill continues to be a primary method for bypass of juvenile salmon at two of the mid-Columbia projects, Wanapum Dam and Priest Rapids Dam, whereas Wells Dam and Rocky Reach Dam are equipped with mechanical bypass systems. At Rock Island Dam, a mix of spill and operating protocols is used to minimize passage through Powerhouse 1 during outmigration of smolts, which results in achieving the level of survival of juvenile salmonids passing the project specified in the HCP for Rock Island Dam. Spillway modifications are part of the approach.

In 1994, FERC ordered sufficient spill at Wanapum and Priest Rapids Dams to pass 70% of outmigrating juvenile salmon during 80% of the migration in the spring and 50% during the summer[9]. Those spill levels are to be

---

[8]More detail is provided in Appendix A of Whitney et al. (1997) as to requirements by FERC, the Northwest Power Planning Council, and NOAA Fisheries (1995 BiOp) for spill and other bypass measures. To those are to be added the requirements of the FCRPS 2000 BiOp, and the Federal court.

[9]FERC Docket No. E-9569-003, Grant County Phase. Order of May 24, 1994.

a)

b)

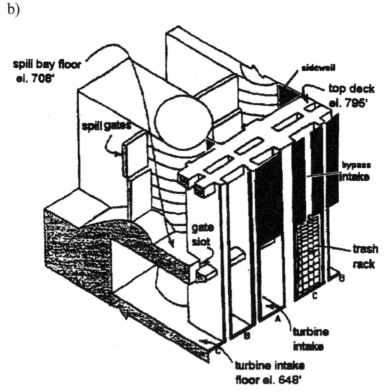

Figure 7.24 Wells Dam (a) photo and (b) diagram, which uses a unique hydrocombine config-uration that is thought to contribute to its success in passing juvenile salmonids with low asso-ciated mortality. Photo from the U.S. Army Corps of Engineers Digital Virtual Library. Diagram from Whitney et al. (1997).

interim measures pending installation of mechanical bypass systems or a contrary order from FERC. Production of gas supersaturation (see discussion below) by these spill amounts has prevented full implementation of the order; consequently spill was limited to 17% in spring and 14% in summer, which in 1994 provided passage for an estimated 50% of the fish in spring and 25% in summer at Wanapum Dam. At Priest Rapids Dam, 50% were passed in the spring and 62% in the summer (Hammond 1994).

In the year 2000, the parties to the mid-Columbia Proceeding arrived at a memorandum of agreement that would modify the FERC order to focus on spill amounts, rather than survival levels because experience had shown that the spill amounts were not achievable within the limits of gas super saturation permitted. Spill levels required by the revised agreement at Wanapum and Priest Rapids Dams are 80% of total river flow in both spring and summer during 95% of the duration of the outmigration. Although not yet approved by FERC, Grant County PUD implemented the agreed upon changes, excepting in the summer of 2001, which was an unusually dry summer and the region was short on power[10]. Implementation details have yet to be determined at the remaining mid-Columbia dams, owned and operated by Douglas County PUD (Wells Dam) and Chelan County PUD (Rocky Reach and Rock Island Dams). Existing HCPs call for specific survivals of juvenile salmonids to be achieved at each project. The PUDs have the option (within limits) of determining what measures will be employed to achieve or mitigate for failure to achieve the survival standards.

With respect to the U.S. Army Corps of Engineers projects in the mainstem Columbia and lower Snake Rivers, a 1989 memorandum of agreement established spill levels to be used in spring and summer as an interim measure at Lower Monumental, Ice Harbor, The Dalles, and John Day Dams, pending the development of solutions to fish passage problems (Fish Passage Managers 1990). The 1995 and 2000 Biological Opinions for endangered Snake River salmon provided additional guidance on required spill levels. The 1995 BiOp required a standard of 80% fish passage during time periods set for spring and summer migrants (NMFS 1995a). Implementation of this standard requires substantial amounts of spill as a supplement to operation of the turbine intake bypass systems. For example, at Lower Granite Dam, to achieve the desired FPE, spill was to be provided at 80% of total instantaneous discharge for 12 hours per day. However, it was found that this level of spill could not be implemented because the total dissolved gas limits would be exceeded when spillway flows reached 60 cfs. Consequently, the 2000 BiOp specifies the required spill amounts project by project (Table 7.2), rather than setting survival standards.

---

[10]Personal communication, Stuart Hammond, Grant County PUD, October, 2003.

Table 7.2 Estimated spill levels and gas caps for FCRPS projects during spring (all) and summer (nontransport projects) (from FCRPS 2000 BiOp).

Project[1]	Estimated Spill Level[2]	Hours	Limiting factor
Lower Granite	60 kcfs	6 pm–6 am	Gas cap
Little Goose	45 kcfs	6 pm–6 am	Gas cap
Lower Monumental	40 kcfs	24 hours	Gas cap
Ice Harbor	100 kcfs (night) 45 kcfs (day)	24 hours	Nighttime–Gas cap Daytime–Adult passage
McNary	120–150 kcfs	6 pm–6 am	Gas cap
John Day	85–160 kcfs/60%[3] (night)	6 pm–6 am[4]	Gas cap/percentage
The Dalles	40% of instant flow	24 hours	Tailrace flow pattern and survival concerns (ongoing studies)
Bonneville	90–150 kcfs (night) 75 kcfs (day)	24 hours	Nighttime–Gas cap Daytime–Adult fallback

[1]Summer spill is curtailed beginning on or about June 20 at the four transport projects (Lower Granite, Little Goose, Lower Monumental, and McNary Dams) due to concerns about low in-river survival rates.

[2]Estimated spill levels shown in the table will increase for some projects as spillway deflector optimization measures are implemented.

[3]The total dissolved gas (TDG) cap at John Day Dam is estimated at 85 to 160 kcfs and the spill cap for tailrace hydraulics is 60%. At project flows up to 300 kcfs, spill discharges will be 60% of instantaneous project flow. Above 300 kcfs project flow, spill discharges will be at the gas cap (up to the hydraulic limit of the powerhouse).

[4]Spill at John Day Dam will be 7:00 pm to 6:00 am (night) and 6:00 am to 7:00 pm (day) between May 15 and July 31.

A recent Federal Court order found the 2004 BiOp to be in violation of the Endangered Species Act and ordered maximum spill at the four lower Snake River projects during outmigration of juvenile fall Chinook salmon, June 20 to August 31, 2005, with production of only enough electricity at each project to provide station service (National Wildlife Federation et al. v National Marine Fisheries Service et al. CV 01-640-RE (Lead Case) CV 05-23-RE (Consolidated Cases). Opinion and Order, Hon J A. Redden, United States District Court, District of Oregon June 10, 2005.) More information is given later in this Chapter.

Spill Effectiveness and Spill Efficiency

Two measures have been used to evaluate the influence of spill amounts on passage of juvenile salmonids: spill effectiveness and spill efficiency. Spill effectiveness is the ratio of the estimated total number of fish using the spillway for passage divided by the estimated total number of fish approaching the dam. Spill efficiency takes into account the volume of water passing in the spillway as a percentage of the total river flow. A long-standing assumption in the region has been that there is a one-to-one relationship between the percentage of river flow that is spilled and the percentage of fish that will be bypassed in spill. This assumption was made in the absence of any data, at a time when technology was not available to acquire the necessary data. The assumption has not held up at any project under close scrutiny but, for lack of better information, is still employed at times.

Studies in the 1980s, using hydroacoustic technology at each of the mid-Columbia projects, revealed that the relationship between the percentage of juvenile salmon passed in spill and the spill volume relative to total river flow is complex and varies from project to project (Biosonics 1983a, 1983b, 1984; Raemhild et al. 1983). As an example of the non-linear relationship often found, at Wanapum Dam in the spring of 1983, nighttime spill of 20% of the instantaneous flow passed an average of about 45% of the fish, while spill of 50% passed 60% of the fish (Biosonics 1983b). On the other hand, at Rocky Reach Dam during the spring of 1983, nighttime spill amounting to 20% of the instantaneous river flow was estimated to pass about 16% of the fish, spill of 50% passed about 30% of the fish, and spill of 80% passed about 55% of the fish (Biosonics 1984).

U. S. Army Corps of Engineers projects in the mainstem Columbia and lower Snake Rivers relied on observation studies to define spill efficiency relationships. In the mainstem Columbia, monitoring studies at John Day Dam focused on developing an overall ratio of percentage fish passage to percentage spill for a range of values from 37% to 66% spill, for the spring and summer seasons, arriving thus at spill effectiveness ratios of 1.3 in 1987, 1.4 in 1989, and 1.1 in 1988 (Magne et al. 1987). The combined data estimate 50% fish passage in 60% spill at John Day Dam. Obviously, spill effectiveness must improve at some spill level beyond the observations, since 100% spill must include 100% of the fish. Therefore, a curve, rather than a single ratio, would be more appropriate for describing spill effectiveness at John Day Dam.

At The Dalles Dam, estimates suggested 30% fish passage in spill at 10% spill and 75% fish passage at 40% spill (Willis 1982). For the lower Snake River Corps dams (Lower Granite, Little Goose, and Lower Monumental), a curve was developed to describe the relationship (Smith et al. 1993, as cited in Ferguson et al. 2003). The U.S. Army Corps of Engineers is continuing studies of fish, spill, and sluiceway passage efficiencies at certain projects, employing radio-tagged juvenile salmonids and hydroacoustic methods.

Measures to Improve Spill Effectiveness

Surface Spill.   Spill effectiveness has been improved at some projects by several means. The standard spill gates in the Columbia River projects are designed to open from the bottom of the spillbay, typically at depths near 50 feet, depending upon reservoir elevation (e.g., 47–58 ft below normal operating pool at John Day Dam) (Giorgi and Stevenson 1995). From a study at John Day Dam, Raymond and Sims (1980) suggested that surface spill would be more effective in passing fish than standard spill. They placed stop logs in the spillbay opening to create surface spill and found that juvenile salmon passing through the bays with surface spill were as likely to pass during the daytime as at night. This was in contrast to samples of juvenile salmon from the turbine intakes, the ceilings of which were located at about the same depth as the bottoms of the unlogged spill bays, which showed a strong peak at night. Results suggested that juvenile salmon approaching the dam delayed sounding to the intakes until after dark, but that they are more readily passed through surface spill.

Giorgi and Stevenson (1995) observed that surface spill remained to be adequately evaluated at U.S. Army Corps of Engineers projects. The regional fisheries managers have great interest in improving spill effectiveness, as increased spill efficiencies and increased juvenile survivals offer both biological benefits (greater fish survivals) and economic incentives (less water needs to be spilled and can therefore be used for power generation). Efforts to improve spill efficiencies have focused primarily on testing the effectiveness of differing spillway configurations. Consequently, improvements are expensive and time consuming, as they are tied to annual tests during the spring juvenile outmigrations.

Examples of some of the actions to evaluate and improve spill effectiveness include attempts to redesign water movements in the forebay toward the spillway, modification of existing ice and trash sluiceways in bypass routes, and modification of the spillway gates themselves. For example, a set of "occlusion plates" that were designed to force water from the surface of the forebay into and through the entrance to the spillway were evaluated at The Dalles Dam in 2000; results were not encouraging and the testing was discontinued. The Corps of Engineers has tested in prototype a removable spillway weir (RSW), a temporary structure placed in front of the spillway that diverts water headed for the gate opening, which is at the bottom, upward into a surface chute, making it unnecessary for the fish to sound in order to pass the dam. At Lower Granite Dam the RSW was shown to reduce passage times of radio tagged fish and also improved spill effectiveness. Plans are being developed for full installation at one or more projects, Ferguson et al. 2003.

At Bonneville Dam, low FGEs of the intake screens require high volumes of spill, particularly in the summer, during passage of subyearling Chinook. Spill rates are maintained between 50,000 and 75,000 ft^3/s during the day to inhibit predation on smolts in the tailrace[11]. At Powerhouse 2, there is a sluice chute, located between the powerhouse and the spillway. When open, the chute has been found to pass significant numbers of juvenile salmonids (Ferguson et al. 2003). It is thought to have good potential as a surface-type collector (see discussion below).

Wanapum and Priest Rapids Dams are fitted with sluiceway spill gates that open from the top. These are smaller spillbays, designed for passage of debris, rather than to assist in control of water elevation in the forebay. However, spill in the sluiceway at Priest Rapids Dam was twice as effective in passing fish as spill in the spillway (Ransom and Malone 1990; McFadden et al. 1992; Ransom and Steig 1995). Results were not as promising at Wanapum Dam, so studies are currently underway to isolate the source of the problem[12]. At Rock Island Dam, when spill was split 50:50 between the shallow (35 feet) and deep (55 feet) spill gates, the shallow spill gates passed 87% of the fish passing in spill and the deep gates only 13% (Ransom et al. 1988).

Spreading Spill Volume. Experience in 1995 at Priest Rapids and Wanapum Dams showed that spreading the spill of a given total volume of water over a 24-hour period doubled the percentage of fish passed in spill, as compared to spilling the same volume over a 12-hour period at night, where it might have been expected to pass the same percentage of fish. In fact, at Priest Rapids Dam in 1995, 17% spill for 24 hours a day for 60 days during the summer achieved 62% fish passage, whereas in the summer of 1994, spill of 40% for 12 hours per night for 34 nights only achieved an estimated 33% fish passage (Hammond 1994). Tests are underway at other mainstem projects to determine to what extent the Priest Rapids results might apply to the hydrosystem as a whole.

Ice and Trash Sluiceways—Effectiveness in Passing Juvenile Salmonids

Ice and trash sluiceways (Figure 7.25) are located at the surface directly above the turbine intakes and positioned well to attract fish approaching the powerhouses. For example, at The Dalles Dam, the ice and trash sluiceway extends the full length of the powerhouse and is designed to skim ice and trash off the water surface and pass it to the tailrace through a channel. Initial testing of the ice and trash sluiceways as juvenile passage routes occurred after juvenile salmon were observed in the sluiceways at Bonneville Dam (Powerhouse 1) and The Dalles Dam (Michimoto and Korn 1969). Efficiency of the sluiceways in diverting juvenile salmon from the turbine intakes was generally in

---

[11]Spill under 50,000 ft³/s creates backeddies and slackwater areas in the tailrace where excessive predation may occur.

[12]Stuart Hammond, Grant County PUD, personal communication, October 7, 2003.

Figure 7.25   Ice and trash sluiceway at Priest Rapids Dam. Photo by R. Whitney.

the neighborhood of 20% to 40%, (Nichols et al. 1978; Willis and Uremovich 1981; Willis 1982). At Bonneville Dam's Powerhouse 2, an ice and trash sluice chute located adjacent to the southernmost turbine was shown to pass an estimated 81% of juvenile salmon passing the powerhouse in the daytime and 30% at night (Magne 1987). In 1998, the Corps began modifying the chute into a surface flow bypass system (a "corner collector") in an effort to improve the effectiveness of the chute in passing juvenile salmonids. Ferguson et al. (2003) report that combined FGE of the chute and screens at turbine units 11 through 13 was 90% in both spring and summer. Considering the powerhouse as a whole, the combined FGE was 90% for spring migrants, and 75% to 80% for summer migrants. A transportation channel 2,800 feet in length has been provided, along with a 500 foot-long outfall channel, which deposits the fish about a quarter mile downstream where a plunge pool has been excavated into the river bottom.

All of the U.S. Corps of Engineers hydroelectric projects in the lower Columbia and Snake Rivers are equipped with ice and trash sluiceways. Given the success of the Bonneville sluiceways in passing juvenile salmonids and the spill efficiencies involved, considerable research effort is currently being directed at testing and refining the use of ice and trash sluiceways at other mainstem projects.

Using marked coho and subsequent regression analysis, Willis (1982) estimated effectiveness of the sluiceway at The Dalles Dam in passing fish at various levels of spill, for example, about 30% fish passage at 10% spill and 75% fish passage at 40% spill. The configuration of The Dalles Dam—where the spillway is at right angles to the natural course of the river, the powerhouse is nearly parallel to the natural course of the river, and the sluiceway is at the downstream end of the powerhouse—contributes to the effectiveness of spill efficiencies at The Dalles Dam. Hydroacoustic studies showed fish were more concentrated in the volume of water entering the ice and trash sluiceway than in water entering the turbines (Nichols and Ransom 1980, 1981; Steig and Johnson 1986). Dawley et al. (1989) estimated survival rates of coho and subyearling Chinook that passed the project by way of the ice and trash sluiceway were 96% and 89% respectively, which were similar to survival estimates of fish passing by way of the spillway (Ferguson et al. 2003).

Higher spill levels reduce the effectiveness of the ice and trash sluiceway in passing juvenile salmonids (Ferguson et al. 2003). For example, doubling the spill percentage reduced sluiceway efficiency by nearly 50%. Substantial predation by northern pikeminnow may occur at the sluiceway outfall and immediately downstream.

Application of Surface Collection at Other Projects

At most of the Corps projects, there have been studies aimed at improving fish guidance through use of surface collector devices, behavioral guidance structures, and simulations of the Wells Dam configuration. In summary, it can be said that while these tests have shown some promise, there has been as yet no breakthrough in design that would substitute for turbine intake screens in combination with provision of spill amounts, as called for in the 2000 BiOp. Further information may be found in Ferguson et al. (2003).

In the lower Snake River, the U.S. Army Corps of Engineers has investigated potential applications of surface collection systems at Corps projects, including several studies of prototype surface collection configurations at Ice Harbor Dam. Three types of surface collectors were tested: (1) vertical slots in front of two turbine intake slots (in conjunction with the ice and trash sluiceway), (2) a sluiceway surface skimming gate, and (3) stop logs that allowed surface spill at two spillbays (Swan et al. 1995). Effectiveness was evaluated by hydroacoustics and radiotelemetry of juveniles. The hydroacoustic study showed that the density of juvenile salmon was greatest in the sluiceway, although more total fish passed in spill because of the high volume of spill (Biosonics 1995).

At Lower Granite Dam, the Corps of Engineers tested a surface bypass collector from 1996 to 1999. The collector partially occluded the upper portion of three of the six turbine intakes. Attraction water entered the collector through four adjustable entrances and was conveyed in a channel to the adjacent spillbay. Fish passage efficiency was approximately 50%. The apparatus is still undergoing tests aimed at improving its effectiveness in passing fish (Ferguson et al. 2003).

Surface collection methods have also been investigated at several mid-Columbia PUD projects. At Rocky Reach Dam, a surface collection device (Figure 7.26) was put in place in 2003 (Peven et al. 1995, 1996). The bypass system consists of 200 cfs of attraction water at the entrance to a structure in the forebay that conveys fish into a flume that penetrates the wall of the dam, crosses the dam above the tailrace, and continues downstream for a distance of about 0.5 miles to the outfall, which is located in an area where water velocities are thought to be sufficient to discourage predatory fishes (Figure 7.27a and 7.27b). In combination with spill at the spillway, the bypass is estimated to provide 93% survival of juvenile yearling Chinook and steelhead, which is the goal specified in the PUD's HCP. Surface collection systems are being investigated at Rock Island and Wanapum Dams; however, results of prototype testing have not been sufficiently promising to warrant full-scale application.

Attracting Juvenile Salmon Directly into Spill for Passage

It is now generally believed that at most projects, the spillway provides the most benign route for passage of juvenile salmonids. Consequently, requirements for specific levels of spill are incorporated into the Federal Columbia River Power

Figure 7.26 Surface collector transported by barge to Rocky Reach Dam for installation. Photo courtesy of Chelan County PUD No. 1, Wenatchee, WA.

a) Panoramic view of dam face and entire juvenile bypass flume and outfall

b) outfall

Figure 7.27 Juvenile bypass system at Rocky Reach Dam, showing (a) a panoramic view of the dam face including the whole juvenile bypass system, and (b) a close-up of the juvenile outfall. Photo courtesy of Kelly Gillin, *The Wenatchee World.*

System's (FCRPS) Biological Opinion (BiOP) the NPCC's Fish Wildlife Program project license requirements by FERC, and most recently by the federal court, as previously mentioned. The requirements are not completely consistent among themselves, probably due to differences in statutory authority.

Survival of Juvenile Salmonids Passing the Dam in Spill

Studies of mortality in spill showed little or no mortality associated directly with the spill passage but, under certain conditions, high estimates of juvenile mortality associated with certain combinations of spill volume occur relative to river flow. Studies of mortality in spill provide 23 estimates (Whitney et al. 1997; Ferguson et al. 2003), in which 8 estimates showed zero mortality in spill and another 9 showed 2% or less mortality. Nevertheless, studies revealed a potential for added mortality from predation below the spillway, particularly at certain combinations of spill volume relative to river flow. One unusually high estimate of 27.5% at Lower Granite Dam was probably associated either with high predation by northern pikeminnow or other adverse conditions below the dam (Muir et al. 1995b).

Spillway design also affects the rate of injury and survival, with freefall being the least injurious (Bell and DeLacy 1972; Stone and Webster Engineering Corp. 1982). Backroll may be created with certain designs and spill levels, which can trap fish in turbulence, adding to the potential for predation and other causes of mortality (Stone and Webster Engineering Corp. 1986). Studies to evaluate spill effectiveness are utilizing radiotelemetry and PIT tag technology to estimate survival of juvenile salmonids in spill and other passage routes (Sheer et al. 1997; Hansel et al. 1998; Holmberg et al. 1998; Skalski et al. 2002).

The Council's Independent Scientific Advisory Board (ISAB) recently reviewed studies of survival in spill conducted by NMFS at The Dalles Dam in the years 1997 to 1999 and concluded that at that project, survival of juvenile salmon in spill is variable from year to year, differs between night and day, and differs between spring and summer, probably being affected by a number of factors, not all of which are understood (ISAB 2000). The Dalles Dam has no turbine intake screens and depends upon the ice and trash sluiceway in conjunction with spill to provide alternate passage routes for juvenile salmonids to achieve the 80% passage objective set by the Council and NOAA Fisheries. Considerable volumes of spill are required to achieve passage goals. Early studies of effectiveness of the ice and trash sluiceway suggested that 75% fish passage in spill could be achieved at 40% spill[13]. An experiment was designed to identify the spill level at which an 80% fish passage goal could be met for all salmonid species (ISAB 2000). However, higher than expected mortalities were observed at certain levels of spill. Studies continue there, where the stilling basin has been identified as a probable source of high mortality at some spill levels.

The Wells Dam Hydrocombine

As noted above, the Wells Dam's surface bypass system and its efficiency served to stimulate interest throughout the basin for the potential of surface bypass systems and relatively small amounts of spill to assist in achieving

---

[13]Spill volume equal to 40% of total volume of flow.

Council and FERC objectives for juvenile survival and passage at mainstem hydroelectric projects. Spill efficiencies take advantage of the unique hydro-combine design of the powerhouse at Wells Dam, where the spillway is located directly above the turbine intakes (Figure 7.24a). This provided a sit-uation in which it was thought that juvenile salmonids, observed to enter the turbines near the ceiling, might be diverted into the spillbays above. Testing of a prototype began in 1983 (Biosonics 1983a) and resulted in modifications using baffles that concentrated the volume of water entering the spillway, so that it reached higher velocity in slots than it would have if spread across the powerhouse (Figure 7.24b). Plumes of water at the spillway entrances appeared across the powerhouse that were apparently more attractive or more readily detected by juvenile salmonids than the evenly distributed flow that occurred when spillways were opened without the baffles. The volume of water required for operation of the bypass varies somewhat depending on river flow and the powerhouse load. In 1995, it ranged between 1.2% and 7.5% of the daily average river flow.

The Wells Dam bypass system was fully installed in 1989. In January 1991, a long-term settlement agreement approved by FERC established a criterion at Wells Dam for bypass of at least 80% of the juvenile salmon for the spring period and at least 70% for the summer. From the resulting studies, the 3-year average bypass effectiveness during both the spring and summer outmigra-tions was estimated to be 89% (Skalski 1993). It is currently the most effec-tive bypass system in the basin. The new standard at Wells Dam is spelled out in the HCP for Wells Dam, an agreement reached for mitigation of effects of development and operation of Wells Dam on survival of salmonids (NOAA Fisheries 2003). The HCP sets a goal of 93% survival of juvenile yearling Chinook and steelhead over 95% of the duration of the outmigration. Actual measured survival was estimated to be 96%.

Supersaturation of Gas due to Spill

A problem encountered with high spill amounts is gas supersaturation (Figure 7.28), leading to a condition in fish similar to the diver's "bends," in which gas bubbles appear in the blood stream and other tissues, which can lead to death (Ebel 1969; Dawley and Ebel 1975; Bouck 1980; Chapter 6). To achieve the survival benefits of spill during dam passage with minimal in-river dam-age from gas bubble disease, a physical and biological monitoring program has been established, which includes both physical and biological criteria for cessation of spill. The U.S. Army Corps of Engineers (1993; Ruffing et al. 1996) has monitored levels of total dissolved gas saturation at near-surface monitoring stations downstream of dams for many years, and the Smolt Monitoring Program and the National Biological Service have monitored downstream migrants for biological signs of gas bubble trauma at smolt

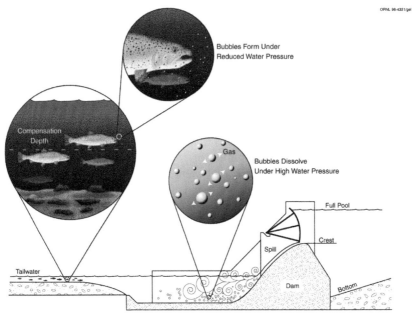

Figure 7.28 Diagram of a mainstem dam showing spill causing gas supersaturation when the spilled water plunges into the stilling basin below the dam under pressure and then affecting fish in the tailwater below.

monitoring stations in dam bypasses since 1994 (McCann 1995; Smolt Monitoring Program 1995), including recommendations from an expert panel convened by the NMFS (Panel on Gas Bubble Disease 1994, 1996; Backman et al. 1996; Schrank and Dawley 1996). While the expert panel recommended that monitoring for signs be augmented by estimates of in-river survival ("reach survival estimates," now possible with PIT tag technology), field research to obtain such estimates is still being developed (Muir et al. 1995a). There have been a few attempts to decipher changed survival due to gas-bubble effects from PIT tag data (Cramer 1996).

The monitoring results have been controversial. Physical monitoring has shown that spill increases gas saturation, both when controlled by the management program and when uncontrolled during major runoff events or unavailability of turbines. Gas saturation at short distances (usually one mile or less) downstream of dams range from about 115% to 120% during controlled spill, but up to about 140% during uncontrolled spill. Even during uncontrolled spill, however, biological monitoring of bubbles in fish at dam bypasses has shown low incidence and severity, much below the biological criterion of 15% incidence in juveniles that would trigger cessation of spill.

On the basis of these monitoring results, risk analyses favoring spill have been prepared by the Fish Passage Center (1995).

Standard methods for measuring and quantifying bubble signs in fish that are clearly related to mortality (or other debilitation) should be useful for routine monitoring (e.g., bubbles in the lateral line, fins, buccal cavity, and gill lamellae). Although a monitoring program for juvenile migrants has been in place at dams for several years (Fish Passage Center, Portland, Oregon, annual reports), the link between gas bubble monitoring data to changes in survival is still unclear. We cannot reliably relate severity of damage or probability of death (survivability) to the presence or absence of specific signs used in monitoring today across a full range of possible effects. Monitoring of juvenile salmonids for gas bubble disease signs in the bypasses of dams is based upon assumptions that have not been substantiated and thus the results may be skewed toward underestimation of effects (Biological Monitoring Inspection Team 1995; Panel on Gas Bubble Disease 1996).

Because high in-river survival of fish is the recovery goal, direct measurement of survival under varying conditions of gas supersaturation would appear to be the most useful source of information for managing total dissolved gas saturation and spill. Methods for obtaining reach survival estimates being developed by Muir et al. (1995a) should help evaluate the effects of gas supersaturation on survival of juvenile salmonids. Preliminary analyses of 1994–1995 PIT-tag survival data by NMFS and the Fish Passage Center (presentation to Council, January 10, 1996) suggested that managed spill yielding gas saturation values generally under 115% did not lower survival.

Reduction of Total Dissolved Gas

One approach to managing spill-induced gas saturation has been to search for mechanisms to lower levels of total dissolved gas during fish emigration. One of the most straightforward is physical modification of spillway exits to incorporate spill deflectors or "flip lips." This was an active program by the Corps of Engineers in the 1970s, temporarily abandoned when spill became less common, but was revived with the BiOp's call for increased spill.

The spill deflector (flip lip) design for the spillway (Figure 7.29) directs the spill in a horizontal direction rather than steeply downward (Smith 1974). At flows of 123 to 169 kcfs, spill deflectors at Little Goose Dam were shown to reduce gas saturation levels downstream by about 10%, relative to levels before the deflectors were installed (pre-installation gas saturation of 128% with spill of 46% to 59% of river flow) (Park et al. 1977). At Lower Monumental and Lower Granite Dams, gas saturation levels were 2% to 8% lower than at Little Goose Dam under the same flow conditions, probably due to the greater depth of the stilling basin below Little Goose Dam and

Figure 7.29 Spill bay outlets at Bonneville Dam showing a "flip lip" installed on the right spill-bay, but absent on the left spillbay. Photo by R. N. Williams.

smaller deflectors there (8 ft in length compared to 12 ft at the others). At McNary Dam, gas saturation was lowered 16% to 20% by installation of spill deflectors (Park et al. 1977).

In a more thorough analysis, Johnsen and Dawley (1974) developed curves showing the relationship of gas saturation levels below the spillway with fore-bay gas levels, spill discharges, water temperatures, tailwater elevations, and effects of deflectors at Bonneville Dam. With forebay gas levels of 110% and tailrace elevations of 24 feet, the deflectors generally reduced gas saturation levels by about 10% (130% reduced to 120%). But at higher discharge rates (and tailrace elevations), the difference lessened to the extent that it appeared the deflectors might be disadvantageous at spill discharges above 14 kcfs per bay.

The demonstrated success of spill deflectors in reducing levels of gas satu-ration led to installation at most projects. Improved devices were recently installed at Ice Harbor, John Day, and Bonneville. In 1996, only Lower Granite Dam was fully equipped with flip-lip spillbays across the spillway. At present, seven of the eight Corps projects in the Snake River and lower Columbia River are equipped with spill deflectors (Bruce 1995; NMFS 1995a). Only The Dalles Dam does not have spill deflectors installed, as juvenile fish passage efforts there have concentrated on passage associated with the ice and trash sluiceway and other options.

The Corps of Engineers is considering other plans for possible structural modifications to spillways and stilling basins for the purpose of abatement of total dissolved gas levels. To date, only the spill deflectors have been implemented.

Mortality of Juvenile Salmonids in Reservoirs

Mortality of juvenile salmon and steelhead occur as emigrating fish move through various routes of passage around and through hydroelectric projects as discussed above. Mortality also occurs in the slackwater reservoirs between projects, where juveniles are lost due to delayed mortality from dam passage and from predation by native and non-native predators (fish, birds, and mammals). Survival of juveniles may also be negatively affected by exposure to increased water temperatures, delays in migration timing, disease opportunities, and increased encounters with predators.

Estimates of juvenile survival through hydrosystem projects and river reaches brought into focus the need to be able to separate direct mortality induced upon juvenile salmon in the turbines themselves from mortality experienced elsewhere, either as an indirect result of turbine passage or other causes, because the solutions will differ. There have been several attempts to separate mortality estimates into components for the reservoir and tailrace; however, estimates are few and results appear highly variable, apparently species and location specific in nature.

For example, Iwamoto et al. (1993) developed an estimate for mortality in the reservoir above Lower Granite Dam in 1993, based on a series of reach survival estimates from a point above Lower Granite Dam to the tailrace at Little Goose Dam, as well as estimates of survival in turbines at both dams. The study produced an estimate of zero mortality for yearling Chinook in the reservoir above Lower Granite Dam in 1993. These results are in contrast to Muir et al. (1995a), who developed an estimate of 42% steelhead smolt mortality from the forebay at Lower Monumental Dam to the tailrace, a surprisingly large number. Unfortunately, there seems to be no estimate of mortality of steelhead in passage through turbines for Lower Monumental Dam. However, even assuming the worst, say 20% mortality in the turbines, the result indicates a high loss of juvenile steelhead in the forebay. Steelhead smolts that are delayed in their outmigration have been observed to residualize and become resident trout (ISAB 2003). Complicating the estimates of mortality of fall Chinook in reservoirs is the recent report that some fish may spend an additional year there before emigrating to sea (Connor et al., 2003). The full ramifications of this finding remain to be explored

Studies indicate that in some instances, losses of juvenile salmon in the forebay and tailrace may exceed the losses in turbines. Poor placement of bypass outfalls can affect juvenile survival in the tailrace, while migration delays in the forebay can result in increased susceptibility to predators. Estimates of survival at the outfalls, especially Bonneville Dam, The Dalles Dam, and John Day Dam, brought out the fact that losses in the tailrace can be substantial due to predation by northern pikeminnow and smallmouth bass.

Predation

Development and operation of the hydroelectric system has produced habitat more favorable to native and non-native predators and coolwater-adapted species than to the native coldwater-adapted salmonid species. Conspicuous among these species is the native piscivorous minnow, the northern pikeminnow (*Ptychocheilus oregonensis*), known until recently as the northern squawfish. Predation by other fish species and by birds, especially gulls and terns on the mainstem, is a well-documented source of mortality for emigrant juvenile salmon in the Columbia River Basin. Direct observations of rates of consumption, and conclusions derived from simulation models, established fish predation as a factor capable of removing a substantial fraction of the annual juvenile emigration (Willis and Ward 1995). It was therefore logical for the Council's Fish and Wildlife Program to consider means of altering predation in ways beneficial to salmon survival.

Prior to implementation, predator reduction as an action to increase survival of emigrants in the Columbia River Basin was extensively discussed over a 2-year period (1988–1989) by a working group of biologists employed by the fisheries agencies and tribes, NPCC, and the hydroelectric industries. One of the primary agents of mortality in reservoirs of the Columbia River was postulated by the Working Group to be predation by piscivorous fishes. The extent to which predation is a documented agent of mortality in juvenile salmonids in the Columbia River system was established by an intensive program of research on predation on juvenile salmon conducted in John Day Reservoir (Thompson 1959; Poe and Riemann 1988; Collis et al. 1995). The Working Group identified predator reduction as one of the few measures within the Fish and Wildlife Program that might immediately reduce mortalities of emigrant and resident juvenile salmonids.

Northern Pikeminnow

The predator reduction program has focused on northern pikeminnow (Figure 7.30a). Pikeminnow were targeted not only because research indicated them to be responsible for the majority of predation on juvenile salmonids in the reservoir behind John Day Dam (Poe and Riemann 1988), but also because other predators were the objects of sport harvesting effort, while northern pikeminnow were not. The states of Idaho, Oregon, and

Figure 7.30   Predators: (a) northern pikeminnow, and (b) Caspian terns.

Washington manage exotic predators such as smallmouth bass (*Micropterus dolomieu*), walleye (*Stizostedion vitreum*), and channel catfish (*Ictalurus punctatus*) as game fish with catch and season limits. The hydrosystem has created habitat favorable to all of these predators, including the native pikeminnow.

The predator reduction program was envisioned in its broad sense to include non-lethal means of reducing access of northern pikeminnow to juvenile salmon in the hydroelectric system, as well as more traditional means of removal by fishing and other lethal means (Poe and Riemann 1988). Modeling studies indicated that annual reduction of the northern pikeminnow population by approximately 15% could reduce the losses of juvenile salmonids by as much as 50% (Beamesderfer and Rieman 1988; Rieman and Beamesderfer 1990; Beamesderfer et al. 1990). The reduction program was seen as a long-term, perhaps continuous, attempt to alter the age composition of the population in favor of the younger, smaller age classes (< 10 inches length) that do not consume juvenile salmonids. Altering the age composition was seen as preferable to eradication efforts, because the northern pikeminnow age structure might be altered without substantially diminishing the reproductive capacity of the population. With sustained northern pikeminnow reproduction, other species of predators, which normally target juvenile northern pikeminnow, would not be forced to switch to juvenile salmonids by declining availability of northern pikeminnow juveniles.

The Working Group also discussed the need to reduce populations of northern pikeminnow in the immediate vicinity of the hydroelectric dams. Very large northern pikeminnow individuals congregate in the forebays and tailraces of the dams, where boat restrictions occur and the general public are prohibited. Angling from the dams and operational procedures such as turbine operating sequences and spill were also identified as possible ways to disrupt intense predation at the dams.

The reduction program was implemented by the fisheries agencies and tribes starting with pilot studies in 1990 (Young 1996). The pilot approaches included (1) paying bounty to members of the public for northern pikeminnow of predaceous size (sport reward fishery), (2) employing net fishers to target northern pikeminnow in the reservoirs, (3) employing professional hook and line anglers to fish in waters adjacent to dams from which the general public is excluded, and (4) fishing with nets near a hatchery outfall. All approaches were initially highly productive except the reservoir net fishing, which was later abandoned.

Reductions in northern pikeminnow populations are not uniformly geographically distributed: Some areas show decreases, whereas others do not. The predator reduction program has demonstrated a sharp decline in the abundance indices of northern pikeminnow in the vicinity of Snake and lower Columbia River dams, and it has demonstrated a shift toward younger, smaller individuals available to private anglers outside the areas of dam influence (Friesen and Ward 1999; Zimmerman and Ward 1999). Northern pikeminnow greater than 11 inches in length are known to be predators on juvenile emigrant salmon. Annual catches and catch per unit efforts by technicians angling below Snake and Columbia River dams declined by about 80% during the 4 years of the program (ending in 1995), but there was no appreciable change in the average size of the individuals caught at dams. Annual catches of private anglers, who were paid for each northern pikeminnow over 11 inches long, did not decline during this period, but the average size of the individual fish in these catches did decline. Average size of the individuals caught by private anglers appeared to be influenced by recruitment from strong-year classes in unimpounded areas below Bonneville Dam and below Priest Rapids Dam.

The sport reward and hatchery outfall fisheries continued to be highly productive as of the 1995 season (Young 1996) and to date. The dam angling projects have seen a sharp drop in catches of northern pikeminnow as of 1995. The program has been instrumental in lowering the rate of northern pikeminnow predation, which would otherwise have been experienced by the juvenile salmon. In the period from 1990 to 1996, the program removed an estimated 1.1 million northern pikeminnow, leading to reduction in predation on juvenile salmonids by 14% to 38% relative to that which occurred prior to implementation of the mitigation program in 1990 (Friesen and Ward 1999).

Caspian Tern

A colony of Caspian terns (Figure 7.30b), which since 1996 have nested on Sand Island below Bonneville Dam, are significant predators on juvenile salmon and steelhead. The seriousness of the problem was first noted when scientists equipped with magnetic sensors detected large numbers of fish tags that had accumulated on the ground among the deposits of bird scat on the island. The number of terns had increased dramatically as a result of new nesting area made available by disposition of dredge spoils by the U.S. Army Corps of Engineers.[14] The terns prefer bare sand and silt as nesting habitat and are discouraged by growth of grass or other vegetation.

In 1997 it was estimated that Caspian terns consumed about 11 million salmon and steelhead smolts, amounting to an estimated 5.4% of the total production of the Columbia and Snake River systems (G. Bisbal, NWPPC staff, personal communication, 29 September 1999). The colony of terms now represents about 70 % of the world's population.

By the year 2000 it was estimated their consumption of juvenile salmon had been reduced to between 3.5 and 7.7 million as a result of efforts to discourage their nesting and by transplantation efforts. Transplantations to areas a distant as San Francisco are being considered. But a new, even more menacing predator, the double crested cormorant, has entered the picture. In 1989 double crested cormorants were estimated to number in the hundreds at Sand Island. Now, it is estimated there are probably 30,000, and still increasing. These larger birds undoubtedly eat more fish, but there are no estimates of the total as yet.

---

[14]Sand Island, the nesting site for the new population of Caspian terns, was created by the U.S. Army Corps of Engineers from dredge spoil, part of which had its origin in the eruption of Mount. St. Helens in 1980. The relocation plan calls for removal of some of the grasses that form cover used by the nesting birds, and plantings on other sand bars to attract the birds.

The Council has called for measures to remove the terns from this location, and the Corps has developed an action plan for implementation in the spring of 2000. The plan became controversial as a segment of the public (and wildlife agencies) expressed concern about the fate of the birds. The result was modification of the plan to emphasize relocation rather than reduction or elimination of the population.

Other birds are known to feed on juvenile salmonids, particularly in the tailraces where fish disoriented from passage through the turbines are vulnerable to capture near the surface. California gulls, ring-billed gulls, double-breasted cormorants, common mergansers, and pelicans have been identified. Nesting bird colonies also have been found at several islands in the McNary Reservoir. Measures to inhibit their feeding activity consist of wires stretched across the tailrace and other harassment. These measures are specified in the HCPs for the mid-Columbia projects.

Proposed, but Largely Untried, Methods for Juvenile Bypass

Spill, turbine intake diversion screens, surface collectors (Wells Dam), and ice and trash sluiceways are the only bypass systems that have proven to be effective for juvenile salmon in the Columbia Basin. Numerous alternatives have been investigated for their potential in directing juvenile salmonids at the dams, including diversion barriers upstream of the dams, a forebay wedge screen, batteries of lights, bubble curtains, electric fields, sound, air lifts to remove fish from gatewells, gatewell conduits without intake screens, and others. These alternative methods were reviewed by the Office of Technology Assessment (1995) and Stone and Webster Engineering Corp. (1986), who concluded that for the most part, these devices have not been accepted by the resource agencies because they have not been shown to divert a high enough percentage of the fish. None was sufficiently effective to justify full-scale or prototype testing in the field for application at large hydroelectric projects.

Some of these proposed methods have met with varying degrees of success for other species in different applications elsewhere, such as at pump intake diversions or irrigation diversions (Office of Technology Assessment 1995). For example, angled louvers have been used effectively at pump intakes and irrigation diversions to divert juvenile salmon and other small fish into alternate channels (e.g., Sacramento River), but high water volumes and velocities preclude their effectiveness in rivers of the Northwest. At a smaller scale, louvers have been used successfully in the Columbia Basin at irrigation diversions, in conjunction with screens, where flows were carefully regulated at low levels and floating debris was sparse (Mighetto and Ebel 1995).

Another bypass concept uses pumps to create attraction flows to direct fish, without completely blocking their path, into a collection device, an enclosure of some kind (e.g., a "Merwin" Trap). Such devices were tested at Pelton Dam on the Deschutes River, Mud Mountain Dam, and Merwin Dam on the Lewis River (Stockley 1959; DeHart 1987). At Baker Lake, Washington, a surface collection device of this type was found to be effective in collecting juvenile sockeye salmon for transportation below the power-house (Wayne 1961; Quistorff 1966). It became a viable solution to the problem of collecting juvenile salmon in the reservoir when a lead net was added to the "gulper" (Cary Feldmann, Puget Sound Power and Light, personal communication). While these approaches have shown some localized effectiveness, they are not widely employed as collection devices.

In some situations, salvage of fish from gatewells has proven to be a worth-while exercise. Since 1980, Grant County PUD has salvaged fish from the gatewells at Wanapum and Priest Rapids Dams on a daily basis during the outmigration. Specially designed nets deployed by mobile cranes from the deck of the powerhouse are used to remove fish that have accumulated. Captured fish are placed into tank trucks, transported below the dam, and released into the tailrace. Approximately 150,000 to 200,000 fish are salvaged at each of the two projects each year in the spring and an additional 30,000 to 50,000 in the summer[15].

Another area of ongoing research has been an attempt to develop turbine designs that are more fish friendly. Justification for this approach comes from studies that showed higher survival of fish passing through turbines when turbine efficiency is higher. Because damage to the machinery is least at high efficiencies, both factors are incentives to operate the machines in the region of their highest efficiency. Currently, the U.S. Army Corps of Engineers is exploring design of more efficient turbines; however, none of these designs is likely to be as benign in effects as diversion by spill, intake screens, or surface collectors. The expected benefits of the fish-friendly turbines to total survival of juveniles passing a project would be slight.

Finally, in the early 1990s, seasonal drawdown of lower Snake and Columbia River reservoirs was proposed as a mitigation tool. The rationale for temporary drawdown focuses primarily on the potential to reduce travel time for emigrants. However, the efficacy of this mitigation measure has not been demonstrated. Since that time, scientists within and outside the region have rejected this approach (NRC 1996). Concentration of salmonid juveniles with predators and loss of shallow water habitats are potential problems with drawdown scenarios.

An extension of the seasonal drawdown approach is the "natural river option," which calls for breaching or bypassing dams and would likely yield conditions and benefits beyond those achieved by drawdown (Figure 7.31).

---

[15]Stuart Hammond, Grant County PUD No. 2, personal communication, October, 2003.

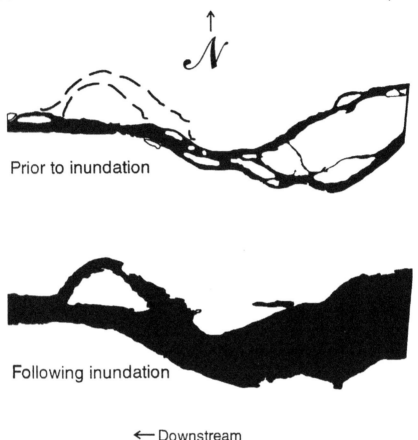

Prior to inundation

Following inundation

← Downstream

Figure 7.31 Natural river drawdown. Silhouettes of the area upstream of John Day Dam on the Columbia River before and after impoundment.

These options to improve ecological conditions and salmon production in the basin have been discussed throughout the region over the last few years and have been evaluated with respect to their biological, as well as social and cultural, benefits and costs. There is insufficient information available to predict with certainty the magnitude of responses to drawdown or breaching that might occur in salmon populations.

Nevertheless, virtually all attempts at analysis predict the greatest and most rapid response by depressed upper basin salmon and steelhead stocks to management scenarios that restore natural river conditions via the breaching or bypassing of the four lower Snake River dams (PATH Scientific Review Panel 1998). The Corps of Engineers (2002) completed an analysis of the feasibility of these alternatives. (See Chapter 12). The political/social response to this proposal has been mostly negative.

Transportation of Juvenile Salmonids

Transportation of juvenile salmon downstream in barges and trucks to avoid mortalities associated with passage at some of the Snake River dams and their reservoirs is one of the techniques employed in an attempt to protect salmon from the harmful effects of the federal Columbia River hydroelectric system. Juvenile salmon emigrants are removed from the turbine intake bypass systems when they arrive at certain Snake and Columbia River hydroelectric dams during their annual migration downriver (Point F, Figure 7.16). Primary collection locations are Lower Granite, Little Goose, and Lower Monumental Dams, with McNary Dam as a secondary collection facility. Juvenile salmonids are diverted from the bypass systems at these dams into holding facilities, from which they are later transferred to the barges (mostly) or trucks (at the beginning and end of the season when few fish are diverted), transported around or through the hydrosystem, and released downstream of Bonneville Dam (Figures 7.32 and 7.33). The proportion of yearling Chinook salmon that is transported from the lower Snake River may be as

Figure 7.32 Transportation of juvenile salmonids by barge around the mainstem hydrosystem projects. Photo from U.S. Army Corps of Engineers Digital Visual Library, website at *http://images.usace.army.mil/photolib.html.*

Figure 7.33 Transportation of juvenile salmonids by truck around the mainstem hydrosystem projects. Photo from U.S. Army Corps of Engineers Digital Visual Library, website at *http://images.usace.army.mil/photolib.html.*

high as 85% (Williams et al. 2003); however, not all species and life history types are equally susceptible to collection, so that the proportion remaining in the river will vary by species and life history type within species.

Effectiveness in collecting fish for transportation depends upon the FGE of the intake screens, which varies among projects according to flow, species, and life history type, as well as state of maturity. In general, FGE is higher for life history types with large juvenile emigrants, such as spring Chinook and steelhead, and lower for life history types with small juvenile emigrants, such as sockeye and fall Chinook. As a consequence, the efficacy of transportation depends heavily on the FGE of the particular turbine intake screens, which are selective as to size of fish, life history type, state of maturity, and other factors. This selectivity of action is a source of concern for some because of the potential for reducing the adaptive capacity of the populations as a whole (Independent Scientific Group [ISG] 2000).

Evaluation of the transport program is based on comparative rates of return of transported and untransported (i.e., "in-river") adults under the assumption that the probability of recapture is the same for all marked fish. A review of NOAA Fisheries studies prior to the recent PIT tag studies showed that transportation increased the survival of yearling spring Chinook to the point of release in 26 out of 36 studies over the years from 1968 to 1995 (ISG 2000). Results with steelhead during the same period indicated improved survival of transported juveniles.

Use of PIT-tagged juvenile salmonids in a single release estimation model (Jolly-Seber) has resolved many of the issues that were raised with the earlier, more limited studies and changed the picture completely (Williams et al. 2003). PIT tag technology allowed researchers to measure the effects of transportation at different points in the life cycle, as well as for routes of passage other than the turbine intake bypass. PIT tag recognition systems have also been installed into juvenile bypass systems at many of the hydrosystem projects that allow individual fish in release groups to be identified as to their passage routes including transported, in-river, and combinations (Williams et al. 2003).

Using the PIT tag results from the years 1994 through 2000, ratios of return rates ranged between 0.55 and 0.63 for the four combinations of species and rearing type (hatchery and wild). On average, return rates of transported juvenile salmonids were lower than those of in-river migrants (Williams et al. 2003). For example, survival rates below Bonneville Dam of wild spring/summer Chinook salmon transported from Lower Granite Dam to below Bonneville Dam were roughly equal to the survival rates of non-transported fish that migrated between Lower Granite Dam and Bonneville Dam, thus showing no advantage for transportation. When results are combined for both species (Chinook and steelhead) and life history types (hatchery reared and wild), the average return rates of transported fish ranged between 0.55 and 0.63; that is, transportation was disadvantageous.

Williams et al. (2003) examined return rates according to timing within the season and found that return rates of transported fish were lowest early in the season and increased later. On the other hand, return rates of in-river migrants showed the opposite trend, with a downward trend as the season progressed. They concluded that early in the season, transportation, which typically requires 1.5 days, rather than the 3 weeks required for in-river migrants, most likely puts the transported juvenile salmonids in the estuary too early, before conditions are right for them in terms of food supply and other ocean conditions.

Snake River Basin spring/summer Chinook have shown a response to transportation that is probably best explained in terms of conditions within the hydroelectric system at the time of transportation. When in-river conditions are clearly adverse, such as the low flow year of 1973, transported fish showed a strongly positive relative rate of adult returns, although both transported and untransported salmon showed extremely low overall survival. In another low flow year (1977), the effects of transportation could not be measured because nearly all of the juvenile salmon marked for the experiment, both transport and control, died before returning as adults[16].

Under passage conditions associated with higher river flows, the responses of relative survivals of spring/summer Chinook to transportation may be equivocal. Consequently, the benefits of transportation would be expected to be smaller during years of high flow, when spill amounts are high.

The 2000 BiOp adopts a "spread the risk" policy, which attempts to produce a balanced approach between in-river migration and transportation during the spring migration, but which during the summer requires maximizing transportation of juvenile Chinook. The 2000 BiOp specifies that during the spring migration, transportation is required at Lower Granite, Little Goose, and Lower Monumental Dams. During the summer migration, transportation is required at the same dams and from McNary Dam on the lower Columbia River. While the 1995 and 1998 BiOps called for collection and transportation from McNary Dam in the spring, the 2000 BiOp suspended that operation when unexpectedly low survivals were measured. Because collection efficiency is affected by spill levels, the BiOp also specifies spill levels and dissolved gas limits for each of the federal projects on the lower Snake and Columbia Rivers.

The possibility that transportation of juvenile salmon might affect their homing ability when they return as adults, is a question that deserves more attention. The data of Chapman et al., 1997, indicate that sockeye smolts transported from Priest Rapids Dam to below Bonneville Dam experienced migration delays and recovery rates different from sockeye transiting in-river. That was not seen with spring Chinook.

Managing Passage of Juvenile Salmonids through the Hydrosystem

Managers are faced with a quandary with respect to in-season decisions on transportation as a tool to decrease juvenile mortalities through the hydrosystem. To maximize the benefits of transportation for those groups of fish that are most susceptible to diversion into the bypass systems, spill should be minimized. But minimizing spill is certain to emphasize the selective effects of transportation, because it is known that the bypass systems favor larger sized juvenile migrants (yearling spring Chinook, coho, and steelhead) and are not as effective for smaller migrants (subyearling fall Chinook and sockeye). On the other hand, the relative benefits of spill for the different stocks of salmon are unknown and are assumed to be equally beneficial for all stocks. In the

---

[16]Results of studies in 2001 will be of particular interest since it was a year of unusually low flow and requirements for spill were relaxed.

face of this quandary, the Council's ISAB recommended a "spread the risk" approach, involving the use of barges as well as in-river measures such as spill to improve survival of the entire spectrum of juvenile salmon and steelhead emigrants across the annual migration period (ISAB 1998). The ISAB recommended against the use of trucks for transportation because the NMFS studies have not demonstrated favorable return rates for trucked fish.

Given the apparent dependence of the survivals of both transported and untransported juvenile salmon on conditions in the hydroelectric system, juvenile transportation alone is unlikely to halt or prevent the continued decline and extirpation of listed species of salmon in the Snake River Basin. While transportation appears to improve the relative survival rates of certain kinds of salmon from the Snake River Basin under certain combinations of dam operations, river flows, and progress of the season, it is selective in its operations, as it removes the mortalities attendant to passage through the hydroelectric system for only a portion of the juvenile salmonids present. Its effects to date have been insufficient to reverse the downward trends in abundance, even for those stocks that it may favor.

Dams as Regulators of Flow

The major effect of full development of the hydroelectric system has been the large-scale regulation of flow made possible by water storage in large upstream reservoirs, available for later release to produce electricity (Chapter 6). As mentioned in Chapter 6, the result has been a shift in the timing of the early spring high flow period to later in the season. Concerns about potential effects of this shift on salmon and steelhead were addressed by identification of a portion of storage to be used to improve their survival through augmentation of flow. In addition to the large-scale effect from storage upstream, each of the dams has an effect as an obstacle and as a regulator of outflow. Outside of the approximately 55 miles of undammed river in the Hanford Reach and about 100 miles in the Snake River below Hells Canyon Dam, dams are spaced in the rivers to the point that their reservoirs extend upstream into the tailraces of the dam next upstream, affecting the head of the turbines there and calling for compensation for the intrusion.

The recognition that the river has been replaced by a system of reservoirs leads us to begin with a focus upon that aspect of flow and its effects on salmon and steelhead, and to move from there to discuss the large-scale mitigation effort that calls for flow augmentation. Reservoirs have slowed and otherwise modified the movement of water to the ocean. The resulting regulation of flow affects migrations of adult as well as juvenile anadromous fishes.

Effects of Flow Regulation on Adult Anadromous Fishes

Construction and operation of dams for hydroelectric power production has led to a change in flow conditions encountered by adult salmon migrating upstream. Formerly, flow was spread to some extent over the entire cross section of the river, with the volume being concentrated in a channel cut by scouring action of the river. At present, flow is usually concentrated through each dam's powerhouse into the tailrace. During times when flow exceeds powerhouse hydraulic capacity, spillways are operated that generally split the flow away from the powerhouse. At some point downstream, water from the powerhouse and spillway meet to continue downstream.

Passage facilities (fish ladders) for adult salmon are normally located at each end of the powerhouse and at the bank opposite the powerhouse. Usually, access to the passage is available at openings across the powerhouse. While this arrangement attempts to simulate natural conditions, it is necessary at times to adjust operations of individual turbines and spill bays to compensate for adverse flow conditions that may be produced at passage entrances under some flows (Dauble and Mueller 1993; Mendel et al. 1994). Careful adjustments of flow and other characteristics of the ladders themselves are required to keep them functioning properly. For example, changes in elevation of the forebay or tailrace require corresponding adjustments in flow through the ladder. Certain levels of spill at some projects may produce conditions in the tailrace that create irregular patterns of flow that can confuse adults attempting to find ladder entrances. Many of these adjustments are now computerized.

Radio tracking of adult salmon and steelhead has identified behavior patterns that may be associated with unfavorable conditions of flow in and around the ladders. A certain percentage of fish may "fall back," that is, they may successfully negotiate the ladder in part or in full and then return downstream either through the ladder, the turbines, or spillway. Energetic costs of these behaviors are not known; however, mortality of upstream-migrating adult salmon and steelhead does not appear to exceed 2% in passing each project reach in the mid-Columbia (NOAA Fisheries 2003) or in the Snake River (Ferguson et al. 2003). Prior to the PIT tag studies, estimates had been in the vicinity of 5% per project (Chapman et al. 1994).

Spawning by fall Chinook still occurs in some open reaches of the mainstem where water velocities, substrate, and percolation are suitable, such as the Hanford Reach, the Snake River below Hells Canyon Dam, and in the tailraces immediately below most Columbia and Snake River mainstem dams (Horner and Bjornn 1979; Dauble and Watson 1990; Garcia et al. 1994).

Observations of fall Chinook spawning in the mainstem indicate that velocity of flow may be more important than water depth in determining location of their spawning (Chambers 1955). Fall Chinook have been observed spawning in water as deep as 30 or 35 feet (Meekin 1967; Chapman et al. 1983; Garcia et al. 1994). However, modification of normal flow results from the practice of "load following" by hydrosystem operators within operating criteria that require maintenance of minimum reservoir elevations.

Extreme fluctuations in flow have been observed at times. For example, load following during low runoff conditions may lead to virtually zero outflow from a project at night, when the powerhouse is shut down to preserve water for use during higher periods of demand for electricity (ISAB 2003). Such operations do not appear to affect adult passage, since adult salmon and steelhead do not normally continue their upstream movements at night (Bjornn et al. 1998); however, spawning, incubation, rearing, emergence, and outmigration of anadromous fishes are affected by these fluctuations. Dewatering of Chinook redds occurred in the tailrace at Chief Joseph Dam (Meekin 1967) and exposed redds and incubating eggs in the Hanford Reach (Watson et al. 1969; Bauersfield 1978; Chapman et al. 1983). Agreements to stabilize flows have been reached that protect spawning fall Chinook in the Hanford Reach and chum salmon spawning below Bonneville Dam.

Irrigation removals affect flow, as do the manipulations by the hydrosystem. The most substantial removal occurs at Grand Coulee Dam where Columbia River water is diverted into the irrigation system leading into the Columbia River Plateau. In the Snake River, irrigation removals have a greater effect on flow than does the hydrosystem there (National Research Council (NRC), 1996). Otherwise, effects of irrigation are primarily in the tributaries. Irrigation removals produce long-term or periodic increase in the extent of the shallow water zone in tributaries, leading to potential for stranding of juveniles or their diversion into the ditches. The region has undertaken an ambitious program for screening of irrigation intakes to prevent entrainment of salmonids. Increases in temperature in the tributaries are a serious problem with no apparent solution.

In a later section we discuss the effects of the hydroelectric system on water temperature and the effects on salmon. Our discussion there focuses upon mainstem Columbia and Snake river effects. Here we focus upon effects in the tributaries, where observed effects are not associated with the hydroelectric system, but with irrigation removals.

In some tributaries, irrigation removals have created problems for passage of adults. For example, low flows at the mouth of the Yakima River below Rosa Dam during the summer months lead to high water temperatures that are a barrier for adult salmon. The problem also occurs in the Okanogan River, where sockeye adults are delayed in their upstream migration by warm water at the mouth of the river (Allen and Meekin, 1980; Major and Mighell 1966; Mullan 1986; Mullan et al. 1986; Mullan et al 1986;) Similar thermal blockages during summer and early fall have been noted for the Chinook and steelhead in the John Day and Umatilla Rivers. In the Umatilla River, water is pumped from the Columbia River upstream into the Umatilla River to augment flows for fish. In combination with a hatchery effort, Chinook salmon have been restored to the Umatilla River, where they had previously been extirpated.

Effects of Flow Regulation on Juvenile Salmonids

Mortalities of Juvenile Salmonids in River Reaches

Studies of juvenile salmonid mortalities are designed to develop project-specific mortality estimates, identify factors causing the mortalities, and develop mortality and survival estimates for the hydrosystem as a whole and its impacts on specific salmon and steelhead stocks.

Studies of juvenile yearling Chinook in the mid-Columbia Reach in the 1980s, prior to installation of bypass systems, found an average of 15% to 16% mortality from one project to the next, including mortality in the turbine, tailrace, and reservoir (Chapman and McKenzie 1980; McKenzie et al. 1982, 1983). System-wide mortality estimates of 20% to 25% per project (reservoir and dam) were derived for the Snake River and lower Columbia River, during the same time period (Raymond 1979; Sims et al. 1984; Berggren and Filardo 1993). With these rates of loss in passing through the hydroelectric system, fewer than half of the fish migrating downstream from the uppermost projects in the Snake River or upper Columbia River would survive to below Bonneville Dam. However, passage improvements through the hydrosystem (described above) have reduced mortalities associated with project passage.

Estimates of survival of yearling Chinook through the hydropower system from above Lower Granite Dam to below Bonneville Dam derived from PIT tagging over the years 1998 to 2003 ranged from 0.267 in the year 2001, a severe drought year during which the requirement for spill as a passage route for juveniles was relaxed, to 0.551 in 2002 (Williams et al. 2003). Without the 2001 estimate, survival has been near 50% on the average, which would produce a per-project estimate of mortality of about 15% in recent years, which might be compared to Raymond's estimate of 20% in the 1970s (Raymond 1979).

Rates of survival of juvenile salmonids differ among the stocks and species. Estimates of survival of subyearling fall Chinook are significantly lower than for yearling Chinook or steelhead: below 50% in the Lower Granite Reach, reservoir to tailrace, in some studies, depending upon the flow pattern and other variables in the particular year (Anderson et al. 2000). Smith et al. (2003) found that survival declined as the season progressed, flow decreased, temperature increased, and water clarity increased, all of which were intercorrelated so no individual factor could be identified as the causative agent.

Steelhead survival estimates from Lower Granite Dam to below Bonneville Dam are somewhat lower than those for yearling Chinook salmon (Williams et al. 2003) The low flows of 2001 were particularly difficult for juvenile steelhead in the Snake River (Williams et al. 2003). Intermittent flows, resulting from load following in response to low base flows in the river, led to periodic inter-ruptions of outflow from the lower Snake River projects. Steelhead that residu-alized (discontinued their downstream migrations) became a common sight and the subject of sport fisheries in some of the reservoirs during 2001 (ISAB 2003).

Radio tagging has also been used to estimate survival through river reaches (Skalski et al. 1998; Hockersmith et al. 2003). Estimates of travel time and survival obtained by radio tagging of yearling Chinook were compared with estimates from PIT tagging from Lower Granite Dam to Bonneville Dam in 1999 and did not differ significantly. Mean survival rate from Little Goose Dam to Lower Monumental Dam was about 0.8, for both groups, and from Lower Monumental Dam to McNary Dam about 0.7 for radio-tagged fish and about 0.8 for PIT-tagged fish.

Effects of Flow Augmentation on Travel Time of Juvenile Salmon

The premise behind flow augmentation is that increased flow will lead to reduced travel time for juvenile salmonids, which in turn would lead to increased rates of survival (Giorgi et al. 2002). In the Snake River, higher flows led to reduced travel times of 5% to 16% for spring Chinook and 6% to 17% for steelhead (discussed in Chapter 6). At flows between 50 kcfs and 100 kcfs, the travel time for Snake River yearling Chinook may be reduced from 16 days to 11 days from Lower Granite Dam to McNary Dam, roughly a one-day reduction in travel time for each 10 kcfs of additional flow. It appears the effect would be about the same for steelhead, reducing travel time from 14 days to 9 days. A relationship between flows and travel time for Snake River subyearling fall (ocean-type) Chinook has been difficult to demonstrate.

In the mid-Columbia Reach, no relationship between flow and travel time was found for yearling Chinook, but appeared to exist for steelhead. Giorgi et al. (1994) found no relationship between travel time, flow, temperature, or release date for fall Chinook in the John Day pool. Those fish could have originated either in the mid-Columbia Reach or the Snake River.

Effects of Flow on Survival of Juvenile Salmon

Sims and Ossiander (1981) plotted 9 years of annual survival estimates of Snake River yearling Chinook against average annual river flow during the season of outmigration. Plots of the Sims and Ossiander data had a peculiar shape that Chapman's "broken stick" model (Chapman et al. 1991) might best describe. In that postulated model, there may be an effect of flow on sur-vival at flows less than 100 kcfs but no discernible effect of flow on survival at flows of 100 kcfs and above. That particular characteristic of the survival data has continued to hold true for yearling Chinook and steelhead[17]. For subyearling Chinook flows are usually low, most often less than 100 kcfs, during the time of their outmigration (late June into August), but relationships of survival and flow have been established (e.g. Berggren 2000).

NOAA Fisheries has continued to estimate survival of yearling Chinook and steelhead in the Snake River using PIT-tagged fish since 1993 (Muir et al. 2001; Smith et al. 2003; Williams et al. 2003). The Fish Passage Center and others have also been involved, particularly with subyearling Chinook (Anderson et al. 2000; Connor et al. 2003). Using PIT tag data from the low flow year 2001, the broken stick issue was resolved. Instead of depending upon one pair of data points for each year representing average annual survival and average flow during outmigration, the PIT tag detections make possible numerous estimates within a year over weekly intervals[18] (Figure 6.17). Analysis estimated a probable "break point" in the data at 96.4 kcfs for yearling Chinook and at 101 kcfs for steelhead. The result was a reasonable correspondence of current PIT tag estimates with the earlier Sims and Ossiander points.

Travel time through the hydrosystem can also affect survival of juveniles with respect to the timing of their entry into the estuary and ocean, which may affect their rate of survival to adult. Studies examining the smolt-to-adult return rate are underway (ISG 2000; NRC 1996; Pearcy 1992; Williams et al. 2003). We discuss this further in the section on transportation.

Juvenile salmon and steelhead migrating out of the system are affected by the volume of flow, presumably as it affects water velocity (Giorgi et al. 2002). Yearling Chinook, steelhead, sockeye, and coho have a life history that normally moves them downstream rapidly in the spring at a time corresponding with the spring freshet. Subyearling Chinook move more slowly through the spring and summer, feeding and resting on the way downstream.

In the Snake River, reduced water movement created by the reservoirs has shifted the average transit time of juvenile subyearling Chinook (ocean-type, fall Chinook) salmon from May and June primarily to late June to early August (NMFS 1999; Giorgi et al. 2002). Prior to construction of the four lower Snake River dams and the Hells Canyon and Brownlee dams, emigrating subyearling Chinook encountered flows in June of about 2,800 to 3,800 m3/s, while flows now range from 850 to 1,300 m^3/s.

Water Budget and Flow Augmentation

The council's decision to provide a water budget to improve survival of juvenile salmonids had its basis in the study by Sims and Ossiander (1981) to which we referred earlier (McConnana 1993).

---

[17]S. Smith of NOAA Fisheries, personal communication to NPCC, December 11, 2002.

[18]S. Smith and J. Williams of NOAA Fisheries, personal communication to NPCC, December 11, 2002.

A primary mitigative activity in the mainstem is the use of water stored in upstream reservoirs to decrease the travel time of juvenile salmonids through the mainstem, with the objective of improving their survival. The premise behind this strategy is that reduced travel time will reduce the opportunity for predation and other time-related mortality sources (Giorgi et al. 2002). This approach has its origin in the Council's 1982 Fish and Wildlife Program, which established a "water budget," a specified volume of water to be used by the region's fish and wildlife agencies and tribes annually from April 15 to June 15, to increase juvenile salmon survival within some limits for power production.

The benefits of flow additions as a measure to improve survival of juvenile salmonids have been questioned from the outset[19]. Upriver interests noted the lack of a clear flow–survival relation associated with flow augmentation in the lower river, while lower river interests cited a need for elevated flows to improve late summer travel time and potentially reduce high temperatures for the benefit of migrating juvenile salmon and steelhead. The critical issue concerns the sufficiency of flow (river discharge) as a predictor of survival for out-migrant salmon and steelhead smolts. The importance of river discharge continues to be argued by the Fish Passage Center, NOAA Fisheries, and others, who relate yearling salmon/steelhead smolt survival estimates to average water travel time, which is a function of river flow (see Figures 6.13–6.17 in Chapter 6, showing the relationship between smolt travel time, water transit time, and smolt survivals, along with the accompanying discussion).

The initial water budget consisted of 4.64 million acre feet (Maf) of water, divided into two portions: 3.45 Maf at Priest Rapids Dam, reflecting flow regulated by storage reservoirs in the Columbia River mainstem to its confluence with the Snake River, and 1.19 Maf at Lower Granite Dam, reflecting flow regulated by storage reservoirs in the upper Snake River Basin. The average annual runoff of the Columbia River at its mouth is about 198 Maf.

In 1992, the Council's Fish and Wildlife Program called for increased volumes of water for flow augmentation (replacing the "water budget" terminology) in the spring months (NPPC 1992). The Council called for an independent scientific evaluation of the available information on effects of flow on survival of juvenile salmon and for experimentally testing a hypothetical relationship between flow, water velocity, fish travel time, and survival.

No such experimental study has been conducted. Difficulties anticipated are problems of controlling the several factors that are covariates with flow, in particular, temperature and turbidity, each of which has been associated with survival in multivariate analyses (Smith et al. 2003). On the other hand, the ISAB has called attention to the fact that flow itself is a variable that can and should be controlled in an experiment that would focus on survival when

---

[19]This resulted in a Congressional hearing by the Senate Subcommittee on Science, Technology, and Space, in Washington DC on June 18, 1996.

base flows are less than 100 kcfs, the break point in survival estimates that we previously discussed (ISAB 2003). Most recently, the ISAB pointed to the opportunity for an experiment in 2005 that could be designed to measure the effects of load following on the survival of juvenile salmonids (ISAB 2005-X)

Implementing the Water Budget

Implementation of the water budget, or flow augmentation, was a complex process that involved close monitoring of sequential passage of juvenile fish from one dam to the next and selective releases of stored water from upstream sites. In response to the Council's designation of a water budget in its 1982 Fish and Wildlife Program, a joint proposal was submitted by the Columbia River tribes and state and federal fishery agencies to establish a Water Budget Center. It was established in 1983 and renamed the Fish Passage Center to better reflect its purpose to monitor juvenile fish passage and coordinate the necessary releases of water with the responsible federal agencies.

Experience in practice showed that base power flows assumed by the Council were not available in the Snake River in low water years, thus compromising the ability to implement the water budget (Olsen et al. 1998). Through 1993, implementation of the water budget had been difficult due to lack of an authoritative management strategy, limited scope, and the absence of firm implementation guidelines (Wood 1993). The NRC concluded that the effectiveness of flow augmentation alternatives had not been demonstrated (NRC 1996). The ISAB pointed to the problems of measuring in the lower Columbia River either the volume of water included in flow augmentation or the small increments of survival that might be expected (ISAB 20004-X). However, with new standards from the 1995 BiOp, continuing into the Biological Opinion of 2000 (FCRPS 2000 BiOp), the flow schedule (Table 7.3) is set according to calendar dates that are intended to encompass the nor-

Table 7.3 Seasonal flow objectives (in kcfs) and planning dates for the mainstem Columbia and Snake Rivers (from 2000 FCRPS BiOp).

Locations	Spring		Summer	
	Dates	Objective	Dates	Objective
Snake at Lower Granite Dam	4/03–6/20	85–100[1]	6/21–8/31	50–55[1]
Columbia at McNary Dam	4/10–6/30	220–260[1]	7/01–8/31	200
Columbia at Priest Rapids Dam	4/10–6/30	135	NA	NA
Columbia at Bonneville Dam	11/1	125–160[2]	NA	NA

[1]Objective varies according to water volume forecasts.
[2]Objective varies based on actual and forecasted water conditions.

mal duration of outmigration of the targeted stocks, runoff, and planning for flood control, irrigation, and power, along with the flow for fish (NPPC 1994). The water for implementation comes from the storage reservoirs upstream, principally in Montana and Canada, and its withdrawal requires drawdown of those reservoirs.

Recovery efforts specified in the Biological Opinion initially focused on flow augmentation in spring and late summer in the mid- and lower reaches of the Snake and Columbia Rivers in an effort to provide flows to reduce travel time of juvenile fall Chinook salmon and to counteract potentially lethal temperatures associated with low flows (NMFS 1995d, 2003). Between 1991 and 1995, the water management strategy changed from predominantly targeting spring migrants to emphasizing summer-migrating fall Chinook salmon (Giorgi et al. 1997). For example, in 1991, 74% of the flow augmentation water was released prior to June 21, but by 1995, only 3% of the augmentation occurred by that date, with the remainder released during the summer period extending through August. Typically, flow augmentation was insufficient to sustain the flow targets specified in the 1995 Biological Opinion. However, flow augmentation increased water velocity through Lower Granite Pool in the spring an average of 3% to 13%, and in the summer an average of 5% to 38%.

Drafting of the storage reservoirs for flow augmentation can lead to adverse effects on resident fishes in those systems (ISAB 1997). For example, loss of the spawning cue associated with the spring freshet is a primary problem for recovery of the endangered Kootenai River sturgeon downstream from Libby Dam (Marotz et al. 1996). Additionally, effects of fluctuations in flow may adversely affect seasonally submerged riparian habitat essential for rearing of juveniles (Coutant 2004). Consequently, the Recovery Plan for these sturgeon (U.S. Fish and Wildlife Service) calls for (1) re-regulation of the outflow from Libby Dam to produce freshet flows in the spring for creation of spawning habitat, and (2) stabilization of daily fluctuations in flows to maintain shallow water habitat for larval recruits. These measures likely would be beneficial to all native fishes in headwater rivers like the Flathead, Kootenai, Clearwater, Clark Fork, Pend Orielle, Upper Columbia, Deschutes, and Willamette, which are regulated by large storage reservoirs.

In 2003, the Council adopted an amendment to its Fish and Wildlife Program aimed at creating a balance between negative impacts on resident fishes due to reservoir drawdown and beneficial effects on downstream-migrating salmonids (NPCC 2003). The amendment and the federal 2000 BiOp for flow augmentation shifted the emphasis of flow augmentation from spring to summer. The amendment would shift the schedule for drawdown of Hungry Horse reservoir to benefit resident fish and fisheries, with the understanding that there would be little effect on rates of survival of juvenile salmonids in the mainstem downstream. A study of the effects on resident fishes in the reservoir was funded. As previously mentioned, the ISAB concluded that effects of the Council's amendment on either flow or survival of juvenile salmon were likely to be smaller than can be measured (ISAB 2004-2).

Mitigation of Effects of Load Following on Survival of Juvenile Salmonids

In the 1970s, it began to be appreciated that, load following at upstream projects during fall Chinook spawning led to exposure of redds and incubating eggs in the Hanford Reach (Watson et al. 1969; Bauersfeld 1978; Chapman et al. 1983). Because of the importance of the Hanford spawning area, studies were conducted which identified flow stabilization measures that could improve spawning success. This resulted in a long-term agreement among the fishery agencies and the power, flood control, and irrigation interests to stabilize flows during spawning, incubation, and emergence of the Hanford Reach fall Chinook (Vernita Bar Settlement Agreement, 1988-2005). In 1999, hourly fluctuations in flow were further stabilized during fry emergence, leading to additional reductions in juvenile mortalities caused by strandings.

Unfortunately in 2001, the Council and NOAA Fisheries relaxed requirements for fish protection because of severe drought conditions, resulting in an estimated 2 million juvenile fall Chinook killed in the Hanford Reach due to strandings in shore areas as flows were reduced periodically (on an hourly basis) to take advantage of short-term demand for power. This is in contrast to losses of 93,000 in 1999, 45,000 in 2000, and 67,000 in 2002 years in which fluctuations in flow were limited. (ISAB 2003).

Daily and hourly flow fluctuations in the Snake River up to Hells Canyon Dam may also affect survival of fall Chinook. The Council's ISAB examined flow patterns in the lower Snake River and concluded that survival of Chinook juveniles during outmigration was inversely related to the frequency, duration, and magnitude of episodes of hourly flow fluctuations brought about by load following (ISAB 2003). An obvious recommendation is to use PIT tag technology to test whether stabilization of flows in the Snake River (when base flows decline to 100 kcfs and below) could have a more beneficial effect on survival of juvenile salmonids than simply adding a volume of water, as presently occurs in flow augmentation (ISAB 2005-3).

Effects of the Hydroelectric System on Water Temperature

Storage and release of water from upstream reservoirs has delayed the timing of peak summer temperatures in the mainstem Columbia River since 1941 (Ebel et al. 1989). The delay amounts to about 30 days at Rock Island Dam, and the effects can be detected at Bonneville Dam (Jaske and Goebel 1967; Jaske and Synoground 1970). Projects downstream of Grand Coulee Dam have little storage capacity and show negligible effects on water temperature.

Water temperatures at the mouth of the Snake River in August and September have historically been a few degrees higher than those in the main-stem Columbia (Roebeck et al. 1954). During late summer of some years, high water temperatures (20° to 22°C) and low dissolved oxygen levels (< 6 mg/L) make conditions marginal for salmon and steelhead in the lower Snake River reservoirs (Bennett et al. 1983). When river water temperatures are greater than 20°C, migrating adults take longer to move upstream (Peery et al. 2003; Stuehrenberg et al. 1978; Chapter 6).

High temperatures in the lower Columbia and Snake Rivers will remain problematic. Heat storage in the mainstem reservoirs will occur, especially during dry, hot years. Coldwater releases from upstream storage dams do not ameliorate high temperatures in the lower Columbia River and cause ecological problems in the river systems downstream of the storage projects, due to negative (potentially lethal) effects of dramatically fluctuating temperatures on fish and the aquatic community (Stanford and Hauer 1992). To ameliorate this effect on resident fishes immediately below Hungry Horse Dam, an outflow structure was installed that allows for release of water from a selected depth (thus temperature) stratum of the reservoir.

Mitigation of Effects on Temperature

Releases from Dworshak Dam (Figure 7.12), a large storage dam on the Clearwater River, are called for in the 2000 BiOp. They are intended to provide cooler water in the Snake River during the summer. Decreases in river temperature at Lower Granite Dam, the uppermost dam encountered by salmonids, are dependent upon base flow in the Snake River relative to volumes released from Dworshak, but may be between 3° and 40°C, while decreases at Ice Harbor Dam, the lowermost dam in the Snake River, may be between 1° and 20°C (Bennett et al. 1997; Karr et al. 1998; Giorgi et al. 2002). Careful coordination is required between flows out of Dworshak Dam and flows from the Snake River to ensure that an appropriate water temperature is obtained to benefit salmon (Anderson 2000; ISAB 2003). Questions remain about the possible effects of such temperature modification on survival of salmonids moving upstream or down.

Summary of Flow Effects

Mitigation Efforts and Their Effectiveness

Stabilization of flows has been the most significant and successful measure undertaken toward normalization of conditions for fish in the Columbia Basin. Thanks to operating agreements that have brought a degree of stability to flow in the Hanford Reach, fall Chinook originating there are arguably

Table 7.4 Estimates of baseline, Habitat Conservation Plan (HCP) goals, and actual estimated survivals of juvenile steelhead and spring-run Chinook salmon in the HCP Reach compared to estimated survival in a hypothetical, free-flowing river. (Source Supporting Document D referred to in the HCPs [Available on NOAA Fisheries web site http://www.salmonrecovery.gov/papers])

Estimated Juvenile UCR Steelhead Survivals

Project	Baseline (Draft QAR 2002)		HCP (93% Juvenile Project Survival)[1]			HCP (Actual Measured)[1,2]		
	Project Survival	Percentage of Free-flowing	Project Survival	Est. Survival Improvement	Percentage of Free-flowing	Project Survival	Est. Survival Improvement	Percentage of Free-flowing
Rock Island	0.878	88.8%	0.927	5.6%	93.8%	0.952	8.4%	96.3%
Rocky Reach	0.871	89.1%	0.927	6.5%	94.9%	0.927	6.5%	94.9%
Wells	0.913	92.8%	0.927	1.6%	94.1%	0.958	4.9%	97.3%
HCP Reach Total	0.698	73.4%	0.798	14.2%	83.9%	0.846	21.1%	88.9%

Estimated Juvenile Spring-run Chinook Salmon Survivals

Project	Baseline (Draft QAR 2002)		HCP (93% Juvenile Project Survival)[1]			HCP (Actual Measured)[1,2]		
	Project Survival	Percentage of Free-flowing	Project Survival	Est. Survival Improvement	Percentage of Free-flowing	Project Survival	Est. Survival Improvement	Percentage of Free-flowing
Rock Island	0.870	87.9%	0.927	6.6%	93.7%	0.952	9.4%	96.2%
Rocky Reach	0.865	88.3%	0.927	7.2%	94.7%	0.927	7.2%	94.7%
Wells	0.890	90.3%	0.927	4.2%	94.1%	0.958	7.6%	97.2%
HCP Reach Total	0.670	70.1%	0.798	19.1%	83.5%	0.846	26.3%	88.5%

[1]Project survival standard applies to 95% of the juvenile migration. The remaining 5% is assumed to pass through turbines. These are minimum estimates, as some juvenile measures (Wells bypass operation and Rocky Reach bypass operations) will cover nearly 100% of the migration of UCR steelhead and/or spring-run Chinook at these projects. Turbine survival estimated at 91.5% (middle of range presented in Table 2-4 of FEIS); pool survival estimated at 96.0% (consistent with draft QAR analysis and similar to pool survival estimates in 2000 FCRPS Biological Opinion).
[2]Wells Project survival estimates represent the average of three juvenile survival studies conducted in accordance with the HCP study requirements and with HCP operations being implemented (covering 95% of the juvenile migrants). Individual estimates were 1998 yearling Chinook salmon (99.7%); 1999 steelhead (94.3%); and 2000 steelhead (94.6%). Rock Island Project estimates represent a single juvenile survival study conducted in accordance with the HCP study requirements (covering 95% of the juvenile migrants) and with HCP operations being implemented. The estimate for 2002 yearling Chinook was 95.6%. Rocky Reach had no survival studies conducted which meet HCP study requirements or operations. UCR, Upper Columbia River; FEIS, Federal Environment Impact Statement; QAR, Quantitative Analysis Report

the healthiest stock of salmonids remaining in the basin. Runs are currently around 200,000 adult fish each year and have been more than 400,000 adults since management efforts began. Stabilization of flows takes place during spawning of adults, incubation of eggs, and emergence and emigration of fry. Hatchery fish contribute variable proportions, generally in the range of 5% to 25% of the run. This effort is matched approximately by the suite of engineering solutions, particularly turbine intake screens, used to reduce juvenile mortality associated with downstream passage through the hydrosystem, although we do not know of an analysis that directly compares these gains.

The estimates of reach survival in the mid-Columbia and in the Snake River are higher in today's system due to improved bypass provisions at all 13 dams (NMFS 1992; NRC 1996). The improvements include construction and modification of turbine intake bypass facilities at the four lower Snake River projects, regular schedules for removal of debris from collection systems, installation of flip lips in spillways to reduce gas supersaturation, changes in turbine operations, requirements for spill, and flow augmentation. In the Snake River, reach survival of juveniles per project was estimated at about 20% in the 1970s (Raymond et al. 1975; McKenzie et al. 1982, 1983; Raymond 1968, 1979; Chapman and McKenzie 1980) and is thought now to be about 10% per project (Iwamoto and Williams 1993; Muir et al. 1999).

Reach survivals have also improved in the mid-Columbia since the studies were conducted in the early 1980s. Wells Dam has a fully functioning bypass system, as well as new turbines with higher efficiency ratings. Other mid-Columbia projects have either installed bypass systems (Rocky Reach), improved spill efficiency (all), or added spill amounts as bypass routes (Rock Island, Wanapum, and Priest Rapids). None of these survival estimates for the mid-Columbia reach extend downstream to include fish in the Hanford Reach.

Passage Goals and Attempts to Achieve Them

HCPs of the three mid-Columbia PUD projects require continued operation of the hydroelectric plants to achieve 91% combined survival of adult and juvenile salmonids through each project reach (Table 7.4). Until adult rates are better defined, goals for survival rates of juvenile salmonids are set at 93%, including reservoir, forebay, dam, and tailrace, thus integrating survival through the various routes that may be provided at the dam. Shortfall in mitigation is to be made up by hatchery production (7%) and habitat improvement measures in tributaries (2%), to be financed by each PUD. In Table 7.4, the reach survivals specified are compared to recently measured survivals. The requirements have been met for Wells Dam, while studies are underway at Rock Island Dam and studies were underway at Rock Island Dam and Rocky Reach Dam in 2004.

Passage goals of the Council, FCRPS 2000 BiOp, and FERC call for 80% fish passage and 95% survival at each dam[20]. This requires combinations of bypass measures unique to each dam. Each dam must provide some level of spill, because of differences in effectiveness of their bypass systems, in order to achieve the acceptable level of safe passage for fish. Even Wells Dam, the most effective system in the basin, operates by diverting fish into spill but uses only 1% to 7% of the river flow.

Spill is a costly alternative from the perspective of the hydroelectric operators. In the period following issuance of the 2000 BiOp, there have been two years when the runoff was unusually low, 2001 and 2005. In both cases hydrosystem operators asked NOAA Fisheries for a variance from the BiOp requirements for spill. The reduction was implemented in 2001 and seemed likely to be in 2005. It appears that the regional policy makers have been willing to forego fish protection when the tradeoffs in cost reach some unspecified level. However, the reader is referred to our previous discussion of the Federal Court order requiring maximum spill at the four lower Snake River projects during the outmigration of juvenile fall Chinook. Federal courts have consistently held that the law(s) require equal consideration be given to power production and restoration of fish and wildlife Most recently, Judge Redden found that "The law is clear that an injunction to protect listed species [listed under the Endangered Species Act] from harm is necessary regardless of economic costs." (National Wildlife Federation et al. v National Marine Fisheries Service et al. CV 01-640-RE (Lead Case) CV 05-23-RE (Consolidated Cases). Opinion and Order, Hon. J A Redden, United States District Court, District of Oregon. June 10, 2005, page 9). The cost of spill, in terms of foregone opportunity, provides motivation to explore improvements to bypass systems. Efforts by the U.S. Army Corps of Engineers and the PUDs are focusing upon surface collector systems and ways of diverting more fish into spill per unit volume of water.

Based on information currently available, the survival goal can probably be achieved (with spill) if the 80% fish passage goal can be achieved without concentrating the fish in the forebays or at outfalls where predators concentrate. Concentration in the forebay can occur if fish are delayed in their migration as a result of power operations. Experience has shown that the effects of predation at bypass outfalls and in the forebays at some projects under certain operating conditions may lead to survival rates lower than 95%, even with 80% fish passage.

---

[20]In its 2003 Mainstem Amendments to the Fish and Wildlife Program, the Council stated that specific measures in the 1994–1995 program that have not been directly superseded remain in effect until the measure is specifically repealed in a subsequent program amendment process (NPCC 2003).

In 1995, an analysis showed that the region's goals for 80% fish passage were not met at any of the Snake River or lower Columbia River projects, except at Ice Harbor Dam where the turbines were out of operation and spill occurred in amounts that passed a high percentage of fish (NMFS 1992; Fish Passage Center 1995). Spill amounts at the projects were limited because of gas saturation levels specified in permits issued by the states under provisions of the Clean Water Act.[21] With the exception of Bonneville Dam (55%–62% passage), all of the lower river projects achieved fish passages in the 70% range. Snake River projects achieved fish passages in the 50% to 60% range. Bypass systems have since been improved and strategies have been modified to focus on components of the juvenile salmonid outmigration that are listed under the Endangered Species Act.

Summary

To summarize the chapter, we return to the quotation from Andrew and Geen (1960) at the start of Chapter 6: *"Although the fish-dam problem has existed for centuries in many countries, no practical solutions have yet been found that afford complete protection for anadromous fish in rivers obstructed and altered by large dams."*
   Our description of efforts that have been undertaken to date has shown that no engineering solution yet developed can afford complete protection for Columbia River anadromous fishes. It has not been possible to eliminate or compensate for the effects of development and operation of the hydrosystem on abundance of anadromous fishes. Upstream movement is still problematic for sturgeon and lamprey, though it is likely that at least partial solutions will be developed. Mainstem spawning and rearing areas of salmon and steelhead have been inundated, resulting in reduced potential for production. Hatcheries have generally fallen short of hopes or expectations that they could compensate for this and other losses (see Chapter 8). Losses of juvenile salmonids in their downstream migrations will continue into the foreseeable future as they encounter the dams. No bypass system yet developed has passed fish with 100% survival. To compensate for those losses experienced at the mainstem dams, mitigation efforts for the hydroelectric system have moved into the tributaries to implement measures that would reduce fish losses that were brought about by other human development.
   Nevertheless, though the mission in the mainstem is not fully accomplished, it is clear that the measures taken to divert fish away from turbine

---

[21]The 2000 BiOp specifies gas saturation levels not to exceed 115% in the forebays of the projects, while the states issue special permits allowing 120% during the spring outmigration period.

intakes and provide safer passage routes have been successful in significantly reducing losses of juvenile salmonids as they migrate to sea. Fish ladders continue to be improved to more effectively pass Salmon, Steelhead, lamprey, and Sturgeon. Absent any rigorous analysis, we can only speculate that these efforts may have prevented extirpation of certain stocks of anadromous fishes, and at least provided a delaying tactic for other stocks while further perfection of operating and other engineering strategies is continued at mainstem dams and in the tributaries. It is clear that society, acting through legislative bodies, has put a high value on maintenance of anadromous fish in the Columbia River, along with operation of the hydroelectric system and other beneficial uses of the river, such as irrigation, flood control, and recreation. Policy makers are in the position of developing an appropriate balance among the various interests that are affected. Optimum solutions will be the rule, and no group will find that its interests are maximized. Those decisions will be further complicated by the requirements to operate within limits established by law. Of particular relevance within the Columbia Basin at the present time are the Endangered Species Act and Treaty Tribal Fishing Rights. The Northwest Power Planning and Conservation Act of 1980 offers the best hope of providing a vehicle, the Northwest Planning and Conservation Council, through which optimum accommodations can be reached.

Literature Cited

Absolon, R. F., D. W. Brege, B. P. Sandford, and D. D. Dey. 1995. Studies to evaluate the effectiveness of extended-length screens at The Dalles Dam, 1994. NSFS/NOAA Northwest Fisheries Science Center. Report to US Army Corps of Engineers, Delivery Order E96930030. Seattle, Washington. 69 p.

Allen, R. L., and T. K. Meekin. 1980. Columbia River sockeye salmon study, 1971–1874. Washington Department of Fisheries. Progress Report, 120. Olympia, Washington. 75 p.

Anderson, J. J. 2000. Heat budget of water flowing through Hells Canyon and the effect of flow augmentation on Snake River water temperature. University of Washington, Seattle. 16 p. (Available on http://www.cqs.washington.edu/papers.)

Anderson, J. J., R. A. Hinrichsen, and C.Van Holmes. 2000, December. Effects of flow augmentation on Snake River fall Chinook. Report provided to NPCC in response to request for comments on mainstem strategies. University of Washington, Seattle. 62 p.

Backman, T. W. H., D. Rondorf, and A. Maule. 1996. Symptoms of gas bubble trauma induced in salmon (*Oncorhynchus* spp.) by total dissolved gas pressure supersaturation in the Snake and Columbia rivers, USA: Collection of preliminary data and development of protocols. Columbia River Inter-Tribal Fish Commission. Draft. Portland, Oregon.

Bates, K. M. 1992. Fishway design guidelines for Pacific salmon. Washington Department of Fish and Wildlife. Working Paper. Olympia, Washington.

Bauersfeld, K. 1978. The effect of daily flow fluctuations on spawning fall chinook in the Columbia River. Washington Department of Fisheries. Technical Report No. 38. Olympia, Washington 32 pp.

Beamesderfer, R. C., and B. E. Rieman. 1988. Predation by resident fish on juvenile salmonids in John Day Reservoir, 1983-86; Part III. Bonneville Power Administration. Final report, Contracts DE-A179-82BP34796 and DE-A179-82BP35097. Portland, Oregon.

Beamesderfer, R. C., B. E. Rieman, L. J. Bledsoe, and S. Vigg. 1990. Management implications of a model of predation by a resident fish on juvenile salmonids migrating through a Columbia River reservoir. *North American Journal of Fisheries Management* 10:290–304.

Bell, M. C., and A. C. DeLacy. 1972. A compendium on the survival of fish passing through spillways and conduits. US Army Corps of Engineers, Portland, Oregon. 121 p.

Bell, M. C., A. C. DeLacy and G. J. Paulik. 1967. A compendium on the success of passage of small fish through turbines. Section I:1–120 in M. C. Bell, 1981. Updated Compendium on the success of passage of small fish through turbines. US Army Corps of Engineers, Contract No. D-35-026-CIVEN-66-C16, and Contract No. DACW-68-76-C-0254.

Bennett, D. H., M.A. Madsen, M.H. Karr 1997. Water temperature characteristics of the Clearwater River, Idaho and Lower Granite, Little Goose, Lower Monumental, and Ice Harbor reservoirs, Lower Snake River, Washington during 1991-1993, with emphasis on upstream water releases. Data volume II, project 14-16-0009-1579. Idaho Department of Fish and Game, Moscow, ID

Berggren, T. J., and M. J. Filardo. 1993. An analysis of variables influencing the migration of juvenile salmonids in the Columbia River basin. *North American Journal of Fisheries Management* 13:48–63.

Berggren, T. J. 2000. Subyearling Chinook survival to Lower Granite Dam vs flow. Memorandum from Fish Passage Center, Portland, OR. October 12, 2000, 2 pp. 9 The Fish Passage Center maintains a web site at http://www.fpc.org/

Bevan, D. E., J. P. Harvilloe, P. K. Bergman, T. C. Bjornn, J. A. Crutchfield, P. C. Klingeman, and J. W. Litchfield. 1993. Snake River Salmon Recovery Plan Recommendations. Draft. NMFS (NOAA Fisheries) appointed this "Snake River Salmon Recovery Team" to develop recommendations for recovery.

Bickford, S. A., and J. R. Skalski. 2000. Reanalysis and interpretation of 25 years of Snake-Columbia river juvenile salmonid survival studies. *North American Journal of Fisheries Management* 20:53–68.

Biological Monitoring Inspection Team. 1995. Research priorities related to gas bubble monitoring needs in the Columbia River Basin. (Biological Monitoring Inspection Team of the Gas Bubble Disease Technical Work Group). National Marine Fisheries Service. Seattle, Washington.

Biosonics. 1983a. Hydroacoustic assessment of downstream migrating salmon and steelhead at Rock Island Dam in 1983. Public Utility District No. 1 of Chelan County. Processed Report. Wenatchee, Washington. 99 p.

Biosonics. 1983b. Hydroacoustic assessment of downstream migrating salmon and steelhead at Wanapum and Priest Rapids Dams in 1983. Public Utility District No. 2 of Grant County. Processed Report. Ephrata, Washington. 41 p.

Biosonics. 1983c. Hydroacoustic monitoring and distribution of downstream migrant salmonids and evaluation of the prototype bypass system at Wells Dam in spring, 1983. Public Utility District No. 1 of Douglas County. Processed Report. East Wenatchee, Washington. 37 p.

Biosonics. 1984. Hydroacoustic assessment of downstream migrating salmon and steelhead at Rocky Reach Dam in 1983. Public Utility District No. 1 of Chelan County. Processed Report. Wenatchee, Washington. 52 p.

Biosonics. 1995, September. Hydroacoustic evaluation of vertical slots, sluiceway and spill at Ice Harbor Dam. Corps of Engineers Meeting, Walla Walla, Washington.

Bjornn, T. C., J. P. Hunt, K. R. Tolotti, P. J. Keniry, and R. R. Ringe. 1998. Effects of zero versus normal flow at night on passage of steelhead in summer and fall. Part VII. Migration of adult salmon and stelhead past dams and through reservoirs in the lower Snake River and into tributaries. 62 p. (Idaho Cooperative Fishery Research Unit, University of Idaho, Moscow, ID 83843)

Bouck, G. R. 1980. Etiology of gas bubble disease. *Transactions of the American Fisheries Society* 109:703–707.

Brookshire, J. R., and D. McKinnon. 1995. 21st century advanced hydropower turbine system. Pages 2003–2008 in J.J. Cassidy, ed. *Waterpower '95. Proceedings of the International Conference on Hydropower.* American Society of Civil Engineers, New York.

Bruce, S. 1995. 1995 migration exceeds expectations. *Salmon Passage Notes, Snake and Columbia River Fish Programs* September.

Burnham, K.P., D. R. Anderson, G. C. White, C. Brownie, and K. H. Pollock. 1987. Design and Analysis Methods for Fish Survival Experiments Based on Release-Recapture. American Fisheries Society. Monograph 5, 437 pp.

Chambers, J. S. 1995. Research relating to study of spawning grounds in natural areas. Washington Department of Fisheries. Annual Report to U.S. Army Corps of Engineers, 1955. Olympia, Washington, p. 88-94.

Chapman, D., C. Arlson, D. Weitcamp, G. Matthews, J. Stevenson, and M. Miller. 1997. Homing in sockeye and Chinook salmon transported around part of their smolt migration route in the Columbia River. *North American Journal of Fisheries Management* 17:101–113.

Chapman, D., A. Giorgi, M. Hill, A. Maule, S. McCutcheon, D. Park, W. Platts, K. Pratt, J. Seeb, L. Seeb, and F. Utter. 1991. Status of Snake River Chinook salmon. Don Chapman Consultants, Boise, Idaho. 520 p.

Chapman, D., A. Giorgi, T. Hillman, D. Deppert, M. Erho, S. Hays, C. Peven, B. Suzumoto, and R. Klinge. 1994. Status of summer/fall Chinook salmon in the Mid-Columbia region. Don Chapman Consultants, Boise, Idaho. 411 p.

Chapman, D. W., and D. McKenzie. 1980. Mid-Columbia River system mortality study. Douglas County Public Utility District No. 1. Report to Mid-Columbia Coordinating Committee. East Wenatchee, Washington. 23 p.

Chapman, D. W., D. E. Weitkamp, T. L. Welsh, and T. H. Schadt. 1983. Effects of minimum flow regimes on fall Chinook spawning at Vernita Bar 1978-1992. Report to Grant County PUD, Ephrata, Washington. Don Chapman Consultants, Boise, Idaho. 123 p.

Collis, K., R. E. Beaty, and B. R. Crain. 1995. Changes in catch rate and diet of northern squawfish associated with the release of hatchery-reared juvenile salmonids in a Columbia River reservoir. *North American Journal of Fisheries Management* 15:346–357.

Connor, W. P., H. L. Burge, J. R. Yearsley, and T. C. Bjornn. 2003. The influence of flow and temperature on survival of wild subyearling fall Chinook salmon in the Snake River, North America. *North American Journal of Fisheries Management* 23:362–375.

Connor, W. P., J. G. Sneva, K. F. Tiffan, R. K. Steinhorst, and D. Ross 2005. Two alternative juvenile life history types for fall Chinook salmon in the Snake River Basin. Transactions of the American Fisheries Society 134 (2):291–304

Coutant, C. C. 2004. A riparian habitat hypothesis for successful reproduction of white sturgeon. Reviews in Fishery Science 12:23-73

Coutant, C. C., and R. R. Whitney. 2000. Fish behavior in relation to passage through hydropower turbines: A review. *Transactions of the American Fisheries Society* 129:351–380.

Cramer, S. P. 1996. Seasonal changes in survival of yearling Chinook smolts emigrating through the Snake River in 1995. S. P. Cramer and Associates. Draft. Gresham, Oregon.

Dauble, D. D., and R. P. Mueller. 1993. Factors affecting the survival of upstream migrant adult salmonids in the Columbia River Basin. Bonneville Power Administration. Recovery Issues for Threatened and Endangered Snake River Salmon, 9 of 11. Portland, Oregon. 72 p.

Dauble, D. D., and D. G. Watson. 1990. Spawning and abundance of fall Chinook salmon (*Oncorhynchus tshawytscha*) in the Hanford Reach of the Columbia River, 1948-1988. U. S. Department of Energy, Pacific Northwest Laboratory. PNL-7289, UC-600. Richland, Washington.

Dawley, E. M., and W. J. Ebel. 1975. Effects of various concentrations of dissolved atmospheric gas on juvenile Chinook salmon and steelhead trout. *Fishery Bulletin* 73:787–796.

Dawley, E. M., L. G. Gilbreath, R. D. Ledgerwood, P. J. Bentley, B. P. Dandford, and M. H. Schiewe. 1989. Survival of subyearling Chinook salmon which have passed through the turbines, bypass system, and tailrace basin of Bonneville Dam Second Powerhouse. NMFS/NOAA Northwest Fisheries Science Center. Report to US Army Corps of Engineers, Portland District, Contract No. DACW57-87-F-0323. Seattle, Washington. 80 p.

DeHart, D. A. 1987. Downstream migrant juvenile salmonid protection systems for low to medium head hydroelectric dams. Ph. D. dissertation, University of Washington, Seattle. 152 p.

Ebel, W. J. 1969. Supersaturation of nitrogen in the Columbia River and its effects on salmon and steelhead trout. *US Fish and Wildlife Service, Fishery Bulletin* 68:1–11.

Ebel, W. J., C. D. Becker, J. W. Mullan, and H. L. Raymond. 1989. The Columbia River: Toward a holistic understanding. Pages 205–219 in D. P. Dodge, ed. *Proceedings of the International Large River Symposium (LARS)*. Special Publication of the *Canadian Journal of Fisheries and Aquatic Sciences*.

Eicher, G. E. 1988. Fish collection, transportation and release in relation to protection at power plants. Pages I-13 and 23 in W. C. Micheletti, ed. *Proceedings: Fish Protection at Steam and Hydroelectric Power Plants*. Electric Power Research Institute (EPRI), Paloalto CA. Document CS/EA/AP-5663-SR.

Federal Columbia River Power System (FCRPS) 2000 Biological Opinion (BiOp) December 21, 2000. Consultation on remand for operation of the Columbia River Power System and 19 Bureau of eclamation Projects in the Columbia basis. NOAA Fisheries, Northwest Region, Seattle, WA.

Ferguson, J. W., R. L. McComas, R. F. Absolon, D. A. Brege, M. H. Gessel, L. G. Gillbreath, B. H. Monk, and G. M. Matthews. 2003, December. Passage of Juvenile and Adult Salmonids at Columbia and Snake River Dams. NOAA Technical Memorandum. 183pp. Available on web site http://www.Salmonrecovery.gov/remand.

Fish Passage Center (FPC). 1995. Summary of the 1995 Spring and Summer Juvenile Passage Season. Fish Passage Center. Portland, Oregon. 32 p.

Fish Passage Managers. 1990. 1989 Fish Passage Managers Annual Report. Fish Passage Center. Report to NPPC, Appendix C. Fish Spill Memorandum of Agreement. Portland, Oregon.

Friesen, T. A., and D. L. Ward. 1999. Management of northern pikeminnow and implications for juvenile salmonid survival in the lower Columbia and Snake rivers. *North American Journal of Fisheries Management* 19:406–420.

Garcia, A. P., W. P. Connor, and R. H. Taylor 1994. Fall Chinook salmon spawning ground surveys in the Snake River. Pages 1-21 in D. W. Rondorf and K. F. Tiffan, eds. Identification of the Spawning, Rearing, and Migratory Requirements of Fall Chinook in the Columbia River Basin Annual Report for 1993 to Bonneville Power Administration, Portland, OR, 157 pp.

General Accounting Office (GAO). 1992. Endangered Species: Past actions taken to assist Columbia River salmon. General Accounting Office. Briefing Report to Congressional Requesters, GAO/RCED-92-173BR. Washington, DC.

GAO. 2002. Columbia River Basin salmon and steelhead: Federal agencies' recovery responsibilities, expenditures and actions. US General Accounting Office. Report to US Senate, GAO-02-612. Washington DC. 86 p.

Gessel, M. H., B. P. Sanford, and D. B. Dey. 1995. Studies to evaluate the effectiveness of extended-length screens at Little Goose Dam, 1994. NMFS/NOAA Northwest Fisheries Science Center. Report to US Army Corps of Engineers, Walla Walla District, Delivery Order E86920164. Seattle, Washington. 33 p.

Giorgi, A. E., D. R. Miller, and B. P. Sanford. 1994. Migratory characteristics of juvenile ocean-type Chinook salmon, *Oncorhynchus tshawytscha*, in John Day Reservoir on the Columbia River. *Fishery Bulletin* 92:872–879.

Giorgi, A. E., M. Miller, and J. Stevenson. 2002. Mainstem Passage Strategies in the Columbia River System: Transportation, Spill and Flow Augmentation. Doc. 2002-3. Northwest Power Planning Council. Portland, Oregon. 97 p.

Giorgi, A. E., J. W. Schlecte, and HDR Engineering. 1997. An evaluation of the effectiveness of flow augmentation in the Snake River, 1991-1995. HDR Engineering, Inc. Report to BPA, DE-AC79-92BP24576. 47 p.

Giorgi, A. E., and J. R. Stevenson. 1995. A review of biological investigations describing smolt passage behavior at Portland District Corps of Engineers Projects: Implications to surface collection systems. Don Chapman Consultants. Report to US Army Corps of Engineers, Boise, Idaho. 33 p.

Giorgi, A. E., and L. Stuehrenberg. 1988. Lower Granite pool and turbine survival study, 1987. Bonneville Power Administration. Annual report, FY 1987. Portland, Oregon. 30 p.

Hammond, S. L. 1994. 1994 Spill program report. Priest Rapids project # 2114. Wanapum and Priest Rapids dams. Grant County Public Utility District No. 2. Report to Federal Regulatory Commission, Ephrata, Washington. 11 p.

Hansel, H. C., R. S. Shively, J. E. Hensleigh, D. D. Liedtke, T. Hatton, R. E. Wardell, R. H. Werheimer, and T. P. Poe. 1998. Movement, distribution, and behavior of radio-tagged sub-yearling Chinook salmon in the forebay of Bonneville Dam, 1998. U.S. Geological Survey, Cook, Washington. Preliminary Report to U.S. Army Corps of Engineers, Portland, Oregon.

Hays, S. G., and K. B. Truscott. 1986. Rocky Reach prototype fish guidance system–1986 developmental testing. Public Utility District of Chelan County. Wenatchee, Washington. 50 p.

Heisey, P.G., D. Mathur, and T. Rineer 1992. A reliable tag-recapture technique for estimating turbine passage survival: application to young-of-the-year American shad. Canadian Journal of Fisheries and Aquatic Sciences 49: 1826-1834

Holmberg, G. S., R. S. Shively, H. S. Hansel, T. L. Martinelli, M. B. Sheer, J. M. Hardiman, B. D. Liedtke, L. S. Blythe, and T. P. Poe. 1998. Movement, distribution, and behavior of radio-tagged juvenile Chinook salmon in John Day, The Dalles, and Bonneville Dam forebays, 1996. US Geological Survey, Cook, Washington. Annual Report of Research to the US Army Corps of Engineers, Portland District, Portland, Oregon.

Hockersmith, E. E., W. D. Muir, S. G. Smith, B. P. Sandford, R. W. Perry, N. S. Adams, and D. W. Rondorf 2003. Comparison of migration rate and survival between radio-tagged and PIT-tagged migrant yearling Chinook salmon in the Snake and Columbia rivers. *North American Journal of Fisheries Management* 23:404–413.

Horner, N. and T. C. Bjornn 1979. Status of Upper Columbia River fall Chinook salmon. Idaho Cooperative Fishery Research Unit University of Idaho, Boise Idaho, 45 pp.

Independent Scientific Advisory Board (ISAB). 1997. Ecological impacts of the Biological Opinion for endangered Snake River salmon on resident fishes in the Hungry Horse and Libby systems in Montana, Idaho, and British Columbia. Northwest Power Planning Council. ISAB Report, 97-3. Portland, Oregon.

ISAB. 1998. Response to the question of the Implementation Team regarding juvenile salmon transportation in the 1998 season. Northwest Power Planning Council and the National Marine Fisheries Service. ISAB Report, 98-2. Portland, Oregon.

ISAB. 2000. Review of studies of fish survival in the spill at The Dalles Dam. Northwest Power Planning Council and the National Marine Fisheries Service. ISAB Report, 2000-1. Portland, Oregon. 18 p.

ISAB. 2003. Review of flow augmentation: Update and clarification. Northwest Power Planning Council, NOAA Fisheries, and the Columbia River Basin Indian Tribes. ISAB Report 2003-1, Portland, Oregon. 67 p.

Independent Scientific Advisory Board (ISAB) 2004. ISAB findings from the Reservoir Operations/Flow Survival Symposium ISAB 2004-2. Available at the web site for the NPCC, http://www.nwcouncil.org/library

Independent Scientific Advisory Board (ISAB) 2005. Recommendation to study effects of load following on juvenile salmon migratory behavior and survival ISAB 2005-3. Available at the web site for the NPCC, http://nwcouncil.org/library

Independent Scientific Group (ISG). 2000. Return to the River: Restoration of Salmonid Fishes in the Columbia River Ecosystem. Northwest Power Planning Council.ISG Report 2000-12. Portland, Oregon. 538 p.

Iwamoto, R. N., W. Muir, B. Sandford, K. McIntyre, D. Frost, J. Williams, S. Smith, and J. Skalski. 1993. Survival estimates for the passage of juvenile Chinook through Snake River dams and reservoirs. NMFS/NOAA Northwest Fisheries Science Center. Report to BPA, Contract No. DE-A179-93BP10891, Project 93-29. Seattle, Washington. 140 p.

Iwamoto, R. N., and J. G. Williams. 1993. Juvenile salmonid passage and survival through turbines. NMFS/NOAA Northwest Fisheries Science Center, jointly with US Army Corps of Engineers, Portland District. Project E86920049. Seattle, Washington. 27 p.

Jaske, R. T., and J. B. Goebel. 1967. Effects of dam construction on temperatures of the Columbia River. *Journal of the American Water Works Association* 59:935–942.

Jaske, R. T., and M. O. Synoground. 1970. Effect of Hanford plant operations on temperature of the Columbia River, 1964 to present. Pacific Northwest Laboratory. BNWL-1345. Richland, Washington.

Johnsen, R. C., and E. M. Dawley. 1974. The effect of spillway flow deflectors at Bonneville Dam on total gas supersaturation and survival of juvenile salmon. NMFS/NOAA. Report to US Army Corps of Engineers, Contract No. DACW-57-74-F-0122. Seattle, Washington. 19 p.

Johnson, G. E. 1995. Fisheries research on phenomena in the forebay of Wells Dam in Spring, 1995 related to the surface flow smolt bypass. Battelle Pacific Northwest Laboratory. Report to U.S. Army Corps of Engineers, Walla Walla District, Contract No. 19478, Task 12. Walla Walla, Washington. 66 p.

Karr, M. H., J. K. Fryer, P. R. Mundy 1998. Snake River water temperature control project phase II: Methods for managing and monitoring water temperatures in relation to salmon in the lower Snake River. Columbia River Intertribal Fish Commission, Portland, OR, and Fisheries and Aquatic Sciences, Lake Oswego, OR.

Koski, C. H., S. W. Petit, J. B. Athearn, and A. L. Heindl. 1986. Fish Transportation Oversight Technical Team Annual Report–FY 1985. Transport operations on the Snake and Columbia Rivers. NMFS/NOAA. Technical Memorandum, NMFS-14. Seattle, Washington. 70 p.

Krcma, R. F. 1985. Preliminary results of fish guiding efficiency of traveling screens at The Dalles Dam. NOAA Fisheries. Northwest Fisheries Science Center.

Krcma, R. F., D. A. Brege, and R. D. Ledgerwood. 1986. Evaluation of the rehabilitated fish collection and passage system at John Day Dam–1985. NMFS/NOAA Northwest Fisheries Science Center. Report to US Army Corps of Engineers, Contract No. DACW57-85-H-0001, Seattle, Washington. 25 p.

Krcma, R. F., C. W. Long, and C. S. Thompson. 1978. Research and development of a fingerling protection system for low head dams–1977. NMFS/NOAA Northwest Fisheries Science Center. Report to US Army Corps of Engineers, Seattle, Washington. 32 p.

Krcma, R. F., W. E. Farr, and C. W. Long. 1980. Research to develop bar screens for guiding juvenile salmonids out of turbine intakes at low head dams on the Columbia and Snake Rivers, 1977–1979. NMFS/NOAA Northwest Fisheries Science Center. Report to US Army Corps of Engineers, Seattle, Washington. 28 p.

Krcma, R. F., D. DeHart, M. Gessel, C. Long, and C. W. Sims. 1982. Evaluation of submersible traveling screens, passage of juvenile salmonids through the ice-trash sluiceway, and cycling of gatewell orifice operations at the Bonneville First Powerhouse, 1981. NMFS/NOAA Northwest Fisheries Center. Report to US Army Corps of Engineers, Portland District, Portland, Oregon, Seattle, Washington.

Krcma, R. W., M. H. Gessel, W. D. Muir, C. S. McCutcheon, L. G. Gilbreath, and B. H. Monk. 1984. Evaluation of the juvenile collection and bypass system at Bonneville Dam –1983. NMFS/NOAA Northwest Fisheries Science Center. Report to US Army Corps of Engineers, Contract No. DACW57-83-F-0315, Seattle, Washington. 68 p.

Krcma, R. F., G. A. Swan, and F. J. Ossiander. 1985. Fish guiding and orifice passage efficiency tests with subyearling Chinook salmon, McNary Dam, 1984. NMFS/NOAA Northwest Fisheries Center. Report to US Army Corps of Engineers, Contract No. DACW68-84-H-0034, Seattle, Washington. 19 p.

Ledgerwood, R. D., E. M. Dawley, L. G. Gilbreath, B. J. Bentley, B. P. Sandford, and M. W. Schiewe. 1991. Relative survival of subyearling Chinook salmon that have passed through the turbines or bypass system of Bonneville Dam Second Powerhouse, 1990. NMFA/NOAA Northwest Fisheries Science Center. Report to US Army Corps of Engineers, Delivery Order E86900104. Seattle, Washington. 90 p.

Lichatowich, J. 1999. *Salmon without rivers*. Island Press, Covelo, California. 317 p.

Long, C. W., F. J. Ossiander, T. E. Ruehle, and G. M. Mathews. 1975. Survival of coho salmon fingerlings passing through operating turbines with and without perforated bulkheads and of steelhead trout fingerlings passing through spillways with and without a flow detector. NMFS/NOAA Northwest Fisheries Science Center. Report to US Army Corps of Engineers, Contract No. DACW68-74-C-0113. 8 p.

Magne, R. A. 1987. Hydroacoustic monitoring of downstream migrant juvenile salmonids at Bonneville Dam, 1987. Pages 209–221 in *Seventh Progress Report. Fish Passage Development and Evluation Program 1984–1990*. US Army Corps of Engineers, Portland District, Operations Division, Fisheries Field Unit, Portland, Oregon.

Magne, R. A., D. R. Bryson, and W. T. Nagy. 1987. Hydroacoustic monitoring of downstream migrant juvenile salmonid passage at John Day Dam in 1984–1985. US Army Corps of Engineers. Portland, Oregon. 29 p.

Magne, R. A., R. J. Stansell, and W. T. Nagy. 1989. Hydroacoustic monitoring of downstream migrant juvenile salmonids at Bonneville Dam, 1988. Pages 257–270 in *Seventh Progress Report: Fish Passage Development and Evaluation Program 1984–1990*. US Army Corps of Engineers, North Pacific Division, Environmental Resource Division. Portland, Oregon

Major, R. L., and J. L. Mighell. 1966. Influence of Rocky Reach Dam and the temperature of the Okanogan River on the upstream migration of sockeye salmon. *Fishery Bulletin* 66:131–147.

Marotz, B. L., C. Althen, B. Lonon, and D. Gustafson. 1996. Model development to establish Integrated Operational Rule Curves for Hungry Horse and Libby Reservoirs–Montana. Bonneville Power Administration. Final Report, DOE/BP-92452-1. Portland, Oregon. 114 p.

McCann, J. 1995. Results of 1995 GBT juvenile salmon monitoring program and 1996 monitoring proposal. Fish Passage Center. Memorandum, Portland, Oregon.

McComas, R. L., B. P. Sandford, and D. B. Dey. 1994. Studies to evaluate the effectiveness of extended-length screens at McNary Dam, 1993. NMFS/NOAA Northwest Fisheries Science Center. Report to US Army Corps of Engineers, Walla Walla District, Delivery Order E86910060, Seattle, Washington. 109 p.

McConnaha, W. E. 1993. History of flow survival studies in the Columbia Basin. April 26, 1993 Memorandum to Donald Bevan, Chairman, NMFS Salmon Recovery Team. Northwest Power Planning Council [Now Northwest Power and Conservation Council}, Portland, OR 7 pp

McFadden, B. D., B. H. Ransom, and B. Schnebly. 1992. Hydroacoustic evaluation of the effectiveness of the sluiceway at Priest Rapids Dam in passing juvenile salmon and steelhead trout during spring and summer 1992. Grant County Public Utility District No. 2. Ephrata, Washington. 77 p.

McKenzie, D., D. Weitkamp, T. Schadt, D. Carlile, and D. Chapman. 1982. 1982 System mortality study. Report to Mid-Columbia Coordinating Committee. Douglas County Public Utility District No. 1. East Wenatchee, Washington. 37 p.

McKenzie, D., D. Carlile, and D. Weitkamp. 1983. 1983 System mortality study. Report to Mid-Columbia Coordinating Committee. Douglas County Public Utility District No. 1. East Wenatchee, Washington. 25 p.

Meekin, T. K. 1967. Observations of exposed fall Chinook redds below Chief Joseph Dam during periods of low flow. Washington Department of Fisheries. Report to Douglas County P.U.D. No. 1, East Wenatchee, WA. 25 pp.

Mendel, G., D. Milks, M. Clizer, and R. Bugert. 1994. Upstream passage and spawning of fall Chinook salmon in the Snake River. Washington Department of Fisheries. Report to BPA, Contract Number DE-BI 79-92 BP60415. Olympia, Washington.

Michimoto, R. T., and L. Korn. 1969. A study of the value of using the ice and trash sluiceway for passing downstream migrant salmonids at Bonneville Dam. Fish Commission of Oregon. Portland, Oregon. 28 p.

Mid-Columbia Coordinating Committee. 1989. Findings of the Mid-Columbia Coordinating Committee for 1988. Douglas County Public Utility District No. 1. Report to Hon. Stephen L. Grossman, Administrative Law Judge for the FERC, MID-COL 90-5. East Wenatchee, Washington. 36 p.

Mighetto, L. M., and W. J. Ebel. 1995. Saving the salmon: A history of the US Army Corps of Engineers' role in the protection of anadromous fish on the Columbia and Snake Rivers. Historical Research Associates. Report to US Army Crops of Engineers, North Pacific Division, Seattle, Washington. 262 p.

Muir, W. D., R. N. Iwamoto, C. P. Paisley, B. P. Sandford, P. A. Ocker, and T. E. Ruehle. 1995a. Relative survival of juvenile Chinook salmon after passage through spillbays and the tailrace at Lower Monumental Dam, 1994. Northwest Fisheries Science Center. Report to US Army Corps of Engineers. Seattle, Washington

Muir, W. D., S. G. Smith, R. N. Iwamoto, D. J. Kamikawa, K. W. McIntyre, E. E. Hockersmith, B. P. Sandford, P. A. Ocker, T. E. Ruehle, and J. G. Williams. 1995b. Survival estimates for the passage of juvenile salmonids through Snake River Dams and Reservoirs, 1994. Bonneville Power Administration. DOE/BP-10891-2, Contract No. DE-A179-93BP10891, COE Walla Walla District Delivery Order E86940119. Portland, Oregon. 187 p.

Muir, W. D., S. G. Smith, J. G. Williams, E. E. Hockersmith, and J. R. Skalski. 1999. Survival estimates for PIT-tagged migrant juvenile Chinook salmon and steelhead in the lower Snake and Columbia rivers, 1993-1998. National Marine Fisheries Service and Northwest Fisheries Science Center. Report to Bonneville Power Administration, DE-A179-93BP10891. Seattle, Washington. 43 p.

Muir, W. D., S. G. Smith, J. G. Williams, and B. P. Sandford. 2001 Survival of juvenile salmonids passing through bypass systems, turbines, and spillways with and without flow deflectors at Snake River dams. *North American Journal of Fisheries Management* 21:135–146.

Mullan, J. W. 1986. Determinants of sockeye salmon abundance in the Columbia River, 1880's–1982: A review and synthesis. US Fish and Wildlife Service. *Biological Report* 86(12). 136p.

Mullan, J. W., M. B. Dell, S. G. Hays, and J. A. McGee. 1986. Some factors affecting fish production in the Mid-Columbia River 1934-1983. US Fish and Wildlife Service. FRI/FAO-86-15. 69 p.

National Marine Fisheries Service (NMFS). 1992. Survival estimates for the passage of juvenile salmonids through dams and reservoirs. Bonneville Power Administration. Seattle, Washington.

NMFS. 1995a. Draft Biological Opinion on reinitiation of consultation on 1994-1998 operation of the federal Columbia River Power System and juvenile transportation program in 1994-

1998. US Army Corps of Engineers, Bonneville Power Administration, Bureau of Reclamation, and National Marine Fisheries Service. Portland, Oregon. 146 p.

NMFS. 1995b. Juvenile Fish Screen Criteria. NOAA Fisheries. Portland, Oregon 97232.

NMFS. 1995d. Proposed Recovery Plan for Snake River Salmon. US Department of Commerce, National Oceanic and Atmospheric Administration. Washington, DC. 387 p.

NMFS. 1999. Salmonid travel time and survival related to flow management in the Columbia River Basin. Draft White Paper. Northwest Fisheries Science Center. Seattle, Washington.

NMFS. 2003. Federal Columbia River Power System Biological Opinion of 2000. NOAA Fisheries, Seattle, Washington.

National Oceanic and Atmospheric Administration (NOAA) Fisheries. 2003. Biological Opinion, Unlisted Species Analysis, and Magnuson-Stevens Fishery Conservation and Management Act Consultation: For Proposed Issuance of a Section 10 Incidental Take Permit to Public Utility District No. 1 of Chelan County for the Rock Island Hydroelectric Project (FERC No. 2145). Anadromous Fish Agreement and Habitat Conservation Plan. NOAA Fisheries, Northwest Region.

National Research Council (NRC). 1996. Upstream: Salmon and society in the Pacific Northwest. Report on the Committee on Protection and Management of Pacific Northwest Anadromous Salmonids for the National Research Council of the National Academy of Sciences. National Academy Press, Washington, DC.

Nichols, D. W., and B. H. Ransom. 1980. Development of The Dalles Dam trash sluiceway as a downstream migrant bypass system for juvenile salmonids. Oregon Department of Fish and Wildlife. Report to US Army Corps of Engineers, Contract No. DACW57-78-COO58. Portland, Oregon. 36 p.

Nichols, D. W., and B. H. Ransom. 1981. Development of The Dalles Dam trash sluiceway as a downstream migrant bypass system, 1981. Oregon Department of Fish and Wildlife. Report to US Army Corps of Engineers, Portland, Oregon. 34 p.

Nichols, D. W., F. R. Young, and C. O. Junge. 1978. Evaluation of The Dalles Dam ice-trash sluiceway as a downstream migrant bypass system during 1977. Oregon Department of Fish and Wildlife. Report to US Army Corps of Engineers, Portland, Oregon. 10 p.

Northwest Power and Conservation Council (NPCC). 2003. Mainstem Amendments to the Columbia River Basin Fish and Wildlife Program. Northwest Power and Conservation Council. Council Doc. 2003-11. 31 p.

Northwest Power Planning Council (NPPC). 1992. Columbia River Basin Fish and Wildlife Program. Document 92-21A. Portland, Oregon.

NPPC. 1994. Columbia River Basin Fish and Wildlife Program. Northwest Power Planning Council. Portland, Oregon. Doc 94-95, 404pp+app.

Office of Technology Assessment. 1995. Fish passage technologies: Protection at hydropower facilities. US Government Printing Office. OTA-ENV-641. Washington, DC. 167 p.

Olsen, D., J. Anderson, R. Zabel, J. Pizzimenti, and K. Malone. 1998, February. The Columbia-Snake River flow targets/augmentation program: A white paper review with recommendations for decision makers. Prepared for Columbia-Snake River Irrigators Assn., Eastern Oregon Irrigators Assn., Idaho Water users Assn., Northwest Irrigation Utilities, and Washington State Water Resources Assn. 41 p.

Page, T. L., R. H. Gray, and D. A. Neitzel. 1976. Fish impingement studies at the Hanford Generating Project (HGP) December 1, 1975 through April, 1976. Battelle, Northwest Laboratory. Report to Washington Public Power Supply System, Contract No. 2311202149. 22 p.

Panel on Gas Bubble Disease. 1994. Report and recommendations. Report to National Marine Fisheries Service, Northwest Fisheries Science Center, Seattle, Washington.

Panel on Gas Bubble Disease. 1996. Summary report. Report to National Marine Fisheries Service, Northwest Fisheries Science Center, Seattle, Washington.

Park, D. L., J. R. Smith, E. Slatick, G. A. Swan, E. M. Dawley, and G. M. Mathews. 1977. Evaluation of fish protective facilities at Little Goose and Lower Granite Dams, and review of nitrogen studies relating to protection of juvenile salmonids in the Columbia and Snake Rivers, 1976. NMFS/NOAA Northwest Fisheries Science Center. Report to US Army Corps of Engineers, Seattle, Washington. 48 p.

PATH Scientific Review Panel. 1998. Conclusions and recommendations from the PATH Weight of Evidence Workshop. ESSA Technologies. Formal Report, Vancouver, British Columbia.

Peery, C. A., T. C. Bjornn, and L. C. Stuehrenberg 2003. Water temperatures and passage of adult salmon and steelhead in the Lower Snake River. Idaho Cooperative Fish and Wildlife Research Unit. University of Idaho, Moscow. Technical Report 2003-2, 79 pp. (Available at web site http://www.cnnuidaho.edu/uiferl/)

Petersen, K. C. 1995. *River of life, channel of death.* Confluence Press, Lewiston, Idaho.

Peven, C. M. 1993. Fish guidance system developmental testing at Rock Island Dam Powerhouse No. 1, spring and summer, 1993. Chelan County Public Utility District No. 1. Wenatchee, Washington. 27 p.

Peven, C. M. 1995. Evaluation of the prototype fish bypass system at Rock Island Dam, 1994. Chelan County Public Utility District No. 1. Wenatchee, Washington.

Peven, C. M. 1996. Study plan for testing of surface collector at Rocky Reach in 1996. Chelan County Public Utility District No. 1. Wenatchee, Washington.

Peven, C. M., and B. G. Keesee. 1992. Rocky Reach fish guidance system developmental testing. Chelan County Public Utility District No. 1. Wenatchee, Washington. 20 p.

Peven, C. M., A. M. Abbott, and B. Bickford. 1995. Biological evaluation of the Rocky Reach surface collector. Chelan County Public Utility District No. 1. Wenatchee, Washington. 38 p.

Piercy, W. G. 1992. Ocean Ecology of North Pacific Salmonids. Washington Sea Grant Program. University of Washington, Seattle.

Poe, T. P., and B. E. Riemann. 1988. Predation by resident fish on juvenile salmonids in John Day Reservoir, Volume I. Bonneville Power Administration. Final Report of Research, Contracts DE-AI79-82BP34796 and DE-AI79-82BP35097. Portland, Oregon.

Quistorff, E. 1966. Floating salmonid smolt collectors at Baker River dams. *Fishery Research Papers* 2:39–52.

Raemhild, G. A., G. E. Johnson, and C. M. Sullivan. 1983. Hydroacoustic studies of downstream migrant salmon and steelhead at Wells Dam in spring, 1984. Douglas County Public Utility District No. 1. Processed Report, East Wenatchee, Washington. 44 p.

Ransom, B. H., and K. M. Malone. 1990. Hydroacoustic evaluation of the sluiceway at Wanapum Dam in passing juvenile salmon and steelhead trout during spring 1990. Hydroacoustic Technology Inc. Report to Grant County Public Utility District No. 2. Ephrata, Washington. 50 p.

Ransom, B. H., G. A. Raemhild, and T. W. Steig. 1988. Hydroacoustic evaluation of deep and shallow spill as a bypass mechanism for downstream migrating salmon and steelhead at Rock Island Dam. Pages I-70–84 in W. C. Micheletti, ed. *Fish Protection at Steam and Hydroelectric Power Plants.*

Ransom, B. H., and T. W. Steig. 1995. Comparison of the effectiveness of surface flow and deep spill for bypassing Pacific salmon smolts (*Oncorhynchus spp.*) at Columbia River basin hydropower dams. Pages 127–280 in J. J. Cassidy, ed. *Waterpower '95. Proceedings of the International Conference on Hydropower.* American Society of Civil Engineers, New York.

Raymond, H. L. 1968. Migration rates of yearling Chinook salmon in relation to flows and impoundments in the Columbia and Snake Rivers. *Transactions of the American Fisheries Society* 97:356–359. Electric Power Research Institute (EPRI), Palo Alto, CA.

Raymond, H. L. 1979. Effects of dams and impoundments on migrations of juvenile Chinook salmon and steelhead from the Snake River, 1966 to 1975. *Transactions of the American Fisheries Society* 108:505–529.

Raymond, H. L., and C. W. Sims. 1980. Assessment of smolt migration and passage enhancement studies for 1979. NMFS/NOAA Northwest Fisheries Science Center. Report to US Army Corps of Engineers, Seattle, Washington. 48 p.

Raymond, H. L., C. W. Sims, R. C. Johnsen, and W. W. Bently. 1975. Effects of power peaking operations on juvenile salmon and steelhead trout migrations, 1974. Northwest Fisheries Science Center, National Marine Fisheries Service. Report to US Army Corps of Engineers, Seattle, Washington.

Rieman, B. E., and R. C. Beamesderfer. 1990. Dynamics of a northern squawfish population and the potential to reduce predation on juvenile salmonids in a Columbia River reservoir. *North American Journal of Fisheries Management* 10:228–241.

Roebeck, G. G., C. Henderson, and R. C. Palange. 1954. Water quality studies on the Columbia River. US Department of Health, Education and Welfare, Public Health Service. Special Report. Washington, DC.

Ruehle, T. E., C. W. Long, and M. H. Gessel. 1978. Laboratory studies of a non-traveling bar screen for guiding juvenile salmonids out of turbine intakes. NMFS/NOAA Northwest Fisheries Science Center. Processed report. Seattle, Washington. 14 p.

Ruffing, F. E., N. A. Flint, and T. P. Miller. 1996. Total dissolved gas (TDG) distribution in the Columbia and Snake rivers. Field data report of transect studies 1995. US Army Corps of Engineers, North Pacific Division. Portland, Oregon.

Salmon and Steelhead Advisory Commission (SSAC). 1984. A New Management Structure for Anadromous Salmon and Steelhead Resources and Fisheries of the Washington and Columbia River Conservation Areas. Report to the United States Secretary of Commerce, Pursuant to The Salmon and Steelhead Enhancement Act of 1980 (16 USC Sections 3301 *et sec*. Publ. 96-561). July, 1984. 72 p.

Schrank, B. P., and E. M. Dawley. 1996. Evaluation of the effects of dissolved gas supersaturation on fish and invertebrates in Priest Rapids Reservoir, and downstream from Bonneville and Ice Harbor dams, 1995. National Marine Fisheries Service. Draft. Seattle, Washington.

Sheer, M. B., G. S. Holmberg, R. S. Shively, T. P. King, C. N. Frost, H. C. Hansel, T. M. Martinelli, and T. P. Poe. 1997. Movement, distribution, and passage behavior of radio-tagged juvenile Chinook salmon in John Day and The Dalles Dam forebays, 1995. Annual Report of Research US Geological Survey, Cook, Washington to US Army Corps of Engineers, Portland District, Portland, Oregon.

Sims, C. W., A. E. Giorgi, R. C. Johnsen, and D. A. Brege. 1984. Migrational characteristics of juvenile salmon and steelhead in the Columbia River basin 1983. US Army Corps of Engineers and National Marine Fisheries Services. Annual. 31 p.

Sims, C. W., and R. Johnson. 1977. Evaluation of the fingerling bypass system at John Day Dam and fingerling bypass outfall at McNary Dam. NMFS/NOAA Northwest and Alaska Fishery Center. Report to US Army Corps of Engineers, Portland, Oregon, Seattle, Washington. 15 p.

Sims, C. W., and F. J. Ossiander. 1981. Migrations of juvenile Chinook slamon and steelhead trout in the Snake River, from 1973 to 1979: A research summary. NMFS/NOAA Northwest and Alaska Fisheries Center. Final Report to US Army Corps of Engineers, Contract No. DACW69-78-C-0038. Portland, Oregon. 31 p.

Skalski, J. R. 1993. Summary of 3-Year bypass efficiency study at Wells Dam. Douglas County Public Utility District No. 1. Processed report. East Wenatchee, Washington. 5 p.

Skalski, J. R., R. L. Townsend, and A. E. Giorgi. 1998. The design and analysis of salmonid tagging studies in the Columbia River Basin: Volume XI, Recommendations on the design and analysis of radiotelemetry studies of salmonid smolts to estimate survival and passage efficiencies. Report to Bonneville Power Administration. Division of Fish and Wildlife. Contract No. DE-B179-90BP02341

Skalski, J. R., R. L. Townsend, J. Lady, A. E. Giorgi, J. R. Stevenson, and R. D. McDonald. 2002. Estimating route-specific passage and survival probabilities at a hydroelectric project from smolt radio-telemetry studies. *Canadian Journal of Fisheries and Aquatic Sciences* 59:1385–1393.

Smith, J. R. 1974. Spillway redesign abates gas supersaturation in Columbia River civil engineering. *American Society of Civil Engineers* 44:70–73.

Smith, S. G., W. D. Muir, E. E. Hockersmith, R. W. Zabel, R. J. Graves, C. V. Ross, W. P. Connor, and B. D. Arnsberg. 2003. Influence of river conditions on survival and travel time of Snake River subyearling fall Chinook salmon. *North American Journal of Fisheries Management* 23:939–961.

Smith, S. G., J. R. Skalski, and A. E. Giorgi. 1993. Statistical evaluation of travel time estimation based on data from freeze-branded Chinook salmon on the Snake River, 1982-1990. Northwest Fisheries Science Center, 2725 Montlake Blvd. E., Seattle, WA 98112. Report to Bonneville Power Administration, Portland, Oregon. Contract DE-B179-91BP35885, 95 p.

Smolt Monitoring Program. 1995. Weekly reports of smolt monitoring results. Fish Passage Center. Portland, Oregon.

Stanford, J. A., and F. R. Hauer. 1992. Mitigating the impacts of stream and lake regulation in the Flathead River catchment, Montana, USA: An ecosystem perspective. *Aquatic Conservation: Marine and Freshwater Ecosystems* 2:35–63.

Stansell, R. J., L. M. Beck, W. T. Nagy, and R. A. Magne. 1991. Hydroacoustic evaluation of juvenile fish passage at The Dalles Dam fish attraction water units in 1990. Pages 315–328 in *Seventh Progress Report: Fish Passage Development and Evaluation Program 1984-1990*. US Army Corps of Engineers, North Pacific Division, Environmental Resources Division, Portland, Oregon.

Stansell, R. J., R. A. Magne, W. T. Nagy, and L. M. Beck. 1990. Hydroacoustic monitoring of downstream migrant juvenile salmonids at Bonneville Dam, 1989. Pages 285–298 in *Seventh Progress Report: Fish Passage Development and Evaluation Program 1984-1990*. US Army Corps of Engineers, North Pacific Division, Environmental Resources Division, Portland, Oregon.

Steig, T. W., and W. R. Johnson. 1986. Hydroacoustic assessment of downstream migrating salmonids at The Dalles Dam in spring and summer, 1985. Bonneville Power Administration. Contract No. DE-AC79-85 BP23174. Portland, Oregon. 56 p.

Stockley, C. 1959. Merwin Dam downstream migrant bypass trap experiments for 1957. Washington Department of Fisheries. Processed report. Olympia, Washington.

Stone and Webster Engineering Corp. 1982. Feasibility study of improvement of fish passage through units 1 and 2 turbines, Rocky Reach Hydroelectric Project. Report to Public Utility District No. 1 of Chelan County, Wenatchee, Washington. 14 p.

Stone and Webster Engineering Corp. 1986. Assessment of downstream migrant fish protection technologies for hydro-electric application. Electric Power Research Institute (EPRI). Report to Electric Power Research Institute (EPRI). AP-4711, Research Project 2694-1. 324 p.

Stuehrenber, L., K. Liscom, and G. Monan. 1979. A study of apparent losses of Chinook salmon and steelhead based on count discrepancies between dams on the Columbia and Snake rivers, 1967-1968, 49 pp. [cited in Ferguson et al. 2003]

Swan, G. A., and W. T. Norman. 1987. Research to improve subyearling Chinook fish guiding efficiency at McNary Dam, 1986. NMFS/NOAA Northwest Fisheries Science Center. Report to US Army Corps of Engineers, Contract No. DACW68-84-H-0034. Seattle, Washington. 34 p.

Swan, G. A., R. F. Krcma, and F. J. Osiander. 1985. Development of an improved fingerling protection system for Lower Granite Dam, 1984. NMFS/NOAA Report to US Army Corps of Engineers, Portland, Oregon, Contract No. DACW68-84-H-0034. Seattle, Washington.

Swan, G. A., R. F. Krcma, and F. J. Osiander. 1986. Continuing studies to improve and evaluate juvenile fish collection at Lower Granite Dam, 1985. NMFS/NOAA Northwest Fisheries Center. Final report to US Army Corps of Engineers, Portland, Oregon, Contract No. DACW68-84-H-0034. Seattle, Washington. 31 p.

Swan, G. A., A. E. Giorgi, T. Coley, and W. T. Norman. 1987. Testing fish guiding efficiency of submersible traveling screens at Little Goose Dam: Is it affected by smoltification levels in yearling Chinook salmon? NMFS/NOAA Northwest Fisheries Science Center. Report to US Army Corps of Engineers, Portland, Oregon, Contract No. DACW68-84-H-0034. Seattle, Washington. 72 p.

Swan, G. A., B. H. Monk, J. G. Williams, and B. P. Sandford. 1990. Fish guidance efficiency of submersible traveling screens at Lower Granite Dam, 1989. NMFS/NOAA Northwest Fisheries Science Center. Report to US Army Corps of Engineers, Contract No. DACW68-84-H-0034. Seattle, Washington. 31 p.

Swan, G. A., R. D. Ledgerwood, B. H. Monk, W. T. Norman, R. F. Krcma, and J. G. Williams. 1992. Summary report on fish guidance efficiency studies at Lower Granite Dam–1984, 1985, 1987, and 1989. Pages 137–143 in *Seventh Progress Report: Fish Passage Development and Evaluation Program 1984-1990*.

Swan, G. A., B. L. Iverson, B. P. Sandford, and M. A. Kaminski. 1995. Radio-telemetry study Ice Harbor Dam. NMFS/NOAA Northwest Fisheries Science Center. Report presented to the US Army Corps of Engineers, Delivery Order E86-95-0113. Walla Walla, Washington.

Thompson, W. F. 1959. An approach to population dynamics of the Pacific red salmon. *Transactions of the American Fisheries Society* 88:206–209.

Thorne, R. E., and E. S. Kuehl. 1989. Evaluation of hydroacoustic techniques for assessment of juvenile fish passage at Bonneville Powerhouse I. Biosonics, Inc. Report to US Army Corps of Engineers, Portland District, Contract No. DACW57-88-C-0057 (Summary: pages 245–255 in US Army Corps of Engineers, North Pacific Division, Environmental Resources Division. 1993. *Seventh Progress Report: Fish Passage Development and Evaluation Program 1984-1990*). Portland, Oregon.

Thorne, R. E., and E. S. Kuehl. 1990. Hydroacoustic evaluation of fish behavioral response to fixed bar screens at Lower Granite Dam in 1989. Pages 299–313 in *Seventh Progress Report: Fish Passage Development and Evaluation Program 1984-1990*. US Army Corps of Engineers, North Pacific Division, Environmental Resources Division.

US Army Corps of Engineers. 1992. Fish bypass projects on the Snake and Columbia rivers. *Fish Passage Notes* Special Digest Edition:1–3.

US Army Corps of Engineers. 1993. Fish passage development and evaluation program 1984-1990. US Army Corps of Engineers, North Pacific Division. Seventh Progress Report. Portland, Oregon. 403 p.

U. S. Army Corps of Engineers 2002. Lower Snake River Juvenile Salmon Migration Feasibility Report/Environmental Impact Statement (Summary) February, 2002 U. S. Army Corps of Engineers, Walla Walla District. 52 pp. More information is available on a web site at http://www.nww.usace.army.mil

U. S. Fish and Wildlife Service (USFWS) 1999. Recovery Plan for the white sturgeon (*Acipenser transmontanus*): Kootenai River population Portland, OR.

Washington Department of Fish and Wildlife and Oregon Department of Fish and Wildlife. 2002. Status report. Columbia River Fish Runs and Fisheries, 1938-2000. Washington Department of Fish and Wildlife and Oregon Department of Fish and Wildlife. Olympia, Washington, and Clackamas, Oregon.

Watson, D. G., C. E. Cushing, C. C. Coutant, and W. L. Templeton. 1969. Effect of Hanford reactor shutdown on Columbia River biota. Proceedings of the Second National Symposium on Radioecology, May 13-17, 1967. Ann Arbor, Michigan. pp. 291-299

Wayne, W. W. 1961. Fish handling facilities for Baker River Project. *Journal of the Power Division, Proceedings of the American Society of Civil Engineers* 87:23–54.

Whitney, R. R., L. D. Calvin, J. Michael W. Erho, and C. C. Coutant. 1997. Downstream passage for salmon at hydroelectric projects in the Columbia River Basin: Development, installation, and evaluation. Northwest Power Planning Council. Technical Report, 97-15. Portland, Oregon. 101 p.

Wik, S. J., and T. Y. Barila. 1990. Evaluation of extended-length screening concept Lower Granite Dam, 1990. Pages 329–337 in *Seventh Progress Report: Fish Passage Development and Evaluation Program 1984-1990*. US Army Corps of Engineers, North Pacific Division, Environmental Resources Division.

Williams, J. G., S. G. Smith, W. D. Muir, B. P. Sandford, S. Achord, R. McNatt, D. M. Marsh, R. W. Zabel, and M. Scheuerell. 2003. Effects of the Federal Columbia River Power System on Salmon Populations. Preliminary Draft White Paper. NOAA Fisheries, Seattle, Washington. 69 p.

Willis, C. F. 1982. Indexing of juvenile salmonids migrating past The Dalles Dam, 1982. Oregon Department of Fish and Wildlife. Report to US Army Corps of Engineers, Contract No. DACW57-78-C-0056. Portland, Oregon. 37 p.

Willis, C. F., and B. L. Uremovich. 1981. Evaluation of the ice and trash sluiceway at Bonneville Dam as a bypass system for juvenile salmonids, 1981. Oregon Department of Fish and Wildlife. Report to NMFS/NOAA, Contract No. 81-ABC-00173. Portland, Oregon. 34 p.

Willis, C. F., and D. L. Ward. 1995. Development of a systemwide predator control program: Stepwise implementation of a predation index, predator control fisheries, and evaluation plan in the Columbia River Basin. Bonneville Power Administration. 1993 Annual Report, Vol. 1, Contract DE-B179-90BP07084. Portland, Oregon.

Wood, C. A. 1993. Implementation and evaluation of the water budget. *Fisheries* 18:6–17.

Wydoski, R. S. and R. R. Whitney 2003. Inland Fishes of Washington. American Fiheries Society, Bethesda, MD and University of Washington Press, Seattle, WA 322 pp.

Young, F. R. 1996. Development of a system-wide predator control program: Stepwise implementation of a predation index, predator control fisheries, and evaluation plan in the Columbia River Basin. Bonneville Power Administration, Fish and Wildlife Program. Annual Report, Project No. 90-077. Portland, Oregon.

Zimmerman, M. P., and D. L. Ward. 1999. Index of predation on juvenile salmonids by northern pikeminnow in the lower Columbia River Basin, 1994-1996. *Transactions of the American Fisheries Society* 128:995–1007.

8

Artificial Production and the Effects of Fish Culture on Native Salmonids

James A. Lichatowich, M.S., Madison S. Powell, Ph.D., Richard N. Williams, Ph.D.

Introduction

History of Artificial Propagation of Salmon in the Columbia Basin
  Harvest and Hatcheries
  Hatcheries and Habitat Stewardship
  Science and Research Enter the Picture

Consequences of Historical Hatchery Programs
  Evaluation of Benefits
    Hatcheries from 1900 to 1980
    Hatcheries from the 1981 Northwest Power Act to the Present
  The Influence of Artificial Propagation on Management Policy
  Results on Uses of Hatcheries

Effects of Hatcheries and Hatchery Fish on Wild Fish
  Interactions Between Wild- and Hatchery-Origin Fish
  A Synthesis of Recent Literature on Wild and Hatchery Fish Interactions
  Genetic Effects of Hatchery Fish on Wild Fish
    Direct Genetic Effects
    Indirect Genetic Effects
  Genetic Changes to Hatchery Stocks

Conclusions and Recommendations about Artificial Propagation
  Conclusions
  Key Critical Uncertainties in Artificial Propagation
  Recommendations
    Risk Assessment and Adaptive Management
    Comprehensive Evaluation of Artificial Propagation
    Link Harvest Management with Hatchery Operations
    The Experimental Nature of Artificial Propagation
    The Need for Regional Decision Documents
    Hatcheries as Genetic Reserves and Refuges

Future Directions for Artificial Propagation in the Columbia River Basin

Literature Cited

". . . fish so abundant that they can be caught without restriction, and serve as cheap food for the people at large, rather than to expend a much larger amount in preventing the people from catching the few that still remain after generations of improvidence." (G. B. Goode's statement on the role of hatcheries)
—G. B. Goode. 1884. Report of the U.S. Commission of Fish and Fisheries (p. 1157).

"The role of hatcheries in restoring threatened and endangered populations of salmon to sustainable levels is one of the most controversial issues in applied ecology. The central issue has been whether such hatcheries can work, or whether, instead, they may actually harm wild populations."
—R. A. Myers et al. 2004. Hatcheries and endangered salmon. Science 303:1980.

Introduction

Artificial propagation is an important tool used in the management of Pacific salmon throughout the Pacific Northwest. The first hatchery was built in 1872 on the McCloud River, a tributary to the upper Sacramento River. From this beginning, hatcheries were rapidly located on streams throughout the region. This chapter focuses on the hatchery program in the Columbia Basin, the largest watershed in the region. The history of the hatchery program in the Columbia Basin parallels the history of artificial propagation in other watersheds in the Northwest. Further, the development of the hatchery program in the Columbia River has influenced and shaped programs in the region's smaller watersheds. In recent years, fish culturists in the Columbia Basin have been experimenting with innovations in the technology (Figure 8.1) that will have a strong influence on the role of artificial propagation outside the basin.

Artificial propagation of Pacific salmon was introduced into the Columbia Basin in 1877, and it has consumed a major portion of the fish management budget in the basin over the intervening years (General Accounting Office [GAO] 1992). By 1898, 26 million salmon fry were being released from hatcheries into the Columbia Basin each year. However, the early hatchery programs were largely ineffective (Columbia Basin Fish and Wildlife Authority 1990); thus, early hatcheries did not significantly increase the number of adult salmon and may actually have contributed to their decline. Nevertheless, with deep roots going back to the beginnings of salmon management, artificial propagation has had a lasting and major influence on fisheries management philosophy and approach. Understanding the growth and evolution of the hatchery program is an important starting point for anyone attempting to understand the current status of salmon, their management, and the possibility for restoration in the Pacific Northwest.

Hatcheries are still a major part of salmon management programs; about 80% of the remnant runs of adult salmon and steelhead entering the Columbia River were hatched and reared in a hatchery (Northwest Power

Figure 8.1 Artificial stream habitat for rearing juvenile Chinook salmon at the Nez Perce Tribal Hatchery near Lewiston, Idaho. Photo by Ron Williams.

Planning Council [NPPC] 1992; National Research Council [NRC] 1996). Between 1981 and 1991, hatcheries consumed 40% of the $1.3 billion spent on salmon restoration in the basin (GAO 1992). Furthermore, about 50% of the increase in salmon production predicted to result from the Council's[1] program is expected to come from artificially propagated fish (Regional Assessment of Supplementation Project 1992; NPPC 1994;). However, three independent panels of fisheries experts—the National Fish Hatchery Review Panel, the NRC, and the Council's Scientific Review Team—have recently called for significant changes in the approach, operation, and expectations from artificial propagation (National Fish Hatchery Review Panel 1994; NRC 1996; Brannon et al. 1999). Whether the region's management institutions are willing or able to act on those recommendations is a major uncertainty.

---

[1]The Council is the Northwest Power and Conservation Council (NPCC), a four-state compact created in 1981 by the Northwest Power Act to oversee both electric power production and fish and wildlife management the Columbia River. See Chapters 1 and 2 for more detailed information on the Council and its role.

In this chapter, we describe the history of the hatchery program in the Columbia Basin. We provide a historical perspective on the benefits of hatchery programs, describe the biological effects of hatcheries on wild populations, and discuss the influence of hatcheries on salmon management. We then discuss the future role of hatcheries, including supplementation and conservation programs.

History of Artificial Propagation of Salmon in the Columbia Basin

Harvest and Hatcheries

By the end of its first decade of existence in 1875, the salmon canning industry in the Columbia Basin had become an important part of the regional economy. Some of the fishermen and cannery owners were aware of the collapse of the Atlantic salmon fishery several decades earlier (e.g., Hume 1893). Consequently, the Oregon legislature petitioned Spencer Baird, the newly appointed head of the U.S. Commission on Fish and Fisheries, for advice on how to maintain the supply of salmon. Baird responded with a letter that correctly identified the threats to the supply of salmon: habitat change, excessive harvest, and dams or other barriers to migration (Baird 1875). However, Baird did not believe that these threats could be avoided through regulations, so he offered an alternative to habitat protection, harvest regulation, and prohibition of dams. The alternative was fish culture, and it was accepted with enthusiasm. Baird offered the region a developer's dream: a technology that he believed would maintain the supply of salmon while simultaneously freeing entrepreneurs to harvest salmon, dam rivers, log riparian areas, and divert water for irrigation with few regulations or restraints.

Hatcheries and Habitat Stewardship

When Spencer Baird recommended fish culture in lieu of stewardship, he had no scientific information to support his recommendation. The first Pacific salmon hatchery began operation just 3 years earlier. Hatcheries were accepted as an alternative to stewardship not on the strength of scientific findings but because they were consistent with the prevailing public policy of laissez-faire access to natural resources. Hatcheries would augment the harvest of massive wild runs of salmon; then, when development degraded habitat, they would mitigate for the losses of naturally produced fish (Figure 8.2).

Livingston Stone, who built the first Pacific salmon hatchery in the headwaters of the Sacramento River, traveled to the Columbia River in 1877 to

Figure 8.2 Historical photo of hatchery worker and incubation trays (Photo courtesy of Christine Moffitt).

help the Oregon and Washington Fish Propagating Company build and operate the first hatchery in the basin (Stone 1879; Hayden 1930). Stone selected a site on the Clackamas River and then built the hatchery buildings and a rack across the river to capture brood fish. By 1928, 15 hatcheries were operating in the basin and a total of 2 billion artificially propagated fry and fingerlings had been released into the river (Figure 8.3).

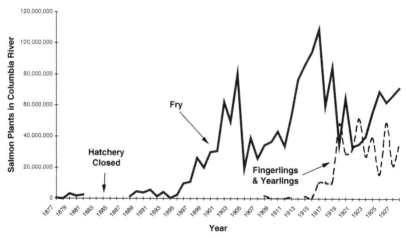

Figure 8.3 Hatchery production of all salmon species in the Columbia River (1877–1928) as the annual numbers of fry, fingerlings, and yearlings released into the river (Cobb 1930).

Figure 8.4 Historical photo of many small boats (Photo courtesy of Christine Moffitt).

Initially, fishermen targeted Chinook salmon (Figure 8.4), especially the spring and summer races, because they made the highest quality canned product and brought the highest prices (Craig and Hacker 1940). The early hatchery program also focused exclusively on the Chinook salmon (Figure 8.5); however, when the abundance and harvest of Chinook salmon began to decline, the fishery switched to other species and the hatcheries followed. Coho salmon and steelhead were propagated in hatcheries beginning about 1900; chum and sockeye salmon were taken into the hatchery program about a decade later (Cobb 1930).

The goals of early fish culture programs were to (1) exert human control over the production of salmon, and (2) maintain a supply of fish for the salmon industry (Goode 1884). The salmon industry supported hatcheries as an alternative to other forms of conservation, such as a restriction in the harvest (DeLoach 1939). These goals, especially human control over production, were based on the belief that natural reproduction was inherently inefficient and wasteful. It was subject to major, uncontrolled sources of mortality, which could be reduced or eliminated through artificial spawning and incubation in a protected environment (Foerster 1936; Hedgepeth 1941). These assumptions are reflected in the hatchery policy of the U.S. Fish Commission, which was to make:

> "... fish so abundant that they can be caught without restriction, and serve as cheap food for the people at large, rather than to expend a much larger amount in preventing the people from catching the few that still remain after generations of improvidence." (Goode 1884, p. 1157)

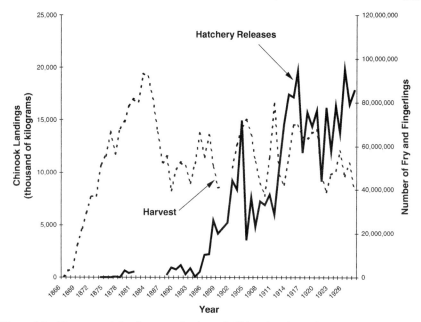

Figure 8.5 Harvest and hatchery production of Chinook salmon in the Columbia River (1866–1928) as the annual numbers of fry, fingerlings, and yearlings released into the river (Cobb 1930; Beiningen 1976).

The belief that protection of incubating eggs in hatcheries would make salmon so abundant that regulations would be unnecessary suggests that carrying capacity or density dependent limits to production were not considered. However, by 1894, after 22 years experience with artificial propagation and few tangible results, the U.S. Fish Commission reduced its expectations for artificial propagation. Marshall McDonald, who succeeded Spencer Baird stated,

". . . we have relied too exclusively upon artificial propagation as a sole and adequate means for maintenance of our fisheries. The artificial impregnation and hatching of fish ova and the planting of fry have been conducted on a stupendous scale. We have been disposed to measure results by quantity rather than quality, to estimate our triumphs by volume rather than potentiality. We have paid too little attention to the necessary conditions to be fulfilled in order to give the largest return for a given expenditure of effort and money." (McDonald 1894, p. 15)

McDonald raised several issues regarding the use of hatcheries, including three important points that are still valid today: (1) a warning regarding an overdependence on hatchery production as a substitute for stewardship, (2) a criticism of evaluations based on the quantity of juveniles released rather than their demonstrated ability to survive to adults, and 3) a recommendation for the need to evaluate the quality of the habitat in receiving waters in watersheds to be stocked with hatchery fish. McDonald's reservations were largely ignored and did not diminish the enthusiasm for artificial propagation.

Science and Research Enter the Picture

The first hatcheries in the Columbia Basin were built less than 20 years after Darwin (1859) published his evolutionary theories. Concepts such as reproductive isolation, natural selection, and local adaptation had not yet become a part of science, much less fisheries management. Salmon from different rivers were believed to be genetically similar (Ricker 1972) and therefore interchangeable; consequently, mass transfers of fish among streams were common. For example, when Bonneville Hatchery was constructed in 1909, one of its chief purposes was to serve as a central clearinghouse for the distribution of salmon eggs throughout the region (Figure 8.6). Eggs were

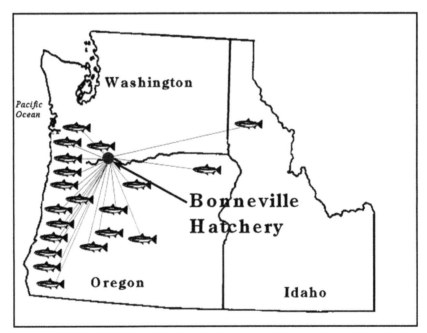

Figure 8.6 Transfers of salmon from Bonneville Hatchery to Washington, Oregon, and Idaho streams from 1909 to 1950. Each line can represent multiple plants (From Lichatowich 1999, page 127).

brought into Bonneville Hatchery from distant rivers and hatcheries, held to the eyed stage, then either the fry were released from Bonneville Hatchery into the Columbia River or the eyed eggs were shipped to hatcheries on other rivers (Figure 8.7). The source stream and the ultimate destination of a group of eggs were rarely the same.

The early history of the hatchery program is marked by extravagant and undocumented claims of hatchery effectiveness. For example, in 1883, George Brown Goode of the U.S. Fish Commission told the International Fisheries Exhibition in London, England, that the Pacific salmon fisheries in the Sacramento and Columbia Rivers were under the complete control of fish culture (Maitland 1884). When Goode made that claim, the only hatchery on the Columbia River had been closed for 2 years (Cobb 1930).

Figure 8.7 Idaho milk can photo. Photo by Christine Moffitt.

By the 1930s, declining harvests and the failure of the hatchery programs to maintain the supply of salmon fueled a growing criticism of the hatchery programs and a call for the development of a scientific approach to propagation (e.g., Culler 1932; Huntsman 1937; Needham 1939). The Fish Culture Division of the American Fisheries Society even questioned the ability of hatcheries to perform the tasks that had been assigned to them (Gottschalk 1942). It was becoming clear that artificial propagation had to be based on science.

Scientific management emphasized the principle of supply and demand, which is best exemplified in the catchable trout program (Bottom 1997). Catchable sized trout are delivered to the stream in the right quantity to meet the demand. The catchable trout program counted on little or no long-term survival of the planted fish. Therefore, the stream, its habitat, carrying capacity, and food gradients were not important considerations for the success of the catchable program (Wood 1953). For anadromous salmonids, the shift during the first half of the 20th century to longer rearing in the hatchery and the release of smolts was in many respects the equivalent to the catchable trout program. As hatchery programs shifted to smolt releases, it diminished the importance of the stream as an integrated ecosystem. The rivers became merely channels to transport smolts to sea (Ortmann et al. 1976). Scientific research focused on prevention and treatment of diseases, development of nutritious feeds, improved physical plant and hatchery infrastructure, and determining the optimal size of juveniles and the timing of their release from hatcheries to improve post-release survival.

Research eventually improved post-release survival of artificially propagated salmon and steelhead, especially after 1960 (Columbia Basin Fish and Wildlife Authority 1990), and reinforced the manager's belief that artificial propagation could compensate for the destruction of habitat in the Columbia River watershed (Schwiebert 1977). By the 1930s, the harvest of salmon in the Columbia Basin was rapidly declining (Figure 8.8), but the 1930s also marked the beginning of a major new threat to the basin's Pacific salmon: the construction of large, mainstem hydroelectric dams (Figure 8.9). Despite the lack of evidence that hatcheries had ever achieved their objective of maintaining the supply of adult salmon, artificial propagation was used to mitigate for the effects of the dams on anadromous salmonids (Lichatowich 1999). Mitigation initiated a massive expansion in the basin's hatchery program that continues today.

The goal of mitigation was not simply to replace the production of salmon and steelhead attributed to habitat losses from development of the hydroelectric system. Managers predicted that genetic selection in the hatchery program would produce strains of steelhead suited to the changing environment of the Columbia River (Ayerst 1977), an argument that is still heard today. Managers still believed that technology would solve the problems of declining salmon

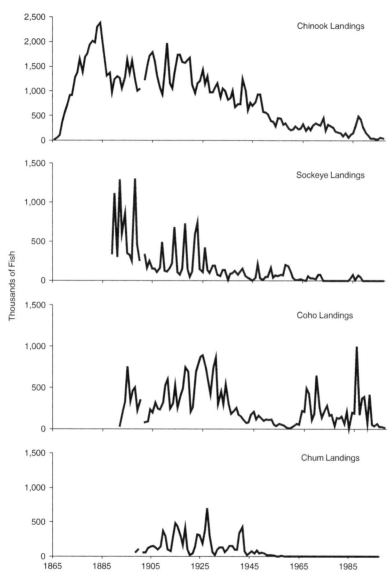

Figure 8.8 Columbia River commercial salmon fishery landings, 1866–1994. Data from Washington Department of Fish and Wildlife and Oregon Department of Fish and Wildlife (1995).

abundance: Through a combination of hatcheries and other technology, such as transportation and spillway deflectors, salmon and steelhead populations would be restored in a few years and ultimately, in the Snake River, would return in numbers greater than had existed before (Ebel 1977).

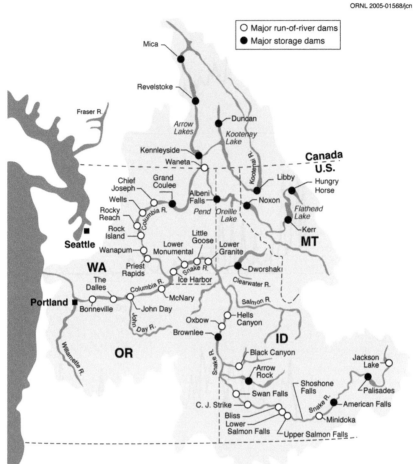

Figure 8.9 Map of the Columbia River Basin, showing major dams and lakes.

Eighteen years later, Chinook salmon, sockeye salmon, and steelhead trout from the Snake River were on the Endangered Species List. The National Fish Hatchery Review Panel (1994) and the NRC (1996) concluded a major revision in the role and objectives of artificial propagation was necessary. In general, the reviews recommended that hatchery programs become integrated into comprehensive ecosystem restoration plans and work toward conservation objectives, rather than focusing on the production of fish for harvest (Flagg et al. 1995b). The Northwest Power Planning Council's Scientific Review Team outlined a comprehensive set of scientific principles and actions needed for hatchery reform (Brannon et al. 1999). Although recommendations for hatchery reform have received increased attention in recent years, it is still an open question whether meaningful change will be implemented.

Consequences of Historical Hatchery Programs

Evaluation of Benefits

Hatcheries from 1900 to 1980

During the first 50 years hatcheries were operated in the Pacific Northwest, claims of success for the hatchery program were based on short-term correlations, weak evidence, or no evidence at all. Early experiments, based on returns of fin-clipped hatchery fish, were poorly designed and executed and did little more than confirm that some of the fish reared at hatcheries returned as adults (Washington Department of Fisheries and Game 1904). In the Columbia Basin, declining or fluctuating catches in spite of an increasing number of fry released from hatcheries (Figure 8.5) discouraged fishery managers (Oregon Department of Fisheries 1908) and led in 1911 to an experiment. The common practice at the time was to release the salmon shortly after hatching and before they started to feed. In the experiment, hatcheries reared small lots of juvenile salmon for several months and released them at larger sizes. The catch increased in 1914, the year managers expected the first returns from their experiment. After 5 successive years of improved catches in the Columbia River, the Oregon Fish and Game Commission announced the success of their experiments:

> "This improved method has now passed the experimental stage, and . . . the Columbia River as a salmon producer has 'come back'. By following the present system, and adding to the capacity of our hatcheries, thereby increasing the output of young fish, there is no reason to doubt but that the annual pack can in time be built up to greater numbers than ever before known in the history of the industry." (Oregon Fish and Game Commission 1919, p. 16).

At the same time, the State of Washington claimed that the increase in harvest beginning in 1914 was due to an increase in production from their hatcheries (Washington Department of Fish and Game 1917). Subsequent review indicated that the claims of hatchery success were premature and the increased catch was not caused by the new methodology (Johnson 1984) and probably had little to do with artificial propagation in Oregon or Washington. Instead, the increase in harvest from 1914 to 1920 was consistent with the pattern of variation in harvest for the previous 20 years (Figure 8.10) and probably resulted from favorable environmental conditions. For example, the 1914 Chinook salmon run into the Umatilla River, which had no hatchery, also increased dramatically (Van Cleve and Ting 1960),

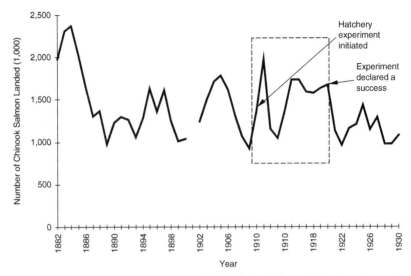

Figure 8.10   The number of Chinook salmon landed in the Columbia River (1882–1930). The data inside the box are discussed in the text (Modified from Lichatowich 1997).

supporting the suggestion that the increase in harvest was a response to natural climatic fluctuations and marine environments.

In 1914, Willis Rich initiated studies of the life history of Chinook salmon that had two practical purposes: (1) to determine the value of hatchery work, and (2) to understand the differences in early life history between spring and fall Chinook (Rich 1920). The latter was important because the spring Chinook were more valuable commercially, and their increase through artificial propagation was an important objective of the industry. To achieve these objectives, Rich initiated several marking experiments at hatcheries in the basin (Rich and Holmes 1929), which were a major improvement over earlier "evaluations," but they did not come close to the standards of experimental design used in later evaluations (e.g., Wahle et al. 1974; Wahle and Vreeland 1978). At the time of Rich's experiments, the institutional infrastructure needed to coordinate coast-wide recovery of marked salmon did not exist.

Based on his observations on the timing of the migration of juvenile Chinook salmon, Rich (1920) concluded that the release of sack fry should be terminated. He recommended that fry be held in the hatchery and released during the natural migration. He also recommended that juveniles be allowed to migrate out of the hatchery ponds on their own volition. One of the more important contributions from Rich's studies was the acquisition of data, which later contributed to his synthesis paper on the importance of stocks or local breeding groups to the maintenance of productive salmon fisheries (Rich 1939).

None of the early studies attempted to evaluate the relative contribution of artificially and naturally propagated salmon, that is, to answer the question "Are hatcheries making a significant contribution to the adult returns to the river?"

Nationally, by the 1920s, biologists were beginning to question the efficacy of fish culture during its first 50 years, and as a result, hatchery programs came under increasing criticism (Wood 1953). The first scientific evaluations of hatchery programs reinforced the growing skepticism. Studies involving yellow-pike perch in Lakes Huron and Michigan (Hile 1936), whitefish in Lake Erie (Van Oosten 1942), and Atlantic salmon in the Penobscot River, Maine (Rounsefell 1947) concluded that artificial propagation was not significantly more efficient than natural production and, in the case of the Atlantic salmon, that hatcheries were not able to prevent a decline in abundance. The lack of rigorous, scientific evaluation of the hatchery programs for Pacific salmon led Cobb (1930) to conclude that artificial propagation was a threat to the continued existence of the Pacific salmon fishery. Cobb was not opposed to artificial propagation, but he believed that managers had to put aside their optimism and stop relying on hatcheries alone to increase or maintain the fishery.

With all the clarity of hindsight, it is now generally recognized that in its early years, artificial propagation had little effect on the abundance of salmon in the Columbia River and elsewhere in the Pacific Northwest (Columbia Basin Fish and Wildlife Authority 1990). Early hatchery programs and practices may have had a detrimental effect on the wild salmonids; however, it is impossible to estimate the impacts of massive stock transfers, stream racking, and the overall mining of eggs from productive, wild populations of salmon, although they may have been considerable (Figure 8.11).

In 1922, the British Columbia Fisheries Commission was concerned about the lack of clear, positive results from its hatchery program for sockeye salmon, so it stipulated that R. E. Foerster carry out a scientific evaluation of artificial propagation. The question addressed by the study was a departure from the norm. The study was designed to evaluate the number of juveniles released from the hatchery and the number of adults that returned. In addition, it was designed to evaluate the benefits of artificial propagation by comparing the contributions from both natural and artificial propagation in a system where they could be monitored (Foerster 1936). The study was carried out at the Cultus Lake sockeye salmon hatchery in the lower Fraser Basin.

The study monitored the contribution of natural and artificial propagation for 10 years. No significant difference in the efficiency of natural and artificial propagation was found. Because the hatchery could incubate only a small fraction of the eggs in the spawning population, the small incremental

Figure 8.11   Photos showing (top) a close-up of salmon eggs and (bottom) a hatchery worker filling incubation trays. Photo from the U.S. Army Corps of Engineers Digital Virtual Library (http://images.usace.army.mil/photolib.html).

increase in adult returns produced by artificial propagation was not worth the expense of the hatchery. Based on this study, British Columbia closed all its sockeye salmon hatcheries (Foerster 1936). Foerster not only conducted one of the earliest scientific evaluations of a hatchery program for Pacific salmon, but he tested the fundamental assumption underlying all salmon hatcheries (artificial propagation was more efficient than natural reproduction) and found it to be false, at least as far as the contemporary hatchery practices for sockeye salmon were concerned. However, Foerster's study only evaluated the difference in survival between natural and artificial propagation of sockeye salmon when the hatchery fish were planted into Cultus Lake or its tributaries as fry or eyed eggs (Foerster 1936).

In 1934, shortly before Foerster completed his study, Salo and Bayliff (1958) started an evaluation of natural and artificial propagation in Minter Creek, a small stream in Puget Sound. They compared the relative survival and contribution of wild and artificially propagated coho salmon. The latter were reared in the hatchery for an extended period before release. The length of hatchery rearing was a major departure from Foerster's study, which evaluated the contribution from fry releases. At the time Salo and Bayliff's study was initiated, most hatcheries released fry with little or no feeding; however, hatchery practices were gradually shifting from fry releases to extended rearing on the assumption that larger, older fish would survive better after release from the hatchery. Like Foerster's study, the Minter Creek evaluation was carried out for several years. The findings, however, differed from Foerster's.

Salo and Bayliff (1958) reported that coho salmon reared in the hatchery for extended periods of 6 to 12 months produced greater adult returns than coho juveniles from an equivalent number of wild spawners. The Minter Creek study showed that under the right hatchery practices, artificial propagation could be more efficient than natural production, and artificially propagated salmon could significantly increase adult production in small populations. However, in the 1940s and 1950s, extended rearing presented hatchery managers with a new set of problems for which they had no clear solutions. Extended rearing required improved disease prevention, treatment, and the development of nutritious feeds.

By the 1940s, individual hatcheries were fin-clipping juvenile salmon in order to evaluate returns to the hatchery from routine production or to evaluate experimental hatchery practices. Often the experiments had too few recoveries to be conclusive. The results of many of those studies were summarized by Wallis (1960).

Beginning in the 1960s, the National Marine Fisheries Service (NMFS) conducted a series of large-scale evaluations of the contribution of Chinook and coho salmon from Columbia River hatcheries to various fisheries in the Northeast Pacific. The evaluation was stimulated by a moratorium on the construction of new hatcheries until it could be demonstrated that such investments were economically justified (Wahle and Vreeland 1978). The 1961 through 1964 broods of juvenile fall Chinook from 13 hatcheries in the Columbia Basin were given special marks (fin clips) before release, so that their contribution to the sport and commercial fisheries could be estimated. Results of the evaluation were positive. The benefit-cost ratio for all hatcheries combined for each of the brood years was 1961, 3.7:1; 1962, 2.0:1; 1963, 7.2:1; and 1964, 3.8:1. The potential catch per 1,000 fish released was 1961,

6.7; 1962, 3.1; 1963, 10.0; and 1964, 6.5. Average survival for all hatcheries combined was 0.7%. Overall, an estimated 14% of the fall Chinook salmon caught in the sport and commercial fisheries from southeast Alaska to northern California originated from the Columbia River hatcheries (Wahle and Vreeland 1978).

The NMFS repeated the fall Chinook evaluation with the 1978 to the 1982 broods. Total survival for all four brood years and all facilities was 0.33% or about half the survival of the earlier study, but the benefit-cost ratio was still positive at 5.7:1. The overall contribution to the fishery was 1.9 adults for each 1,000 juveniles released (Vreeland 1988). The NMFS used a similar approach to evaluate the contribution made to the west coast fisheries by the 1965 and 1966 broods of coho salmon. Juvenile coho salmon from 20 hatcheries in the Columbia Basin were marked for the study. Recoveries were monitored from British Columbia to California. Coho salmon from Columbia River hatcheries made up about 16% of the total catch in the sampling area. The catch from both brood years combined was 55 adults for each 1,000 smolts released for a benefit-cost ratio of 7.0:1 (Wahle et al. 1974). These results prompted additional investment in artificial production programs.

Hatcheries from the 1981 Northwest Power Act to the Present

Evaluation of hatchery programs continues to be a difficult and controversial, but necessary, task. Artificial production programs consume about 40% of the annual fish and wildlife expenditures in the Columbia River Basin (GAO 1992, 2002) and in the Council's Fish and Wildlife Program (Independent Scientific Review Panel 1999). Expenditures of this magnitude deserve to be justified by performance-based evaluations, something that has not occurred at a system level and only rarely at a facility or program level. A complete evaluation of a hatchery or group of salmon hatcheries should address three basic questions:

1. Do the salmon and steelhead of hatchery origin contribute to the fisheries and/or escapement, and is the economic value of that contribution greater than the cost to produce it?
2. Is the level of contribution consistent with the purpose or objective of the hatchery? For example, if a hatchery is intended to replace natural production lost due to habitat degradation, is the hatchery, in fact, replacing the lost production?
3. Do artificially propagated fish add to existing natural production, or do they replace it? That is, does the hatchery operation generate a cost to natural production through mixed stock fisheries, domestication, and genetic introgression?

The NMFS evaluations of the early 1980s were well designed and executed, but they only addressed the first question. That was a serious omission. From a historical perspective, it is clear that artificial propagation has failed to replace natural production lost due to habitat degradation. In addition, hatcheries have caused direct and indirect costs to the existing natural production (Flagg et al. 1995b; Utter et al. 1995).

As the NMFS study demonstrated, coho smolts released from Columbia River hatcheries achieved high levels of survival in the late 1960s and early 1970s. Although some biologists recognized that favorable ocean conditions contributed to the improved production, managers largely credited hatcheries for the improved harvests which, *"while most encouraging, was not unplanned, nor unexpected"* (Oregon Fish Commission 1964).

Columbia River coho salmon are a major contributor to the Oregon Production Index (OPI), which is a measure of the abundance of coho salmon south of Illwaco, Washington (Oregon Department of Fish and Wildlife 1982). The hatchery and wild stocks of coho salmon from the Columbia River are managed as part of the OPI. The history of ocean harvest of coho salmon in the OPI illustrates the need for more comprehensive evaluations of hatchery programs. It is now understood that the pattern of production with lows from the 1930s to the 1950s, followed by a period of high production in the 1960s and 1970s and another trough in the 1980s and 1990s (Figures 8.12 and 8.13), largely reflects population responses to

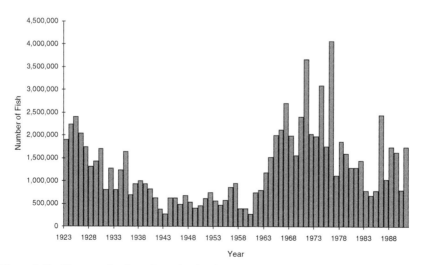

Figure 8.12 Harvest of coho salmon in the Oregon Production Index (OPI) (Lichatowich 1997).

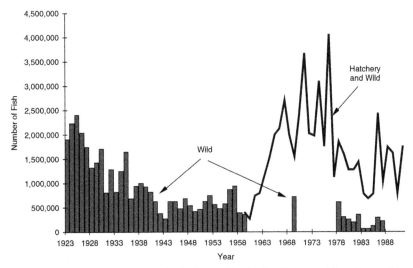

Figure 8.13 Harvest of coho salmon in the Oregon Production Index partitioned into wild and hatchery fish. Solid bars are catch of wild coho salmon. All coho salmon are assumed to be wild before 1960 (Lichatowich 1997).

changing ocean conditions and climate patterns, rather than only to the release of hatchery-reared coho (Nickelson 1986; Lichatowich 1996).

Prior to 1960, most of the coho salmon harvested in the OPI were naturally produced (Oregon Department of Fish and Wildlife 1982). After 1960, artificially propagated salmon made up an increasing proportion of the catch. Unfortunately, the separate contributions of hatchery and wild coho salmon to the OPI ocean harvest were monitored in only 11 of the years between 1960 and 1992 (Figure 8.13). What appears to be a recovery in the 1960s and 1970s was dominated by artificially propagated coho salmon; wild fish showed little sign of recovery. Harvest targeted on the dominant hatchery component of the OPI had significant impact on the natural production of Oregon's coastal and lower Columbia River coho stocks. The mixed stock (hatchery plus wild) fishery in the OPI consistently overharvested the wild coastal stocks of coho salmon. Of 55 coastal stocks of coho identified by the Oregon Department of Fish and Wildlife, 41 were classified as depressed (Nickelson et al. 1992), and between 1981 and 1991, escapement goals were met in only 3 of the 11 years (Pacific Fishery Management Council 1992).

Wild coho salmon from the lower Columbia River, which were also part of the OPI, are largely extinct, although remnant populations may still exist in the Clackamas, Hood, and Klickitat Rivers. High harvest rates on the mixed hatchery and wild stocks, which often exceeded 90%, were exacerbated by hatchery practices. Flagg et al. (1995b) identified the following hatchery practices that contributed to the decline and extirpation of coho salmon in the lower Columbia River: (1) selection for early spawners, (2) fry stocking that exceeded carrying capacity, (3) planting fry that were larger than their wild counterparts, (4) interhatchery stock transfers, and (5) excessive harvest in the fisheries targeting mixed hatchery and wild stocks.

The overly optimistic expectations and a tradition of inadequate evaluation have extended to the present. Recently, several key studies and reviews have addressed critical aspects of artificial production (Integrated Hatchery Operations Team 1994; Montgomery Watson 1996; Brannon et al. 1999; NPPC 1999; NPCC 2003). However, these analyses were unable to reach a bottom-line evaluation (i.e., egg-to-adult survival rates, as well as costs), due to inadequacies in the available databases. Thus, in spite of these new and more sharply focused review efforts, after 120 years in which hatcheries have been a primary management tool in the basin, there has never been a comprehensive evaluation of the program.

The Influence of Artificial Propagation on Management Policy

Perhaps the most important legacy of the hatchery program throughout its 128-year history has been its influence on management policy rather than any direct contribution to the various fisheries. Belief in the success of artificial propagation, which was largely unsubstantiated prior to 1960, made compromise leading to habitat destruction and overharvest easier to accept (Hilborn 1992; Lichatowich 1999). As suggested in the U.S. Fish Commission's hatchery policy, fish culture was viewed as an alternative to other forms of management, such as effective harvest regulation or habitat conservation. In addition, hatcheries were viewed as a means of compensating for production lost through habitat degradation. If hatcheries could compensate for lost and degraded habitat, managers could afford to give habitat protection and restoration a lower priority, which they did. By 1932, 50% of the best spawning and rearing habitat in the Columbia Basin had been lost or severely degraded (Oregon Fish Commission 1933). This loss, and the loss of habitat that continued after 1930, is at least in part, the legacy of overoptimism regarding the effectiveness of artificial propagation.

In 1930, John Cobb, Dean of the College of Fisheries at the University of Washington, listed artificial propagation as one of the threats to the fishing industry for Pacific salmon.

"In some sections an almost idolatrous faith in the efficacy of artificial culture of fish for replenishing the ravages of man and animals is manifested, and nothing has done more harm than the prevalence of such an idea. While it is an exceedingly difficult thing to prove, the consensus of opinion is that artificial culture does considerable good, yet the very fact that this can not be conclusively proved ought to be a warning to all concerned not to put blind faith in it alone." (Cobb 1930, p. 493)

Hatcheries influenced management policy in two important ways: First, from the late 1800s and to the 1970s, management institutions were willing to trade habitat for hatchery programs. The result was a massive shrinkage in habitat and in the natural production base and increasing dependence on a large, expensive hatchery program, which could only maintain salmon and steelhead at a fraction of their historical abundance. Second, management agencies are now forced to provide major emphasis and allocate resources to the restoration of those degraded habitats in an attempt to enhance the depleted base of natural production.

Artificial propagation's influence on policy has continued to the present, which is clearly shown in the distribution of expenditures for salmon protection and restoration in the Columbia River prior to 1980. Less than 1% of the funds was spent on habitat, whereas 43% of the expenditures went to the hatchery program (Figure 8.14; GAO 1992). In recent years, the situation has

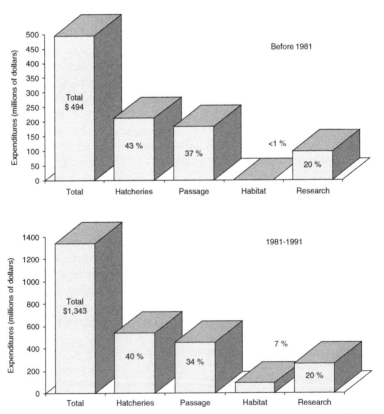

Figure 8.14 Distribution of expenditures for salmon restoration in the Columbia River prior to 1981 and in 1981–1991 (General Accounting Office 1992).

improved, but expenditures on habitat are still only 6% of the total; hatchery expenditures are 40% of the total (GAO 1993, 2002).

For the past two decades, salmon management has been changing. From a program almost entirely devoted to hatchery production and harvest regulation, management is shifting toward a greater concern for natural production. In recent years, Idaho, Oregon, and Washington have conducted extensive surveys of the status of naturally reproducing stocks of salmon and steelhead (Washington Department of Fisheries et al. 1993; Kostow 1995; Lutch et al. 2003). Hatchery programs are being designed to minimize their impact on natural production, and new programs are subject to extensive monitoring (Bowles and Leitzinger 1991; Messmer et al. 1992; Lutch et al. 2003; Hesse et al. 2004). Harvests are severely restricted to protect weak natural stocks, and biologists are recommending that hatchery programs be revised to include conservation objectives, instead of merely supplying fish for harvest (White et al. 1995; Flagg et al. 1995a; Flagg and Nash 1999; Flagg et al. 2000; Flagg and Manhnken 2000; Berejikian 2001). Which direction an emerging new role for artificial propagation will take is hard to predict; however, biologist Gary Meffe has outlined one approach that has merit:

> ". . . a management strategy that has as a centerpiece artificial propagation and restocking of a species that has declined as the result of environmental degradation and overexploitation, without correcting the causes for the decline, is not facing biological reality. Salmonid management based largely on hatchery production, with no overt and large scale ecosystem-level recovery program is doomed to failure. Not only does it fail to address the real causes of salmonid decline, but it may actually exacerbate the problem and accelerate the extinction process." (Meffe 1992, p. 351)

Results on Uses of Hatcheries

After nearly 130 years of salmon management based largely on the assumption that artificial propagation could replace natural production in the Columbia Basin and the development of a massive system of hatcheries, it is instructive to note that the most productive stock in the basin is the fall Chinook population that spawns naturally in the free flowing Hanford Reach of the mainstem Columbia (see Figures 3.4 and 3.12–3.14). In the context of the entire history of the hatchery program, and the history of salmon management in the basin, the hatchery program has failed to meet its objectives of offsetting (i.e., mitigating) habitat losses and degradation and maintaining salmon and steelhead abundance near historical levels. In 1994, the smallest number of salmon and steelhead entered the Columbia River since counts began in 1938, and by 1938, salmon production was already far below historical levels. Even the increases in abundance in recent years (1999–2004) are far below historical levels of abundance. Artificial propagation of salmon did not maintain salmon production. The early optimism that predicted hatcheries would make up for overharvest and habitat degradation has given way to the reality of depletion, closed fisheries, and a fragmented ecosystem in which natural production is severely restricted. Today the dominance of hatcheries in management programs is being questioned (Hilborn and Walters 1992; Washington and Koziol 1993; NRC 1996). New roles for hatcheries and guidelines for their operation are being developed or proposed (White et al. 1995; Flagg and Nash 1999; Waples 1999; Flagg et al. 2000; Berejikian 2001; Williams et al. 2003; Pollard and Flagg 2004). However, in the past, the hatchery program has been slow to adopt change. For example, by 1939, fish culturists recognized that the stock concept in Pacific salmon meant interhatchery transfers were detrimental (Oregon Fish Commission 1939). Fifty-six years later, Flagg et al. (1995b) were still recommending that hatcheries restrict that practice.

The total release of hatchery-reared salmonids grew from 79 million in 1960 to about 200 million in 1987—a level where it has remained ever since (+ 10%; Figure 8.15). The number of adult salmon and steelhead returning to the Columbia River has shown a steadily decreasing trend from 1960 through the 1990s, notwithstanding substantial interannual variation. Prior to 1960, most of the adult salmon and steelhead entering the Columbia Basin were naturally produced (Columbia Basin Fish and Wildlife Authority 1990; NPPC 1999); however, over the past three decades, the proportion of hatchery-reared fish in the adult population has grown to about 80% (Figures 8.16 and 8.17; NPPC 1992, 1999; Williams et al. 2003).

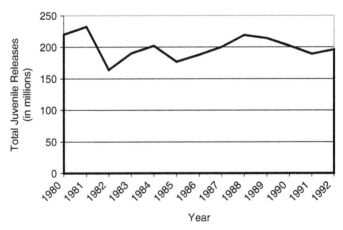

Figure 8.15 Total releases of Juvenile anadromous salmonid into the Columbia Basin, 1980–1992 (From Pacific States Marine Fisheries Commission http://www.streamnet.org.).

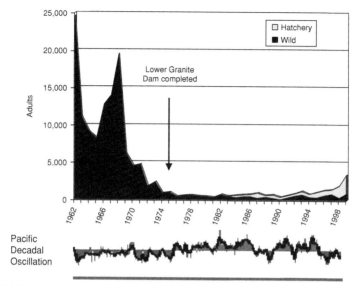

Figure 8.16 Annual returns of wild and hatchery fall Chinook to the Snake River Basin, 1962–1999, showing relationship to the Pacific Decadal Oscillation (PDO) Index and the date when Lower Granite Dam, the last of the four lower Snake River dams, was completed. Data from Williams et al. 2003. When the PDO index is blue, production of Columbia River stocks is usually good. When the PDO index is red, production of Columbia River stocks is usually poor.

These two figures show the mixed results of hatchery programs. For steelhead, it could be argued that in recent decades the hatchery program has accomplished its objective—hatchery production has more than replaced natural production circa 1960 lost through habitat degradation. The same argument cannot be made for fall Chinook. The success or failure of the hatchery program is a complex picture. The hatchery program since 1960 contains some successes; in other cases, hatcheries have failed to reach mitigation goals, and hatchery practices have been directly linked to depleted natural populations. Individual examples of success notwithstanding, overall in the Columbia Basin, artificial propagation has fallen for short of maintaining the supply of salmon.

The hatchery program for coho salmon contributed to the depletion of wild coho populations in tributaries below Bonneville Dam. Flagg et al. (1995b) identified factors related to the hatchery program that contributed to

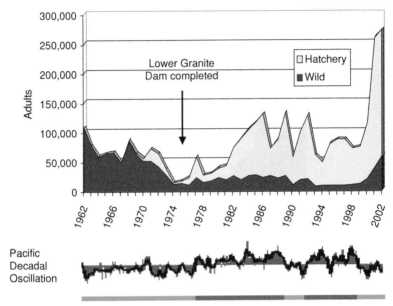

Figure 8.17   Annual returns of wild and hatchery summer steelhead to the Snake River Basin, 1962–2002, showing relationship to the Pacific Decadal Oscillation Index and the date when Lower Granite Dam, the last of the four Lower Snake River dams, was completed. From Williams et al. 2003. When the PDO index is blue, production of Columbia River stocks is usually good. When the PDO index is red, production of Columbia River stocks is usually poor.

the decline in natural production of coho salmon in the lower Columbia River.

In the upper Salmon River, hatchery mitigation has not replaced or maintained natural production lost due to smolt mortality, especially at the four lower Snake River Dams; however, it has slowed the decline of total production (Bowles 1993). In the South Fork of the Salmon River, hatchery mitigation has increased total production (Bowles 1993); however, returns through the 1980s and 1990s showed a pattern of steady decline, particularly for wild stocks, to the point where many were judged to be on the brink of extirpation (Thurow et al. 1997, 2000). That situation has eased somewhat since 1999, with sharply increased returns of both wild and hatchery-origin salmon and steelhead the last few years. The increase is primarily attributed to increased ocean survival for Columbia River stocks and is associated with a shift in the Pacific Decadal Oscillation (PDO cycle; see the extended discussion of this phenomenon in Chapter 10). In particular, many wild stocks showed dramatic increases in return numbers, although hatchery-origin fish still account for about 80% of the returning adults.

In the upper Columbia River, the present geographic distributions and genetic population structures of fall-run Chinook salmon reflect stock transfers and hatchery operations carried out between 1939 and 1943 under the Grand Coulee Fish Maintenance Project (GCFMP). The GCFMP intercepted upstream migratory salmonids at Rock Island Dam near Wenatchee, Washington, from 1939 through 1943 for relocation in tributaries downstream of Grand Coulee Dam. In this 5-year period, almost all adult spring-run and summer-fall-run Chinook, regardless of original destination, were either confined to restricted areas for natural reproduction or used in hatchery operations (Utter et al. 1995). This large-scale program of interceptions, stock transfers, and stock mixing permanently altered the salmon populations in the Upper Columbia River and provided the foundation for their present population structures.

In contrast, the Warm Springs National Fish Hatchery in the Deschutes River subbasin has successfully increased production in that river without adversely affecting wild stock production (Olson et al. 1995). This program appears to be an example of the effective use of adaptive management.

These examples suggest that the results of artificial propagation in the Columbia River since 1930, and especially after 1960, have been mixed. Unfortunately, the lack of a meaningful comprehensive evaluation dose not permit a determination and detailed description of the net effects of artificial propagation. Given the current state of the salmon and steelhead in the basin, it would be straightforward to conclude that in its 128-year history, the net effect of hatcheries has been negative.

In recent years, there has been increasing interest in an all-inclusive audit spanning hatchery operations in the Pacific Northwest. The Northwest Power and Conservation Council (NPCC) has recently completed a draft of such an audit, the Artificial Production Review and Evaluation (APRE; NPCC 2003). This review is based upon the process developed by the Hatchery Scientific Review Group in Washington state. Hatchery programs under the APRE were judged to be successful if they met four critical conditions: (1) The hatchery must produce a healthy and viable population. (2) It must make a sustainable contribution of adult returns to conservation and/or harvest. (3) Its potential effects on wild and native populations and the environment must be understood. (4) It must collect, record, evaluate, and disseminate information pertaining to the first three conditions so that decision makers may be informed about the benefits and risks of the program relative to other means of achieving similar conservation goals. The 2003 APRE effort is but the initial audit step in a more comprehensive and longer-term review effort led by the NPCC to evaluate and reform artificial production in the Columbia River Basin. Time will tell whether this latest effort will lead to substantive hatchery reform and to integration of artificial and natural production objectives within the basin.

Effects of Hatcheries and Hatchery Fish on Wild Fish

Even though stocking of artificially produced fish has long been recognized as a tool in fisheries management, the demographic, ecological, and genetic hazards of such use upon wild populations has been slower to come to light. Miller and Kapuscinski (2003) and Waples and Drake (2004) provide reviews of risk-benefit considerations as well as guidelines for the use of artificially produced fish.

Interactions Between Wild- and Hatchery-Origin Fish

With the magnitude of hatchery releases in the Pacific Northwest (e.g., greater than 80% of outmigrating smolts in the Columbia River are of hatchery origin), interactions between wild and hatchery fish occur at all life history stages. Understanding the ecological, genetic, and demographic impacts of the interactions between wild and hatchery fish is a critical need in order to effectively manage the potentially conflicting objectives of increasing overall salmon and steelhead abundance without causing further declines to the remaining naturally reproducing salmonid stocks.

A Synthesis of Recent Literature on Wild and Hatchery Fish Interactions

In desiring to integrate production between wild natural-origin fish and hatchery-origin fish, the single biggest hurdle faced by hatchery managers is the need to produce fish that are the functional equivalents of wild fish (ecologically, genetically, behaviorally, and reproductively). Recommendations from several reviews have helped guide hatchery reforms needed to reach this goal (National Fish Hatchery Review Panel 1994; White et al. 1995; NRC 1996; Brannon et al. 1999; Williams et al. 2003). With over 125 years of experience in culturing salmonids, today's hatchery programs are capable of producing very large numbers of juvenile salmon and steelhead for release into the natural environment. For example, approximately 1.2 billion juvenile salmon are released each year from U.S. west coast hatcheries, including 200 million salmon released into the Columbia River (Mahnken et al. 1998; Flagg et al.

2000). When environmental conditions are favorable, these releases can return large numbers of adult fish, as evidenced by the largest returns of Chinook and steelhead in several decades to the Snake River Basin in 2001 and 2002.

Unfortunately recent evidence suggest that the performance of hatchery-origin fish in the natural environment often falls significantly short of that of wild fish (Reisenbichler and Rubin 1999; Einum and Fleming 2001; Fleming and Petersson 2001; Chilcote 2003). These and other reviews (Nickelson et al. 1986; Jonsson et al. 1991; McKinnell et al. 1994) have led some authors to caution that supplementation programs designed to boost natural production may actually pose a threat to the long-term persistence of the target wild stock through the inhibition of local adaptation and eventually of natural production (Fleming and Petersson 2001; Utter 2002), as well as negative interspecific interactions (Levin and Williams 2002).

Fleming and Petersson (2001) reviewed and evaluated the ability of hatchery-origin salmonids to breed and contribute to the natural productivity of wild populations based on their own and others' studies of coho salmon (*O. kisutch*), Atlantic salmon (*Salmo salar*), and brown trout (*S. trutta*) in Europe and North America. They compared fitness-related behavioral traits and spawning performance among wild and hatchery males (overt aggression, courting, and number of spawnings) and among wild and hatchery females (overt aggression, number of nests, and breeding success). For males, the majority of comparisons (12 of 17) showed decreased performance by the hatchery-origin fish as compared to the wild fish, while for females, 6 of 15 comparisons showed decreased performance by the hatchery-origin fish. In no instance did hatchery-produced fish show increased performance over wild fish.

Based on their review, Fleming and Petersson (2001) concluded that current hatchery practices appear to result in competitively and reproductively inferior fish. Recent results for Pacific Northwest research efforts on Chinook, coho, and steelhead are consistent with these conclusions, generally showing reduced performance for hatchery-origin fish as compared to wild fish (Berejikian et al. 1996, 1997, 1999, 2001a, 2001b; Kostow et al. 2003). The conservation and management implications of these results are profound, as they suggest there may be inherent system-level limitations on the use of hatchery-origin fish to rebuild depressed native wild salmonid stocks, most of which are Endangered Species Act (ESA)-listed stocks, if it is not technically possible to artificially rear hatchery-origin fish that are wild fish equivalents. Nevertheless, the Pacific Northwest is currently making a concerted effort and investment into hatchery reform, focused on the technical aspects of improving aquaculture practices (Berejikian 2001). These include various modifications of the structural and rearing environments of the hatchery for juvenile salmonids with subsequent testing for improved juvenile post-release survival and improved adult reproductive performance compared to wild-origin fish.

Ultimately, however, the central uncertainty for the supplementation and captive brood approaches is whether we can produce salmonids that are, for all practical purposes, wild fish. The premise that we can successfully do this needs to be rigorously examined and tested. Currently, a number of intensive research studies tied to answering critical uncertainties about captive brood-stock technology are underway throughout the Pacific Northwest and are likely to provide insights into our ability to achieve this goal (Berejikian et al. 1996, 1997, 1999, 2001a, 2001b). Data to this point show clearly that improvements in hatchery protocols and operational conservation practices, such as those outlined by Brannon et al. (1999), can improve post-release survival of juvenile salmonids. However, in spite of these performance improvements, there continues to be a disparity between hatchery and wild fish performance (Chilcote 2003; Kostow et al. 2003).

These results suggest, as Fleming and Petersson (2001) note, that the use of hatchery-origin fish, even under our current best conservation hatchery operations, may be inimical to the sustained rebuilding of depressed wild stocks, due to the reproductive and performance inferiority of hatchery-origin fish compared to wild fish. Utter (2002), in discussing genetic interactions among wild and cultured salmon in British Columbia, offers a similar caution, noting that continued large-scale releases of hatchery fish threaten wild fish and are inconsistent with British Columbia's Wild Salmon Policy's principles to optimize sustainable benefits and conserve wild salmon.

Genetic Effects of Hatchery Fish on Wild Fish

Genetic interactions between hatchery-origin fish and wild-origin fish fall into two classes: direct effects and indirect effects. Direct effects are those resulting from interbreeding between hatchery fish or non-native fish with wild fish, while indirect genetic effects result from the ecological and behavioral interactions between wild and hatchery fish that occur without direct genetic exchange. Nevertheless, the indirect genetic interactions have genetic, and therefore fitness, consequences.

Direct Genetic Effects

Direct genetic effects are those that result from hybridization of cultured fish with wild fish. The effects of such interactions are generally negative and usually result in reduced fitness in the wild population, due to the breakup of various co-adapted gene complexes that are linked to local adaptation, performance, and fitness in the local population. Progeny of such matings usually suffer increased mortality and lowered reproductive success as compared to progeny of native wild fish (Leary et al. 1995; Fleming and Petersson 2001). Numerous studies document losses of within- or among-population genetic variability as a result of genetic interactions between hatchery and wild fish (Allendorf and Ryman 1987; Currens et al. 1990; Williams et al. 1996; Einum and Fleming 2001; Utter 2002). Increasing numbers of studies are documenting decreased levels of reproductive performance by hatchery fish as compared to wild fish (Chilcote 2003; Kostow et al. 2003).

Loss of within-population variation is usually linked to small effective population size (Ne), where allelic diversity can be lost through drift or sampling error. Generally, wild populations are not affected by this process, unless their numbers reach very low levels, such as experienced by many Idaho salmon stocks in the 1990s; however, considerable data exist documenting the debilitating effect of small Ne on hatchery populations.

In those few instances where hatchery and wild fish populations are similar genetically and slightly inbred, heterosis, or F1 hybrid vigor may occur. However, as genetic differences between the hatchery and wild stocks increase (typically measured by genetic distance), the more likely it is that outbreeding depression will occur and lead to reduced fitness in the F1 hybrids (Figure 8.18). Recombination in the F2, and subsequent generations, is likely to reduce fitness even further (Emlen 1991; Waples 1991).

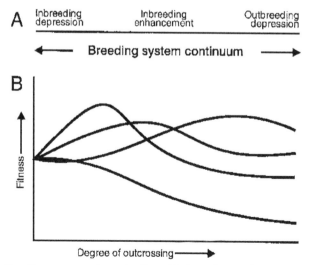

Figure 8.18 Breeding systems and genetic effects. (A) A continuum of breeding systems that, at opposite extremes, can lead to inbreeding or outbreeding depressions. (B) Several possible forms of the relationships between fitness and the degree of outcrossing (Allendorf and Waples 1996).

Reductions in among-population genetic variances can occur where a single broodstock is used over a wide geographic area (Reisenbichler and Phelps 1989), such as occurred with the Carson spring Chinook stock and the Skamania steelhead stock, or where substantial numbers of cultured fish have strayed into natural populations, as has occurred for steelhead in the Deschutes River (Chilcote 1998, 2003) and for Atlantic salmon in Norwegian rivers due to net pen escapees (Hindar et al. 1991; Gausen 1993; Heggberget et al. 1993; Fleming and Petersson 2001;). Reductions in reproductive fitness are most likely the result of genetic interactions between hatchery and wild fish (Hindar et al. 1991; Waples 1991). Such reductions in fitness are due to outbreeding depression (Figure 8.18), where two genetically dissimilar individuals (or stocks) interbreed.

Indirect Genetic Effects

Indirect genetic effects result from the ecological and behavioral interactions between wild and hatchery fish that occur without direct genetic exchange; however, the interactions have genetic, and therefore fitness, consequences (Waples 1991). Any factor that causes a reduction in population size can have an indirect effect on the genetic structure of wild fish populations, as well as increasing the risk of local extinction of that population through stochastic environmental perturbations (Soule 1987; Lande 1988). Factors that can adversely affect population size include competition, hatchery stocking densities that exceed carrying capacity, increased physiological stress associated with agonistic encounters, predation, disease, harvest of hatchery fish (underharvest increases opportunities for hatchery fish to stray or to breed with wild fish; overharvest also harvests wild stock and reduces its population size), and altered selection regimes. There is a substantial body of literature, which is not reviewed here, that documents the interactions between wild and hatchery fish. For example, Fausch (1988), Kaeriyama and Edpalina (2004), and Utter (2003) present reviews of competitive interactions between introduced and native fishes in stream systems. Some of these factors can have profound effects on genetic variability and population viability. An extreme example that illustrates some of the negative consequences that can result from large-scale, indirect interactions of hatchery-raised fish and wild fish occurred in Norwegian Atlantic salmon populations. Heggberget et al. (1993) note that disease transfer from farmed fish into native fish, after a catastrophic release of net pen fish, led to the complete extirpation of more than 30 native populations. Many of these factors alter the selection regimes faced by populations, which can shift the population's genetic and phenotypic attributes, as well as numerical abundance.

Effects of "Directional or Non-Random Selection"

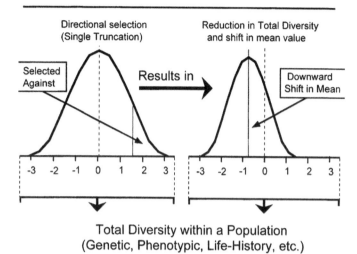

**Total Diversity within a Population
(Genetic, Phenotypic, Life-History, etc.)**

Figure 8.19 Directional selection happens when selection occurs for a character value other than the mean. A typical example of this type of selection is the effect that fishing pressure using size selective nets have in selectively harvesting larger fish, causing the mean size of fish in the run to decrease.

Character values, such as run timing, size at maturity, and so forth, are typically distributed in normal frequency distributions that are bell-shaped. Directional selection happens when selection occurs for a character value other than the mean (Figure 8.19). A typical example of this type of selection is the effect that fishing pressure (and various types of nets and gear) have in selectively harvesting larger fish, causing the mean or average size of fish in the run to decrease.

Stabilizing or truncating selection happens when selection occurs specifically for the mean of a character, which reduces overall variation (Figure 8.20). Management actions that focus on mean values may promote selection of this type. An example of this is the reduction observed in the number of wild smolts emigrating during the late summer and fall in the Columbia River because smolt protection programs have focused on the mean emigration time for all smolts combined (mid-April to mid-June). Smolts emigrating during this time period are favored by circumstances related to human development, while those outmigrating in the early spring or the late summer and fall months appear to have been disadvantaged and in some cases eliminated.

Effects of "Managing for the Mean"

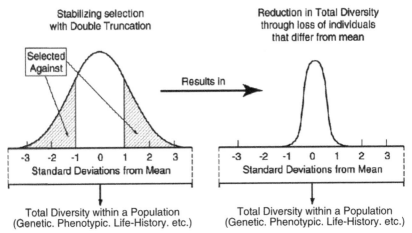

Figure 8.20 Stabilizing or truncating selection happens when selection occurs specifically for the mean character, which will act to reduce overall variation.

Genetic Changes to Hatchery Stocks

Genetic changes, and the potential for such changes, have been well documented for hatchery stocks. Reductions in overall levels of genetic variability, usually due to small effective breeding population size associated with hatchery practices, and concomitant reduced fitness attributed to inbreeding depression have been reported in some hatchery stocks (Allendorf and Phelps 1980; Leary et al. 1985; Allendorf and Ryman 1987; Waples and Smouse 1990) but not all hatchery stocks (Utter et al. 1989). Genetic changes in hatchery stocks can also be attributed to artificial selection or domestication selection.

Directed or inadvertent artificial selection can occur in the hatchery environment (Waples 1991; Reisenbichler 1995). A well-known example of this is the common advancement of time of spawning in hatchery strains of rainbow trout and timing of spawning migrations of Pacific salmon and steelhead trout that occurs from a greater than representative contribution to spawning populations from early maturing fish.

Domestication selection is natural selection occurring within the hatchery environment, whereby fish that perform better in the hatchery environment have a selective advantage (Reisenbichler and McIntyre 1977). In general, domestication selection results in increased fitness in the hatchery environment but decreased fitness under wild conditions (Campton 1995).

Conclusions and Recommendations about Artificial Propagation

Conclusions

1. Artificial propagation has failed to achieve the objective of replacing natural production lost because of habitat degradation in the basin (1).[2]
2. Hatcheries have been successful at preserving some of the genetic legacy, which would otherwise have been lost from salmon populations formerly occupying presently, severely degraded or occluded habitats (1).
3. Belief in the efficacy of artificial propagation led to disproportionate budgets for habitat protection and restoration (2).
4. In the 128-year history of the artificial propagation in the Columbia Basin, the program has never been subjected to a comprehensive evaluation (1).
5. The ecological, behavioral, and energetic interactions of hatchery fish with native species (including wild salmon) and fish assemblages of the Columbia River ecosystem have not been evaluated (1). In the operation of hatcheries, those interactions are generally assumed to be inconsequential, benign, or outside the concern or responsibility of the hatchery program managers (5).
6. The extent to which the artificial propagation program has implemented relevant research, particularly where the interaction between natural and artificially propagated fish is concerned, has been slow (3 historical, but 2 recently).
7. Hatchery operations including broodstock selection, interbasin transfers, and release practices have contributed to the decline of natural production and loss of locally adapted stocks in the basin (2).
8. Management of fisheries on mixed hatchery and wild stocks has contributed to the decline of natural production in the Columbia Basin (2).
9. Because of the declining natural production in the Columbia Basin, those fisheries that still harvest Columbia River salmon are largely supported by the hatchery program (1).
10. Hatchery practices are one of the factors that have altered the genetic structure of stocks in the basin (1).
11. Hatchery programs have contributed to the protection of genetic diversity in some life history types (1).

---

[2]The numbers in parentheses represents Levels of Proof that indicate the amount of scientific support for the implied assumption as follows: (1) Thoroughly established, accepted, peer-reviewed empirical evidence in its favor; (2) Strong weight of evidence in support, but not fully conclusive; (3) Theoretical support with some evidence from experiments or observation; (4) Speculative, but with little empirical support; and (5) Misleading or demonstrably wrong, based on good evidence to the contrary.

Key Critical Uncertainties in Artificial Propagation

Several key uncertainties remain to be addressed in the use of artificial production technologies for Columbia River salmon and steelhead populations.
The first uncertainty stems from the question "Can we successfully integrate
natural and artificial production systems in the same basin to achieve sustainable long-term productivity?" Integration includes the question "Can we
artificially produce salmonids that are the genetic and ecological equivalents
of wild fish in the target stream?" A second major uncertainty associated
with the use of supplementation is the condition of the habitat that will
receive the juvenile salmon; that is, is the habitat capable of supporting
salmon at levels of survival that will bring about restoration?

Recommendations

Artificial production is one of the cornerstones of the region's and the
Council's fisheries management and restoration plan and of the federally
directed recovery plans for ESA-listed stocks. We offer the following observations and recommendations to provide guidance on central concerns in the
application of artificial propagation to the management and recovery of
Pacific Northwest salmon and steelhead.

Risk Assessment and Adaptive Management

Because each application of aquaculture to the conservation of a local population of a life history type represents a new technology in the recovery of
endangered or depleted stocks, there are uncertainties associated with them.
Any use of artificial propagation to restore depleted salmon populations
should be preceded by an assessment of the risks, and applications must be
accompanied by a well-designed and adequately funded monitoring and
evaluation (M&E) program, the key link to a successful adaptive management approach.

Comprehensive Evaluation of Artificial Propagation

There remains a need for a valid comprehensive evaluation of the artificial
propagation in the Pacific Northwest. The evaluation should cover the
entire 128-year history of the program and include direct and indirect,
positive and negative effects. The comprehensive evaluation should also
include an assessment of the adequacy of existing monitoring to answer
ecological questions. The Council's recently completed APRE (NPCC

2003) is a first step toward such a comprehensive review. The APRE came out of the congressionally mandated Artificial Production Review (NPPC 1999) and audited hatchery programs and facilities in the Columbia River Basin and made preliminary recommendations on reform of hatchery practices.

There is also a need for a separate comprehensive evaluation of the mitigation hatcheries in the Pacific Northwest. What were their objectives? Did they achieve their objectives, and if not, why not? Are their original mitigation goals still relevant and appropriate?

As a next step in the comprehensive evaluation, the objectives of each hatchery need to be evaluated and redefined if necessary. The objectives should be established within the contexts of the subbasin where the hatchery operates, paying particular attention to the linkages between salmonids and their habitats, and the potential for metapopulation rebuilding. The hatchery operations should be integrated into the total production system and should assist in the recovery efforts in the subbasin. The hatchery's objectives need to be integrated and defined by the rebuilding objectives of the subbasin. The objectives should consider non-target species and the existence of metapopulation structure of the target species.

Finally, there are three questions that need to be answered in any evaluation of the hatchery program: (1) Do the artificially propagated fish contribute to the fishery and/or escapement and is the economic benefit of that contribution greater than its cost? (2) Has the program achieved its objective; that is, has it replaced lost natural production if it is a mitigation hatchery? (3) Has the operation of the hatchery incurred costs to natural production? The first and the third questions are related in that a meaningful cost-benefit analysis should include ecological costs.

Link Harvest Management with Hatchery Operations

The region needs to develop an interim policy regarding the operation and harvest management of production from each hatchery where monitoring has been inadequate to complete a comprehensive evaluation. The interim policy should be designed to minimize the ecological costs of the hatchery until the evaluation can be carried out.

The Experimental Nature of Artificial Propagation

Artificial propagation must be treated as an experiment, with hypotheses related to uncertainties, experimental design, analysis, and integration of results with available knowledge consistent with the adaptive management provisions of the Council's Fish and Wildlife Program.

The Need for Regional Decision Documents

Decisions about when, how, and at what scale to employ artificial propagation for salmonids need to be made through a standardized, consistent, and transparent process throughout the Pacific Northwest. A "decision tree" should be described that explains what biological and environmental considerations lead to a decision to bring a population into artificial or captive propagation, and what biological and/or environmental factors will drive decisions about continuing or discontinuing the program. Are there "trigger points" or metrics that drive decisions about implementation of the program? Decisions to employ supplementation programs should also incorporate metapopulation theory into the planning stages.

Hatcheries as Genetic Reserves and Refuges

Existing hatchery populations should be protected and carefully evaluated to identify their genetic legacy and its potential role in rebuilding populations and metapopulation structure. In many production and supplementation programs, broodstock were developed from wild returns. As founders to these hatchery populations, they contain varying amounts of genetic variation also observed (at least at one point in time) in their wild counterparts. In conservation programs such as Redfish Lake (Idaho) sockeye (Flagg et al. 2004; Powell and Faler 2004), the genetic constitution of the broodstock contains all remaining variation within the population. Thus, hatchery populations should neither be lumped with nor separated from wild populations as a whole. Instead, careful genetic evaluation of each program should be undertaken.

Though it is generally considered that conservation populations can act as "refuges" for remaining genetic variation, the inclusion or exclusion of other hatchery populations as part of ESA-listed species will remain a controversial issue. Specific language must be adopted which addresses the issue of the genetic legacy within supplementation and production populations.

Future Directions for Artificial Propagation in the Columbia River Basin

At any given level, human intervention, such as artificial propagation to manipulate biological processes, poses some associated risk to the existing environment and species therein. Moreover, the success of such intervention is ultimately tied to both freshwater and saltwater environmental conditions. Though these truths have been borne out over the past 100 years of artificial production in the Columbia Basin, we are just beginning to take note of them (Independent Scientific Advisory Board 2003; NPCC 2003). Use of artificial production to aid in the recovery and restoration of trout and salmon within the basin must employ this knowledge and incorporate it into both existing operations and future plans.

How is this best achieved amidst the uncertainties of the long-term efficacy of hatchery programs? The answer is by recognizing artificial production simply as a suite of tools and part of the repertoire in an overall comprehensive conservation framework, a framework that extends past geopolitical boundaries and past any single operational hierarchy within an agency. This does not infer that more hatcheries are needed to support conservation efforts, but rather that more attention is needed to incorporate artificial production programs into larger, landscape-scale processes. We are just beginning to understand that salmon and steelhead life history patterns likely operate on larger geographic scales than previously thought (Brannon et al. 2004; Hendry and Stearns 2004 and references therein). Thus, conservation efforts and the coordinated use of artificial production must also expand to incorporate these spatial scales (Williams et al. 2003).

A larger spatial framework may include watersheds where conservation, supplementation, and/or production are coordinated yet remain separated hierarchically. An example of this may be found within the Salmon River, where wild steelhead preserves are present in the Middle Fork and South Fork, while the Main Salmon operates as a supplementation project, and the Little Salmon and Rapid Rivers contain production programs.

The future of artificial production must not only consider itself in a broader context, but also recognize a tangential yet important issue in the Pacific Northwest, that is, population growth. Lackey (2001) suggests rapid population growth and human expansion in the Pacific Northwest over the next decade will have a profound effect on our ability to sustain populations and habitats for salmon, resulting in the potential for a much greater need for intervention requiring artificial production to reduce demographic risk to these populations. Such potential will necessitate a greater and accelerated need for comprehensive monitoring and evaluation of hatchery programs. Thus, the challenges we will face in the future of the Pacific Northwest demand that we understand how to use artificial production wisely to achieve societal benefits, to rebuild depressed salmon and steelhead populations, and to avoid doing inadvertent harm to the region's remaining wild salmon and steelhead populations. Only by critically examining our past record of artificial production activities can we hope to gain such understanding. Perhaps along with that understanding will come the wisdom needed to successfully integrate the artificial and natural production regimes.

Literature Cited

Allendorf, F. W., and S. R. Phelps. 1980. Loss of genetic variation in a hatchery stock of cut-throat trout. *Transactions of the American Fisheries Society* 109:537–543.

Allendorf, F. W., and N. Ryman. 1987. Genetic management of hatchery stocks. Pages 141–160 in N. Ryman and F. Utter, eds. *Population genetics and fishery management.* University of Washington Press, Seattle.

Allendorf, F. W., and R. S. Waples. 1996. Conservation and genetics of salmonid fishes. Pages 238–280 in J. C. Avise, ed. *Conservation genetics: Case histories from nature.* Chapman and Hall, New York.

Ayerst, J. D. 1977. The role of hatcheries in rebuilding steelhead runs of the Columbia River system. Pages 84–88 in E. Schwiebert, ed. *Columbia River salmon and steelhead, March 5-6, 1976.* American Fisheries Society, Washington, DC.

Baird, S. 1875, March 3. Salmon fisheries of Oregon. *Oregonian*, Portland, Oregon.

Beiningen, K. T. 1976. Fish Runs, Report E. *Investigative Reports of Columbia River Fisheries Project.* Pacific Northwest Regional Commission, Portland, Oregon.

Berejikian, B. A. 2001. Release of captively-reared adult salmon for use in recovery. *World Aquaculture* 32:63–65.

Berejikian, B. A., S. B. Mathews, and T. P. Quinn. 1996. Effects of hatchery and wild ancestry and rearing environments on the development of agonistic behavior in steelhead trout (*Oncorhynchus mykiss*) fry. *Canadian Journal of Fisheries and Aquatic Sciences* 53:2004–2014.

Berejikian, B. A., E. P. Tezak, L. Park, S. L. Schroder, E. P. Beall, and E. LaHood. 2001a. Male dominance and spawning behavior of captively-reared and wild coho salmon (*Oncorhynchus kisutch*). *Canadian Journal of Fisheries and Aquatic Sciences* 58:804-810.

Berejikian, B. A., E. P. Tezak, and S. L. Schroder. 2001b. Reproductive behavior and breeding success of captively-reared Chinook salmon (*Oncorhynchus tshawytscha*). *North American Journal of Fisheries Management* 21:255–260.

Berejikian, B. A., E. P. Tezak, S. L. Schroder, T. A. Flagg, and C. M. Knudsen. 1999. Competitive differences between newly emerged offspring of captively reared and wild coho salmon (*Oncorhynchus kisutch*). *Transactions of the American Fisheries Society* 128:832–839.

Berejikian, B. A., E. P. Tezak, S. L. Schroder, C. M. Knudsen, and J. J. Hard. 1997. Reproductive behavioral interactions between wild and captively-reared coho salmon (*Oncorhynchus kisutch*). *ICES Journal of Marine Science* 54:1040–1050.

Bottom, D. L. 1997. To till the water: A history of ideas in fisheries conservation. Pages 569–597 in D. Strouder, P. Bission, and R. Naiman, eds. *Pacific salmon and their ecosystems: Status and future options.* Chapman and Hall, New York.

Bowles, E. 1993. Operation of compensation hatcheries within a conservation framework. Idaho Department of Fish and Wildlife, Boise, Idaho.

Bowles, E., and E. Leitzinger. 1991. Salmon supplementation studies in Idaho Rivers (Idaho supplementation studies), Experimental design. USDOE, Bonneville Power Administration. Project 89-098. Portland, Oregon. 132 p.

Brannon, E. L., K. P. Currens, D. Goodman, J. A. Lichatowich, W. E. McConnaha, B. E. Riddell, and R. N. Williams. 1999. Review of salmonid artificial production in the Columbia River basin as a scientific basis for Columbia River production programs. Northwest Power Planning Council, Portland, Oregon. 77 p.

Brannon, E. L., M. S. Powell, T. P. Quinn, and A. Talbot. 2004. Population structure of Columbia River Basin Chinook salmon and steelhead trout. *Reviews in Fisheries Science* 12:99–232.

Campton, D. E. 1995. Genetic effects of hatchery fish on wild populations of Pacific salmon and steelhead: What do we really know? *American Fisheries Society Symposium: Uses and effects of cultured fishes in aquatic ecosystems* 15:337–353.

Chilcote, M. 1998. Conservation status of steelhead in Oregon. Oregon Department of Fish and Wildlife. Information Reports, Number 98-3. Portland, Oregon.

Chilcote, M. W. 2003. Relationship between natural productivity and the frequency of wild fish in mixed spawning populations of wild and hatchery steelhead (*Oncorhynchus mykiss*). *Canadian Journal of Fisheries and Aquatic Sciences* 60:1057–1067.

Cobb, J. N. 1930. Pacific salmon fisheries. Bureau of Fisheries. Document No. 1092. Washington, DC.

Columbia Basin Fish and Wildlife Authority. 1990. Review of the history, development, and management of anadromous fish production facilities in the Columbia River Basin. Columbia Basin Fish and Wildlife Authority and US Fish and Wildlife Service. Portland, Oregon. 52 p.

Craig, J. A., and R. L. Hacker. 1940. The history and development of the fisheries of the Columbia River. *US Bureau of Fisheries Bulletin* 32:133–216.

Culler, C. F. 1932. Progress in fish culture. *Transactions of the American Fisheries Society* 62:114–118.

Currens, K. P., C. B. Schreck, and H. W. Li. 1990. Allozyme and morphological divergence of rainbow trout (*Oncorhynchus mykiss*) above and below waterfalls in the Deschutes River, Oregon. *Copeia* 1990:730–746.

Darwin, C. 1859. *The origin of species by means of natural selection.* The Modern Library, New York.

DeLoach, D. B. 1939. The salmon canning industry. *Oregon State Monographs* Economic Studies No. 1.

Ebel, W. J. 1977. Panel 2: Fish passage problems and solutions to major passage problems. Pages 33–39 in E. Schwiebert, ed. *Columbia River salmon and steelhead, March 5-6, 1976.* American Fisheries Society, Washington, DC.

Einum, S., and I. A. Fleming. 2001. Implications of stocking: Ecological interactions between wild and released salmonids. *Nordic Journal of Freshwater Research* 75:56–70.

Emlen, J. M. 1991. Heterosis and outbreeding depression: A multi-locus model and an application to salmon production. *Fisheries Research* 12:187–212.

Fausch, K. D. 1988. Tests of competition between native and introduced salmonids in streams: What have we learned? *Canadian Journal of Fisheries and Aquatic Sciences* 45:2238–2246.

Flagg, T. A., and C. V. W. Mahnken. 2000. Endangered species recovery: Captive broodstocks to aid recovery of endangered salmon stocks. Pages 290–292 in *Encyclopedia of aquaculture.* John Wiley and Sons, New York.

Flagg, T. A., C. V. W. Mahnken, and K. A. Johnson. 1995a. Captive broodstocks for the recovery of Snake River sockeye salmon. *American Fisheries Society Symposium: Uses and effects of cultured fishes in aqautic ecosystems* 15:81–90.

Flagg, T. A., D. J. Maynard, and C. V. W. Mahnken. 2000. Conservation hatcheries. Pages 174–176 in *Encyclopedia of aquaculture.* John Wiley and Sons, New York.

Flagg, T. A., W. C. McAuley, P. A. Kline, M. S. Powell, D. Taki and J. C. Gislason. 2004. Application of captive broodstocks to preservation of ESA-listed stocks of Pacific salmon: Redfish Lake sockeye case example. *American Fisheries Society Symposium: Propogated fish in resource management* 44:383–396.

Flagg, T. A., and C. E. Nash, eds. 1999. A conceptual framework for conservation hatchery strategies for Pacific salmonids. US Department of Commerce. NOAA Tech. Memo, NMFS-NWFSC-38. Washington, DC. 48 p.

Flagg, T. A., F. W. Waknitz, D. J. Maynard, G. B. Milner, and C. V. W. Mahkhen. 1995b. The effect of hatcheries on native coho salmon populations in the lower Columbia River. *American Fisheries Society Symposium: Uses and effects of cultured fishes in aquatic ecosystems* 15:366–375.

Fleming, I. A., and E. Petersson. 2001. The ability of released hatchery salmonids to breed and contribute to the natural productivity of wild populations. *Nordic Journal of Freshwater Research* 75:71–98.

Foerster, R. E. 1936. Sockeye salmon propagation in British Columbia. Biological Board of Canada. Bulletin, No. 53. Ontario, Canada.

Gausen, D. 1993. The Norwegian gene bank programme for Atlantic Salmon (*Salmo salar*). Pages 181–188 in J. G. Cloud and G. H. Thorgaard, eds. *Genetic conservation of salmonid fishes*. Plenum Press, New York.

General Accounting Office (GAO). 1992. Endangered species: Past actions taken to assist Columbia River salmon. US General Accounting Office. Report to Congressional Requesters, GAO/RCED-92-173BR. Washington, DC.

GAO. 1993. Endangered species: Potential economic costs of further protection for Columbia River salmon. US General Accounting Office. Report to Congressional Requesters, GAO/RCED-93-41. Washington, DC. 34 p.

GAO. 2002. Columbia River Basin salmon and steelhead: Federal agencies' recovery responsibilities, expenditures and actions. US General Accounting Office. Report to the U.S. Senate, GAO-02-612. Washington DC. 86 p.

Gharrett, A. J., and W. W. Smoker. 1991. Two generations of hybrids between even- and odd-year pink salmon (*Oncorhynchus gorbuscha*): A test for outbreeding depression? *Canadian Journal of Fisheries and Aquatic Sciences* 48:1744–1749.

Goode, G. B. 1884. The Status of the US Fish Commission in 1884. Part XLI in Part XII. Report of the Commission of Fish and Fisheries. Washington, DC.

Gottschalk, J. 1942. Report of the Division of Fish Culture. *Transactions of the American Fisheries Society, Seventy-first Annual Meeting, August 25-26, 1941.*

Hayden, M. V. 1930. History of the salmon industry of Oregon. Master's thesis, University of Oregon, Eugene.

Hedgepeth, J. W. 1941. Livingston Stone and fish culture in California. *California Fish and Game* 27:126.

Heggberget, T. G., B. O. Johnsen, K. Hindar, B. Jonsson, L. P. Hansen, N. A. Hvidsten, and A. J. Jensen. 1993. Interactions between wild and cultured Atlantic salmon: A review of the Norwegian experience. *Fisheries Research* 18:123–146.

Hendry, A. P. and S. C. Stearns. 2004. *Evolution illuminated: Salmon and their relatives*. Oxford University Press, Oxford, United Kingdom. 510 p.

Hesse, J. A., J. R. Harbeck, and R. W. Carmichael. 2004. Monitoring and evaluation plan for the Northeast Oregon Hatchery Imnaha and Grande Ronde subbasin spring Chinook salmon. Department of Fisheries Resources Management, Nez Perce Tribe. Lapwai, Idaho. 144 p.

Hilborn, R. 1992. Hatcheries and the future of salmon in the Northwest. *Fisheries* 17:5–8.

Hilborn, R., and C. J. Walters. 1992. *Quantitative fisheries stock assessment: Choice, dynamics and uncertainty*. Chapman and Hall, New York.

Hile, R. 1936. The increase in the abundance of the yellow pike-perch, (*Stizostedion vitreum* (Mitchill)), in lakes Huron and Michigan, in relation to the artificial propagation of the species. Pages 143–159 in *The collected papers of Ralphe Hile, 1928-73*. US Department of Interior, Fish and Wildlife Service, Washington, DC.

Hindar, K., N. Ryman, and F. Utter. 1991. Genetic effects of cultured fish on natural fish populations. *Canadian Journal of Fisheries and Aquatic Sciences* 48:945–957.

Hume, R. D. 1893. *Salmon of the Pacific coast*. Schmidt Label & Lithographic Co., San Francisco, California.

Huntsman, A. G. 1937. Races and homing of salmon. *Science* 85:477–478.

Independent Scientific Advisory Board. 2003. Review of salmon and steelhead supplementation. Northwest Power Planning Council. Report ISAB 2003-03. Portland, Oregon. 229 pp.

Independent Scientific Review Panel. 1999. Review of the Columbia River Basin Fish and Wildlife Program as directed by the 1996 amendment to the Power Act. Northwest Power Planning Council. Annual Report, ISRP 98-2. Portland, Oregon.

Integrated Hatchery Operations Team. 1994. Policies and procedures for Columbia Basin anadromous salmonid hatcheries. Bonneville Power Administration. DOE/BP-2432. Portland, Oregon.

Johnson, S. L. 1984. Freshwater environmental problems and coho production in Oregon. Oregon Department of Fish and Wildlife. Information Report, 84-11. Corvallis, Oregon.

Jonsson, B., N. Jonsson, and L. P. Hansen. 1991. Differences in life history and migratory behavior between wild and hatchery-reared Atlantic salmon in nature. *Aquaculture* 98:69–78.

Kaeriyama, M., and R. R. Edpalina. 2004. Evaluation of the biological interaction between wild and hatchery population for sustainable fisheries management of Pacific salmon. Pages 260–306 in K. M. Leber, S. Kitada, H. L. Blankenship, and T. Svasand, eds. *Stock enhancement and sea ranching: Development, pitfalls, and opportunities.* Blackwell Publishing, Oxford, United Kingdom.

Kostow, K. 1995. Biennial report of the status of wild fish in Oregon. Oregon Department of Fish and Wildlife, Portland, Oregon.

Kostow, K. E., A. R. Marshall, and S. R. Phelps. 2003. Naturally spawning hatchery steelhead contribute to smolt production but experience low reproductive success. *Transactions of the American Fisheries Society* 132:780–790.

Lackey, R. T. 2001. Policy conundrum: Restoring wild salmon to the Pacific Northwest. In *Proceedings of the Biennial Conference of the International Institute of Fisheries Economics and Trade* July 10-14, 2000, Corvallis, Oregon.

Lande, R. 1988. Genetics and demography in biological conservation. *Science* 241:1455–1460.

Leary, R. F., F. W. Allendorf, and K. L. Knudsen. 1985. Developmental instability as an indicator of related genetic variation in hatchery trout. *Transactions of the American Fisheries Society* 114:230–235.

Leary, R. F., F. W. Allendorf, and G. K. Sage. 1995. Hybridization and introgression between introduced and native fish. *American Fisheries Society Symposium: Uses and effects of cultured fishes in aquatic ecosystems* 15:91–101.

Levin, P. S., and J. G. Williams. 2002. Interspecific effects of artificially propagated fish: An additional conservation risk for salmon. *Conservation Biology* 16:1581–1587.

Lichatowich, J. 1997. Evaluating the performance of salmon management institutions: The importance of performance measures, temporal scales and production cycles. Pages 69–87 in D. J. Stouder, P. A. Bisson, and R. J. Naiman, eds. *Pacific salmon and their ecosystems.* Chapman and Hall, New York.

Lichatowich, J. 1999. *Salmon without rivers: A history of the Pacific salmon crisis.* Island Press, Washington, DC.

Lichatowich, J., and J. D. McIntyre. 1987. Use of hatcheries in the management of anadromous salmonids. *American Fisheries Society Symposium* 1:131–136.

Lutch, J., C. Beasley, and K. Steinhorst. 2003. Evaluation and statistical review of Idaho Supplementation Studies: 1991-2001. March 200. Bonneville Power Administration. IDFG Report, Project Number 89-098, Contract Number DE_B179-89BP01466. Portland, Oregon.

Mahnken, C., G. Ruggerone, W. Waknitz, and T. Flagg. 1998. A historical perspective on salmonid production from Pacific Rim hatcheries. *North Pacific Anadromous Fisheries Commission Bulletin* 1:38–53.

Maitland, J. R. G. 1884. The culture of salmonidae and the acclimatization of fish. *The Fisheries Exhibition Literature, International Fisheries Exhibition, London, 1883.* William Clowes and Sons, London.

McDonald, M. 1894. The salmon fisheries of the Columbia River basin. Pages 3–18 in *Report of the Commissioner of Fish and Fisheries Investigations in the Columbia River Basin in Regard to the Salmon Fisheries.* Senate Miscellaneous Document No. 200, 53rd Congress 2d Session, and House Miscellaneous Document No. 86, 53d Congress, 3d Session.

McGurrin, J., D. Ubert, and D. Duff. 1995. Use of cultured salmonids in the federal aid in sport fish restoration program. *American Fisheries Society Symposium: Uses and effects of cultured fishes in aquatic ecosystems* 15:12–15.

McKinnell, S., H. Lundquist, and H. Johansson. 1994. Biological characteristics of the upstream migration of naturally and hatchery-reared Baltic salmon, *Salmo salar L. Aquaculture and Fisheries Management* 25(Suppl. 2):45–63.

Meffe, G. K. 1992. Techno-arrogance and halfway technologies: Salmon hatcheries on the Pacific coast of North America. *Conservation Biology* 6:350–354.

Messmer, R. T., R. W. Carmichael, M. W. Flesher, and T. A. Whitesel. 1992. Evaluation of Lower Snake River Compensation Plan facilities in Oregon. Oregon Department of Fish and Wildlife. Portland, Oregon.

Miller, L. M. and A. R. Kapuscinski. 2003. Genetic guidleines for hatchery supplementation programs. Pages 329–355 in E. M. Hallerman ed. *Population genetics: Principles and applications for fisheries scientists.* American Fisheries Society, Bethesda, Maryland.

Montgomery Watson. 1996. Hatchery evaluation report: Bonneville Hatchery–URB Fall Chinook. Bonneville Power Administration. BPA 95AC49468. Portland, Oregon.

Myers, R. A., S. A. Levin, R. Lande, F. C. James, W. W. Murdoch, and R. T. Paine. 2004. Hatcheries and endangered salmon. *Science* 303:1980.

National Fish Hatchery Review Panel. 1994. U. S. Fish and Wildlife Service National Fish Hatchery Review. The Conservation Fund, The National Fish and Wildlife Foundation. Report. Arlington, Virginia.

National Research Council (NRC). 1996. Upstream: Salmon and society in the Pacific Northwest. Report on the Committee on Protection and Management of Pacific Northwest Anadromous Salmonids for the National Research Council of the National Academy of Sciences. National Academy Press, Washington DC.

Needham, P. R. 1939. Natural propagation versus artificial propagation in relation to angling. Pages 326–331 in *Fourth North American Wildlife Conference, February 13-15.* American Wildlife Institute, Washington, DC.

Nickelson, T. E. 1986. Influences of upwelling, ocean temperature, and smolt abundance on marine survival of coho salmon (*Oncorhynchus kisutch*) in the Oregon production area. *Canadian Journal of Fisheries and Aquatic Sciences* 43:527–535.

Nickelson, T. E., M. F. Solazzi, and S. L. Johnson. 1986. Use of hatchery coho salmon (*Oncorhynchus kisutch*) presmolts to rebuild wild populations in Oregon coastal streams. *Canadian Journal of Fisheries and Aquatic Sciences* 43:2443–2449.

Nickelson, T. E., J. W. Nicholas, A. M. McGie, R. B. Lindsay, D. L. Bottom, R. J. Kaiser, and S. E. Jacobs. 1992. Status of anadromous salmonids in Oregon Coastal basins. Oregon Department of Fish and Wildlife, Portland, Oregon.

Northwest Power Planning Council (NPPC). 1987. 1987 Columbia River Basin Fish and Wildlife Program. Portland, Oregon. 246 p.

Northwest Power and Conservation Council (NPCC). 2003. Artificial production review and evaluation. Draft basin-level report, Document 2003-17. Portland Oregon.

NPPC. 1992. Columbia River Basin Fish and Wildlife Program 92-21A. Portland, Oregon.

NPPC. 1994. Columbia River Basin Fish and Wildlife Program 94-55. Portland, Oregon.

NPPC. 1999. Artificial Production Review: Report and recommendations of the Northwest Power Planning Council. Northwest Power Planning Council. NPPC 99-15. Portland, Oregon. 245 p.

Olson, D. E., B. C. Cates, and D. H. Diggs. 1995. Use of a national fish hatchery to complement wild salmon and steelhead production in an Oregon stream. *American Fisheries Society Symposium: Uses and effects of cultured fishes in aqautic ecosystems* 15:317–328.

Oregon Department of Fish and Wildlife. 1982. Comprehensive plan for production and management of Oregon's anadromous salmon and trout. Part II: Coho salmon plan. Portland, Oregon.

Oregon Department of Fish and Wildlife and Washington Department of Fish and Wildlife. 1995. Columbia River Fish runs and fisheries. Portland, Oregon. 291 p.

Oregon Department of Fisheries. 1908. Annual report to the 25th legislative session. Salem, Oregon.

Oregon Fish Commission. 1933. Biennial report of the Fish Commission of the State of Oregon to the Governor and the 37th legislative assembly, 1933. Salem, Oregon.

Oregon Fish Commission. 1939. Biennial report to the 40th legislative assembly. Salem, Oregon.

Oregon Fish Commission. 1964. Biennial report, July 1, 1962 -June 30, 1964, to the Governor and the 53rd legislative assembly. Salem, Oregon.

Oregon Fish and Game Commission. 1919. Biennial Report of the Fish and Game Commission of the State of Oregon to the Governor and the thirteenth legislative assembly. Portland Oregon.

Ortmann, D. W., F. Cleaver, and K. R. Higgs. 1976. Artificial propagation. Pacific Northwest Regional Commission, Portland, Oregon.

Pacific Fishery Management Council. 1992. Review of the 1991 Ocean Salmon Fisheries. Pacific Fishery Management Council, Portland, Oregon.

Philipp, D. P., and J. E. Clausen. 1995. Fitness and performance differences between two stocks of largemouth bass from different river drainages within Illinois. *American Fisheries Society Symposium: Uses and effects of cultured fishes in aquatic ecosystems* 15:236–243.

Philipp, D. A., J. M. Epifanio, and M. J. Jennings. 1993. Conservation genetics and current stocking practices. *Fisheries* 18:14–16.

Pollard, H. A. and T. A. Flagg. 2004. Guidelines for use of captive broodstocks in recovery efforts of Pacific salmonids. *American Fisheries Society Symposium: Propogated fish in resource management* 44:329–341.

Powell, M. S. and J. C. Faler. 2004. Geneitc analysis of Redfish Lake sockeye (*Oncorhynchus nerka*). Completion Report. Bonneville Power Administration, Portland Oregon. 79 pp.

Regional Assessment of Supplementation Project. 1992. Supplementation in the Columbia basin. Bonneville Power Administration. Report, Contract DE-AC06-75L01830. Portland, Oregon.

Reisenbichler, R., and S. R. Phelps. 1989. Genetic variation in steelhead (*Salmo gairdnerii*) from the North Coast of Washington. *Canadian Journal of Fisheries and Aquatic Sciences* 46:66–73.

Reisenbichler, R. R. 1995. Questions and partial answers about supplementation – genetic differences between hatchery fish and wild fish. Pages 1–18 in *Proceedings of the Columbia River Anadromous Salmonid Rehabilitation Symposium*. Richland, Washington.

Reisenbichler, R. R., and J. D. McIntyre. 1977. Genetic differences in growth and survival of juvenile hatchery and wild steelhead trout, *Salmo gairdneri*. *Journal of the Fisheries Research Board of Canada* 34:123–128.

Reisenbichler, R. R., and S. P. Rubin. 1999. Genetic changes from artificial propagation of Pacific salmon affect the productivity and viability of supplemented populations. *ICES Journal of Marine Science* 56:459–466.

Rich, W. H. 1920. Early history and seaward migration of Chinook salmon in the Columbia and Sacramento Rivers. *Bulletin of the US Bureau of Fisheries* 37.

Rich, W. H. 1939. Fishery problems raised by the development of water resources. Pages 176–181 in *Dams and the problems of migratory fishes*. Fish Commission of the State of Oregon. Stanford University, Palo Alto, CA.

Rich, W. H., and H. B. Holmes. 1929. Experiments in marking young Chinook salmon on the Columbia River, 1916 to 1927. *Bulletin of the US Bureau of Fisheries*, Document No. 1047.

Ricker, W. E. 1972. Hereditary and environmental factors affecting certain salmonid populations. Pages 19–160 in R. C. Simon and P. A. Larkin, eds. *The stock concept in Pacific salmon.* University of British Columbia, Vancouver.

Rounsefell, G. A. 1947. The effect of natural and artificial propagation in maintaining a run of Atlantic salmon in the Penobscot River. *Transactions of the American Fisheries Society* 74:188–208.

Salo, E. O., and W. H. Bayliff. 1958. Artificial and natural production of silver salmon, *Oncorhynchus kisutch,* at Minter Creek, Washington. Washington Department of Fisheries. Research Bulletin, 4. Olympia, Washington.

Schwiebert, E. 1977. Some notes on the symposium. Pages 1–12 in E. Schwiebert, ed. *Columbia River Salmon and Steelhead, March 5-6, 1976.* American Fisheries Society, Washington, DC.

Soule, M. E., ed. 1987. *Viable populations for conservation.* Cambridge University Press, New York.

Stone, L. 1879. Report of operations at the salmon-hatching station on the Clackamas River, Oregon, in 1877. *Report of the Commissioner for 1877.* US Commission of Fish and Fisheries, Washington, DC.

Thurow, R. F., D. C. Lee, and B. E. Rieman. 1997. Distribution and status of seven native salmonids in the interior Columbia River basin and portions of the Klamath River and Great basins. *North American Journal of Fisheries Management* 17:1094–1110.

Thurow, R. F., D. C. Lee, and B. E. Rieman. 2000. Status and distribution of Chinook salmon and steelhead in the interior Columbia River basin and portions of the Klamath River basin. Pages 133–160 in E. C. S. E. Knudsen, D. Mac Donald, J. Williams, and D. Reiser, eds. *Sustainable fisheries management: Pacific salmon.* CRC Press, Boca Raton.

Utter, F. M. 2002. Kissing cousins: Genetic interactions between wild and cultured salmon. Pages 119–135 in B. Harvey and M. MacDuffee, eds. *Ghost runs: The future of wild salmon on the North and Central coasts of British Columbia.* Rainforest Conservation Society, Victoria, British Columbia.

Utter, F. M., 2003. Genetic impacts of fish introductions. Pages 357–378 in E. M. Hallerman ed. *Population genetics: Principles and applications for fisheries management.* American Fisheries Society, Bethesda, Maryland.

Utter, F., G. Milner, G. Stahl, and D. Teel. 1989. Genetic population structure of Chinook salmon, *Oncorhynchus tshawytscha,* in the Pacific Northwest. *Fishery Bulletin* 87:239–264.

Utter, F. M., D. W. Chapman, and A. R. Marshall. 1995. Genetic population structure and history of Chinook salmon of the upper Columbia River. *Evolution and the aquatic ecosystem: Defining unique units in population conservation* 17:149–168.

Van Cleve, R., and R. Ting. 1960. The condition of salmon stocks in the John Day, Umatilla, Walla Walla, Grande Ronde, and Imnaha Rivers as reported by various fisheries agencies. Department of Oceanography, University of Washington. Seattle.

Van Oosten, J. 1942. Relationship between the plantings of fry and production of whitefish in Lake Erie. Pages 119–121 in *American Fisheries Society 71st Annual Meeting.* St. Louis, Missouri.

Vreeland, R. R. 1988. Evaluation of the contribution of fall Chinook salmon reared at Columbia River hatcheries to the Pacific salmon fisheries. Bonneville Power Administration, Report No. DOE/BP-39638-4. Portland, Oregon.

Wahle, R. J., and R. R. Vreeland. 1978. Bioeconomic contribution of Columbia River hatchery fall Chinook salmon, 1961 through 1964 broods, to the Pacific salmon fisheries. *Fisheries Bulletin* 76:179–208.

Wahle, R. J., R. R. Vreeland, and R. H. Lander. 1974. Bioeconomic contribution of Columbia River hatchery coho salmon, 1965 and 1966 broods, to the Pacific salmon fisheries. *Fisheries Bulletin* 72:139–169.

Wallis, J. 1960. Recommended time, size and age for release of hatchery reared salmon and steelhead trout. Fish Commission of Oregon. Clackamas, Oregon.

Waples, R. S. 1991. Genetic interactions between hatchery and wild salmonids: Lessons from the Pacific Northwest. *Canadian Journal of Fisheries and Aquatic Sciences* 48:124–133.

Waples, R. S. 1999. Dispelling some myths about hatcheries. *Fisheries* 24:12–21.

Waples, R. S., and J. Drake. 2004. Risk/benefit considerations for marine stock enhancement: a Pacific salmon perspective. Pages 260–306 in K. M. Leber, S. Kitada, H. L. Blankenship, and T. Svasand, eds. *Stock enhancement and sea ranching: Development, pitfalls, and opportunities.* Blackwell Publishing, Oxford, United Kingdom.

Waples, R. S., and P. E. Smouse. 1990. Gametic disequilibrium analysis as a means of idenitfying mixtures of salmon populations. *American Fisheries Society Symposium* 7:439–458.

Washington, P. M., and A. M. Koziol. 1993. Overview of the interactions and environmental impacts of hatchery practices on natural and artificial stocks of salmonids. *Fisheries Research* 18:105–122.

Washington Department of Fish and Game. 1917. Twenty-sixth and twenty-seventh annual reports of the State Fish Commission to the Governor of the State of Washington. Olympia, Washington.

Washington Department of Fisheries, Washington Department of Wildlife, and Western Washington Treaty Indian Tribes. 1993. 1992 Washington State salmon and steelhead stock inventory. Olympia, Washington.

Washington Department of Fisheries and Game. 1904. Fourteenth and fifteenth annual report of the State Fish Commissioner to the Governor of the State of Washington. Seattle, Washington.

White, R., J. R. Karr, and W. Nehlsen. 1995. Better roles for fish stocking in aquatic resource management. *American Fisheries Society Symposium: Uses and effects of cultured fishes in aqautic ecosystems* 15:527–547.

Williams, R. N., D. K. Shiozawa, J. E. Carter, and R. F. Leary. 1996. Genetic detection of putative hybridization between native and introduced rainbow trout populations of the Upper Snake River. *Transactions of the American Fisheries Society* 125:387–401.

Williams, R. N., J. A. Lichatowich, P. R. Mundy, and M. Powell. 2003, September 4. Integrating artificial production with salmonid life history, genetic, and ecosystem diversity: A landscape perspective. Issue Paper for Trout Unlimited, West Coast Conservation Office, Portland, Oregon.

Wood, E. M. 1953. A century of American fish culture, 1853-1953. *The Progressive Fish Culturist* 15:147–160.

9

Harvest Management

Phillip R. Mundy, Ph.D.

Introduction

Historical Context
  Introduction to the History of the Chinook Fishery
  Prior to 1866
  First Era: 1866–1883
  Second Era: 1884–1920
  Third Era: 1921–1931
  Fourth Era: 1932–1952
  Fifth Era: 1953–Present

Problems with the Old Paradigm
  Introduction
  Lack of Escapement-based Harvest Management
  Failure to Fully Account for Fishing Mortality

Conclusions for Harvest

A New Pacific Salmon Harvest Management Paradigm
  Introduction to the New Paradigm
  The Extended Salmon Fishery Model
  A Sustainable Salmon Fishery Policy
  Principles of Sustainable Salmon Fishing
    Protect Wild Salmon and Their Habitats
    Maintain Spawning Escapements
    Harvest with Caution
    Establish and Apply an Effective Management System
    Maintain Public Support

Conclusions for Harvest and Harvest Management
  Conclusions (and Levels of Proof)
  Recommendations for Harvest and Harvest Management

Literature Cited

> "The way in which the Chinook salmon runs have held up under the excessive exploitation and a constant reduction in the available spawning area is remarkable."
> —Willis H. Rich, 1941. The present state of the Columbia River salmon resources. In *Sixth Pacific Science Congress*. Fishery Commission of the State of Oregon, Berkeley (p. 429).

Introduction

Harvest management of salmon is a complex mixture of science, politics, economics, and human psychology that is intended to mold human behavior along lines that enable the long-term persistence of salmon (Mundy 1996). In the Columbia River Basin, harvest management has not been able to provide for the long-term persistence of wild salmon, due to both lack of support from governmental institutions (Lichatowich 1999; Mundy 1999) and the inability of harvest managers to understand, let alone to measure, the many different sources of anthropogenic mortality that occur over a very large geographic range throughout the life cycle. As is the case for many other exploited fish species in other parts of the world (Alverson 2002), governmental institutions in the Columbia River Basin historically have subordinated conservation interests of even very lucrative commercial species and populations to other societal objectives, such as agriculture and hydroelectric power production (White 1996; Taylor 1999). At the same time, harvest managers have based their operational decisions on information derived from adult harvests in the river proper, passage of adults at hydroelectric dams, and levels of hatchery production, while largely ignoring other parts of the life cycle and attributes of wild populations.

As a consequence of the lack of harvest management, the once abundant Columbia River salmon populations, which supported annual harvests in excess of one million Chinook salmon for 57 consecutive years (1876–1931) (Figures 9.1 and 9.2), have been reduced to a handful of arguably self-sustaining wild stocks, a somewhat larger number of federally protected spawning populations, and an even larger number of hatchery-supported stocks (see Chapter 8). Despite the hatchery efforts, Columbia River commercial, subsistence, and sports fisheries combined have landed more than 500,000 Chinook (10 million pounds) in only one year (1988) of the last 50 (1954–2003) (Oregon Department of Fish and Wildlife [ODFW] 2002; Columbia River Compact 2003b, 2003c, 2003d).

Current harvest management practices offer little to no respite from historical practices. After an 8-year hiatus (1993–2000), during which combined commercial and sports landings of spring Chinook salmon did not exceed 1,000 adults, spring Chinook salmon fisheries returned to the Columbia River in 2001 and persisted through 2003 in response to a bonanza of hatchery returns (Columbia River Compact 2003b, 2003c).

The re-opening of salmon fisheries in 2001 to 2003, despite the presence of small wild salmon populations listed as threatened or endangered by the federal government in the harvest areas, serves to underscore the main thesis of this chapter.

The combination of institutional neglect of internationally recognized conservation principles and the lack of full accounting for anthropogenic

Figure 9.1 Historical scenes of salmon harvesting including seining by hand or with horses, a day's catch, and a salmon cannery. Photos from the Oregon Public Library Historical Photo Collections from the period of 1879–1909. Photographer unknown.

sources of mortality continue to leave the future of wild salmon populations in the Columbia River Basin in doubt. A new harvest management paradigm that respects internationally recognized principles of sustainable harvest, and which includes an accounting of all major sources of mortality throughout the life cycle, is required for wild salmon populations to survive, and to once again be a meaningful part of human commerce and culture.

The current (2001–2003) fisheries, targeting all types of Chinook salmon from the Columbia River, serve to introduce the old salmon harvest management paradigm and why it needs to be replaced. The willingness to sacrifice vulnerable wild salmon stocks in order to harvest the bountiful hatchery returns of 2001 to 2003, and especially 2002, follows a long-established harvest management formula that has frequently led to disaster for conservation of wild salmon stocks in the Columbia River and elsewhere in the Pacific Northwest (Bottom 1996; Lichatowich 1999; Mundy 1999; Taylor 1999; see also Chapter 8). Very large returns of adults to Columbia River salmon hatcheries have once again motivated a harvest management institution, in this case the Columbia River Compact, to sanction fishing in waters known to contain threatened and endangered wild salmon populations. For example,

Figure 9.2   Fish wheel and harvest boat on the lower Columbia River. Fish wheel systems were so efficient at harvesting upriver migrating Chinook salmon that they were eventually banned. Photos from the Oregon Public Library Historical Photo Collections from the period of 1879–1909. Photographer unknown.

the Interim Management Agreement governing salmon fisheries in the Columbia River covering the time period 2001 to 2003 permits mortalities to salmon populations listed under the Endangered Species Act. Permissible impacts to listed upriver spring Chinook in 2003 were 2% for non-Indian fisheries and 10% for treaty Indian fisheries (Columbia River Compact 2003a). Unfortunately, the actual impact of the fisheries operated under this agreement on the number of listed salmon that are able to spawn successfully is not known until several weeks to several months after the fisheries have closed, if then.

As illustrated by the 2001 to 2003 salmon fisheries agreements (Columbia River Compact 2003a), the old harvest management paradigm needs to be replaced because it risks even critically damaged wild stocks in the process of harvesting hatchery stocks without actually knowing how the fisheries will impact future production. In so doing, it takes into account only a few somewhat flawed estimates of anthropogenic sources of mortality. A new harvest management paradigm will be fully developed below; however, the reader first needs to be grounded in the history of the Columbia River Basin salmon fisheries and production and then to acquire an understanding of the most pressing problems of the current harvest management paradigm.

Historical Context

Introduction to the History of the Chinook Fishery

The intertwined processes of destruction of the habitat base and harvest beyond the productive capacities of the stocks (overfishing) may have started with the first human contact, as has been the apparent history of so many other natural resources throughout the world (Jackson et al. 2001). In any event, written historical records support the idea that loss of Columbia River Chinook as a viable resource capable of supporting the regional economy, subsistence users, and the recreational experience was a very long process that was well underway by the end of the 19th century (Mundy 1996; Taylor 1999). Salmon harvest managers were in the unfortunate position of working with increasingly damaged and often declining Chinook populations from about 1884 onward, as some appeared to have understood at the time (see Willis Rich quote at beginning of chapter; Craig and Hacker 1940). From before the beginning of the 20th century, salmon harvest management was shoveling sand against the inexorable tide created by agricultural and industrial forces. For as long as written records have been kept, agriculture and industry have been converting the Columbia from the largest Chinook salmon producing river the world has ever seen into an "organic machine" capable of mass production of electricity, food, and minerals (White 1996).

Although the Columbia River historically produced commercial quantities of five salmon species—chum, coho, sockeye, steelhead, and Chinook—the Columbia River has always been legendary for its Chinook. Hence, the fishery for the icon of wild salmon bearing rivers in North America must be the bellwether for tracing the history of mortalities due to habitat loss and fishing. Mundy (1996) divided the history of the Chinook fishery into five eras with starting years of 1866, 1884, 1921, 1932, and 1953, based on the 5-year moving average of the annual landings (Figure 9.3). Some caveats need to be offered on the interpretation of landing observations before the history may be told.

While the annual landing observation is a reliable means of describing the history of a salmon fishery, it is not a consistent estimator of fish population size (Quinn and DeRiso 1999). Annual landings change for reasons that have little to do with changes in the numbers of salmon available to be harvested, such as collapse of markets, labor disputes, and even misidentification of the species involved. It is also important to keep in mind that landings do not account for all fish killed by fishing, as some fraction of the population is killed by fishing gear and then lost before it can be secured. As a consequence, care must be taken in making inferences about changes in the size of the population based on changes in landings.

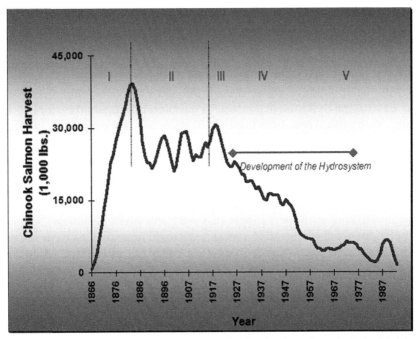

Figure 9.3 Five-year moving average of landings of Chinook salmon from the Columbia River in millions of pounds, 1866 to 1993. Starting about 1950, the proportion of landings of hatchery origin fish increased to become predominant after about 1980. Averages based on landings after 1993 are not presented because they are not even arguably representative of the status of wild Chinook salmon populations, as the fishery was primarily targeted on hatchery stocks, particularly in the spring and summer.

An additional caveat is that landings as total weight is to be preferred over numbers of Chinook because of the practice of excluding 3-year-old Chinook, known as jacks, from the count of numbers landed and because the size at age changes annually, so that a number of Chinook this year may weigh much more (or less) than the same number weighed last year. Numbers reported here are converted from weight by assuming a 20-pound (9.08 kg) average individual weight, unless otherwise noted. Note that Chapman (1986) used 10.45 kg for spring/summer Chinook. Because the historical landings are reported in pounds, and because the landings here include fall Chinook, which may be smaller than spring/summer Chinook, the 20-pound per Chinook figure was chosen for convenience in converting number of pounds to number of individuals. When numbers landed represent actual counts, a literature citation for the source is given.

A final caveat is that the annual Chinook landings (Figure 9.3) do not provide wild stock composition and hatchery contribution information, even though these significant attributes have changed substantially over the history of the Chinook fisheries. Stock composition of Chinook is usually based on the time of year the adults enter freshwater to spawn: spring, summer, and fall (see Chapter 4 for additional discussion). Stock composition of landings is significant for management because it is associated with important biological differences such as location of spawning grounds, tolerance for water temperatures on spawning grounds, and age at saltwater entry (Taylor 1990).

In considering the landings data, the reader should bear in mind that the dominant stock composition changed from wild summer stocks (1883) to wild spring-fall stocks (1920) to wild fall stock (1931) to wild fall-hatchery fall (1952) to hatchery spring-fall (present) (Craig and Hacker 1940; Lichatowich 1999; Van Hyning 1973; Thompson 1951, as cited in Chapman 1986; Chapter 8). These five eras of stock composition ending 1883, 1920, 1931, 1952, and present also correspond to the five eras defined by the 5-year moving average of landings, discussed below.

Prior to 1866

The history of the Chinook salmon resources prior to the imperfect image captured by the commercial fishery starting in 1866 is uncertain. Although one would expect that Chinook populations were very robust and minimally exploited, it should be noted that indigenous peoples in other parts of the world were apparently able to substantially alter, and to possibly irreversibly damage, the resources on which they subsisted (Jackson et al. 2001). Craig and Hacker (1940) estimated annual aboriginal harvest at 18.2 million pounds of Chinook (about 900,000 individuals), which is remarkably close to the 5-year moving average minimum yield of one million individuals that was sustained by the commercial fishery from 1876 to 1931 (Figure 9.1). Somewhat higher levels of aboriginal salmon harvests were estimated by Schalk (1986), based on estimates of aboriginal population size; however, many of the aboriginal peoples had perished in epidemics beginning shortly after the first sustained western contact in the early 19th century, so pre-contact estimates are necessarily imprecise.

First Era: 1866–1883

By the end of the first era (1866–1883), annual landings were increasing toward the 40 million pound level (2 million fish), based on apparent growth in effort, addition of gear types, refinement within gear types, increasing skill of the harvesters, and development of canning technology, among other factors (Lichatowich 1999; Taylor 1999). During the 1870s, overfishing of the salmon runs in the lower river by the commercial fishery may have limited subsistence use of salmon by Native American harvesters (Simms 1877). Species identification and mixing of species in reporting of landings cast some doubt on the overall accuracy of observations (Craig and Hacker 1940), and it is possible that the largest reported annual landing in 1883 could have included other species, such as steelhead. Note that the annual landing figure for 1883 was included in 5-year moving averages for each of the years, from 1883 to 1887 (Figure 9.3).

Second Era: 1884–1920

Annual landings at the end of the second era were on the order of 30 million pounds (1.5 million fish). From 1884 until the end of the second era in 1920, the fishery was working at an apparent annual equilibrium landings level on the order of 25 million pounds (1.25 million fish). Although the total Chinook landings oscillated about 1.25 million individuals, the stock composition of the landings was changing, with availability of late spring and early summer runs declining, and exploitation of early spring and late summer (fall) Chinook increasing to make up the difference (Thompson 1951, as cited in Chapman 1986).

The gradual decline of the 5-year moving average to lower levels during the second era, starting about 1884, appears also to have been influenced by declining salmon markets, reduced fishing effort, and substantial loss and degradation of spawning and rearing habitat from activities such as splash dam logging. Hydroelectric dam building started in 1888. A number of authors point to overfishing as a factor in the decline of Columbia River Chinook during and before the second era. The declines in nominal landings per unit effort of spring and summer Chinook between 1876 and 1919 documented by William Francis Thompson strongly support the concept of declines in actual Chinook population size during this time period (Thompson 1951). Chapman (1986) cited Thompson as one basis for the conclusion that overfishing in the late 19th and early 20th centuries contributed to the decline of Chinook.

Although the overall landing observations do not actually support the finding, salmon fishery historian Anthony Netboy (1974) related the widely held belief that Chinook were overfished, and in "radical decline," following the 1885 season (pp. 282–283). His reference to radical decline may have been to a specific stock that was the mainstay of the first era fisheries, the "June hogs," which were replaced in the fishery by earlier (spring) and later (fall) running stocks by about 1920.

Then as now, upriver biologists were looking to the lower river fisheries to explain sharp declines in salmon (particularly spring Chinook and sockeye) returns to the Yakima River in Washington state (McDonald 1894). Note that harvest rates or escapements from the commercial fisheries of the lower Columbia River were not available to McDonald (1894).

Third Era: 1921–1931

In 1921, the 5-year moving average of Chinook landings fell below an annual harvest of 30 million pounds (1.5 million Chinook). The fall below 30 million pounds is taken here as the point of origin for the long-term slide into extirpation that continues today. In 1922, it appears that the cumulative effects of habitat loss and degradation, combined with ineffective governmental institutions and harvest management regimes, placed wild salmon population numbers below the critical point of replacement. Additionally, this is the first era in which commercial harvest of Chinook in coastal ocean troll fisheries (hook and line) would have contributed consistently to apparent declines in harvest as measured in the river (see Van Hyning 1973). Declines in marine productivity as a result of climate cycles may have also exacerbated the effects of habitat loss and overfishing (Ware and Thompson 1991; Mantua et al. 1997; Hare et al. 1999).

Starting in the third era and continuing until now, it appears habitat degradation accelerated to increasingly reduce salmon productivity, which would have caused harvest managers of the time to consistently overestimate the sustainable levels of annual harvest, leading to overfishing. Subsequently, habitat loss and overfishing combined to cause long-lasting reductions in landings (Craig and Hacker 1940; Rich 1941; Van Hyning 1973). The rate of decline from 1921 to 1931 was similar to the rate of the first notable decline of 1884 to 1900, and followed three sharp rebounds (Figure 9.3). Real reductions in Chinook population levels are likely to have occurred in the third era, because increased fishing effort was used to harvest fewer fish (Craig and Hacker 1940).

Fourth Era: 1932–1952

The fourth era was clearly a time of negative changes for the Chinook fishery. Five-year moving average of landings crossed below the 20 million pound mark (1 million Chinook) in 1932. Rock Island Dam, the first of the large main river "low" dams with fish passage went into operation in 1933, followed in 1938 by another low dam which is closest to the mouth of the Columbia River, Bonneville, and then by the first main river "high" dam, Grand Coulee, in 1941. High dams lack salmon passage, which effectively eliminates the productive capacity of habitats upriver from them. Beginning in 1938, levels of harvest in the ocean and river combined to create overfish-ing (Van Hyning 1973). Note that Van Hyning's (1973) approach used indi-rect measures of the effects of habitat degradation, as well as landings and levels of fishing effort.

The 1930s was the last time when salmon fisheries could be said to be a mainstay of the local economies. About 3,800 harvesters landed fish that put $10 million annually into the local economies. It was recognized at the time that overfishing was only one of the problems causing a decline in landings, and it was generally acknowledged that maintenance of spawning and rear-ing habitats was important to the success of conservation efforts. However, it is interesting to note that Johnson et al. (1948) did not express particular con-cern about trends in escapement of Chinook as of 1935. Such concerns emerged in the literature during the 1950s, especially with respect to spring Chinook (Thompson 1951).

In addition to overfishing, all of the usual suspects—human population growth, logging, mining, hydroelectric power, and flood control and naviga-tional improvements—were invoked by authorities to explain declines in salmon landings that were increasingly coming from fall Chinook stocks (Craig and Hacker 1940; Rich 1941). Netboy (1974) listed the presence of over 300 dams in 1948 in support of his thesis that habitat loss was a factor in decline of salmon during this era (p. 285). The combination of overfishing and habitat loss contributed to a marked downturn in the 5-year moving average of landings throughout most of this era (Figure 9.3).

Fifth Era: 1953–Present

At the start of the current, or fifth, era, the 5-year moving average Chinook landings fell below the 10 million pound mark (500,000 Chinook, estimated), not to return to date. The Columbia River commercial harvest of Chinook in 1988 was 10.54 million pounds (489,000 individuals, actual), but the 5-year moving average in 1988 was lower, due to lower landing figures before and after 1988. The combined sports and commercial harvests of Chinook exceeded 500,000 in 1988. The present era experienced a fivefold increase over the fourth era in dam building and operations. Starting with McNary Dam operations in 1953, 15 Columbia and Snake River dams were built and put into operation from 1953 to 1975. Attempts to produce salmon in hatcheries increased sharply after World War II to slow the decline in landings during the current era (Chapter 8).

The recent 2001 to 2003 upswing in Chinook total returns, not shown in Figure 9.3, would not register very strongly on the 5-year moving average of landings for two reasons. First of all, harvests were relatively modest in these years, even though estimated total returns (landings plus escapement) topped one million (mostly hatchery fish) in both years. Secondly, the landings were at such low levels during the 1990s that the large returns of 2001 to 2003 would have to have been harvested at high rates to reach levels of landings that would change the 5-year moving average substantially upward. Due to federal restrictions imposed under authority of the Endangered Species Act, such high harvest rates were not applied. For these reasons, the years 2001 to 2003 are classified as part of the era that began in 1953, which continues to be marked by some of the smallest landings in the history of the Columbia River Chinook fishery.

Problems with the Old Paradigm

Introduction

The paradox of small harvests in the midst of million-salmon runs is the ultimate legacy of the old salmon harvest management paradigm. Threatened and endangered wild salmon populations are buried in a tidal wave of hatchery returns while all harvesters sit largely idle and Treaty Indian fishing rights go unfulfilled. The problems with the old paradigm are many, but there are two that are particularly serious.

For any harvest management regime to be effective, it must be supported by governmental institutions that consistently bring regulation and enforcement in line with its principles (National Research Council [NRC] 1999). Leaving the issue of institutional support aside, there are two fatal flaws evident in the current principles of the Columbia River salmon management paradigm (i.e., Columbia River Compact 2003a, 2003c, 2003d) that would prevent it from succeeding under any circumstances. First, counts of salmon passing upriver of certain key hydroelectric dams, such as Bonneville, McNary, and Lower Granite, are erroneously interpreted as counts or estimates of spawning escapements for the purposes of harvest control. Such observations cannot be considered counts or consistent estimators of spawning escapements because these observations aggregate hundreds of spawning sites with inherently different characteristics, occur hundreds of miles from some spawning sites, and are made weeks to months before actual spawning occurs. The confusion of dam passage counts with spawning escapements has left harvest managers with no consistent estimator of future stock production that can be applied as a criterion on which to base levels of harvest during the course of fishing. In sustainable salmon fishing, harvest decisions are based on an estimate of the impact of levels of harvest on future production.

The second fatal flaw in the present Columbia River salmon harvest management paradigm is a lack of accounting for all significant sources of mortality. The lack of accounting is best illustrated by the steadfast refusal to consider that all types of Chinook salmon, including spring, are subject to anthropogenic mortalities in the ocean, including fishing mortality. Despite strong circumstantial evidence to the contrary, harvest managers continue to assume that spring Chinook are not killed by fisheries or other human activities in the ocean. As explained below, in the absence of studies to confirm or discount the available evidence, it is reasonable and precautionary (Garcia 1994) to assume that spring Chinook could be killed in substantial numbers by ocean fisheries, even though they rarely are reported as landings.

Lack of Escapement-based Harvest Management

Harvest management of Columbia River Chinook populations has failed to reverse the declines in salmon runs because the two principal harvest control entities do not provide harvest regulations which explicitly provide for salmon spawning escapements to individual tributaries; that is, they do not manage according to the productive capacities of the individual stocks (Paulik 1969). Salmon harvest regulations under the Columbia River Fisheries Management Plan (CRFMP) (U.S. Federal District Court, Portland, Oregon), is implemented through the Columbia River Compact (2003a) by state and tribal fisheries managers. The management actions of the CRFMP, as coordinated through the Pacific Fishery Management Council (PFMC), Portland, Oregon (PFMC 1992) and Pacific Salmon Commission (PSC) (Jensen 1986; PSC 1993) provide for aggregated spawning escapements to large river counting sites, such as hydroelectric dams, and not to tributary spawning grounds (PFMC 1992).

In addition to the lack of watershed-specific spawning escapement objectives, the harvest management of Columbia River Chinook salmon under the freshwater Columbia River Compact and its marine coordinating entities, the PFMC and the PSC, fails to account for escapements of earlier life cycle stages from the gantlet of fisheries encountered during migration (e.g., see Figure 9.4 and discussion below). Put simply, the impacts of PSC and PFMC marine harvests on the overwhelming majority of naturally spawning Columbia River Basin salmon stocks are not directly measured, and the estimates of impacts are based on the assumption that hatchery stocks behave identically to wild stocks, which is virtually impossible to test with current technologies. Prior to about 1985, only hatchery salmon were routinely annually tagged with coded wire in numbers sufficient to permit their identification in the landings of ocean fisheries. After 1985, routine annual tagging programs were extended to some natural stocks, so that contributions of several naturally spawning Columbia River Chinook stocks to ocean salmon landings can now be measured. Unfortunately, contributions to ocean landings of most naturally spawning salmon populations are not actually

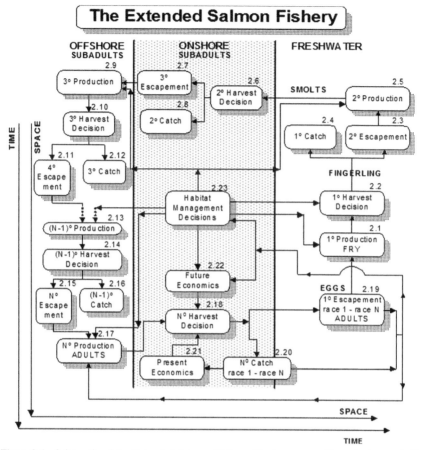

Figure 9.4 Schematic of the Extended Salmon Fishery. The term, races (demes), denotes individual spawning aggregates, and 1° Production (first degree salmon production), denotes the first life cycle stage beyond the egg (fry) (Mundy 1998 after Mundy and Mathisen 1981).

measured, but estimated from landings of other, mostly hatchery, stocks that have been tagged with coded wire.

Many salmon hatchery stocks, and some naturally spawning stocks, such as the Columbia River Hanford Reach fall Chinook, are tagged as juveniles with coded wire and fin clipped so that they can be identified in samples of fishery landings. Although hatchery stocks may be appropriate biological entities from which to infer the impacts of PSC fisheries on some naturally spawning populations, it seems unlikely that hatchery salmon stocks can be valid proxies for each and every natural salmon population of concern. For example, annual variations in oceanic distribution and migratory timings of life cycle stages are but two attributes for which differences between hatchery and natural populations could render any indices of fishing mortality (proportional to escapement), which are based on hatchery populations, invalid for naturally spawning populations. Further, even for a population that is tagged, if the life cycle is such that the landings of the individuals of legal size are out of proportion to the actual catches (total killed by fishing) from the population, then the indices of fishing mortality, and understanding of escapements of individual stocks from ocean fisheries, will also be invalid.

Failure to Fully Account for Fishing Mortality

The second shortcoming, the disparity between catch and landings in the PSC salmon fisheries and PFMC non-PSC salmon and groundfish fisheries (Table 9.1) is relevant to the Columbia River salmon recovery because mortalities due to catch and bycatch of Columbia River Chinook salmon populations in PSC and PFMC fisheries have not been routinely estimated. Catch is a measure of the number of salmon actually killed, whereas landings measure the number of salmon actually kept on the vessel. Landings is a fraction

Table 9.1 Annual estimates of total landings, incidental catch by fishery category; shaker, legal, sublegal, total catch, total incidental catch, total incidental catch per total landing, I/L, and percent of incidental catch in the total catch, of number of Chinook salmon in adult equivalents, for all Pacific Salmon Commission fisheries, 1979–1992. Computed from data on pages K–1 through K–14 of Pacific Salmon Commission (1992).

					Type of Fishery			
	Retention		Non-Retention			Incidental Mortality		
Year	Landed	Shaker	Legal	Sublegal	Catch	Incident	Index I/L	Percent
1979	2365600	301995	0	0	2667595	301995	0.127661	11.32
1980	2251730	294866	0	0	2546596	294866	0.130951	11.58
1981	2189445	303828	4076	3032	2500381	310936	0.142016	12.44
1982	2287289	368901	23770	18315	2698275	410986	0.179683	15.23
1983	2205210	352261	29489	22839	2609799	404589	0.183470	15.50
1984	2186297	337119	31160	23640	2578216	391919	0.179262	15.20
1985	1851845	233542	41140	57518	2184045	332200	0.179389	15.21
1986	1926438	276115	27723	35470	2265746	339308	0.176132	14.98
1987	2050465	304586	57044	62858	2474953	424488	0.207020	17.15
1988	2114972	291768	34880	66431	2508051	393079	0.185855	15.67
1989	1741698	274492	42939	50345	2109474	367776	0.211159	17.43
1990	1740361	300181	36512	52113	2129167	388806	0.223405	18.26
1991	1584825	314182	49235	61889	2010131	425306	0.268361	21.16
1992	1583080	358163	62216	70382	2073841	490761	0.310004	23.66

of the number of fish killed in any fishery, and the difference between catch and landings is often called bycatch, or incidental mortality. In the PSC hook and line fisheries, a regulation requires the release of fish under a minimum size limit (shakers), and in some hook and line and net fisheries, no Chinook salmon of any size are allowed to be retained (Chinook non-retention, legal sized and sublegal sized; Table 9.1). Not all of the fish so released are expected to live, so the reported number of salmon landed is necessarily an underestimate of the number of salmon actually killed.

To appreciate the magnitude of the potential impacts of ocean fisheries on salmon stocks of concern, and the disparities between catch and landings, some estimates are available from the 1992 report of the PSC Joint Chinook Technical Committee. For example, in 1992 it is estimated that the PSC sports fishery in the Strait of Georgia caught the equivalent of 233,509 adult Chinook salmon; however, the number reported landed for this fishery, also in adult equivalents, was only 126,922 (Pacific Salmon Commission 1993). Also in 1992, the PSC Chinook troll fishery in southeast Alaska reported landing the equivalent of 142,076 adult Chinook, but the total catch in this fishery was estimated to be the equivalent of 276,310 adult Chinook (Pacific Salmon Commission 1993). Note that the disparity between annual catch and landings figures will vary by fishery due to changes in the number of small salmon available to be caught. In the aggregate, annual incidental harvests of Chinook in PSC Chinook salmon fisheries from 1979 to 1992 ran from 294,866 to 490,761, which represented incidental harvests of 11% to 24% of total catch, with ratios of landings to incidental harvest ranging from approximately 9:1 to 3:1, as measured in adult equivalents. The 1979 to 1992 time trend in percentage incidentally harvested Chinook in PSC fisheries is decidedly positive.

Although Columbia River stream-type Chinook (spring Chinook and Snake River summer Chinook) are only a very small proportion of PSC Chinook landings (based on recoveries of coded wire tags applied to hatchery populations), the proportion of these populations represented in the PSC Chinook catch is unknown, as is also the case for non-PSC fisheries under the PFMC. Measurements of the stock composition by fishery of the PSC and PFMC Chinook catches have not been taken. Juvenile Chinook salmon, including spring Chinook salmon originating in the Columbia River Basin, are known from tagging studies to be available for harvest in the areas of some of the present PSC fisheries.

Given that the combined Canadian and U.S. PSC fisheries caught, but did not land, the equivalent of at least 294,866 to as many as 490,761 adult Chinook between 1979 and 1992 (Table 9.1), if the Columbia River stream-type Chinook constitute even 0.5% of these incidental harvests, the annual loss in adult equivalents to the Columbia River Basin would be between 1,500 and 3,000. Any such estimate of actual impacts of PSC fisheries on Columbia

River stream-type adult returns is necessarily speculative, due to the lack of stock composition data, and the impact of each fishery could be expected to vary substantially. The salmon bycatch in PFMC fisheries beyond the jurisdiction of the PSC for salmon and groundfish would add to the potential impacts of ocean fisheries on Columbia River, which are not presently being addressed by assessment programs. However, given only the estimated magnitude of PSC incidental Chinook harvests, the lack of stock composition information is a matter of serious concern to recovery of these types of salmon in the Columbia River Basin. Please note that the comments on PSC fisheries offered here do not apply to those salmon fisheries under the jurisdiction of the Fraser River Panel, which is a distinctly different management regime.

The relevance of incidental mortalities to recovery of listed Columbia River stream-type (spring/summer) Chinook is that the ocean fishery with the largest average catch had no stock identification based on sampling of the mortalities. For a very substantial portion, 12% to 24% in the years 1979 to 1992 (Table 9.2), of the total annual number of adult equivalent Chinook salmon taken by PSC ocean fisheries, there is no routine biological sampling at all. Hence, any assumption that the ocean fisheries have no effect on the various spawning populations of Columbia River spring/summer Chinook stocks is not based on data and therefore it is certainly not precautionary (Garcia 1994).

Data on the ocean distributions of spring/summer Chinook juveniles (Alverson et al. 1994) indicate that these stocks could be impacted by ocean

Table 9.2 Estimates of the percentage of Chinook salmon that die after being returned to the water by fishery type. Extracted from Alverson (1994).

Bycatch Species	Area	Fishery	Percent Mortality	
			Low/avg.	Upper
Chinook	North Pacific	Purse seine	50	90
Chinook	North Pacific	Troll	20	30
Chinook	Washington	Gillnet	2	28
Chinook, large	SE Alaska	Purse seine	24	
Chinook, medium	SE Alaska	Purse seine	68	
Chinook, small	SE Alaska	Purse seine	60	
Chinook, illegal	SE Alaska	Purse seine	50	90
Chinook, legal	GOA	Troll	25	
Chinook, legal	GOA	Troll, coho	20	
Chinook, legal	SE Alaska	Troll	8	13
Chinook, sublegal	GOA	Troll	28	
Chinook, sublegal	GOA	Troll, coho	24	
Chinook, sublegal	SE Alaska	Troll	19	28

GOA, Gulf of Alaska.

fisheries (Table 9.2). The likelihood of such impacts would depend on the timing and location of the fishery in relation to the ocean distribution of Chinook juveniles.

The known ocean distributions of juvenile Chinook should put them at risk of being caught by any fishery operating in the right place at the right time. When small salmon are returned to the water, a certain percentage of them are expected to die (Table 9.2). Hence, it is reasonable to expect ocean fisheries could substantially affect that spring/summer Chinook.

The working hypothesis is that juvenile stream-type (yearling emigrant) Chinook originating in the Columbia River Basin can be killed in ocean fisheries without being detected by current sampling programs. Present Compact, PFMC, and PSC management programs assume that juvenile stream-type Chinook occur at the same frequency in the incidental mortality (Table 9.2) as do the older, larger, legal-sized Chinook that are sampled in current programs (see Table 9.3). Such assumptions in the absence of data to support them are not prudent when working with recovery of listed salmon

Table 9.3 Number of years of marine fishery recoveries of coded wire tags by hatchery for stream-type (spring) Chinook salmon produced in the Snake River Basin. Fishery recoveries are from 1977 to 1996. For hatcheries producing more than one stock, the stock source is noted.

Hatchery	Fishery	Years
Dworshak	SEAK ocean troll	1
	OR ocean troll	1
Kooskia	CA ocean sport	1
Lookingglass	Imnaha River	
	CA ocean troll	1
	WA ocean troll	3
	WA ocean sport	1
	Non-Imnaha	
	SEAK ocean gill net	1
	No. BC ocean troll	1
	So. BC ocean troll	2
	OR ocean sport	1
	OR ocean troll	2
	CA ocean sport	1
Lyons Ferry	Tucannon River	
	No. BC ocean troll	1
	So BC mixed net	1
	WA treaty troll	2
Rapid River	No. BC ocean troll	1
Sawtooth	No. BC ocean troll	1
	Gulf AK groundfish	1

species. It is accurate to say that spring Chinook of harvestable size are not often sampled from the landings of ocean fisheries, but the landings are widely geographically distributed from the coast of California to the Gulf of Alaska (Table 9.3). Unfortunately, data on fishery landings of salmon of harvestable size do not necessarily apply to understanding the effect of those fisheries on salmon of less than legal size.

Consequently, present information about the occurrence of legal-sized stream-type (spring) Chinook in the incidental harvests of the ocean fisheries (Table 9.2) does not support the present assumption of little or no impact on the smaller salmon. The degree to which ocean fisheries may affect stream-type (spring/summer) Chinook of all sizes has not been systematically studied. There are substantial gaps in understanding the stock composition of salmon that are killed but not landed in ocean fisheries. Available research on ocean distribution of spring Chinook indicates that it is possible for juvenile spring Chinook to be killed by ocean fisheries. Coded wire tag recoveries of legal-sized spring Chinook (Table 9.3) confirm that spring Chinook are widely, if infrequently, harvested in ocean fisheries. A discussion of the data and literature follows.

Juvenile stream-type (spring) Chinook are found feeding in coastal ocean waters in April through October. Spring/summer juveniles are more oceanic than fall Chinook, which are more prominent in inside marine waters such as Puget Sound and the Inside Passage (Milne 1957; Hartt 1980; Hartt and Dell 1986; Healey 1991). There is a review of ocean distribution of juvenile Chinook (Healey 1991) that summarizes the geographically extensive catches of juvenile Chinook in purse-seining studies along the Pacific coast from the mouth of the Columbia River to Bristol Bay (Hartt 1980; Hartt and Dell 1986). The ocean distribution of juvenile Chinook was based on 3,073 sets during the 15-year period, 1956 to 1970. The seining was conducted from April to October, but concentrated in the summer months. Most young Chinook were caught during the first year of ocean life along the coast.

Catches were greatest from Columbia River to southeast Alaska, with smaller catches being made to the north and west. No Chinook were captured in the first year of ocean life in the central Gulf of Alaska or Bering Sea. Only 253 Chinook were caught in their first year of ocean life over the 15-year period, but peak catches were from June to August, and Chinook appeared in the catches of the north later in the year than in the south. All but eight of the first ocean year Chinook were stream type, and six of those were captured off Cape Flattery. From Cape Flattery northward, the first ocean year Chinook are overwhelmingly stream type, whereas those from inside waters were predominantly ocean type. First-year ocean-type Chinook are more common in coastal ocean waters south of Cape Flattery (Fisher et al. 1983, 1984; Miller et al. 1983).

Conclusions for Harvest

Based on the preponderance of evidence and experience of the past 100 years, the key points relevant to understanding the relationship among harvest, habitat, and salmon productivity are as follows.

1. Lack of effective harvest controls played a role in the decline and extirpation of Pacific salmon populations. Unlimited exploitation incurred by the combination of fishing with other natural resource extraction activities on salmon contributed to reductions in the production of salmon in the Columbia River Basin.
2. Traditional harvest management, through imposition of limits only on exploitation in directed salmon fisheries, has not been sufficient to allow salmon populations of the Columbia River to persist.
3. Traditional harvest management actions will not compensate for losses due to human activities other than directed harvest because estimates of salmon production from habitats that are constantly declining in productivity will always be too high. Overfishing results when estimates of harvestable surplus are too high. A new harvest management paradigm is needed which will take habitat productivity into account.

A New Pacific Salmon Harvest Management Paradigm

Introduction to the New Paradigm

The new paradigm is based on internationally recognized principles of the conservation of exploited animal populations, as articulated in the Sustainable Salmon Fishery Policy adopted by the Alaska Board of Fisheries in 2000. Before the new paradigm can be appreciated, it must be visualized. A visual model, the Extended Salmon Fishery Model, is an essential introduction to the complexity of the problem of managing a salmon population over a migratory pathway that is inscribed within an area of one million square miles, or more.

The Extended Salmon Fishery Model

The Extended Salmon Fishery Model (Figure 9.4; Mundy 1998) was originally developed as a much simpler cycle for Bristol Bay sockeye by Mundy and Mathisen (1981); however, it is applicable to any salmon-producing area and combination of salmon species. It demonstrates in graphic detail the sequence of harvest decisions that occurs throughout the life cycle of the salmon and the consequences that each decision must have for the abundance of the next life cycle stage.

The life cycle of the salmon and the mortalities experienced as a result of human decisions are schematically illustrated (Fig. 9.4). Most of the boxes represent a series of harvest decisions (boxes 2.2, 2.6, in Figure 9.4) and consequent escapement (survivors) and catch (mortalities) at every life cycle stage (see boxes 2.3 & 2.4, 2.7 & 2.8). At the center of the new harvest management paradigm are habitat management decisions (see box 2.23) that drive production at every life cycle stage. Salmon harvest management is therefore to a large extent salmon habitat management.

To build a new harvest management paradigm for the Columbia River Basin requires understanding the two fundamentals: the salmon life cycle and salmon habitats. The salmon life cycle is deceptively simple in concept: egg, fry, fingerling, smolt, subadult, adult. The complexity is added by the habitat, which encompasses an enormous amount of geography with each full turn of the life cycle, potentially thousands of linear miles from headwaters in mountains to the North Pacific Ocean and back. It also requires recognizing that the series of habitats through which the salmon pass from egg to adult each take a toll in mortality. Understanding how the life cycle interacts with the habitats to produce each year class is the essence of the new harvest management paradigm.

The application of the Extended Salmon Fishery Model to the Columbia River is direct. Salmon are harvested by many different human activities in the Columbia River Basin other than fishing (see especially Chapters 6–8). Intentional, or directed, harvest of adults and immature salmon for commercial, subsistence, ceremonial, and recreational purposes are the best understood among human sources of mortalities because records of commercial harvest date from 1866 and estimates of abundance have been produced by a number of authors and institutions (Craig and Hacker 1940; Chapman et al. 1991, 1994; National Marine Fisheries Service [NMFS] 1995; NRC 1996; Schoonmaker et al. 2003).

Nonetheless, directed harvest is by no means the predominant source of mortality for Columbia River Basin salmon. Unintentional or incidental harvest of salmon occurs in those activities that are not intended to capture the salmon species or life history stage that is taken. Incidental harvest of Columbia River salmon occurs in marine and freshwater fisheries for other species of fish, during salmon fisheries targeted on older life history stages of salmon, in the production of electricity at hydroelectric dams (Chapters 6 and 7), during and after logging operations, during and after irrigation withdrawals, during land development operations such as road and real estate building, and during and after some types of mining operations (Chapter 5).

Applying adult harvest observations as if they were the only significant source of mortality is problematic for a number of reasons. Harvest observations have long been applied to understand if the productive capacity of the populations was being exceeded (Ricker 1954; Beverton and Holt 1957). Directed harvest estimates are useful in proper context because in theory, and in practice, it is possible to harvest at a rate high enough to diminish a salmon population's spawning potential and to cause it to be extirpated (Cushing 1983). It was also assumed that the health of the salmon populations could be assured through appropriate limitations on directed harvests (Mundy 1985). Such assumptions were clearly in error.

In the Columbia River Basin, it is clear that directed harvest is only one of many sources of mortality, and it follows that all sources of mortality should be accounted for in order to permit the persistence of the salmon. In practice, all human induced mortalities are measured to the extent possible, with all remaining sources, such as predation by marine mammals, being attributed to natural mortality. Very early on, Ricker (1958) examined the effects of a fluctuating environment (variable mortalities induced at early life stages) on the productive capabilities of fish stocks, which pointed to problems associated with conventional harvest management. Clearly, traditional harvest management, which seeks only to control directed sources of fishing mortality (Ricker and Smith 1975), is not sufficient to provide for the sustainable production of the Columbia River Basin's salmon. However, the principles of sustainable harvest management (Beverton and Holt 1957; Cushing 1983) need to be carried forward in framing a new harvest management paradigm, which is appropriate to the persistence of the full diversity of species and life history types of the basin's salmon.

The Extended Salmon Fishery Model (Fig. 9.4) makes the point that harvest management in the Columbia River and elsewhere must be concerned with a great deal more than the number of adults in the catch (Box 2.20 in Figure 9.4) and on the spawning grounds (Box 2.19). Consequently, under the new harvest management paradigm, the minimum information necessary to implement these principles under effective Pacific salmon harvest management (Fried and Yuen 1987; Hilborn 1987; Walters and Collie 1988; Eggers 1992; McAllister and Peterson 1992;) goes well beyond the information required to achieve these objectives under the old single species management paradigm.

Information requirements are more intensive because the assumptions permitted by productive, stable habitat are no longer valid, because the sources of mortality are numerous, and because harvests by humans are often not identified as such. In this paradigm, the inadvertent taking of salmon by humans is recognized as incidental harvest. Salmon are inadvertently taken by other human activities during the course of the salmon's life cycle, by activities such as logging, road building, agricultural cultivation and irrigation, many kinds of pollution, hydroelectric power generation, fishing for other species, and by directed fishing for the same, and for other life cycle stages of salmon.

A Sustainable Salmon Fishery Policy

The experience of 150 years of salmon fishing in the Columbia River Basin points to the absolute necessity for a new harvest management paradigm to sustain salmon populations and salmon fisheries. A Pacific salmon harvest management paradigm appropriate to the Columbia River Basin was adopted by the Alaska Board of Fisheries in 2000 after a rigorous process of research, scientific peer review, and public involvement. The research phase (Mundy 1998) pointed to internationally recognized principles of conservation (Eggers 1993; Food and Agriculture Organization [FAO] 1995; Olver et al. 1995; Starnes et al. 1995; Mangel et al. 1996; Naiman and Stouder 1996; Mooney 1998; NRC 1996, 1999; Alverson 2002).

Principles of Sustainable Salmon Fishing

Internationally recognized principles of sustainable fishery management, as applied to salmon, have been adapted to salmon fisheries in the form of five basic principles (Mundy 1998; Alaska Department of Fish and Game 2000). When expanded by the rich body of scientific literature and regulatory experience applicable to each principle, the five principles can be applied as criteria of success to analyze any salmon fishery (Mundy 1998). The five principles are:

- **Principle I.** Protect wild salmon and its habitat in order to maintain resource productivity
- **Principle II.** Maintain spawning escapements within ranges necessary to conserve and protect potential salmon production and to maintain normal ecosystem functioning
- **Principle III.** Harvest salmon in a manner consistent with the degree of uncertainty regarding the status and biology of the resource
- **Principle IV.** Establish and apply an effective management system to control human activities that affect salmon
- **Principle V.** Maintain public support and involvement for sustained use and protection of salmon resources

I. Protect Wild Salmon and Their Habitats

Principle I. Protect wild salmon and its habitat in order to maintain resource productivity. The goal of conservation for sustainable use is to secure present and future options by maintaining biological diversity at genetic, species, population, and ecosystem levels; as a general rule, neither the managed resource nor any other components of the ecosystem should be perturbed beyond natural boundaries of variation (FAO 1995; Olver et al. 1995; Starnes et al. 1995; Mangel et al. 1996). The point of sustainable salmon management activities is to keep the full range of salmon resources productive to the full extent possible (Paulik et al. 1967). Protection of salmon production in the short-term takes the form of limiting harvests to allow adults to reach the spawning grounds (Mundy 1985). Protecting the salmon production in the long-term means protecting the spawning and rearing habitats, including the entire salmon bearing ecosystem, from degradation (Nehlsen et al. 1991; Huntington et al. 1996; Myers et al. 1998). Habitat protection takes the form of land use planning and regulation, including regulating natural resource extraction activities.

As Rich (1941) and Craig and Hacker (1940) wrote more than half a century ago, understanding habitat is essential to sustaining salmon production. For understanding habitat, it is essential to develop "a framework for integrating predictable and observable features of flowing water systems with the physical geomorphologic environment" (Vannote et al. 1980, p. 135). The hydrology and geomorphology of the watersheds (Leopold et al. 1964), as well as the consequences of riparian vegetation for salmon production (Volk et al. 2003), need to be part of Pacific salmon harvest management for salmon originating in all types of habitats, but it is especially important for conservation of stocks originating from damaged habitat.

Protection of spawning and rearing habitat (Figure 9.5) is likely to be as important to long-term sustainable salmon production as harvest control is

Figure 9.5 Spawning and rearing habitat for summer and spring Chinook in Bear Valley Creek, Middle Fork Salmon drainage, central Idaho. Photo by R. N. Williams.

to short-term sustainable salmon production (Walters 1996). Some of the best-known Alaskan salmon populations have survived decades of apparent overharvest (Cooley 1963) to prosper under escapement goal management (Mundy 1996). It is probable that an important factor enabling these Alaskan salmon populations to rebound from the effects of a prolonged period of uncontrolled harvest was the health of their spawning and rearing habitats (Mundy 1996). On the other hand, experience from the Pacific Northwest has demonstrated that salmon populations are unlikely to recover from habitat degradation and loss even with total closures to fish harvest (Kostow 1996; Mills et al. 1996).

Principle I emphasizes that the limits on salmon exploitation rates appropriate to conservation are ultimately dependent on the productive capacity of the habitat from which the populations originate and on objectives for the magnitude and geographic distribution of spawners. Hence, salmon harvest managers need to look at the effects of degradation of the habitat on which spawners and juveniles depend for survival (Figures 9.6 and 9.7). The long-term persistence of all species of salmon throughout their ranges is dependent on the implementation of a salmon harvest management paradigm that applies exploitation rates consistent with the status of the salmon-bearing ecosystems (i.e., the ecosystem's production capability).

Understanding the interactions and dependencies among harvest, the health of habitat, and the productivities of salmon populations is essential to improving our abilities to identify and implement salmon restoration efforts (Ricker 1954). In examining the Columbia River Basin, habitat alteration (loss and degradation) and unlimited fishing emerge as parallel companions of the initial decline in population numbers of the principal commercial salmon species (see Chapter 5 on freshwater habitat). The evolution of harvest management protection for naturally spawning Columbia River Basin salmon was restrained by increases in hatchery production during the 1960s (Chapter 8). The large numbers of hatchery salmon drove the public policy process to sanction intensive fisheries on mixtures of hatchery and natural salmon that obscured the downward trends in production of the natural salmon populations. At present, continuing habitat losses and ineffective harvest regulation are probable causes for the continuing failure of Columbia River Chinook salmon. Therefore, an effective harvest management paradigm cannot be developed outside of an ecosystem context (NRC 1999).

In addition to direct harvests impacts, protection of wild salmon needs to take into account the indirect impacts of reducing the diversity of phenotypes in the population, which can impact factors important to basic productivity, such as average number of eggs per female (Miller 1957; Ricker 1981; Russell 1931, cited in Cushing 1983; Beatty 1992). Overfishing reduces

Figure 9.6 Habitat degradation to a spring Chinook spawning tributary (McIntyre Creek) in the upper Grande Ronde Basin caused by grazing impacts and loss of the riparian vegetation. Photo by R. N. Williams.

the production of salmon by reducing or eliminating the populations that have adapted to the habitat types and environmental conditions of the basin (Ricker 1972; Riddell and Legget 1981; Thorpe 1995). As it has developed from the experience of the last three generations of fisheries scientists, and as harvest regulations increasingly reflect, protection of wild salmon populations ought to protect the productive capacity of salmon runs by pursuing the reasonable and essential objective of protecting the genetic diversity of Pacific salmon populations on which production ultimately depends (Paulik 1969; Lande and Barrowclough 1987).

Figure 9.7 Two-person gold-dredging operation in the Clearwater drainage in central Idaho. Photo by R. N. Williams.

II. Maintain Spawning Escapements

Principle II. Maintain spawning escapements within ranges necessary to conserve and protect potential salmon production and to maintain normal ecosystem functioning. The concept of the escapement goal is fundamental to the practice of successful salmon fishery management (Mundy 1982; NRC 1996; Eggers 1993; Walters 1996). The escapement goal is the annual number of adults, or a range of values, that the management entity intends to successfully spawn within a designated watershed. Having a recognized numerical objective for seeding of the salmon spawning grounds (Figure 9.8) is fundamental because it is a clear measure of the success of a management program. Without a tangible measure of success at the watershed level, the management program has no means of judging the effects of its actions in terms of future salmon production, and no means to determine the urgency of any particular course of action.

Although the existence of an escapement goal and its monitoring program at the watershed level is a necessary condition for an effective sustainable salmon management program, alone it is not sufficient. To be sufficient, escapement goals and their monitoring programs need to be combined

Figure 9.8   Multiple spawning redds of summer Chinook, Johnson Creek, Salmon River drainage in central Idaho. Orange areas in the gravel beds indicate the location of active salmon redds. Photo by R. N. Williams.

with the other elements identified in the Principles, including habitat protection and monitoring, stock identification, and juvenile assessments (see Chapter 11).

A clear lesson from the history of Pacific salmon management is that setting an escapement objective, and meeting it, is at least as important as having an escapement objective that is selected to optimize some utility, such as long-term average landings. The NRC recommended an escapement goal approach to "reduce the risk of continued loss of salmon populations and production" (NRC 1996, p. 295). As a step toward escapement goal management, the NRC recommended the minimum sustainable escapement (MSE) be defined as some minimum viable population size below "optimum escapement." A number of spawners less than MSE would serve as a danger signal of management system failure. The MSE may be seen as a particular case of the more general approach of Eggers (1993, Table A-9). In studies of the effects of failing to meet escapement objectives under conditions of normal variability in management errors and stock size, Eggers (1993) found that a fairly broad range of escapements was expected to deliver future salmon production near maximum sustained yield. In many localities, the Alaska

Department of Fish and Game now manages under guidelines that prescribe a range of escapements as the management objective.

Explicit recognition of the role of habitat in determining salmon productivity is essential to adapt the basic approaches to harvest regimes of the salmon fisheries of the Fraser River, Canada (Roos 1991) and Bristol Bay, Alaska (Mundy and Mathisen 1981; Eggers 1992) to the new paradigm. The measure of performance for conservation needs to be extended beyond traditional measures of success, such as sustainable yield for a population of a single species, to include measures of ecological diversity (Pielou 1969) and of ecological processes, especially biogeochemical cycles involving marine-derived nutrients (Mathisen 1972; Kyle et al. 1988; Stockner et al. 2000; Naiman et al. 2002; Murota 2003). Attempts are increasingly being made to move from single-species management toward multi-species approaches, and toward incorporating many more elements of the ecosystem in fisheries management (Fujita et al. 1998; Hofman and Powell 1998; Jarre-Teichmmann 1998).

The concept of stock recruitment, which holds that future spawning stock size is to some extent dependent on present spawning stock size (Ricker 1954; Cushing 1983; Hilborn and Walters 1992; Walters 1986), needs to be enlarged to include other indicators of the status of the ecosystem (Ricker 1958). Although the relation between present and future spawning stock size can be highly variable for healthy salmon populations, the understanding of physical limits on future population growth posed by the number of eggs per female becomes extremely critical at the low population sizes common to salmon in the contiguous United States. Enlarging this concept will require new models to be developed that explicitly incorporate the role of habitat in determining salmon productivity.

It is essential for harvest managers to find ways to establish salmon spawning escapements objectives for a watershed based on analyses of habitat-based watershed attributes in addition to historical time series of the numbers of salmon spawning in the watershed. The parameters of a stock-recruitment function appropriate to effective harvest management in an ecosystem context should include information on habitat quality and quantity. Quantitative data on riparian vegetation and streambed condition in relation to surveys of spawning adults and rearing juveniles are generally lacking. Such information can be drawn from functions of the density and species composition of riparian vegetation, the percentage of fine sediment in the spawning substrate, the abundance of critical life history stages of at least one prey species and one predator species, and the abundance of one species utilized as an alternative by the salmon's predators. If it were possible to explicitly include one or more of the preceding habitat variables in the salmon stock-recruitment function, it would remind harvest managers of salmon originating from areas of high human population density of the ephemeral nature of the productive capacity of the environment.

III. Harvest with Caution

Principle III. Harvest salmon in a manner consistent with the degree of uncertainty regarding the status and biology of the resource. It is now becoming widely recognized that limitations on information and lack of enforcement should sharply limit the ways in which sustainable natural resource extraction may be conducted (FAO 1995; Mangel et al. 1996). If it is reasonable to assume that major uncertainties in how ecological systems respond to management actions will always be with us, then the way in which resource management decisions are made needs to reflect the uncertainties (Ludwig et al. 1993). Making decisions on the use of the salmon resource consistent with the degree of uncertainty requires judicious collection and use of the information from the salmon fisheries and their habitats. Decision making under uncertainty further requires management actions to use the information collected to adapt to changing climate and population levels in ways that provide for sustainable salmon populations (Walters 1986). Harvest and habitat management programs need to be structured to advise decision makers on the chances of success (or failure) associated with an action, and on how to respond to each success (or failure) with additional management actions. This iterative process is known as adaptive management (Holling 1978).

In the Columbia River Basin, migratory behaviors such as timing in fishing areas and rates of migrations are particularly important to understanding risk to listed salmon populations because they determine the proportion of the listed population that is vulnerable to harvest during any fishing event. Migratory behaviors must be measured in order to calculate risk because they are known to vary annually within populations, and among populations within years in Chinook and other salmon species (i.e., Mundy 1985). In addition, migratory behaviors in Chinook are known to vary in concert with climate variables (Mundy 1982), and they can be influenced by changes in river flow and hydroelectric operations, which are also related to variations in climate. Finally, migratory behaviors must be measured in order to calculate meaningful estimates of risk from harvest because differences in migratory behaviors among populations can cause them to be aggregated in time with respect to any given harvest area. Such schooling behavior during migration would negate the assumption that listed species are randomly distributed with respect to hatchery salmon in the waters of a fishing area at an instant in time. The assumption of random distribution of salmon populations is critical to the risk calculation for listed spring Chinook in the Columbia River in 2003 (Columbia River Compact 2003b).

Unfortunately, measurement of critically important migratory behaviors is problematic for listed spring Columbia River Chinook populations because there are too few of them to allow capture and marking. Even if migratory behaviors could be measured for listed species, the supposition that all listed wild Chinook populations are experiencing the elevated return rates of the hatchery brood years contributing to the 2001 to 2003 Columbia River adult spring Chinook migrations would not qualify as precautionary (FAO 1995). In an area as large as the Columbia River Basin, which was estimated to have more than 8,700 km of spring Chinook spawning and rearing habitats of widely varying qualities as of 1975 (see Mundy 1996), it is hardly reasonable to expect all populations to have the same survival rates.

As embodied in the new harvest management paradigm (Principle II), measures of escapement are reasonable and prudent harvest control (Mundy 1985) indicators on which to base harvest management decisions for listed species of Columbia River salmon. On the other hand, the harvest rate estimates applied to landings by the Columbia River Compact (2003b) are not reliable harvest control indicators because of unmeasured variability introduced by variations in migratory behavior and by the distances and lag times between dam count observations and the spawning event. Harvest rate-landing approaches applied in the absence of an understanding of what proportion of the population is vulnerable to harvest cannot provide a mathematically valid estimate of losses (risk) to the listed species due to harvest (Quinn and DeRiso 1999). In this case, any given number landed or presumed killed cannot serve as a valid harvest control indicator (Mundy 1985) because landings cannot reliably estimate spawning escapement, so the impact of any given harvest action on the future of the listed species is unknown.

IV. Establish and Apply an Effective Management System

Principle IV. Establish and apply an effective management system to control human activities that affect salmon. Protecting salmon populations and habitat, providing escapements, and limiting harvests (Principles I–III) all require an effective management system for implementation (FAO 1995; Starnes et al. 1995; Mangel et al. 1996). Effective management systems that provide long-term protection for the salmon also provide substantial net economic benefits to society (Eggers 1992). Knowledge needs to be translated into actions in the form of effective fishing regulations, enforcement of fishing regulations, habitat protection regulations, and enforcement of habitat protection regulations. As is the case with any public process, the management system that helps to sustain salmon production needs to be subject to periodic performance evaluations that measure how well objectives are being met.

Effective harvest management embodies zero-sum mortality allocation (Principle I), escapement goals (Principle II), and geographic gradients in fishing mortalities (Principle III), and it promotes public support through information and public involvement (Principle V). Implementing zero-sum mortality allocation requires tracking all significant sources of mortality (Figure 9.4) in order to be able to know where and when it is necessary to decrease mortality when mortality in one or more other sources increases. Establishing expectations among concerned parties that survivals be held within specified zero-sum bounds at each step in the life cycle (Figure 9.4) is crucial to building public support (Principle V). It is particularly important to publicly establish the concept that increases in "natural" mortalities require corresponding decreases in human mortalities. Spreading the consequences across as many different sectors of the economy as possible lowers the impact of conservation on an individual sector (Figure 9.4).

Effective management strategies are designed to provide adequate spawning escapements to all spawning grounds (Principle II) and to accurately measure the attainment of these goals on an annual basis. Without monitoring, there is no effective harvest management of salmon, because salmon harvest management depends upon information (Walters 1986). Escapement goals under effective harvest management are quantifiable objectives, by locality and life history type, for spawning numbers, habitat, and associated species. Escapement goals must be accompanied by monitoring programs in order to be meaningful.

Effective management strategies also are designed to decrease fishing mortalities in proportion to the fishery's distance from the spawning grounds (Mundy 1996) in order to deal with increasing uncertainty regarding stock composition of catches (Principle III). Such a concept is especially important to minimize risks for salmon populations that have low overall productivity, such as those with damaged spawning and rearing habitats. Given current technologies, the accuracy of stock identification is roughly inversely proportional to the distance of the fishery from the spawning grounds. Thus, in order to maintain an effective harvest management regime in the face of high uncertainty in stock identification means very low harvest rates.

Specifically, the ideal situation is to use stock identification to effectively limit fishing mortality to a level that permits persistence of the smallest identifiable stock, also called a "deme" or population (NRC 1996). In practice, fisheries management agencies are usually able to protect only "stocks" that have been defined as some identifiable aggregate of local spawning populations. The number of populations in the pragmatic stock definitions might be more a function of logistic considerations and the amount of funding available for monitoring than of biological considerations. It is now essential to effective salmon fishery management that the definition of a stock consider the biological criteria engendered by Endangered Species Act definitions of stock, such as the Evolutionarily Significant Unit (ESU) (Waples 1991; Mundy et al. 1995; NRC 1995).

In the face of mounting losses of Pacific salmon populations (Nehlsen et al. 1991; Forest Ecosystem Management Assessment Team 1993), fisheries for mixtures of salmon stocks could be curtailed or eliminated in an attempt to protect damaged salmon populations, including federally threatened or endangered species. Widespread losses of fishing opportunities might be necessary, unless it is possible to identify and define successful concepts and approaches to sustainable salmon harvest management within the context of effective salmon fishery management.

The effective approach to implementing the new harvest management paradigm in situations where mixtures of salmon populations occur is informed mixed stock fishing. Informed mixed stock fishing uses information about migratory pathways, migratory timing of different populations, and other differences among salmon populations to determine the impact of fishing on the individual stocks. In an informed mixed stock fishery, catches taken in mixed stock areas can be assigned to their stock of origin in a process known as stock identification. Ideally, stock separation is quantitative with proportions of each stock in the harvest being estimated. In cases when only presence or absence of a stock in the harvest of a locality can be determined reliably, stock identification can still serve a useful purpose by determining whether fishing at that locality needs to be prohibited in the interest of protecting the stock whose presence has been ascertained.

Effective management works best when humans expect to share in both the consequences and benefits inherent in the exploitation of a natural resource (Principle V). The benefits of harvest are well understood to be food and money; however, lack of food and loss of income need to be understood as the result of failing to achieve escapement goals. Two of the world's largest sockeye salmon runs, Fraser River (Canada) and Bristol Bay (Alaska), are intact today due to the virtual elimination of harvest during critically low periods of abundance (Roos 1991; Mundy 1996).

In the Columbia River Basin, the directed salmon fishery is the only human harvest sector that has operational guidelines that are directly linked to escapement objectives, however flawed the measurement of the attainment of those objectives may be. Chinook salmon harvests were drastically reduced during the 1990s, even though the opportunity to harvest large numbers of hatchery returns has marred that record in 2001 to 2003. Other sectors of the economy that inflict substantial mortalities on migratory and resident salmon, such as the hydroelectric system, barge transportation, agricultural irrigation, and the timber industry, are not directly linked to escapement objectives, and there are no routine estimates of the effects of their operations on the attainment of salmon escapement objectives. Although the hydroelectric industry is limited in power production to some extent by concerns over salmon survivals, hard escapement objectives that might require some level of instability in the power grid are certainly not in place. Until there is a harvest management regime in the Columbia River Basin that subordinates the management of risks to salmon from all sources of anthropogenic mortality to escapement objectives, wild salmon populations are unlikely to recover.

V. Maintain Public Support

Principle V. Maintain public support and involvement for sustained use and protection of salmon resources. In addition to suitable habitat, spawning escapements, and effective management, the active involvement of a broad cross section of the public is essential to sustainable salmon fishing. The cardinal test of the health of the public contribution toward sustaining salmon fishing is the presence of a governmental process that incorporates appropriate mechanisms for resolution of disputes. The mechanisms are expected to include an open and fair public involvement process that addresses management and harvest allocation decisions and an allocation of the conservation burden for salmon across all consumptive user groups. An additional test of the adequacy of the governmental process to support sustainable salmon fishing is the extent to which it provides adequately funded public information and education programs for the public. The public needs to be informed concerning salmon habitat requirements, salmon habitat threats, the value of salmon and habitat to the public and the ecosystem, natural variability and population dynamics, the value of salmon to other fish and wildlife, current status of relevant fish stocks and fisheries, and the nature of the public involvement process (Figure 9.9).

Figure 9.9 Angler with a wild spring Chinook on the South Fork of the Salmon River in central Idaho. Sport angling and harvest has occurred on the river in recent years, due to increasing run size and large hatchery returns after a 20-year closure. Wild fish, like the one pictured, must be released, while adipose fin clipped hatchery fish may be harvested. Photo by R. Howell.

The degree to which management contributes to the success of the public involvement process provides two key additional tests of the public involvement process. Within the process, effective management provides for dissemination of results of management actions and monitoring to all interested parties in a timely fashion. In this context, management is expected to promote public understanding of the proportion of mortality inflicted on each stock by each consumptive user group.

Conclusions for Harvest and Harvest Management

Harvest regulation is a sufficient means of protecting and increasing salmon production only in the presence of functional spawning, rearing, and migratory habitats. The old harvest management paradigm has failed to consider the relation of salmon abundance to other components of the ecosystem, which are connected by the life cycle of the salmon.

Conclusions (and Levels of Proof)

1. Directed (intentional) and incidental (unintentional) harvest of Columbia River Basin (CRB) salmon has occurred in the absence of knowledge of harvest impacts on the abundances and viabilities of the majority of the individual native spawning populations. Viability means having a reasonable probability of survival within an arbitrary time horizon. (1)
2. Harvest rates on native spawning populations of CRB salmon from incidental and direct sources have increased since development of the Columbia River Basin by western civilization in the early 19th century. (3)
3. Both directed and incidental harvests exert levels of mortality on salmon spawning populations that are large enough to influence their annual abundances and viabilities. (2)
4. Harvest, both incidental and intentional, has contributed to the decline in abundance of CRB salmon and it is a factor limiting their recovery. However, harvest restrictions in the absence of habitat restoration are not sufficient to permit recovery. (2)
5. Interactions between mortality associated with habitat degradation (incidental harvest) and directed harvests by fisheries have led to the extirpation of many CRB salmon populations. (3)
6. All Columbia River stocks, with the possible exception of Hanford fall Chinook, are at such low levels that harvest in the ocean will have to be very low or non-existent to allow the habitat restoration proposed herein to have a reasonable chance to succeed. (2)

Recommendations for Harvest and Harvest Management

1. Harvest management needs to recognize the relation of salmon abundance to other components of the ecosystem, which are connected by the life cycle of the salmon.
2. Management Implications. A prime example of the success in sustainable salmon conservation brought about by implementation of the contemporary salmon harvest management paradigm can be found in the best known salmon fishery in Alaska. Many times in the past 100 years, Bristol Bay, Alaska, has been the world's largest salmon fishery. Since at least as early as 1961, the management program has attempted to provide a minimum level of spawning escapement to each of the watersheds. Spawner counts are made near the outlets of the nursery lakes and verified by qualitative aerial surveys of distributions of the spawners within each watershed (Fried and Yuen 1987). Informed mixed stock harvest is achieved by restricting the fishery to relatively small marine harvest areas near shore, and by timing harvest periods to correspond to the timing of adult returns to the various watersheds. Informed mixed stock harvest in conjunction with the application of the precautionary principle to harvest decisions, firm sustained yield management objectives, and the maintenance of pristine spawning and rearing habitats, have all contributed to the populations' abilities to sustain unprecedented levels of harvest during the 1980s and 1990s.

   Conclusions: Sustained yield management of a salmon population, or deme (see NRC 1995, 1996), needs to be based on numerical spawning escapement goals at the watershed level that represent both the productive capacities of the habitats for the salmon population and all related salmon populations, geographic gradients in fishing mortality appropriate to the nature of the stock composition information for each fishery, and a zero-sum mortality allocation across all fisheries.
3. Implement the five principles of the Sustainable Salmon Fishery Policy and regularly evaluate fisheries management using the principles as criteria of the attainment of effective sustainable salmon management.

Literature Cited

Alaska Department of Fish and Game. 2000. Sustainable Salmon Fishery Policy. Alaska Department of Fish and Game, Juneau, Alaska.

Alverson, D. L. 2002. Factors influencing the scope and quality of science and management decisions (The good, the bad and the ugly). *Fish and Fisheries* 3:3–19.

Alverson, D. L., M. H. Freeberg, S. A. Murawski, and J. G. Pope. 1994. A global assessment of fisheries bycatch and discards. FAO Fisheries Technical Paper No. 339. Food and Agriculture Organization of the United Nations, Rome.

Beatty, R. E. 1992. Changes in size and age at maturity of Columbia River upriver bright fall Chinook salmon (*Oncorhynchus tshawytscha*): Implications for stock fitness, commercial value, and management. Master's thesis, Oregon State University, Corvallis. 270 p.

Beverton, R. J. H., and S. J. Holt. 1957. *On the dynamics of exploited fish populations*. Ministry of Marine Fisheries and Ministry of Agricultural Fishing and Food, London, United Kingdom.

Bottom, D. L. 1996. To till the water: A history of ideas in fisheries. Pages 569–597 in R. J. Naiman and D. Stouder, eds. *Pacific salmon and their ecosystems: Status and future options*. Chapman and Hall, New York.

Chapman, D., A. Giorgi, M. Hill, A. Maule, S. McCutcheon, D. Park, W. Platts, K. Pratt, J. Seeb, L. Seeb, and F. Utter. 1991. Status of Snake River Chinook salmon. Don Chapman Consultants, Boise, Idaho. 520 p.

Chapman, D., A. Giorgi, T. Hillman, D. Deppert, M. Erho, S. Hays, C. Peven, B. Suzumoto, and R. Klinge. 1994. Status of summer/fall Chinook salmon in the mid-Columbia region. Don Chapman Consultants, Boise, Idaho. 411 p.

Chapman, D. W. 1986. Salmon and steelhead abundance in the Columbia River in the nineteenth century. *Transactions of the American Fisheries Society* 115:662–670.

Columbia River Compact. 2003a, July 3. Joint staff report: Winter/spring summary fact sheet. Oregon Department of Fish and Wildlife and Washington Department of Fish and Wildlife.

Columbia River Compact. 2003b, March 25. Calculated below Bonneville: Recreational spring Chinook effort, catch, and impacts to ESA listed stocks for 2003. Washington Department of Fish and Wildlife.

Columbia River Compact. 2003c, January 23. Joint staff report concerning commercial seasons for spring Chinook, steelhead, sturgeon, shad, smelt, and other species and miscellaneous regulations for 2003. Joint Columbia River Management Staff, Oregon Department of Fish and Wildlife, and Washington Department of Fish and Wildlife.

Columbia River Compact. 2003d, July 16. Joint staff report concerning 2003 fall in-river commercial harvest of Columbia River fall Chinook salmon, summer steelhead, coho salmon, chum salmon, and sturgeon. Joint Columbia River Management Staff, Oregon Department of Fish and Wildlife, and Washington Department of Fish and Wildlife.

Cooley, R. A. 1963. *Politics and conservation: The decline of the Alaska salmon*. Harper and Row Publishers, New York.

Craig, J. A., and R. L. Hacker. 1940. The history and development of the fisheries of the Columbia River. *U.S. Bureau of Fisheries Bulletin* 32:133–216.

Cushing, D. H. 1983. *Key papers on fish populations*. IRL Press Limited, Washington, DC.

Eggers, D. M. 1992. The costs and benefits of the management program for natural sockeye salmon stocks in Bristol Bay, Alaska. *Fisheries Research* 14:159–177.

Eggers, D. M. 1993. Robust harvest policies for Pacific salmon fisheries. Pages 85–106 in *Proceedings of the International Symposium on Management Strategies for Exploited Fish Populations*. Report No. 93-02. University of Alaska Sea Grant College Program, Fairbanks, Alaska.

Food and Agriculture Organization. 1995. *Code of conduct for responsible fisheries*. Food and Agriculture Organization of the United Nations, Rome.

Forest Ecosystem Management Assessment Team. 1993. Forest ecosystem management: An ecological, economic, and social assessment. Interagency Supplemental Environmental Impact Statement Team, Portland, Oregon.

Fisher, J. P., W. G. Pearcy, and A. W. Chung. 1983. Studies of juvenile salmonids off the Oregon and Washington coast, 1982. Oregon State University. Oregon State University Collections in Oceanography Cruise Report, 83-2. Corvallis, Oregon.

Fisher, J. P., W. G. Pearcy, and A. W. Chung. 1984. Studies of juvenile salmonids off the Oregon and Washington coast, 1983. Sea Grant College Program. Oregon State University Collections in Oceanography Cruise Report, ORESU-T-85-004. Corvallis, Oregon.

Fried, S. M., and H. J. Yuen. 1987. Forecasting sockeye salmon (*Oncorhynchus nerka*) returns to Bristol Bay, Alaska: A review and critique of methods. *Canadian Special Publication in Fisheries and Aquatic Sciences* 45:850–855.

Fujita, R. M., T. Foran, et al. 1998. Innovative approaches for fostering conservation in marine fisheries. *Ecological Applications* 8(1 Suppl):139–150.

Garcia, S. M. 1994. The precautionary principle: Its implications in capture fisheries management. *Ocean and Coastal Management* 22:99–109.

Hare, S. R., N. J. Mantua, and R. C. Francis. 1999. Inverse production regimes: Alaska and west coast Pacific salmon. *Fisheries* 24(1):6–14.

Hartt, A. C. 1980. Juvenile salmonids in the oceanic ecosystem: The critical first summer. Pages 25–57 in W. J. McNeill and D. C. Himsworth, eds. *Salmonid ecosystems of the North Pacific*. Oregon State University Press, Corvallis, Oregon.

Hartt, A. C., and M. B. Dell. 1986. Early oceanic migrations and growth of of juvenile Pacific salmon and steelhead trout. *International North Pacific Fisheries Commission Bulletin* 46:1–105.

Healey, M. C. 1991. Life history of Chinook salmon. Pages 313–393 in C. Groot and L. Margolis, eds. *Pacific salmon life histories*. University of British Columbia Press, Vancouver.

Hilborn, R. 1987. Living with uncertainty in resource management. *North American Journal of Fisheries Management* 7:1–5.

Hilborn, R., and C. J. Walters. 1992. *Quantitative fisheries stock assessment: Choice, dynamics and uncertainty*. Chapman and Hall, New York.

Hofman, E. E., and T. M. Powell 1998. Environmental variability effects on marine fisheries: four case histories. *Ecological Applications* 8(1 Suppl):S23–S32.

Holling, C. S. 1978. *Adaptive environmental assessment and management*. John Wiley and Sons, New York.

Huntington, C., W. Nehlsen, et al. 1996. A survey of healthy native stocks of anadromous salmonids in the Pacific Northwest and California. *Fisheries* 21(3):6–14.

Jackson, J. B., M. X. Kirby, W. H. Berger, K. A. Bjorndal, L. W. Botsford, B. J. Bourque, R. H. Bradbury, R. Cooke, J. Erlandson, J. A. Estes, T. P. Hughes, S. Kidwell, C. B. Lange, H. S. Lenihan, J. M. Pandolfi, C. H. Peterson, R. S. Steneck, M. J. Tegner, and R. R. Warner. 2001. Historical overfishing and the recent collapse of coastal ecosystems. *Science* 293:629–638.

Jarre-Teichmmann, A. 1998. The potential role of mass balance models for the management of upwelling ecosystems. *Ecological Applications* 8(1 Suppl):S93–S103.

Jensen, T. C. 1986. The United States-Canada Pacific salmon interception treaty: An historical and legal overview. *Environmental Law* 16:363–422.

Johnson, D. R., W. M. Chapman, and R. W. Schoning. 1948. The effects on salmon populations of the partial elimination of fixed fishing gear on the Columbia River in 1935. Oregon Fish Commission. Contribution No. 11. Portland, Oregon.

Kostow, K. 1996. The status of salmon and steelhead in Oregon. Pages 145–178 in R. J. Naiman and D. Stouder, eds. *Pacific salmon and their ecosystems: Status and future options*. Chapman and Hall, New York.

Kyle, G. B., J. P. Koenings, and B. M. Barrett. 1988. Density-dependent, trophic level responses to an introduced run of sockeye salmon (*Oncorhynchus nerka*) at Frazer Lake, Kodiak Island, Alaska. *Canadian Journal of Fisheries and Aquatic Sciences* 45:856–867.

Lande, R., and G. F. Barrowclough. 1987. Effective population size, genetic variation, and their use in population management. Pages 87–124 in M. E. Soule, ed. *Viable populations for conservation*. Cambridge University Press, New York.

Leopold, L. B., M. G. Wolman, and J. P. Miller. 1964. *Fluvial processes in geomorphology*. Dover Publications, New York.

Lichatowich, J. 1999. *Salmon without rivers: A history of the Pacific salmon crisis*. Island Press, Washington, DC.

Ludwig, D., R. Hilborn, et al. 1993. Uncertainty, resource exploitation, and conservation: Lessons from history. *Science* 260:17–36.

Mangel, M., L. M. Talbot, et al. 1996. Principles for the conservation of wild living resources. *Ecological Applications* 6(2):338–362.

Mantua, N. J.; S. R. Hare, Y. Zhang, J. M. Wallace, and R. C. Francis. 1997. A Pacific inter-decadal climate oscillation with impacts on salmon production. *Bulletin of the American Meteorological Society* 78(6):1069–1079.

Mathisen, O. A. 1972. Biogenic enrichment of sockeye salmon and stock productivity. *Verhandlungen Internationale Vereinigung fur Theoretische und Angewandte Limnologie* 18:1089–1095.

McAllister, M. K., and R. M. Peterson. 1992. Experimental design in the management of fisheries: A review. *North American Journal of Fisheries Management* 12:1–18.

McDonald, M. 1894. The salmon fisheries of the Columbia River basin. Pages 3–18 in Report of the Commissioner of Fish and Fisheries Investigations in the Columbia River Basin in Regard to the Salmon Fisheries. Senate Miscellaneous Document No. 200, 53rd Congress 2d Session, and House Miscellaneous Document No. 86, 53d Congress, 3d Session.

Miller, D. R., J. G. Williams, and C. W. Sims. 1983. Distribution, abundance, and growth of juvenile salmonids off the coast of Oregon and Washington, summer of 1980. *Fisheries Research* 2:1–17.

Miller, R. B. 1957. Have the genetic patterns of fishes been altered by introductions or selective fishing? *Journal of the Fisheries Research Board of Canada* 14:797–806.

Mills, T. J., D. R. McEwan, et al. 1996. California salmon and steelhead: Beyond the crossroads Pages 91–112 in R. J. Naiman and D. Stouder, eds. *Pacific salmon and their ecosystems: Status and future options.* Chapman and Hall, New York.

Milne, D. J. 1957. Recent British Columbia spring and coho salmon tagging experiments, and a comparison with those conducted from 1925 to 1930. *Fisheries Research Board of Canada* Bulletin No. 113:55.

Mooney, H. A. 1998. *Ecosystem management for sustainable marine fisheries.* Ecological Society of America, Washington, DC.

Mundy, N. M. 1999. The institutional context of natural resources management: A case study of fisheries management in the Columbia River basin of Oregon and Washington. Ph.D. dissertation, Portland State University, Portland, Oregon.

Mundy, P.R. 1982. Computation of migratory timing statistics for adult Chinook salmon in the Yukon River, Alaska, and their relevance to fisheries management. *North American Journal of Fisheries Management* 4:359–370.

Mundy, P. R. 1985. Harvest control systems for commercial marine fisheries management: Theory and practice. Pages 1–34 in P. R. Mundy, T. J. Quinn, and R. B. Deriso, eds. *Fisheries dynamics, harvest Management and sampling.* Washington Sea Grant Technical Report 85-1. University of Washington, Seattle.

Mundy, P. R. 1996. The role of harvest management in determining the status and future of Pacific salmon populations: Shaping human behavior to enable the persistence of salmon. Pages 315–330 in R. J. Naiman and D. Stouder, eds. *Pacific salmon and their ecosystems: Status and future options.* Chapman and Hall, New York.

Mundy, P. R. 1998. Principles and criteria for sustainable salmon management. Alaska Department of Fish and Game. Contract Report No. IHP-98-045. Juneau, Alaska.

Mundy, P. R., and O. A. Mathisen. 1981. Abundance estimation in a feedback control system applied to the management of a commercial salmon fishery. Pages 81–98 in K. B. Haley, ed. *Applied operations research in fishing.* Plenum Press, New York.

Mundy, P. R., T. W. H. Backman, and J. Berkson. 1995. Experiences from the Columbia River. *American Fisheries Society Symposium: Defining conservation units for endangered species* 17:28–38.

Murota, T. 2003. The marine nutrient shadow: A global comparison of anadromous salmon fishery and guano occurrence. *American Fisheries Society Symposium: Nutrients in salmonid ecosystems: Sustaining production and biodiversity* 34:17–31.

Myers, J. M., R. G. Kope, et al. 1998. Status review of Chinook salmon from Washington, Idaho, Oregon, and California. US Department of Commerce, National Oceanic and Atmospheric Administration, Seattle, Washington.

Naiman, R. J. and D. Stouder 1996. *Pacific salmon and their ecosystems: Status and future options.* Chapman and Hall, New York.

Naiman, R, J., R. E. Bilby, D. E. Schindler, and J. M. Helfield. 2002. Pacific salmon, nutrients and the dynamics of freshwater and riparian ecosystems. *Ecosystems* 5:399–417.

National Marine Fisheries Service (NMFS). 1995. Proposed Recovery Plan for Snake River Salmon. US Department of Commerce, National Oceanic and Atmospheric Administration, Washington. 387 p.

National Research Council (NRC). 1995. *Science and the Endangered Species Act.* National Academy Press, Washington, DC.

NRC. 1996. *Upstream: Salmon and society in the Pacific Northwest.* Report on the Committee on Protection and Management of Pacific Northwest Anadromous Salmonids for the National Research Council of the National Academy of Sciences. National Academy Press, Washington, DC.

NRC. 1999. *Sustaining marine fisheries.* Commission on Geosciences, Environment and Resources (CGER) and the Ocean Studies Board (OSB) for the National Research Council of the National Academy of Sciences. National Academy Press, Washington, DC.

Nehlsen, W., J. E. Williams, and J. A. Lichatowich. 1991. Pacific salmon at the crossroads: Stocks at risk from California, Oregon, Idaho, and Washington. *Fisheries* 16:4–21.

Netboy, A. 1974. *The salmon: Their fight for survival.* Houghton Mifflin, Boston.

Olver, C. H., B. J. Shuter, and C. K. Minns 1995. Toward a definition of conservation principles for fisheries management. *Canadian Journal of Fisheries and Aquatic Sciences* 52:1584–1594.

Oregon Department of Fish and Wildlife (ODFW). 2002. Status Report: Columbia River Fish Runs and Fisheries 1938-2000. Tables Only. Washington Department of Fish and Wildlife and the Oregon Department of Fish and Wildlife, Portland, Oregon.

Pacific Fishery Management Council (PFMC). 1992. Review of the 1991 Ocean Salmon Fisheries. Pacific Fishery Management Council, Portland, Oregon.

Pacific Salmon Commission. 1992. Joint Chinook Technical Committee Annual Report. Pacific Salmon Commission. Report TCCHINOOK (92)-2. Vancouver, British Columbia.

Pacific Salmon Commission. 1993. Joint Chinook Technical Committee Annual Report. Pacific Salmon Commission. Report TCCHINOOK (93)-2. Vancouver, British Columbia.

Paulik, G. J. 1969. Computer simulation models for fisheries research, management, and teaching. *Transactions of the American Fisheries Society* 98:551–559.

Paulik, G. J., A. S. Hourston, et al. 1967. Exploitation of multiple stocks by a common fishery. *Journal of Fisheries Research Board of Canada* 24: 2527–2537.

Pielou, E. C. 1969. *An introduction to mathematical ecology.* John Wiley and Sons, New York.

Quinn, T. J. and R. B. DeRiso. 1999. *Quantitative fish dynamics.* Oxford, United Kingdom, Oxford University Press.

Rich, W. H. 1941. The present state of the Columbia River salmon resources. Pages 425–430 in *Sixth Pacific Science Congress.* Fish Commission of State of Oregon. University of California, Berkeley.

Ricker, W. E. 1954. Stock and recruitment. *Journal of the Fisheries Research Board of Canada* 11:559–623.

Ricker, W. E. 1958. *Handbook of computations for biological statistics of fish populations.* Fisheries Research Board of Canada, Ottawa.

Ricker, W. E. 1972. Hereditary and environmental factors affecting certain salmonid populations. Pages 19–160 in R. C. Simon and P. A. Larkin, eds. *The stock concept in Pacific salmon.* University of British Columbia, Vancouver.

Ricker, W. E. 1973. Linear regressions in fishery research. *Journal of the Fisheries Research Board of Canada* 30:409–434.

Ricker, W. E. 1981. Changes in the average size and age of Pacific salmon. *Canadian Journal of Fisheries and Aquatic Sciences* 38:1636–1656.

Ricker, W. E., and H. D. Smith. 1975. A revised interpretation of the history of the Skeena River sockeye salmon (*Oncorhynchus nerka*). *Journal of the Fisheries Research Board of Canada* 32:1369–1381.

Riddell, B. E., and W. C. Legget. 1981. Evidence for an adaptive basis variation in body morphology and time of downstream migration juvenile Atlantic salmon (*Salmo salar*). *Canadian Journal of Fisheries and Aquatic Sciences* 38:308–320.

Roos, J. F. 1991. *Restoring Fraser River salmon.* Pacific Salmon Commission, Vancouver, British Columbia.

Schalk, R. F. 1986. Estimating salmon and steelhead usage in the Columbia Basin before 1850: The anthropological perspective. *Northwest Environmental Journal* 2:1–29.

Schoonmaker, P. K., T. Gresh, J. Lichatowich, and H. D. Radtke. 2003. Past and present Pacific salmon abundance: Bioregional estimates for key life history stages. *American Fisheries Society Symposium: Nutrients in salmonid ecosystems: Sustaining production and biodiversity* 34:33–40.

Simms, J. 1877. Letter to the Commissioner of Indian Affairs from J. Simms, U.S. Indian Agent, Colville Agency.

Starnes, L. B., G. C. Jiminez, et al. 1995. North American Fisheries Policy. *Fisheries* 20(4):6–9.

Stockner, J. G.; E. Rydin, and P. Hyenstrand. 2000. Cultural oligotrophication: Causes and consequences for fisheries resources. *Fisheries* 25(5):7–14.

Taylor, E. B. 1990. Environmental correlates of life-history variation in juvenile Chinook salmon, *Oncorhynchus tshawytscha* (Walbaum). *Journal of Fish Biology* 37:1–17.

Taylor III, J. E. 1999. *Making salmon: An environmental history of the Northwest fisheries crisis.* University of Washington Press, Seattle.

Thompson, W. F. 1951. An outline for salmon research in Alaska. Fisheries Research Institute, University of Washington. Circular No. 18. Seattle.

Thorpe, J. E. 1995. Impacts of fishing on genetic structure of salmonid populations. Pages 67–80 in J. G. Cloud and G. H. Thorgaard, eds. *Genetic conservation of salmonid fishes.* Plenum Press, New York.

Van Hyning, J. M. 1973. Factors affecting the abundance of fall Chinook salmon in the Columbia River. *Research Reports of the Fish Commission of Oregon* 4.

Vannote, R. L., G. W. Minshall, K. W. Cummins, J. R. Sedell, and C. E. Cushing. 1980. The river continuum concept. *Canadian Journal of Fisheries and Aquatic Sciences* 37:130–137.

Volk, C. J., P. M. Kiffney, and R. L. Edmonds. 2003. Role of riparian red alder in the nutrient dynamics of coastal streams of the Olympic Peninsula, Washington, USA. *American Fisheries Society Symposium: Nutrients in salmonid ecosystems: Sustaining production and biodiversity* 34:213–225.

Walters, C. J. 1986. *Adaptive management of renewable resources.* MacMillan Publishing Company, New York.

Walters, C. 1996. Information requirements for salmon management. Pages 61–68 in R. J. Naiman and D. Stouder *Pacific salmon and their ecosystems: Status and future options.* Chapman and Hall, New York.

Walters, C. J., and J. S. Collie. 1988. Is research on environmental factors useful to fisheries management? *Canadian Journal of Fisheries and Aquatic Sciences* 45:1848–1954.

Waples, R. S. 1991. Definition of "species" under the endangered species act: Application to Pacific salmon. National Marine Fisheries Service, Seattle, Washington. 29 p.

Ware, D. M., and R. E. Thompson. 1991. Link between long-term variability in upwelling and fish production in the northeast Pacific Ocean. *Canadian Journal of Fisheries and Aquatic Sciences* 48:2296–2306.

White, R. 1996. *The organic machine: The remaking of the Columbia River.* Hill and Wang Publishers, New York.

Return to the River

10

The Estuary, Plume, and Marine Environments

Daniel L. Bottom, M.S., Brian E Riddell, Ph.D.,
James A. Lichatowich, M.S.

The Columbia River Estuary . . . Transition from River to Ocean
  Estuarine Influence on Salmonid Life Histories
  Estuarine Support of Juvenile Salmon
  Human Influences
    Physical Habitats and Processes
    The Estuarine Food Web
    Salmon Life Histories and Hatchery Effects
  Conclusions
The Columbia River Plume . . . and Local Upwelling
  Riverine Influence on Coastal Waters
  The Coastal Upwelling System
The Pacific Ocean
  The Ocean and Coastal Currents
  The El Niño-Southern Oscillation Cycle and Influences on Salmon Production
  Summary
Scales of Ocean-Climate Variability and Salmon Production
The Ocean Environment and Salmon Production in the Columbia River
Salmon Management in a Sea of Change
Conclusions and Recommendations
  Conclusions
  Critical Uncertainties
  Recommendations
Literature Cited
Appendix: Ocean-Climate Indices

    "Some might be tempted to attribute all changes in salmon abundance to changes in ocean conditions and to conclude that management related to rivers is therefore unimportant. However, because all human effects on salmon are reductions in the total production that environment allows, management interventions are more important when the ocean environment reduces natural production than when ocean conditions are more favorable." (p. 361)
    —National Research Council. 1996.

This quotation from *Upstream* (NRC 1996) exemplifies one of the fundamental changes in the past 20 years about the sources of variation in salmon production. Previously, biologists considered the ocean environment for Pacific salmon to be relatively stable in its productive capacity at levels that were substantially greater than those utilized by the extant number of salmon produced from freshwater environments. Changes in salmon abundance were attributed primarily to poor habitat conditions in freshwater. These ideas were formalized in theoretical population models, which emphasized the role of density-dependent mortality during egg and early juvenile stages, and in hatchery programs, which assumed that annual production would be increased by eliminating various causes of freshwater mortality (Lichatowich et al. 1996; Bottom 1997). This was a basic premise to the widespread expansion of artificial production of salmon throughout the north Pacific Ocean (Lichatowich 1999). Given the extensive development of the Columbia River as a hydrosystem, the extent of habitat changes, and the use of hatcheries to mitigate for the loss of juvenile salmon production, it is not surprising that most of this book would be directed to consideration of those impacts and how to restore production. However, the life of Pacific salmonids is largely spent in marine environments and recent history has demonstrated that those environments can be strong determinants of salmon production. Understanding the effect of ocean conditions and associating these with changes in the freshwater environments will be increasingly important as agencies consider the "normative river concept" and attempt to assess the results from restoration activities in the Columbia Basin.

Major changes in annual returns of Pacific salmon in the Columbia River and among species and major river systems throughout the north Pacific provide compelling evidence for an effect of ocean conditions on salmon production. The time scales of changes can be much more rapid than most impacts on freshwater habitats, and scientists are increasingly providing evidence of cycles in ocean productivity. Regional fluctuations in fish populations have been linked to large-scale climatic changes at interannual (McClain and Thomas 1983), interdecadal (Mantua et al. 1997), and multicentennial (Finney et al. 2002) periods. At interannual scales, for example, strong El Niño conditions in the tropics are associated with major changes in marine fauna throughout the Northeast Pacific, including northern range extensions for marine, fishes, birds, and plankton (McClain and Thomas 1983; Pearcy et al. 1985; Mysak 1986); reduced reproductive success of Oregon seabirds (Graybill and Hodder 1985); changes in the migration routes of adult sockeye salmon (*Oncorhynchus nerka*) returning to the Fraser River in British Columbia (Wickett 1967; McClain and Thomas 1983); and reduced size, fecundity, and survival of adult coho salmon (*O. kisutch*) off Oregon (Johnson 1988). Although the specific mechanisms are poorly understood, these results underscore the importance of the larger oceanic and atmospheric system within which the Columbia River Basin and its migratory stocks of salmon are embedded.

The extent of annual changes in marine survival of Chinook (*O. tshawytscha*) and coho salmon has been tracked since the early 1970s using small coded-wire tags embedded in the snout of juveniles and subsequently recovered in fisheries and spawning areas (tagging program described in Johnson 1990). Coronado and Hilborn (1998a, 1998b) provide an extensive review of these tagging programs over time and by geographic area. These data have been updated through recoveries in 2000 (data provided by Hilborn and Magnusson, University of Washington, personal communication). The results are presented for representative Columbia River coho and fall Chinook stocks to demonstrate the annual variations in survival and synchrony between the stocks over time (Figure 10.1). The magnitude of the between-year variation is substantial with coho survival rates for individual stocks varying from less than 0.25% to over 8% and fall Chinook survivals ranging from less than 0.10% to almost 6% (a range of over 100-fold). Variation in survival over these ranges can mean returns to the Columbia River of a few thousand fish to a few hundred thousand fish, but that variation may have had little to do with freshwater conditions.

This chapter will consider three marine environments that influence the production of salmonids in the Columbia Basin: the tidal river downstream of Bonneville Dam and the estuary (Figure 10.2), the Columbia River plume and coastal upwelling system, and the north Pacific Ocean. The ecological state of the lower river and estuary (Figure 10.3) has been compromised by extensive habitat alteration from human activities in those areas and in the upper river. The plume is directly influenced by the change in seasonal flows and regulated discharges from the upriver hydrosystem. Beyond the plume, alterations within the Columbia Basin are not likely to influence the ocean, but the opposite is clearly not true. Ocean conditions and changing climate can directly affect salmon survival rates and returning abundances of salmon, and indirectly affect abiotic factors in the basin. These large-scale environmental effects are beyond what we can "fix" or regulate, but resource managers in the Columbia Basin should account for them when assessing production trends, developing annual management plans, and developing future conservation and recovery actions.

The Columbia River Estuary . . . Transition from River to Ocean

Although the term "estuary" is defined as a semi-enclosed body of water where freshwater and saltwater mix, in common usage and in our discussion here, it may be considered the entire portion of a river that is influenced by

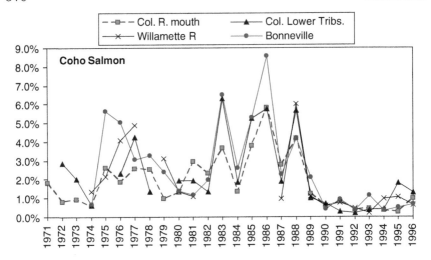

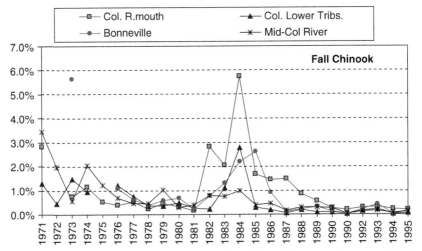

Figure 10.1 Marine survival rates (% survival of tagged juveniles to recruitment to fisheries, methods described by Coronado and Hilborn 1998a, 1998b) for groups of coho and fall Chinook released in the lower portions of the Columbia River, 1971–1996 spawning years. Col. R. mouth = hatcheries in tributaries to the Columbia estuary, Col. Lower Tribs. = tributaries downstream from Bonneville Dam but upstream of the estuary, Willamette R. = any hatcheries within the Willamette River, Bonneville = hatchery complex at Bonneville Dam. Mid-Col. River = bright fall Chinook released in the Hanford Reach and mid-Columbia River.

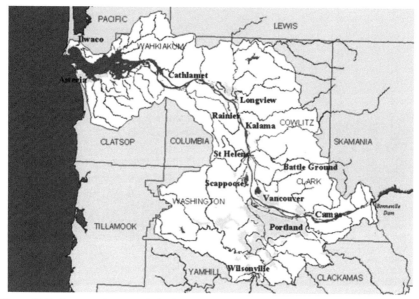

Figure 10.2 Large-scale map of the lower Columbia River mainstem and estuary showing major tributaries and population centers. Map provided by the Columbia Basin Fish and Wildlife Authority.

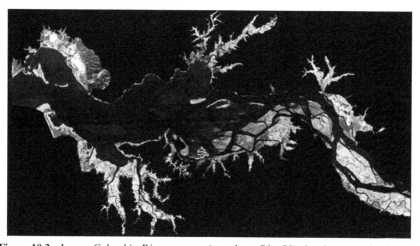

Figure 10.3 Lower Columbia River estuary (mouth to Rkm75) showing extensive islands, shorelines, and shallow areas important to juvenile outmigrating salmonids along the southern border of the estuary and along tributary mouths. Tidewater extends upstream on the mainstem Columbia River to Rkm 233 at Bonneville Dam. Photo from Lower Columbia River Estuary Partnership (LCREP), 2001. Landsat 7 imagery acquired 24 March 2000.

Figure 10.4   The town of Astoria lies on the south shore of the Columbia River in Oregon at the mouth of the river. The town was one of the early west coast settlements and has a rich history in whaling and fishing. Photo by Mike Pinney (*www.photobymike.com*).

ocean tides—in the case of the Columbia River, from the mouth at Astoria (Figure 10.4) to Bonneville Dam (Figure 10.2). Twice during the migrations of Pacific salmon, they undergo extensive physiological changes to make the transition between radically different environments: from freshwater to saltwater as juveniles and the reverse as maturing adults. Although this important transition takes place in the estuary, this part of the salmon's ecosystem has received little attention (Simenstad et al. 1982). Because the estuary is the terminus of the river, it is also where the cumulative impacts of upriver actions all focus (Simenstad et al. 1992), including potential adverse effects of pollution, changes in biological and non-biological input, and alteration of seasonal flow patterns. The estuary is critical habitat that can constrain total salmon production, particularly that of the more estuarine-dependent species such as chum (*O. keta*) and fall Chinook salmon. In a recent review of the influence of estuarine conditions on the Columbia River Fish and Wildlife Program, the Independent Scientific Advisory Board (ISAB) concluded that "All of the investment and effort in the Fish and Wildlife Program flow through this unique environment, but interaction of change in the estuary with projects of the Fish and Wildlife Program, and their combined effect, has basically been ignored." (ISAB 2000, p. iii). Yet the estuary's role in meeting the objectives of the Fish and Wildlife Program has been largely ignored.

Earlier in this book, salmon habitat was equated to beads on a string. Beads were the places where salmon carried out important parts of their life cycle such as spawning, rearing, holding, or avoiding predation; and the string was equated to migration corridors giving salmon access to those places. The estuary is a critical part of the string, a migratory corridor connecting riverine and oceanic habitats, and a place (bead) where some juvenile salmon may rear for extended periods prior to their migration to the sea.

Estuarine Influence on Salmonid Life Histories

Salmon have developed a variety of strategies to utilize the estuary and move between freshwater and marine areas. Juvenile sockeye and coho salmon generally spend a limited time in the estuary and move quickly from the riverine to the marine environment (Groot and Margolis 1991; Pearcy 1992). Pink (*O. gorbuscha*) and chum salmon use the estuary for spawning, as well as an early rearing phase that may last from days to weeks (Pearcy 1992). Chinook salmon display a variety of estuarine strategies. Reimers (1973) identified five different juvenile Chinook life history strategies in the Sixes River based on size and timing of entry into the estuary. Schluchter and Lichatowich (1977) observed seven patterns in the use of the estuary by juvenile Chinook in the Rogue River.

Variations in juvenile salmon life history may be linked to the diversity of habitat opportunities throughout a river basin, including those within the estuary (Healey 1991; Healey and Prince 1995; see also Chapters 3 and 6). Many studies indicate that the patterns of salmonid movement and habitat use within estuaries are size-related. Small subyearlings (Chinook and chum salmon fry) often occupy shallow, near-shore habitats, including salt marshes, tidal creeks, and intertidal flats (Levy and Northcote 1982; Myers and Horton 1982; Simenstad et al. 1982; Levings et al. 1986). As subyearling salmon grow to fingerling and smolt stages, their distribution typically shifts toward deeper habitats farther from the shoreline (Healey 1982, 1991; Myers and Horton 1982). McCabe et al. (1986) reported that subyearling Chinook in shallow intertidal habitats of the Columbia River estuary were smaller than subyearlings captured in deeper pelagic areas (see shallow water habitats in Figure 10.3). Similarly, a 1980–1981 survey of the estuary found that most yearling Chinook salmon occupied deeper channel sites rather than intertidal sites near shore (Bottom et al. 1984).

Differences in estuarine rearing and migration strategies can be distinguished at least at the level of the two major life history types. Subyearling or ocean-type Chinook salmon enter the estuary gradually as part of their protracted downstream rearing and growth phase (see detailed discussion of juvenile migration patterns in Chapter 6). They spend several weeks to months in the estuary prior to a long marine migration (Bottom et al. 1984; Healey 1991). The yearling or stream-type life history, on the other hand, appears to spend little time in the estuary (Healey 1991; Pearcy 1992).

Rich's (1920) historic survey of Columbia River Chinook salmon described variation in juvenile life history far beyond the simple ocean- versus stream-type dichotomy. For example, juvenile Chinook salmon occupied the estuary 12 months of the year, including large numbers of fry that dispersed into estuarine habitats soon after emergence. A recent analysis of Rich's results (Bottom et al. in press; Burke 2004) concluded that at least five variants of subyearling (ocean-type) life history existed in the early decades of the 20th century, including juveniles that reared primarily in the estuary and others that exhibited a combination of riverine- and estuarine-resident behaviors. The diversity of juvenile life histories within the estuary may reflect the complex geographic structure of Chinook salmon populations throughout the vast Columbia River Basin. For example, particular life history patterns of juvenile Chinook salmon in the Nanaimo (Carl and Healey 1984) and Salmon River (Bottom et al. 2005) estuaries have been linked to discrete spawning areas within each basin.

Estuarine Support of Juvenile Salmon

Simenstad et al. (1982) hypothesized three important functions for estuaries that enhance the growth and survival of Pacific salmon:

1. *Physiological Transition Zone*: Salmonids undergoing physiological change may benefit from the gradual change from freshwater to saltwater within the estuary.
2. *Predator Avoidance*: Although the abundance of predators is higher in the estuary than upriver, juvenile salmon may disperse into estuarine habitats that offer protection from predators. In addition, during the period of juvenile residence in estuaries, turbidity is higher, reducing predator efficiency.
3. *Optimum Foraging Conditions*: The size, distribution, and density of many prey organisms in the estuary appear to be optimal for juvenile salmon.

The Columbia River plume, which is the freshwater lens that extends into the near-shore ocean, could be considered an extension of the estuary and may afford similar benefits as rearing habitat. For example, recent studies by National Oceanic and Atmospheric Administration (NOAA) Fisheries suggest that the Columbia River plume and frontal zone may offer productive feeding habitat for juvenile salmon (Schabetsberger et al. 2003). Density gradients concentrate important zooplankton prey resources of young salmon at the frontal margins of the Columbia River plume (e.g., Fresh et al. in press), and small juveniles appear to prefer plume and frontal habitats to marine conditions further offshore. Thus, changes in the hydrograph of the Columbia River, which have altered the size and structure of the plume during the spring and summer, could affect important habitat functions as many juvenile salmonids exit the river mouth (Pearcy 1992).

The Columbia River plume also may limit predation pressure by distributing juvenile salmon offshore away from predators near the shoreline (Pearcy 1992). Recent observations indicate that the small Chinook and coho salmon are distributed in surface waters (< 12m depth) within the plume (Emmett et al. 2004) and that their distribution extends further offshore during peak spring flows than in low-flow years, when the area of the Columbia River plume is substantially reduced (Fresh et al. in press). In addition, laboratory studies imply that high turbidities, as often occur in estuarine and river-plume environments, may be sufficient to reduce vulnerability of young salmon to piscivorous fish and seabirds without affecting their own foraging success (De Robertis et al. 2003).

Given the number of potential predators in the estuary and in the nearshore ocean, the foraging conditions found in the natural estuarine food webs may be the most important function of estuarine habitats for salmon. Rapid growth in the estuary allows the juvenile salmon to grow out of their vulnerability to predators (Simenstad et al. 1982). For that reason, changes in the estuary that impact the food web for juvenile Pacific salmon or their opportunity to access productive estuarine habitats could represent an important constraint on production and recovery in the Columbia Basin (Figure 10.5).

The few studies of salmon food habits in the Columbia River estuary have sampled primarily near the main ship channel and in open-water areas, where various epibenthic crustaceans, including the amphipod *Corophium* spp., are often among the most important prey items. Despite their acknowledged importance to salmon in many other estuaries (e.g., Levy and Northcote 1982; Healey 1991), shallow wetlands of the Columbia had never been surveyed before 2002, when NOAA Fisheries initiated a study of selected emergent, shrub, and forested wetlands in the Cathlamet Bay region (Figure 10.6). Surveys to date reveal consistent use of wetland channels by small ocean-type Chinook salmon during March through August, and juvenile chum salmon for a short period, primarily in April (Roegner et al. 2004). The dominant prey of marsh-rearing Chinook are emergent chironomid insects, which are likely consumed as the pupae move to the water's surface and emerge as adults (Lott 2004). The results indicate that vegetated wetlands play an important role in the production of insect prey, which are heavily consumed by small subyearling salmon before entering deeper, open-water habitats.

Figure 10.5 Sampling juvenile salmonid use of estuarine marsh habitats using a trapnet. Photo by Dan Bottom, NOAA Fisheries.

Human Influences

Changes in the extent and nature of the Columbia River estuary (Figures 10.2 and 10.3) that affect the quality of habitat for salmonids are due to a combination of natural and human-caused factors. These may involve local changes within the estuary or external tidal or riverine influences that shape estuarine conditions. During the last century, the most significant human effects on the estuary and its support of juvenile salmon have included (1) physical modification of habitat and habitat-forming processes, and (2) biological changes, including shifts in estuarine food webs and in salmon populations and life histories. These results reflect changes both within and outside the estuary.

Physical Habitats and Processes

Natural aging of the estuary results from the interplay between accretion of sediments derived from upriver areas and the gradual rise in sea level since the last glaciation (Day et al. 1989). The process of estuary filling is accelerated by attached vegetation that acts to trap and stabilize sediments (Day et al. 1989).

Figure 10.6   The emergent wetlands of Cathlamet Bay, including Russian Island shown above, are dissected by a dendritic network of shallow channels, where many subyearling (ocean-type) salmon rear and grow before moving offshore to deeper habitats. Although Cathlamet Bay wetlands have expanded due to sediment accretion (Elliott 2004), the majority of historic emergent and forested wetlands of the Columbia River estuary have been lost to dyking, filling, and shoreline development (Thomas 1983). Photo by Susan Hinton, NOAA Fisheries.

In the Columbia River estuary, an emergent grass, Lyngby's Sedge (*Carex lyngbei*), is among the most important plant species in this regard (Thomas 1983). Accretion of sediment causes a gradual uplifting of the estuary that is countered to some extent by the continuing increase in sea level since the last glaciation. As the estuary builds up, marsh is gradually converted to willow and spruce swamp. Swamp-dominated flood plain is the end product of the estuarine successional process in the Columbia River (Thomas 1983).

In addition to the slow filling of the estuary over time, natural processes act to move and modify the estuary continuously, often in a very rapid and dramatic fashion. The result is a highly dynamic physical structure. Sherwood et al. (1990) analyzed early navigational charts and noted profound changes in the river entrance from year to year. The pre-development river mouth was characterized by shifting shoals, sandbars, and channels forming ebb and flood tidal deltas. Prior to dredging and maintenance, the navigable channel over the tidal delta varied from a single, relatively deep channel in some years to two or more shallow channels in other years (Sherwood et al. 1990).

Although the processes of erosion and deposition are ongoing, the overall rate of natural aging is generally slow. Relative to the rate of change caused by human actions, natural aging of the Columbia estuary has likely been insignificant over the last century (Thomas 1983). Most of the human alteration of the estuary results from attempts to stabilize and simplify a naturally dynamic and complex environment.

Early activities in the estuary attempted to stabilize the navigation channel. Jetties were constructed on the north and south shores of the river (Figure 10.7) to hold a channel in place, while a regular dredging program deepened the channel (Sherwood et al. 1990). Material dredged from the main channel was deposited in shallow water areas. Many of these areas were subsequently removed from the estuary by dykes.

Dredging, filling, and the construction of dykes (Figure 10.8) resulted in important changes in the morphology of the estuary (Thomas 1983; Sherwood et al. 1990). Total volume of the lower estuary from the mouth to Puget Island (Rkm 75) has declined by about 12% since 1868 (Sherwood et al. 1990). Thomas (1983) estimated that 40% of the original estuarine area in this region has been converted to developed flood plain (Figures 10.9 and 10.10).

Figure 10.7 North jetty at the entrance to the Columbia River estuary. The north jetty was completed in 1917, more than 20 years after the south jetty, to further stabilize the river mouth and a navigation channel from the ocean to the estuary. Since the 1970s, a minimum 40-foot-deep navigation channel has been maintained by dredging 170 km of tidal river from the estuary entrance to the city of Portland, Oregon. Construction is planned in 2005 to further deepen the entire navigational channel by 3 feet. Photo from Lower Columbia River Estuary Workshop, 2003.

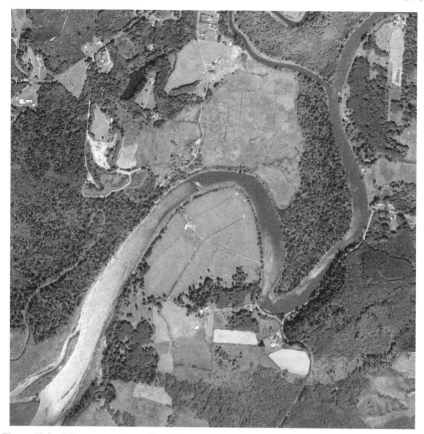

Figure 10.8 Like many other tributaries of the lower Columbia River estuary, much of the lower Grays River has been dyked to eliminate tidal inundation and reclaim low-lying wetlands for agricultural use. Tideland dyking and filling have reduced the surface area of the estuary and eliminated productive shallow-water rearing habitat for juvenile salmon. Photo from Columbia River Estuary Study Taskforce, Astoria, Oregon.

Overall, development since the mid-19th century has resulted in a loss of 77% of the tidal swamps, 62% of the tidal marshes, and 7% of the tidal flats in the lower estuary (Thomas 1983). Dykes, roads, and shoreline development have similarly altered the flood plain throughout the vast tidal freshwater reach above Puget Island to Bonneville Dam, but habitat changes in this region have not yet been quantified (Bottom et al. in press).

In the mid-1970s, with the completion of the upper basin hydrosystem storage projects (see Chapter 7), flow patterns were altered considerably, changing the seasonal input of freshwater to the estuary. In particular, the strength of the spring freshet was appreciably diminished when upriver

Figure 10.9 This earthen dyke along the banks of the Lewis and Clark River, a tributary to Youngs Bay, protects a low-lying pasture (on the right) from tidal inundation. Photo from Columbia River Estuary Study Taskforce, Astoria, Oregon.

storage projects were used to shift water into the winter (Figure 10.11). The result has been a general flattening of the seasonal hydrograph (Figure 10.12). For example, in January (1969–1999), the current average observed flow is 42.5% greater than the estimated adjusted flows (Figure 10.13). Adjusted flows are estimated natural flows expected in the absence of storage (Orem 1968; Sherwood et al. 1990). In addition, the timing of the spring freshet has been moved forward about one month. The biological impacts of these changes have not been studied; however, changes in physical parameters with potential biological impacts have been significant. Changes in estuarine bathymetry and flow have altered the extent and pattern of salinity intrusions into the river and have increased stratification and reduced mixing (Sherwood et al. 1990). Ebbesmeyer and Tangborn (1992) present evidence that the shift of spring flow into the winter has altered sea surface salinities along a large part of the North American coast.

Flow regulation has nearly eliminated access by juvenile salmon to the historic tidal flood plain, which likely provided productive rearing habitat and off-channel refugia during high-flow periods. Whereas periods of overbank flow were common before 1900, flow regulation and water withdrawals have allowed only a few significant overbank events since 1948 (Figure 10.14)

Figure 10.10   The Highway 101 fill and tidegate severely restrict tidal exchange at the mouth of the Chinook River, which enters Baker Bay on the Washington shore of the Columbia River estuary near Rkm 10. Remnant marsh habitat is visible in the lower half of the photo (below the highway crossing), but most historic wetlands of the lower Chinook River have been dyked for agriculture. A restoration plan has been developed for the Chinook River watershed and estuary. Photo from Columbia River Estuary Study Taskforce, Astoria, Oregon.1

(Bottom et al. in press). The season when overbank flow is most likely has also shifted from spring to winter, because winter floods in the western sub-basin (rather than spring freshets in the interior subbasin) are now the major source of high-flow events. Dykes and other changes have increased the historical bank-full flow level below Vancouver from approximately 18,000 m^3s^{-1} to approximately 24,000 m^3s^{-1}. However, even at historical bank-full levels, flow manipulations would have prevented most bank-full events (Bottom et al. in press).

Kukulka and Jay (2003) estimated that dykes and flow regulation combined have reduced the total area of shallow-water habitat (defined as 0.2 to 2.0m depth) by 62% in the reach between Skamokawa (Rkm 80) and Beaver (Rkm 145). Not only has potential habitat been removed, but the remaining shallow habitat has been displaced to a lower elevation relative to historical levels because river stage has been reduced during the spring season (Kulkulka and Jay 2003). If downstream migration and estuarine residency of small subyearling salmon is linked to shallow estuarine habitat, then

Figure 10.11 Dworshak Dam on the North Fork of the Clearwater River in north central Idaho. This high dam is typical of upper basin storage reservoirs where elevation differences between the forebay and tailrace are greater than 300 feet. Photo from US Army Corp Digital Visual Library, website at *http://images.usace.army.mil/photolib.html*.

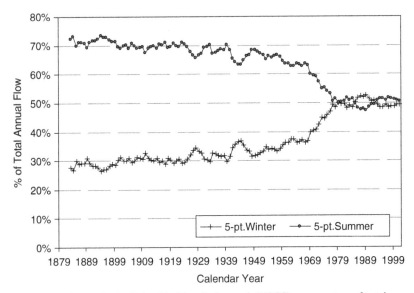

Figure 10.12 Changes in the Columbia River hydrograph (USGS) as percentage of total annual runoff at The Dalles, Oregon, 1879–2001. Summer flow = April–September. Five-point moving averages used to smooth trends in annual data; the first data point is for the average of 1879–1883 and plotted as 1883.

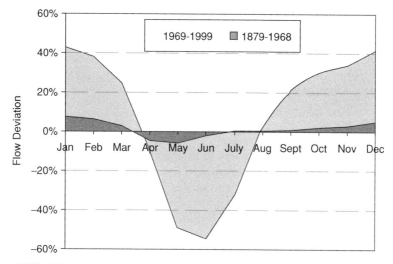

Figure 10.13 Change in monthly flow patterns comparing years 1879–1968 (no to low storage capacity) to current storage conditions (1969–1999). Graph from ISAB (2000, Figure 4, p. 11). Plotted values are (Observed flow minus Adjusted flow/Observed flow) from data provided by Dr. D. Jay and P. Naik (Oregon Graduate Institute, Oregon Health and Science University). Deviations in cubic meters per second of flow.

removal of a large proportion of the tidal flood plain could limit expression of some ocean-type life histories (Bottom et al. in press).

Development also has changed the circulation patterns and increased the shoaling rates in the estuary. Sediment input to the estuary has declined due to the altered hydrograph, but the estuary is now a more effective sediment trap. An analysis of marsh-vegetation change at Russian Island in the Cathlamet Bay region of the estuary reveals a contemporary shift toward late

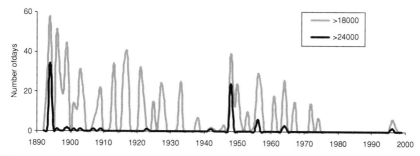

Figure 10.14 The incidence of flows above 18,000 m³s⁻¹ (the pre-1900 estimated bank-full flow level) and above 24,000 m³s⁻¹ (the present bank-full flow level). The present bank-full flow level has only been exceeded a few times since 1948 (from Bottom et al. in press).

successional stages—high marsh and scrub-shrub vegetation types—consistent with effects of sediment accretion and reduced disturbance from flow regulation (Elliott 2004). Although the Columbia River is characterized as a highly energetic system, it has been changing as a result of development and is now similar to more developed and less energetic estuaries throughout the world (Sherwood et al. 1990).

The Estuarine Food Web

Internal changes to the estuary and in the watershed above it have altered the food web in ways that are detrimental to Pacific salmon. The estuarine food webs that support juvenile salmon are detritus-based (Bottom et al. 1984; Salo 1991). However, in the Columbia estuary, the detritus-based food web has undergone an important shift in response to development. Macro-detritus derived from emergent marsh vegetation has undergone a dramatic reduction due to the loss of shallow-water habitat. The loss of those production areas reduced emergent plant production by 82% (Sherwood et al. 1990). Sherwood et al. (1990) estimate that, prior to development, the standing crop of organisms that feed on the macro-detritus would have been 12 times the current standing crop. Since those organisms are prominent prey of juvenile salmonids, it is not unreasonable to assume that a reduction in the food web supported by macro-detritus has had a negative impact on the Pacific salmon. However, Sherwood et al. (1990) could not provide empirical evidence of a linkage between the food web and the status of salmon in the Columbia Basin. Bottom et al. (1984) found that the index of feeding intensity (IFI, i.e., total weight of the stomach contents expressed as a percentage of the fish's total weight) was lower in subyearling Chinook salmon from the Columbia estuary compared with the IFI for Chinook salmon in either the Fraser (British Columbia) or Sixes (Oregon) estuaries.

A food web based on macro-detritus characterized by shallow water, benthic consumers has largely been replaced by a food web composed of deepwater, benthic, and pelagic consumers. Primary production of pelagic algae has increased in the impounded stretches of the mainstem Columbia and Snake Rivers and the resulting micro-detritus input to the estuary is nearly equivalent to the macro-detritus removed from the estuary by dykes (Sherwood et al. 1990). The new food web is favored by fishes such as Pacific herring, smelts, and the non-native American shad (Sherwood et al. 1990). Shad has become one of the dominant species in the Columbia River estuary (Bottom and Jones 1990).

At least 16 non-indigenous fishes and 4 invertebrate species have been introduced into the Columbia River estuary (Weitkamp 1994), with uncertain effects on estuarine food chains. Since its appearance in the estuary in 1938, the Asian bivalve, *Corbicula fluminea*, has expanded its distribution, including recently documented occurrences in the lower mainstem reservoirs and in the Willamette and tributary basins (Wentz et al. 1998). *C. fluminea* distribution is restricted primarily to tidal freshwater regions of the Columbia River estuary, but it also extends into brackish habitats. Biomass estimates as high as 10,000 mg carbon m^{-2} have been measured in some subsidiary channels (Holton et al. 1984; Simenstad et al. 1984). Although *C. fluminea* commonly occurs in association with important juvenile salmonid prey in channel habitats (McCabe et al. 1997, 1998), ecological interactions among *C. fluminea* and salmonid prey species have not been investigated (Bottom et al. in press).

An introduced calanoid copepod, *Pseudodiaptomus inopinus*, first appeared in the Columbia River estuary after 1980 and has become abundant since then (Cordell et al. 1992; Cordell and Morrison 1996), raising concerns about potential impacts on native calanoid copepods and associated estuarine food chains. Estuarine dynamics favoring the trapping of particles and zooplankton in the region of the Estuarine Turbidity Maximum (ETM) have likely increased since the coordination of river-flow regulation by the hydropower system in the mid-1960s (Jay et al. 1990). As noted above, however, salmon may derive few direct benefits from increased pelagic production in the ETM compared with the historic macrodetrital food sources that were once distributed throughout the lower estuary (Bottom et al. in press).

Another major change in the estuarine food web involves the effects of piscivorous marine birds on outmigrating salmon stocks. Since the early 1990s, abundance of Caspian terns (*Sterna caspia*) has reached unprecedented numbers through the establishment of nesting colonies on artificial dredge-spoil islands in the lower Columbia River estuary (Figure 10.15). Following the arrival of approximately 1,000 pairs of terns from Willapa Bay in 1984, colonies in the Columbia River estuary rapidly expanded to 9,000 to 10,000 pairs, concentrated at Rice Island (Fresh et al. in press). During 1999 and 2000, respectively, juvenile salmonids accounted for 77% and 90% of the diet of Rice Island terns.

A successful relocation of Caspian terns to East Sand Island near the estuary mouth has reduced the impact on outmigrating salmon since 1999. Compared to the Rice Island colony, the diet of East Sand Island terns consists of a greater proportion of marine forage fishes and a smaller salmonid component, ranging from 46% to 33% during the period 1999 to 2001 (Collis et al. 2002; Roby et al. 2003). Following the relocation of the Rice Island tern colony, total salmonid mortalities declined from an estimated 11.7 million in 1999 to approximately 5.9 million in 2001 (Collis et al. 2001b).

Research results suggest that effects of bird predation in the estuary are greatest on salmon and steelhead smolts with a stream-type life history (Fresh et al. in press). This conclusion is supported by the proportions of various salmonid stocks represented among tags detected on spoil islands

Figure 10.15 By 1999, as many as 10,000 pairs of Caspian terns were nesting on Rice Island. Two years later, a successful relocation of the colony to East Sand Island near the estuary mouth had reduced the estimated loss of juvenile salmon to marine birds by approximately half. Photo by Brad Ryan, NOAA Fisheries.

used by Caspian tern colonies. For example, steelhead (*O. mykiss*) smolts appear most vulnerable. In 1998 and 1999, respectively, 2.8% and 11.7% of the wild steelhead smolts and 5.4% and 13.4% of the hatchery steelhead smolts detected at Bonneville Dam were later detected at the Rice Island bird colony (Collis et al. 2001a). In 1998, 0.5% of wild yearling Chinook salmon and 1.6% of hatchery yearling Chinook salmon identified at Bonneville Dam subsequently were detected at Rice Island. Total tag recoveries at Rice Island and East Sand Island from 1998 to 2000 indicate a higher predation of steelhead (0.6% to 8.1% East Sand Island and 1.3% to 1.6% Rice Island) than of Chinook salmon (0.2% to 2.0% East Sand Island and 0.6% to 1.6% Rice Island) (Ryan et al. 2003). The greater vulnerability of stream-type than ocean-type salmonids may reflect their estuarine migration period, which coincides with the peak period of tern nesting and chick production during May and June (Fresh et al. in press; Roby et al. 2003).

While complex changes in estuarine food webs may be important factors influencing salmonid growth and survival, many such effects are directly tied to physical changes upriver and in the estuary (Bottom et al. in press). For example, the apparent shift from macro-detrital to micro-detrital food chains stems from dykes and filling of intertidal wetlands and the creation of deep reservoirs behind mainstem dams. Explosion of the American shad population may reflect the creation of habitat opportunities (i.e., mainstem reservoirs) that favor pelagic-feeding fishes, an enhanced micro-detrital food web, and the increased trapping efficiency of the ETM. Rather than a major new salmonid competitor, shad abundance thus may be seen as a symptom of physical habitat change. Similarly, the sudden appearance of large numbers of Caspian terns in the lower estuary is linked to new habitat opportunities (i.e., dredge-spoil islands) created along the route of out-migrating salmonid smolts. Concentrated releases of hatchery salmon and steelhead smolts that migrate past these islands are perhaps another contributing factor. The underlying physical mechanisms for many biological changes in the estuary suggest that restoration efforts must target the ultimate causes rather than the proximal symptoms of salmon decline (Bottom et al. in press).

Salmon Life Histories and Hatchery Effects

Limited contemporary data for the Columbia River estuary suggest that juvenile Chinook life histories may have become more uniform than those that were present during the early 20th century (Rich 1920; Bottom et al. in press-b; Burke 2004). In particular, recent patterns suggest a greater proportion of yearling and subyearling migrants that reside primarily in natal streams (including hatchery-reared juveniles) and/or main rivers and a lower proportion of juveniles with estuarine-resident life histories (Bottom et al. in press; Burke 2004). These results are indicated by a narrower size range and migration period among juveniles today than described by Rich (1920).

The simplification of juvenile life histories, if correct, is consistent with many changes that have affected conditions for salmon in the estuary and upriver. For example, loss of tidal wetlands may limit rearing opportunities for fry and fingerling migrants whose estuarine growth and survival may be directly linked to availability of shallow, near-shore habitat (Levy and Northcote 1982, Myers and Horton 1982; Simenstad et al. 1982; Levings et al. 1986). Furthermore, loss of mainstem spawning habitat following dam construction and increased water temperatures in the lower sections of some subbasins have eliminated ocean-type life histories from many Chinook populations (see Chapter 3). These effects could limit the proportion of subyearling migrants that might otherwise make use of rearing opportunities in the estuary.

Among the most significant influences on contemporary rearing patterns of juvenile Chinook salmon is the extensive use of artificial propagation in the Columbia River Basin. Hatchery production narrows the range of sizes and time periods of downstream migration and concentrates use of the estuary in ways that may reduce survival of juvenile salmonids. The release of large numbers of hatchery-reared fish over a short time interval could create a density barrier in the river (Royal 1972) or the estuary and near-shore ocean (Oregon Department of Fish and Wildlife 1982).

The relationship between massive hatchery releases and a density-dependent limitation in the survival of juvenile coho salmon in the near-shore ocean and estuary has been the subject of at least 10 studies (Gunsolus 1978; Oregon Department of Fish and Wildlife 1982; Clark and McCarl 1983; McCarl and Rettig 1983; McGie 1984; Nickelson and Lichatowich 1983; Peterman and Routledge 1983; Nickelson 1986; Emlen et al. 1990; Peterman 1989). The studies produced contradictory findings, so the issue is still not resolved (Lichatowich 1993).

Hatchery releases and subsequent downstream migration of juveniles often differ from the natural emigration of wild Pacific salmon. Some of those differences include the mass releases from hatcheries instead of the sequential movement of juveniles from individual tributaries, the pulsed movement of hatchery fish (all fish from a hatchery released at the same time) compared with the natural migration over a longer time interval, and hatchery juveniles that are usually larger than their wild counterparts. Hatchery fish may be the wrong size or they may arrive at the wrong time to exploit the food resources of the estuary (Simenstad et al. 1982). For example, wild Chinook salmon from the Lewis River, Washington, are in healthy condition (Washington Department of Fisheries and Washington Department of Wildlife 1993) and it is one of the largest and most stable populations in the Columbia Basin (McIsaac 1990). One of the reasons for the success of this stock might be the timing of juvenile migration through the estuary. The migration of fall Chinook salmon from Lewis River peaks 2 months after all other salmonids and is later than the other subyearling fall Chinook stocks (McIsaac 1990).

Today, patterns of estuarine habitat use by Chinook salmon are dominated by hatchery juveniles whose times and sizes of migration to the estuary are dictated by hatchery rearing and release schedules. For example, Dawley et al. (1986) found that peak Chinook salmon catches at Jones Beach (Rkm 75) were composed primarily of hatchery-reared juveniles and corresponded to the timing of hatchery releases. McIsaac (1990) found that wild Lewis River juveniles reared in the hatchery arrived at Jones Beach 3 to 5 days after release regardless of the time of year. Thus, it may be difficult to understand the true contribution of the estuary to wild salmon based on present-day patterns of migration and habitat use that are driven largely by artificial propagation.

Conclusions[1]

1. The estuary provides a variety of alternative rearing habitats and contributes to the diversity of juvenile life histories in Columbia River salmon populations (2).
2. Development activities in the estuary and in the river have altered the historical estuarine habitat opportunities and food webs in ways that is likely to negatively affect the survival, growth, and diversity of juvenile salmonids (2).
3. Regional consideration of the biological impacts of flow modification in the Columbia River as being limited to areas above Bonneville Dam ignore the potential impact of these alterations on the physical and biological nature of the estuary (2).
4. Other changes in the physical estuarine processes have significantly altered physical and biological processes in the estuary and may impact salmon production, but additional research is needed to quantify and document the linkages (3).
5. Hatchery operations may have altered the patterns of estuary usage by salmonids and further reduced the survival and growth of hatchery-produced and naturally produced salmonids (3).
6. While historic changes in estuarine habitats and food webs may affect populations throughout the Columbia River Basin, the Fish and Wildlife Program has largely ignored the importance of the estuary as a rearing environment for juvenile salmon (2).

The Columbia River Plume . . . and Local Upwelling

Riverine Influence on Coastal Waters

The Columbia River plume influences the distribution of nutrients, salinity, and the upwelling front off Washington and Oregon. Changes in the river hydrograph associated with flow regulation may significantly impact coastal ecosystems. Discharge from the Columbia River is the dominant source of freshwater runoff to the Washington and Oregon coast, particularly during the late spring and early summer. Both the Columbia River and the Fraser River are point sources of high nitrate, phosphate, and silicate near shore in winter and summer (Landry et al. 1989). The low-salinity surface water of the plume represents an offshore extension of the estuary that varies seasonally

---

[1]Levels of Proof are shown in the parentheses after each conclusion and indicate the amount of scientific support for the implied assumption as follows: (1) Thoroughly established, accepted, peer-reviewed empirical evidence in its favor; (2) Strong weight of evidence in support, but not fully conclusive; (3) Theoretical support with some evidence from experiments or observation; (4) Speculative, but with little empirical support; and (5) Misleading or demonstrably wrong, based on good evidence to the contrary.

in its location along the coast. During winter when surface currents are predominantly northward, the Columbia River plume forms a low-salinity tongue of cold water near the Washington coastline to the north (Landry et al. 1989). During the spring/summer regime, low-salinity water from the Columbia River is located offshore and southward off Oregon. The plume can extend beyond Cape Mendocino, California, and its effects are even visible past San Francisco. Measurements in July 1961 reported the maximum depth of the plume as 2 meters off the Columbia River mouth and 0.5 meters off of Cape Blanco (Huyer, 1983). As a result of the Columbia River plume, variability in surface salinity is much greater in the Pacific Northwest than off California or in the subarctic region (Landry et al. 1989).

The Columbia River plume influences surface density gradients and the cross-shelf properties of coastal waters, which may affect patterns of biological production and biomass. Specifically, the plume can retard offshore transport during upwelling, particularly when maximum river flow occurs (e.g., June). The zone of upwelling influence can be most narrow off northern Oregon where the Columbia River plume forms a partial barrier to the offshore movement of surface water (Huyer 1983). Interaction between upwelling intensity and the volume of flow from the Columbia River affect the location of the upwelling front and, therefore, the distribution of chlorophyll and zooplankton biomass. During strong upwelling the Columbia River plume is advected far offshore. Changes in the distribution of the upwelling front may not only influence environmental conditions for emigrating juveniles but may be important to the movements of adult salmon. Coho salmon, for example, prefer temperatures between 11° and 14°C, which are intermediate between the offshore ocean water (15°–17°C) and upwelled water at the coast (8°–10°C) (Smith 1983). Short-term changes in temperature and feeding conditions that concentrate or disperse fish, in turn, create significant variations in salmon catch rates and landings (Nickelson et al. 1992).

The region of the Columbia River plume is a summertime spawning area for an endemic subpopulation of northern anchovy (Bakun 1993). Local stability of the water column and circulation characteristics associated with the plume during the summer may provide the conditions needed to support larval production. A local minimum in wind velocity and upwelling intensity (< 500 m3/s3) minimize offshore transport while the low-salinity lens of the plume maintains vertical stability and reduces turbulence. Furthermore, the density gradient at the interface of the plume and higher salinity surface waters may provide a counterclockwise circulation that would benefit retention of larvae and other organisms (Bakun 1993). Because such convergence zones tend to concentrate larvae and food particles, they are often important areas of secondary production. Fish and plankton surveys conducted inside and outside the Columbia River plume in 2001–2002 appear to substantiate the importance of the plume and frontal region as habitat: Zooplankton abundance was concentrated at the plume front, and salmon abundance was generally greater inside the plume and frontal zone than in the ocean water beyond (Fresh et al. in press).

Ebbesmeyer and Tangborn (1993) conclude that impoundment of summer flows and releases during the winter by Columbia River dams have altered sea surface salinities from California to Alaska. In terms of the seasonal transition in coastal currents, this shift in the hydrograph results in a decrease in the volume of Columbia River water transported off the Oregon coast during the summer and an increase off Washington in the winter. In the last 60 years, salinity has decreased approximately 1.0 ppt over a distance of 500 km to the north and increased 0.6 ppt over the same distance to the south (Ebbesmeyer and Tangborn 1993). The influence of the plume on other physical and biological properties—for example, temperature, nutrients, density gradients, and the upwelling front—suggests that regulation of Columbia River flows may significantly affect coastal ecosystems of the California Current and subarctic region.

The Coastal Upwelling System

The upwelling system of the California Current, which extends along the west coast of the United States, has been the subject of extensive physical and biological research. Since 1949, large-scale systematic surveys have been conducted off California, primarily south of San Francisco, as part of the California Cooperative Fisheries Investigations (CALCOFI) (Huyer 1983). Detailed small-scale studies of the coastal upwelling system were completed off central Oregon in the 1960s and early 1970s (Peterson et al. 1979; Small and Menzies 1981; Smith 1983). The shorter time frames and local scales of most early research off Washington and Oregon are not directly comparable to the larger interannual scales of information collected off California. Several reviewers (Huyer 1983; Strub et al. 1987a; Landry et al. 1989) later synthesized a variety of data sets to better understand the regional features of the Washington and Oregon coastal ocean.

Research results indicate considerable physical variability that may be important to the life history and production of salmon stocks. Depending on the specific time and location of emigration, local populations may enter very different ocean environments. For example, when coho salmon production collapsed off Oregon in the late 1970s, biologists first considered whether upwelling conditions in near-shore coastal waters might explain variations in marine survival. The successful prediction of adult returns from the previous year's run of jacks (2-year-old males) implied that conditions during the first few months in the ocean were most critical. Researchers initially focused attention on the upwelling process along the Oregon coast (Gunsolus 1978;

Scarnecchia 1981), which was known to increase nutrient levels and biological productivity at about the time that salmon smolts entered the ocean. Nickelson (1986) found a positive correlation between the survival percentage of hatchery coho salmon released and average upwelling conditions in the spring and summer. The results further suggested a threshold response to upwelling levels: In years of "strong" upwelling, survival of hatchery coho averaged 8% survival of smolts to adults compared with only 3.4% during "weak" upwelling years. However, in the years following the 1976 collapse, the correlation between salmon survival and upwelling indices has not held, suggesting that upwelling alone is not sufficient to explain variation in coho salmon production (Jamir et al. 1994; Koslow et al. 2002; Logerwell et al. 2003). Most recently, Lawson et al. (2004) also related freshwater environmental factors to interannual variation in coho survival in the Pacific Northwest and demonstrated correlations between freshwater and marine environmental conditions.

The following characteristics of the coastal ocean and local upwelling system appear to play an important role in the marine survival of salmon:

1. *Variations in the intensity and frequency of upwelling events influence biological production and the recruitment of pelagic marine fishes.*

    Small and Menzies (1981) reported differences in the distribution of chlorophyll biomass and its productivity under different upwelling conditions off Oregon. During weak or intermittent periods of upwelling, the band of maximum chlorophyll was located against the coast and had very high concentrations. Productivity of chlorophyll bands during periods of relaxation between upwelling events could be twice that of the strong upwelling state and often 20 times that in the surrounding water. Peterson et al. (1979) found that very high concentrations of zooplankton off Oregon occurred shoreward of the upwelling front (the sharp interface between upwelled water and the warmer ocean water displaced offshore) and were carried below the pycnocline (density gradient) when upwelling relaxed. However, patterns of abundance of zooplankton populations varied by species.

    The most favorable upwelling conditions for fish production also likely vary by species. Lasker (1978) found that physical factors associated with upwelling affected the survival of anchovy (*Engraulis mordax*) larvae and explained variations in year-class strength. Successful year classes were associated with calm periods between upwelling events that supported the production of favored prey species. Cury and Roy (1989) found evidence that successful recruitment of pelagic fishes depended on winds that were strong enough to promote upwelling but sufficiently calm to prevent turbulent mixing that disperses concentrations of food required for larval survival. Cushing (1995) further notes that northern anchovy and sardine

(*Sardinops sagax*) may have developed different survival strategies for upwelling systems: Anchovy grow more slowly and can tolerate periods of low food availability and intermittent periods of stronger upwelling; sardine seem to grow more rapidly and favor a weaker but more persistent upwelling state. Yet both species appear to avoid spawning locations of the strongest upwelling. Such nonlinear relationships raise questions about the apparent threshold level of upwelling associated with juvenile coho salmon survival in the 1960s and 1970s (e.g., Nickelson, 1986) or the absence of a relationship between upwelling indices and salmon survival since the mid-1980s (Logerwell et al. 2003).

2. *Geographic variations in coastal currents and upwelling affect patterns of biological production off Washington and Oregon. Such variations may be important to the survival and adaptations of salmon populations originating from different river systems and following different migratory paths.*

The gradient in atmospheric pressure that produces southward winds along the Pacific coast varies with location and with seasonal and daily changes, creating geographic and temporal variation in winds, currents, and the strength of coastal upwelling. South of about 40° N latitude (approximately Cape Mendocino, California), winds are southward throughout the year, while north of this location, winds are northward, and therefore, unfavorable for upwelling during the winter months. The average intensity of upwelling is relatively weak northward from the central Oregon coast, but upwelling off the narrow Oregon continental shelf is generally stronger than off Washington and more evenly distributed throughout the summer (Landry et al. 1989). Maximum upwelling off Washington occurs in June, one or two months earlier than along the Oregon coast. South of Coos Bay, coastal currents show considerable short-term variability, while a smoother seasonal cycle is apparent in currents from the central Oregon coast northward (Strub et al. 1987b). Complex bathymetry and the orientation of the shoreline also result in considerable local variation in the intensity of upwelling (Huyer, 1983), with uncertain but potentially significant effects on local salmon stocks.

From geographic differences in winds, currents, bathymetry, and upwelling, Bottom et al. (1989) classified three major physical regions of the continental margin north of Cape Mendocino, California: (1) A Washington coastal region south to the mouth of the Columbia River, (2) a northern Oregon coastal region (south of the Columbia River to Cape Blanco), and (3) a southern Oregon and northern California region south to Cape Mendocino. The discontinuity in winds and currents at Cape Blanco is particularly noteworthy. The zone of upwelling and increased nutrients is wider south of Cape Blanco than along the central and northern Oregon coast, and influence of the Columbia River plume is reduced. Summer winds and upwelling are stronger and more variable than in regions to the north. Furthermore, strong offshore flow much greater than is explained by typical upwelling processes may have an important influence on the transport of phytoplankton biomass and could explain large-scale patterns of zooplankton in areas of the California Current (Abbott and Zion 1987). It is interesting that the ocean migration patterns of coastal Chinook stocks also show a discontinuity at Cape Blanco: Stocks from Elk River (located on the south side of Cape Blanco) and northward appear to rear in waters from Oregon to Alaska; stocks south of Elk River generally rear off southern Oregon and northern California (Nicholas and Hankin 1988).

3. *The coastal ocean off Washington and Oregon exhibits distinct winter and summer regimes. The shift to the summer upwelling regime occurs suddenly, and the specific timing varies between years. While areas of coastal upwelling involve local scale events, the transition to a coastal upwelling regime is regulated by large-scale atmospheric conditions.*

The annual northward migration and strengthening of the atmospheric high-pressure system in the North Pacific causes a shift in wind direction that produces the transition from a winter to a spring/summer regime in the coastal ocean off Washington and Oregon (Huyer 1983). In the winter, coastal currents over the shelf are northward, sea levels are high, and downwelling occurs. Summer conditions are characterized by reduced sea levels, southward mean surface currents over a northward undercurrent, and a strong density gradient across the continental shelf (Strub et al. 1987b). Southward winds and the resulting offshore flow raise cold, nutrient-rich water at the surface along the west coast of the United States. Strub et al. (1987b) report that the spring transition in sea level, currents, and temperatures is driven by the large-scale wind system at scales of 500 to 2,000 km at latitudes north of approximately 37° N. The zone of active upwelling is generally restricted to a narrow coastal band (about 10–25 km) but the affected region can be much broader. The response of the coastal system to southward winds is very rapid. A single upwelling event of a few days' duration, typically in March or April, may be sufficient to cause the shift to the spring/summer regime (Huyer, 1983). Thus, timing of the onset of the transition relative to the period of smolt migration may be important to the survival of juvenile salmon (Pearcy 1992).

The Pacific Ocean

The Ocean and Coastal Currents

Both zoogeographic patterns and fluctuations of plankton biomass in the California Current point to large-scale processes that are not fully explained by upwelling. The "classical view" of the eastern boundary regions of the world's oceans has generally assumed that local upwelling is the major factor controlling pelagic production (Bernal and McGowan 1981). But over the last two decades, new evidence indicates that the productivity of the California Current is not entirely regulated by internal processes, but may be substantially influenced by input from outside the system.

The California Current is a transition zone between subarctic and subtropical water masses and the freshwater systems that enter the ocean along its landward boundary (Figure 10.16). Unlike the large semi-enclosed gyres that circulate in the Central and North Pacific, the California Current is a relatively open system affected by annual fluctuations in currents that contribute water of varying properties from adjacent water masses. After traversing eastward across the North Pacific, the Subarctic Current splits into the northward flowing and counterclockwise Alaskan Gyre and the southward-flowing California Current. During the upwelling season, the California Current carries cold nutrient-rich water from the subarctic Pacific along the west coast. When upwelling subsides in the fall and the downwelling season returns, the northward-flowing California Undercurrent (Davidson Current) appears at the surface and carries warm equatorial water inshore (Favorite et al. 1976).

In the 1960s, biogeographers discovered a close association between the major water masses of the Pacific Ocean as characterized by temperature and salinity profiles (Sverdrup et al. 1942) and the boundaries of large biotic provinces of the pelagic ocean as defined by the distributions of planktonic and nektonic species. North of the equator, Johnson and Brinton (1962)

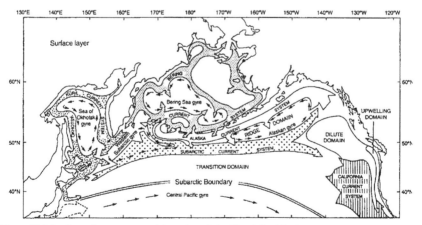

Figure 10.16 Oceanic currents and domains of the North Pacific Ocean north of the subarctic boundary. From Groot and Margolis (1991) as redrawn from Favorite et al. (1976).

identified three major biotic provinces of the Pacific Ocean: A Subarctic assemblage associated with the nutrient-rich waters roughly north of 40° N latitude, a Central Pacific faunal group corresponding to the oligotrophic waters of the Central Pacific gyre, and a group of Transition Zone species occupying the boundary between these two groups along the east–west path of the Subarctic Current and West Wind Drift. Because these biological provinces correspond generally with the boundaries of large semi-enclosed ocean gyres, McGowan (1971, 1974) suggested that they represent discrete, functional ecosystems.

A major exception to these patterns is, however, the California Current system where a small number of coastal species endemic to the region co-occur with a larger mixture of subarctic, subtropic, and equatorial species, many near the peripheries of their distributional range (Johnson and Brinton 1962; McGowan 1971, 1974). Researchers inferred from these results that remote physical factors controlling the input of water and species from other regions may be more important determinants of species composition and abundance in the California Current than biological interactions such as competition and predation (Bernal 1981; Bernal and McGowan 1981). For example, Wickett (1967) first reported that annual concentrations of zooplankton off southern California vary directly and concentrations in the western Bering Sea vary inversely with the southward transport of water at the divergence of the California Current and the Alaskan gyre (see Figure 10.16).

Local upwelling may play a somewhat greater role in interannual variability off the Washington and Oregon Coast than off California. Unlike California (Chelton et al. 1982), monthly anomalies of temperature and salinity off Washington and Oregon in the summer are negatively correlated (Landry et al. 1989), which is an indicator of the upward advection of cold, high-salinity water during upwelling (as opposed to lower salinity water transported from the north). Monthly nutrient (nitrate) anomalies along the midshelf of Washington are also positively correlated with temperature and with upwelling. Landry et al. (1989) conclude that interannual scales of variability off Washington and Oregon are probably influenced by both regional and global scale processes. A global influence is suggested by a consistent pattern of temperature anomalies throughout the California Current and subarctic regions and by the influence of El Niño (Mysak 1986) events in the eastern tropical Pacific. It is likely that a gradient of factors affect biological production along the California Current as evidenced by the north–south pattern in the variability of winds, currents, and upwelling; the latitudinal cline in the relative proportions of subarctic, transitional, and equatorial species (Chelton et al. 1982); the north–south gradient in the amount of Columbia River water found along the Oregon coast during the summer; and the southward decline in the relative proportion of protected inland bay and estuarine habitat from British Columbia to California (Nickelson and Lichatowich 1983).

Interannual variations along this California Current ecotone create special challenges for southern salmonid stocks, which are generally less productive in Washington and Oregon compared with areas located nearer the center of their range (Fredin 1980). Fulton and LeBrasseur (1985) defined a Subarctic Boundary based on interannual variations in the distribution of mean zoo-plankton biomass (Figure 10.17). They reported a large area between Cape Mendocino and the Queen Charlotte Islands where the transition between high and low biomass varied widely between extreme "cold" and "warm" years (e.g., warm during strong El Niño events). They hypothesized that in years of strong southward advection of cold water, the larger zooplankton

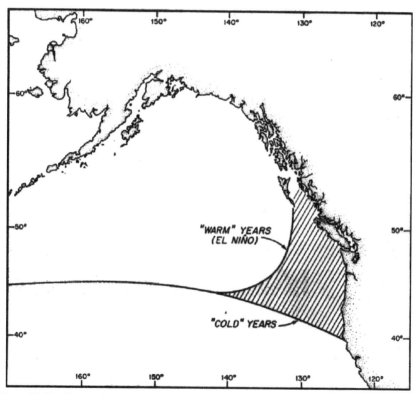

Figure 10.17   Changes in the subarctic boundary based on interannual variations in the distribution of mean zooplankton biomass. The shaded area is between Cape Mendocino, California, and the Queen Charlotte Island, British Columbia, where the transition between high and low biomasses varies widely between extreme "cold" and "warm" years. From Fulton and LeBrasseur (1985).

characteristic of the subarctic water mass may provide a better source of food for juvenile pink salmon than the smaller species otherwise typical of the California Current. As noted above, the strength of southward advection changes not only during El Niño events. Interannual variations in the sub-arctic boundary; the location of the divergence of the California Current; and associated changes in temperature, zooplankton, or other conditions may be particularly important to the survival of salmon at their southern-most distributions.

Although advection may exert a direct physical influence on survival and variability of some pelagic fishes (e.g., Sinclair et al. 1985), the influence of advective processes on year-to-year salmon survival is not well determined. Unlike larval marine fishes, salmon are free swimming when they enter the ocean, but due to small sizes their distribution may be affected by the strength of surface currents. Pearcy (1992) reports that juvenile salmon from Oregon and Washington generally swim northward against the current. However, during May and June soon after they entered the ocean, juvenile coho off Oregon were captured south of the area of ocean entrance, sug-gesting a southward advection of the smallest fish during their first few weeks in the ocean. Later in the summer, when currents were weaker and fish were larger, most young salmon were caught north of their point of ocean entry. The fact that year-class strength of coho salmon may be decided sometime within the first few weeks in the ocean (Pearcy and Fisher 1988) suggests that early survival conditions, perhaps not far from the point of ocean entry, may be important.

The surveys of salmon distribution off Washington and Oregon in the early 1980s (Pearcy and Fisher 1988, 1990) may not, however, be representa-tive of the movements of local salmon stocks under variable current condi-tions. These surveys were completed after wild populations of salmon had been reduced to a small proportion of the total quantity in the region and during a period of relatively warm ocean conditions, poor upwelling, and reduced flow of the California Current. Wild stocks may or may not exhibit these same patterns, and movements could change under conditions of strong upwelling and increased southward transport. Ocean surveys by the National Marine Fisheries Service and Oregon State University since 1998 have begun to provide new information about salmonid distributions in the region from northern Washington to northern California (Brodeur et al. 2003, 2004). These results, which may encompass a shift to a cooler and more productive climatic regime (Chavez et al. 2003; Peterson and Schwing 2003), will provide a useful comparison of salmonid distribution patterns to those observed during the 1980s (Pearcy and Fisher 1988, 1990).

Most assessments of ocean survival of salmon continue to focus on rela-tionships to biological productivity in the coastal ocean. Yet in contrast to the classical view of local upwelling as the sole driver of pelagic production, recent analyses account for both local and regional scale processes that may jointly affect salmon during their marine life history. For example, Koslow et al. (2002) found correlations between coho salmon survival and coastal conditions in the spring when juvenile salmon first enter the ocean and during the winter preceding adult spawning migrations. Their results suggest that a series of independent physical factors likely work in concert to regulate biological productivity and affect salmon survival, including upwelling, sea surface temperature, wind mixing, stratification of the mixed layer, and the strength of the southward-flowing California Current.

Building on this and other literature on the marine survival of Northwest coho salmon, Logerwell et al. (2003) proposed a conceptual model that incorporates environmental conditions before, during, and after salmon smolts enter the coastal ocean. Their results defined four independent factors that may benefit salmon survival: (1) cool sea surface temperature (SST) during the winter preceding smolt migration, a variable that may influence water column stability (and primary production) the following spring; (2) an early transition from winter downwelling to the spring upwelling season, which may increase food production when salmon arrive in the ocean; (3) low spring sea level, an indicator of southward winds, offshore transport from upwelling, and increased southward flow of the North Pacific Current; and (4) cool SST during the winter after smolt migration, which could be an indicator of improved feeding conditions or reduced predation. A predictive model incorporating these four factors explained 75% of the variability in coho salmon survival in the Oregon Production Index area (northern California to southern Washington) from 1969 to 2000 (Logerwell, et al. 2003).

The El Niño-Southern Oscillation Cycle and Influences on Salmon Production

While the above describes the "big picture" of physical processes along California, Oregon, and Washington, significant annual variations are also associated with the El Niño-Southern Oscillation and related current and climate impacts. Until 30 years ago, El Niño was believed to be the result of local changes in the winds that produced upwelling along the coasts of Peru and Ecuador (Mann and Lazier 1991). Oceanographers later discovered that this upwelling system was part of a higher level of organization involving global winds and ocean dynamics across the entire Pacific Ocean basin. They concluded that the upwelling system is a component of the global heat budget such that the physical and biological characteristics of coastal systems change as the thermal budget of the ocean and atmosphere is disturbed (Barber 1988). Barber (1988) describes this single interconnected system, which is structured by the El Niño-Southern Oscillation (ENSO) cycle in the tropical Pacific, as the "basinwide ocean ecosystem." Within this large ecosystem, habitats continually shift, producing opposing regions of abundance and scarcity with the displacement of entire water masses and changes in the thermal structure of particular locales (Sharp 1991).

The ENSO cycle is reviewed by Barber (1988), Enfield (1989), and Mysak (1986) and is only briefly summarized here. El Niño originally referred to a warm southward current off the coast of Ecuador and Peru that generally begins around Christmas and persists for about 3 months. In more recent years, the term has been applied to periods of exceptionally strong warming that usually begin around January, last more than one season, and cause economic crisis associated with mortality of pelagic fish and guano birds (Mysak 1986). El Niño is one part of a basinwide oscillation in the atmospheric pressure gradient of the equatorial Pacific known as the southern oscillation. The oscillation refers to shifts between the South Pacific high pressure system and the Indonesian low pressure system that cause changes in the westward trade winds along the equator. The interaction of the trade winds and mid-latitude westerlies with the ocean creates a slope in the sea level and in temperature, density, and nutrient gradients across the ocean basin (Barber 1988).

El Niño occurs when a critical break point is reached causing a sudden "flip" in the system, which otherwise maintains higher productivity in the eastern portion of the tropical Pacific due to the shallow thermal structure and upwelling of nutrients. Increasing instability in the east–west thermal gradient results as the trade winds continue to increase the volume of warm surface water in the west, causing the warm pool to expand eastward together with the region of atmospheric heating. As the associated low pressure system also migrates eastward, weakening and reversals in the trade winds produce internal waves that cause warm surface water to rush into the eastern basin. These waves and the migration of warm water deepen the thermocline, so that upwelling is less effective in raising cool water to the surface. Further warming and migration of the zone of atmospheric heating eastward finally produces the sustained low productivity state of El Niño (Barber 1988). The eastern upwelling region then takes on the physical and biological characteristics of the less productive western basin.

El Niño may directly affect the upwelling system, thermal structure, and biotic assemblage of the California Current. At the height of El Niño, the pool of warm water accumulating in the eastern ocean drains toward the poles influencing conditions in the northeast Pacific and the Southern Ocean. The California Undercurrent, which carries warm water northward along the west coast of the United States, thus may serve as a "release valve" for the build-up of heat in the tropical Pacific, and may be involved in the resetting of the ENSO cycle to the cold (La Niña) phase (Barber 1988). Following the mature phase of El Niño in the winter and spring, southward flow of the

California Current is reduced (Chelton et al. 1982), and the strength of northward flow in the California Undercurrent is increased (McClain and Thomas 1983; Mysak 1986). Responses to El Niño along the west coast of the United States may include elevated sea levels and sea surface temperatures, increased thermocline depths, and the northward expansion of the ranges of southern species (McClain and Thomas 1983).

Climatologists and oceanographers have described abrupt shifts in the predominant patterns of atmospheric circulation, oceanic currents, and thermal regimes that may be linked to conditions in the tropical Pacific and often follow strong El Niño events. Included in these patterns are extended periods of eastward migration and intensification of the North Pacific Aleutian low pressure system during the winter half of the year (Trenberth 1990). The intensification of the Aleutian Low causes a southward migration of storm tracks, anomalous southerly winds and warming along the west coast of North America and Alaska, and anomalous northerly winds and cold temperatures in the Central North Pacific region (between Japan and 160° N) (Trenberth 1990; Ebbesmeyer et al. 1991). In the Northeast Pacific, it is associated with a rise in sea level and ocean surface temperature; reduced flow of the California Current (Mann and Lazier 1991); and reduced precipitation, increased river temperatures, and low stream flow conditions in Oregon (Greenland 1994).

Johnson (1988) summarized the direct effects of the very strong 1982–1983 El Niño on Oregon coastal and Columbia River stocks of salmon. Both adult and juvenile salmon experienced increased mortality. Higher mortality was indicated by returns of adult coho salmon to the Oregon Production Area and tule fall Chinook to the lower Columbia-Bonneville pool area that were much lower than the preseason predictions. Mean sizes of Chinook and coho salmon that survived El Niño were much smaller than average, and fecundity of female coho salmon also was reduced. Unlike Chinook stocks off southern Oregon and locally distributed stocks from the Columbia River, northward-migrating populations from the Columbia River showed little or no decline in abundance during the El Niño (Johnson 1988).

Similar effects on coho salmon production may have occurred during a strong 1957–1958 El Niño (McGie 1984; Pearcy 1992). The mean weight of returning adults was low in 1959 (McGie 1984), and total ocean landings in 1960 from smolts that entered the ocean the previous year declined to its lowest level since 1917. Anomalously high water temperatures from 1957 to 1960 probably indicate that the relatively strong upwelling during this period was not effective in raising cold, nutrient-rich water above a deepened thermocline (Pearcy 1992). Decreases in zooplankton biomass and abundance of larval fish in the California Current coincided with a depressed thermocline during the 1958 to 1960 warming (McGowan et al. 1998).

From scale analyses of survivors returning to Tenmile Lakes, Oregon, Bottom (1985) reported little interannual variability in the relative growth of juvenile coho salmon over a wide range of upwelling conditions and salmon survival rates (among 13 brood years of salmon sampled between 1954 and 1981). A major exception to these results was the larger than average growth rates among those juveniles that entered the ocean during the 1983 El Niño and survived to return as adults in 1984. These results are consistent with the hypothesis of a brood failure during 1983, which might have caused better than average growth rates among the survivors if ocean habitats were not seeded to their capacity (Isles 1980; Lichatowich 1993). Survival of smolts entering the ocean during El Niño was very poor and stock density was likely quite low as indicated by the return of 2-year-old coho jacks in the fall of 1983 (Johnson 1988). The direct mortality of adults during this strong El Niño (Johnson 1988) suggests a different scale, habitat, and mechanism of population regulation than the control of juvenile survival, which occurs soon after smolts enter the coastal ocean (Nickelson 1986; Pearcy 1992).

The frequency and intensity of El Niño events also exhibit patterns of variation. Through a reconstruction of El Niño occurrences over the last 450 years, Quinn et al. (1987) note that intervals between strong and very strong events have averaged close to 10 years, but may range from 4- or 5-year intervals to as high as 14 to 20 years. El Niños classified as "very strong" events are relatively rare and have occurred with a frequency of 14 to 63 years. Thus, the 15 years between the most recent very-strong El Niño events (1982–1997) is near the high end of the frequency spectrum for the last 450 years.

Decadal or longer climatic changes are indicated by extended periods of unusually strong El Niño activity. Examples include the periods 1701 to 1728, 1812 to 1832, 1864 to 1891, and 1925 to 1932 (Quinn et al. 1987). Several studies note that El Niño events became more frequent and intense during the extended period of climatic change that followed the shift in the Aleutian low pressure system in 1976–1977 (Guilderson and Schrag 1998). For example, four major El Niños were recorded between 1981 and 1998 with only two major intervening cold (La Niña) events, including the return to cold conditions following the very strong 1997–1998 El Niño (Kumar et al. 1994; McPhaden 1999). Changes in the tropical Pacific have raised questions about whether the general warming trend after 1976 might have influenced the frequency of ENSO cycles and whether the increased heat itself could be an early sign of global warming from greenhouse gases (Kerr 1994; Kumar et al. 1994).

Summary

In summary, local responses to remote atmospheric and oceanic disturbances support the concept of an interconnected basinwide ecosystem in which the background conditions for different regions of the Pacific Ocean continually shift in response to the global heat budget (Barber 1988). It is within this shifting background of oceanic and atmospheric conditions that local and regional scales of processes are embedded. Basinwide forcing produces different responses among regions based on the distribution of atmospheric pressure gradients and their influence on local winds, currents, upwelling, ocean thermal structure, and precipitation patterns. Thus, for example, an increase in the intensity and extent of the wintertime Aleutian Low tends to cause cooling in the western subarctic Pacific at the same time it is warming the eastern subarctic Pacific.

Within the California Current, remote forces regulate the thermal structure and, through advective processes, determine the along-shore distribution of nutrients and the location of the subarctic boundary (Figure 10.17). Three types of forcing mechanisms may be involved in periods of warming in the California Current: (1) depression of the thermocline, strengthening of the California Undercurrent, and decreased effectiveness of upwelling during El Niño events; (2) decreased southward advection from the subarctic divergence into the California Current and increased northward advection into the Alaskan gyre related to the strengthening of the wintertime Aleutian low pressure system; and (3) increased periods of downwelling, decreased intensity of upwelling, and changes in the onset of the spring transition caused by shifts in the regional wind field also associated with patterns of the Aleutian Low (Norton et al. 1985). The specific responses along the Washington or Oregon coast involve the interaction of many scales of variability with direct effects on salmon growth and survival.

Scales of Ocean-Climate Variability and Salmon Production

In recent years, interdecadal variations in fish populations have been traced to large-scale climatic changes influencing oceanic regimes. Perhaps most dramatic was the analysis of synchronous trends in sardine (*Sardinops* sp.) abundance (as indicated by harvest) from three widely separated regions of the Pacific Ocean basin: California, Japan, and Chile (Figure 10.18) (Kawasaki 1983). The specific relationships explaining the 40-year cycle in abundance was not clear, but all three stocks appear to track variations in mean surface air temperatures in the northern hemisphere.

It is now generally accepted that patterns in salmon abundance are also linked to patterns in ocean-climate conditions. These conditions can be relatively stable for many years but then shift abruptly from one pattern to another (see Beamish 1995) with associated changes in salmon production (Beamish et al. 1999). Beamish and Bouillon (1993) first documented synchronous trends in pink, chum, and sockeye salmon abundance estimated

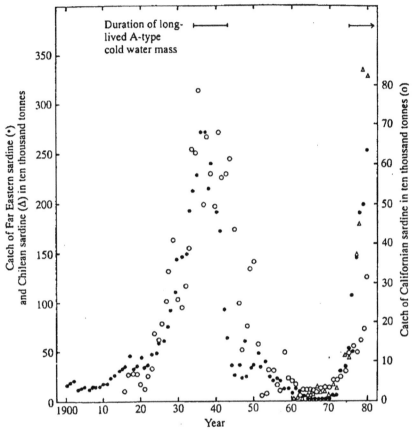

Figure 10.18 Trends in sardine (*Sardinops* sp.) abundance, as indicated by harvest in 10,000 tonnes, from three widely separated regions of the Pacific Ocean basin: California, Japan, and Chile. From Kawasaki (1983).

from the combined annual harvests in U.S., Canadian, Japanese, and Russian fisheries. These trends, as well as abundance of copepods (sampled at Ocean Station P, 50°N, 145°W), were associated with an Aleutian Low Pressure Index[2] (ALPI) (see Beamish et al. 1997 or *http://www.pac.dfo-mpo. gc.ca/sci/sa-mfpd/climate/clm_indx_alpi.htm*). Combined all-nation harvest for all salmon species averaged 673,100 t from the mid-1920s to the early 1940s and reached a peak of 837,400 t in 1939. After a period of low catch

---

[2]A number of measures or indices of ocean-climate variation have been developed; a sample of them is presented in an appendix to this chapter to demonstrate their patterns and the similarities between these indices. The indices plotted include the PDO, AFI, and PNI.

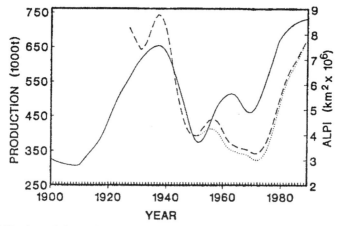

Figure 10.19   Annual changes in the combined annual harvests (landings) of all salmon species in U.S., Canadian, Japanese, and Russian fisheries, 1900 to 1990, in thousand metric tons (solid line) compared to annual changes in climate using the Aleutian Low Pressure Index (broken line). From Beamish and Bouillon (1993).

from the mid-1940s to a minimum in 1974, production again climbed to nearly 720,000 t in 1985. These patterns generally follow trends in the ALPI (Figure 10.19).

A profound shift in climatic regime of the North Pacific in 1976–1977 (Ebbesmeyer et al. 1991) was associated with the strongest Aleutian Low since 1940–1941. In addition to corresponding increases in salmon abundance, this shift is implicated in the almost doubling of chlorophyll a in the Central North Pacific north of Hawaii (Venrick et al. 1987), a doubling of summer zooplankton abundance in the Alaskan gyre between 1956 and 1962 and 1980 and 1989 (Brodeur and Ware 1992), simultaneous increases in the abundance of a variety of non-salmonid fishes in various regions of the North Pacific (Beamish 1993), and increases in prey availability for marine birds and mammals (Francis et al. 1998).

Through time series analysis, Francis and Hare (1994) and Francis et al. (1998) described multidecadal variations in salmon production associated with sudden changes in atmospheric conditions of the North Pacific. These findings were further supported by Mantua et al. (1997), who described a recurrent interdecadal pattern of climate across the Pacific Basin—the Pacific (inter) Decadal Oscillation (PDO)—that shifts between alternative climatic regimes every 20 to 30 years. From Alaska to California, the PDO affects sea surface and land temperatures, sea level pressure, and stream flow but appears independent of the ENSO (Francis et al. 1998). In retrospect, four major oceanic/atmospheric regimes have been identified that closely track patterns of the PDO: 1900 to 1924, 1925 to 1946, 1947 to 1976, and 1977 through the late 1990s (Francis et al. 1998). The regimes beginning in 1925 and in 1977 were associated with periods of high salmon abundance in the Gulf of Alaska. Variations in the harvest of coho and Chinook salmon from Washington and Oregon also show interdecadal patterns, but these fluctuate out of phase with the more northerly stocks of pink, chum, and sockeye salmon in the Gulf of Alaska (Figure 10.20; Francis 1993; Hare et al. 1999; also see Mueter et al. 2002a). McGowan et al. (1998) have argued, however, that the physical evidence for interdecadal changes in the strength of the California Current is lacking.

One hypothesis for these opposing patterns is that atmospheric forcing may influence the position of the subarctic divergence and relative flows into the Alaska and California Currents: periods of a strong Aleutian Low and increased northward flows into the Alaska Current may reduce southward flows into the California Current and vice versa, with inverse effects on the productivities of each region (Wickett 1967). More recently, Gargett (1997) proposed a mechanistic framework that associated the strength of the Aleutian pressure system with water column stability and biological processes that could explain these opposing patterns. Gargett concludes her paper with an important observation that such mechanistic bases are preferable to correlation studies since they provide testable hypotheses for study.

Evidence of a recent shift in climatic regime in late 1998 was suggested by the reversal in sign of the PDO from positive to negative and associated

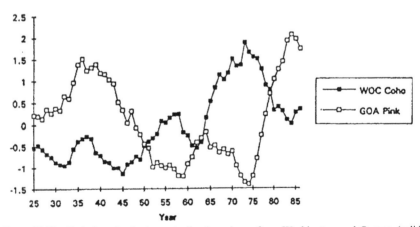

Figure 10.20 Variations in the harvest of coho salmon from Washington and Oregon (solid squares) also show interdecadal patterns, but these fluctuate out of phase with the more northerly stocks of pink salmon in the Gulf of Alaska (open squares). From Francis and Sibley (1991).

ecosystem changes in the Northeast Pacific, including a shift from warm- to cold-water zooplankton species and increased abundance of forage fishes and salmon stocks (Chavez et al. 2003; Peterson and Schwing 2003).

Chavez et al. (2003) describe coherent biological changes that occur over periods of about 50 years. They characterize a "warm sardine regime" and a "cool anchovy regime" for the Pacific Basin ecosystem. Associated with these regimes are shifts in sea surface temperature and thermocline slope throughout the Pacific Basin similar to those of El Niño and La Niña, but over longer time scales. In the Northeast Pacific, these regimes are associated with intensification (sardine) and relaxation (anchovy) of the Aleutian Low, a weaker (sardine) or stronger California Current (anchovy), and opposing patterns of salmon abundance between Alaska (increased during sardine) and the Pacific Northwest (decreased during sardine). Beginning in 1999, dramatic increases in ocean chlorophyll off California and in baitfish (including anchovy), zooplankton, and salmon populations off Oregon and Washington suggested a possible return to a cool anchovy regime (Chavez et al. 2003; Emmett 2003; Peterson and Schwing 2003). Since 2002, however, warm ocean conditions in the Northeast Pacific, a strengthening of the Aleutian Low, and a positive PDO imply that the pronouncements of a return to a persistent (e.g., interdecadal) cool-water regime may have been premature (e.g., see *http://www.pac.dfo-mpo.gc.ca/sci/ psarc/OSRs/OceanStatus2003.pdf*).

Observed patterns of coho salmon production from the Pacific Northwest coincide with coherent ecological changes throughout the region of the California Current. Trends in average commercial harvest of coho salmon in Oregon, for example, appear to follow a similar pattern to the combined biomass of dominant pelagic species (hake, anchovy, and California sardine) estimated from fish scales deposited in anaerobic marine sediments off California (Smith 1978; Lichatowich 1993). Ware and Thomson (1991) identified a 40- to 60-year cycle in wind and upwelling conditions that may influence long-period fluctuations in pelagic fish biomass off southern California. They report an extended period of relaxation in upwelling and primary production between 1916 and 1942, which also coincides with the decline in total fish biomass. However, abundances of California sardine this century are out-of-phase with the trends in combined biomasses of pelagic species including sardine (Lluch-Belda et al. 1992). Sardine populations off California were abundant before the 1950s, very low in abundance between 1950 and the late 1970s, and increased from the late 1970s to the mid-1980s. Both the period of peak sardine production this century and the recent recovery beginning in the late 1970s coincide with periods of relaxation of upwelling, which, as noted above, tend to favor strong recruitment of sardine (Bakun 1990; Lluch-Belda et al. 1992).

These same large-scale climatic conditions that influence the ocean environment of salmon also affect the quality of their freshwater habitats.

For example, in a review of the climate on the H. J. Andrews (HJA) Experimental Forest in western Oregon, Greenland (1994) found correlations between various indices of atmospheric circulation with local temperature and precipitation. The results indicate that during periods of a strong Aleutian Low, storms are pushed north of Oregon, causing relatively dry weather during the winter and raising January air temperatures due to the southwesterly flow of warm air into the region. These patterns are also associated with El Niño events. During many El Niño years, winter water year precipitation on the HJA Forest is low, and annual mean temperatures are high, while in La Niña years, winter precipitation increases, particularly a year later, and annual temperatures are below average.

These results underscore the importance of geographic heterogeneity to salmon production in a shifting climate. While interdecadal regimes affect vast areas, the degree of independence between freshwater and ocean conditions may be important to salmon resilience. In western Oregon, stream and ocean conditions affecting salmon survival tend to oscillate in phase. That is, during periods of warm ocean conditions and reduced biological productivity, freshwater habitat conditions may also decline due to reduced stream flows and increasing river temperatures. These effects suggest a kind of "double jeopardy" for salmon stocks caused by a synchrony of mortality factors that involve more than one stage of life history (e.g., see Lawson et al. 2004). It is also possible that in other regions or among diverse watersheds within large basins like the Columbia River, ocean and river conditions for salmon survival are not in phase so that the effects of large-scale climatic change may be dampened. For example, after 1980, during a favorable regime for salmon survival in the Gulf of Alaska, annual precipitation on the coast of British Columbia was above average. On the other hand, discharge of the large Fraser River declined during the same period due to reduced snowpack in the interior of British Columbia (Beamish 1993). These varying degrees of "connectedness" between the environments supporting different salmon life stages illustrate the importance of stock diversity and habitat heterogeneity to dampen the otherwise synchronous effects of large-scale climatic change.

The above discussion pertains largely to large-scale spatial and long-term temporal variability, with the exception of the shorter-term ENSO events along the eastern Pacific coasts. However, recent studies have identified that marine fish and populations of Pacific salmon covary in interannual production at much smaller spatial scales (500–1000 km) (Myers et al. 1997; Mueter et al. 2002b). The covariation has been associated with localized marine environmental parameters, but as noted above, and reported by Lawson et al. (2004), these parameters may also be correlated with environmental factors in freshwater. The Lawson et al. report even indicates that seasonal scales of variability may have to be accounted for to fully understand the sources of variation in survival of salmon populations. Further, evidence suggests that interspecific competition (Bugaev et al. 2001; Ruggerone et al. 2003; Ruggerone and Goetz 2004) and intraspecific competition (Sweeting et al. 2003) exist in the marine environments and may vary with ocean production regimes (Kaeriyama et al. 2004). While concerns have been expressed about competition in the past, few studies have been able to demonstrate any affect, even though billions of artificially produced salmonids are released into the north Pacific Ocean annually (Noakes et al. 2000). Evidence of competition now creates interesting hypotheses about the implications of large numbers of salmon produced in hatcheries (see Chapter 8) and their interaction with local populations (Levin et al. 2001; Levin and Williams 2002) and between geographic regions.

The Ocean Environment and Salmon Production in the Columbia River

Until recently, changes in salmon abundance were attributed primarily to poor habitat conditions in freshwater. These ideas were formalized in theoretical population models that emphasized the role of density-dependent mortality during egg and early juvenile stages, and in hatchery programs, which assumed that annual production would be increased by eliminating various causes of freshwater mortality (Lichatowich et al. 1996; Bottom 1997; Chapter 8). The first serious challenge to these assumptions came after 1976 when abundance of Oregon coho salmon precipitously declined despite continued increases in production of hatchery juveniles. In the last few decades though, oceanographers have described dramatic changes in marine fish assemblages and food chains that have important implications for salmon conservation. From analysis of fish scales deposited in anaerobic marine sediments off southern California, researchers documented large fluctuations in abundance and shifts in the dominance of pelagic species that occurred well before intensive fisheries had any impact on fish stocks (Soutar and Isaacs 1974; Soutar and Isaacs 1969; Smith 1978). Regional fluctuations in fish populations have been linked to large-scale climate changes at interannual (McClain and Thomas 1983), interdecadal (Mantua et al. 1997), and multi-centennial (Finney et al. 2002) periods.

Variability in marine climate and ecosystems raise fundamental issues for salmon conservation. First, although fishery managers cannot control environmental variations in the ocean, this does not mean that they can afford to ignore them. For example, in the 1960s and early 1970s, the assumption that hatcheries were responsible for increases in adult returns led to continued growth in hatchery programs and in ocean fisheries during a period of unusually favorable conditions for marine survival of salmon. As noted above, abundance of salmon later collapsed after the ocean returned to less productive conditions. The failure to account for natural variability in marine conditions may lead to faulty conclusions about the response (or lack of response) of salmon to hatchery practices, fishery quotas, habitat restoration efforts, or other management prescriptions (Lawson 1993). To avoid choices that undermine conservation, restoration programs in the Columbia River Basin must account for the full "background" (freshwater, estuary, and marine) of environmental change upon which management actions are superimposed.

Second, changes in the ocean environment raise important questions about the appropriate spatial and temporal scales for observation, and indicators of biological response and change. Traditionally, salmon management programs have emphasized year-to-year variations in adult abundance and, in studies of environmental factors, the spatial scales of freshwater habitats, stream reaches, or river systems (meters to tens of kilometers). But changes in salmon production involve processes that vary over periods of decades and centuries and extend to distances of thousands of kilometers during migrations. Unlike the well-defined and relatively restricted boundaries of streams and watersheds, oceans are highly "open" systems in which physical and biological properties are linked across vast distances. Oceanic and atmospheric influences on salmon production therefore involve multiple spatial and temporal scales of variability, and appropriate indicators of changes affecting salmon may include many species other than salmon. Salmon are members of complex marine communities. Significant shifts in the distribution of predators or in the structure of food chains may be important factors in the dynamics of salmon populations, whose ocean distributions, physical environments, and biotic interactions are at least partially predetermined by the migratory route they must follow to and from their home streams.

Third, large-scale changes in ocean climate and productivity undermine simple distinctions between density-dependent and density-independent factors that may inhibit understanding of population regulation. If fluctuations in the ocean environment involve qualitative shifts in the structure of entire communities and in the physical distribution of species, food, and nutrients, how do we separate density-dependent from density-independent effects? In the case of salmon, a changing background of current or upwelling conditions each year of smolt migration may create new sets of potential "winners" and "losers" among different salmon stocks based on when and where they enter the ocean and tend to migrate. Understanding how physical and biological processes interact to influence the relative performance of geographically discrete stocks is more important than arbitrarily characterizing trends in salmon abundance as either density-dependent or density-independent.

In summary, the fact that salmon production may be regulated by conditions in the ocean argues for a broader management perspective incorporating the entire riverine-estuarine-marine habitat continuum of salmon life cycles.

Salmon Management in a Sea of Change

The Pacific Ocean and atmosphere do not move toward a steady state condition but continually shift in response to changes in the global heat budget. The risk of sudden changes in production regimes and shifts of ecological conditions within different oceanic regions can dramatically alter both freshwater and marine conditions for salmon and call into question management programs that emphasize constancy of the natural environment. Local salmon populations may encounter different combinations of conditions each year they enter the coastal ocean, as determined by the basinwide climatic regime and its interactions with regional and local scales of variation. Simply stated, conservation programs that were designed under one climatic regime may not be appropriate under another. For example, the same levels of artificial propagation established during optimum ocean conditions may not be appropriate following a shift to a low productivity state (Beamish and Bouillon 1993). Genetic or ecological risks of hatchery programs on wild salmonids might increase disproportionately during poor survival conditions, due to intensified harvest pressures or increased interactions with hatchery fish.

Temporal shifts in the ocean environment also undermine the assumptions of traditional analytical models that are often used to establish management goals (i.e., a desired number of fish to reproduce) and set harvest levels in fisheries. The maximum sustained yield concept is based on a logistic growth curve developed from animal populations held under a constant food supply and environmental conditions (Botkin 1990). Such models assume that abundance of a salmon population is regulated primarily by density-dependent factors during early life stages in freshwater. In practice, production relationships rely on multiple years of observation to show an "average" relationship between parental number of spawning fish and the number of progeny produced. Thus, rather than trying to understand the effects of environmental change on populations, traditional models assume that change is insignificant by averaging conditions over the period of observation (Cushing 1995). In a system that oscillates unpredictably between different climatic states, and where physical changes continually reset ecological conditions and regulate ocean survival of salmon smolts, these theoretical population models may offer little practical guidance for long-term conservation.

Within the Columbia River Basin, salmon management has been designed around efforts to create a stable river system for the benefit of various economic uses. The idea that hatcheries could increase salmon production by eliminating variability ("limiting factors") in the freshwater environment was consistent with efforts to regulate the river (discussed in detail in Chapter 5). One important result of regulating the river with dams and controlling salmon production in hatcheries has been the narrowing of salmon life histories to conform to the rigid conditions imposed by the management system itself. Hatchery programs (Chapter 8) replace diverse riverine habitats with a constant rearing environment and replace diverse native stocks with a more uniform hatchery "product." By dampening seasonal fluctuations in the hydrograph, dam operations (Chapters 6 and 7) also reduce the diversity of freshwater habitats and the variety of flow conditions represented in the river. Furthermore, release strategies for hatchery salmon are programmed to fit the scheduled releases of water through the dams. Survival advantage is given to those fish that migrate downstream according to the operations of bypass, spill, and transportation systems. Unlike historical patterns of migration, which maximized use of freshwater habitats throughout the year and varied the time of ocean entry through a wide array of migratory behaviors, river operations concentrate the migrations of salmon through narrow windows of opportunity prescribed by the management system.

This approach fails to account for the effects of environmental control on other ecosystems or life stages of salmon downstream. Regulation of river flows eliminates salmonid access to historic tidal floodplain habitats in the estuary (Bottom et al. in press; Kukulka and Jay 2003) and directly controls the density and dynamics of offshore waters of the coastal ocean (Ebbesmeyer and Tangborn 1993) with unknown consequences for natural production processes. Similarly, control of salmon life histories in the river may directly regulate the subsequent survival of salmon in the estuary and ocean. A variable ocean requires flexibility for anadromous species to successfully respond to a wide array of potential conditions. This flexibility may be particularly important in the highly variable environment of the California Current ecotone, which encompasses the southern edge of a shifting subarctic boundary and the distributional limit of subarctic salmonids. Different times of migration, for example, may be advantageous in different years depending on the onset of the spring/summer transition, the distribution of the Columbia River plume, the timing and location of upwelling episodes, or the northward extent of warming caused by occasional strong ENSO events in the tropics.

A critical point for future salmon management is that the performance of salmon in the estuary or ocean is not independent of the selection processes in freshwater. Management manipulations that alter population structure, life histories, or habitat diversity during the freshwater phase of life history may directly influence the capacity of salmon to withstand natural fluctuations in the estuary and ocean. Efforts to control variability in freshwater may unwittingly eliminate behaviors that buffer salmon production in an unstable marine environment. *By the 1970s, about two-thirds of the coho salmon smolts produced in the Oregon Production Area were derived from only two stocks of fish released from numerous hatcheries in the Columbia River Basin!* Across Oregon, the replacement of a diversity of wild populations and life history patterns with relatively few hatchery-produced stocks may have depressed survival rates of smolts during poor ocean conditions and increased interannual variability of returning adult salmon.

Finally, variability in the ocean underscores the importance of life history diversity to the production and recovery of salmon. Diverse life histories minimize the risk of brood failure in an uncertain ocean environment because not all individuals behave uniformly. Slightly different migration times, for example, may be advantageous in different years depending upon the exact timing of the shift in along-shore surface currents (e.g., the spring transition), the specific location of the Columbia River plume, the location and timing of upwelling events, or the distribution of predators along the coast. These and other conditions in the ocean can vary markedly between years. The diversity of life history characteristics within and among populations may determine the number of individuals that successfully make the transition from riverine to oceanic life. Habitat quality, flow conditions, or other factors that affect mortality during early life history may subsequently limit the range of sizes or times of emigration among surviving juveniles. Loss of life history diversity in the Columbia River may thus limit the capacity of salmon to survive variable conditions at sea, as well as to persist in the freshwater environment. Further, the location of Columbia River populations toward the southern range of Pacific salmon species and their sensitivity to changes in the coastal ocean raise concerns about the effects of global climate change on salmon production. Melack et al. (1997) and Neitzel et al. (1991) discuss effects of global warming on inland waters of the Columbia River Basin, and Mote et al. (2003) consider broader impacts on water, salmon, and forests in the Pacific Northwest (see also Chapter 12). Significant ecological changes in the ocean environment could also be critical to local salmon production. While we cannot predict the extent of future change, these possibilities underscore the importance of maintaining habitat complexity and stock diversity in freshwater to help buffer potential effects of global warming on salmon survival and recovery within the Columbia River.

Clearly, the identification of such temporal and spatial interactions is essential to properly assess the impact of development within the basin versus limitations imposed from outside the basin, and for reasonable expectations of salmon production and recovery over time. Pulwarty and Redmond (1997) have already presented a very similar conclusion: "*To allow for salmon recovery it is therefore necessary to develop freshwater recovery measures that minimize vulnerability to unanticipated periods of poor ocean conditions, in addition to mitigating anthropogenic stresses. (p. 394)*" These authors also provide two important comments worthy of completing this chapter: Uncertainty in managing such large ecological systems will be inescapable and large, and beyond the technical issues, resource managers must be willing to act to implement change. Accounting for marine environmental conditions within a full life-cycle framework and within the normative river concept will require substantial change from how salmon recovery in the Columbia Basin has been viewed until now.

Conclusions and Recommendations

Conclusions

1. Global and regional scale processes in the ocean and atmosphere can regulate the productivity of local marine, estuarine, and freshwater habitats for salmon. Although managers cannot control these processes, natural variability must be understood to correctly interpret the response of salmon to management actions in the Columbia Basin.
2. The North Pacific Ocean oscillates on an interdecadal time scale between alternate climatic states associated with changes in the Aleutian low pressure system. Years when the winter Aleutian Low is strong and centered in the eastern North Pacific are associated with high mean sea levels, increased ocean surface temperatures, and increased stratification of the upper water column in the northeast Pacific; increased southwesterly winds and downwelling off Oregon and Washington; and reduced precipitation levels, low stream flows, and increased water temperatures in Oregon streams. Conversely, periods characterized by a weak winter Aleutian Low centered in the western North Pacific are associated with reduced surface temperatures and stratification in the California Current, enhanced westerly winds and upwelling in the Northeast Pacific, and increased rainfall and stream flows in Oregon.
3. Because salmon migrations are tied to major ocean circulation systems and because the life cycles of salmon are shorter than the interdecadal periods of large-scale climatic change, abundance of salmon "tracks" large-scale shifts in climatic regime. The specific mechanisms of this tracking are poorly understood.
4. Salmon abundances in the California Current region (off Washington, Oregon, and California) and in the Central North Pacific Ocean domain (off British Columbia and Alaska) respond in opposite ways to shifts in climatic regime. During periods of a strong Aleutian Low, zooplankton and salmon production generally increase in the Central North Pacific and decrease in the California Current, suggesting geographically distinct mechanisms of aquatic production.
5. Stocks with different life history traits and ocean migration patterns may be favored under (or differentially tolerant of) different combinations of climatic regime and local habitat characteristics. Such differences

afford stability to salmon species over multiple scales of environmental variability.

6. Together, landscape modifications, construction of dams, overfishing, and hatchery programs have simplified the geographic mosaic of habitat conditions in the Columbia River Basin and reduced the variety of salmon life histories formerly associated with this mosaic. Such changes limit the capacity of salmon to adapt to periodic shifts in large-scale atmospheric and oceanic conditions.

7. The cumulative effects of human disturbance on salmonid ecosystems may not become fully apparent until severe climatic stresses trigger a dramatic response. Such interactions may be particularly severe in the Columbia Basin. Although climatic fluctuations may be a proximate factor in regional salmon decline, the ultimate causes may involve a longer history of change affecting species and population resilience into the future. Conservative standards of salmon protection may be necessary even during a high productivity state in order to ensure preservation of diversity needed to withstand subsequent productivity troughs.

8. Shifts in oceanic regime involve substantial changes in the distribution of species, the structure of marine food chains, and the physical processes of biological production. Anticipating such change and understanding its effects on salmon production in the Columbia Basin will require long-term monitoring of ecological indicators other than just the abundance of salmon.

9. Increased salmon abundance coincident with periods of improved ocean conditions should not be interpreted as evidence that recovery efforts in the Columbia River Basin have suddenly succeeded. The baseline for evaluating current population trends is the entire history of salmon records (Pauly 1995), including previous favorable climatic regimes, rather than the most recent poor-survival years of the last strong Aleutian Low (1978–1998). Moreover, indicators of salmon recovery must include the spatial distribution and diversity of stocks and not simply measures of salmon abundance (McElhaney et al. 2000).

Critical Uncertainties

1. Studies of salmon, forage fish populations, and zooplankton off the Oregon and Washington coasts since 1998 are providing new information about physical and biotic interactions of the Columbia River plume, upwelling system, and the northern California Current ecosystem (e.g., Brodeur et al. 2003, 2004; Peterson and Schwing 2003). Continued long-term monitoring will be important to understand their impact on salmon production and annual variation in survival rates. Such

understanding is needed if hydropower and fish management programs are to adapt to natural variations to assist recovery of Columbia River salmon.

2. The genetic basis for existing life history diversity within the Columbia Basin and potential differences between hatchery and wild salmon are largely unknown. Methods of genetic analysis now provide new opportunity to study genetic versus environmental sources of variation and to identify origins of juvenile salmon and steelhead sampled in the ocean (e.g., Brodeur et al. 2004; Teel et al. 2003). It is important to determine whether genetic variation remains to adjust to future conditions or whether we are limited to phenotypic responses within remaining and limited genetic backgrounds.

3. The ecological interactions between salmon species, hatchery versus wild production within species, and between geographic regions may be significant during regimes of poor ocean productivity. Given the extent of ecological change within the Columbia Basin and magnitude of artificial production of salmonids, this topic merits much more directed research. Further, there is increasing evidence worldwide that ocean fisheries can have a destabilizing influence on marine food chains.

4. The risks of global warming are potentially great for Columbia Basin salmon due to the sensitivity of southern salmon stocks to climate-related shifts in the position of the subarctic boundary, the strength of the California Current, the intensity of coastal upwelling, and the frequency and intensity of El Niño events. While the potential effects of global warming on ocean circulation patterns are poorly understood, the implications for salmon restoration efforts throughout the Pacific Northwest are tremendous.

Recommendations

1. Research on the uncertainties listed above should be encouraged because salmon management must link freshwater habitats and ecosystems to ocean properties and patterns if our alternative conceptual foundation that builds around the salmon life history ecosystem is adopted.

2. Research on effects of ocean conditions on productivity of salmon must to be integrated with estuarine and riverine research.

3. A conceptual foundation involving the total life cycle of Columbia Basin salmonids must be adopted to properly assess the impact of development within the basin versus limitations imposed from outside the basin, and for reasonable expectations of salmon production and recovery over time.

Acknowledgements

The authors wish to thank Dr. R. Hilborn and Mr. Arni Magnusson, University of Washington, for providing the coded-wire tag survival data needed to update Figures 10.1 and 10.6 of this chapter. Dr. N. Mantua, University of Washington, provided permission to use the Pacific Decadal Oscillation data from the website *http://tao.atmos.washington.edu/pdo*.

Literature Cited

Abbott, M. R., and P. M. Zion. 1987. Spatial and temporal variability of phytoplankton pigment off northern California during Coastal Ocean Dynamics Experiment 1. *Journal of Geophysical Research* 92:1745–1755.

Anderson, J. J. 2000. Decadal climate cycles and declining Columbia River salmon. Pages 467–484 in E. E. Knudsen, C. R. Steward, D. MacDonal, J. E. Williams, and D. W. Reiser, eds. *Sustainable fisheries management: Pacific salmon*. CRC Press, New York.

Bakun, A. 1990. Global climate change and intensification of coastal ocean upwelling. *Science* 247:198–207.

Bakun, A. 1993. The California current, Benguela current, and southwestern Atlantic shelf ecosystems: A comparative approach to identifying factors regulating biomass yields. Pages 199–221 in K. Sherman, L. M. Alexander, and B. D. Gold, eds. *Large marine ecosystems: Stress, mitigation, and sustainability*. American Association for the Advancement of Science Press, Washington, DC.

Barber, R. T. 1988. Ocean basin ecosystems. Pages 171–193 in L. R. Pomeroy and J. J. Alberts, eds. *Concepts of ecosystem ecology: A comparative view*. Springer-Verlag, New York.

Beamish, R. J. 1993. Climate and exceptional fish production off the west coast of North America. *Canadian Journal of Fisheries and Aquatic Sciences* 50:2270–2291.

Beamish, R. J., ed. 1995. Climate change and northern fish populations. *Canadian Special Publication of Fisheries and Aquatic Sciences* 121.

Beamish, R. J., and D. R. Bouillon. 1993. Pacific salmon production trends in relation to climate. *Canadian Journal of Fisheries and Aquatic Sciences* 50:1002–1016.

Beamish, R. J., C. E. Neville, and A. J. Cass. 1997. Production of Fraser River sockeye salmon (*Oncorhynchus nerka*) in relation to decadal-scale changes in the climate and the ocean. *Canadian Journal of Fisheries and Aquatic Sciences* 54:543–554.

Beamish, R. J., D. J. Noakes, G. A. McFarlane, L. Klyashtorin, V. V. Ivanov, and V. Kurashov. 1999. The regime concept and natural trends in the production of Pacific salmon. *Canadian Journal of Fisheries and Aquatic Sciences* 56:516–526.

Bernal, P. A. 1981. A review of low-frequency response of the pelagic ecosystem in the California current. *California Cooperative Fisheries Investigations* 22:49–62.

Bernal, P. A., and J. A. McGowan. 1981. Advection and upwelling in the California current. Pages 381–389 in F. A. Richards, ed. *Coastal Upwelling*. American Geophysical Union, Washington, DC.

Botkin, D. B. 1990. *Discordant harmonies: A new ecology for the twenty-first century*. Oxford University Press, New York.

Bottom, D. L. 1985. Research and development of Oregon's coastal salmon stocks. Oregon Department of Fish and Wildlife. Progress Report, Project No. AFC-127, 1 October 1984 to 30 September 1985. Portland, Oregon.

Bottom, D. L. 1997. To till the water: A history of ideas in fisheries conservation. Pages 569–597 in R. J. Naiman and D. Stouder, eds. *Pacific salmon and their ecosystems: Status and future options.* Chapman and Hall, New York.

Bottom D. L. and K. K. Jones. 1990. Species composition, distribution, and invertebrate prey of fish assemblages in the Columbia River Estuary. *Progress in Oceanography* 25:243–270.

Bottom, D. L., K. K. Jones, and M. J. Herring. 1984. Fishes of the Columbia River Estuary. Oregon Department of Fish and Wildlife. Final Report on the Fish Work Unit of the Columbia River Estuary Data Development Program, Portland, Oregon.

Bottom, D. L., K. K. Jones, J. D. Rodgers, and R. F. Brown. 1989. Management of living marine resources: A research plan for the Washington and Oregon continental margin. National Coastal Resources Research and Development Institute. Publication No. NCRI-T-89-004. Newport, Oregon.

Bottom, D. L., K. K. Jones, T. J. Cornwell, A. Gray, and C. A. Simenstad. 2005. Patterns of Chinook salmon migration and residency in the Salmon River estuary (Oregon). *Estuarine Coastal and Shelf Science,* 64:79–93.

Bottom, D. L., C. A. Simenstad, A. M. Baptista, D. A. Jay, J. Burke, K. K. Jones, E. Casillas, and M. H. Schiewe. In press. Salmon at river's end: The role of the estuary in the decline and recovery of Columbia River salmon. NOAA Technical Memorandum NMFS-NWFSC-XX.

Brodeur, R. D., and D. M. Ware. 1992. Interannual and interdecadal changes in zooplankton biomass in the subarctic Pacific Ocean. *Fisheries Oceanography* 1:32–38.

Brodeur, R. D., K. W. Myers, and J. H. Helle. 2003. Research conducted by the United States on the early ocean life history of Pacific salmon. *North Pacific Anadromous Fish Commission Bulletin* 3:89–131.

Brodeur, R. D., J. P. Fisher, D. J. Teel, E. Casillas, R. L. Emmett, and T. M. Miller. 2004. Juvenile salmonid distribution, growth, condition, origin, and environmental and species associations in the Northern California Current. *Fishery Bulletin* 102:25–46.

Bugaev, V. F., D. W. Welch, M. M. Selifonov, L. E. Grachev, and J. P. Eveson. 2001. Influence of the marine abundance of pink (*Oncorhynchus gorbuscha*) and sockeye salmon (*O. nerka*) on growth of Ozernaya River sockeye. *Fisheries Oceanography* 10(1):26–32.

Burke, J. L. 2004. Life histories of juvenile Chinook salmon in the Columbia River Estuary, 1916 to present. Master's thesis, Oregon State University, Corvallis. 88 pp.

Carl, C. M., and M. C. Healey. 1984. Differences in enzyme frequency and body morphology among three juvenile life history types of Chinook salmon (*Oncorynchus tshawytscha*) in the Nanaimo River, British Columbia. *Canadian Journal of Fisheries and Aquatic Sciences* 41:1070–1077.

Chavez, F. P., J. Ryan, S. E. Lluch-Cota, and C. Miguel Niquen. 2003. From anchovies to sardines and back: Multidecadal change in the Pacific Ocean. *Science* 299:217–221.

Chelton, D. B., P. A. Bernal, and J. A. McGowan. 1982. Large-scale interannual physical and biological interaction in the California current. *Journal of Marine Research* 40:1095–1125.

Clark, J., and B. McCarl. 1983. An investigation of the relationship between Oregon coho salmon (*Oncorhynchus kisutch*) hatchery releases and adult production utilizing law of the minimum regression. *Canadian Journal of Fisheries and Aquatic Sciences* 40:516–523.

Collis, K., D. D. Roby, D. P. Craig, S. L. Adamany, J. Y. Adkins, and D. E. Lyons. 2002. Colony size and diet composition of piscivorous waterbirds on the lower Columbia River: Implications for losses of juvenile salmonids to avian predation. *Transactions of the American Fisheries Society* 131:537–550.

Collis, K., D. D. Roby, D. P. Craig, B. A. Ryan, and R. D. Ledgerwood. 2001a. Colonial waterbird predation on juvenile salmonids tagged with passive integrated transponders in the Columbia River estuary: Vulnerability of different salmonid species, stocks and rearing types. *Transactions of the American Fisheries Society* 130:385–396.

Collis, K., D. D. Roby, D. E. Lyons, R. M. Suryan, M. Antolos, S. K. Anderson, A. M. Meyers, and M. Hawbecker. 2001b. Caspian Tern Research on the Lower Columbia River, Final 2001 Summary. Columbia Bird Research. (*www.columbiabirdresearch.org*)

Cordell, J. R., and S. M. Morrison. 1996. The invasive Asian copepod Pseudodiaptomus inopinus in Oregon, Washington, and British Columbia estuaries. *Estuaries* 19:629–638.

Cordell, J. R., C. A. Simenstad, and C. A. Morgan. 1992. Establishment of the Asian calanoid copepod Pseudodiaptomus inopinus in the Columbia River estuary. *Journal of Crustacean Biology* 12:260–269.

Coronado, C., and R. Hilborn. 1998a. Spatial and temporal factors affecting survival in coho and fall Chinook salmon in the Pacific northwest. *Bull. Mar. Sci.* 62(2):409–425.

Coronado, C., and R. Hilborn. 1998b. Spatial and temporal factors affecting survival in coho salmon (*Oncorhynchus kisutch*) in the Pacific northwest. *Canadian Journal of Fisheries and Aquatic Sciences* 55:2067–2077.

Cury, P., and C. Roy. 1989. Optimal environmental window and pelagic fish recruitment success in upwelling areas. *Canadian Journal of Fisheries and Aquatic Sciences* 46:670–680.

Cushing, D. 1995. *Population production and regulation in the sea: A fisheries perspective.* Cambridge University Press, Cambridge, United Kingdom.

Dawley, E. M., R. D. Ledgerwood, T. H. Blahm, C. W. Sims, J. T. Durkin, R. A. Kirn, A. E. Rankis, G. E. Monan, and F. J. Ossiander. 1986. Migrational characteristics, biological observations, and relative survival of juvenile salmonids entering the Columbia River estuary. Report to Bonneville Power Administration, Contract DE-A179-84BP-39652. 256 p.

Day, J., John W., C. A. S. Hall, W. M. Kemp, and A. Yáñez-Arancibia. 1989. *Estuarine ecology.* John Wiley and Sons, New York.

De Robertis, A., C. H. Ryer, A. Veloza, and R. D. Brodeur. 2003. Differential effects of turbidity on prey consumption of piscivorous and planktivorous fish. *Canadian Journal of Fisheries and Aquatic Sciences* 60:1517–1526.

Ebbesmeyer, C. C., D. R. Cayan, D. R. McClain, F. H. Nichols, D. H. Peterson, and K. T. Redmond. 1991. 1976 step in Pacific climate: Forty environmental changes between 1968-1975 and 1977-1984. Pages 115–126 in J. L. Betancourt and V. L. Tharp, eds. *Proceedings of the Seventh Annual Pacific Climate (PACLIM) Workshop, April, 1990.* California Department of Water Resources.

Ebbesmeyer, C. C., and R. M. Strickland. 1995. *Oyster condition and climate: Evidence from Willapa Bay.* Publication WSG-MR 95-02. Washington Sea Grant Program, University of Washington, Seattle. 11 p.

Ebbesmeyer, C. C., and W. Tangborn. 1992. Linkage of reservoir, coast, and strait dynamics, 1936-1990: Columbia River basin, Washington coast, and Juan de Fuca Strait. Pages 288–299 in *Interdisciplinary approaches in hydrology and hydrogeology.* American Institute of Hydrology, St. Paul, Minnesota.

Ebbesmeyer, C. C., and W. Tangborn. 1993. Great Pacific surface salinity trends caused by diverting the Columbia River between seasons (Manuscript). Evans-Hamilton, Seattle, Washington.

Elliot, C. 2004. Tidal emergent plant communities, Russian Island, Columbia River Estuary. Master's thesis, University of Washington, Seattle. 87 p.

Emlen, J. M., R. R. Reisenbichler, A. M. McGie, and T. E. Nickelson. 1990. Density-dependence at sea for coho salmon (*Oncorhynchus kisutch*). *Journal of the Fisheries Research Board of Canada* 47:1765–1772.

Emmett, R. 2003. Recent changes in the abundance of northern anchovy (*Engraulis mordax*) off the Pacific Northwest, tracking a regime shift? *Pices Press* 11(2):20–21, 26.

Emmett, R. L., R. D. Brodeur, and P. M. Orton. 2004. The vertical distribution of juvenile salmon (*Oncorhynchus* spp.) and associated fishes in the Columbia River plume. *Fisheries Oceanography* 13(6):392–402.

Enfield, D. B. 1989. El Niño past and present. *Review in Geophysics* 27:159–187.

Favorite, F., A. J. Dodimead, and K. Nasu. 1976. Oceanography of the subarctic Pacific region, 1960-71. International North Pacific Fisheries Commission. Bulletin No. 33. Vancouver, British Columbia.

Finney, B. P., I. Gregory-Eaves, M. S. V. Douglas, and J. P. Smol. 2002. Fisheries productivity in the northeast Pacific Ocean over the past 2,200 years. *Nature* (London) 416:729–733.

Francis, R. C. 1993. Climate change and salmon production in the North Pacific Ocean. Pages 33-43 in K. T. Redmond and V. J. Tharp, eds. *Proceedings of the Ninth Annual Pacific Climate (PACLIM) Workshop, April 21–24, 1992.* California Department of Water Resources.

Francis, R. C., and S. R. Hare. 1994. Decadal-scale regime shifts in the large marine ecosystems of the northeast Pacific: A case for historical science. *Fisheries Oceanography* 3:279–291.

Francis, R. C., S. R. Hare, A. B. Hollowed, and W. S. Wooster. 1998. Effects of interdecadal climate variability on the oceanic ecosystems of the NE Pacific. *Fisheries Oceanography* 7(1):1–21.

Francis, R. C., and T. H. Sibley. 1991. Climate change and fisheries: What are the real issues? *The Northwest Environmental Journal* 7:295–307.

Fredin, R. A. 1980. Trends in North Pacific salmon fisheries. Pages 59–119 in W. J. McNeil and D. C. Himsworth, eds. *Salmonid ecosystems of the North Pacific.* Sea Grant College Program, Oregon State University, Corvallis.

Fresh, K. L., E. Casillas, L. Johnson, and D. L. Bottom. In press. Role of the estuary in the recovery of Columbia River basin salmon and steelhead: An evaluation of the effects of selected factors on salmonid population viability. NOAA Technical Memorandum NMFS-NWFSC-XX.

Fulton, J. D., and R. J. LeBrasseur. 1985. Interannual shifting of the subarctic boundary and some of the biotic effects on juvenile salmonids. Pages 237–247 in W. S. Wooster and D. L. Fluharty, eds. *El Niño north: Niño effects in the Eastern Subarctic Pacific Ocean.* Washington Sea Grant Program, University of Washington, Seattle.

Gargett, A. E. 1997. The optimal stability 'window': a mechanism underlying decadal fluctuations in North Pacific salmon stocks? Fisheries Oceanography 6:109-117.

Graybill, M., and J. Hodder. 1985. Effects of the 1982-83 El Niño on reproduction of six species of seabirds in Oregon. Pages 205–210 in W. S. Wooster and D. L. Fluharty, eds. *El Niño north: Niño effects in the Eastern Subarctic Pacific.* Washington Sea Grant Program, University of Washington, Seattle.

Greenland, D. 1994. The Pacific Northwest regional context of the climate of the H. J. Andrews Experimental Forest. *Northwest Science* 69:81–96.

Groot, C., and L. Margolis, eds. 1991. *Pacific salmon life histories.* University of British Columbia, Vancouver, British Columbia.

Guilderson, T. P., and D. P. Schrag. 1998. Abrupt shift in subsurface temperatures in the tropical Pacific associated with changes in El Niño. *Science* 281:240–243.

Gunsolus, R. T. 1978. The status of Oregon coho and recommendations for managing the production, harvest, and escapement of wild and hatchery-reared stocks (Manuscript). Oregon Department of Fish and Wildlife, Clackamas, Oregon.

Hare, S. R., and R. C. Francis. 1995. Climate change and salmon production in the northeast Pacific Ocean. *Canadian Special Publication of Fisheries and Aquatic Sciences* 121:357–372.

Hare, S. R., N. J. Mantua, and R. C. Francis. 1999. Inverse production regimes: Alaska and West Coast Pacific Salmon. *Fisheries* 24:6–14.

Healey, M. C. 1982. Juvenile Pacific salmon in estuaries: the life support system. Pages 315–341 in V. S. Kennedy, ed. *Estuarine comparisons.* Academic Press, New York.

Healey, M. C. 1991. Life history of Chinook salmon. Pages 313–393 in C. Groot and L. Margolis, eds. *Pacific salmon life histories.* University of British Columbia Press, Vancouver.

Healey, M. C., and A. Prince. 1995. Scales of variation in life history tactics of Pacific salmon and the conservation of phenotype and genotype. *American Fisheries Society Symposium* 17:176–184.

Holton, R. L., D. L. Higley, M. A. Brzezinski, K. K. Jones, and S. L. Wilson. 1984. Benthic infauna of the Columbia River estuary. Columbia River Estuary Data Development Program, Astoria, Oregon.

Huyer, A. 1983. Coastal upwelling in the California Current system. *Progress in Oceanography* 12:259–284.

Independent Scientific Advisory Board (ISAB). 2000. The Columbia River estuary and the Columbia River Basin Fish and Wildlife Program. ISAB 2000-5. Northwest Power Planning Council, 851 SW 6th Ave., Suite 1100, Portland, Oregon 97204.

Isles, T. D. 1980. Environmental pressure and intra- and inter-year-class competition as determinants of year-class size. *Rapports et Proces-verbaux des Reunions* 177:315–331.

Jamir, T. V., A. Huyer, W. Pearcy, and J. Fisher. 1994. The influence of environmental factors on the marine survival of Oregon hatchery coho (*Oncorhynchus kisutch*). Pages 115–138 in M. Keefe, ed. *Salmon ecosystem restoration: Myth and reality, 1994 Northeast Pacific Chinook and Coho Salmon Workshop*. Oregon Chapter American Fisheries Society, Corvallis, Oregon.

Jay, D. A., B. S. Giese, and C. R. Sherwood. 1990. Energetics and sedimentary processes in the Columbia River estuary. *Progress in Oceanography* 25:157–174.

Johnson, M. W., and E. Brinton. 1962. Biological species, water-masses and currents. Pages 381–414 in M. N. Hill, ed. *The sea*. Interscience Publishers, New York.

Johnson, S. L. 1988. The effects of the 1983 El Niño on Oregon's coho (*Oncorhynchus kisutch*) and Chinook salmon (*O. tshawtscha*) salmon. *Fisheries Research* 6:105–123.

Johnson, J. K. 1990. Regional overview of coded wire tagging of anadromous salmon and steelhead in northwest America. *American Fisheries Society Symposium* 7:782–816.

Kaeriyama, M., M. Nakamura, R. Edpalina, J. R. Bower, H. Yamaguchi, R. V. Walker, and K. W. Myers. 2004. Change in feeding ecology and trophic dynamics of Pacific salmon (Oncorhynchus spp.) in the central Gulf of Alaska in relation to climate events. *Fisheries Oceanography* 13(3):197–206.

Kawasaki, T. 1983. Why do some pelagic fishes have wide fluctuations in their numbers? *FAO Fish Report* 291:1065–1080.

Kerr, R. A. 1994. Did the tropical Pacific drive the world's warming? *Science* 266:544–545.

Koslow, J. A., A. J. Hobday, and G. W. Boehlert. 2002. Climate variability and marine survival of coho salmon (Oncorhynchus kisutch) in the Oregon production area. *Fisheries Oceanography* 11(2):65–77.

Kukulka, T., and D. A. Jay. 2003. Impacts of Columbia River discharge on salmonid habitat II. Changes in shallow-water habitat. *Journal of Geophysical Research* 108(C9), 3294 doi 10.1029/2003JC001829.

Kumar, A., A. Leetmaa, and M. Ji. 1994. Simulations of atmospheric variability induced by sea surface temperatures and implications for global warming. *Science* 266:632–634.

Landry, M. R., J. R. Postel, W. K. Peterson, and J. Newman. 1989. Broad-scale distributional patterns of hydrographic variables on the Washington/Oregon shelf. Pages 1–40 in M. R. Landry and B. M. Hickey, eds. *Coastal oceanography of Washington and Oregon*. Elsevier, Amsterdam.

Lasker, R. 1978. The relation between oceanographic conditions and larval anchovy food in the California Current: Identification of factors leading to recruitment failure. *Rapports et Proces-verbaux des Reunions* 173:212–230.

Lawson, P. W. 1993. Cycles in ocean productivity, trends in habitat quality, and restoration of salmon runs in Oregon. *Fisheries* 18:6–10.

Lawson, P. W., E. A. Logerwell, N. J. Mantua, R. C. Francis, and V. N. Agostini. 2004. Environmental factors influencing freshwater survival and smolt production in Pacific Northwest coho salmon (*Oncorhynchus kisutch*). *Canadian Journal of Fisheries and Aquatic Sciences* 61:360–373.

Levin, P. S., and J. G. Williams. 2002. Interspecific effects of artificially propagated fish: an additional conservation risk for salmon. *Cons. Biol.* 16(6):1–8.

Levin, P. S., R. W. Zabel, and J. G. Williams. 2001. The road to extinction is paved with good intentions: negative associations of fish hatcheries with threatened salmon. Proc. Royal Soc. London, Series B 286:1–6.

Levings, C. D., C. D. McAllister, and B. D. Chang. 1986. Differential use of the Campbell River estuary, British Columbia, by wild and hatchery-reared juvenile Chinook salmon (*Oncorhynchus tshawytscha*). *Canadian Journal of Fisheries and Aquatic Sciences* 43(7):1386–1397.

Levy, D. A., and T. G. Northcote. 1982. Juvenile salmon residency in a marsh area of the Fraser River estuary. *Canadian Journal of Fisheries and Aquatic Sciences* 39:270–276.

Lichatowich, J. 1993. Ocean carrying capacity: Recovery issues for threatened and endangered Snake River Salmon. Bonneville Power Administration, US Department of Energy. Technical Report, DOE/BP-99654-6. Portland, Oregon. 25 p.

Lichatowich, J. 1999. *Salmon without rivers: A history of the Pacific salmon crisis.* Island Press, Washington, DC.

Lichatowich, J. A., L. Mobrand, and T. Vogel. 1996. A history of frameworks used in the management of Columbia River Chinook salmon (Draft). Bonneville Power Administration, Portland, Oregon. 117 p.

Logerwell, E. A., N. Mantua, P. W. Lawson, R. C. Francis, and V. N. Agostini. 2003. Tracking environmental processes in the coastal zone for understanding and predicting Oregon coho (*Oncorhynchus kisutch*) marine survival. *Fisheries Oceanography* 12(6):554–568.

Lott, M. A. 2004. Habitat-specific feeding ecology of ocean-type juvenile Chinook salmon in the Lower Columbia River estuary. Master's thesis, University of Washington, Seattle. 110 p.

Lluch-Belda, D., S. Hernandez-Vasquez, D. B. Lluch-Cota, and C. A. Salinas-Zavala. 1992. The recovery of the California sardine as related to global change. *California Cooperative Fisheries Investigations Report* 33:50–59.

Mann, K. H., and J. R. N. Lazier. 1991. *Dynamics of marine ecosystems: Biological-physical interactions in the oceans.* Blackwell Scientific Publications, Boston.

Mantua, N. J., S. R. Hare, Y. Zhang, J. M. Wallace, and R. C. Francis. 1997. A Pacific inter-decadal climate oscillation with impacts on salmon production. *Bulletin of the American Meteorological Society* 78:1069–1079.

McCabe, G. T., Jr., R. L. Emmett, W. D. Muir, and T. H. Blahm. 1986. Utilization of the Columbia River estuary by subyearling Chinook salmon. *Northwest Science* 60(2):113–124.

McCabe, G. T., Jr., S. A. Hinton, and R. L. Emmett. 1998. Benthic invertebrates and sediment characteristics in a shallow navigational channel of the lower Columbia River, before and after dredging. *Northwest Science* 72:116–126.

McCabe, G. T., Jr., S. A. Hinton, R. L. Emmett, and B. P. Sandford. 1997. Benthic invertebrates and sediment characteristics in main channel habitats in the lower Columbia River. *Northwest Science* 71:45–55.

McCarl, B. A., and T. B. Rettig. 1983. Influence of hatchery smolt releases on adult salmon production and its variability. *Journal of the Fisheries Research Board of Canada* 40:1880–1886.

McClain, D. R., and D. H. Thomas. 1983. Year-to-year fluctuations of the California Countercurrent and effects on marine organisms. *California Cooperative Fisheries Investigations Report* 24:165–181.

McElhaney, P., M. H. Ruckelhaus, M. J. Ford, T. C. Wainwright, and E. P. Bjorkstedt. 2000. Viable salmon populations and the recovery of evolutionary significant units. US Department of Commerce. NOAA Technical Memorandum NMFS-NWFSX-42. Seattle, Washington. 156 p.

McFarlane, G. A., J. R. King, and R. J. Beamish. 2000. Have there been recent changes in climate? Ask the fish. *Progress in Oceanography* 47(2000):147–169.

McGie, A. M. 1984. Commentary: Evidence for density dependence among coho salmon stocks in the Oregon Production Index Area. Pages 37–49 in W. G. Pearcy, ed. *The influence of ocean conditions on the production of salmonids in the North Pacific.* Oregon State University Sea Grant College Program, ORESU-W-83-001, Corvallis, Oregon.

McGowan, J. A. 1971. Oceanic biogeography of the Pacific. Pages 3–74 in B. M. Funnell and W. R. Riedel, eds. *The micropppaleontology of oceans.* Cambridge University Press, Cambridge, Massachuetts.

McGowan, J. A. 1974. The nature of oceanic ecosystems. Pages 9–28 in C. B. Miller, ed. *The biology of the oceanic Pacific*. Oregon State University Press, Corvallis.

McGowan, J. A., D. R. Cayan, and L. M. Dorman. 1998. Climate-ocean variability and ecosystem response in the Northeast Pacific. *Science* 281:210–217.

McIsaac, D. O. 1990. Factors affecting the abundance of 1977-79 brood wild fall Chinook salmon (*Oncoryhnchus tshawytscha*) in the Lewis River, Washington. Ph. D. dissertation, University of Washington, Seattle.

McPhaden, M. J. 1999. Genesis and evolution of the 1997-98 El Niño. *Science* 283:950–954.

Melack, J. M., J. Dozier, C. R. Goldman, D. Greenland, A. M. Milner, and R. J. Naiman. 1997. Effects of climate change on inland waters of the Pacific coastal mountains and western Great Basin of North America. *Hydrological Processes* 11:971–992.

Mote, P. W., E.A. Parson, A. F. Hamlet, W. S. Keeton, D. Lettenmaier, N. Mantua, E. L. Miles, D. W. Peterson, D. L. Peterson, R. Slaughter, and A. K. Snover. 2003. Preparing for climate change: the water, salmon, and forests of the Pacific Northwest. *Climatic Change* 61:45–88.

Mueter, F. J., R. M. Peterman, and B. J. Pyper. 2002a. Opposite effects of ocean temperature on survival rates of 120 stocks of Pacific salmon (Oncorhynchus spp.) in northern and southern areas. *Canadian Journal of Fisheries and Aquatic Sciences* 59:456–463.

Mueter, F. J., D. M. Ware, and R. M. Peterman. 2002b. Spatial correlation patterns in coastal environmental variables and survival rates of salmon in the north-east Pacific Ocean. *Fisheries Oceanography* 11(2):1–14.

Myers, K. W., and H. F. Horton. 1982. Temporal use of an Oregon estuary by hatchery and wild juvenile salmon. Pages 388–392 in V. S. Kennedy, ed. *Estuarine comparisons*. Academic Press, New York.

Myers, R. A., G. Mertz, and J. Bridson. 1997. Spatial scales of interannual recruitment variations of marine, anadromous, and freshwater fish. *Canadian Journal of Fisheries and Aquatic Sciences* 54:1400–1407.

Mysak, L. A. 1986. El Niño, interannual variability and fisheries in the northeast Pacific Ocean. *Canadian Journal of Fisheries and Aquatic Sciences* 43:464–497.

National Research Council (NRC). 1996. *Upstream: Salmon and society in the Pacific Northwest*. Report on the Committee on Protection and Management of Pacific Northwest Anadromous Salmonids for the National Research Council of the National Academy of Sciences. National Academy Press, Washington DC.

Neitzel, D. A., M. J. Scott, S. A. Shankle, and J. C. Chatters. 1991. The effect of climate change on stream environments: the salmonid resource of the Columbia River Basin. *Northwest Environmental Journal* 7:271–293.

Nicholas, J. W., and D. G. Hankin. 1988. Chinook salmon populations in Oregon coastal river basins: Description of life histories and assessment of recent trends in run strengths. Oregon Department of Fish and Wildlife. Information Reports (Fish), 88-1. Portland, Oregon.

Nickelson, T. E. 1986. Influences of upwelling, ocean temperature, and smolt abundance on marine survival of coho salmon (*Oncorhynchus kisutch*) in the Oregon production area. *Canadian Journal of Fisheries and Aquatic Sciences* 43:527–535.

Nickelson, T. E., and J. A. Lichatowich. 1983. The influence of the marine environment on the interannual variation in coho salmon abundance: An overview. Pages 24–36 in W. G. Pearcy, ed. *The influence of ocean conditions on the production of salmonids in the North Pacific*. Oregon State University, Sea Grant ORESU-W-83-001, Corvallis, Oregon.

Nickelson, T. E., J. W. Nicholas, A. M. McGie, R. B. Lindsay, D. L. Bottom, R. J. Kaiser, and S. E. Jacobs. 1992. Status of anadromous salmonids in Oregon Coastal basins. Oregon Department of Fish and Wildlife. Portland, Oregon.

Noakes, D. J., R. J. Beamish, R. Sweeting, and J. King. 2000. Changing the Balance: Interactions between hatchery and wild Pacific coho salmon in the presence of regime shifts. *North Pacific Anadromous Fish Commission Bulletin* 2:155–163.

Norton, J., D. McClain, R. Brainard, and D. Husby. 1985. The 1982-83 El Niño event off Baja and Alta California and its ocean climate context. Pages 44–72 in W. S. Wooster and D. L. Fluharty, eds. *El Niño north: Niño effects in the Eastern Subarctic Pacific Ocean.* Washington Sea Grant Program, University of Washington, Seattle.

Oregon Department of Fish and Wildlife. 1982. Comprehensive plan for production and management of Oregon's anadromous salmon and trout. Part II: Coho salmon plan. Portland, Oregon.

Orem, H. J. 1968. Discharge in the lower Columbia River basin, 1928-1965. USGS Circular 550. 24p.

Pauly, D. 1995. Anecdotes and the shifting baseline syndrome of fisheries. Trends in Ecoloy and Evolution 10(10): 430.

Pearcy, W. G. 1992. *Ocean ecology of North Pacific salmonids.* University of Washington Press, Seattle.

Pearcy, W. G., J. Fisher, R. Brodeur, and S. Johnson. 1985. Effects of the 1983 El Niño on coastal nekton off Oregon and Washington. Pages 188–204 in W. S. Wooster and D. L. Fluharty, eds. *El Niño north: Niño Effects in the Eastern Subarctic Pacific.* Washington Sea Grant Program, University of Washington, Seattle.

Pearcy, W. G., and J. P. Fisher. 1988. Migrations of coho salmon, *Oncorhynchus kisutch*, during their first summer in the ocean. *Fishery Bulletin* 86:173–195.

Pearcy, W. G., and J. P. Fisher. 1990. Distribution and abundance of juvenile salmonids off Oregon and Washington, 1981-1985. National Oceanic and Atmospheric Administration. Technical Report, NMFS 93.

Peterman, R. M. 1989. Application of statistical power analysis to the Oregon coho (*Oncorhynchus kisutch*) problem. *Canadian Journal of Fisheries and Aquatic Sciences* 46:1183–1187.

Peterman, R. M., and R. D. Routledge. 1983. Experimental management of Oregon coho salmon (*Oncorhynchus kisutch*): Designing for yield of information. *Canadian Journal of Fisheries and Aquatic Sciences* 40:1212–1223.

Peterson, W. T., C. B. Miller, and A. Hutchinson. 1979. Zonation and maintenance of copepod populations in the Oregon upwelling zone. *Deep-Sea Research* 26A:467–494.

Peterson, W. T., and F. B. Schwing. 2003. A new climate regime in northeast Pacific ecosystems. *Geophysical Research Letters* 30(17), 1896, doi:10.1029/2003GL017528, 2003.

Pulwarty, R. S., and K. T. Redmond. 1997. Climate and salmon restoration in the Columbia River basin: the role and usability of seasonal forecasts. *Bulletin of the American Meteorological Society* 78(3):381–397.

Quinn, W. H., V. T. Neal, and S. E. A. de Mayolo. 1987. El Niño occurrences over the past four and a half centuries. *Journal Geophysical Research* 92:14,449–14,461.

Reimers, P. E. 1973. The length of residence of juvenile fall Chinook salmon in Sixes River, Oregon. *Research Reports of the Fish Commission of Oregon* 4:3–43.

Rich, W. H. 1920. Early history and seaward migration of Chinook salmon in the Columbia and Sacramento Rivers. *U.S. Bur. Fish., Bull.* 37:2–73.

Roby, D. D., D. E. Lyons, D. P. Craig, K. Collis, and G. H. Visser. 2003. Quantifying the effects of predators on endangered species using a bioenergetics approach: Caspian terns and juvenile salmonids in the Columbia River estuary. *Canadian Journal of Zoology* 81:250–265.

Roegner, G. C., D. L. Bottom, A. Baptista, S. Hinton, C. A. Simenstad, E. Casillas, and K. Jones. 2004. Estuarine habitat and juvenile salmon—Current and historical linkages in the lower Columbia River and estuary, 2003. Report of Research by the Fish Ecology Division, NOAA Fisheries, Northwest Fisheries Science Center to Portland District US Army Corps of Engineers. Northwest Fisheries Science Center, Seattle, Washington.

Royal, L. A. 1972. An examination of the anadromous trout program of the Washington State Game Department. Washington State Department of Game. Olympia, Washington. 176 p.

Ruggerone, G. T., and F. A. Goetz. 2004. Survival of Puget Sound Chinook salmon (*Oncorhynchus tshawytscha*) in response to climate-induced competition with pink salmon (*Oncorhynchus gorbuscha*). *Canadian Journal of Fisheries and Aquatic Sciences* 61:1756–1770.

Ruggerone, G. T., M. Zimmermann, K. W. Myers, J. L. Nielsen, and D. E. Rogers. 2003. Competition between Asian pink salmon (*Oncorhynchus gorbuscha*) and Alaskan sockeye salmon (*O. nerka*) in the North Pacific Ocean. *Fisheries Oceanography* 12(3):209–219.

Ryan, B. A., S. G. Smith, J. M. Butzerin, and J. W. Ferguson. 2003. Relative vulnerability to avian predation of juvenile salmonids tagged with Passive Integrated Transponders in the Columbia River Estuary, 1998–2000. *Transactions of the American Fisheries Society* 132:275–288.

Salo, E. O. 1991. Life history of chum salmon (*O. keta*). Pages 231–310 in C. Groot and L. Margolis, eds. *Pacific salmon life histories*. University of British Columbia Press, Vancouver.

Scarnecchia, D. L. 1981. Effects of streamflow and upwelling on yield of wild coho salmon (*Oncorhynchus kisutch*) in Oregon. *Canadian Journal of Fisheries and Aquatic Sciences* 38:471–475.

Schabetsberger, R., C. A. Morgan, R. D. Brodeur, C. L. Potts, W. T. Peterson, and R. L. Emmett. 2003. Prey selectivity and diel feeding chronology of juvenile Chinook (*Oncorhynchus tshawystcha*) and coho (*O. kisutch*) salmon in the Columbia River plume. *Fisheries Oceanography* 12:523–540.

Schluchter, M., and J. A. Lichatowich. 1977. Juvenile life histories of Rogue River spring Chinook salmon *Oncorhynchus tshawytscha* (Walbaum), as determined from scale analsysis. Oregon Department of Fish and Wildlife. Information Report Series, Fisheries No. 77-5. Corvallis, Oregon.

Sharp, G. D. 1991. Climate and fisheries: Cause and effect–A system review. Pages 239–258 in T. Kawasaki, S. Tanaka, Y. Toba, and A. Taniguchi, eds. *Long-term variability of pelagic fish populations and their environment*. Pergamon Press, Oxford, United Kingdom.

Sherwood, C. P., D. A. Jay, R. B. Harvey, P. Hamilton, and C. A. Simenstad. 1990. Historical changes in the Columbia River estuary. *Progressive Oceanography* 25:299–352.

Simenstad, C. A., K. L. Fresh, and E. O. Salo. 1982. The role of Puget Sound and Washington coastal estuaries in the life history of Pacific salmon: An unappreciated function. Pages 343–364 in V. S. Kennedy, ed. *Estuarine comparisons*. Academic Press, New York.

Simenstad, C. A., D. Jay, C. D. McIntire, W. Nehlsen, C. R. Sherwood, and L. F. Small. 1984. The Dynamics of the Columbia River Estuarine Ecosystem, Vols. I and II. Columbia River Estuary Data Development Program, Astoria, Oregon.

Simenstad, C. A., D. A. Jay, and C. R. Sherwood. 1992. Impacts of watershed management on land-margin ecosystems: The Columbia River estuary. Pages 266–306 in R. J. Naiman, ed. *Watershed management: Balancing sustainability and environmental change*. Springer-Verlag, New York.

Sinclair, M., M. J. Tremblay, and P. Bernal. 1985. El Niño events and variability in a Pacific mackerel (*Scomber japonicus*) survival index: Support for Hjort's second hypothesis. *Canadian Journal of Fisheries and Aquatic Sciences* 42:602–608.

Small, L. F., and D. W. Menzies. 1981. Patterns of primary productivity and biomass in a coastal upwelling region. *Deep-Sea Research* 28A:123–149.

Smith, P. E. 1978. Biological effects of ocean variability: Time and space scales of biological response. *Rapports et Proces-verbaux des Reunions* 173:117–127.

Smith, R. L. 1983. Physical features of coastal upwelling systems. Washington Sea Grant Program, University of Washington. Technical Report, WSG 83-2. Seattle, Washington.

Soutar, A., and J. D. Isaacs. 1969. History of fish populations inferred from fish scales in anaerobic sediments off California. *California Cooperative Oceanic Fisheries Investigations Report* 13:63–70.

Soutar, A., and J. D. Isaacs. 1974. Abundance of pelagic fish during the 19th and 20th centuries as recorded in anaerobic sediment off the Californias. *Fishery Bulletin* 72:257–273.

Strub, P. T., J. S. Allen, A. Huyer, and R. L. Smith. 1987a. Large-scale structure of the spring transition in the coastal ocean off western North America. *Journal of Geophysical Research* 92(c2):1527–1544.

Strub, P. T., J. S. Allen, A. Huyer, R. L. Smith, and R. C. Beardsley. 1987b. Seasonal cycles of currents, temperatures, winds and sea level over the northeast Pacific continental shelf: 35°N to 48°N. *Journal of Geophysical Research* 92:1507–1526.

Sverdrup, H. U., M. W. Johnson, and R. H. Fleming. 1942. *The oceans: Their physics, chemistry and general biology.* Prentice Hall, Englewood Cliffs, New Jersey.

Sweeting, R. M., R. J. Beamish, D. J. Noakes, and C. M. Neville. 2003. Replacement of wild coho salmon by hatchery-reared coho salmon in the Strait of Georgia over the past three decades. *North American Journal of Fisheries Management* 23:492–502.

Teel, D. J., D. M. Van Doornik, D. R. Kuligowski, and W. S. Grant. 2003. Genetic analysis of juvenile coho salmon (*Oncorhynchus kisutch*) off Oregon and Washington reveals few Columbia River wild fish. *Fishery Bulletin* 101:640–652.

Thomas, D. W. 1983. Changes in Columbia River estuary habitat types over the past century. Columbia River Estuary Data Development Program. Astoria, Oregon.

Trenberth, K. E. 1990. Recent observed interdecadal climate changes in the northern hemisphere. *Bulletin of the American Meteorological Society* 71:988–993.

Venrick, E. L., J. A. McGowan, D. R. Cayan, and T. L. Hayward. 1987. Climate and chlorophyll a: long-term trends in the Central North Pacific Ocean. *Science* 238:70–72.

Ware, D. M., and R. E. Thompson. 1991. Link between long-term variability in upwelling and fish production in the northeast Pacific Ocean. *Canadian Journal of Fisheries and Aquatic Sciences* 48:2296–2306.

Washington Department of Fisheries and Washington Department of Wildlife. 1993. 1992 Washington State salmon and steelhead stock inventory Appendix 3: Columbia River stocks. Washington Department of Fisheries. Olympia, Washington. 580 p.

Weitkamp, L. A. 1994. A review of the effects of dams on the Columbia River estuarine environment, with special reference to salmonids. Report to Bonneville Power Administration, PO Box 3621, Portland, OR 97208. DE-A179-93BP99021, 148 p.

Wentz, D. A., I. R. Waite, and F. A. Rinella. 1998. Comparison of streambed sediment and aquatic biota as media for characterizing trace elements and organochlorine compounds in the Willamette Basin, Oregon. *Environmental Monitoring and Assessment* 51:673–693.

Wickett, W. P. 1967. Ekman transport and zooplankton concentration in the North Pacific Ocean. *Journal of the Fisheries Research Board of Canada* 24:581–594.

Zhang, Y., J. M. Wallace, and D. S. Battisti. 1997. ENSO-like interdecadal variability: 1900-93. *Journal of Climate* 10:1004–1020.

Appendix: Ocean-Climate Indices

A number of measures or indices have been developed to monitor changes in ocean condition; a few are presented here to demonstrate the patterns in ocean conditions and the similarities between these indices. The indices plotted include the Pacific Decadal Oscillation index, the Atmospheric Forcing Index, and the Pacific Northwest Index. References to the source papers and websites are provided.

(a) *The Pacific Decadal Oscillation (PDO):* This index was presented by Mantua et al. (1997) and Zhang et al. (1997). The index is the first principal component of monthly sea surface temperature anomalies in the Pacific Ocean north of 20° N. The index is associated with two phases: a positive phase associated with warming of surface waters in the northeast Pacific and cooling in the central and western North Pacific, and a negative phase with the opposite temperature pattern. The data presented in Figure 10.21 is the annual average of the monthly values maintained and provided at a University of Washington website: *http://tao.atmos.washington.edu/pdo.*

This site describes the PDO as *"a long-lived El Niño-like pattern of Pacific climate variability. While the two climate oscillations have similar spatial climate fingerprints, they have very different behavior in time. Fisheries scientist Steven Hare coined the term 'Pacific Decadal Oscillation' (PDO) in 1996 while*

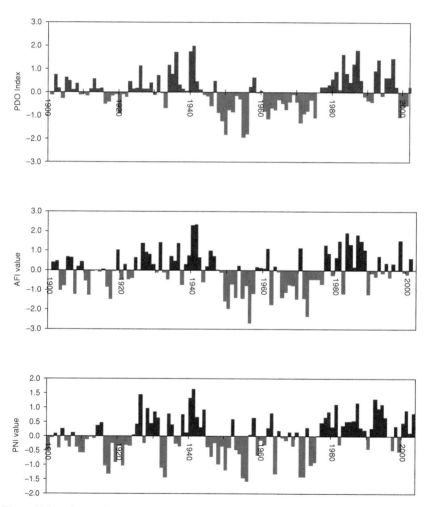

Figure 10.21 Comparison of the three indices described above. PDO and AFI are indices of changes in ocean conditions, and the Pacific Northwest Index (PNI) characterizes Pacific Northwest climate patterns in both coastal waters and freshwater habitats.

researching connections between Alaska salmon production cycles and Pacific climate. . . . Two main characteristics distinguish PDO from El Niño/Southern Oscillation (ENSO): first, 20th century PDO 'events' persisted for 20 to 30 years, while typical ENSO events persisted for 6 to 18 months; second, the climatic fingerprints of the PDO are most visible in the North Pacific/North American sector, while secondary signatures exist in the tropics–the opposite is true for ENSO. Several independent studies find evidence for just two full PDO cycles in the past century: 'cool' PDO regimes prevailed from 1890-1924 and again from 1947-1976, while 'warm' PDO regimes dominated from 1925-1946 and from 1977 through (at least) the mid-1990's."

"Major changes in northeast Pacific marine ecosystems have been correlated with phase changes in the PDO; warm eras have seen enhanced coastal ocean biological productivity in Alaska and inhibited productivity off the west coast of the contiguous United States, while cold PDO eras have seen the opposite north-south pattern of marine ecosystem productivity."

Only yearly average PDO values have been presented for comparison with these indices.

(b) *The Atmospheric Forcing Index (AFI):* This index was presented by McFarlane et al. (2000) and is a composite of three other indices. The AFI utilizes standardized scores of the first principal components of an analysis on the Aleutian Low Pressure Index, Pacific Interdecadal Oscillation Index, and the northwesterly atmospheric circulation anomalies for the North Pacific (December–March). Positive values represent intense Aleutian Lows, above average frequency of westerly and southwesterly winds, cooling of sea surface temperatures in the Central North Pacific, and warming within North American coastal waters. (see *http://pac.dfo-mpo.gc.ca/sci/sa-mfpd/downloads/afi.txt*)

c) *The Pacific Northwest Index (PNI):* This index was presented by Ebbesmeyer and Strickland (1995) and is a terrestrial index described as being useful in studying climate effects on salmon productivity trends (Anderson 2000). The PNI is another composite index *"that characterizes Pacific Northwest climate patterns in both coastal waters and freshwater habitats. . . . A composite climate index is an effective measurement because many environmental parameters in the Northwest are statistically related to one another; consequently, they may be combined to furnish a broad-scale under-standing of the state of the Pacific Northwest environment. The PNI uses three parameters: 1) air temperature at Olga in the San Juan Islands, averaged annually from daily data; 2) total precipitation at Cedar Lake in the Cascade Mountains; and 3) snowpack depth at Paradise on Mount Rainier on March 15 of each year. For each parameter, annual values were normalized by subtracting the annual value from the average of all years and dividing by the standard deviation. The sign of the normalized temperature is reversed because low tempera-ture corresponds with high precipitation and snow. Finally, the three variables are averaged yearly giving a relative indicator of the variations in climate. Years with positive values of the PNI are warmer and drier than average and those with negative values are cooler and wetter than average"* (see *http://www.cbr.washington.edu/data/pni_data.html*).

The coincidence of patterns is apparent among indices. The PDO pattern was described above, but the other indices follow very similar patterns. The important question though is how these changes in ocean-climate conditions influence salmon production in the Columbia Basin. Several early publications link the patterns to trends in salmon production in large geographic areas, notable examples being Beamish and Boullion (1993), Hare and Francis (1995), and Mantua et al. (1997). While the evidence is strong for marine environmental effects on large spatial scales, the patterns over time vary between geographic areas, and effects are more difficult to identify at more localized spatial scales. Anderson (2000) describes the importance of recognizing these patterns and the interaction of ocean-climate regimes with anthropogenic developments within the Columbia and Snake Rivers. Clearly, the identification of this interaction is essential to properly assess the impact of development within the basin and for reasonable expectations of salmon production over time. As Anderson (2000) suggests, *"Beyond the issue of rebuilding damaged populations, an understanding of climate cycles is essential for long term sustainable management of fisheries resources. Since climate cycles have such a large impact on stock productivity, an improved understanding of the cycles and their coupling to fisheries should eventually improve our ability to sustain fisheries over periods of scarcity and abundance"* (p. 481).

Return to the River

11

Monitoring and Evaluation: Salmon Restoration in the Columbia River Basin

Lyman L. McDonald, Ph.D., Charles C. Coutant, Ph.D.,
Lyle D. Calvin, Ph.D., Richard N. Williams, Ph.D.

Introduction
Background
  Historical Record
  Hydrosystem Development
  The Northwest Power Act
  Endangered Species Act—Section 7 Consultation: Biological Opinions
System-wide Monitoring and Evaluation in the Columbia Basin
  Current Efforts
  Types of Monitoring Needed
    Implementation Monitoring
    Trend (or Change) Monitoring (Tier 1 Monitoring)
    Statistical Monitoring (Tier 2 Monitoring)
    Research Monitoring (Tier 3 Monitoring)
    Effectiveness Monitoring
Recommendations for Monitoring and Evaluation
  Evaluation of Monitoring Efforts
  Monitoring and Adaptive Management
  Role of Databases in Monitoring and Evaluation
  Role of Index Sites
  Major Monitoring and Evaluation Exercises
Summary
Literature Cited

> *"That in view of the lack of definite assurance as to the degree of success to be anticipated from the plan as proposed, its experimental character should be recognized; and it follows that the adoption of the plan for trial should not be understood as implying an indefinite commitment to its support, but only for so long as the results may reasonably appear to justify its continuance." (p. VI)*

—Calkins, R. D., H. F. Durand, and W. H. Rich. 1939. Report of the Board of Consultants on the fish problems of the upper Columbia River. Sections 1 and 2. Stanford University, California.

Introduction

Monitoring and evaluation has been a long-standing issue of concern in the United States (Bernhardt et al. 2005) and the Columbia River Basin, as well as for the Fish and Wildlife Program (FWP) of the Northwest Power and Conservation Council (NPCC). The importance of the monitoring and evaluation functions has been recognized for some time (Calkins et al. 1939), including the necessary linkage of monitoring and evaluation with adaptive management (Figure 11.1; Lee 1993; Volkman and McConnaha 1993; McConnaha and Pacquet 1996). Although implied from the earliest planning under the Northwest Power Act, a monitoring and evaluation role was made explicit in the 1987 amended program with inclusion of a System Monitoring and Evaluation Program to track progress of the FWP in achieving the NPCC's goals of doubling the runs of salmon and steelhead in the basin. While substantial progress has occurred on refining monitoring and evaluation needs at the project-specific level, the region continues to struggle with how to assess programs at higher levels of organization. System-level monitoring and evaluation of subbasins and the Columbia Basin remain elusive goals (Monitoring and Evaluation Group [MEG] 1988; ISRP 1997, 1998, 1999, 2000).

A collaborative effort is needed to develop an effective monitoring and evaluation program for the Columbia River Basin (Columbia Basin Fish and Wildlife Authority [CBFWA] 2003). Monitoring of survival and movement of juveniles and adults through the hydropower system on the mainstem of the Columbia and Snake Rivers and monitoring of harvest of adult fish are reasonably well coordinated among federal, state, and tribal agencies. (See the World Wide Web sites of the Fish Passage Center, *www.fpc.org*, for monitoring of movement and survival of juveniles and adults through the hydropower system and the Pacific States Marine Fisheries Commission, *www.psmfc.org*, for information on harvest.) However, there are urgent needs in monitoring and evaluation of trends and changes in habitat (tributary and mainstem) and the interactions of hatchery and naturally produced fish. In this chapter, we will concentrate primarily on monitoring and evaluation of habitat and interactions of hatchery and naturally produced fish. Currently, state, federal, and tribal agencies often utilize different procedures for selection of study sites in the field, different data collection methods, and different formats for electronic storage of data. These differences can lead to intractable problems for summary of results across boundaries of watersheds, subbasins, and institutions. Nevertheless, considerable funds and manpower are utilized in these individual efforts. A cooperative monitoring and evaluation program should provide better information at more reasonable economic costs.

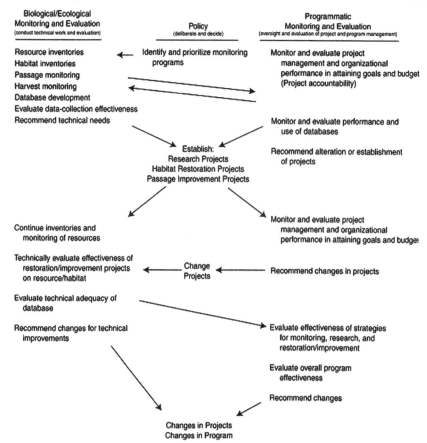

Biological/Ecological
Monitoring and Evaluation
(conduct technical work and evaluation)

Policy
(deliberate and decide)

Programmatic
Monitoring and Evaluation
(oversight and evaluation of project and program management)

Resource inventories
Habitat inventories
Passage monitoring
Harvest monitoring
Database development
Evaluate data-collection effectiveness
Recommend technical needs

Identify and prioritize monitoring
programs

Monitor and evaluate project
management and organizational
performance in attaining goals and budget
(Project accountability)

Monitor and evaluate performance and
use of databases

Establish:
Research Projects
Habitat Restoration Projects
Passage Improvement Projects

Recommend alteration or establishment
of projects

Continue inventories and
monitoring of resources

Monitor and evaluate project
management and organizational
performance in attaining goals and budget

Technically evaluate effectiveness of
restoration/improvement projects
on resource/habitat

Change
Projects

Recommend changes in projects

Evaluate technical adequacy of
database

Recommend changes for technical
improvements

Evaluate effectiveness of strategies
for monitoring, research, and
restoration/improvement

Evaluate overall program
effectiveness

Recommend changes

Changes in Projects
Changes in Program

Figure 11.1 Relationships among three aspects of monitoring and evaluation: (a) technical work of biological—ecological monitoring and data analysis, (b) oversight of programs (programmatic monitoring and evaluation), and (c) policy deliberation and decisions regarding projects to conduct and adaptive management based on the results. All three are important for an effective monitoring program.

"Research" is often mentioned in the same breath with monitoring and evaluation. We choose to separate relatively short-term research projects from long-term ecological monitoring and evaluation, because there are fundamental differences and objectives. We discuss some of these differences later in the chapter.

Subbasin and system-wide monitoring and evaluation of fish status is needed to address requirements of biological opinions (BiOps) and recovery plans for threatened and endangered populations as well as for healthy stocks of anadromous and resident species. There is a need to document, integrate, and make available existing monitoring data that bear on the problem of evaluating the status of fishes and their responses to management actions in the Columbia Basin. Agencies in the basin need to collaboratively design improved monitoring and evaluation methods that will fill information gaps and provide better answers to questions in the future.

Background

Historical Record

The earliest surveys in the Columbia River Basin were conducted by Evermann (1895); however, the first systematic fisheries and stream habitat surveys in the Columbia River were conducted by the federal Bureau of Fisheries (now NOAA Fisheries; previously the National Marine Fisheries Service [NMFS]). The Bureau of Fisheries conducted fisheries and stream habitat surveys in parts of the basin from 1934 to 1942. The surveys were intended to cover streams in the Columbia River Basin, which provided, or once had provided, spawning (Figure 11.2) and rearing habitat (Figure 11.3) for salmon and steelhead, to evaluate the streams' condition, availability, and

Figure 11.2 Chinook salmon on spawning habitat, Johnson Creek of the Salmon River, Idaho. Redds (spawning nests) are denoted by the lighter colored gravel areas in the foreground and background. Photo by R. N. Williams.

Figure 11.3   Spring Chinook rearing habitat in the upper Imnaha River, Oregon. Photo by R. N. Williams.

usefulness for migration, breeding, and rearing of migratory fishes (Rich 1948). Most of the quantitative records of those surveys has been lost. Surviving material consists of summaries or brief, qualitative accounts (Rich 1948; Bryant 1949; Bryant and Parkhurst 1950; Parkhurst 1950a, 1950b, 1950c). Despite their brevity, these summaries have formed the basis for esti-mating habitat losses and conditions in the Columbia River Basin (Fulton 1968, 1970; Thompson 1976; Sedell and Luchessa 1981; Northwest Power Planning Council [NPPC] 1986).

Field notebooks from the early fishery surveys were discovered in the mid-1990s. The data are now archived and stored in the Forest Science

DataBank at Oregon State University and have been published as exact repli-
cates of the originals as part of the FWP. The habitat surveys included the
Clearwater, Salmon, Weiser, and Payette River Basins (McIntosh et al.
1995a), the Cowlitz River Basin (McIntosh et al. 1995b), the Umatilla,
Tucannon, Asotin, and Grande Ronde River Basins (McIntosh et al. 1995c),
the Willamette River Basin (McIntosh et al. 1995d), and the Yakima River
basin (McIntosh et al. 1995e). These records, as noted by the compilers of the
present publications, are the earliest and most comprehensive documentation
available of the condition and extent of anadromous fish habitat before
hydropower development in the Columbia River Basin. They are unique
because they are the only long-term data set that quantifies fish habitat in a
manner that is replicable over time. Other surveys, such as Thompson and
Haas (1960), inventoried extensive areas, but in a manner that was mostly
qualitative. Knowledge of past and present quantity and quality of habitat
for anadromous fishes is essential to evaluating our efforts to enhance fish
populations. Habitat condition has to be recognized as a key element in monitoring and evaluating progress toward the NPCC's restoration goals and
recovery of populations listed under the Endangered Species Act.

   The data sets include detailed information on the character of the watershed and station, marginal vegetation and extent of erosion, elevations and
slopes, observed flows and fluctuations, water and air temperatures, pool and
riffle characteristics, character of the bottom, areas available that were suitable and unsuitable for spawning, obstructions, diversions, pollution, fish
observations (redds, run sizes and timing, juvenile rearing), non-salmonid
fish observed, extent of sport fishing, and miscellaneous field observations
and opinions of the surveyors.

   The next comprehensive stock surveys were compiled through the initial
efforts at subbasin planning in the Columbia River Basin in the late 1980s
and early 1990s. The planning process led to summaries of tributary stocks
of salmonids in subbasins throughout the Columbia River Basin. Efforts
were driven by the NPCC's FWP and overseen by the Columbia River
Coordinated Information System (now StreamNet). Draft reports were published in 1992, and the material is stored in retrievable electronic form at the
StreamNet offices (*www.streamnet.org*) (Hymer et al. 1992a, 1992b; Kiefer
et al. 1992; Olsen et al. 1992a, 1992b, 1992c, 1992d). The stock summaries of
major tributaries are a valuable record of the information that is available.
Many of the stocks for which information has been compiled have not been
systematically monitored, but have scattered records.

   The most recent updating of these surveys occurred in the eastern portion
of the Columbia River Basin through the Interior Columbia Basin
Ecosystem Management Process (Quigley et al. 1996), which conducted
comprehensive surveys of salmonid stocks and habitats (Figure 11.4).
Currently, the Northwest Power and Conservation Council, Portland,

Figure 11.4 Chinook salmon migrating to spawning grounds on Johnson Creek of the Salmon River, Idaho. The Johnson Creek summer Chinook population is one of the region's index stocks, where population abundance is tracked in a defined portion of the river annually for Tier 1 (trend) monitoring. Photo by L. L. McDonald.

Oregon, has instigated a new effort at subbasin planning (NPCC 2003). The NPCC has funded a project under the CBFWA to coordinate monitoring and evaluation in the Columbia Basin (CBFWA 2003). Federal Agencies (i.e., the "Action Agencies": Bonneville Power Administration, U.S. Army Corps of Engineers, and Bureau of Reclamation) and NOAA Fisheries have collaborated on a plan for research, monitoring, and evaluation to meet requirements of the BiOps in the Columbia Basin (Action Agencies 2003). Finally, at the time of this writing, a system-level cooperative monitoring and evaluation program in the Pacific Northwest is being proposed and is in the early stages of development. The effort includes biologists from concerned federal, state, and tribal agencies. It remains to be seen if these collaborative efforts can be successful in developing protocols for the inventory, monitoring, and assessment of the status of salmonids stocks and habitats that lead to coordinated management plans and an improved system-wide monitoring and evaluation program with common data collection and storage procedures.

Hydrosystem Development

The development of the Columbia River hydropower system in the Pacific Northwest began in the 1930s under a program of regional cooperation to meet the needs of electric power production, land reclamation, flood control, navigation, recreation, and other river uses (Chapter 6; NPCC 2003). Finally, by the late 1970s at the end of the major dam construction period, it became clear that the region's prosperity, which resulted in large measure from inexpensive hydropower from the federal dams, had extracted a price on fish and wildlife in the Columbia River Basin. Just a century earlier, for example, between 10 million and 16 million salmon returned to the Columbia each year (NPPC 1986). By the late-1970s, however, there were only about 2.5 million salmon and most (approximately 80%) were of hatchery origin.

The Northwest Power Act

On December 5, 1980, President Carter signed the Pacific Northwest Electric Power Planning and Conservation Act into law as Public Law 96-501. The intention of the act is to protect the fish and wildlife that had been impacted over the years by the construction and operation of hydropower dams. The Northwest Power Act established a four-state compact, the Northwest Power and Conservation Council (Oregon, Washington, Idaho, and Montana), and required the NPCC to prepare a program to protect, mitigate, and enhance Columbia River Basin fish and wildlife, and related spawning grounds and habitat, that had been affected by hydroelectric development, and to review the program at least every 5 years. The NPCC's 2000 FWP established a basin-wide vision for fish and wildlife. Ultimately, the program will be implemented through subbasin plans developed locally in the 53 tributary subbasins of the Columbia Basin. Major emphases in the FWP are on improvements in passage for migrating juvenile and adult salmonids, improvements in freshwater habitat, and on the use of artificial production to offset production losses due to passage mortalities or to blocked access to historical spawning areas (e.g., Columbia River above Grand Coulee Dam and the North Fork Clearwater above Dworshak Dam). An amendment to the Act (1996) formed the Independent Scientific Review Panel (ISRP) to review the projects proposed for funding for their scientific merit, consistency with the FWP, and provisions for monitoring to evaluate the success of projects in providing benefits for fish and wildlife. One of the benefits of this process has been increased attention and rigor for monitoring and evaluation at the project level. Plans are being formed for cooperative subbasin and system-wide monitoring and evaluation; however, little concrete progress has occurred to date.

Endangered Species Act—Section 7 Consultation: Biological Opinions

The Endangered Species Act (ESA) (16 USC 1531-1544), amended in 1988, establishes a national program for the conservation of threatened and endangered species of fish, wildlife, and plants and the habitat on which they depend. Section 7(a)(2) of the ESA requires federal agencies to consult with USFWS and NMFS, as appropriate, to ensure that their actions are not likely to jeopardize the continued existence of endangered or threatened species or to adversely modify or destroy their designated critical habitats. These consultations resulted in Biological Opinions (NMFS 2000; USFWS 2000) for the Federal Columbia River Power System.

The BiOps clearly recognize the central and critical role that subbasin and system-level research, monitoring, and evaluation (RME) must play in assessing the regional progress toward ESA-mandated recovery goals for listed species. Reasonable and Prudent Alternatives (RPAs) in the NMFS BiOps call for RME that will establish the results of management actions, undertaken by the Action Agencies, that are intended to recover listed species.

System-wide Monitoring and Evaluation in the Columbia Basin

The NPCC selects projects for funding in its FWP and requires monitoring and evaluation to show benefits of fish and wildlife and the overall success of its program. The 2000 BiOp (NMFS 2000) issued by NOAA Fisheries is being contested in court, but at the time of this writing, it remains one of the primary documents for guidance for recovery of species listed under the ESA.

The project review structure imposed on the NPCC and BPA by the 1996 amendment to the Power Act, and the formation of the ISRP as a central technical review group, provided a forum through which issues concerning monitoring and evaluation at both project and programmatic levels are starting to be addressed in a disciplined manner. Through the process of annual project proposal reviews (ISRP 1997, 1998, 1999) and the more recent provincial reviews (e.g., ISRP 2002), the ISRP has iteratively reviewed ongoing and new project proposals and provided the region with advice on different kinds of monitoring needs and their appropriate data needs and evaluations. Together, the FWP, the BiOps, and the 1996 amendment to the Power Act are the driving forces behind current attempts to establish a cooperative system-wide monitoring and evaluation program in the Columbia Basin.

Current Efforts

Development of a system-wide monitoring and evaluation program is presently in a formative stage and will require the integration of three relatively new initiatives. The first of these is through a FWP project administered by the CBFWA. The CBFWA has broad representation from state and tribal agencies and has established a workgroup including federal agencies to accomplish the project. Specifically, fisheries scientists and fisheries managers are working together with biometricians to:

1. Document, integrate, and make available existing monitoring data that bear on the problem of evaluating the status of salmon, steelhead, bull trout, and other species of regional importance across the U.S. portion of the Columbia Basin (i.e., system-wide).
2. Work collaboratively to critically assess the strengths and weaknesses of existing monitoring and evaluation methods for answering key questions regarding both stock status and responses to management actions.
3. Work collaboratively to design improved monitoring and evaluation methods that will fill information gaps and provide better answers to these questions in the future.

The second initiative is occurring through the federal Action Agencies, which have proposed a draft RME Plan (Action Agencies 2003). The RME Plan describes six principal components:

1. Population and Environmental Status Monitoring
2. Action Effectiveness Research
3. Critical Uncertainty Research
4. Project Implementation Monitoring
5. Data Management
6. Regional Coordination

Finally, a cooperative monitoring and evaluation program in the Pacific Northwest is being proposed by an *ad hoc* partnership of biologists from concerned federal, state, and tribal agencies under the name Pacific Northwest Aquatic Monitoring Partnership (PNAMP). This group recognizes that government agencies and other organizations use a variety of different monitoring efforts and that such monitoring efforts have typically included little or no coordination with other agencies. However, new questions are now being asked that are best answered at large-scale landscape levels. This will necessitate coordination across traditional lines. A draft plan called The Partnership

Plan was issued in January 2004 (PNAMP 2004). The partnership is built on four principles:

1. The purpose of monitoring is to coordinate important scientific information needed to inform public policy and resource management decisions.
2. Cooperative monitoring enhances efficiencies and effectiveness of the efforts.
3. Environmental monitoring must be scientifically sound.
4. Monitoring data must be accessible to all on a timely basis.

The momentum to develop and implement a cooperative subbasin and system-wide monitoring and evaluation programs for the Columbia Basin seems to be building at the time of this writing. We hope that these efforts yield improved data in the future for inventory, monitoring, and assessment of fish stocks and their habitat.

Types of Monitoring Needed

NOAA Fisheries' 2000 BiOp introduced nomenclature for the types of research and monitoring judged necessary for documentation of successful (or unsuccessful) recovery of listed populations of anadromous fish (NMFS 2000). Their three tiers of research and monitoring provide a useful framework for discussion of the complex array of studies and data collection procedures that exist.

Implementation Monitoring

Implementation monitoring is the monitoring of task completion in a specific project. This might include, for example, documenting the miles of stream fenced, number of culverts removed, completion of reports, irrigation diversions maintained, implementation of an experiment, and so on. Implementation monitoring has often been the only monitoring included in projects conducted in the Columbia Basin. While implementation monitoring results must be collected by specific projects, sound science (and review criteria from the 1996 Power Act amendment) requires that project results also be measured in terms of benefits to fish and wildlife.

Trend (or Change) Monitoring (Tier 1 Monitoring)

The purpose for Tier 1 monitoring is to provide long-term, daily (yearly), mundane, dull, economical, and repeatable data with enough accuracy and

Figure 11.5   Temporary trap for spawning chum salmon on Hamilton Spring, a small tributary below Bonneville Dam. Photo by L. L. McDonald.

precision to detect trend or change in the face of background noise. Examples of Tier 1 trend monitoring include estimation of harvest using coded wire tags, counts of adults passing mainstem dams or tributary weirs (Figures 11.5 and 11.6), numbers of juveniles released from a hatchery, digitized habitat layers in GIS, aerial photographs, and so forth. Tier 1 (trend or change) monitoring, which is not necessarily expensive or time consuming, obtains repeated measurements over time, usually representing a single spatial unit (e.g., a watershed or subbasin), with a view to quantifying trends or changes over time on the unit and perhaps documenting direct effects of a project. Study sites can range from site-specific locations to river reach or watershed level. Trends or changes must be distinguished from background noise.

Tier 1 monitoring can involve a low level of monitoring on individual project sites, such as a fenced enclosure and photo time series to document grazing effects (Figure 11.7), or on a large area consisting of subbasins or the entire Columbia Basin. For example, aerial photography (Figure 11.8) or data layers in a GIS would be used for long-term monitoring of trend or change in riparian and other terrestrial habitat over time in large subbasins.

Figure 11.6 Permanent trap for counting adult and juvenile steelhead and Chinook on the Imnaha River, a tributary in the Grande Ronde River system that is managed as a wild steelhead index stream. Photo by R. N. Williams.

Figure 11.7 Photo showing effect of elimination of livestock grazing on riparian habitat by fencing of Foley Creek, a tributary of the Deschutes River, Oregon. Photo by L. L. McDonald.

In general, Tier 1 monitoring does not establish the cause of observed trends or changes and does not provide probabilistic statistical inferences to larger areas or longer time periods. When trends or changes are detected, then relatively short-term experimental research projects can be developed to help explain why the trend or change occurred. For example, Tier 1 monitoring counts of adults passing both the Bonneville Dam and The Dalles Dam might indicate that in 2025 there was a 40% drop in the number expected to pass The Dalles. Why? Although Long-term Tier 1 monitoring has indicated the change, more concentrated, relatively short-term research is needed to answer this question. The results of Tier 1 monitoring are often not of much value until a significant period of time has passed to establish a "baseline," perhaps, 10 to 15 years or longer.

If Tier 1 monitoring is replicated on similar observational studies over time and space, compelling evidence for general conclusions can be obtained. In this inductive sense, Tier 1 monitoring does support "research on cause of effects." However, direct conclusions on causes of effects are lacking in most Tier 1 monitoring. Most often, the study is unreplicated, and the conclusion is that a trend or change has occurred (this is, in a sense, "research"), but the Tier 1 monitoring data do not provide the cause for a trend or change. At this point, Tier 1 monitoring has done its job, but it is often unfairly criticized because although decision makers know a trend or change occurred, they do not know why it occurred.

Statistical Monitoring (Tier 2 Monitoring)

Tier 2 (statistical) monitoring allows statistical inferences to be made on parameters in a study area for a certain length of time as measured by certain data collection protocols (i.e., The Methods in a report). These conclusions apply to areas larger than the study sites and to time intervals between data collection periods. The inferences require probabilistic selection of study sites and/or repeated visits over time. Again, Tier 2 monitoring by itself does not establish the cause of observed effects and may not be of much value for 10 to 15 years.

Tier 2 statistical monitoring is research in the sense that probabilistic conclusions are drawn about areas or time periods not measured, for example, estimates of the number of Chinook redds in the Wenatchee River Basin in 2025 based on counts in a probabilistic sample of sites from the subbasin. In this regard, Tier 1 and Tier 2 monitoring are different. Otherwise, the same limitations exist on learning why trends or changes occurred. The cause of an effect detected by Tier 1 or Tier 2 monitoring is elusive unless more data are gathered from similar observational studies replicated in time and space or from a "true" experiment with random assignment of treatments to experimental unit.

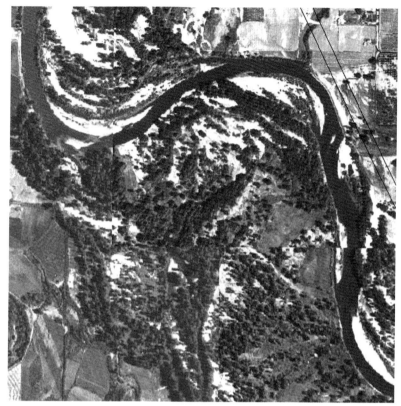

Figure 11.8   Aerial photograph time series showing cottonwood tree replacement in the river flood plain of the Yakima River (RM 96). First photo is from 1947 and second is from 2002. Photos courtesy of the Yakima County Geographic Information Services.

Individual projects, which might lend themselves most easily to Tier 1 monitoring, can support larger Tier 2 statistical monitoring projects by using the same coordinated methods to select study sites and field data collection methods. Most large projects should implement sampling designs that allow Tier 2 statistical monitoring or contribute data to statistical monitoring. Tier 2 statistical monitoring is usually required for estimation of parameters such as number of spawners in the escapement to a subbasin, juvenile production in a subbasin, acres of noxious weed present, and so on.

An example of the difference between Tier 1 (trend) and Tier 2 (statistical) monitoring can be seen in the Oregon Plan for Salmon and Watersheds Monitoring Program. The Oregon Plan (Maleki and Riggers 2000) was implemented successfully to estimate coho distribution and abundance in Oregon coastal systems by applying a rigorous design for probabilistic site

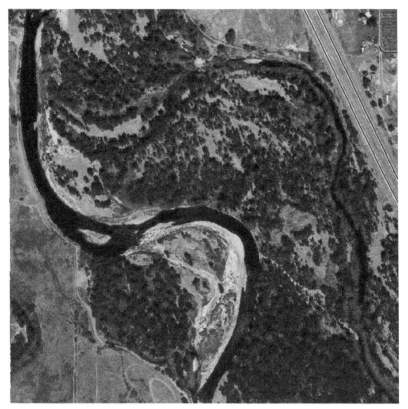

Figure 11.8   *(Continued)*

selection to answer key monitoring questions. Efforts are presently underway by NOAA Fisheries to develop Tier 2 statistical monitoring for status and trend of salmonids and habitat over large subbasins (Wenatchee, John Day, and one of the subbasins in the upper Salmon) in the Columbia Basin. The *ad hoc* partnership has endorsed Tier 2 statistical monitoring for estimation of certain fish population and habitat parameters.

Research Monitoring (Tier 3 Monitoring)

Tier 3 research monitoring is for those projects or groups of projects whose objectives include establishment of mechanistic links between management actions and salmon or other fish or wildlife population response. Bisbal (2001) defines this level of effort as *effects* or *response monitoring*, that is, the repeated measurement of environmental variables to detect changes caused by external influences. The key phrases here are "establishment of mechanistic links" and "detect changes caused by external influences." Tier 3 research monitoring requires the use of "true" experimental designs incorporating treatments and controls randomly assigned to study sites. Generally, Tier 3 research monitoring is for a relatively short time period (e.g., 3–5 years), and the results qualify for publication in the refereed scientific literature.

Examples of Tier 3 monitoring would include projects to evaluate (1) the effects of different levels of fertilization on growth and survival of juvenile salmonids with streams selected randomly for reference and treatment; (2) the survival rates of adult salmonids caught and released from tangle nets; (3) the survival rates of juveniles migrating past a dam at different levels of spill and turbine passage; (4) the swimming ability of lamprey during upstream migration; and (5) the effectiveness of various land restoration or management techniques.

Large-scale non-randomized observational studies that involve treatment-control, before-after, or before-after-control-impact (BACI) designs fall under Tier 1 or 2 monitoring and do not establish cause-and-effect relationships as in Tier 3 research monitoring. A good example in the Columbia Basin is the Idaho Supplementation Study (ISS) on Chinook salmon (Lutch et al. 2003). As initially envisioned, the ISS would randomly decide which member of a pair of matched streams would receive supplementation by hatchery fish. However, this randomization proved to be infeasible, and subjective judgment was used to determine which streams were supplemented. The end result is an observational study that does not establish cause-and-effect relationships. However, with a large number of replications, as in the ISS, compelling evidence for general conclusions based on regression-correlation type analyses can be obtained.

Effectiveness Monitoring

The 2000 BiOp called for assessing the effectiveness of tributary habitat actions (NMFS 2000). This level of monitoring falls under Tier 3 Research Monitoring, not long-term biological monitoring. The objective is to measure environmental parameters to determine whether management actions were effective in creating a desired outcome at either the project (stream reach) or watershed scale. It relies on an experimental design approach utilizing random assignment of treatments and controls to study sites.

It is very difficult at the watershed or higher level to conduct true effectiveness monitoring, because ecological processes occurring upstream or upslope from the project influence both habitat and fish population parameters. In addition, downstream effects (e.g., ocean conditions, passage mortality at dams, harvest, interactions with hatchery fish, etc.) influence the viability of anadromous populations. This reality limits the ability to determine cause-and-effect effectiveness at larger spatial scales. There is general agreement that a regional network of Intensively Monitored Watersheds (IMW) is needed to evaluate restoration projects, programs, and policies at the watershed or subbasin scale (ISAB 2003; PNAMP 2004). Effectiveness monitoring at these large scales attempts to answer the question "Does the collective effect of restoration actions result in improved watershed condition?" (PNAMP 2004).

Understanding the effect of habitat conditions on salmon population performance requires replicated observational studies or intensive research level experiments to be conducted at large spatial and long temporal scales. Very few evaluation efforts for tributary habitat that have been implemented to date in the Columbia River basin meet these criteria. Considerable expense and effort are required to establish studies or experiments that can enhance our understanding of habitat-population relationships and thus provide a sound basis for the development of tributary habitat restoration efforts (Bernhardt et al. 2005).

The basic premise of IMW is that cause-effect relationships in complex systems can best be understood by concentrating monitoring and research efforts at a few locations. Closely spaced measurements in space and time are often required to develop a thorough understanding of the processes responsible for habitat or fish population response to a management action. Concentration of effort can focus sufficient resources and research expertise to begin to tease apart some of the complex interactions governing system response to restoration activities (Figures 11.9, 11.10, and 11.11).

Recommendations for Monitoring and Evaluation

It is not easy to condense the advice given by the various government agencies to a simple set of recommendations on research and monitoring of fish populations and habitat in large watersheds, subbasins, or the Columbia Basin. Further, the situations in different parts of a large area, for example, a subbasin, are likely to require different approaches. For example, evaluation of effectiveness of habitat actions on forest lands might be integrated with the U.S. Forest Service monitoring procedures, while evaluation on private lands may require development of survey procedures. We believe the following four points contain the essential elements for development of an appropriate large-scale RME plan.

First, develop a sound Tier 1 trend (or change) monitoring procedure based on remote sensing, photography, and data layers in a GIS. Landscape changes in terrestrial and aquatic habitat and land use should be monitored for the smallest units possible. Accuracy and precision of data layers in the

Figure 11.9 Smolt trap (screw trap) for monitoring of numbers of juvenile salmon and steelhead migrating down the Secesh River of the Salmon River, Idaho. Photo by R. N. Williams.

Figure 11.10 Weir with automated adult fish counting facilities maintained by the Nez Perce Tribe on the Secesh River, a branch of the Salmon River in Idaho. Photo by L. L. McDonald.

Figure 11.11 Automated data collection equipment for fish counting facilities maintained by the Nez Perce Tribe on the Secesh River, a branch of the Salmon River, Idaho. Photo by L. L. McDonald.

GIS should be evaluated using "blind" classification of randomly selected units by on-the-ground verification during field visits.

Second, cooperate with Columbia Basin-wide attempts to develop common Tier 2 probabilistic (statistical) site selection procedures for population and habitat status monitoring (Action Agencies 2003; PNAMP 2004). Use common protocols for on-the-ground or remotely sensed data collection. In so far as possible, measurement of indicator variables should be collocated on the same sites. Status monitoring plans are being developed by the Action Agencies for implementation of the EPA EMAP probabilistic selection of aquatic sites in pilot projects in the Wenatchee, John Day, and Salmon subbasins (Action Agencies 2003). Subbasin planning required by the NPCC is a unique opportunity to promote the collection of research and monitoring data with common methods throughout the entire Columbia Basin. Status of fish and wildlife populations and habitat would be evaluated in a long-term biological monitoring program.

Third, as data are obtained on a wildlife or fish population and habitat status monitoring program, develop empirical models for prediction of current abundance or presence–absence of focal species. Potential predictor variables include not only physical habitat variables (flow, temperature, etc.) but also measures of habitat recovery actions that are currently in place or are implemented in the future. Use the empirical models to evaluate the relative importance of physical factors and habitat improvements and to predict abundance or presence-absence throughout major sections of the subbasin. If adequate coverage exists with current study sites, it may be advisable to conduct initial analyses on current data. However, a shift to probabilistically selected sites should be made as soon as possible to avoid inherent biases in subjectively selected and non-collocated study sites.

Fourth, make best professional judgment based on the Tier 1 and 2 monitoring as to whether any new research in the spirit of the IMW approach should be instigated. This step can be based on expert systems and existing data if adequate coverage of the watershed or subbasin exists.

We judge that the approach in these four steps is the most likely to accomplish successful large-scale, long-term monitoring and evaluation programs. An extensive long-term status monitoring program identifies important and unexplained trends and changes; that is, it identifies the intensive research that, if conducted, would explain the "why." Tier 1 trend monitoring by remotely sensing procedures and Tier 2 statistical monitoring provide indications of trend and change in indicator variables, but the "why" of certain trends and changes is usually not well understood. For example, the status monitoring may indicate that a major and unexpected increase in juvenile fish production occurred in a watershed with high summer water temperature and low flow during the period 2010 to 2020. Why? A population of bull trout is detected in an area where current knowledge and logic indicate they should not exist. Why?

We do not recommend an intensive research project to explain "why changes occurred" on every habitat improvement project, but rather periodic economical monitoring on individual projects to indicate benefits to fish and wildlife. In general, individual projects should depend on large coordinated monitoring programs to establish changes and trends in fish populations and habitat on a larger scale and the relationships of changes and trends to actions intended to improve habitat.

Evaluation of Monitoring Efforts

It is important to separate evaluation of long-term Tier 1 or 2 ecological monitoring (standard, everyday, dull, every-year data collection from large areas and over long time periods) from evaluation of Tier 3 research. Evaluation is an important part of all three processes, and there must be a perceived need and clear procedure for analysis of data being collected. If there is a perceived need and clear procedure for analysis of data being collected, plans for evaluation of the data are less important for Tier 1 and Tier

2 monitoring than for Tier 3, because Tier 1 and Tier 2 data should have a long shelf life (in the range of 50 to 100 years minimum). In fact, the methods for evaluation of Tier 1 and 2 data in the future, say 2025, probably have not been conceived. On the other hand, Tier 3 research data (e.g., effectiveness monitoring) are often for relatively short-term evaluation of specific research projects and must have well-defined plans for analysis and evaluation. Tier 3 data are also likely to have a shorter effective shelf life than either Tier 1 or Tier 2 data.

A part of evaluation is the reporting of analyses and summary statistics from Tier 1, 2, or 3 monitoring. Technical review groups in the Columbia Basin, including the ISRP, have been insistent that research proposals report or reference past achievements and that annual and final reports be issued on time and made available to the region. Review groups also have consistently placed an emphasis on peer-reviewed publication of project results wherever possible.

When the first attempts are made to draw conclusions, it may emerge that the data that have been collected up to that time will prove inadequate or excessive for the purpose. This will naturally lead to suggestions for revisions of the design (or even revision of the list of variables that are measured). The possibility of such iterative false starts should motivate all players in this enterprise to attempt to crystallize their statistical analysis techniques and decision rules as soon as is feasible. When statistical procedures and decision rules are defined in advance, it is technically possible to analyze proposed designs and determine in advance the probability that the design will deliver conclusive results after some stated period of time. This gives added depth to the concept of planning and is much encouraged (ISAB/ISRP 2004).

Monitoring and Adaptive Management

Knowledge of the management actions that would be sufficient to achieve recovery goals in the NPCC's FWP is tremendously important to the eventual success and cost of the recovery effort. Because present knowledge is not sufficient to determine all details of an implementation plan, there is need for some element of adaptive management in the recovery efforts themselves.

The adaptive management approach (Lee 1993; Volkmann and McConnaha 1993) offers the region a means to integrate new knowledge and experimentation into the applied effort of salmon recovery and maintenance of the Columbia River ecosystem. There is a fine balance to be struck in drafting a plan that has sufficient flexibility to accommodate a realistic need for ongoing fine tuning, but which still is concrete and specific enough to provide meaningful guidance.

It is impossible to evaluate the effects of management actions and make appropriate adaptive changes without some data, that is, some research, monitoring, and evaluation. The design and analysis of observational studies, and design and operations of the required monitoring, may constitute a fair fraction of the recommended investment of the resources of projects and may occupy a fair fraction of the available talent, for these are demanding problems.

Role of Databases in Monitoring and Evaluation

It is imperative that each data collection activity result in verified electronic data and metadata that are quickly available for use, and that each database system be housed in an organization that can guarantee data integrity, maintenance, access, and documentation (ISAB/ISRP 2004). It is necessary to define methods and procedures of quality assurance/quality control, data transfer and entry, data access rules, metadata standards, and methods to link data for queries. At a minimum, an effective data management plan must guarantee that each data collection activity takes responsibility for getting its data recorded in some electronic database system in a timely manner, where every data entry will be recorded in defined electronically interpretable fields (i.e., *NOT* in comment fields):

- What variable was measured (this description should include the medium in which it was measured)
- What value was obtained
- What method was used for the measurement (referring to a common manual of approved methods, and the associated quality control and quality assurance procedures)
- The location where the sample was taken or the measurement was made, as appropriate
- The date and time when the sample was taken or the measurement was made, as appropriate
- The study design type (ISAB/ISRP 2004).

It is particularly critical that metadata (methods and data collection procedures) be archived in a database structure that maintains the association between original data and their pertinent metadata (ISRP 2000; Figure 11.12).

Role of Index Sites

Index sites subjectively selected are of limited value in Tier 2 monitoring. Use of index areas is Tier 1 monitoring with no ability to make inductive

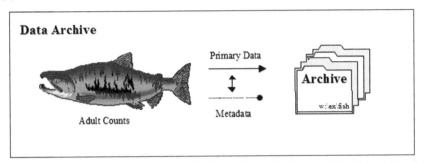

Figure 11.12 Primary data and their associated metadata need to be stored in a data archive. In a distributed data system, the data archive has a unique Internet address through which its contents can be accessed. Figure from ISRP Report 2000-3, 2000.

(statistical) inferences beyond the specific sites measured. One compromise in the design of Tier 2 probabilistic sampling designs is to keep the "good old" index sites, but select new sites by probabilistic procedures. This approach yields a 100% sample from the stratum "index sites" and a probabilistic sample from the rest of the area. Statistical inference would involve estimation of a parameter on the non-index sites combined with data from the index sites.

In general, we caution against the use of index sites and recommend that a general protocol for probabilistic selection of aquatic sites be developed. One possibility is that old and new sites be measured for the indicator variables for a few years, say 5 to 10 years, during which time correlation and regression relationship can be developed between the two sets of sites. We endorse the current plans (Action Agencies 2003; PNAMP 2004) in their use of probabilistic sampling plans for aquatic populations and habitat. The EPA EMAP strategy for probabilistic selection of study sites appears to be the procedure that will be adopted throughout the Columbia Basin.

Major Monitoring and Evaluation Exercises

In this chapter, we have concentrated on the current philosophy of monitoring rather than on who has been monitoring what and for what purpose. Monitoring of nearly every conceivable type is being carried out in the Columbia River Basin, from checking the effectiveness of habitat modifications to major stock assessment monitoring. Nonetheless, certain recent and current monitoring and evaluation exercises deserve mention.

The Northwest embarked on a massive analytical effort to use historical monitoring information when it initiated in 1995 a multi-agency process termed PATH (Plan for Analyzing and Testing Hypotheses; Marmorek and Parnell 1995). This project attempted to synthesize many years of monitoring data (especially of index stocks) in the context of models for fish migration, survival and population dynamics. The results have been published in many outlets (e.g., Deriso et al. 2001; Marmorek and Peters 1998; Petrosky et al. 2001; Schaller et al. 1999) and continue to guide salmon management and the collection of relevant monitoring data. The extensive analysis and evaluation showed that, despite a long history of monitoring, the data were generally insufficient to resolve major technical issues and establish cause-and-effect relationships. Additional monitoring approaches were needed.

A different analytical framework for both using and guiding monitoring data has been no less influential. This is the conservation-biology-based modeling of salmon populations by staff of the Northwest Fisheries Science Center (e.g., Kareiva et al. 2000). This approach requires population-specific data on fish numbers so that rates of population change can be calculated and projected into the future. To accommodate such population-level evaluations, it has been necessary to delve deeply into the genetic makeup of populations and population aggregates such as Evolutionarily Significant Units in the context of the U.S. Endangered Species Act (e.g., Interior Columbia Basin Technical Recovery Team 2003; Myers et al. 2003). The appropriate populations and how to monitor them are still being resolved by various technical recovery teams.

Perhaps the most intensive monitoring effort currently underway is the measurement and documentation of smolt survival in passage through the hydropower system. This effort became possible with development of the Passive Integrated Transponder (PIT) tag, a radio-frequency identification tag the size of a grain of rice that can be implanted in a small fish and allow it to be individually identified (Prentice 1990). This tag began its use experimentally in about 1987, and its use has progressed to true monitoring status recently with installation of tag detectors for migrating anadromous fish and for returning adults at most Columbia River mainstem dams. The high value of PIT tags for monitoring and evaluating salmon populations is exemplified by periodic white papers published by the Northwest Fisheries Science Center of NOAA Fisheries and available on their website (e.g., Williams et al. 2003). Since inception of PIT tagging, about 20 organizations have tagged and monitored smolt movements and survival in the Columbia River Basin, with over 7 million smolts tagged and monitored. The work has progressed from research on techniques and monitoring approaches (e.g., Muir et al. 1995) to full-scale and routine monitoring (e.g., Zabel et al. 2001). The work has necessitated the development and use of new statistical tools (Skalski 1998). The great success in monitoring survival of smolts during downstream passage is now leading to monitoring of the returning adults and the calculation of smolt-to-adult return ratios for discrete fish stocks.

Summary

Although routine monitoring of the numbers of upstream- and downstream-migrating salmonids and of selected index populations has a long history in the Columbia River Basin, recent evaluation exercises have shown the data to be insufficient for recovery planning. At the same time, it is apparent that the effectiveness of many habitat improvements and alterations of physical structures at dams have been poorly monitored and evaluated. The concept of monitoring encompasses several levels of data collection, while monitoring often blends with efforts more correctly classified as research. Research projects often become simplified and standardized over time and morph into monitoring projects, causing some confusion over objectives. Layered on these technical considerations is an increasing demand by the public and policy makers for fiscal and functional accountability in salmon restoration programs (i.e., they want success or failure documented).

As a result of these attributes, there has been a reconsideration of research, monitoring, and evaluation activities recently. The activities have been better categorized so that those undertaking them can more fully understand their objectives in relation to other potential objectives. New statistical designs have been developed and implemented, for example, stratified, randomized designs and advanced mark-recapture designs. Specific plans are being made to monitor and evaluate the results of actions prescribed in biological opinions and recovery plans. Overall, we seem to be entering into a new era of monitoring, in which the documentation of population and habitat trends comes closer to equaling the effort spent on remediation projects.

Literature Cited

Action Agencies. 2003. Research, Monitoring & Evaluation Plan for the NOAA-Fisheries 2000 Federal Columbia River Power System Biological Opinion. C. Jordan, J. Geiselman, M. Newsom, J. Athearn, eds. Available at *http://www.nwr.noaa.gov/1hydrop/hydroweb/fedrec.htm*. Bonneville Power Administration, Portland, Oregon.

Bernhardt, E. S. et al. 2005. Synthesizing U.S. river restoration efforts. *Science* 308:636–637.

Bisbal, G. A. 2001. Conceptual design of monitoring and evaluation plans for fish and wildlife in the Columbia River ecosystem. *Environmental Management* 28:433–453.

Bryant, F. G. 1949. A survey of the Columbia River and its tributaries with special reference to its fishery resources. No. 2. Washington streams from the mouth of the Columbia River to and including the Klickitat River (Area I). US Fish and Wildlife Service. Special Scientific Report, No. 62.

Bryant, F. G., and Z. E. Parkhurst. 1950. Survey of the Columbia River and its tributaries, Number 4, Area III. Washington streams from the Klickitat and Snake Rivers to Grand Coulee Dam. US Fish and Wildlife Service. Special Scientific Report–Fisheries No. 37.

Calkins, R. D., H. F. Durand, and W. H. Rich. 1939. Report of the Board of Consultants on the fish problems of the upper Columbia River. Sections 1 and 2. Stanford University, Palo Alto, California.

Columbia Basin Fish and Wildlife Authority (CBFWA). 2003. Proposal entitled "Collaborative, Systemwide Monitoring and Evaluation Program" submitted to the Northwest Power and Conservation Council. Available at *http://www.cbfwa.org/*. Columbia Basin Fish and Wildlife Authority, Portland, Oregon.

Deriso, R. B., D. R. Marmorek, and I. J. Parnell. 2001. Retrospective patterns of differential mortality and common year-effects experienced by spring and summer Chinook salmon (*Oncorhynchus tschawytscha*) of the Columbia River. *Canadian Journal of Fisheries and Aquatic Sciences* 58:2419–2430.

Evermann, B. W. 1895. A preliminary report upon salmon investigations in Idaho in 1894. *Bulletin of the US Fisheries Commission* 15:253–284.

Fulton, L. A. 1968. Spawning areas and abundance of Chinook salmon (*Oncorhynchus tshawytscha*) in the Columbia River Basin–Past and present. USDI, Fish and Wildlife Service. Special Scientific Report–Fisheries No. 571.

Fulton, L. A. 1970. Spawning areas and abundance of steelhead trout and coho, sockeye, and chum salmon in the Columbia River Basin–Past and present. USDC, NOAA, and NMFS. Special Scientific Report–Fisheries No. 618.

Hymer, J., et al. 1992a. Stock summary reports for Columbia River anadromous salmonids. Volume III: Washington below the Snake River. Columbia River Coordinated Information System, Columbia River Inter-Tribal Fish Commission. Portland, Oregon.

Hymer, J., et al. 1992b. Stock summary reports for Columbia River anadromous salmonids. Volume IV: Washington Upper Columbia and Snake River. Columbia River Coordinated Information System, Columbia River Inter-Tribal Fish Commission. Portland, Oregon.

ISAB. 2003. *Review of strategies for recovering tributary habitat* Independent Scientific Advisory Board Report to the Northwest Power and Conservation Council, the National Marine Fisheries Service, and the Columbia River Basin Indian Tribes, 851 SW 6th Avenue, Suite 1100 Portland, Oregon.

Independent Scientific Advisory Board and Independent Scientific Review Panel (ISAB/ISRP). 2004. Review of the "Draft Research, Monitoring & Evaluation Plan for the NOAA-Fisheries 2000 Federal Columbia River Power System Biological Opinion." Joint Independent Scientific Advisory Board and Independent Scientific Review Panel Report to the Northwest Power and Conservation Council. Report 2004-1.

Independent Scientific Review Panel (ISRP). 1997. Review of the Columbia River Basin Fish and Wildlife Program as directed by the 1996 amendment to the Power Act. Independent Scientific Review Panel Report to the Northwest Power Planning Council. Annual Report, ISRP 97-1. Portland, Oregon.

ISRP. 1998. Review of the Columbia River Basin Fish and Wildlife Program as directed by the 1996 amendment to the Power Act. Independent Scientific Review Panel Report to the Northwest Power Planning Council. Annual Report, ISRP 98-1. Portland, Oregon.

ISRP. 1999. Review of the Columbia River Basin Fish and Wildlife Program as directed by the 1996 amendment to the Power Act. Independent Scientific Review Panel Report to the Northwest Power Planning Council. Annual Report, ISRP 98-2. Portland, Oregon.

ISRP. 2000. Review of databases funded through the Columbia River Basin Fish and Wildlife Program. Independent Scientific Review Panel Report to the Northwest Power Planning Council. Annual Report, ISRP 2000-3. Portland, Oregon.

ISRP. 2002. Final Review of Fiscal Year 2003 Proposals for the Upper and Middle Snake, Columbia Cascade, and Lower Columbia and Estuary Provinces. Independent Scientific Review Panel Report to the Northwest Power Planning Council. Report No. ISRP 2002-11. Portland Oregon.

Interior Columbia Basin Technical Recovery Team. 2003. Independent populations of Chinook, steelhead, and sockeye for listed Evolutionarily Significant Units within the Interior Columbia River Domain (Draft). Northwest Fisheries Science Center, National Marine Fisheries Service, Seattle, Washington.

Kareiva, P., M. Marvier, and M. McClure. 2000. Recovery and management options for spring/summer Chinook salmon in the Columbia River basin. *Science* 290:977–979.

Kiefer, S. W., M. Rowe, and K. Hatch. 1992. Stock summary reports for Columbia River anadromous salmonids. Volume V: Idaho. Bonneville Power Administration. Report for the Coordinated Information System, Portland, Oregon.

Lee, K. 1993. *Compass and gyroscope: Integrating science and politics for the environment.* Island Press, Washington, DC.

Lutch, J., C. Beasley, and K. Steinhorst. 2003, March. Evaluation and statistical review of Idaho Supplementation Studies: 1991-2001. IDFG Report. BPA Project Number 89-098, Contract Number DE_B179-89BP01466.

Maleki, S. M., and B. L. K. Riggers. 2000. Watershed Restoration Inventory. Monitoring Program Report to the Oregon Plan for Salmon and Watersheds. Governor's Natural Resources Office, Salem, Oregon.

Marmorek, D., and I. Parnell. 1995. Plan for analyzing and testing hypotheses (PATH): Information package for Workshop I. Design of retrospective analyses to test key hypotheses of importance to management decisions on endangered and threatened Columbia River salmon stocks. ESSA Technologies, Ltd., Vancouver, British Columbia.

Marmorek, D. R., and C. N. Peters, eds. 1998. Plan for analyzing and testing hypotheses (PATH): Preliminary decision analysis report on Snake River spring/summer Chinook. ESSA Technologies, Ltd., Vancouver, British Columbia.

McConnaha, W. E., and P. J. Pacquet. 1996. Adaptive strategies for the management of ecosystems: the Columbia River experience. *American Fisheries Society Symposium* 16:410–421.

Mcintosh, B. A., S. E. Clarke, and J. R. Sedell. 1995a. Summary report for Bureau of Fisheries stream habitat surveys: Clearwater, Salmon, Weiser, and Payette River basins, 1934-1942. Bonneville Power Administration. DOE/BP-02246-2. Portland, Oregon.

McIntosh, B. A., S. E. Clarke, and J. R. Sedell. 1995b. Summary report for Bureau of Fisheries stream habitat surveys: Cowlitz River Basin, 1934-1942. Bonneville Power Administration. DOE/BP-02246-4. Portland, Oregon.

McIntosh, B. A., S. E. Clarke, and J. R. Sedell. 1995c. Summary report for Bureau of Fisheries stream habitat surveys: Umatilla, Tucannon, Asotin, and Grande Ronde River basins, 1934-1942. Bonneville Power Administration. DOE/BP-02246-1. Portland, Oregon.

McIntosh, B. A., S. E. Clarke, and J. R. Sedell. 1995d. Summary report for Bureau of Fisheries stream habitat surveys: Willamette River basin, 1934-1942. Bonneville Power Administration. DOE/BP-02246-3. Portland, Oregon.

McIntosh, B. A., S. E. Clarke, and J. R. Sedell. 1995e. Summary report for Bureau of Fisheries stream habitat surveys: Yakima River basin, 1934-1942. Bonneville Power Administration. DOE/BP-02246-5. Portland, Oregon.

Monitoring and Evaluation Group (MEG). 1988. Recommendations for a program to monitor and evaluate the Fish and Wildlife Program of the Northwest Power Planning Council. Northwest Power Planning Council. Final Report, Portland, Oregon.

Muir, W. D. et al. 1995. Survival estimates for the passage of juvenile salmonids through Snake River dams and reservoirs, 1994. Bonneville Power Administration. DOE/BP-10891-2. Portland, Oregon.

Myers, J., C. Busack, D. Rawling, and A. Marshall. 2003. Historical population structure of Willamette and lower Columbia River basin Pacific salmonids. Northwest Fisheries Science Center, National Marine Fisheries Service, Seattle, Washington.

National Marine Fisheries Service (NMFS). 2000. Final FCRPS (Federal Columbia River Power System) Biological Opinion. National Marine Fisheries Service, Northwest Fisheries Science Center, and NOAA. Endangered Species Act–Section 7 Consultation. Seattle, Washington.

Northwest Power and Conservation Council (NPCC). 2003. Guidelines for final formatting and submission of subbasin plans. *http://www.nwcouncil.org/fw/subbasinplanning/.* Northwest Power and Conservation Council, Portland, Oregon.

Northwest Power Planning Council (NPPC). 1986. Council Staff Compilation of Information on Salmon and Steelhead Losses in the Columbia River Basin. Northwest Power Planning Council, Portland, Oregon.

Olsen, E., K. Hatch, P. Pierce, and M. McLean. 1992a. Columbia River Coordinated Information System Project Phase II: Stock summary reports for Columbia River anadromous salmonids, Volume I: Oregon below Bonneville Dam (Draft). Columbia River Coordinated Information System Project, Portland, Oregon.

Olsen, E., K. Hatch, P. Pierce, and M. McLean. 1992b. Columbia River Coordinated Information System Project Phase II: Stock summary reports for Columbia River anadromous salmonids, Volume II: Oregon above Bonneville Dam (Draft). Columbia River Coordinated Information System Project, Portland, Oregon.

Olsen, E., K. Hatch, P. Pierce, and M. McLean. 1992c. Columbia River Coordinated Information System Project Phase II: Stock summary reports for Columbia River anadromous salmonids, Volume III: Washington below the Snake River (Draft). Columbia River Coordinated Information System Project, Portland, Oregon.

Olsen, E., K. Hatch, P. Pierce, and M. McLean. 1992d. Columbia River Coordinated Information System Project Phase II: Stock summary reports for Columbia River anadromous salmonids, Volume IV: Washington Upper Columbia and Snake River (Draft). Columbia River Coordinated Information System Project, Portland, Oregon.

Pacific Northwest Aquatic Monitoring Partnership (PNAMP). 2004. Recommendations for Coordinating State, Federal, and Tribal Watershed and Salmon Monitoring Programs in the Pacific Northwest. Portland, Oregon. Available at *http://www.reo.gov/pnamp/*. Columbia Basin Fish and Wildlife Authority, Portland, Oregon.

Parkhurst, Z. E. 1950a. Survey of the Columbia River and Its Tributaries, Part 6, Area V. Snake River from the Mouth through the Grande Ronde River. US Fish and Wildlife Service. Special Scientific Report–Fisheries No. 39.

Parkhurst, Z. E. 1950b. Survey of the Columbia River and Its Tributaries, Part 7, Area VI. Snake River from above the Grande Ronde River through the Payette River. US Fish and Wildlife Service. Special Scientific Report–Fisheries No. 40.

Parkhurst, Z. E. 1950c. Survey of the Columbia River and Its Tributaries, Part 8, Area VIII. Snake River above Payette River to Upper Salmon Falls. US Fish and Wildlife Service. Special Scientific Report–Fisheries No. 57.

Petrosky, C. E., H. A. Schaller, and P. Budy. 2001. Productivity and survival rate trends in the freshwater spawning and rearing stage of Snake River Chinook salmon (*Oncorhynchus tschawytscha*). *Canadian Journal of Fisheries and Aquatic Sciences* 58:1196–1207.

Prentice, E. F. 1990. PIT-tag monitoring system for hydroelectric dams and fish hatcheries. *American Fisheries Society Symposium* 7:323–334.

Quigley, T. M., R. W. Haynes, and R. T. Graham. 1996. Integrated scientific assessment for ecosystem management in the Interior Columbia Basin and portions of the Klamath and Great Basins. US Department of Agriculture, Forest Service, Pacific Northwest Research Station. General Technical Report, PNW-GTR-382. Portland, Oregon.

Rich, W. H. 1948. A survey of the Columbia River and its tributaries with special reference to the management of its fishery resources. US Fish and Wildlife Service. Special Scientific Report–No. 51.

Schaller, H. A., C. E. Petrosky, and O. P. Langness. 1999. Contrasting patterns of productivity and survival rates for stream-type Chinook salmon (Oncorhynchus tschawytscha) populations of the Snake and Columbia rivers. *Canadian Journal of Fisheries and Aquatic Sciences* 56:1031–1045.

Sedell, J. R., and K. J. Luchessa. 1981. Using the historical record as an aid to salmonid habitat enhancement. Pages 210–223 in N. B. Armantrout, ed. *Acquisition and utilization of aquatic habitat inventory information.* American Fisheries Society, Bethesda, Maryland.

Skalski, J. R. 1998. Estimating season-wide survival rates of outmigrating salmon smolt in the Snake River, Washington. *Canadian Journal of Fisheries and Aquatic Sciences* 55:761–769.

Thompson, K. 1976. Columbia Basin Fisheries: Past, Present and Future. Pacific Northwest Regional Commission. Columbia River Fisheries Project Report. Portland, Oregon.

Thompson, R. N., and J. B. Haas. 1960. Environmental Survey Pertaining to Salmon and Steelhead in Certain Rivers of Eastern Oregon and the Willamette River and Its Tributaries. Fish Commission of Oregon, Research Division. Clackamas, Oregon.

USFWS. 2000. BIOLOGICAL OPINION: Effects to Listed Species from Operations of the Federal Columbia River Power System. Endangered Species Act–Section 7 Consultation. US Fish and Wildlife Service, Portland, Oregon.

Williams, J. G., et al. 2004. Effects of the Federal Columbia River Power System on salmon populations. Northwest Fisheries Science Center, National Marine Fisheries Service, Seattle, Washington.

Volkman, J. M., and W. E. McConnaha. 1993. Through a glass, darkly: Columbia River Salmon, The Endangered Species Act, and adaptive management. *Environmental Law* 23:1249–1272.

Zabel, R. W., S. G. Smith, W. D. Muir, D. M. Marsh, J. G. Williams, and J. R. Skalski. 2001. Survival estimates for the passage of spring-migrating juvenile salmonids through Snake and Columbia River dams and reservoirs, 2001. Report to Bonneville Power Administration, Portland, Oregon.

Return to the River

12

Federal and State Approaches to Salmon Recovery at the Millennium

Peter A. Bisson, Ph.D., James A. Lichatowich, M.S., William J. Liss, Ph.D., Daniel Goodman, Ph.D., Charles C. Coutant, Ph.D., Lyman L. McDonald, Ph.D., Dennis Lettenmeier, Ph.D., Eric J. Loudenslager, Ph.D., Richard N. Williams, Ph.D.

Introduction
Are the Plans Likely to Succeed?
  Are the Plans an Adequate Response to the Salmon Recovery Problem?
    Mainstem Habitat and Fish Passage
    Tributary Habitat
    Hatcheries
    Harvest
    Models, Monitoring, and Evaluation
    Climate, Hydrology, and Water Resources
    Institutional Arrangements
  A Significant Change or the Status Quo?
    Mainstem Habitat and Fish Passage
    Tributary Habitat
    Hatcheries
    Harvest
    Models, Monitoring, and Evaluation
    Climate, Hydrology, and Water Resources
    Institutional Arrangements
  Are Linkages among the Proposed Strategies Adequate?
    Mainstem Habitat and Fish Passage
    Tributary Habitat
    Hatcheries
    Harvest
    Models, Monitoring, and Evaluation
    Climate, Hydrology, and Water Resources
    Institutional Arrangements

Discussion
  Data Gaps
  Conceptual Gaps
    Hatchery Reform
    Climate and Demographic Trends
    Tributary Habitat
    Harvest
  Integration
  Implementation
Summary
  Addendum
Literature Cited

Introduction

This chapter reviews approaches to salmon recovery put forward in four recent documents: (1) the Four Northwest States Governors' Plan (Recommendations of the Governors of Idaho, Montana, Oregon, and Washington for Protecting and Restoring Columbia River Fish and Wildlife and Preserving the Benefits of the Columbia River Power System, July 2000, updated in June 2003, and here termed the *Governors' Plan*); (2) the Northwest Power Planning Council's Plan (2000 Columbia River Basin Fish and Wildlife Program, November 30, 2000, here termed the *Fish and Wildlife Program*); (3) the National Marine Fisheries Service's Biological Opinion (Final 2000 Federal Columbia River Power System [FCRPS] Biological Opinion, December 21, 2000, here termed the *BiOp*); and (4) the Federal Caucus Plan (Conservation of Columbia Basin Fish, December 21, 2000, here termed the *Basinwide Recovery Strategy* or *All-H Paper*). The BiOp is the federal action plan for recovering threatened and endangered salmon listed under the Endangered Species Act (ESA) in the Columbia River Basin. It contains specified progress milestones to be achieved by 2003, 2005, 2008, and 2010, including both habitat-based and population-based evidence of recovery.

The salmon recovery documents are available online and can be viewed at the following websites:

Governors' Plan
http://www.nwcouncil.org/library/2003/4governors.pdf
Fish and Wildlife Program
http://www.nwcouncil.org/library/2000/2000-19/
Biological Opinion
http://www.nwr.noaa.gov/1hydrop/hydroweb/docs/Final/2000Biop.html

Basinwide Recovery Strategy
http://www.salmonrecovery.gov/strategy.shtml

There are other planning documents in the Columbia River Basin that are significant and important but which are not reviewed here. These include tribal salmon recovery strategies, for example, Spirit of the Salmon *Wy-Kan-Ush-Mi-Wa-Kish-Wit*, June 15, 1995, draft; the Forest Service and Bureau of Land Management's Interior Columbia Basin Ecosystem Management Plan (Interior Columbia Basin Final Environmental Impact Statement and Proposed Decision, December 2000, usually referenced by the acronym ICBEMP); the Fish and Wildlife Service's Biological Opinion on hydrosystem operations affecting bull trout and Kootenai River white sturgeon (December 21, 2000), as well as various agency strategy reports. Some of these documents are quite large and contain details about specific programs, implementation strategies, and monitoring.

Our objective is to examine the four major planning documents that form the core of the region's salmon recovery approach and answer the general question "Do they collectively outline salmon recovery strategies that are likely to have a high probability of success?" and to determine their compatibility with the normative river approach described in this book. We considered the scientific foundations of different elements of the papers, including the familiar "H" categories: Habitat, Harvest, Hatcheries, and Hydroelectric operations, as well as modeling and monitoring programs, climate and ocean conditions, and institutional arrangements. We do not review individual projects but rather examine the overall approaches expressed in the documents in light of current scientific information.

We approach our task by asking three specific questions of each document relative to factors considered important in salmon recovery: mainstem habitat, tributary habitat, hatcheries, harvest, monitoring, climate change, and institutional arrangements. The three questions are designed to assess each document relative to its (1) scientific adequacy and likelihood of success, (2) consistency and type of approach, and (3) linkages to each of the four Hs. The answers to these questions provided the basis of our assessment of the plans' collective likelihood of success.

Are the Plans Likely to Succeed?

Are the Plans an Adequate Response to the Salmon Recovery Problem?

Questions that need to be addressed are whether the documents in aggregate represent an adequate response to the salmon recovery problem and whether they describe courses of action that are likely to achieve regional goals such as ESA delisting, harvestable fish, and sustainable ecosystems.

Mainstem Habitat and Fish Passage

None of the recovery documents recommends removal or breaching of the four dams on the lower Snake River, but they all recognize issues of mainstem habitat structure and function more than did earlier plans. Some aspects of the recovery strategies represent an evolution in thinking about mainstem habitat and fish passage, whereas other proposed actions represent a continuation of previous recovery programs. The Governors' Plan is essentially a statement of social and political support for restoration activities. Next in level of detail is the Fish and Wildlife Program, which is largely a statement of principles, goals, and general strategies. The Basinwide Recovery Strategy further outlines proposals that are restated and elaborated in a different format in the BiOp, which provides more details about hydroelectric operations than any other aspect of salmon recovery (Figure 12.1). The strategies have a salmon life cycle orientation, which represents a more realistic approach than that which existed in some of the older recovery plans.

The success of the new plans for managing the mainstem Columbia River hydrosystem depends on the level of implementation. The old plans had some action items that were ecologically oriented, but these items were often ignored in practice (Independent Scientific Review Panel [ISRP] 1997, 1999). For mainstem restoration efforts to be successful, feasibility studies would be required and proposed actions deemed feasible before full-scale implementation. This has not often been done. In contrast, traditional measures (e.g., flow regulation, screens and juvenile bypass systems at dams, smolt transportation) have the momentum of past history that makes it likely they will be continued and even expanded

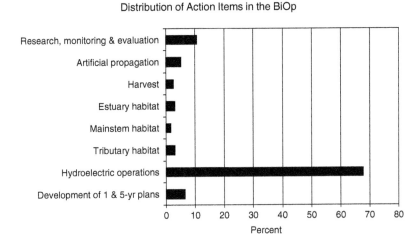

Figure 12.1   The distribution of action items specified by the December 21, 2000, BiOp among various salmon restoration categories.

incrementally, perhaps to the detriment of newer, ecologically oriented initiatives. For example, the barge transportation program for smolt migration is still advocated in the BiOp, despite adopting an overall recovery strategy that moves toward natural ecological processes (either directly or through engineering designs that mimic nature)—a direction the Independent Scientific Group (ISG) termed moving toward more normative conditions (ISG 1999).

Tributary Habitat

The Governors' Plan and the Fish and Wildlife Program outline conceptual or procedural approaches to tributary habitat restoration but do not explicitly consider limiting factors or habitat performance measures. The BiOp and the Basinwide Recovery Strategy, on the other hand, contain a much more substantive discussion of tributary habitat. The process for recovering tributary habitat relies heavily on a combination of modeling, interagency cooperation, and landscape assessments at the subbasin level. Progress toward meeting regional tributary habitat goals depends on (1) landscape modeling efforts being able to generate useful first-level subbasin assessments, (2) action agencies being able to agree upon robust and ecologically meaningful sets of performance standards, (3) adequate habitat and fish population monitoring, and (4) participating organizations learning from past restoration failures and new scientific information. A breakdown in any of these steps will significantly delay implementation of landscape-based restoration (Independent Scientific Advisory Board [ISAB] 2003b). Given current institutional monitoring programs and the extended time period required for many habitat recovery actions to become fully effective (Figure 12.2), monitoring will probably not provide *quantitative* answers about the success of many tributary habitat restoration projects in meeting the goals of the BiOp within 10 years.

Hatcheries

Little evidence is provided that hatchery reform measures proposed in an earlier artificial production review will be implemented or that they will improve salmon recovery. The BiOp and Basinwide Recovery Strategy do not outline a quantitative way to assess the extent to which hatchery programs impact ESA-listed species, or to what extent the artificial propagation of one species could affect the recovery prospects of another species (e.g., Levin and Williams 2002). Since the magnitude of hatchery impacts on naturally spawning salmon is unknown, benefits derived from reducing the impacts of hatchery fish on native stocks are also unknown. Similarly, evaluations of existing supplementation and captive rearing programs provided in the documents are inadequate to determine if these activities can make significant contributions to recovery of listed species (ISAB 2003a).

Figure 12.2   Tributary habitat improvement measures such as riparian revegetation and cattle exclusion often take decades to achieve full recovery. Marks Creek near Prineville, Oregon. Photo by P. Bisson.

Harvest

The documents do not directly address harvest but rather assume that changes in harvest management that evolved over the past 10 to 20 years will continue, depending on the status of stocks. Those trends have included substantial reductions in the total rates of exploitation on naturally spawning salmon populations, fisheries responsive to changes in abundance, and management of total fishing mortality, that is, catch plus associated incidental mortality. In general, the agencies preparing the recovery documents do not regulate Columbia River fisheries, with the notable exception of the National Marine Fisheries Service (NMFS) or the U.S. Fish and Wildlife Services' (USFWS) responsibility for jeopardy evaluations of ESA-listed fishes.

Consequently, there are no guarantees that harvest rate reductions on weak stocks assumed in the assessments will continue. If the productivity of the natural populations remains depressed in years with large returns of hatchery fish, as seen in 2000–2003 in the Snake River Basin, harvest rates of wild stocks will likely increase. This would not be an "allocated" harvest, but rather harvest due to mortality associated with bycatch and catch-and-release.

Each document supports the expansion of selective fishing techniques and, in particular, the development of mass-mark selective fisheries in which marked hatchery fish would be retained and unmarked wild fish would be released. The implicit assumption in mass-mark selective fisheries is that the total mortality of naturally produced salmon associated with catch-and-release selective fisheries is less than the mortality in complete retention fisheries and that the resulting harvest rate of wild stocks is sustainable. Intuitively this assumption seems obvious, but it has not been adequately tested.

A related concern is the potential impact of mass-mark selective fisheries on the Coded Wire Tag (CWT) program. The CWT program is essential to the estimation of total exploitation rates in ocean and freshwater fisheries by age and stock (natural and hatchery) and is currently the only means to measure these parameters (Figure 12.3). At this time, it is uncertain whether the viability of the CWT program can be maintained if widespread mass-mark

Figure 12.3 The Coded Wire Tag (CWT) program permits the identification of different stocks of salmon in ocean fisheries, thus permitting managers to target strong stocks and protect weak stocks. Photo by T. Quinn.

(e.g., adipose clip) selective fisheries are implemented. The loss of information may be unacceptable to other fishery management processes outside the Columbia River system. For example, the Pacific Salmon Treaty requires that each country assess the aggregate exploitation rates over all Chinook salmon fisheries to ensure that this aggregate value is less than a maximum value stated in the treaty agreement. The adoption of mass-mark selective fisheries as a harvest tool may also ignore potential ecological interactions between hatchery-produced and naturally produced fish. Mass marking hatchery fish and developing selective fisheries to utilize this production provides a powerful incentive for maintaining large-scale production of hatchery fish. However, what is not considered in this strategy is the potential for ecosystem effects associated with the continued release of large numbers of hatchery-produced fish.

Models, Monitoring, and Evaluation

The documents do not present very much detail with respect to monitoring. They identify the need for successful long-term monitoring programs, but only general suggestions for monitoring are given. Determining monitoring success, that is, the ability to detect the effects of management actions, will depend on details of statistical design and the intended scale of monitoring efforts. Successful implementation requires a high level of cooperation, including give and take by all concerned—state and federal agencies, tribes, and private organizations. The recovery documents do not contain enough details to provide assurance that monitoring plans will be up to the difficult task of tracking recovery actions in the Columbia River Basin. Whether sufficient details about monitoring programs will be presented in subbasin plans is not yet known.

Climate, Hydrology, and Water Resources

The salmon recovery documents do not fully consider the interactions of climate and hydrology as they affect the managed Columbia River hydrosystem. Specifically, the influence of climatic variability (Figure 12.4) at a variety of time scales on hydrologic variability and salmon life histories (felt primarily through variations in sea surface temperature, which also affect ocean survival of salmon) is ignored. Additionally, the documents do not address the potential effects of long-term climate changes, which are likely to result in permanent alteration of the patterns of winter snow accumulation throughout the Columbia Basin (Figure 12.5), and consequently, the timing and amount of seasonal discharge. Implications for management of the reservoir system with an earlier spring freshet and reduced summer flows are likewise not addressed.

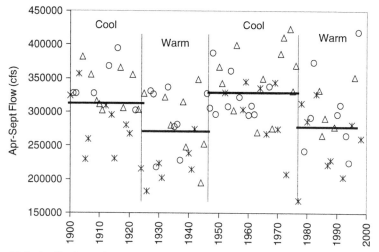

Figure 12.4 April—September average naturalized stream flow for the Columbia River at The Dalles. Solid horizontal lines are Pacific Decadal Oscillation (PDO) phase averages, crosses denote El Niño years, circles are La Niña years, triangles are El Niño-Southern Oscillation (ENSO) neutral years. Figure adapted from Hamlet and Lettenmaier (1999a).

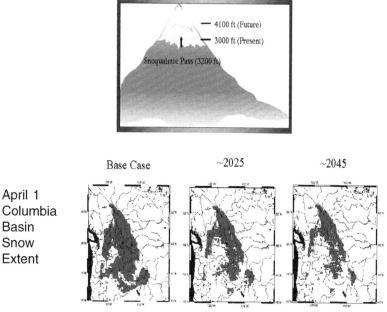

Figure 12.5 Mean Columbia River Basin April 1 snow extent for base case (current climate), year 2025 and year 2045. Figure adapted from Hamlet and Lettenmaier (1999b).

Institutional Arrangements

While the recovery documents attempt to define the problems and identify desirable future conditions, and in some cases suggest measures to determine whether those conditions have been attained, they provide little guidance on how institutions can function more effectively to promote ecosystem recovery. With the exception a general discussion in the Governors' Plan, there is no treatment in any of the documents of probable trends in human population and economic growth, and of the impacts these trends could have on the Columbia River Basin. In addition, there is almost no discussion of the very complex institutional structure existing within the Columbia River Basin or of the ways in which elements of that structure might facilitate, impede, or otherwise influence planning processes and implementation of recovery actions. Previous reviews (e.g., National Research Council 1996) have repeatedly pointed to the Columbia Basin's complex institutional structure and the resulting fragmented jurisdictional authority as one of the major barriers to salmon recovery efforts.

A Significant Change or the Status Quo?

Do the salmon recovery documents, in aggregate, represent a significant change in the status quo or a continuation of past efforts? Do they provide a consistent course of action?

Mainstem Habitat and Fish Passage

The documents represent a change in the status quo, but probably not a major change. The principles, justifications, and the specific recovery actions for the mainstem Columbia and Snake Rivers in the documents are ecological and life cycle based, in contrast to technologically based engineering fixes stressed earlier (e.g., specified flows and dam modifications). Many past efforts are also continued, but with a decidedly more ecological and life-cycle emphasis, for example, an emphasis on compatible surface bypasses and weirs for fish passage at dams (Figure 12.6) rather than on artificial screening systems. However, there are still vestiges of obsolete salmon biology in the BiOp. For example, in one section it is stated that habitat features such as riparian vegetation, food, and rearing space are not needed by certain stocks in the Columbia River mainstem. Such inaccurate statements have been used to justify further simplification of the mainstem based on a narrow view of protecting a few prominent stocks, when a return to natural habitat complexity is needed for the entire assemblage of diverse stocks and species.

Recognizing that each document has its own specific objective, audience, and level of detail, the documents seem quite consistent about protecting and

Figure 12.6 Fish bypass systems can be complex and expensive, but they do provide an opportunity to track the downstream movements of PIT-tagged fish. The automated detection system shown here from Bonneville Dam separates marked from unmarked fish for further examination. Photo by P. Bisson.

restoring habitat values in the mainstem and providing fish passage opportunities that better match natural migration needs. The declaration of a flow emergency in early 2001 that vacated the terms of the BiOp and other recovery plans, however, raises important issues. What criteria are used to determine when fish recovery actions must give way to the need for hydropower? Some unwritten threshold was clearly exceeded in 2001, resulting in management actions that included passing all water through turbines (as opposed to allowing part of the flow to pass over the dams), and capturing all smolts at the upstream dams and transporting them downstream in barges (as opposed to allowing them to move downstream on their own volition). It would be useful to know what these criteria were, or alternatively, for the documents to establish a process for creating criteria explicitly for use in the future when emergencies arise. The documents in aggregate are silent on such emergency criteria from the fish recovery perspective.

Tributary Habitat

The greatest difference between the approaches outlined in the four documents and past tributary habitat restoration efforts is that current strategies place much more emphasis on formulating landscape-based recovery actions, although relatively few details are given. This means that tributary restoration projects will, in principle, be chosen on the basis of their overall contribution to recovery within the context of salmon life history needs (preferred habitats, seasonal movements) and the extent of habitat alteration (locations of habitat bottlenecks or high-quality refugia) within a tributary subbasin of interest. Whether implementation of tributary restoration efforts under the strategies represents a continuation of past programs or a real change in the status quo depends on their ability to take advantage of recent improvements in knowledge of salmon life history and watershed processes. This requires agencies to achieve a level of communication and coordination that is unprecedented for such a large area, particularly because their institutional mandates may occasionally promote actions that are contradictory. Habitat performance standards that transcend water quality hazard thresholds, the traditional approach, and take dynamic watershed processes into account are appropriate for this approach (ISAB 2003b). Subbasin plans that demonstrate linkages between programs and integrate the entire freshwater life cycle of salmon, rather than serving as a collection of individual restoration projects, will be most effective.

The BiOp, Basinwide Recovery Strategy, and Fish and Wildlife Program outline a consistent course of action for tributary habitat restoration, and all three documents utilize independent scientific peer review to provide external quality checks. The Governors' Plan appears to be least consistent with the other documents. Although the section on habitat in the Governors' Plan is brief, it is clear that it endorses restoration efforts designed by local organizations such as watershed councils, municipal and county authorities, and conservation groups, as opposed to being designed by large federal agencies. The Governors' Plan, however, makes no mention of scientific peer review of local plans. There is thus a tendency for the documents to endorse traditional bioengineering approaches to habitat restoration (Figure 12.7).

Hatcheries

The salmon recovery documents tend to advocate a continuation of past hatchery efforts. With regard to artificial production, this may not be an entirely unavoidable or undesirable approach, at least in the short term. Abandoning or substantially curtailing artificial production, concluding that it is a failure, will only happen once stakeholders have exhausted their efforts to make it succeed. Even though all four documents recognize a need for hatchery reform, none of them describes specific reforms or provides guidance on how to implement them.

All four documents explicitly recognize genetic and ecological risks associated with artificial production and acknowledge that past and some current programs continue with high-risk practices. The reports deal with that issue simply by calling for implementation of reforms outlined in earlier reviews of artificial production programs (ISRP 1997; Brannon et al. 1999), and as a

Figure 12.7 Tributary habitat restoration, showing an engineered stream channel, mid-stream rock weirs, and streamside vegetation planting in Asotin Creek, a tributary of the Clearwater River that supports steelhead and spring Chinook spawning populations. While such projects are commonplace, few are peer-reviewed or monitored. Photo by R. Williams.

consequence, they assume that the risks will be minimized (see Chapter 8 on artificial production for additional detail). Unfortunately, most reviews lack details about what reforms are needed or how reforms are to be implemented and subsequently monitored (Figure 12.8). This lack of detail is a result of several data gaps. First, there are no clear statements of different types of risks associated with different hatchery practices, for example, selective breeding changing the makeup of originally native-source broodstock versus the use of non-native broodstock. In addition, a thorough evaluation of the evidence for genetic and ecological risks is lacking—at least for salmonid fishes (ISAB 2003a). Concerns over risk are acknowledged in a vague sense, but priorities for dealing with those concerns are not detailed in practice or policy. A clear explanation of the genetic and ecological issues, as well as an assessment of the relevant evidence, would provide helpful guidance for hatchery operations in the Columbia River Basin.

Harvest

The documents propose both a continuation of the status quo and some new harvest initiatives. There is a general assumption that reduced harvest

Figure 12.8 Many recovery documents are unclear with regard to the specific steps needed to reform hatchery operations. Bonneville Hatchery. Photo by P. Bisson.

impacts will be continued, although the large salmon returns in the years 2001 to 2003 saw both new fisheries and extended seasons. The implementation of more selective fishing techniques is an emerging policy and is a significant change from the status quo. In the Columbia River, selective fisheries have involved taking hatchery fish marked with an adipose fin clip and releasing naturally spawned, unmarked fish (Figure 12.9). This policy has several issues associated with it that are not fully discussed in the four documents. In general, however, there is a high degree of concordance among the four documents. They include strategies for maintaining conservative harvest levels and honoring treaty obligations.

Although harvest strategies are generally consistent among the recovery documents, the topic of harvest definitely receives the least attention of the Hs in recovery planning. This likely follows from two factors: (1) Each document notes major reductions in exploitation of salmon stocks, to the extent that additional reductions are assumed to have marginal value to most populations, and (2) the organizations responsible for the documents have a direct role in annual harvest management plans. As summarized in the Basinwide Recovery Strategy, regulation of ocean harvest occurs pursuant to the Magnuson-Stevens Fisheries Management and Conservation Act and the Pacific Salmon Treaty, and management of in-river harvest occurs under the

Figure 12.9 Upper: hatchery steelhead marked with an adipose fin clip. Note erosion of the dorsal fin commonly associated with hatchery rearing. Lower: unmarked naturally spawned steelhead with intact fins. Photos by R. Williams.

auspices of the federal court in *U.S. v. Oregon*. In addition, NMFS or USFWS must authorize any harvest of ESA-listed fish. Even the new Pacific Salmon Treaty agreement was reviewed by the NMFS under Section 7 for consistency with the ESA. However, while the treatment of harvest may be limited, the importance of harvest management to salmon recovery is paramount for the immediate short-term survival of populations, and management policies reflect the degree of change that has already occurred in the fisheries and fishing communities. As stated in the Governors' Plan, "*Salmon fishing has decreased to a level that represents a mere fraction of what once occurred.*" Maintaining these reductions in harvest impacts is a core assumption of recovery planning.

Models, Monitoring, and Evaluation

With respect to harvest, hatcheries, and hydropower, the strategies in the documents advocate a continuation and evolution of past efforts in monitoring, evaluation, and modeling. Increased emphasis is placed on monitoring and evaluation of management actions for improvement of tributary habitat for anadromous species and on the effects of hatchery-produced fish on naturally spawning stocks. Also, added emphasis is placed on monitoring reproductive performance of wild stocks throughout the Columbia River Basin, that is, monitoring "fish coming in and fish going out" of natural production areas.

The four documents are consistent in their call for aggressive monitoring and evaluation of management actions aimed at recovering threatened or endangered salmonid populations and supporting a sufficient abundance of salmon to allow for sustainable harvests. Emphasis is placed on (1) continuing the present monitoring programs for the harvest and hydropower systems and adding the tiered monitoring efforts of the hierarchical plan identified in the BiOp and Basinwide Recovery Strategy for threatened and endangered fish species (including resident fishes) and their habitats; (2) increasing monitoring of the effectiveness of projects in the Fish and Wildlife Program consistent with monitoring recommendations in the BiOp and Basinwide Recovery Strategy and recent programmatic recommendations by the Council's Independent Scientific Review Panel (ISRP 2001, 2002); (3) implementing the recommendations for monitoring of hatcheries called for in a recent review of performance standards for artificial production (ISAB 2000); and (4) re-evaluating the harvest and hydropower monitoring programs as needed.

Climate, Hydrology, and Water Resources

The four documents do not represent a meaningful change in the status quo. The inherent assumption in them is that future climate conditions will resemble the past. This assumption, which underlies essentially all water resource design and management in the Columbia River Basin, is now being called into question. The four documents are consistent in that they deal, in one way or another, with flow and flow augmentation issues in the Columbia River mainstem and major tributaries. Hence, even if not explicitly recognized, the use of historical flow observations to determine effects of different flow management options implies an assumption about climate. None of the documents explicitly considers the implications of proposed climate changes on hydrosystem performance, particularly in low water years that are the basis for annual projections of the minimum available power.

Institutional Arrangements

While agency plans have grown in detail and complexity, the primary focus of the documents is on desired future conditions of the Columbia River Basin, largely ignoring the institutional arrangements that have led to the current situation. Proposed recovery strategies, for the most part, rest on the assumption that top-down planning (at least, at the subbasin scale), informed by science, can restore productive salmon ecosystems. The Governors' Plan departs from this assumption somewhat by proposing more authority for local planning inputs, but it presents little evidence that local planning will lead to a dramatic change in the status quo resulting in effective salmon recovery actions. The four documents are likewise, and somewhat understandably, inconsistent with respect to institutional arrangements. Each document has different goals and was inspired by somewhat different problems. Assignment of organizational responsibility for planning, key participants in identifying restoration priorities, and responsibility for monitoring and evaluation differs among the reports.

Are Linkages among the Proposed Strategies Adequate?

Among the various documents and plans, are linkages among strategies for dealing with the four Hs adequately identified?

Mainstem Habitat and Fish Passage

For the most part, the four documents do not describe how management of mainstem salmon habitat and fish passage will be adjusted for changes in management actions with respect to the other Hs. Cross-linking of items in subsequent drafts of the reports can be useful for coordination within the Columbia River Basin. Implementation of certain mainstem-related actions from the BiOp might not be seen in isolation (or worse, as competitors with other strategies), but as responding to an element of the Basinwide Recovery Strategy, a principle or strategy in the Council's Fish and Wildlife Program, and a general mandate in the Governors' Plan. Although it might be seen as a bookkeeping exercise, such cross-referencing (including referencing the Tribal Plan and Interior Columbia Basin Ecosystem Management Plan) could be the first step toward a mutually accepted, integrated, regional recovery plan.

Tributary Habitat

Tributary habitat recommendations in the documents, in general, are inadequately linked with recovery strategies in the other Hs. The Fish and Wildlife Program does, however, stress the need for supplementation to be linked to watershed condition and to be integrated into subbasin planning. The consequence of putting more water back into tributaries (one of the goals of tributary habitat recovery) is not clearly related to mainstem habitat management or to water quality issues such as temperature and dissolved gas. The important ecological role of salmon carcasses as vectors of marine-derived nutrients in salmon-producing watersheds is not adequately linked to harvest and escapement levels in most of the documents. Changes in habitat restoration tactics are not related to climate shifts or disturbance agents such as droughts, floods, or wildfires.

Hatcheries

The connection between hatchery production and harvest level is recognized; however, the problems of developing selective and terminal fisheries are not adequately considered. Coordination of habitat restoration and supplementation is acknowledged, but how subbasin planning and habitat modeling will inform decisions on where and how much supplementation is warranted is not clear. The cumulative effects of hydroelectric operations and other habitat changes (e.g., water withdrawals) on mainstem habitats and how these alterations limit the effectiveness of artificial production deserve increased attention. A climatic regime shift producing favorable ocean conditions and abundant returns of hatchery salmon similar to those of 2001 to 2003 creates pressure for increased harvest levels. How that pressure might be addressed in the context of conserving wild stocks is not articulated.

Harvest

Harvest is only one source of mortality in the life cycle of salmon populations, and life cycle analyses are appropriate means to integrate harvest mortality with other sources of mortality. More in-depth consideration of two issues could have strengthened the discussions of harvest. First, the level of harvest that can be sustained by a stock is determined by its productivity in the existing environment and the size of the spawning population. To assess the appropriateness of harvest that can take place and still achieve recovery requires establishing spawning escapement goals for each production unit (group of spawning populations), predicting adult returns in the next generation, and a management plan for harvesting surplus returns or imposing harvest restraints to increase the spawning population sizes. The Fish and Wildlife Program calls for the development of these production plans (in the subbasin planning process), but the empirical basis for these assessments in the subbasins is often quite limited. Second, promotion of mass-mark selective fisheries without considering the potential ecological consequences of such a harvest approach may have significant risks. Mass-marking artificially produced salmon and promoting selective fisheries to utilize marked salmon provide an incentive for maintaining large-scale production of hatchery fish, which has significant implications for the long-term fitness of naturally spawning fish in the same watershed. What is not adequately considered in the documents, however, is the potential for population impacts associated with the continued release of large numbers of hatchery-produced fish on wild stocks (Levin and Williams 2002; ISAB 2003a).

Models, Monitoring, and Evaluation

The four documents identify the need for monitoring and evaluation of certain important linkages, such as the effect of naturally spawning hatchery fish on wild populations and the effect of habitat improvement on carrying capacity. Examples of specific monitoring needs that have not received adequate planning include monitoring the effect of selective fisheries on wild stocks, monitoring the survival rates of salmon in the ocean, and developing a long-term plan for monitoring the survival of juvenile downstream migrants. In fairness to the documents, it is unrealistic to expect them to propose monitoring and evaluating all linkages among the Hs everywhere. Given that limitation, the importance of identifying monitoring plans that respond to priority needs is all the more apparent.

Climate, Hydrology, and Water Resources

Linkages between climate, hydrology, and water resources and the various Hs are not adequately identified. Some documents (notably the Basinwide Recovery Strategy) do mention the role of climate cycles on ocean survival, but flow variation is ignored in management planning except for changes in smolt transportation options during low flow years.

Institutional Arrangements

The four documents do not identify specific improvements in institutional coordination within the Columbia River Basin that would make actions involving each of the Hs more integrated and effective. A widely held view in the Columbia River Basin is that scientific research will identify and resolve key uncertainties, and that once the necessary knowledge is obtained, effective decisions will become apparent. There are at least two difficulties with this belief. First, although scientific knowledge is always desirable and provides insight into unanswered questions, it invariably gives rise to new issues and consequently new uncertainties. Second, even if all the necessary data existed, it is not clear from the four documents that the institutional framework is adequate to utilize that information in ways appropriate to make and successfully implement decisions on salmon recovery that reflect current and best scientific understanding. The region's institutions may simply be developing salmon recovery plans that are consistent with the current organizational framework. Insufficient attention has been devoted to improving the way institutions incorporate scientific information into recovery strategies, and to ways in which coordination of efforts undertaken by different organizations to improve each of the Hs can be made more effective (see Lee 1993).

Discussion

Two major positive trends distinguish the strategies in these documents from previous recovery plans: (1) They tend to reflect a functional ecosystem approach to salmon recovery, and (2) they make use of quantitative models to assess recovery actions, determine jeopardy, and to evaluate management alternatives. With regard to ecosystem health, the current documents emphasize landscape-based approaches and attempt to direct recovery actions at major components of salmon habitat in the Columbia River Basin. In contrast to previous plans, they address recovery of the estuary, tributary habitat, and features of mainstem habitat beyond water temperature, flow, and gas saturation. The documents propose a watershed planning process that tailors recovery actions to natural biophysical conditions of subbasins and provinces. The documents also acknowledge the importance of using natural conditions as a guide for restoration.

With regard to the use of models, the recovery documents place more reliance on mathematical simulation than previous plans. Extinction risk models developed by NMFS were used to assess jeopardy in the BiOp, and the Cumulative Risk Initiative model was used to assess impacts at different stages of the life cycle and to establish reasonable recovery alternatives. The Salmonid Watershed Assessment Model, another NMFS model, is intended for use in developing recovery actions for tributary habitat. The Northwest Power and Conservation Council recommends the Ecosystem Diagnosis and Treatment (EDT) model to evaluate recovery strategies at the provincial and subbasin scales.

Although we believe the overall answer to the question of whether the four documents will lead collectively to salmon recovery actions that have a high chance of succeeding is probably "no," we do not wish to diminish the scientifically sound recommendations contained in each of them. We reach this conclusion for reasons that hinge on data gaps, conceptual gaps, program integration, and implementation of recovery actions. While the strategies outlined in the documents offer some real advances in the science of salmon recovery, particularly with adoption of an ecosystem perspective and better use of models, important scientific data necessary to resolve critical uncertainties still have not been obtained. Shortcomings in program integration and implementation, inadequately addressed in the documents, are particularly troublesome because of the lack of clear institutional arrangements to carry out the programs. While implementation is not strictly a science issue, failure to clearly specify how recovery strategies would be achieved is a problem these documents share with many previous Columbia River Basin salmon plans.

Data Gaps

One of the fundamental shortcomings of salmon recovery planning in the Columbia River Basin has been the failure of management organizations to establish historical population and environmental databases. As a result, current recovery efforts rest on geographically limited data of varying quality and applicability. If reliable data collection protocols are established, future comparisons to current data will have difficulty discerning whether population trends are due to real changes caused by management actions, changes in the environment unrelated to management actions, or whether they just reflect the inaccuracy of historical estimates. This problem did not originate with the present generation of recovery documents, and in fact reasonable improvements in plans for future monitoring and evaluation are specified in some of them. This is a situation, perhaps unfair, in which it will be difficult to assess the effects of proposed management actions and know whether to continue or change them, because baseline data are inadequate. To assume that monitoring strategies can be implemented in time to assess real changes in the 10-year time frame proposed in the 2000 BiOp is probably unrealistic.

Conceptual Gaps

The salmon recovery documents too often fail to address important issues in a really meaningful way. Several examples are noteworthy.

Hatchery Reform

Although all of the documents acknowledge the need for hatchery reform for a variety of reasons, it is not clear from them what they mean by hatchery reform or how it should be implemented. This is a significant gap between concept and application. The four documents do not map a detailed strategy for reducing risks of harmful interactions between wild and hatchery fish, but instead they defer to the Artificial Production Review and recommendations by Brannon et al. (1999). The documents assume that supplementation will succeed in rebuilding populations and that artificial production will mitigate loss of naturally produced fish to habitat destruction—two frequent but unverified assumptions (ISAB 2003a).

Climate and Demographic Trends

Even though the documents acknowledge that environmental variation must be taken into account, they do not appear to be sensitive to the types of environmental variations that are systematic, that is, constitute probable trends. Two such variations seem especially relevant to salmon recovery.

Climate Change. If current forecasts of climate change are correct, it is quite likely that hydrologic runoff patterns in the region will change (Figure 12.5), probably with negative implications for recovery efforts. The documents appear to assume that the Columbia River Basin will remain within the range of climatic variations observed over the last century. They do not specify alternative actions if this assumption proves to be incorrect.

Human Demographic Changes. If current forecasts of human demographic trends are correct, increasing stress will be put on the Columbia River Basin's natural resources and perhaps even more importantly, on the power demands of the hydroelectric system. This trend will have implications for any recovery program. We found few if any attempts to reconcile salmon restoration efforts with regional strategies for future population growth and development.

Tributary Habitat

The documents tend to lack a strong conceptual foundation for determining desired habitat conditions in a watershed, estimating the productive capacity of watersheds for salmonids, and evaluating restoration alternatives. Natural disturbances, usually viewed as undesirable, but in reality important for long-term salmon habitat creation and maintenance (ISAB 2003b), are acknowledged in some of the documents. However, there is little indication of how managing the effects of natural disturbances such as wildfires and floods would figure in restoration programs.

Harvest

The documents do not provide a conceptual basis for the establishment of escapement goals for each production unit, prediction of adult returns, and plans for how harvest levels factor into conservation and recovery goals. All of the documents support selective fisheries based on retention of marked fish, but there are potential conflicts between such fisheries and the region's coded wire tag program that has been fundamental to the management and conservation of wild stocks.

Integration

To be truly effective, recovery actions integrated in a way that strategically addresses problems occurring throughout the salmon life cycle are necessary. Plans can only be as effective as the weakest link in the chain of management decisions that influence life stage survival. Too often the four documents do not adequately consider interactions between policies that affect different salmon life stages in the context of the various Hs. For example, the potential interaction between habitat rehabilitation projects and population supplementation is not adequately addressed, nor is the potential effect of harvest on nutrient levels in salmon spawning and rearing areas discussed.

Implementation

There are several very difficult issues with implementing the proposed actions. The level of institutional cooperation between state agencies, tribes, federal agencies, and private landowners needed to achieve salmon recovery in the Columbia River Basin is unprecedented. This point is emphasized strongly in the most recent version of the Governors' Plan, which is critical of the lack of coordination between the subbasin planning process and recovery plans being formulated by the NOAA Fisheries Technical Recovery Teams (TRTs). Fully implementing the proposed actions would require a level of cooperation that has never before been achieved, and the documents do not explain how this cooperation would be facilitated. In particular, the recovery documents reject mainstem dam breaching in favor of aggressive tributary habitat restoration, but how coordination will occur between public, private, and tribal land managers to provide habitat improvement is inadequately addressed.

Details of recovery actions and implementation strategies are often lacking. In many instances, the four documents present "plans to do planning." They assume that details will be worked out sometime in the future in spite of the fact that it has not been possible to work them out effectively in the past. The documents do not provide explicit strategies for dealing effectively with limited knowledge and high uncertainty in an adaptive management context. Some management decisions in the past have been to postpone potential recovery actions pending future scientific findings and verified population responses, but critical data gaps remain. Not acting is a decision that places the burden of proof on organizations attempting to conserve the resource (i.e., to demonstrate that a significant improvement would be likely). Developing explicit strategies for dealing with high levels of uncertainty is a painful but necessary process.

None of the documents adequately explain the procedures and circumstances that would trigger a departure from their recommendations. In the winter of 2000–2001, the Columbia River Basin experienced the most severe drought conditions since 1977, and many of the action items in the BiOp pertaining to operation of the hydrosystem were modified to accommodate the need for electricity production, sometimes to the possible detriment of juvenile salmonid outmigration. This included downstream barge transportation of all smolts collected at the Snake River dams and elimination of spill over the dams in favor of power generation. However, the documents do not describe how or when such extraordinary circumstances might cause a departure from stated restoration strategies. Nowhere are environmental thresholds identified that would lead to significant changes in management actions, including abandonment of existing plans. Such a lack of specificity underscores our concern that these four documents may not have the collective strength to serve as a clear, detailed, and robust blueprint for salmon recovery in the Columbia River Basin.

Summary

The salmon recent recovery documents provide federal and state strategies for salmon recovery in the Columbia River Basin over the next decade. They vary in their scientific content, ranging from the technically detailed BiOp to the more general policy- and process-oriented Governors' Plan. The purpose of this review was not to provide a thorough appraisal of the science contained in each of those documents individually, but to address the question "Do these four documents collectively outline salmon recovery strategies that are likely to have a high probability of success?" Overall, we believe the answer to the question is "probably not," unless state and federal organizations follow up these reports with strategies and actions that address the significant deficiencies in past recovery efforts and provide a more explicit recovery blueprint.

Taken together, the four papers represent a realistic assessment of the problems facing salmon recovery, and there is consistency in many of the kinds of recovery actions proposed in the documents. However, the strategies often lack details about how various recovery actions would be implemented, with the exception of actions related to mainstem passage. There is no doubt that the proposed strategies would result in some beneficial results, but the status of many wild stocks has become grave. Recovery documents containing *explicit and quantified details* are needed so that their sufficiency can be evaluated. We believe the four documents, collectively, fall short of providing this detail. Furthermore, the documents propose actions that mix ecological recovery (the approach advocated in this book) with approaches that involve artificial substitution and mitigation. While this is unavoidable in a river basin as heavily developed as the Columbia River, we feel a clear and well-coordinated strategy is needed that lays out a rationale for when and where different types of restoration should be used.

Recovering salmon in the Columbia River Basin will be an enormous undertaking, and the four documents represent a serious effort by state and federal organizations to develop a regional salmon strategy. Most of the scientific underpinnings of the documents are consistent with current ecological beliefs. We found relatively few instances in which they are clearly based on outdated science. But many passages in the documents appear to be works-in-progress in which details, hopefully, will emerge from subbasin assessments, experimental management, operational reforms, and research and monitoring. These details, of course, ultimately determine the successes or failures of the strategies contained in the reports. We hope some of the ideas and suggestions in this chapter and the book will be helpful to scientists and policy makers as recovery actions continue to evolve and monitoring programs are put in place.

Addendum

Since the majority of this chapter was written, there have been two important court decisions that have strongly affected federal salmon recovery policy. In the first case, U.S. District Court judge Michael R. Hogan ruled on September 12, 2001, in *Alsea Valley Alliance v. Evans* that NOAA Fisheries could not split an Evolutionarily Significant Unit (ESU) into two components—hatchery and wild fish—when making a determination of whether the ESU deserved to be listed under the ESA. The decision suspended the listing of Oregon coast coho salmon and potentially affected the listing status of 23 out of the 25 ESA-listed west coast salmon and steelhead. NOAA

Fisheries announced that it would review each of the ESUs that included fish reared in hatcheries, and on June 3, 2004, published its proposed hatchery listing policy, the *Federal Register* (*69 Fed. Reg. 31354*), followed by a re-evaluation of listed Pacific salmon ESUs on June 14, 2004 (*69 Fed. Reg. 33102*). Although under the new policy, hatchery and wild fish are included in jeopardy determinations, overall changes in listed ESUs are relatively minor. NOAA Fisheries proposed a re-listing of Oregon coast coho salmon and further proposed to list lower Columbia River coho salmon as Threatened. Two ESUs (Sacramento winter Chinook salmon and Upper Columbia River steelhead) were proposed for Threatened listing, an improvement from their previously Endangered status, and one ESU (California central coast coho) were proposed for Endangered, a change from their Threatened listing status. The Hogan decision has not resulted in major changes in ESA listing actions, but the precedent has been set to include the status of both hatchery and wild fish in the listing and delisting process.

However, NOAA Fisheries' new policy of including both hatchery and wild fish when considering ESUs for listing has sparked considerable scientific debate. The ISAB (2002) recommended that surplus hatchery fish not be allowed to spawn in the wild, arguing that this could lead to both ecological and genetic harm to wild salmon populations. A group of scientists formerly members of NOAA Fisheries' Recovery Science Review Panel (Myers et al. 2004) argued that hatchery salmon should not be included with wild fish when determining listing status. Both groups pointed out the scientific evidence for loss of genetic fitness associated with mixing hatchery and wild fish. Recently, a group of hatchery advocates (Brannon et al. 2004) defended the use of hatchery salmon in rebuilding wild populations, arguing that properly managed supplementation programs should be used to assist the recovery of wild runs. Readers are referred to Chapter 8 in this book to learn more about the evidence for and against the efficacy of supplementation.

The second important court decision occurred on May 7, 2003, when U.S. District Court Judge James A. Redden ruled in *National Wildlife Federation v. NMFS* that the 2000 Federal Columbia River Hydropower System BiOp did not meet the requirements of the ESA. The BiOp was referred back to NOAA Fisheries, which was given a year to revise it. Judge Redden found that the BiOp was inadequate because it relied heavily on non-federal actions outside the mainstem Columbia and Snake River hydrosystem (e.g., tributary habitat improvements) to mitigate ESA-listed salmon and steelhead losses, and there was little certainty that such mitigation would be successful. The judge stated that the BiOp was arbitrary and capricious because it was improper for NOAA to rely on both federal basin-wide actions encompassing the Columbia Basin as well as the dams and off-site mitigation actions such as those directed at hatchery operations and habitat improvement that had not undergone Section 7 consultation under the ESA. Additionally, the judge stated that non-federal basin-wide, off-site mitigation actions "are not reasonably certain to occur."

On September 9, 2004, NOAA Fisheries and the federal action agencies (U.S. Army Corps of Engineers, Bureau of Reclamation, and Bonneville Power Administration) released their draft responses to Judge Redden's remand order. In its "Draft Revised 2000 BiOp," also referenced as the "2004 Draft BiOp" on the *www.salmonrecovery.gov* website, the agencies proposed "actions in the operation of the dams that taken as a whole will not jeopardize the existence of the protected stocks." Thus, the question of whether to remove or breach certain federal Columbia River hydrosystem dams was removed from consideration provided suitable modifications were implemented. This policy position departed from the 2000 BiOp, which considered dam removal a possibility if specified salmon recovery goals were not met according to a 10-year schedule. As this addendum is being written, it is uncertain whether Judge Redden will accept the revised BiOp, and in any event advocates of dam removal have promised to litigate. On May 26, 2005, Judge Redden ruled that the 2004 BiOp was legally flawed, and as this addendum is being written the future of the BiOp is uncertain.

The two court cases illustrate how quickly federal policy for salmon recovery can change. In both instances, federal agencies shifted important positions (including hatchery and wild fish in an ESU; consideration of dam removal) that had strongly influenced governance of the Columbia River Basin during the 1990s. We note that both policy shifts appear to be at odds with some of the concepts discussed in this book. These new policies have been highly controversial and in all likelihood will catalyze further court battles. The use of "sound science" or "best science" will be an important part of the debates. While there is no scientific certainty in the optimum path to recovering salmon in a river basin as large or as complex as the Columbia, we hope the "normative" restoration principles identified here will assist agencies and all other stakeholders in making the best use of current scientific information.

Literature Cited

Brannon, E. L., D. A. Amend, M. A. Cronin, J. E. Lannan, S. LaPatra, W. J. McNeil, R. E. Noble, C. E. Smith, A. J. Talbot, G. A. Wedemeyer, and H. Westers. 2004. The controversy about salmon hatcheries. *Fisheries* 29(9):12–31.

Brannon, E. L., K. P. Currens, D. Goodman, J. A. Lichatowich, W. E. McConnaha, B. E. Riddell, and R. N. Williams. 1999. Review of salmonid artificial production in the Columbia River basin as a scientific basis for Columbia River production programs. Northwest Power Planning Council, Portland, Oregon. 77 p.

Hamlet, A. F., and D. P. Lettenmaier. 1999a. Columbia River streamflow forecasting based on ENSO and PDO climate signals. *ASCE Journal of Water Research, Planning and Management* 125(6):333-341.

Hamlet, A. F., and D. P. Lettenmaier. 1999b. Effects of climate change on hydrology and water resources in the Columbia River Basin. *American Water Resources Association* 35(6):1597–1623.

Independent Scientific Advisory Board (ISAB). 2000. Recommendations for the design of hatchery monitoring programs and the organization of data systems. Report ISAB 2000-4, Northwest Power Planning Council, Portland, Oregon.

ISAB. 2002. Hatchery surpluses in the Pacific Northwest. *Fisheries* 27(12):16–27.

ISAB. 2003a. Review of salmon and steelhead supplementation. Report ISAB 2003-3, Northwest Power Planning Council, Portland, Oregon.

ISAB. 2003b. A review of strategies for recovering tributary habitat. Report ISAB 2003-2, Northwest Power Planning Council, Portland, Oregon.

Independent Scientific Group (ISG). 1999. Return to the river: Scientific issues in the restoration of salmonid fishes in the Columbia River. *Fisheries* 24(3):10–19.

Independent Scientific Review Panel (ISRP). 1997. Review of the Columbia River Basin Fish and Wildlife Program as directed by the 1996 amendment to the Northwest Power Act. Report ISRP 97-1, Northwest Power Planning Council, Portland, Oregon.

ISRP. 1999. Review of the Columbia River Basin Fish and Wildlife Program for Fiscal Year 2000 as directed by the 1996 amendment to the Northwest Power Act. Report ISRP 99-2, Northwest Power Planning Council, Portland, Oregon.

ISRP. 2001. Final Review of Fiscal Year 2002 Project Proposals for the Mountain Snake and Blue Mountain Provinces. Report ISRP 2001-12A, Northwest Power Planning Council, Portland, Oregon.

ISRP. 2002. ISRP Final Review of Fiscal Year 2003 Proposals for the Upper and Middle Snake, Columbia Cascade and Lower Columbia and Estuary Provinces. Report ISRP 2002-2, Northwest Power Planning Council, Portland, Oregon.

Lee, K. N. 1993. *Compass and gyroscope: Integrating science and politics for the environment.* Island Press, Washington, DC.

Levin, P. S., and J. G. Williams. 2002. Interspecific effects of artifically propagated fish: An additional conservation risk for salmon. *Conservation Biology* 16(6):1581–1587.

Myers, R. A., S. A. Levin, R. Lande, F. C. James, W. W. Murdoch, and R. T. Paine. 2004. Hatcheries and endangered salmon. Policy Forum. *Science* 303:1980.

13

Return to the River: Strategies for Salmon Restoration in the Columbia River Basin

Richard N. Williams, Ph.D., Jack A. Stanford, Ph.D., James A. Lichatowich, M.S., William J. Liss, Ph.D., Charles C. Coutant, Ph.D., Willis E. McConnaha, Ph.D., Richard R. Whitney, Ph.D., Phillip R. Mundy, Ph.D., Peter A. Bisson, Ph.D., Madison S. Powell, Ph.D.

Introduction
  Development of a Regional Fish and Wildlife Program
Salmonid Restoration in Regulated Rivers
Improving Conditions for Salmon
  Lessons from the Hanford Reach
  Salmon and Habitat in the Mid-Columbia and Lower Snake Rivers
  Lessons from the Salmon Themselves
Conclusions and Recommendations
  Need for an Explicit Conceptual Foundation
  Need for an Integrated Approach
  Manage for Biological Diversity
  Protect and Restore Habitat
  Reduce Sources of Mortality
  Improve Effectiveness of Mitigation Actions
  Manage Considering Ocean and Estuary Conditions
  Establish Salmonid Reserves
Return to the River: The Challenge Ahead
Literature Cited

> "Fundamentally, the salmon's decline has been the consequence of a vision based on flawed assumptions and unchallenged myths—a vision that guided the relationship between salmon and humans for the past 150 years. We assumed we could control the biological productivity of salmon and 'improve' upon natural processes that we didn't even try to understand." (p. 8)
> —Jim Lichatowich. 1999. *Salmon without Rivers: A History of the Pacific Salmon Crisis.* Island Press, Washington, DC.

Introduction

The Columbia River today is a great "organic machine" (White 1995) that dominates the economy of the Pacific Northwest. Even though natural attributes remain—for example, salmon production in Washington State's Hanford Reach, the only unimpounded reach of the mainstem Columbia River—the Columbia and Snake River mainstems are dominated by technological operations supporting the region's economy (e.g., hydropower production, irrigation systems, flood control, commercial barging). Operation of the river via the hydropower system is driven largely by economic considerations of water usage in the basin and constrains conservation and restoration efforts for anadromous and resident salmonid fishes (Snake River Salmon Recovery Team 1993; Petersen 1995; NRC 1996; ISG 1999; NOAA 2004).

During more than a century of development in the Columbia River Basin (Figure 13.1), the region attempted to provide technological solutions for losses of salmon habitat and reductions in salmon survival, first through hatcheries (Figures 13.2 and 13.3) and fish ladders (Figure 13.4), then later through installation of screens at turbine intakes (Figure 13.5) and irrigation diversion screening (Figures 13.6 and 13.7), and finally barging and trucking of juvenile fish around the dams (Figures 13.8–13.10) (ISG 1998, 1999; Lichatowich 1999). The total amount of money spent maintaining and restoring salmon in the Columbia River Basin is difficult to determine, but it exceeds 3 billion dollars over the last two decades (General Accounting Office 1992, 2002). Despite these efforts, anadromous salmonids have continued to decline from their historical abundance (Figure 13.11). Total returns of cultured and wild anadromous salmonids reached an all-time low in 1995 of 750,000 fish (WDFW and ODFW 1996; 2001), although improved ocean conditions 2000 to 2004 brought increased adult returns to levels not seen in several decades. The returns are disproportionately composed of hatchery-raised fish. The total returns, and especially returns of wild fish, are still far short of historical numbers.

Prior to Euro-American development in the basin, the Columbia River may have supported more than 200 anadromous stocks, which returned 7 to 30 million adult salmon and steelhead to the river annually (Chapman 1986; NPPC 1986; Nehlsen et al. 1991). By the late-1990s, only nine salmon and steelhead stocks were considered healthy: Lewis River (Washington) and Hanford Reach (Washington) fall Chinook, Lake Wenatchee and Lake Osoyoos (Washington) sockeye, and five summer steelhead stocks in the John Day River (Oregon) (Huntington et al. 1996; Mullan et al. 1992). Recent increased adult returns, attributable largely to improved ocean conditions associated with the long-term productivity cycles in the Pacific Ocean (described in Chapter 10), have improved the abundance and status of many individual populations and stocks (Figure 13.12); however, extensive analyses

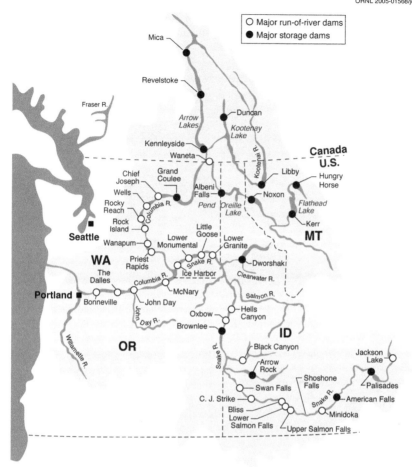

ORNL 2005-01568/jcr

Figure 13.1   Map of the Columbia River Basin, showing major dams and lakes.

equivalent to the Huntington et al. (1996) study or NOAA Fisheries population status reviews (Busby et al. 1996; Gustafson et al. 1997; Johnson et al. 1997; Myers et al. 1998) have not been conducted within the past several years. Monitoring population abundance in years to come will determine whether these increases contributed to stock recoveries or were only momentary gains.

A consequence of the declines in salmon and steelhead has been a proliferation of legal challenges and endangered species listings and petitions. Presently, 13 "evolutionary significant units" comprising four species of salmon and steelhead that spawn in the Columbia River or its tributaries

Figure 13.2 Photo of an upper Columbia River mitigation hatchery near Chelan, Washington, showing the proximity of the hatchery to the mainstem Columbia River and the constrained nature of the canyon geography. Photo by R. N. Williams.

Figure 13.3 Sorting of coho salmon prior to artificial spawning at Chiwawa Hatchery in central Washington. Photo by R. N. Williams.

Figure 13.4 The north (Washington shoreline) fish ladder and spillbays at John Day Dam looking south to the Oregon shore. Photo from U.S. Army Corps Digital Visual Library, website at *http://images.usace.army.mil/photolib.html.*

Figure 13.5 Installation of a turbine screen at McNary Dam. Photo from U.S. Army Corps Digital Visual Library, website at *http://images.usace.army.mil/photolib.html.*

Figure 13.6 Irrigation diversion dam in a middle reach of Fifteenmile Creek near The Dalles, Oregon. Photo by R. N. Williams.

Figure 13.7 A rotating drum screen located on the water diversion ditch created by the irrigation dam on Fifteenmile Creek shown in Figure 13.6. The rotating drum screen passes water down the irrigation ditch but diverts any entrained juvenile and adult fish back to the creek. Photo by R. N. Williams.

Figure 13.8 Loading juvenile salmon and steelhead captured in the bypass systems of Columbia and Snake River hydroelectric projects into a fish barge for transportation downriver and release below Bonneville Dam, the lowermost dam in the system. Photo from U.S. Army Corps Digital Visual Library, website at *http://images.usace.army.mil/photolib.html*.

have been listed as threatened or endangered under the Endangered Species Act (ESA). These include Snake River fall Chinook, Snake River spring/summer Chinook, Snake River sockeye, Snake River steelhead, upper Columbia River spring Chinook, upper Columbia River steelhead, middle Columbia River steelhead, lower Columbia River spring Chinook, lower Columbia River coho, lower Columbia River chum, lower Columbia River steelhead, upper Willamette River spring Chinook, and upper Willamette River steelhead. The federal role and responsibility for regional salmon recovery is complex and rapidly changing (see more detailed description in Chapter 12).

Figure 13.9 A U.S. Army Corps of Engineers' fish barge used in "Operation Fish Run" to transport juvenile salmon and steelhead around Columbia and Snake River hydroelectric projects. Photo from U.S. Army Corps Digital Visual Library, website at *http://images.usace. army.mil/photolib.html*.

Figure 13.10 Idaho Department of Fish and Game tanker truck used to transport juvenile salmon and steelhead. Photo by R. N. Williams.

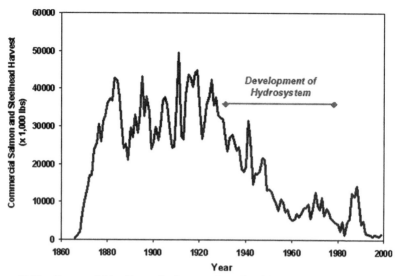

Figure 13.11 Commercial landings of salmon and steelhead in thousands of pounds in the Columbia River, 1866 to 2000. Figure courtesy of Washington Department of Fish and Wildlife and Oregon Department of Fish and Wildlife (2001).

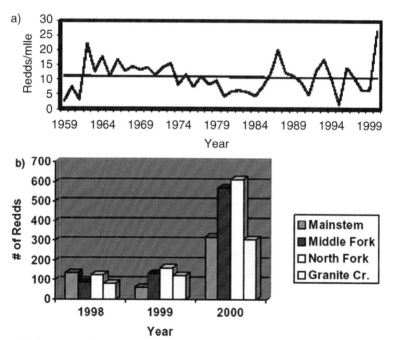

Figure 13.12 Status and trends of spring Chinook based on redd counts (a) in the John Day River overall from 1959 to 1999, and (b) from 1998 to 2000 for the John Day mainstem and the North Fork, Middle Fork, and Granite Creek tributaries. Graphs from the John Day Subbasin Management Plan (Columbia-Blue Mountain Resource Conservation and Development Area 2004).

Development of a Regional Fish and Wildlife Program

Since the early 1980s, salmon restoration has been approached regionally through implementation of the Columbia River Basin Fish and Wildlife Program[1] of the Northwest Power Planning Council (now called the Northwest Power and Conservation Council). The Northwest Power Act directed the Council to develop a plan to "protect, mitigate, and enhance" the fish and wildlife as affected by the Columbia River Basin hydroelectric system. The Fish and Wildlife Program was based on recommendations submitted by state fish and wildlife managers, Native American tribes, federal agencies, and other stakeholders. Those recommendations were solicited, compiled, and discussed by the Council in public hearings before being adopted into the program. Consequently, the Fish and Wildlife Program, in its early manifestations, represented a collection of individual, reactive measures (proposed by a diverse constituency), rather than a cohesive program derived from a single *a priori* conceptual framework. Thus, these recommendations did not share a common scientific understanding of the physical and biological components of the Columbia River watershed and the ways those components interact to form a salmonid-producing ecosystem.

While the Fish and Wildlife Program actions to date represent a good faith effort by the Council and the region's fisheries managers to recover salmonids, those efforts have failed so far to stem the decline of wild salmonids in the basin, notwithstanding increases in adult returns from 2000 to 2004 when ocean conditions improved and salmonid marine survivals increased. Salmon returns have declined from almost 2.5 million annually in the early 1980s to less than 1 million returning adults in the late 1990s, most of which (> 80%) are of hatchery origin. Wild fish abundance is approximately 1% of historical pre-development abundance (NRC 1996). The Council's current fish and wildlife program aims to achieve a modest return of salmon and steelhead to the Columbia Basin of 5 million fish.

Recent iterations of the Council's Fish and Wildlife Program (NPPC 1997, 1998, 2000, 2004) include an integrated framework (NPPC 1997) and scientific principles (NPPC 1998) for fish and wildlife restoration, at least in part due to consistent critical advice by the Council's independent scientific advisory groups (ISG 1993, 1998, 1999; ISRP 1997, 1998, 1999). The Council's Fish and Wildlife Program, in concert with NOAA Fisheries Biological Opinions (NMFS 1995, 2000; NOAA Fisheries 2004), emphasize actions to increase survival of salmon and steelhead in the lower Snake River (i.e., downstream from Hells Canyon

---

[1]Enactment of the Pacific Northwest Electric Power Planning and Conservation Act by Congress in 1980 (hereafter the Northwest Power Act) formed the Northwest Power Planning Council (renamed the Northwest Planning and Conservation Council in 2003) and directed it to develop a regional fish and wildlife plan.

Dam, Idaho/Oregon, which is a barrier to upstream adult migration), the middle and lower reaches of the mainstem Columbia River (i.e., downstream from Chief Joseph Dam, Washington), and their tributaries (Figure 13.1). Major actions implemented so far include the following:

1. Modifying mainstem dam operations and facilities to improve upstream and downstream passage of adults and juveniles (Chapter 7)
2. Coordinating river operations to enhance spring and summer flows aimed at improving smolt survival (Chapters 6 and 7)
3. Reducing smolt predators (Chapter 7)
4. Constructing and operating hatcheries (Chapter 8)
5. Modifying existing artificial production operations, including supplementing naturally reproducing populations (Chapter 8)
6. Implementing best management practices for land use activities (Chapter 5)
7. Screening irrigation diversions (Chapter 5)
8. Improving habitat and other measures as well as research and monitoring designed to answer critical recovery questions (see Chapter 5 on habitat and Chapter 11 on monitoring).

Our collective review of the management of the Columbia River and its salmonid populations led to the conclusion that management over the course of the Council's Fish and Wildlife Programs (1982–2000) and the NOAA Fisheries Biological Opinions (1995, 2000, 2004) have been based on a conceptual foundation (i.e., a belief) that natural ecological processes comprising a healthy salmonid ecosystem can, to a large degree, be replaced, circumvented, simplified, and controlled by humans, while the production of anadromous salmonids is maintained or even enhanced (Chapter 2). In a review of the Pacific Northwest's use of salmon hatcheries, Meffe (1992) identified this approach to restoration (which he called "techno-arrogance") as the driver behind the region's reliance on large-scale hatchery technology to rebuild depleted salmon runs.

Salmonid Restoration in Regulated Rivers

Our alternative conceptual foundation (Chapter 3 specifically and Chapters 4–6 supporting) explicitly recognizes the Columbia River as a natural-cultural system, in which human development and its consequences are an integral part of the ecosystem. At the same time, the conceptual foundation also recognizes the critical function of natural biophysical processes in the creation and maintenance of healthy salmon habitat and fulfillment of life history functions (Chapters 3 and 5). The formation and maintenance of complex and interconnected habitats is fundamental to the expression of life history diversity and the spreading of the risk of mortality in variable environments and, ultimately, to the realization of sustainable production (described more fully in Chapter 3). Human development in the Columbia Basin has weakened or eliminated the natural habitat forming and maintenance processes (Chapter 5), which together with inappropriate hatchery practices (Chapter 8) and overharvest (Chapter 9), have caused the depletion and extinction of salmon populations (Petersen 1995; NRC 1996; ISG 1998, 1999; Naiman and Bilby 1998; Stanford et al. 2005a). Thus, in highly developed natural-cultural ecosystems like the Columbia, there is an inescapable tension between the benefits derived from development and the costs of that development in terms of lost goods and services naturally produced by a healthy ecosystem (e.g., salmon and clean water) (Miller 1997; Blumm et al. 1998; Wood 1998; Lackey 1999; Lichatowich 1999). We recognize this tension between development and salmon production in our conceptual foundation.

It is not possible to return the Columbia River system to a completely natural state (i.e., the "historical" river) in order to achieve salmon restoration. However, maintaining the current approach to salmon restoration will not achieve the Council's salmon restoration goals of returning approximately 5 million adult fish to the basin annually and is likely to continue the present trends of declining wild salmon abundance, local population extinctions, and proliferating ESA listings. A major conclusion embedded in our alternative conceptual framework (Chapter 3) is the need to restore a greater degree of "naturalness" to the river than exists today (see also Poff et al. 1997; Gregory et al. 2001; Stanford et al. 2005). With historical (i.e., pristine) conditions not attainable, what standard of naturalness is appropriate? An acceptable level of naturalness rests somewhere between the current developed state and a completely natural river. The ecological and biophysical attributes of the pre-development river represent the norms or standards under which salmon in the Pacific Northwest evolved. Management actions that restore or improve these attributes will also improve ecological conditions for salmon and aid in their restoration. Some examples of natural and artificial attributes that suggest possible management actions to improve conditions for salmon are shown in Table 13.1.

We believe an ecosystem with a mix of natural and cultural features such as the Columbia River can still sustain all life stages of a diversity of salmonid populations (ISG 1998, 1999). The region will have to improve ecological conditions in the river system before sustained salmon recovery is possible. This is a major change in approach to salmon recovery from the current approach, which has emphasized activities and actions that circumvented the natural ecological attributes of the river, that is, attempting to restore salmon without restoring the natural river functions. In our dynamic ecosystem view, a cornerstone of salmonid restoration is protection and restoration of the natural processes that create and change habitat conditions (e.g., Reeves et al. 1995; Stanford et al. 1996; Beechie and Bolton 1999; ISAB 2003). Restoration of habitat-forming processes is crucial for re-establishing and

Table 13.1 Examples of natural and artificial conditions and approaches to salmon restoration.

Natural	Artificial
Natural spawning and rearing	Artificial propagation and rearing in man-made structures; population relocations or stock transfers
Unimpeded passage to and from spawning and rearing sites	Migrations blocked or hindered by anthropogenic factors such as instream structures (dams and other migration barriers), water withdrawals, water pollution, or unfavorable flows; artificial migration pathways that don't mimic natural features
Flow regimes produced by local and regional climates, unencumbered by regulation	Regulated flow regimes in which natural patterns of seasonal and diurnal discharge do not occur and characteristics of naturally flowing water are absent or limited
Riverine habitats formed and maintained by natural processes through the interactions between flowing water and the surrounding landscape	Replacement of free-flowing river channels with impoundments; substitution of artificial habitats for habitats formed by natural disturbance processes
Community interactions dominated by species with which native salmonids co-evolved	Introductions of non-native plants and animals, including other game fishes, which have altered survival, growth, and behavior of native salmonids
Survival rates that permit enough adults to return so that (1) naturally spawning populations are capable of sustaining and rebuilding themselves, (2) sufficient numbers exist to repopulate favorable but currently vacant habitats, and (3) sufficient marine-derived nutrients are returned to maintain aquatic and riparian productivity	Anthropogenic mortality, including harvest, is sufficiently high that (1) populations are incapable of sustaining or rebuilding themselves, (2) there are insufficient adults and juveniles to recolonize favorable habitats and interbreed with other locally reproducing populations, and (3) not enough nutrients are returned to maintain aquatic and terrestrial food webs dependent on salmon carcasses

sustaining diverse salmonid populations and life histories (Healey and Prince 1995; Reeves et al. 1996; Ebersole et al. 1995). Knowing how to restore natural processes is exceedingly challenging in a landscape that has been subject to more than 100 years of human development and is highly fragmented, dominated by the inertia of established land and fish management practices, and in which public sentiment may not favor large-scale changes (e.g., reduced fire suppression and a more natural fire regime, re-regulation of rivers to achieve a more natural hydrograph) (Stanford et al. 1996; Miller 1997; Poff et al. 1997).

Along with protection and restoration of core habitats and habitat-forming processes, protection of core salmon and steelhead populations and their associated habitats should be of the highest priority (McElhany et al. 2000; Williams 2001). The Hanford stock of fall Chinook in central Washington and the free-flowing river habitat where they spawn are a particularly potent example of a core stock and its critical riverine habitat. Both should receive the utmost protection possible.

Improving Conditions for Salmon

Lessons from the Hanford Reach

Two important factors emerge from an examination of the Hanford Reach fall Chinook stock, one of only two truly robust fall Chinook population in the Columbia River Basin (the other being the healthy, but much smaller, Lewis River fall Chinook). First, the Hanford population originates in a series of linked habitats amid a free-flowing river section that provide suitable adult spawning habitat, successful incubation of eggs, and various juvenile rearing areas that are immediately adjacent to, or within a few kilometers downstream of, the spawning areas (Geist and Dauble 1998) (Figures 13.13 and 13.14). Second, fisheries regulations are appropriate to maintain adequate escapement of spawners to the Hanford Reach area.

While the 50-mile Hanford Reach is unimpounded and the Columbia River flows freely through this remarkable section, flows coming into the Hanford Reach are regulated by a number of upstream dams. In recent years, the Hanford Reach fall Chinook population has benefited from agreements to regulate river flows out of these upstream dams that take into consideration the spawning and rearing needs of the Hanford Chinook population. As a result of an agreement reached among affected parties (the three mid-Columbia public utility districts, the Bonneville Power Administration, the U.S. Army Corps of Engineers, the states of Washington and Oregon, and certain Treaty Tribes), flows are maintained while Chinook salmon are spawning to provide a stable boundary within which the redds are constructed. Following spawning, flows are maintained to protect the incubating eggs from dewatering. Finally, after the fry emerge from the redds, flows are stabilized to prevent stranding of juveniles, prior to their movement downstream (ISAB 1998). It is clear that the population has responded positively to these measures that have improved the functionality of the existing habitat. In addition, hatcheries at Priest Rapids and Ringold have also contributed to returning adults. The Chinook juveniles from the Hanford Reach are typical fall (ocean-type) Chinook, beginning and completing their migration to the sea during the first year after hatching. They move downstream

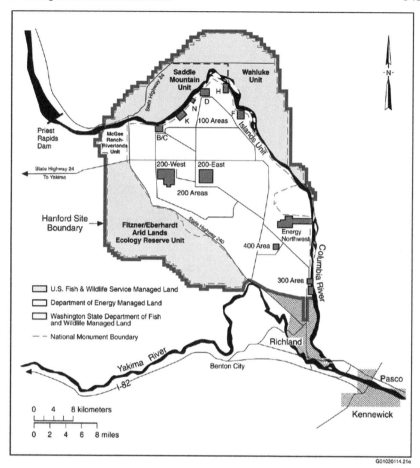

Figure 13.13 Map of the Hanford Reach in the Middle Columbia River showing its position adjacent to the Yakima River. Figure from U.S. Fish and Wildlife Service, *http://hanfordreach. fws.gov/fac.html.*

slowly, feeding and taking advantage of the excellent rearing habitat found in the lower portions of the Hanford Reach. Fall Chinook do not guide well with turbine intake screens. Like sockeye salmon, they are probably subject to higher mortality than spring Chinook or steelhead passage past the dams. Nevertheless, Hanford Reach fall Chinook are captured in ocean fisheries off the coasts of Washington, British Columbia, Alaska, and in the Columbia River. To date, passage mortalities on juveniles and these fisheries have not prevented attainment of the escapement goals in the Columbia River, although there are other stocks of fall Chinook that have not fared as well.

Figure 13.14   Beach seining for juvenile fall Chinook at Hanford Reach, showing typical shoreline habitat within the Hanford Reach National Monument. Photo courtesy of D. Geist.

Salmon and Habitat in the Mid-Columbia and Lower Snake Rivers

Areas in the mid-Columbia, adjacent to the Hanford Reach and extending downstream to below The Dalles, were historically highly productive spawning and rearing habitats for salmon, particularly fall Chinook (Figure 13.15). The forested upper reaches of the Snake River Basin and its many tributaries (the Boise, Payette, Weiser, Owyhee, Bruneau, Salmon, and Clearwater Rivers) produced the majority of the Columbia River Basin's spring Chinook salmon. The lower Snake River (the site of the four lower Snake River dams) (Figure 13.1) is largely constrained through a high desert canyon, yet also was a historically productive river reach for fall Chinook salmon. The Columbia River, from its confluence with the Snake River in eastern Washington, downstream to below The Dalles, was characterized by a shallow river gradient replete with complex and diverse riverine habitats that were used extensively by spawning fall Chinook salmon, including the large and storied "June hogs" and today's "upriver brights." The entire area was also used as rearing habitat for juvenile fall Chinook and as resting and feeding habitat for outmigrating spring Chinook and steelhead from the upper basin (Chapter 6).

While the mid-Columbia area near Washington's Tri-Cities (Richland, Kennewick, and Pasco) and John Day and McNary Dams is extensively developed for hydroelectric generation, barge transportation of inland goods, and large-scale pump-fed crop irrigation, the adjacent Hanford Reach holds great

Figure 13.15 View of the mid-Columbia River near Vantage, Washington, showing the broad valley and shorelines habitats typical of the mid-Columbia region. Photo by Mike Pinney, *www.photobymike.com.*

potential for restoration of salmon-bearing habitat and increasing salmon abundance in the basin. Nevertheless, large-scale restoration of salmon habitat in the mid-Columbia area would necessitate major modification of the Columbia River federal hydroelectric system, something not currently considered in the 2004 NOAA Fisheries Biological Opinion (NOAA 2004), nor supported by the Northwest Regional Congressional delegation. An earlier version of the Biological Opinion (NMFS 2000) had decision points where modification of the hydrosystem, including the possibility of partial removal of the four lower Snake River dams, would have been considered if explicit restoration steps did not achieve numerical targets by 2008. Recent federal actions and the new 2004 Biological Opinion removed any consideration of dam removal or major modification to the hydrosystem from the federal planning process (Chapter 12); however in May 2005, the 2004 Biological Opinion was rejected by the federal courts after determining that it violated the Endangered Species Act. In June 2005, Judge Redden ordered summer spill to occur at selected Snake and Columbia River dams after noting that "irreparable harm results to listed species as a result of the action agencies' implementation of the updated proposed action" (Judge James A. Redden, p. 9; Injunctive Order and Opinion for Case CV 01-640-RE, June 10th, 2005).

In spite of the current federal position constraining modification of the Snake or Columbia River hydrosystem, several recent studies have examined the importance of geomorphic features in large rivers (Geist and Dauble 1998) and assessed the impacts of development and operation of the Columbia and Snake River hydroelectric system on mainstem riverine processes and salmon habitats–primarily on fall Chinook spawning and rearing habitats (Battelle's Pacific Northwest Division and U.S. Geological Survey 2000; Dauble et al. 2003).

Among the findings from the study are that only about 13% and 58% of the historical mainstem Columbia and Snake Rivers, respectively, are still riverine in nature (as opposed to the lacustrine nature of the impounded reaches). The largest loss of riverine habitat in the Columbia River occurred downstream of the Snake River confluence where only about 3% of the historical riverine habitat between the confluence and Bonneville Dam still exists, mostly in the tailraces downstream of hydroelectric projects (Figure 13.16). Other recent studies have shown remnant fall Chinook populations utilizing these specific habitats for spawning (Garcia et al. 1994). In the upper Snake River, nearly 70% of the historical mainstem riverine habitat still remains; however, it lies upstream of Hells Canyon Dam (Figure 13.1) and is no longer accessible to anadromous salmonids.

Dauble et al.'s (2003) analysis of historical spawning areas for fall Chinook was coupled with model-based analysis of river reach geomorphology and concluded that historical fall Chinook spawning areas were primarily associ-

Figure 13.16 The Snake River immediately below Hells Canyon Dam on the Idaho-Oregon border, showing fall Chinook spawning habitat. Photo by R. N. Williams.

ated with the wide alluvial flood plains that were once common in the main-stem Columbia and Snake Rivers (Battelle's Pacific Northwest Division and U.S. Geological Survey 2000; Dauble et al. 2003). From the analysis, they identified three river reaches with the highest potential for restoration of riverine processes: (1) the Columbia River upstream of John Day Dam, (2) the Columbia-Snake-Yakima River confluence, and (3) the lower Snake River upstream of Little Goose Dam.

The John Day pool lies immediately downstream of McNary Dam. The upper portion of the pool also lies downstream of the confluence of the Snake and Columbia Rivers and contains what was formerly a large alluvial reach that served as a highly productive area for mainstem spawning Chinook salmon populations. Populations in this area may have been linked together into a regional metapopulation. The large mainstem spawning population may have served as the core of the metapopulation, and it may have stabilized Chinook salmon production in the mid-Columbia and its tributaries. Restoration and revitalization of the upper John Day pool as a free-flowing river segment might assist in the re-establishment of Chinook salmon production and metapopulation structure through straying and dispersal from the upstream Hanford Reach Chinook. Dauble et al. (2003) examined three different river management options for the John Day reservoir: (1) natural river drawdown, (2) spillway crest drawdown, or (3) normal pool conditions; and found that up to 53%, 22%, and 3% of the total reservoir area would be potentially suitable as spawning area for fall Chinook, respectively (Figure 13.17). Because of the shallow river gradient in the Columbia River below the confluence with the Snake River, the John Day Dam pool is one of the longest reservoirs on the Columbia River at more than 70 miles in length. As suggested by one scenario in the Dauble et al. (2003) study, drawing the John Day Dam pool elevation down 40 feet to spillway crest would reduce the reservoir length by half, exposing about 35 miles of river that was inundated when the dam was built. The resulting river would be shallow and braided: ideal spawning and juvenile rearing habitat for fall Chinook (e.g., Figure 13.18).

In the lower Snake River, Dauble et al. (2003) concluded that the majority (74%) of the 266-km study area was classified as alluvial or partially alluvial habitat. They estimated that approximately 55% of the study area may have been suitable as fall Chinook spawning habitat prior to hydroelectric development. Of particular interest is the river section between Little Goose Dam and Lower Granite Dam, in which 87% of the lineal river distance was predicted to be suitable fall Chinook spawning habitat.

Lessons from the Salmon Themselves

While it is important to look at historical distribution and abundance patterns of salmon in the basin as one means of identifying potential restoration

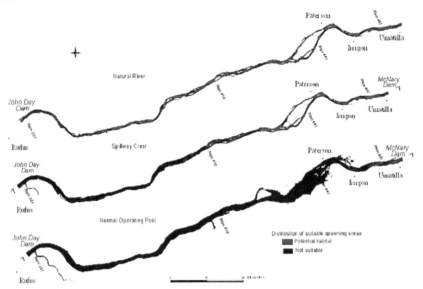

Figure 13.17 Distribution of suitable (red) and unsuitable (blue) fall Chinook spawning habitat in the John Day pool area based on water depth and velocity for three operational scenarios: natural river (top), spillway crest drawdown (middle), and normal operating pool (bottom). Figure from draft final report by Battelle's Pacific Northwest Division and the U.S. Geological Survey (2000), courtesy of D. Dauble.

sites and opportunities, we also need to look closely at current distribution and abundance patterns, as these indicate how salmon are adapting to and using the present system. One of the most interesting distribution patterns to emerge from recent surveys of the Columbia and Snake River mainstems is the persistent use of tailwater areas below hydrosystem projects by fall Chinook salmon as spawning areas (Garcia et al. 1994; Fish Passage Center website, *www.fpc.org*). Remnant fall Chinook populations have been observed below nearly all projects in the mainstem Columbia and Snake Rivers, attesting to the dispersal ability of fall Chinook, as well as to their ability to find and colonize suitable available spawning habitats. We view this response by fall Chinook to these habitats as evidence of the effectiveness of the normative river concept (Chapter 3) as a restoration strategy for increasing salmon abundance and productivity.

The most striking example of fall Chinook use of tailwater habitats immediately below a hydroelectric project occurs below Bonneville Dam around Pierce and Ives Islands. Lower Columbia River chum salmon, an ESA-listed threatened species, have also used the area for spawning since the 1960s, or earlier. The first recorded stream survey of the area (Hamilton Slough on the north bank of the Columbia) occurred in November 1967 by Washington Department of Fish and Wildlife personnel, where they counted 63 chum

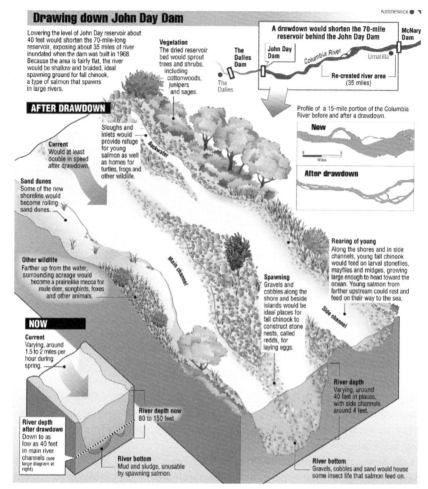

Figure 13.18 An example of how ecological conditions would be improved by restoring free-flowing river conditions to the upper John Day pool area through a permanent drawdown of John Day Reservoir to spillway crest. Graphic design provided by Molly Swisher and Jonathan Brinckman, *Oregonian* staff.

salmon adults. Another survey in the fall of 1976 noted 13 chum salmon and 75 redds at the upstream end of Pierce Island. In spite of these records, the area was not systematically surveyed until the late 1990s.

In November 1993, fall Chinook were observed in the area by WDFW personnel. Surveys in December 1994 counted more than 150 spawning fall Chinook salmon. More extensive surveys in 1997 and 1998 counted over 1,000 adult fall Chinook in the Pierce and Ives Islands and the Hamilton slough area. Figure 13.19 shows the distribution of fall Chinook and chum salmon redds in the Hamilton Slough area in late November 1999. Genetic

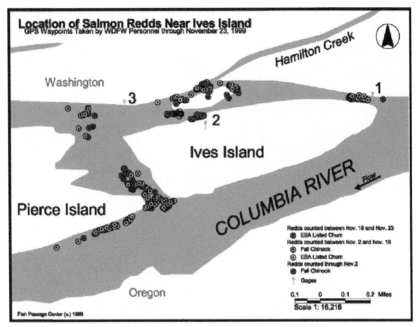

Figure 13.19 Map showing chum and fall Chinook salmon redds in normative habitats surrounding Ives and Pierce Islands immediately below Bonneville Dam, fall spawning season 1999. Map courtesy of the Fish Passage Center.

analysis of fall Chinook in this area by WDFW suggests they are derived from remnant lower Columbia River fall Chinook through natural colonization and growth, rather than being derived from mid-Columbia or upper-river fall Chinook salmon.

Low flows from Bonneville Dam in October 1997 (117 kcfs) threatened to dewater the spawning area. Further research and monitoring indicate that dewatering of redds could occur at flows below approximately 150 kcfs. Presently, Bonneville Dam operations are coordinated with spawning area water level sensors through the Fish Passage Center's monitoring program to avoid dewatering redds of both chum and fall Chinook salmon in the Hamilton Slough area.

Conclusions and Recommendations

Salmon restoration in the Columbia River is based on the prevailing belief that the primary problem for anadromous fish is mortality associated with juvenile passage through the mainstem dams and reservoirs. The traditional solution involves a combination of hatchery technology (to maximize the number of smolts produced), flow augmentation, and juvenile transportation via barges to move them as rapidly and efficiently as possible past the dams. This strategy is reflected in restoration expenditures (General Accounting Office 1992, 2002; Bernhardt et al. 2005) and in the measures supported by management agencies and tribes for implementation (ISRP 1997, 1998, 1999; CBFWA 2004).

The region, through its policy representatives and the evaluative processes described above, must decide how far it is willing to restore the river based on its economic, cultural, and ecological values. If the region concludes it cannot or is unwilling to improve the ecological conditions needed to achieve the Council's current salmon recovery goals, then those goals must be changed. The challenge before the region is to reach agreement on the extent to which the numerous social and biophysical constraints on the Columbia River can be relaxed or removed. Defining what the river must be and moving the ecosystem to that point is the only way to bring about salmon recovery and to achieve the Fish and Wildlife Program's salmon restoration goals.

Unfortunately, the restoration program based on the current set of assumptions has failed to curtail the decline of wild salmonid fishes. Moreover, it may be actively interfering with conservation efforts for resident fishes or other management goals in headwater areas not accessible to salmon; for example, eutrophication controls in Flathead Lake are negated by discharges from Hungry Horse Reservoir made to accommodate late summer smolt movement in the lower Columbia River (Stanford and Hauer 1992).

Need for an Explicit Conceptual Foundation

The lack of progress towards salmon recovery goals in the Columbia Basin has been linked to restoration programs derived from a conceptual foundation that sought to circumvent *important ecological processes*. Recovery of anadromous salmonids in the basin needs to be centered around an explicitly defined conceptual foundation based on ecological principles. We have consistently recommended that the region adopt an explicitly defined conceptual foundation that is based on ecological principles (see Chapter 3). Without a fundamental change in our approach to salmon restoration, more extinctions of salmon populations are likely and progress toward the rebuilding goal is unlikely. Temporary increases in some populations may occur in response to fluctuations in ocean conditions, and small increases may result from large-scale use of technology such as hatcheries, but the overall downward trend in returns of wild fish that has occurred throughout this century will likely continue without a fundamental change in approach.

The most recent (2000, 2004) Fish and Wildlife Program drafted by the Northwest Power and Conservation Council is a major move toward the conceptual foundation and the programmatic approach we have been recommending; however, more needs to be done.

Need for an Integrated Approach

The potential social, economic, and biological trade-offs that are needed to improve the ecological conditions in the Columbia River are not fully known; nor have they been subject to open discussion and debate. Identifying and quantifying those trade-offs where possible is a high priority. Although uncertainty exists regarding our restoration approach, it offers an opportunity to move from the continued pattern of decline and to boost recovery of salmon and the goals of regional fisheries management and recovery plans (IEAB 1999).

A rigorous program of evaluation, monitoring, research, and adaptive management derived from the appropriate conceptual foundation will be required. An approach based on the re-establishment of more natural riverine processes, combined with an implementation program governed by the principles of adaptive management, offers the best hope for preventing large-scale extinction of salmon in the basin. This approach might be tested at the subbasin level as a first step (Hill and Platts 1998). Recent versions of the Council's program (NPPC 2000, 2004) have stressed ecological principles, multi-species actions, and increased coordination and integration.

Manage for Biological Diversity

Biological diversity is the cornerstone for long-term population viability among species inhabiting stochastic environments (Frankham et al. 2002). The Pacific Northwest region needs to explicitly recognize that salmonid fishes in the Columbia River exist naturally as aggregates of populations (Brannon et al. 2004), likely organized as metapopulations (described in Chapter 4). Thus, managing for life history and genetic diversity among salmonid populations is essential for the long-term persistence of these species. The results of such actions will likewise manifest themselves in increased survival and total production within these areas. This, in turn, further reduces the cumulative impacts from deterministic effects which pose hazards to the populations' longevity and reduces the effect of unforeseen catastrophies; that is, strong, healthy, interconnected populations are better able to weather demographic and environmental stochasticity (Gilpin 1996; Frankham et al. 2002). Although much of the natural diversity of salmonid fishes has been lost (Nehlsen et al. 1991; Huntington et al. 1996), we must assume that salmonids retain the capacity to re-express life history and population diversity if opportunities for access to suitable habitat are provided (Healey 1994). As habitats improve in the Columbia Basin, metapopulation structure will likely develop from the natural expansion of remaining wild core populations (e.g., fall Chinook in the Hanford Reach) by processes variously described as metapopulation "rescue effects" (Stacey and Taper 1997).

Protect and Restore Habitat

Freshwater habitat for all life history stages must be protected and restored, with a special emphasis on key alluvial river reaches and lakes. Protecting healthy habitat (Figure 13.20), restoring degraded habitat, and providing access for salmonids to diverse habitats should be management priorities (Chapter 5). This approach will permit the re-expression of phenotypic diversity in salmonid populations.

At least three generalized actions could begin to rebuild habitat quantity and quality of the mainstem and tributaries:

Figure 13.20 The Wind River Canyon in southwestern Washington. This section of the Wind River is home to a remnant but depressed wild steelhead population. Photo by R. N. Williams.

1. Re-regulate flows to restore the spring high-water peak, including occasional, especially high flood peaks, that will revitalize the mosaic of habitats in alluvial riverine reaches.
2. Re-regulate flows to stabilize daily fluctuations in flow (caused by the practice of "power peaking") to allow food web development in shallow water habitats and reduce juvenile mortalities via stranding.
3. Provide incentives for watershed planning that emphasize riparian and upland land use activities that support natural interactions between land and water, and insist on empirical evaluation of effectiveness of management practices (Figure 13.21).

Reduce Sources of Mortality

Reduce sources of mortality in the mainstem of the Columbia and Snake Rivers and improve effectiveness of mitigation activities within the hydroelectric system (Chapters 6 and 7). These goals include managing stocks with a more complete understanding of their migratory behavior and how this behavior is affected by various modes of river regulation. Mitigation meas-

Figure 13.21 The lower Lemhi River is a highly productive tributary of the upper Snake River in central Idaho, where water and land incentives have encouraged local ranchers to protect important spawning and rearing habitat for Chinook salmon and resident rainbow trout. Photo by R. N. Williams.

ures for dams should be directed toward increasing natural riverine processes and functions needed by salmon for spawning and rearing.

We identified four specific areas or activities that would improve the survival of salmon in the mainstems of the Columbia and Snake Rivers:

1. Couple seasonality of flow with spill rates over the dams that efficiently bypasses juveniles and adults around mainstem dams and behaviorally cue (rather than physically flush) the juveniles through the mainstem.
2. Reduce mortality from gas bubble trauma with field research on causes of the problem and installation of devices that reduce nitrogen gas supersaturation.
3. Transport (barge) juvenile salmon around mainstem dams only if all life history types are included, if the currently perceived benefits of transportation are real for all life history stages, and if it is clear that natural habitats in the mainstems cannot be restored.
4. Restore mainstem, floodplain, and estuary habitats to more natural conditions where possible, which will reduce predation rates on migrating juvenile salmon and provide more rearing and resting habitat.

Improve Effectiveness of Mitigation Actions

The Pacific Northwest region needs a better strategy to reduce inadvertent negative impacts and improve the effectiveness of mitigation actions associated with habitat restoration, artificial propagation, and harvest management. Planning and implementation of mitigation measures will be more effective if they occur within the context of an explicitly defined conceptual foundation and the normative river concept. Measures should be evaluated for effectiveness in reaching stated objectives. Habitat restoration in both the mainstem and tributaries must receive high priority (Chapter 5). Restoration efforts should be directed at providing the habitat opportunities that historically supported salmonids in their natural state and the connectivity to those habitats (Healey 1994).

Artificial propagation (Chapter 8) must be viewed as an experiment to be implemented within an adaptive management framework (NRC 1996), rather than a proven technique to increase the number of fish for harvest or accelerate the recovery of wild stocks. It will be difficult to determine if it is possible to integrate hatchery operations with natural production in the basin (Scientific Review Group 1999; Myers et al. 2005); monitoring and evaluation should be designed to make that determination. Resolving that issue has been avoided for too long. The role and scale of artificial production at the subbasin level should be consistent with the rebuilding goals for natural production. Supplementation goals emphasizing harvest should be secondary to population restoration and self-sufficiency. Monitoring, and especially

evaluation, remains inadequate to address the many outstanding uncertainties associated with the use of artificial propagation (ISAB 2003).

Appropriate harvest control is necessary for successful salmon conservation (Chapter 9), with full accounting for harvest (both direct and indirect) to
ensure the persistence of salmon populations. With degraded habitats,
reduced life history diversity, and reduced abundance, it is essential to
account for all sources of mortality in all localities and to control harvest to
levels consistent with those other sources of mortality and with salmon
recovery. Harvest regulation should take into consideration the keystone role
that salmon carcasses play in riverine food webs.

Manage Considering Ocean and Estuary Conditions

Estuary and ocean dynamics are now recognized as major controlling factors
of salmon productivity (Chapter 10). This will require adjustments in management actions for all other aspects of the life cycle under human control,
such as harvest, hatchery operations, river and tributary habitats, and
hydrosystem operations. Prudent management actions should increase or
maintain biodiversity in salmon populations to help them cope with the
effects of a fluctuating marine environment. A better understanding of estuarine and oceanic food webs is needed. Much recent work has been conducted in this area (Bottom et al. in press), yet our understanding and ability
to maintain salmon productivity in the estuary is limited.

Estuarine habitats and the near-shore Columbia River plume can be
improved by pollution abatement and continuing enhancement of the spring
freshet associated with the restoration of more normal flow regimes (Cury and
Roy 1989; Bottom and Jones 1990; Lawson 1993). Numbers of smolts released
from hatcheries should take ocean productivity into account; it may be prudent
to limit releases during periods of low ocean survival and growth (Francis
1997) or to time hatchery releases to avoid attracting large numbers of predators during smolt migration periods. Management actions affecting freshwater
parts of the salmon's life cycle should emphasize the linkages between habitat
and biological diversity, as a biologically diverse suite of salmon and steelhead
populations are likely to be buffered against fluctuating ocean conditions
(Francis 1993; Bisbal and McConnaha 1998; Hilborn et al. 2003).

Establish Salmonid Reserves

It is critical to protect remaining viable, naturally spawning salmon and steelhead populations and to restore habitats with the potential to re-establish core
populations at strategic locations within the basin. One way to accomplish this
would be to reconsider the concept of salmonid reserves. Reserves protect habitats that currently support remaining viable core populations (Figure 13.22),
and they could serve as foci for rebuilding salmonid abundance and metapopu-

Figure 13.22 Granite Creek is an important tributary in the John Day River system for spring Chinook spawning and rearing. The John Day system is significant, as it is the only major sub-basin in the Columbia and Snake River systems where artificial production of salmon and steelhead has not occurred. Of the seven healthy salmon and steelhead populations identified by Huntington et al. (1996) in the Columbia River Basin, five of them occurred in the John Day subbasin. Photo by R. N. Williams.

lation structure throughout the Columbia Basin. We believe priority should be given to considering the confluence of the Snake and Columbia Rivers, including the Hanford Reach, as a potential reserve area (Williams 2001).

The concept of salmon reserves has been discussed by salmon managers for over 100 years, including at least four recommendations for the inclusion of reserves in the Columbia Basin (Rahr et al. 1998). Moreover, promulgations by the Clinton administration regarding their "Roadless Area Conservation Rule" states that 237,000 sq. km of inventoried roadless areas (IRAs) would remain roadless and be protected from timber extraction and other activities under the rule (USDA Forest Service 2000). Loucks et al. (2003) concluded 77% of these IRAs have outstanding potential to conserve threatened, endangered, or imperiled species and that the conservation of such areas would have far-reaching effects. The U.S. Forest Service Roadless Area Final Environmental Impact Statement identified IRAs as having a direct or indirect influence on critical habitat for more than 30 threatened or endangered species, including most trout and salmon (USDA Forest Service

2000). In spite of its long history of discussion, no salmon reserves have ever been implemented in the basin. The Hanford Reach, a roughly 75 km long portion of the mid-Columbia River, is the only remaining large, undammed mainstem river segment, and it contains the largest natural spawning population of fall Chinook in the Columbia Basin above Bonneville Dam. Over the last two decades, naturally spawning Hanford Reach fall Chinook have continued to be productive while other stocks have declined. These fish exhibit characteristics of a core population both in their resiliency, being the only remaining mainstem Chinook salmon population of significance, and their contribution to spawning populations elsewhere in the basin. (Marked individuals have been recovered at other mid-Columbia and Snake river sites; see Figure 3.16.) The Hanford Chinook stock likely has remained productive because of the lack of dams in this river section and the maintenance of necessary ecological processes and functions through the re-regulation of flows during the Chinook spawning season.

The Hanford Reach also provides an excellent case study of how using normative river principles (in this case, flow management and protection of riparian areas) can maintain a healthy salmon population in spite of other human development. Success to date has been achieved as a result of restricting operations of the hydropower system to stabilize flows in the Reach during spawning, incubation, and emergence of fall Chinook. Additionally, land surrounding the Reach is isolated from human activities, as it lies within the Federal Hanford Nuclear Reservation where management actions, other than those associated with development of atomic resources, have not been permitted to date. The Hanford Reach is now part of the Hanford Reach National Monument, which has a series of management units based on their "ecological values," which includes spawning areas for salmon. The U.S. Fish and Wildlife Service is currently going through a Draft Comprehensive Conservation Plan/Environmental Impact Statement (*http://hanfordreach.fws.gov/fac.html*) for the Monument which will, in theory, result in a management plan.

The viability of naturally spawning Chinook in the Hanford Reach provides a template that may be replicated elsewhere in the Columbia Basin. Therefore, adequate protection of both the habitat function of the Reach and the fall Chinook that spawn there are of the highest priority (Geist 1995; Whidden 1996; Williams 2001). While continuing to protect the natural features of the Hanford Reach area, the region should search for other candidate areas in the Columbia and Snake Rivers where spawning and rearing habitat can be protected or restored, and natural population and metapopulation structure re-established. A likely candidate is the approximately 50-mile stretch of undammed Snake River between Hells Canyon Dam and the upper reaches of the reservoir behind Lower Granite Dam, where rapid daily fluctuations in flow are still permitted. The John Day River, which currently supports healthy Chinook and steelhead populations, is another candidate area for special management attention (Figure 13.22).

Return to the River: The Challenge Ahead

"There is no course of action for society to select that will reverse the apparent decline of wild salmon that is not socially disruptive and economically expensive."
—*Robert T. Lackey. 1999. Salmon policy: Science, society, restoration, and reality. Renewable Resources Journal 17(2):6–16.*

Returning the river to a more natural state runs counter to the management philosophy that has guided salmon restoration in the Columbia River Basin for much of the 20th century. For this reason, restoration or improvement of ecological conditions will require an examination of the values that underlie Columbia River management (Miller 1997; Wood 1998). However, the conceptual foundation outlined above (and in Chapters 3–6) provides a scientific basis for that debate. In the recent past, failure of the scientific community to resolve key restoration issues was often used to justify maintaining the status quo and avoid the necessary public debate over the social and economic costs of salmon recovery. However, expecting scientists to agree on each of the key questions is an unrealistic assumption. The healthy exercise of scientific debate should not be used as an excuse to hold progress hostage to the unattainable goal of perfect scientific consensus.

Maintaining the historical fisheries management approach is unlikely to result in significant improvement in the status of Pacific salmon in the Columbia River. It is more likely to result in further declines, extirpations, and extinctions. If the region is genuine in its desire to restore Pacific salmon in the Columbia Basin, continuing the status quo is not an option (ISG 1998, 1999). While a more natural river can be made somewhat compatible with other uses of the river, it cannot be achieved without significant changes in the way the river is managed. The 2004 Biological Opinion (BiOp) for the Federal Columbia River Power System Operations did not require the U.S. Army Corps of Engineers or the Bureau of Reclamation to significantly change the current hydroelectric operations. It called on river operators to make relatively minor, albeit expensive, modifications that left the currently altered flow regime in place (see Chapter 12). In May 2005, Judge Redden of the US District Court in Portland, Oregon invalidated the 20004 BiOp on the grounds that it was legally flawed. The opinion was the latest in a series of decisions issued by judges from the Portland District Court since 1994, addressing the 1993, 1995, 2000, and 2004 Biological Opinions issued by the National Marine Fisheries Service (now NOAA Fisheries). The Biological Opinions have attempted to balance the economic, commercial, and recreational interests in the Pacific Northwest, served by the ongoing operations

ORNL-DWG 96M-1217

Return to the Normative[1] River Depends on Location and Prior Development

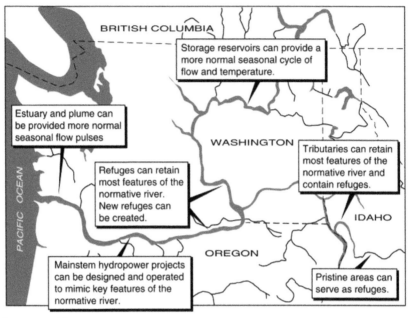

[1] the norm of Standard for Measure

Figure 13.23 Opportunities to preserve, restore, or enhance ecological conditions favorable to salmon and steelhead exist in the Columbia River Basin, although individual opportunities may be constrained by the impacts of prior development or by the social and cultural value derived from the development.

of the Columbia and Snake River hydroelectric system, with the conservation of salmon species listed under the Endangered Species Act.

In invalidating the 2004 Biological Opinion, Judge Redden described the BiOp as having a jeopardy analysis that ignored the reality of past, present, and future effects of federal actions on listed species. Consequently, he judged that NOAA Fisheries' interpretation of jeopardy and their proposed mitigation measures conflicted with the structure, purpose, and policy behind the Endangered Species Act. In June 2005, Judge Redden further offered injunctive relief against NOAA Fisheries and the 2004 BiOp by requiring summer spill from some of the Columbia and Snake River hydrosystem projects to aid outmigrating juvenile salmonids. The fact that (for now) salmon recovery efforts are being directed by the Portland Oregon US District Court, rather than by state, federal, and Tribal fisheries managers, underscores the long-term failure of the fisheries management community, as well as our appointed and elected policy-makers, in the Pacific Northwest to deal adequately with the decline and endangerment of Pacific Northwest salmon and steelhead stocks.

The first step in developing a Scientifically-sound restoration program for salmon is to clearly articulate the conditions needed for salmon relative to the region's salmon recovery goals. The next step is to determine what changes in the federal hydropower system and other uses of the river are needed to achieve these conditions (Figure 13.23). The difficult job of debating cost and benefits of salmon restoration is the next step. Significant changes will, in many cases, require painful decisions, perhaps even congressionally mandated alteration of federal hydrosystem project operations. Other lesser changes might limit, but not eliminate, the region's ability to use the Columbia River as a navigation corridor and to supply some irrigation needs.

This volume and other recent reviews of the salmon problem (NRC 1996; Stouder et al. 1997) provide an improved scientific foundation for salmon recovery. Consequently, the biggest challenge facing the region is not the biological uncertainties associated with salmon recovery efforts, but whether the policy makers are willing to face the difficult task of significantly changing the status quo. Restoration of fish and wildlife in the Columbia River Basin will require difficult decisions and will continue to test whether the region's policy makers and elected officials can find the political will necessary to endorse and implement a scientifically sound salmon recovery program (Figure 13.24).

Figure 13.24 Rainbow over the lower Columbia River Gorge, symbolizing the hope that all in the region have that salmon and steelhead populations can be maintained in the Columbia and Snake Rivers while the cultural, societal, and economic benefits derived from the river and its salmon can also be maintained. Photo by Mike Pinney, *www.photobymike.com*.

Literature Cited

Battelle's Pacific Northwest Division and US Geological Survey. 2000. Assessment of the impacts of development and operation of the Columbia River hydroelectric system on mainstem riverine processes and salmon habitats (Draft Final Report). Bonneville Power Administration. Portland, Oregon.

Beechie, T. and S. Bolton. 1999. An approach to restoring salmonid habitat-forming processes in Pacific Northwest watersheds. *Fisheries* 4:6–15.

Bisbal, G. A., and W. E. McConnaha. 1998. Consideration of ocean conditions in the management of salmon. *Canadian Journal of Fisheries and Aquatic Sciences* 55:2178–2186.

Blumm, M. C., L. J. Lucas, D. B. Miller, D. J. Rohlf, and G. H. Spain. 1998. Saving Snake River water and salmon simultaneously: The biological, economic, and legal case for breaching the Lower Snake River dams, lowering John Day Reservoir, and restoring natural river flows. *Environmental Law* 28:101–153.

Bottom, D. L., and K. K. Jones. 1990. Species composition, distribution, and invertebrate prey of fish assemblages in the Columbia River estuary. *Progressive Oceanography* 25:243–270.

Bottom, D. L., C. A. Simenstad, J. Burke, A. M. Baptista, D. A. Jay, K. K. Jones, E. Casillas, and M. H. Schiewe. In press. Salmon at river's end: The role of the estuary in the decline and recovery of Columbia River salmon. NOAA Technical Memorandum.

Brannon E. L., M. S. Powell, T. P. Quinn, and A. Talbot. 2004. Population structure of Columbia Basin Chinook salmon and steelhead trout. *Reviews in Fisheries Science* 12:99–232.

Busby P. J., T. C. Wainwright, G. J .Bryant, L. J. Lierheimer, R. S. Waples, F. W. Waknitz and I. V. Lagomarsino. 1996. Status review of west coast steelhead from Washington, Idaho, Oregon and California. NOAA Technical Memorandum NMFS-NWFSC-27.

Columbia Fish and Wildlife Authority (CBFWA). 2004. FY 2001-2003 Provincial Review Implementation Report. Columbia Fish and Wildlife Authority. Portland, Oregon. 299 p.

Chapman, D. W. 1986. Salmon and steelhead abundance in the Columbia River in the nineteenth century. *Transactions of the American Fisheries Society* 115:662–670.

Columbia-Blue Mountain Resource Conservation and Development Area. 2004, May 28. John Day Draft Subbasin Plan. Prepared for the Northwest Power and Conservation Council.

Cury, P., and C. Roy. 1989. Optimal environmental window and pelagic fish recruitment success in upwelling areas. *Canadian Journal of Fisheries and Aquatic Sciences* 46:670–680.

Drawdown Regional Economic Workgroup. 1999. Lower Snake River juvenile salmon migration feasibility study, draft economic analysis project study plan (PSP). US Army Corps of Engineers. Walla Walla, Washington.

Dauble, D. D., T. P. Hanrahan, D. R. Geist, and M. J. Parsley. 2003. Impacts of the Columbia River hydroelectric system on main-stem habitats of fall Chinook salmon. *North American Journal of Fisheries Management* 23:641–659.

Ebersole, J. L. W. J. Liss, and C. F. Frissell. 1997. Restoration of stream habitats in the western United States: Restoration as reexpression of habitat capacity. *Environmental Management* 21:1–14.

Francis, R. C. 1993. Climate change and salmon production in the North Pacific Ocean. Pages 33–43 in K. T. Redmond and V. J. Tharp, eds. *Ninth Annual Pacific Climate (PACLIM) Workshop, April 21-24, 1992*. California Department of Water Resources.

Francis, R. C. 1997. Managing resources with incomplete information: Making the best of a bad situation. Pages 513–524 in D. J. Stouder, P. A. Bisson, and R. J. Naiman, eds. *Pacific salmon and their ecosystems*. Chapman and Hall, New York.

Frankham, R., J. D. Ballou, and D. A. Briscoe. 2002. *Introduction to conservation genetics*. Cambridge University Press, Cambridge, United Kingdom. 617 p.

Garcia, A. P., W. P. Connor, and R. H. Taylor. 1994. Fall Chinook salmon spawning ground surveys in the Snake River. Annual Report for 1993. Pages 1–21 in D. W. Rondorf and K. F.

Tiffan, eds. *Identification of the spawning, rearing, and migratory requirements of fall Chinook salmon in the Columbia River Basin*. Bonneville Power Administration. Portland, Oregon.

Geist, D. R. 1995. The Hanford Reach: What do we stand to lose? *Illahee* 11:130–141.

Geist, D. R., and D. D. Dauble. 1998. Redd site selection and spawning habitat use by fall Chinook salmon: The importance of geomorphic features in large rivers. *Environmental Management* 22:655–669.

General Accounting Office. 1992. Endangered Species: Past actions taken to assist Columbia River salmon. General Accounting Office. Briefing Report to Congressional Requesters, GAO/RCED-92-173BR. Washington, DC.

General Accounting Office. 2002. Columbia River Basin salmon and steelhead: Federal agencies' recovery responsibilities, expenditures and actions. General Accounting Office. Report to US Senate, GAO 02-612. Washington, DC.

Gilpin, M. E. 1996. Spatial structure and population vulnerability. Pages 124–158 in M. E. Soule, ed. *Viable populations for conservation*. Cambridge University Press, Cambridge, United Kingdom.

Gregory, S. V., L. R. Ashkenas, and P. Minear. 2001. Application of analysis of historical channel change in the restoration of large rivers. *Verh. Internat. Verein. Limnol* 27:4077–4086.

Gustafson, R. G., T. C. Wainwright, G. A. Winans, F. W. Waknitz, L. T. Parker, and R. S. Waples. 1997. Status Review of sockeye salmon from Washington and Oregon. NOAA Technical Memorandum NMFS-NWFSC-33.

Hauer, F. R., C. N. Dahm, G. A. Lamberti, and J. A. Stanford. 2003. Landscapes and ecological variability of rivers in North America: Factors affecting restoration strategies. Pages 81–105 in P. A. Bisson, ed. *Strategies for restoring river ecosystems: Sources of variability and uncertainty in natural and managed systems*. American Fisheries Society, Bethesda, Maryland.

Healey, M. C. 1994. Variation in the life history characteristics of Chinook salmon and its relevance to conservation of the Sacramento winter run of Chinook salmon. *Conservation Biology* 8:876–877.

Healey, M. C., and A. Prince. 1995. Scales of variation in life history tactics of Pacific salmon and the conservation of phenotype and genotype. *Evolution and the aquatic ecosystem: Defining unique units in population conservation* 17:176–184.

Hilborn, R., T. P. Quinn, D. E. Schindler, and D. E. Rogers. 2003. Biocomplexity and fisheries sustainability. *Proceedings of the National Academy of Sciences* 100(11):6564–6568.

Hill, M. T., and W. S. Platts. 1998. Ecosystem restoration: A case study in the Owens River Gorge, California. *Fisheries* 23:18–27.

Huntington, C., W. Nehlsen, and J. Bowers. 1996. A survey of healthy native stocks of anadromous salmonids in the Pacific Northwest and California. *Fisheries* 21:6–14.

Independent Economic Advisory Board (IEAB). 1999. River Economics: Evaluating Trade-offs in Columbia River Basin Fish and Wildlife Programs and Policies. Northwest Power Planning Council. IEAB 99-1. Portland, Oregon.

Independent Scientific Advisory Board (ISAB). 1998. Recommendation for stable flows in the Hanford Reach during the time when juvenile fall Chinook are present each spring. Northwest Power Planning Council and the National Marine Fisheries Service. ISAB Report, 98-5. Portland, Oregon.

ISAB. 2003. Review of salmon and steelhead supplementation. Northwest Power Planning Council. Report ISAB 2003-03. Portland, Oregon. 229 p.

Independent Scientific Group (ISG). 1993. Critical uncertainties in the Columbia River Basin Fish and Wildlife Program. Bonneville Power Administration. Report to Policy Review Group, SRG 93-3. Portland, Oregon. 17 p.

ISG. 1998. Return to the river: An ecological vision for the recovery of the Columbia River salmon. *Environmental Law* 28:503–518.

ISG. 1999. Scientific issues in the restoration of salmonid fishes in the Columbia River. *Fisheries* 24:10–19.

Independent Scientific Review Panel (ISRP). 1997. Review of the Columbia River Basin Fish and Wildlife Program as directed by the 1996 amendment to the Power Act. Northwest Power Planning Council. Annual Report, ISRP 97-1. Portland, Oregon.

ISRP. 1998. Review of the Columbia River Basin Fish and Wildlife Program as directed by the 1996 amendment to the Power Act. Northwest Power Planning Council. Annual Report, ISRP 98-1. Portland, Oregon.

ISRP. 1999. Review of the Columbia River Basin Fish and Wildlife Program as directed by the 1996 amendment to the Power Act. Northwest Power Planning Council. Annual Report, ISRP 98-2. Portland, Oregon.

Johnson, O. W., W. S. Grant, R. G. Kope, K. Neely, F. W. Waknitz, and R. S. Waples. 1997. Status review of chum salmon from Washington, Oregon, and California. NOAA Technical Memorandum, NMFS-NWFCS-32.

Lackey, R. T. 1999. Salmon policy: Science, society, restoration, and reality. *Renewable Resources Journal* 17:6–16.

Lawson, P. W. 1993. Cycles in ocean productivity, trends in habitat quality, and restoration of salmon runs in Oregon. *Fisheries* 18:6–10.

Lichatowich, J. 1999. *Salmon without rivers: A history of the Pacific salmon crisis.* Island Press, Washington, D.C.

Loucks, C., N. Brown, A. Loucks, and K. Cesareo. 2003. USDA Forest Service roadless areas: Potential biodiversity conservation reserves. *Conservation Ecology* 7(2):5. Available at *http://www.consecol.org/vol7/iss2/art5*

McElhany, P., M. H. Ruckelshaus, M. J. Ford, T. C. Wainwright, and E. P. Bjorkstedt. 2000. Viable salmonid populations and the recovery of evolutionarily significant units. US Department of Commerce. NOAA Tech. Memo. NMFS-NWFSC-42. 156 p.

Meffe, G. K. 1992. Techno-arrogance and halfway technologies: Salmon hatcheries on the Pacific coast of North America. *Conservation Biology* 6:350–354.

Miller, D. B. 1997. Of dams and salmon in the Columbia/Snake Basin: Did you ever have to make up your mind? *Rivers: Studies in the Science, Environmental Policy, and Law of Instream Flow* 6:69–79.

Mullan, J. W., A. Rockhold, and C. R. Chrisman. 1992. Life histories and precocity of Chinook salmon in the Mid-Columbia River. *Progressive Fish-Culturist* 54:25–28.

Myers, J. M., R. G. Kope, G. J. Bryant, D. Teel, L. J. Lierheimar, T. C. Wainwright, W. S. Grant, F. W. Walknitz, K. Neely, S. T. Lindley, and R. S. Waples. 1998. Status review of Chinook salmon from Washington, Idaho, Oregon, and California. NOAA Technical Memorandum, NMFS-NWFSC-35.

Myers, R. A., S. A. Levin, R. Lande, F. C. James, W. W. Murdoch, and R. T. Paine. 2005. Hatcheries and endangered salmon. Science 303:1980.

Naiman, R. J., and R. E. Bilby, eds. 1998. *River ecology and management: Lessons from the Pacific Coastal Ecoregion.* Springer-Verlag, New York.

National Marine Fisheries Service (NMFS). 1995. Draft Biological Opinion on reinitiation of consultation on 1994-1998 operation of the federal Columbia River Power System and juvenile transportation program in 1994-1998. US Army Corps of Engineers, Bonneville Power Administration, Bureau of Reclamation, and National Marine Fisheries Service. Portland, Oregon. 146 p.

NMFS. 2000. Final FCRPS (Federal Columbia River Power System) Biological Opinion. Northwest Fisheries Science Center, NOAA. Endangered Species Act–Section 7 Consultation. Seattle, Washington.

National Oceanic and Atmospheric Administration (NOAA) Fisheries, Northwest Fisheries Science Center. 2004. Final FCRPS (Federal Columbia River Power System) Biological

Opinion. Northwest Fisheries Science Center, NOAA. Endangered Species Act–Section 7 Consultation. Seattle, Washington.

National Research Council (NRC). 1996. Upstream: Salmon and society in the Pacific Northwest. Report on the Committee on Protection and Management of Pacific Northwest Anadromous Salmonids for the National Research Council of the National Academy of Sciences. National Academy Press, Washington DC.

Nehlsen, W., J. E. Williams, and J. A. Lichatowich. 1991. Pacific salmon at the crossroads: Stocks at risk from California, Oregon, Idaho, and Washington. *Fisheries* 16:4–21.

Northwest Power Planning Council (NPPC). 1986. Council Staff Compilation of Information on Salmon and Steelhead Losses in the Columbia River Basin. Northwest Power Planning Council. Portland, Oregon.

NPPC. 1997. An integrated framework for fish and wildlife management in the Columbia River. Northwest Power Planning Council. NPPC 97-2. Portland, Oregon.

NPPC. 1998. A proposed scientific foundation for the restoration of fish and wildlife in the Columbia River. Northwest Power Planning Council. NPPC 98-16. Portland, Oregon.

NPPC. 2000. Columbia River Fish and Wildlife Program. Northwest Power Planning Council. Portland, Oregon.

NPPC. 2004. Columbia River Fish and Wildlife Program. Northwest Power Planning Council. Portland, Oregon.

Petersen, K. C. 1995. *River of life, channel of death: Fish and dams of the lower Snake*. Confluence Press, Lewiston, Idaho.

Poff, N. L., J. D. Allan, M. B. Bain, J. R. Karr, K. L. Prestegaard, B. D. Richter, R. E. Sparks, and J. C. Stromber. 1997. The natural flow regime. *Bioscience* 47:769–784.

Rahr, G. R., J. A. Lichatowich, R. Hubley, and S. M. Whidden. 1998. Sanctuaries for native salmon: a conservation strategy for the 21st century. *Fisheries* 23:6–8.

Reeves, G. H., L. E. Benda, K. M. Burnett, P. A. Bisson, and J. R. Sedell. 1995. A disturbance-based approach to maintaining and restoring freshwater habitats of evolutionarily significant units of anadromous salmon in the Pacific Northwest. Pages 334–339 in J. L. Nielsen, ed. *Evolution and the aquatic ecosystem*. American Fisheries Society Symposium 17, Bethesda, Maryland.

Scientific Review Group (SRG): Brannon, E. L., K. P. Currens, D. Goodman, J. A. Lichatowich, W. E. McConnaha, B. E. Riddell, and R. N. Williams. 1999. Review of salmonid artificial production in the Columbia River basin as a scientific basis for Columbia River production programs. Northwest Power Planning Council. Portland, Oregon. 77 p.

Snake River Salmon Recovery Team. 1993. Snake River salmon recovery plan recommendations (Draft). NMFS/NOAA. Portland, Oregon. 364 p.

Stacey, P. B., and M. L. Taper. 1997. Migration within metapopulations: The impact upon local population dynamics. Pages 267–291 in I. A. Hanski and M. E. Gilpin, eds. *Metapopulation biology: Ecology, genetics, and evolution*. Academic Press, San Diego, California.

Stanford, J. A., and F. R. Hauer. 1992. Mitigating the impacts of stream and lake regulation in the Flathead River catchment, Montana, USA: An ecosystem perspective. *Aquatic Conservation: Marine and Freshwater Ecosystems* 2:35–63.

Stanford, J. A., F. R. Hauer, S. V. Gregory, and E. B. Snyder. 2005a. The Columbia River, pp. 591–653. In: Benke, A. C. and C. E. Cushing, eds. *Rivers of North America*. Academic Press/Elsevier, Boston, Massachusetts. 1168 p.

Stanford, J. A., M. S. Lorang, and F. R. Hauer. 2005b. The shifting habitat mosaic of river ecosystems. *Verh. Internat. Verein. Limnol.* 29(1):123–136.

Stanford, J. A., J. V. Ward, W. J. Liss, C. A. Frissell, R. N. Williams, J. A. Lichatowich, and C. C. Coutant. 1996. A general protocol for restoration of regulated rivers. *Regulated Rivers* 12:391–413.

Stouder, D. J., P. A. Bisson, and R. J. Naiman, eds. 1997. *Pacific salmon and their ecosystems: Status and future options.* Chapman and Hall, New York.

USDA Forest Service. 2000. Forest Service, roadless area conservation, final environmental impact statement. USDA Forest Service, Washington, DC.

Washington Department of Fish and Wildlife and Oregon Department of Fish and Wildlife. 1996. Status report. Columbia River fish runs and fisheries, 1938-1998. Washington Department of Fish and Wildlife and Oregon Department of Fish and Wildlife. Olympia, Washington, and Clackamas, Oregon.

Washington Department of Fish and Wildlife and Oregon Department of Fish and Wildlife. 2001. Status report. Columbia River fish runs and fisheries, 1938-2000. Washington Department of Fish and Wildlife and Oregon Department of Fish and Wildlife. Olympia, Washington, and Clackamas, Oregon.

Whidden, S. M. 1996. The Hanford Reach: Protecting the Columbia's last safe haven for salmon. *Environmental Law* 26:265–297.

White, R. 1995. *The organic machine: The remaking of the Columbia River.* Hill and Wang, New York.

Williams, R. N. 2001. Refugia-based conservation strategies: Providing safe havens in managed river systems. Pages 59–63 in Oregon Trout, ed. *Oregon salmon: Essays on the state of the fish at the turn of the millennium.* Oregon Trout, Portland, Oregon.

Wood, M. C. 1998. Reclaiming the natural rivers: The Endangered Species Act as applied to endangered river ecosystems. *Arizona Law Review* 40:197–286.

Subject Index

Page numbers followed by f indicate figures; page numbers followed by t indicate tables

A

Acipenser medirostris (green sturgeon), 155–157
Acipenser transmontanus (white sturgeon), 155–157
Adaptation
  and geographic variations, 533
  local, 104–105
  opportunities for, 186
  and run timing, 116, 118
Adaptive management, 443
  introduction of, 47
  limited success of, 47
  monitoring, 592
  risk assessment in, 452
Adipose fin clips, 614, 615f
Advective processes, 538
Agriculture, 9, 485
  and estuary development, 519f
  and harvest management, 466
  and management of natural resources, 33
  and riverine ecosystems, 208–210, 209f
Alaska
  Bristol Bay, 70–73, 499
  and sockeye salmon, 139
  stream-type life histories in, 67
Alaska Board of Fisheries, Sustainable Salmon Fishery Policy of, 483
Alaska Current, 546
Alaskan gyre, 543
Aleutian Low, 541, 542, 543, 554
  and California Current, 546
  and climatic shift, 545
  and El Niño events, 548
  intensification and relaxation of, 547
Aleutian Low Pressure Index (ALPI), 544–545
All-H Paper, 602, 603

Alluvial flood plains, 177, 178f. *See also* Flood plains
  beaver activity in, 201
  occurrence of, 184
Alluvial river ecosystems, 62f, 194f
Alsea River, 290
Alsea Valley Alliance v. Evans, 625
American Fisheries Society, Division of Fish Culture of, 426
Analytical models, traditional, 551. *See also* Models
Anchovy *(Engraulis mordax)*
  in Columbia River plume, 530
  and sea surface temperatures, 547
  and upwelling events, 532–533
Aquaculture management programs, 444
Army Corps of Engineers, U.S., 17, 401
  gas levels monitored by, 375
  and ice and trash sluiceways, 366
  "Operation Fish Run" of, 636f
  and spill levels, 361
  spillway modification by, 375
  and surface collection systems, 367
  and turbine intake screens, 347
Artificial migration corridor, 10
Artificial production, 556, 612 *See also* Hatcheries
  and human population growth, 455
  intention of, 53
  investment in, 434
  and ocean environment, 508
  reliance on, 52
Artificial Production Review and Evaluation (APRE), 443–444, 452–453
Artificial propagation, 33, 424 *See also* Hatcheries, 451
  adaptive management framework for, 655–656
  British Columbia study of, 431–433
  of Chinook salmon, 132

Artificial propagation (*Cont.*)
  of coho salmon, 134, 136
  in Columbia River Basin, 528
  comprehensive evaluation of, 452–453
  density dependent limits to, 423
  experimental nature of, 453
  history of, 420–429
  influence on management of, 434–439
  introduced to Columbia Basin, 418
  recommendations for, 452–454
  results of, 443
  of salmon in hatcheries, 42
  success of, 454–455
  technological innovation in, 418, 419f
  as threat, 431, 437
  uncertainties in, 452
Artificial selection, in hatchery stocks, 450
Asian bivalve (*Corbicula fluminea*),
    524–525
Asotin subbasin, habitat restoration in, 23f
Astoria, WA, 512, 512f
Asynchrony, beneficial effects of, 91
Atlantic salmon (*Salmo salar*)
  of hatchery-origin vs. wild, 445
  and hatchery programs, 431
Atmospheric system, 508–509. *See also*
    Climate
  adaptations to changes in, 555
  global, 554
Auger Falls, and Chinook salmon spawning,
    126

B

Baird, Spencer, 420
Baker Lake, bypass methods tested at, 381
Balloon tags, mortality studies with,
    344–345
Barge transportation, 605
  conditions for, 655
  for juveniles, 383–386, 383f
  loading juvenile salmon for, 635f, 636f
  as technological solution, 630, 635f, 636f
Basinwide Recovery Strategy, 602, 603, 604,
    614
  climate in, 619
  and hatcheries, 605
  mainstem-related actions in, 617
  monitoring and evaluation in, 616
  tributary habitats in, 605, 612
Bathymetry, estuarine, 520
Bear Valley Creek, 487f

Beaver Army Terminal, discharge statistics
    for, 175t
Beaver trapping, and riverine ecosystem,
    200–201
Bed load, 177
Before-after-control-impact (BACI) designs,
    587
Belloni, Robert, 19
Benthic consumers, in food web, 524
Bering Sea, sockeye salmon in, 139
"Best management practices" (BMPs)
  of Fish and Wildlife Program, 17
  uncertainty about, 235
Biodiversity, 101. *See also* Diversity
  management for, 652
  and river regulation, 231
  of stocks, 107
Biological Opinion (BiOp), 2000, 335, 579,
    602, 624
  action items specified by, 604f
  and ESA requirements, 626
  on flow augmentation, 395
  and gas saturation levels, 400
  and hatcheries, 605
  on hydrosystems effects, 645
  mainstem-related actions in, 617
  monitoring and evaluation in, 616
  obsolete salmon biology in, 610–611
  passage goals of, 400–401
  performance standards of, 342
  recovery efforts in, 395
  and spill levels, 361–362
  "spread the risk" policy of, 386–387
  tributary habitats in, 605, 612
  turbine intake screens in, 367
Biological Opinion (BiOp), 2004
  and dam removal, 645
  earlier version of, 645
  limitations of, 659
Biological Opinion (BiOp), 2005, 657
Biological rationale, for NPCC measures, 39
Body size. *See also* Size
  and life history, 69
  and probability of predation, 72
Boldt, George, 19
Bonneville Dam, 10f, 57f, 214, 259, 272, 332,
    473, 526, 646, 658
  adult fish ladder at, 341f
  Bradford Island fish ladder at, 341
  bypass facilities at, 286, 292, 343
  construction of, 344

electricity-generating turbines in, 329f
FGE at, 351t
flow objectives for, 394t
juvenile bypass at, 355f
low flows from, 650
monitoring at, 584
navigational locks at, 342f
and passage goals, 400
salmon counts at, 475
sluiceways at, 365–366
spill at, 359, 362t, 364
statistics for, 179t
and survival estimates, 391
tailrace losses in, 376
traveling screens at, 347–348
turbine intake screens at, 350, 355
Bonneville Hatchery, transfers of salmon
    from, 424–425, 424f
Bonneville Power Administration (BPA), 4,
    37, 288
funds collected by, 15
mandate for, 15–16
power lines controlled by, 333
scientific reviews conducted by, 199
BPA. See Bonneville Power Administration
Breeding groups, in hatchery programs, 431
Breeding system continuum, 447, 447f
Bristol Bay, Alaska
biocomplexity of, 73
fishery, 499
salmon run, 496
sockeye salmon in, 70–73
British Columbia, 67
and sockeye salmon, 139, 140
sockeye smolt migration in, 289
British Columbia Fisheries Commission,
    artificial propagation study of,
    431–433
Broken stick model, for smolt travel times,
    296–298, 299, 300f, 313
Brook trout, proliferation of non-native,
    220–221, 221f, 222
Brownlee Dam, 179t, 257, 262
Brownlee Reservoir, 272
Brown trout (S. trutta), of hatchery-origin
    vs. wild, 445
Bubbles. See also Gas
biological monitoring of, 372–373
measuring and qualifying, 373
Bull charr (Salvelinus confluentus), 152,
    153f

distribution of, 154
under ESA, 155
genetic structure of, 153–154
and habitat degradation, 143–144
propagation efforts for, 154
as species of special concern, 153
Bull trout (Salvelinus confluentes), 18. See
    also Bull charr
bounties placed on, 53
on ESA listing, 334
and food web changes, 234
genetic structure of, 145
and habitat degradation, 143–144
hydrosystems affecting, 603
Bureau of Fisheries, surveys conducted by,
    574
Bureau of Reclamation, System Operation
    Review of, 218
Bursts turbulent, 282f
Burst swimming, 186
Bypass provisions, 179
and gas supersaturation, 371–373, 372f
ice and trash sluiceways, 365–367, 366f
proposed, 380–382, 382f
for smolt, 359, 361–362
Bypass systems. See also Spillways
at Bonneville Dam, 355f
criteria for successful, 355
improvements to, 401
at Rocky Reach Dam, 368, 369f
surface, 357–358, 370–371
turbine intake, 345
Wells Dam, 360f, 370–371

C

Calanoid copepods (Pseudodiaptomus
    inopinus), in Columbia River estuary, 525
California Cooperative Fisheries
    Investigations (CALCOFI), 531
California Current, 534
and Aleutian pressure system, 546
and coastal species, 536
interannual variations along, 537–538
periods of warming in, 543
plankton biomass in, 534–535
and salmon production, 547
as transition zone, 535, 535f
upwelling system of, 531–534
California Undercurrent, 540–541
Canadian fisheries, salmon harvests of, 544,
    545f

Cannery, Columbia River, 130
Canning industry, salmon, 8, 420
Canyon reaches, 177, 178f
Cape Blanco, 533
Cape Flattery, 482
Cape Mendocino, 533
Carson spring Chinook stock, 448
Caspian terns *(Sterna caspia)*, 377f
  in Columbia River estuary, 525, 526f
  relocation of, 525, 526f
  as salmon predators, 379–380
Catchable trout program, 426
Catchment basin, environmental diversity of,
  176
Cathlamet Bay
  late successional stages of, 523–524
  wetlands of, 517f
CBFWA, monitoring and evaluation of, 577
Celilo Falls, 33f
Central Pacific gyre, 536
Change monitoring, 581–584
Channel banks, logging and, 203
Channel catfish *(Ictalurus punctatus)*, 377
Channels, and tidal deltas, 517
Chief Joseph Dam, 10, 126, 250, 258, 338
  construction of, 326
  statistics for, 179t
Chinook fishery, history of, 469–471
Chinook salmon *(O. tshawytscha)*, 124f. *See
    also* Fall Chinook; Spring Chinook
  commercial salmon fishery landings of,
    134f
  dendrogram of, 117f
  distribution of, 123–124, 125–126, 127f,
    128f
  in early hatchery programs, 422, 423f
  effects of dams on, 256–257
  effects of transportation on, 386
  and El Niño, 541
  ESA listing for, 18, 132, 132t, 334, 428,
    635
  in estuarine environment, 513
  evolutionary history of, 124
  extinctions of, 109
  fall spawning of, 76, 77
  feeding habits of, 223
  genetic structure of, 114–118, 115f, 116f,
    117f, 119f, 124–125
  geographic organization of, 79f, 80f,
    110–111, 129
  and habitat degradation, 78, 106
  at Hanford Reach, 83, 84f, 85, 642

  harvest summary of, 130–131, 131f
  life history diversity in, 129–130
  in the lower Snake River, 331f
  marine survival of, 509, 510f
  marsh-rearing, 515
  migration patterns of, 69, 80
  ocean distributions of, 481
  and ocean fisheries, 481, 482
  ocean-type, 80, 260
  populations structure of, 118, 119f
  propagation efforts for, 132
  rearing of, 259, 260
  run size for, 5, 7
  in Snake River, 292–293, 313
  spawning areas of, 82f, 126, 127f, 128f,
    577f
  spawning patterns for, 389
  stream-type, 260
  synchronous decline of, 110
  temperature requirements of, 192–193,
    218
  in Umatilla River, 429–430
Chinook subyearlings
  distribution of, 268
  in mid-Columbia Reach, 293
  migration times for, 268–269, 269f
  shoreline habitat for, 269–270
Chinook yearlings, 279–280
  in estuary, 287
  and flow structure, 281–285, 282f–284f
  in mid-Columbia Reach, 293
  in reservoirs, 285–287
  travel time of, 298f
  and water transit time, 296–297, 299f
Chironomids, 223
Chiwawa Hatchery, 46f, 632f
Chlorophyll biomass, 532
Chum salmon *(O. keta)*, 137f
  commercial landings of, 134f
  distributions of, 137
  ESA listing for, 138, 635
  in estuarine environment, 513
  genetic structure of, 113, 136–137
  harvest summary of, 137
  in hatchery programs, 422
  migration of, 184
  propagation efforts for, 138
  rearing of, 259
  regional differentiation in, 114
  run size for, 7
  size of, 69
Clackamas hatchery, 31

Clackamas River, 12, 134
Clark, William, 1–2, 19, 32, 190
Clark Fork River, westslope cutthroat trout in, 151
Clean Water Act, 210, 402
Clearwater National Forest, 220
Clearwater River, 214, 215f, 233, 257
  drainage, 490f
  Dworshak Dam on, 339f
  instream mining on, 207f
  sockeye salmon in, 118
  westslope cutthroat trout in, 151, 152
Clearwater River Basin
  discharge statistics for, 175t
  roadless catchments of, 199
Climate. *See also* California Current; Temperatures
  effect on fish populations of, 543, 549
  impact of logging on, 204
  and migratory behavior, 493
  and ocean environment, 508
  in recovery plans, 616–617, 619
  in recovery programs, 608, 609f
  and salmon harvest, 473
Climate change, 622
Clinton administration, 657
Coast, upwelling system of, 531–534. *See also* Upwelling system
Cobb, John, 437
Coded Wire Tag (CWT) program, 607–608
Coeur d'Alene River, westslope cutthroat trout in, 151
Coho salmon *(O. kisutch)*, 133f
  apparent recovery of, 134f, 135f, 136
  coastal stocks, 436
  distribution of, 132–134
  and El Niño, 541
  ESA listing for, 136, 635
  in estuarine environment, 513
  evolutionary history of, 133
  extinction of, 18, 133, 134
  genetic structure of, 133
  harvest of, 134–136, 135f, 435–436, 435f
  hatchery-origin vs. wild, 445
  in hatchery programs, 422, 441
  in hatchery study, 433
  marine survival of, 509, 510f
  migration of, 288–289
  Oregon coastal, 87
  propagatiion efforts for, 136
  rearing of, 259
  run size for, 7

sorting of, 632f
spawning areas for, 133
temperatures preferred by, 530
wild, 436
Coldwater releases, 397
Collection systems
  at McNary Dam, 383
  removal of debris from, 400
  surface, 368, 368f, 369f
Colorado River, 253
Columbia Plateau, 177
Columbia River
  characteristics of, 5–7
  development of, 52, 630, 631f
  discharge statistics for, 175t
  early history of, 630
  ESA listing for, 18
  free-flowing riverine reaches of, 11
  as integrated system, 16
  location of, 6f
  mainstems of, 176f
  physiography of, 174–180
  predevelopment, 177
  riverine *versus* lacustrine nature of, 646
  salmon in, 5–15
  salmon management in, 15–16
Columbia River Basin, 53, 54f
  boundaries of, 6f
  dams on, 327, 328
  decline of salmon in, 8–9
  environmental diversity of, 176
  human development of, 78
  impact of human activity on, 198–199
  and institutional arrangements, 610
  in late 19th century, 3
  map of, 9f
  native culture in, 2
  salmonid habitat in, 328f
  snow extent for, 609f
Columbia River Basin Fish and Wildlife Program, 638
Columbia River Basin hydrosystem. *See also* Hydrosystem
  Bonneville Dam, 10f
  dams, 10
  and Fish and Wildlife Program, 17
  fish ladders in, 12, 13f
  and salmonid ecosystem, 42–44, 43f
Columbia River Compact, 467, 468, 476
  harvest rate estimates of, 494
  management program of, 481

Columbia River Coordinated Information
    System, 576
Columbia River Fisheries Management Plan
    (CRFMP), 476
Columbia River Gorge, 661
Columbia River Inter-Tribal Fish
    Commission, 140
Columbia River management
    failure of, 659–660
    values of, 659
Columbia-Snake River Basin, history of, 313
Commercial fishery, in Columbia River, 8.
    See also Landings, commercial
Conceptual foundation
    defined, 29, 30–32
    derivation of, 32
    for fish stocks, 104
    importance of, 30–31
    for salmon management, 32–35
    of salmon recovery, 36–47
Conceptual framework, of Fish and Wildlife
    Program, 38–47
Congress, U.S.
    and mitigation measures, 333
    NPPC set up by, 36, 37
Congressional delegation, Northwest
    Regional, 645
Connectivity
    importance of, 90
    and life history patterns, 66
Conservation. See also Fish conservation
    and fluctuations in marine climate, 549
    and institutional neglect, 466–467
    internationally recognized principles of,
        483
    and marine environment, 65
    and societal objectives, 466
    of stocks, 106–108
    for sustainable use, 486
Cool anchovy regime, 547
Core populations
    concept of, 75
    of fall chinook in Hanford Reach, 85,
        86f
    and metapopulations, 91
    protection of, 91
Corophium spp., 228–229
    in food chain, 227
    as salmon prey, 515
Corps of Discovery, 1–2
Counting

    automated data collection equipment for,
        600f
    permanent trap for, 583f
    weir for, 589f
Cowlitz Basin, discharge statistics for, 175t
Cowlitz River, 251
Creston National Fish Hatchery, 154
Critical habitats, 183–184. See also Habitat
    loss
Cropland agriculture, and riverine
    ecosystems, 208–210, 209f. See also
    Agriculture
Cultus Lake sockeye salmon hatchery, 431
Cutthroat trout, 148t
    coastal, 149, 150–151, 150f
    distribution of, 143, 144t, 148, 150–151
    under ESA, 151
    evolutionary divergence in, 148–149
    and food web changes, 234
    genetic structure of, 145, 148–149
    and habitat degradation, 143
    propagation efforts for, 150
    westslope, 149, 151–152, 152f

D

The Dalles Dam, 3f, 272, 286, 292
    bypass provisions of, 370
    effectiveness of sluiceway at, 367
    FGE at, 351t, 352
    monitoring at, 584
    Native American petroglyph near, 332f
    naturalized stream flow at, 609f
    sluiceways at, 365
    spill deflectors at, 374
    spill effectiveness at, 363, 364
    spill levels at, 361–362
    statistics for, 179t
    tailrace losses in, 376
    turbine intake screens at, 350, 355
    turbine passage at, 292
Dalles Reservoir, sturgeon in, 156, 157
Dam forebays, smolts in, 309. See also
    Forebays
Dam obstacle hypothesis, of transit time,
    311–312
Dam passage counts, vs. spawning
    escapements, 475
Dams
    on Columbia River, 174, 175t, 176, 176t
    of Columbia River Basin, 251f, 327f, 328,
        428f, 631f

construction of, 9, 80
effects on anadromous fishes of, 253
effects on spawning of, 256–259
extensive regulation by, 177
federal, 330, 341
and FGE, 351t
and food chain changes, 527
high, 473, 522f
irrigation diversion, 634f
low-head, 230
mitigation measures for, 655
as obstacles for anadromous fish, 338
    in adult upstream passage, 338–342,
       340f–342f
    bypass provisions, 345–347
    in downstream passage of juveniles,
       342–343
    and juvenile management, 386–387
    mortalities associated with, 343–345,
       343f
    reservoir mortality, 375–380
    spill and surface bypass systems,
       357–375
    in transportation of juveniles,
       383–386, 383f, 384f
    at turbine intakes, 347–357
as obstacles to salmon migration,
    253–256
as regulators of flow
    effects on adult anadromous fishes,
       387–390
    effects on juveniles of, 390–392
    and mitigation efforts, 397,
       398t–399t
    and water temperatures, 396
and riverine ecosystems, 211–212
typical Columbia River, 252f
typical hydroelectric, 254–255, 255f
Dams, beaver, 200–201, 202f
DART, 307
Data archive, 594f
Databases, in recovery planning, 593, 594f
Data sets, in conceptual foundation, 31
Decision tree, 454
Deme, defined, 102
Demographic changes, human, 622. *See also*
    Population growth
Deposition, in estuarine environment, 518
Descaling
  in bypass systems, 353–354
  standards defining, 353

Deschutes River, 443, 448
  bull charr in, 154
  and juvenile migration, 69, 81t
Dexter Dam, 288
Diel patterns, 272, 312
  of Chinook salmon, 263f
  and fish distribution, 213–214
  weekly, 264f
Dipnetting, of Native American salmon,
    2f
Directional selection, 449, 449f
Dispersal
  metapopulation processes of, 108–109
  of salmonid populations, 108
Displacement, studies on, 312
Diurnal behavior
  in steelhead emigration, 290
  studies of, 312
Diversions, and riverine ecosystems,
    218–220, 219f. *See also* Irrigation
Diversity
  biological, 101, 107, 231, 652
  in evolutionary theory, 104
  in fishery management, 92
  habitat, 229–231
  importance of, 158
  of Pacific salmon populations, 489
  and run timing, 116, 118
  of salmonids, 101–102
  stock, 313
Domestication selection, in hatchery stocks,
    450
Drawdowns
  and feeding behavior, 279
  as mitigation tool, 381
Dredge mining, 206, 206f, 207f
Dredging program, 518
Drought, 624
  fisheries management during, 108
  and fish protection, 396
  impact on juveniles of, 260
Dry year, and flow strategies, 314
Dworshak Dam, 257, 338, 339f, 397, 522f
Dworshak Reservoir, 214, 215f
Dykes. *See also* Estuaries
  in Columbia River estuary, 518, 519f,
    520f
  and flood plain, 519, 521f
  and food chain changes, 527
  and shallow-water habitat, 521
  tideland, 519f

E

Eagle Creek, 258
East Sand Island, 525, 526, 526f
Ecology
  of regulated rivers, 229
  of regulated streams, 212
  and salmon recovery, 651
Ecosystem approach, 56
Ecosystem Diagnosis and Treatment (EDT)
    model, 620
Ecosystem management, 35
Ecosystem Management Process, 576
Ecosystems
  alluvial river, 62f, 63–64, 64f, 194f
  basinwide ocean, 540
  beavers in, 200–201, 202f
  and California Current, 555
  and climatic stress, 555
  coastal, 531
  Columbia River, 220
  dynamic nature of, 53, 87–89
  and harvest management, 499
  human alteration of, 198
  normative concept of, 56–59, 57, 87–88
  Pacific Basin, 547
  population diversity in, 70
  role of lamprey in, 158
  of salmon, 35
  and salmon production, 24, 25f
  social dimensions of, 88
  static view of, 85–86
  and technology, 20–22
  tributary, 200
Ecosystems, riverine
  beaver trapping in, 200–201
  effect of dams on, 211–212
  effect of human activities on, 197
  flood pulses in, 225
  food web in, 223–225
  human alteration of, 221
  impact of diversions on, 218–220, 219f
  impact of reservoirs on, 212–218
Eel River, cutthroat trout in, 150
Effectiveness monitoring, 587–588
Eggs, salmon
  incubation of, 45f
  mining of, 431, 432f
Egg size, and habitat conditions, 72
Electricity, provided by Columbia River
    dams, 328, 329f
El Niño, 59, 65

  direct effects of, 541
  impact of, 508, 536
  and naturalized stream flow, 609f
  occurrence of, 540
  tracking of, 542
  and winter precipitation, 548
El Niño Southern Oscillation (ENSO), 105,
    314, 539–540, 609f
Endangered Species Act (ESA; 1973), 3, 15,
    104, 334, 402, 403, 579
  bull charr under, 155
  Chinook salmon under, 132, 132t
  chum salmon under, 138
  coastal cutthroat trout under, 151
  and coho salmon, 136
  and court decisions, 625–626
  and habitat restoration, 233
  and Northwest Power Act, 18
  pink salmon under, 142
  provisions of, 18
  sockeye salmon under, 141
  and steelhead trout, 147, 147t
  sturgeon under, 157
Entiat River, coho spawning in, 133
Environment. *See also* Ecosystems; Habitat
  and agents of change, 24
  and conventional harvest management,
    485
  human interactions with, 4
  marine,65, 507 (*see also* Oceans)
  ocean, 549–550
Environmental Protection Agency(EPA),
    U.S., 211
Erosion, in estuarine environment, 518
Escapement. *See also* Spawning escapements
  goal management for, 488, 490–491
  measures of, 494
  and MSE, 491
Estuaries
  Chinook yearlings in, 287
  defined, 509, 512
  food web of, 524
  invertebrate species of, 227, 228f
  natural aging of, 516–517, 518
  recommendations for, 556
  and salmon management, 656
  subyearling migrations in, 274–275
Estuarine Turbidity Maximum (ETM), 525
Estuarinization, 227
Estuary, Columbia River, 509
  human influences on, 516–529

of Lower Columbia River, 511f
maps of, 511f
Estuary environment, 509, 510f
changes in, 529
and hydrosystem storage projects,
519–520, 522f
juvenile salmon in, 513, 514–515, 516f
rearing habitats in, 529
and salmonid life histories, 513–514
Euro-Americans, and natural resources, 32
Evaluation. *See also* Monitoring
exercises, 594–595
of habitat condition, 576
importance of, 572
recommendations for, 588, 590–595
reconsideration of, 596
in recovery plans, 608, 616, 619
role of databases in, 593
system-wide, 579–588
three aspects of, 573f
Evolutionarily Significant Units (ESUs),
15
and ESA, 495
listings for, 631, 635
steelhead as, 120, 122
in stock conservation, 107
Experiments, marking, 430
Exploitation rates, salmon, 488
Extended Salmon Fishery Model, 477f,
483–485
application of, 484
information requirements of, 485
Extinction
and colonization, 76
metapopulation processes of, 108–109
risk of, 7
spreading of risk for, 70
of stock, 106
vulnerability to, 109

F

Fall Chinook salmon
annual returns for, 440–441, 441f
compared with spring Chinook, 111,
129–130
ESA listing for, 633
Hanford Reach, 83, 84f, 85, 109, 174, 276,
477, 642, 658
in John Day pool, 648f
in Snake River, 276–278, 331f
spawning areas of, 64f, 646

spawning of, 76, 77, 388, 439, 648–650,
649f
Feasibility, for mainstem restoration efforts,
604
Federal agencies, 638. *See also specific
agencies*
Federal Caucus Plan, 602
Federal Columbia River Power System
(FCRPS), 334
Federal Columbia River Power System
(FCRPS) BiOp, 2000. *See* Biological
Opinion 2000
Federal Endangered Species Act. *See*
Endangered Species Act
Federal Energy Regulatory Commission
(FERC), 333
passage goals of, 400, 402
and spills as bypass measures, 359
Federal Hanford Nuclear Reservation,
658
Federal Register (69 Fed. Reg. 31354), 626
Feeding areas, main river, 228f
Feeding patterns. *See also* Food web;
Nutrients
changes in, 222
and hydropower development, 227
F1 hybrid vigor, 447
Fifteenmile Creek, 634f
Fin-clipping, 433
Fingerlings, in estuarine environment, 513
Fish, around Columbia River plume, 530
Fish and Wildlife Program, NPCC, 602, 651.
See also Biological Opinion
actions implemented by, 639
early manifestations of, 638
harvest in, 619
integrated framework of, 638
and mainstem habitat, 604
mainstem-related actions in, 617
monitoring and evaluation in, 616
scientific advisory groups of, 638
tributary habitats in, 605, 612, 618
Fish and Wildlife Service, U.S., Draft
Comprehensive Conservation
Plan/Environmental Impact Statement
of, 658
Fish behavior, and FPE, 357
Fish bypasses, development of, 262. *See also*
Bypass systems
Fish Commission, U.S., hatchery policy of,
422, 423

Fish conservation. *See also* Conservation
central principles for, 55–56
habitats in, 59
static view in, 87
strongholds in, 82–83
Fish culture, 420. *See also* Artificial
propagation
goals of, 33–34, 422
in management policy, 437
Fish-dam problem, 253
Fisheries
Canadian, 544, 545f
Russian, 544, 545f
stock concept in, 102
Fishery landings, commercial salmon, 426,
427f. *See also* Landings
Fishery management
changing paradigm for, 21–22
and conservation principles, 55–56
and Fish and Wildlife Program, 22
and judicial decisions, 19
and local populations, 103
ocean ignored by, 65
and political agendas, 92
stock concept in, 108
technological improvements in, 52
traditional perspective in, 52–54
Fish guidance efficiency (FGE), 345–346
comparisons for, 351t
and juvenile transportation, 383–384
and turbine intake screens, 347, 351, 352
Fishing rights, of Indian tribes, 19
Fishing techniques, selective, 614
Fish ladders, 254
at Bonneville Dam, 341f
criteria for, 342
flow regulation for, 387–388
at Ice Harbor Dam, 340f
at Little Goose Dam, 46f
requirement for, 339
as technological solution, 630, 633f
of typical dam, 343f
Fish orientation hypothesis, of travel time,
308–311
Fish passage. *See also* Bypass systems
facilities
failure of, 262
lack of, 10
in recovery plans, 610–611, 611f, 617
Fish Passage Center (FPC)
annual planning of, 254

monitoring program of, 650
risk analyses of, 373
Fish passage efficiency (FPE), 345, 346, 357
Fish scales, analysis of, 549
Fish stranding, with fluctuating flows, 304
Fish wheel systems, 468f
Fish and Wildlife Program, NPPC, 4, 335,
572
adaptive management provisions of, 453
background on, 35–38
conceptual framework of, 38–47
and estuarine conditions, 512
first, 17, 334
fish passage efficiency standard of, 347
subbasin planning of, 337
water budget established by, 393
Fish and Wildlife Service, U.S. (USFWS), 18
Fitness. *See also* Genetics
and degree of outcrossing, 447f
and domestication selection, 450
Flathead Basin, discharge statistics for, 175t
Flathead Lake
non-native salmonids in, 181
toxic pollution of, 211
Flathead River
North Fork of, 178f
Nyack Flood Plain of, 195f
westslope cutthroat trout in, 152
Flip lips, 373, 374, 374f, 398
Flood control, 174, 330, 334, 472
Flooding
annual cycles, 224
and habitat diversity, 229
loss of riparian, 226–227
spring, 189f
Flood plains
alluvial, 177, 178f, 184, 201
of Columbia basin, 63
competition for segments of, 196
developed, 518, 519, 520f
logging and, 203
and spawning areas, 646
tree replacement in, 585f–586f
Flood surge, 282
Flow augmentation, in recovery plans, 616
Flow dynamics, in migration, 265
Flow provisions, of hydroelectric system,
342
Flow rates, 39
average, 312
and body size, 69

dams as regulators of
    effects on adult anadromous fishes,
        387–389
    effects on juveniles, 390–392
and feeding behavior, 279
and fish orientation, 311
and hydroelectric system, 253
impact of flooding on, 227
and load following, 258–259
and migration, 294
and rearing of juveniles, 259–260
and salmonid habitats, 63
and salmon migration, 254, 255f
and spawning patterns, 73, 256
study of, 235
and swimming behavior, 275–276
and travel time, 301–312
and wood accumulations, 190
and WTT, 284f
Flow strategy, and stock diversity, 314
Fluctuating flow hypothesis, of travel time,
        302–308
Flume experiments, 274
Foerster, R.E., 431–433
Food chains, and salmon production, 550
Food production, changes in, 222
Food webs
    detritus-based, 524
    estuarine, 524–527, 529
    hydrosystem alterations of, 225–229
    riverine, 223–225
Foraging conditions, in estuaries, 514
Forebays
    fluctuating flows in, 307–308
    gas levels in, 374
    hourly water elevations in, 305f
    juvenile losses in, 375–376
    rates of mortality in, 335–336
Forest Science DataBank at Oregon State
        University, 575–576
Forest Service Roadless Area Final
        Environment Impact Statement, U.S.,
        658
Four Northwest States Governors' Plan,
        602
    harvest in, 615–616
    implementation of, 623
    institutional arrangements in, 610, 617
    and mainstem habitat, 604
    mainstem-related actions in, 617
    tributary habitats in, 605, 612

Fraser River, 253
    restoration, 104
    salmon run, 496
Fraser River Panel, 480
Freshwater environment
    and marine environment, 65
    variability in, 552
    vulnerability of, 181
Fry, in hatchery programs, 430–431
Fyke net array, for estimating juvenile
        turbine bypass mortality, 349f
Fyke net screens, smolted fish caught in,
        286

G

Gas
    reduction of total dissolved, 373–375,
        374f
    supersaturation of, 361, 371–373, 372f,
        398
Gas bubble disease, 373
Gas bubble trauma, 655
Gas caps, estimated, 362t
Gene flow, and straying, 187
General Accounting Office (GAO), 332
Genetic diversity
    and juvenile migration patterns, 125
    and population growth, 90
Genetic drift, and metapopulation concept,
        74
Genetic introgression, 144
Genetic reserves, hatcheries as, 454
Genetics
    of Chinook salmon, 111
    of hatchery stocks, 450
    of hatchery-wild fish interactions
        direct, 446–448
        indirect, 448–449
    and variations in life histories, 66
Genetic structure
    of anadromous salmonid populations,
        111–122
    of bull charr, 153–154
    of bull trout, 145
    of Chinook salmon, 114–118, 115f, 116f,
        117f, 119f, 124–125
    of chum salmon, 136–137
    of coho salmon, 133
    of cutthroat trout, 145, 148–149
    of pink salmon, 142
    of rainbow trout, 145–146

Genetic structure (*Cont.*)
 of salmonid populations, 112–114, 113f,
  115f
 of sockeye salmon, 139–140
 of steelhead trout, 120–123, 121f, 122f
 substructuring, 114
Genus *Oncorhynchus*, 101f
Geography, and genetic structure, 114, 115,
  115f, 116
Global heat budget, and Pacific Ocean, 65,
  551
Global warming, risks of, 556
Gold-dredging, two-person, 500f
Goode, George Brown, 425
Grand Coulee Dam, 3, 10, 214, 250, 254,
  339, 473
 construction of, 326
 installation of turbine fan in, 346f
 statistics for, 179t
 and temperature changes, 217f
Grand Coulee Fish Maintenance Program
  (GCFMP), 333, 443
Grande Ronde Basin
 discharge statistics for, 175t
 and placer mining, 208
Grande Ronde River, 233
 Chinook salmon spawning in, 130
 coho spawning in, 133
 and juvenile migration, 69
Grande Ronde Stream, temperature changes
  in, 215
Granite Creek, 657f
Gravel-bed rivers, 61, 223–224
Grays River, 519*f*
Grazing
 and riverine ecosystems, 204–205, 205f
 and riverine habitat, 188
Green sturgeon, 155–157
Groundwater upwelling, and riverine
  habitat, 193–196, 194f, 195f
Growth, temperature dependence of, 190,
  193. *See also* Size

H

Habitat Conservation Plans (HCPs), 337
Habitat degradation, 24, 440, 489f
 causes of, 196–200, 234
 consequences of, 196–200
 and harvest management, 473, 498
 indirect measures of, 474
 and non-native fishes, 234

 and population growth, 455
 recovery from, 488
Habitat destruction
 and artificial propagation, 426
 in historical context, 469
Habitat diversity, and river regulation,
  229–231
Habitat-forming processes, restoration of, 89
Habitat loss
 and Columbia River fisheries, 220–222
 freshwater, 174
 largest, 646
 and non-native fishes, 220
 technological solutions for, 630
Habitats, salmonid, 105–106. *See also*
  Ecosystems; Riverine habitat
 accessible, 327, 328f
 artificial stream, 419f
 and body size, 69
 chain of interconnected, 66
 critical, 279
 diversity of, 59, 61
 effects of river regulation on, 229–232
 estuarine, 528 (*see also* Estuaries)
 events and processes controlling, 60t
 in fish conservation, 59–61
 genetic diversity and, 108
 and harvest management, 483
 and harvest success, 492
 healthy, 653, 653f
 and human development, 640
 for juveniles, 270f
 and life cycle, 484
 and life history diversity, 70–73
 in lower Snake River, 644
 mainstem, 604–605
 maintaining diversity of, 75
 in mid-Columbia River, 644
 monitoring changes in, 572
 native salmonids, 56
 and population diversity, 70
 protection of, 91, 486–490, 487, 653–654
 rearing, 575f
 recovery measures for, 591, 603
 reduction of freshwater, 552
 reservoirs as, 213
 restoration of, 87, 233, 653–654, 654f
 riparian, 583ff
 shallow-water, 521
 shoreline, 645f
 spatial and temporal variation in, 61–64

spawning, 574f
stronghold, 83
tributary, 78, 605, 606f
Hamilton Slough, 648
Hamilton Spring, 582f
Hanford Generating Plant, 356
Hanford Reach, 11, 81, 177, 179, 180f, 221,
    224, 232, 250, 251, 252f, 287, 326–327,
    439, 630, 643f
  alluvial reaches in, 63–64, 64f
  chinook salmon in, 174, 226
  fall Chinook in, 477, 642
  fish stranding at, 304
  and food web components, 235
  healthy stocks in, 7
  juveniles at, 644f
  lessons from, 642–643
  management in, 276–278
  as potential reserve, 655
  resilience of, 236
  salmonid stock in, 397
  salmon spawning at, 126
  sockeye salmon in, 289
  spawning in, 257
  as stronghold for fall Chinook, 83, 84f, 85
  subyearling Chinook movement through,
    270–271, 271f
  uniqueness of, 656
Hanford Reservation, atomic wastes at, 251
Harvest boat, 468f
Harvest management, 466
  caution in, 493–494
  and ecosystem context, 488
  effective, 494
  extended salmon fishery model in, 477f,
    483
  in historical context, 466, 467f, 469–475
  and impact on future production, 475
  by indigenous peoples, 471
  lack of escapement-based, 476–478
  and mortality accounting, 478–482, 478t
  new paradigm for, 468, 483, 484
  and ocean fluctuations, 551
  old paradigm for, 475
  recommendations for, 499
  regulation in, 498
  in 1930s, 474
  sustainable, 485
  Sustainable Salmon Fishery Policy in,
    486
  traditional, 483

Harvests, salmon
  decline of, 426, 427f
  incidental, 484
  informed mixed stock, 499
  interdecadal patterns of, 546, 546f
  1914 to 1920 increase in, 429, 430f
  ocean, 498
  and recovery operations, 603
  in recovery plans, 606–608, 613–616,
    618–619, 623
  reduction of, 466, 496
  and salmon productivity, 483
  and spawning populations, 498
Hatcheries. See also Artificial propagation
  annual releases of, 444–445
  benefit-cost ratio for, 433–434
  at Bonneville dam, 14f
  Chinook salmon in, 422, 423f
  in Columbia Basin, 3
  as core populations, 77
  early experiments in, 429
  early issues raised in, 423–424
  egg incubation at, 45f
  evaluation of, 434–435
  first Pacific salmon, 420–421, 421f
  as genetic reserves and refuges, 454
  and habitat stewardship, 420–424
  harvest and, 420–424
  history of, 420–424, 421f, 422f
  inferior fish from, 445
  and juvenile losses, 402
  management model for, 34
  and management policy, 437–439
  matings in, 46f
  on McCloud River, 418
  mistaken practices of, 441–442
  mitigation, 453
  practices causing decline in, 436–437
  Priest Rapids, 85
  production of, 421f
  and recovery operations, 603
  in recovery plans, 605, 612–613, 614f, 618
  reforms for, 444
  in salmon management programs,
    418–419
  as technological solution, 630, 632f
  technology in, 332, 639
  total releases of, 440–441, 440f
Hatcheries in Columbia River
  history of, 12–13
  management of, 13–14

Hatchery-origin fish, 446
  first generation, 77
  genetic effects on wild fish of, 446–449
  genetics changes to, 450
  interaction of wild fish with, 445–446
  juveniles, 445–446
  mass marking, 608
  survival of, 185
  system-level limitations, 445
Hatchery programs
  under APRE, 443–444
  benefits derived from
    after 1981, 434–437
    from 1900 to 1980, 429–434
  for coho salmon, 441
  criticism of, 431
  early, 418
  evaluation of, 453
  extended rearing in, 433
  genetic selection in, 426
  and harvest management, 453
  historical effectiveness of, 425–426
  and juvenile survival, 528
  long-term efficacy of, 455
  monitoring of, 439
  national reviews of, 428
  resistance to change of, 440
  results on use of, 439–444
  smolt releases of, 426
Hatchery reform, 622
Hatchery Scientific Review Group, 443
Hazards, anthropogenic, and local
  adaptation, 105
Headwater areas, of Columbia River Basin,
  327f
Heat storage, in mainstem reservoirs,
  396–397
Hell's Canyon, 12f
  salmon production in, 11
  water velocity in, 284
Hell's Canyon Dam, 10, 76, 126, 250, 254,
  257, 272, 278, 338, 340, 646, 659
  construction of, 326, 327
  statistics for, 179t
Herring, in Columbia River estuary, 524
Heterosis, 447
H.J. Andrews (HJA) Experimental Forest, 548
Hogan decision, 625, 626
Hoh River
  characteristics of, 189f
  steelhead spawning in, 188f
Holmes, Harlan, 344

Home stream theory, debate over, 31, 107
Hood River
  coho salmon in, 134
  coho spawning in, 133
Hot spots, of bioproduction, 195, 195f
Human development, and normative
  ecosystem, 58
Hume, R.D., 102
Hungry Horse Dam, outflow structure
  installed at, 395
Hungry Horse/Flathead Lake Basin, 259
Hungry Horse Reservoir, drawdown
  schedule for, 395
Hunting rights, of Indian tribes, 19
Hybridization
  of cultured fish with wild fish, 446–447
  in trout populations, 144
Hydraulic patterns, studies on, 312
Hydroacoustics, 352, 363, 367
Hydrocombine design, at Wells Dam, 360f,
  370–371
Hydroelectric dams
  and artificial propagation, 426
  introduction of, 472
  and recovery operations, 603
Hydroelectric power generation, 174, 485
Hydroelectric system. *See also* Dams
  ameliorating effects of, 337–338
  Columbia River Basin, 326, 327f
  development of, 250–253
  effects of
    and rates of mortality, 335–336
    on water temperature, 396–397
  and harvest management, 466
  lasting adverse effects of, 326
  mitigation activities within, 654–655
  and predation, 376–380
  typical dam in, 254–255, 255f
Hydrograph
  annual changes, 53, 55f
  changes in the Columbia River, 520, 522f
Hydrology, in recovery plans, 608, 609f,
  616–617, 619
Hydropower, management of, 334
Hydrosystem
  alterations of food webs in, 225–229
  changes in flow patterns caused by, 520,
    523f
  and Chinook salmon harvest, 131f
  effects on life history diversity of, 313–314
  managing passage of juvenile through,
    386

operation of, 630
and salmon decline, 578
and salmonid mortality, 335, 336f, 337f
travel time through, 392
Hyperheic zone, flow patterns in, 63
Hyporheic zone, defined, 62

I

Ice and trash sluiceways, 365–367, 366f
Ice Harbor Dam, 86f, 297f
    discharge statistics for, 175t
    FGE at, 351t
    fish ladder at, 340f
    and passage goals, 402
    spill deflectors at, 374
    spill levels at, 361–362
    statistics for, 179t
    surface collection configurations at, 367
    turbine intake screens at, 355
Ice Harbor pool, 306
Ice Harbor Reservoir, 305
Ichthyofauna, 220
Idaho Supplementation Study (ISS), 587
Identification studies, 140
Iliamna Lake, 71
Imnaha River, 575f, 583f
    Chinook salmon spawning in, 130
    spring Chinook spawning in, 127f
Impingement rates, 353
Implementation monitoring, 581
Incubation of salmonid eggs
    habitats for, 72
    occurrence, 180
Independent Scientific Advisory Board
    (ISAB), NPPC, 45, 87, 370, 512
Independent Scientific Review Panel (ISRP),
    45, 578, 579, 616
Index sites, in recovery planning, 593–594
Indian tribes. *See also* Native Americans
    ceremonial harvests of, 140
    federal treaties with, 15, 19
Information
    in conceptual foundation, 30
    ineffective transfer of, 24
Informed mixed stock fishing, 496
Insects
    impact of hydroelectric development on,
        226
    in shoreline feeding, 277
Institutional arrangements, in recovery
    plans, 610, 617, 619–620

Intensively Monitored Watersheds (IMW),
    588
Interim Management Agreement, 468
Interior Columbia Basin Ecosystem
    Management Plan (ICBEMP), 603
Intermountain valley development, 232
International Pacific Salmon Fisheries
    Commission, 253
Invertebrates, estuarine, 235
Irrigation, 485. *See also* Dams
    hydroelectric projects for, 328, 330f
    management of, 334
    and riverine ecosystems, 208–210, 210f
Irrigation diversion
    dam for, 634f
    as technological solution, 630, 634f
Isolation, population. *See also* Genetics
    and extinction, 110
    and habitat fragmentation, 129
    of individual sockeye populations, 140
    and metapopulation concept, 74
Ives Island, 648, 649, 650f

J

Japanese fisheries, salmon harvests of, 544,
    545f
Jetties, in Columbia River estuary, 518,
    518f
John Day Basin, 198
    discharge statistics for, 175t
    dredge mining sites in, 206, 207f
John Day Dam, 43f, 126, 177, 263, 267, 272,
    285, 286, 294, 644
    FGE at, 351t
    fish ladder at, 633f
    monitoring studies at, 363
    mortality rates at, 353
    and natural river drawdown, 381, 382f
    passage pattern at, 289
    sockeye smolts at, 289
    spillbays at, 633f
    spill deflectors at, 374
    spill effectiveness at, 364
    spill levels at, 361–362
    statistics for, 179t
    steelhead passage at, 292
    tailrace losses in, 376
    turbine intake screens at, 355
John Day pool, 391, 647, 648f
John Day Reservoir, 76, 223, 268, 273,
    293

John Day Reservoir (*Cont.*)
  juvenile predators in, 376
  management options for, 647
  permanent drawdown of, 649f
  sturgeon in, 156, 157
John Day River, 233, 389, 637f, 657f, 659
  bull charr in, 154
  and Chinook salmon spawning, 126
  coho spawning in, 133
  dredge mine reclamation site on, 207f
  fluoride effluent in, 211
  healthy stocks in, 7
  juveniles in, 81t
  late summer temperatures of, 192
  as spawning habitat, 21f
  westslope cutthroat trout, 151
John Day subbasin, 657f
Johnson Creek, 577f
  Chinook spawning in, 128f
  multiple spawning of, 491f
  summer chinook in, 183f
Jones Beach
  Columbia River estuary at, 288
  hatchery-reared juveniles at, 527
Jordan, David Starr, 31
"June hogs," 472, 644
Juveniles. *See also* Bypass provisions;
    Migration, juvenile
  barge transportation of, 635f, 636f
  bypass of
      and mortality and descaling, 353
      provisions for, 345–347
      turbine intake screens for, 350–356
  changing life histories of, 527
  Chinook, 124
  in coastal waters, 482
  collection facilities for, 179t
  daily migration cycles of, 263–264, 263f, 264f
  diversion of
      proposed methods, 380–382, 382f
      reduction of total dissolved gas,
          373–375, 374f
      into spills, 368–369
      into spills and bypass systems, 357
      and supersaturation of gas due to
          spill, 371–373, 372f
      surface collection for, 367
      and survival, 369–370
      at turbine intakes, 347–357
      at Wells Dam, 360f, 370–371
  downstream migration of, 260–262,
      342–343

effects of flow regulation on, 390–392
effects of load following on, 396–397
and El Niño, 542
in estuarine environment, 513, 514–515,
    516f
in fluctuating flows, 213
and food web, 225–226
habitats for, 81t, 182–186, 270f
at Hanford Reach, 644f
from Hanford Reach, 642
and hydrodynamics, 295
and hydropower system, 44
identifying, 31
life history requirements of, 301
management of, 386–387
migration of, 67
nutritional status of, 228
ocean distributions of, 481
rearing of, 259–260
in reservoirs, 214
sockeye, 289
solutions for passage of, 332
stream-type, 82
survival of, 12
temperature requirements of, 218
transportation of, 383–386, 383f, 384f
and water budget, 392–396

K

Kamchatka Peninsula, 139
Kelly Creek, westslope cutthroat trout in,
    152
Kerr Dam, 259
Klamath River, bull charr in, 153
Klickitat River, coho salmon in, 134
Kokanee salmon, 181, 187
Kootenai River, 259
  rainbow trout in, 146
  westslope cutthroat trout in, 151
Kootenai River white sturgeon, (*Acipenser
    transmontanus*), 18, 157, 395

L

Lake Chelan drainage, westslope cutthroat
    trout in, 151
Lake Osoyoos, healthy stocks in, 7
Lake Roosevelt, 214
Lakes
  of Columbia River Basin, 251f, 428f, 631f
  dams on, 174, 176
  migrations into, 61

Lake systems
    population structure within, 71
    and spawner counts, 73
Lampetra tridentata (Pacific lamprey),
    157–158
Landings. *See also* Harvest management
    annual estimates of, 478t
    annual observation of, 469, 470–471
    of chum salmon, 134f
    data for, 471
    decline in, 474
    disparities between catch and, 479
    equilibrium for, 472
    five-year moving average for, 470–471,
        470f, 474–475
    measurement of ocean, 476–477
    reporting of, 472
Landings, commercial
    of Chinook salmon, 134f
    from 1866 to 2000, 35f, 637f
    and hydrosystem development, 8, 8f
    of sockeye salmon, 134f
    of steelhead trout, 8, 8f, 35f
Landscape-based recovery actions, 611–612
Landscaped-based restoration, 605
La Niña, 542, 548, 609f
Lava flows, 177
Lava rock, in Columbia River basin, 63
Lemhi River, 69, 654f
Lewis, Meriwether, 1–2, 19, 32, 190
Lewis and Clark River, 520
Lewis River, 528
    bull charr in, 154
    healthy stocks in, 7
Lewiston Dam, 280
Libby Dam, 259
Life cycles
    and habitats, 61
    of salmon, 477f, 484
Life histories, 53
    of Chinook salmon, 129–130
    decline of ocean-type, 79–80
    diversity of, 90, 234, 556
    ecosystem approach to, 56
    effect of hydrosystem development on,
        313–314
    effects of temperatures on, 193
    estuarine influence on, 513–514
    habitats in, 61
    juvenile, 527
    ocean-type, 66–67
    and population diversity, 66

salmonid, 105–106
steelhead, 123
stream-type, 66–67
studies of, 430
and variability in ocean, 553
Little Goose Dam, 43f, 46f, 305
    FGE at, 351t
    gas saturation levels at, 373–374
    juvenile collection at, 383
    mortality rates at, 354
    spill effectiveness at, 362–363
    spill levels for, 362t
    statistics for, 179t
    and survival estimates, 391
    transportation required at, 386
    turbine intake screens at, 350, 355
Little Goose pool, 304, 306, 307
Little Goose Reservoir, 216f, 305
    food limitations in, 228
    juvenile fall chinook in, 277
Livestock
    elimination of grazing of, 583f
    impacts on streams of, 205
Load following
    and Chinook spawning, 258
    defined, 253
    and survival of juvenile salmonids,
        396–397
Load rejection, 258
Local adaptation, 424. *See also* Adaptive
        management
Logging, 9, 474, 485
    and riverine ecosystem, 201, 203–204
    and riverine habitat, 188
Loss, estimates of, 344–345. *See also*
        Mortalities
Lotic (running water) spawners, 180–181
Louvers, angled, 380
Lower Columbia River Fishery
        Development Program (LCRFDP), 104
Lower Columbia River Gorge, 57f
Lower Granite Dam, 10, 11f, 272, 280, 281,
        297f, 305, 306, 311, 659
    descaling rates at, 353
    FGE at, 351t
    flow objectives for, 394t
    gas saturation levels at, 373–374
    juvenile collection at, 383
    in the lower Snake River, 331f
    salmon counts at, 475
    spill effectiveness at, 363
    spill levels for, 362t

Lower Granite Dam (*Cont.*)
  statistics for, 179t
  surface bypass collection at, 368
  and survival estimates, 391
  transportation required at, 386
  turbine intake screens at, 350, 354, 355
Lower Granite Dam Reservoir, 216f, 257, 278, 291, 292
  experimental drawdown of, 279
  food limitations in, 228
  juvenile fall chinook in, 277
  mortality in, 375
Lower Granite pool, 304, 307
Lower Monumental Dam, 305, 339f
  daily flow fluctuations at, 303f
  FGE at, 351t
  gas saturation levels at, 373–374
  juvenile collection at, 383
  spill at, 358f
  spill effectiveness at, 363
  spill levels at, 361–362
  statistics for, 179t
  and survival estimates, 390
  transportation required at, 386
  turbine intake screens at, 355
Lower Monumental Dam forebay, mortality from, 375
Lower Monumental pool, 306
Lower Monumental Reservoir, 305
Lower Snake River Compensation Plan, 333

M

Magnuson-Stevens Fisheries Management Conservation Act, 614
Mainstem habitat, in recovery plans, 604–605, 610–611, 617
Mainstems, of Columbia and Snake Rivers, 176f
Management. *See also* Harvest management; Salmon management
  adaptive, 47, 443, 452, 592
  commodities-driven concept in, 4
  of human activities affecting salmon, 494–496
Management programs. *See also* Harvest management
  and marine environment, 65
  misplaced priorities in, 199–200
Marine environment. *See* Oceans; Pacific Ocean
Marking experiments, 430

Marshes. *See also* Estuaries
  and estuary modification, 517, 519
  food web in, 524
Mass-mark selective fisheries, 607
McDonald, Marshall, 423
McIntyre Creek, 489f
McNary Dam, 76, 177, 232, 267, 286, 294, 474, 644, 647
  descaling rates at, 353
  FGE at, 351, 351t
  flow objectives for, 394t
  gas saturation at, 374
  juvenile bypass at, 353
  juvenile collection at, 383
  mortality rates at, 354
  salmon counts at, 473
  spill levels for, 362t
  statistics for, 179t
  and survival estimates, 390
  transportation required at, 386
  turbine intake screens at, 348f, 349, 350, 355
McNary pool, 273
McNary Reservoir, 277
Measures of conceptual framework, potential inadequacy of, 39
Meffe, Gary, 439
M&E program, 452
Merwin Dam, bypass methods tested at, 380
Metapopulations, 647
  concept of, 74
  core satellite, 75
  extinction of, 74–75
  of fall Chinook spawning, 129
  first order, 119f, 120, 122, 122f
  and habitat improvement, 652–653
  human impacts on, 110
  non-native fishes of, 222
  persistence of, 76
  protection and restoration of, 77
  salmonid, 74, 108–109
  steelhead populations as, 120, 122
Metapopulation theory
  application of, 123
  chinook salmon in, 118
  value of, 78
Methow Basin, discharge statistics for, 175t
Methow River, 219f
  chinook salmon in, 111
  coho spawning in, 133
Mica Dam, 179
Microhabitat system, 60t

Mid-Columbia River, salmon habitat in, 644–647

Migration. *See also* Travel time
changed feeding patterns in, 222
and connectivity, 90
downstream, 260–262
    active vs. passive, 265–268, 266f
    of coho salmon, 288–289
    daily cycles, 263–264, 263f, 264f
    and environmental variables, 294
    flow dynamics in, 265
    of sockeye salmon, 289–290
    of steelhead trout, 290–292
    of subyearling Chinook, 268
    and surface orientation, 262
    of yearling Chinook, 279
effects of flow on
    in mid-Columbia Reach, 293
    in Snake River, 292–293
estuarine, 513, 526
flow dynamics in, 265
flushing mode of, 291
and hydrologic studies, 311
measurement of behavior during, 493–494
obstacles to, 253–256
and ocean currents, 554
passive, 296
passive vs. active, 278–279
physiological changes during, 512
and rising water flows, 281
spiraling of, 261f
and survival, 296–300
swimming behavior during, 267
travel time in, 262
upstream and downstream conditions for, 337f
and upwelling system, 534
and water velocities, 281

Migration, adult, 67
annual losses during, 479–480, 480t
habitat for, 186–187
and hydroelectric system, 339

Migration, juvenile
of Chinook salmon, 125
and hydroelectric system, 255–256, 342–343
and life histories, 67–69, 68f
long-distance, 185–186
mortality associated with, 336, 336f, 337f
in riverine habitat, 182–186
of sockeye salmon, 139
stream flow and, 67f, 269f

timing of, 527

Milk can, Idaho, 425f

Minimum sustainable escapement (MSE), 491

Mining, 9
gold-dredging, 490f
placer, 208
and riverine ecosystems, 206–208, 206f, 207f
and salmon landings, 474

Minter Creek study, 433

Mitchell Act, 333

Mitigation, 330
alternative passage routes for, 347
for dams, 655–656
effectiveness of, 397–398, 398t–399t
of effects of hydroelectric systems on temperature, 369–397
of effects of load following, 396–397
engineering in, 333
goal of, 426
hatchery, 442
legal authorities for, 333–335
and predator reduction program, 378
purpose of, 335

Mitigation hatchery, 632f

Models, in recovery plans, 608, 616, 619, 620

Monitoring
and adaptive management, 592
change, 581–584
coordination of programs for, 572
effectiveness, 587–588
of escapement goals, 490–491
evaluation of, 591–592
exercises, 594–595
of habitat condition, 576
historical record for, 574–577
implementation, 581
importance of, 572
long-term, 582, 585f–586f, 591
planning for, 631
recommendations for, 588, 590–595
reconsideration of, 596
in recovery plans, 608, 616, 619
of research, 586–587
research and, 572–573
role of databases in, 593
statistical, 584–586
system-wide, 577, 579–588
three aspects of, 573f
trend (or change), 581–584
types of, 581–588

Monitoring and evaluation program,
    cooperative, 580
Mortalities, salmonid
    accounting for, 475–476
    in bypass systems, 353–354
    at dams, 343–345, 343f
    in effective management, 495
    estimates for, 344–345, 480t
    failure to account for, 478–482
    and human decisions, 477f, 484
    and hybridization, 447
    identification of sources of, 336, 336f
    implementing zero-sum allocation for,
        494–495
    of juveniles
        in reservoirs, 375–380
        in river reaches, 390–392
        in spills, 370
    and passage through turbines, 345
    public understanding of, 498
    reducing sources of, 654–655
    sources of, 336f, 484, 485
    and tern relocation, 525
Mud Mountain Dam, bypass methods
    tested at, 380

N

Natal spawning areas, straying from,
    186–187. See also Spawning
National Biological Service, 371–372
National Fish Hatchery Review Panel, 419,
    428
National Marine Fisheries Service (NMFS),
    17, 118, 141, 143, 296, 604
    Biological Opinion of, 602
    hatchery studies of, 433–435
    ocean surveys of, 538
National Oceanic and Atmospheric
    Administration (NOAA), 18, 334, 335
    on Columbia River plume, 514
    Fisheries Biological Opinions of, 638, 639
    population status reviews of, 631
    on screen bypass systems, 347
    survival estimates of, 392
National Research Council (NRC), 23, 419,
    428
National Wildlife Federation vs. NMFS,
    626–627
Native Americans, 638
    archaeological sites of, 330f, 332f
    conceptual foundations of, 32
    fishing rights of, 53
    historic fishing grounds of, 330f, 331f
    Pacific lamprey prized by, 158
Natural-cultural system, Columbia River as,
    639–640
Naturalness, acceptable level for, 640
Natural population, compared with
    hatchery fish, 477–478. See also Wild
    fish
Natural processes, in fishery management,
    92
Natural resource management, politics and,
    38
Natural resources
    in Native American societies, 32
    and Resource Conservation Ethic, 33
"Natural river option," 381, 382f
Natural selection, 424. See also Genetics
Navigation lock, of typical dam, 343f
Netboy, Anthony, 472
Nez Perce Tribal Hatchery, 419f
Niche segregation, 190
Nighttime behavior
    during migration cycles, 264
    swimming, 276
Nitrates
    in Columbia River plume, 529
    and local upwellings, 536
NMFS. See National Marine Fisheries
    Service
NOAA. See National Oceanic and
    Atmospheric Administration
Non-native species
    and habitat degradation, 234
    and river regulation, 231
"Normative" restoration principles, 627
"Normative river concept," 60f, 508, 603
Northern pikeminnow (Ptychocheilus
    oregonensis), 376–379, 377f
North Fork, of Flathead River, 178f
North Pacific, climatic regime of, 545. See
    also Pacific Ocean
North Pacific Aleutian low pressure system,
    541, 542, 543. See also Aleutian Low
Northwest Electric Power Conservation
    Council, 333
Northwest Electric Power Planning and
    Conservation Council. See Northwest
    Power and Conservation Council
Northwest Power Act (1980), 14, 403, 572,
    638
    intention of, 578
    1996 amendment to, 579

objectives of, 24
provisions of, 16–19
requirements of, 20
Northwest Power and Conservation Council
    (NPCC), 4, 14, 333, 419, 1, 602, 638
    APRE of, 443, 452–453
    conservation program of, 21f
    current guidelines, 37
    design of, 16
    established, 578
    Fish and Wildlife Program of, 4, 403, 572
        passage goals of, 400, 402
        scientific support for assumptions of,
            42
        subbasin planning of, 337
    goals of, 20
    and Indian tribes, 19
    limitations of, 37
    mandate of, 36
    monitoring and evaluation of, 577
    prevailing conceptual foundation of, 36
    program of, 16–17
    purpose, 15
    recovery goals of, 58–59
    Scientific Review Team of, 419, 428–429
    subbasin planning of, 576–577, 590
Northwest Power Planning and
    Conservation Act (1980). See
    Northwest Power Act
Norwegian Atlantic salmon
    interactions of hatchery-raised fish with,
        448
    migrating behavior of, 281
NPCC. See Northwest Power and
    Conservation Council
Nutrients. See also Feeding patterns
    and California Current, 543
    excessive, 210–211
    and local upwellings, 536
Nutritional status
    evaluation of, 229
    of juveniles, 228–229
Nyack Flood Plain, of Flathead River,
    195f

O

Observation studies, 363
Ocean conditions, 56, 65
    changes in, 34
    effect on salmon spawning of, 72–73
    and freshwater habitat, 548
    and salmon stocks, 630, 656
    shifts in, 53
    temporal cycles of, 59
Ocean currents, and salmon migrations, 554.
    See also California Current
Ocean fisheries
    Hanford Reach fall Chinook in, 643
    potential impacts of, 479, 480–481, 480f,
        482
    and salmon mortality, 474
Oceans, 534–539 See also Pacific Ocean
    adaptations to changes in, 555
    coastal, 534
    and coho salmon harvest, 435–436, 436f
    currents of, 534–539
    global, 554
    juvenile Chinook in, 482
    migration into, 61
    as open systems, 550
    and salmonid populations, 508, 538, 550
    survival in, 13
    temporal shifts in, 551
Office of Technology Assessment, 380
Offshore fishing, shift to, 135
Okanagan River, 254, 289
Okanagan Stream, temperature changes in,
    215
Oncorhynchus, occurrence of, 112
Oncorhynchus clarki clarki (coastal cutthroat
    trout), 150–151, 150f
Oncorhynchus clarki (cutthroat trout),
    148–152
Oncorhynchus clarki lewisi (westslope
    cutthroat trout), 151–152
Oncorhynchus gorbuscha (pink salmon), 141
Oncorhynchus keta (chum salmon), 136–138
Oncorhynchus kisutch (coho salmon),
    132–136
Oncorhynchus mykiss (rainbow and steelhead
    trout), 120, 121f, 145–147
Oncorhynchus nerka (sockeye salmon),
    138–140
Oncorhynchus tshawytscha (Chinook
    salmon), 123–132
Operation Fish Run, 636f
Oregon
    coasts, 67
    spawning localities in, 82f
Oregon and Washington Fish Propagating
    Co., 421
Oregon Fish and Game Commission, 429
Oregon Plan for Salmon and Watersheds
    Monitoring Program, 585

Oregon Production Index (OPI), 135, 135f,
    435–436, 435f, 539
Oregon State University, ocean surveys of,
    538
"Organic machine," Columbia River as, 469,
    630
Oscillations (seiches), 304, 306. *See also*
    Pacific Decadal Oscillation
  in reservoirs, 302
  and water movement, 305
Osoyoos Lake, 141
  juveniles from, 289
  sockeye salmon in, 140
  sockeye salmon spawning in, 138, 139
Outfall bypass. *See also* Bypass systems
  for juveniles, 355f
  mortality associated with, 355
Outmigration, 266f. *See also* Migration
  and flow regulation, 396
  yearling, 266f
Overbank events, reduction in, 520–521,
    523f
Overfishing, 469
  and decline of Chinook, 472
  and landing reductions, 473
  results of, 488–489
  during 1870s, 471–472
Overharvesting, 488. *See also* Harvest
    management
  of coho salmon, 134
  and investment in hatcheries, 440
Owyhee Dam, discharge statistics for, 175t
Owyhee River, late summer temperatures of,
    192
Oxbow Dam, statistics for, 179t

P

Pacific Decadal Oscillation (PDO), 105, 440,
    545–546
  and naturalized stream flow, 609f
  and salmon survival, 314
  and shift in climatic regime, 547
Pacific Decadal Oscillation (PDO) index, 72,
    73
Pacific Fishery Management Council
    (PFMC), 476, 481
Pacific lamprey, 70
  distribution of, 157–158
  life history of, 157–158
  propagation of, 158
  status of, 157–158

Pacific Northwest Aquatic Monitoring
    Partnership (PNAMP), 580–581
Pacific Northwest Electric Power Planning
    and Conservation Act (1980), 333, 334
Pacific Ocean
  currents of, 534–539 (*See also* California
    Current; Oceans)
  El Niño Southern Oscillation Cycle in,
    539–542
  and global heat budget, 65
Pacific salmon
  local populations of, 103
  population classification for, 107
  stock concept in, 102–105
  stock structure of, 104
Pacific Salmon Commission (PSC), 476, 481
Pacific salmon species
  bull charr, 152–155
  Chinook salmon, 123–132
  chum salmon, 136–138
  coho salmon, 132–136
  Columbia River trout and char, 142–145
  cutthroat trout, 148–152
  pink salmon, 141
  rainbow and steelhead trout, 145–147
  sockeye salmon, 138–140
  status of, 123
Pacific Salmon Treaty, 140, 608, 614, 615
Palmer, Gov. Joel, 19
Parafluvial zone, 230f
Passage facilities, 254. *See also* Bypass
    systems
  flow regulation for, 388
  goals of, 400
  technology for, 332
Passive displacement, 267
Passive Integrated Transponder (PIT)
    tagging, 595
  at Lower Granite Dam, 281
  to measure effects of transportation, 385
  for measuring travel time, 292
  methodology, 352
  of smolts, 300f
  to study mortality, 345
  for survival estimates, 372, 392, 396
  of wild steelhead, 291
PATH (Plan for Analyzing and Testing
    Hypotheses), 594
PDO. *See* Pacific Decadal Oscillation
Pelagic consumers
  in Columbia River estuary, 527

in food web, 524
Pelton Dam, bypass methods tested at, 381
Phosphates, in Columbia River plume, 529
Phylopatry, 102, 104
Pierce Island, 648, 649, 650f
Pikeminnow, 376–379, 377f
Pinchot, Gifford, 33
Pink salmon *(O. gorbuscha)*, 142f
  distribution of, 142
  under ESA, 142
  in estuarine environment, 513
  evolutionary history of, 114, 141
  genetic structure of, 142
  migration of, 184
  run size for, 7
PIT. *See* Passive Integrated Transformer tagging
Placer mining, 208
Plankton surveys, around Columbia River plume, 530. *See also* Zooplankton
Plume, Columbia River, 509, 514
  effects of, 529–531
  and surface density gradients, 530
Plunge pool, 366
Politics, and scientific innovation, 38
Pollution
  and estuary environment, 512
  inadvertent taking of salmon by, 485
  sources of, 210–211
Pool/riffle system, 60t
Population growth
  and ecosystems, 88
  and environmental conditions, 68f
  human, 474
  and salmon habitat, 455
  salmon restoration and, 622
Populations, salmonid. *See also* Salmon populations
  continuum of richness for, 103t
  defined, 102
  diversity of, 53
  ecosystem approach to diversity of, 56
  factors affecting size of, 448
  human impacts on structure of, 78–82
  linkages among, 109
  and metapopulation concept, 108
  observed differences between, 106
Power Act, 333, 334
Powerhouse
  bypass systems, 347

of typical dam, 343f
Predation
  in Columbia River estuary, 525
  in Columbia River plume, 515
Predator reduction program, 377–378
Predators
  caspian tern, 377f, 379–380
  and estuarine environment, 514
  northern pikeminnow, 376–379, 377f
  and salmon production, 550
Preservation, opportunities for salmon, 660, 660f
Priest Rapids Dam, 177, 232, 257, 295
  bar screen on, 350f
  bypass system at, 356, 361
  flow objectives for, 394t
  fyke net array at, 349f
  hatcheries at, 642
  regulation of, 304
  spill effectiveness at, 365
  statistics for, 179t
  tank trucks transport at, 381
  turbine intake screens at, 349
Priest Rapids hatchery, 271
Prince William Sound, cutthroat trout in, 150
Propellers, to enhance migration, 311
Proposed Recovery Plan for Endangered Kootenai River sturgeon, 259
Protection, salmon, expenditures for, 438–439, 438f
Public
  and fish conservation, 53
  and sustainable salmon fishing, 497–498
Public Law 96-503, 578
Public policy, development of, 38
Public utility districts (PUDs), 333, 334
  fish passage projects of, 340–341
  and survival standards, 361
Pumps, to enhance migration, 311

Q

Quantitative models, 620
Quartz Lake, 232

R

Radio-tagged fish
  to estimate survival, 391
  mortality studies with, 344–345
  tracking of, 341

Radio-tagged fish (*Cont.*)
   and unfavorable flow conditions, 388
Radiotelemetry, 280, 286, 351, 367
   of steelhead smolts, 291
   of yearling migration, 311–312
Rainbow trout, 120, 146*f*, 281. *See also*
      Steelhead
   anadromous and resident forms of, 142
   coastal and inland forms, 145
   distribution of, 143, 144t
   distributions of, 146
   evolutionary history of, 145
   genetic structure of, 145–146
   and habitat degradation, 143, 147
   hatchery, 146, 147
   propagation efforts for, 146–147
   rapid downstream movement of, 296
Rainfall, seasonal, 177. *See also* Climate
Reach system, 60t
Rearing habitat, 575*f*
   Hanford Reach, 643
   impact of hydroelectric system on, 645
   and inundation, 402
   protection of, 487–488, 487f
Reasonable and Prudent Alternatives
      (RPAs), 579
Recolonization
   and connectivity, 90
   metapopulation processes of, 108–109
Recovery, salmon
   conceptual foundation of, 36–47, 649
   dynamic ecosystem view in, 640
   federal policy for, 627
   freshwater measures for, 553
   functional ecosystem approach to, 620
   goals for, 651
   indicators of, 555
   integrated approach to, 652
   relevance of incidental mortalities to, 480
   scientifically sound program for, 661
   of spring Chinook salmon, 481
Recovery documents, 602, 603
   conceptual gaps in, 621
   data gaps in, 621
   implementation of, 623–624
   integration of, 623
   and quantified details, 625
   variation in, 624
Redband trout, under ESA, 147
Redd (salmon nests) counts
   for Chinook salmon, 130
   for spring Chinook, 637f

Redds (salmon nests)
   identification of, 574f
   location of, 257
   multiple spawning of, 491f
Redfish Lake, 140, 141, 289
Refuges, hatcheries as, 454
Refugia, protection of cool water, 232
Regional decision documents, need for, 454
Regional stochasticity, 109–110
Reproduction. *See also* Spawning
   and habitat quality, 180–181
   and life history patterns, 66
Reproductive isolation, 424. *See also*
      Genetics
Reproductive success
   and genetic interactions, 448
   in hybridization, 447
Research
   general conclusions of, 196–197
   in hatchery programs, 426
   on ice and trash sluiceways, 366–367
   on impact of logging, 203
   monitoring associated with, 572–573
   reconsideration of, 596
Research, monitoring, and evaluation
      (RME), 579, 580
Research monitoring, 586–587
Reserves, establishing salmonid, 656–659
Reservoir habitat, 230
Reservoirs. *See also specific reservoir*
   on Columbia River, 177
   estuarine invertebrate species in, 227, 228f
   for flow augmentation, 395
   mortality of juveniles, 375–380
   rates of mortality in, 335–336
   and riverine ecosystems, 212–218
   smolt travel time in, 309
   Snake River, 262
   storage capacity of, 251
   subyearling migrations in, 272–274
   turbulence in, 310–311, 310f
   typical Columbia River, 252f
   unstable hydraulics of, 304–307, 305f, 306f
   yearling Chinook in, 285–287
Resource Conservation Ethic, 33
Resting/feeding, studies on, 312
Restoration, fish
   mitigation approach to, 330, 332–333
   negative feedback loop in, 200
   whole-river, 233
Restoration, habitat, 235, 651–654, 654f
   and non-native species, 236

recommendations for, 236–238
Restoration, salmon, 54
  in Columbia Basin, 3
  ecosystem approach to, 56–59
  progress in, 59
  role of hatcheries in, 44
Restoration, salmonid, 486, 638
  alternative conceptual framework for, 640
  approaches to, 641, 641t
  cost and benefits of, 660
  ecological processes in, 20
  economics of, 438–439, 438f, 630
  failure of, 14
  and global warming, 556
  habitat in, 655
  needed changes for, 22
  opportunities for, 660, 660f
  in regulated rivers, 639–642, 641t
  response to, 24
  in static view of ecosystems, 85–86
  traditional solution for, 650–651
  in Umatilla River, 390
Return rates, for transported fish, 385
Rice Island, tern colonies at, 526
Rich, Willis, 430
Ringold, hatcheries at, 642
Riparian vegetation
  and harvest management, 492
  and thermal regime, 197
Riparian zone, defined, 62
Risk assessment
  and adaptive management, 452
  spreading of
    for Bristol Bay salmon, 73
    and population diversity, 70
River channels, shapes of, 63
River discharge, 393
  and migration rates, 296, 297f
  and water transit time, 297f
River flow, 301 See also Flow rates
  and fish survival, 300–301
  within-day variation in, 303
Riverine ecosystems
  beaver trapping in, 200–201
  direct human alteration of, 221
Riverine habitat
  high-quality
    for adult migration, 186–187
    and groundwater upwelling, 193–196, 194f, 195f
    influence of wood on, 188–190, 189f

juveniles in, 182–186
    for reproduction, 180–181
    and temperature patterns, 190–193, 191f
  impact of human activities on
    beaver trapping, 200–201
    excessive nutrients, 210–211
    grazing, 204–205, 205f
    irrigation and cropland agriculture, 208–210, 209f, 210f
    logging, 201, 203–204
    mining, 206–208, 206f, 207f
    stream regulation, 211–220
    toxic pollutants, 210–211
Riverine zone, defined, 62
River-riparian interface (ecotone), 187
Rivers, geomorphic features of, 645. See also Columbia River
Road densities, and spawning strongholds, 83
Roadless Area Conservation Rule, 657
Roads
  building, 485
  and flood plan, 519, 521f
  and habitat loss, 234
  and riverine ecosystems, 203
  and riverine habitat, 199
  and salmonid populations, 204
Rock Creek, and Chinook salmon spawning, 126
Rock Island Dam, 10, 214, 250, 254, 295, 473
  bypass system at, 356, 361
  construction of, 344
  spill effectiveness at, 365
  statistics for, 179t
  surface collection at, 368, 368f, 369f
  temperature cycles at, 217, 217f
Rocky Reach Dam
  bypass system at, 356, 361
  nighttime spill at, 363
  power lines from, 329f
  statistics for, 179t
  surface collection at, 368, 368f, 369f
  traveling screen at, 348f
Rogue River, timing for smolts in, 215
Rosa Dam, 254
Rotation drum screen, 634f
Runoff, for Columbia Basin, 177
Run timing
  of chum salmon, 136–137
  and genetic structure, 115–116

Russian fisheries, salmon harvests of, 544,
   545f

S

Sack fry, in hatchery programs, 430–431
St. Joe River, westslope cutthroat trout in,
   151, 152
Salinity profiles, of Pacific Ocean, 535
Salmo, occurrence of, 112
Salmon. *See also* Pacific salmon; *specific*
      species
   adaptation of, 647
   in Columbia River, 5–15
   commercial fishing of, 2–3
   commercial landings of, 8, 8f, 35f,
      637f
   decline in numbers of, 3, 4, 578, 638
   and ecological conditions, 25f
   ecosystem, 35
   high fecundity of, 34
   marine survival of, 531
   prior to European settlement, 2
   production goals for, 58f
   species of, 469
   spiraling migration of, 261f
   threats to, 420
   trucking transport for, 636f
   upstream passage for, 341
Salmon conservation
   marine environment in, 65
   metapopulation theory in, 77
Salmon depletion
   assumptions associated with, 40t–41t
   causes of, 39t, 42
   solutions to problem, 39t
Salmon fisheries
   extended, 477f
   re-opening of, 466
Salmon Fishery Policy, sustainable, 484. *See*
      also Sustainable salmon fishing
Salmonid conservation
   dynamic approach to, 53–54
   metapopulation theory in, 74
Salmonid populations
   decline in, 630
   evaluation of, 8
   genetic structure of, 111–122
   habitats for, 59, 61, 61t
   status of, 197
   sustaining, 90–91
Salmonid restoration, natural processes in,
   89–90. *See also* Restoration

Salmonids
   distribution of, 61
   diversity of, 100, 101–102, 158
   evolution of, 100–101, 101*f*
   genetic structures of, 112–114, 113f, 115f
   genetic variation in, 112
   habitats for, 62–63, 105–106
   introduced, 181
   life histories of, 105–106
   metapopulations of, 74
   mortality of, 211
   non-nature vs. native, 144, 144t
   phylopatry in, 104
   populations structure of, 108–111
Salmonid Watershed Assessment Model, 620
Salmon management. *See also* Harvest
      management
   adaptive, 25
   change in, 553–554
   in Columbia River, 15–16
   conceptual foundation for, 32–35
   and estuarine environment, 552
   and marine fluctuations, 541
   stock structure in, 104
   uncertainty in, 553
Salmon populations
   decline of, 14
   effect of habitat conditions on, 588
   at Hanford Reach, 642
Salmon production
   and Aleutian Low, 554
   and climatic changes, 543–544
   and environmental shifts, 68f
   multidecadal variations in, 545
   and ocean environment, 549–550
   regional goals for, 58f
   and shifts in oceanic regime, 551
Salmon River, 233, 258, 291, 497f, 574f
   and Chinook salmon spawning, 126
   importance of, 77
   Middle Fork, 280
   westslope cutthroat trout, 151
   westslope cutthroat trout in, 152
Salvelinus confluentus (bull charr), 152–155
Sandbars, in tidal deltas, 517
Sand Island, 379
Sardines (*Sardinops* sp.)
   harvest of, 543, 544*f*
   and sea surface temperatures, 547
   and upwelling systems, 533
Sawmill, early, 201
Schooling behavior, during migration, 493

Scientific management, 426
Scientific method, and adaptive
    management, 47
Scientific Review Team,, of NPCC, 419,
    428–429
Scooting, 186
Scouring flows, 217
Screening systems, 12. *See also* Turbine
    intake screens
Screw trap, 589f
Sea levels, and salmon survival, 539
Searching behavior, 288
Sea surface, and PDO, 545
Sea surface salinities
    alteration in, 531
    impact of hydrosystem on, 520
Sea surface temperature (SST)
    and Aleutian Low, 554
    and biological changes, 547
    and salmon survival, 539
Secesh River, 589f, 590f
Sediment trap, Columbia River estuary as,
    523
Segment system, 60t
Seiches (oscillations), 304, 305
    flows induced by, 306–307, 306f
    and powerhouse discharges, 302
Seining, beach, 644f
Serial Discontinuity Concept (SDC), 231
Shad, in Columbia River estuary, 524, 527
Shoaling rates, 523
Shoals, shifting, 517
Shoreline development, and flood plan, 519,
    521f
Shoreline distribution, for subyearling
    juveniles, 273
Shoshone Falls, 5, 6, 7f, 126
Silicates, in Columbia River plume, 529
Sink populations, 76
Sixes River, 513
Size, fish
    of juveniles, 265–266
    migration and, 275
    and predation, 378
Skamania steelhead stock, 448
Smallmouth bass *(Micropterus dolomieu)*,
    376–377
Smelts, in Columbia River estuary, 524
Smoltification, 277
    explained, 266
    and migration, 293
    and outmigration, 295

process of, 280
and travel times, 286–287
Smolt Monitoring Program, 228, 371–372
Smolts
    in estuarine environment, 511
    factors affecting, 295
    migrating steelhead, 291
    migration of, 308–309
    protection programs for, 449
    in Snake River, 277
    spills as bypass for, 359, 361–362
    steelhead, 375, 526
Smolt trap, 589*f*
Snake River, 257, 273, 281, 287, 294
    average velocity in, 295
    Chinook salmon in, 111, 211, 279
    and Chinook salmon spawning, 126, 128f,
        130, 331f
    cutthroat trout in, 148
    dams on, 11, 281, 286
    descaling rates at, 353
    estuarinization of, 227
    fall run fish in, 314
    Ice Harbor Dam, 340f
    juvenile salmon migrants in, 12
    lower, 644, 647
    mainstems of, 176f
    management in, 276–278
    migration in, 280, 285
    reservoirs on, 214, 215f
    riverine vs. lacustrine nature of, 646
    smolts in, 277
    spring Chinook salmon in, 313
    subyearlings in, 272
    temperature changes in, 217
Snake River Basin, 327, 607
Snake River Canyon reach, 278
Snake River Salmon Recovery, 357
Sockeye salmon *(O. nerka)*, 138f
    commercial salmon fishery landings of,
        134f
    distribution of, 138, 139, 140
    on Endangered Species List, 428
    ESA listing for, 18, 141, 633
    in estuarine environment, 513
    genetic structure of, 139–140
    harvest summary of, 140
    in hatchery programs, 422
    in mid-Columbia Reach, 294
    migration of, 184, 289–290
    propagation efforts for, 141
    rearing of juveniles, 259–260

Sockeye salmon (*Cont.*)
  run size for, 7
  sustainability of, 70–72
  and temperature changes, 190, 191f
Spawning
  effects of dams on, 256–259
  fall vs. spring, 181
  habitats for, 72
  impact of harvests on, 476
  impact of hydroelectric system on, 645
  and inundation, 402
  surveys of, 257
  temporary trap for, 582f
  vulnerability during, 187
Spawning escapements
  effective management of, 495
  maintaining, 490–492
  objectives for, 492
Spawning grounds
  of Chinook salmon, 577f
  surveys of, 140
Spawning habitat
  groundwater-influenced, 194
  protection of, 487–488, 487f
  reduction of mainstem, 11
Special interests, and NPPC, 37
Species, introduced, 220
Spill deflectors, 373, 374, 374f
Spill efficiency, determination of, 363
Spill levels
  establishing, 361
  estimated, 362t
Spills
  attracting juvenile salmon into, 368–369
  at Bonneville Dam, 358f
  effectiveness of, 363–365
  juvenile salmonids diverted into, 357–359,
    358f
  at Lower Monumental Dam, 358f
  and seasonality of flow, 655
  for smolt bypass, 359, 361–362
  supersaturation of gas due to, 371–373,
    372f
Spillways, 400
  design of, 370
  economics of, 400
  and flow rates, 254, 255f
  spreading spill volume, 365
  survivals of juveniles in, 369–370
  of typical dam, 343f
Spiraling, 271
Spiraling migrations, 261f

Spirit of the Salmon *Wy-Kan-Ush-Mi-Wa-
    Kish-Wit*, 603
Spokane Basin, discharge statistics for, 175t
Spokane River
  coho spawning in, 133
  westslope cutthroat trout, 151
Sport reward fishery, 378, 379
Spring Chinook salmon
  ESA listing for, 334, 635
  migration of, 279–280
  and ocean fisheries, 481, 482
  rearing habitat of, 575f
  in Snake River, 313
  trends of, 637f
  in Willamette River, 287–288
Spring freshets
  impact of hydrosystem on, 520
  and reservoir system, 608
Spring/summer Chinook
  effects of transportation on, 385
  ESA listing for, 635
Stabilizing selection, 450, 450f. *See also*
    Genetics
Stage waves, 282, 312
Statistical monitoring, 584–586
Steady state thinking, 65
Steelhead trout *(O. mykiss)*, 143f, 147f. *See
    also* Rainbow trout
  annual returns of, 440–441, 442f
  commercial landings of, 8, 8f, 35f
  decline of, 14
  declining populations of, 3, 4
  downstream migration, 260–262
  under ESA, 147, 147t
  ESA listing, 147, 147t
  ESA listing for, 18, 334, 428, 635
  genetic structure of, 120–123, 121f, 122f
  in hatchery programs, 422
  in Hoh River, 188f
  in mid-Columbia Reach, 294
  migration of, 290–292
  population cluster for, 122f
  population structure of, 120, 122f
  rearing of, 259
  run size for, 7
  in Snake River, 292–293
  Snake River Basin, 122
  survival estimates for, 391
  travel time for, 298f
  trucking transport for, 636f
  upstream passage for, 341
  and water transit time, 296, 300f

Stevens, Gov. Isaac, 19
Stevens' treaty tribes, 20
Stewardship. *See also* Management
 habitat, 420–424
 hatchery production as substitute for, 423
 streamside, 232
Stock concept, 101–102
 elements of, 104
 in fisheries, 102
 in Pacific salmon, 102–105
Stock identification, and fishing mortality, 495
Stocks
 comprehensive surveys of, 576
 conservation, 106–108
 decline in numbers of, 630
 definition of, 102, 107, 495
 genetic changes to, 450
 in hatchery programs, 430–431
 healthy, 630
 massive transfers of, 431
 recruitment concept for, 492
 wild salmon, 467
Stone, Livingston, 420
Stone and Webster Engineering
  Corporation, 380
Storage-to-flow ratios, 214
Straying, of salmonid populations, 108. *See
  also* Genetics
Streambed condition, and harvest
  management, 492
Stream drift, chironomids in, 224–225
StreamNet, 576
Streams
 habitat changes in, 60t
 racking of, 431
Stress levels, and fish mortality, 354
Strongholds, as restoration focus, 83
Sturgeon, 155f
 distribution of, 155–156
 under ESA, 157
 FCRPS 2000 Biological Opinion on, 394
 fish and Wildlife Service's Biological
  Opinion on hydrosystems affecting,
  603
 and flooded riparian habitat, 225
 harvest of, 156–157
 propagation of, 157
 spawning patterns of, 156
 status of, 155–157
Subarctic Current, 536
Subbasins

 habitat connectivity in, 81
 planning for, 590
Submerged traveling screen (STS), 347–348,
  348f
Subyearling Chinook
 distribution of, 268
 in mid-Columbia Reach, 293
 migration times for, 268–269, 269f
 shoreline habitat for, 269–270
Subyearlings
 annual movement of, 275
 in Columbia River estuary, 527
 in estuaries, 274–275
 in estuarine environment, 513, 514
 experimental research on, 275–276
 management risks for, 278–279
 in reservoirs, 272–274
 in Snake River, 272
 spiraling migration of, 261f
Summer Chinook salmon
 in Johnson Creek, 183f
 migration of, 280, 281–282
 multiple spawning of, 491f
Supreme Court, U.S., 19
Surface bypass systems, objective of,
  357–358
Surface collection, application of, 367–368,
  368f, 369f
Surface elevation changes, 306–307, 306f
Surface orientation, and migration, 262
Surface spill, 358, 358f
Surges, studies on, 312
Surveys
 around Columbia River plume, 530
 comprehensive stock, 576
 earliest, 574
 field notebooks from, 575–576
Survival
 coho, 532
 dam obstacle hypothesis of, 311–312
 effects of flow on, 392
 fish orientation hypothesis of, 308–311
 and fluctuating flow hypothesis, 302–308
 and gas bubble disease, 373
 and geographic variations, 533
 of juveniles
  in river reaches, 390–391
  in spills, 369–370
  transported, 385
 and life history patterns, 66
 of marine salmon, 509, 510f, 532
 migration and, 296–300

Survival (*Cont.*)
  monitoring, 595
  ocean, 538–539
  reach, 400
  steelhead, 391
  and temperature modification, 397
  and turbine intake screens, 354, 356
  and upwelling conditions, 532
  velocity hypothesis of, 301–302
  and water transit time, 299*f*
Sustainability
  and diversity, 70–73
  of salmonid populations, 90–91
Sustainable Salmon Fishery Policy, of
      Alaska Board of Fisheries, 483
Sustainable salmon fishing
  effective management in, 494–496
  harvesting with caution, 493–494
  maintaining public support for, 497–498
  maintaining spawning escapements,
      490–492
  principles of, 486
  protecting wild salmon, 486–490
Sustainable salmon management, 499
Swamps, and estuary modification, 517, 519
Swan Falls, 76
Swan River, 204
Swan River Basin, 199
Swimming behavior
  in flume experiments, 274
  during migration, 267
  subyearling, 275–276
Synchrony
  and human development, 110
  in salmonid fishes, 74–75
  and salmonid restoration, 91
System Monitoring and Evaluation
      Program, 572

T

Tagging experiments, 103
Tagging programs. *See also* Passive
      Integrated Transponder (PIT) tagging
  CWT, 607–608
  review of, 509
Tailraces
  downstream of hydroelectric projects, 646,
      646f
  juvenile losses in, 375–376
  rates of mortality in, 336
  spawning in, 256

Tailwater habitats, Chinook use of, 648
Taneum Creek, irrigation diversion dam on,
      330f
Technical Recovery Teams (TRTs), of
      NOAA, 623
"Techno-arrogance," 639
Technology
  and ecological process, 20–22
  hatcheries, 44
  persistent belief in, 426–427
Telemetry studies
  flushing seen on, 280
  of migration rates, 284
  of wandering smolts, 309
Teleosts, 100–101
Temperatures
  effects of dams on, 214, 215f
  effects of hydroelectric system on, 396
  and fish distribution, 213–214
  and groundwater upwelling, 195
  and irrigation diversion, 209
  and life histories, 82
  of Pacific Ocean, 535
  and PDO, 545
  and reservoirs, 214, 215f
  restoring, 217
  and salmon abundance, 197
  and sardine abundance, 543, 544f
  study of, 235
  summer, 184–185
  and water quality, 190–193, 191f
  and zooplankton biomass, 537f
Tenmile Lakes, 542
Thermocline, depressed, 541
Thompson, William Francis, 66, 472
Toeslope surfaces, logging and, 203
Trade winds, 540
Travel time
  calculating, 285
  compared with water transit time, 297f
  dam obstacle hypothesis of, 311–312
  effects of flow on, 301–312
  fish orientation hypothesis of, 308–311
  and flow augmentation, 391
  and fluctuating flow hypothesis, 302–308
  in Lower Granite Reservoir, 292
  smolt, 309, 310f
  and smoltification, 286–287
  velocity hypothesis of, 301–302
  water/fish movement in, 294–295
Treaty Tribal Fishing Rights, 403

Trend monitoring, 581–584
Tributaries
  dams and, 250, 251
  far-from-ocean, 258
  impact of human activity on, 232
  regulation of, 179
  and temperature patterns, 192
  unsuitable temperatures for, 217
  whole-river model for, 236
Tributary habitats, in recovery planning,
  605, 606f, 611–612, 613f, 618, 622–623
Trigger points, 454
Troll fisheries. coastal ocean, 473
Trout. See Cutthroat trout; Steelhead
  trout
Trucking
  as technological solution, 630, 635f,
    636f
  for transporting juveniles, 382–386
Truncating selection, 450, 450f. See also
  Genetics
Tucannon River
  fish habitat in, 197–198, 198f
  juveniles in, 81t
Tumwater Canyon, 254
Turbine intake screens, 345, 347–355, 348f,
  350f, 351t
  application of, 355–356
  cleaning of, 347
  effectiveness of
    for bypass of juveniles, 350–352
    bypass outfall, 355
    conduit to tailrace, 352–353
    stress measured during passage,
      354–355
  extended length bar screens, 350, 351
  installation of, 633f
  principle of operation for, 343f, 347,
    348f
  as technological solution, 630, 633f
Turbines
  design research for, 381
  diverting juveniles from, 342
  "fish friendly," 345
  and fish mortality, 344–345
  at Grand Coulee Dam, 346f
  horizontally oriented bulb, 344
  Kaplan-type, 344
  operations, 400
Turbulence, effect on smolt migration of,
  309–311, 310f

U

Umatilla River, 254, 390
  Chinook salmon in, 429–430
  and Chinook salmon spawning, 126
  juveniles in, 81t
Umatilla Stream, temperature changes in,
  215
Uncertainties
  about best management practices, 235
  critical, 555–556
"Upriver brights," 644
Upwelling influence, zone of, 530
Upwelling systems
  coastal, 531–534
  and El Niño, 540
  in global heat budget, 539
  and pelagic production, 538
  and sardine production, 547
  and smolt migration, 550
  in summer, 534
  and variability, 536
Urbanization, 9
U.S. Fish and Wildlife Service (USFWS),
  143, 344, 606
U.S. v. Oregon, 615

V

Variability
  in freshwater, 552
  seasonal scales of, 548–549
Varial zone, 213, 217
Variation
  in fishery management, 92
  loss of within-population, 447
  in marine survival, 509
Vegetation
  flooded riparian, 225, 226
  submerged riparian, 224
Velocities. See also Water velocity
  factors affecting, 295–296
  formula for estimating, 302
Velocity hypothesis, of travel time,
  301–302
Vernita Bar, 64f, 86f
Vernita Bar Agreement, 258, 304, 396
Viability, and harvest management,
  498
Vortices
  and migration rates, 283f
  studies on, 312

W

Walla Walla River, 126, 277
Walleye *(Stizostedion vitreum)*, 377
Wanapum Dam
    bypass system at, 356, 359, 363
    spill effectiveness at, 365
    statistics for, 179t
    surface collection at, 368, 368f, 369f
    tank trucks transport at, 381
    turbine intake screens at, 348
Warm sardine regime, 547
Warm Springs National Fish Hatchery, 443
Washington State, coasts of, 67
Water
    from irrigated fields, 209
    in recovery plans, 608, 609f, 616–617, 619
Water budget
    defined, 393–394
    and flow augmentation, 392–394, 394t
    implementing, 394–396, 394t
    terminology, 393–394
Water displacement times, 312
Waterfalls, 186
Water/fish movement, principles of, 294–295
Watersheds
    and genetic structuring in salmonids, 114
    human development in, 110
    life history variation within, 69–70
    and resource management, 220
Water transit time (WTT). *See also* Travel
        time
    and Chinook travel time, 296–301,
        297f–300f
    and flow rates, 284f
    and smolt survival, 299f
Water velocity, 281 *See also* Flow rates
    and migration rates, 281, 283f
    turbulent burst, 282f
    waves, 282, 282f, 284
Weir, with counting facilities, 589f
Weiser River, 258
Wells Dam, 262, 295, 367
    bypass system at, 356, 361, 400, 401
    hydrocombine at, 360f, 370–371, 372f
    spillway at, 357, 360f
    statistics for, 179t
Wenaha River, Chinook salmon spawning
        in, 130
Wenatchee Lake, 7, 138, 139, 141, 289
Wenatchee River, 254, 289
    Chinook salmon in, 111

coho spawning in, 133
    sockeye salmon in, 140
Wenatchee River Basin, Chinook redds in,
        584
Westerlies, 540
Westslope cutthroat trout *(O. clarki lewisi)*,
        18, 151–152
West Wind Drift, 536
Wetlands, 515 *See also* Estuaries
    of Cathlamet Bay, 517f
    and food chain changes, 527
    loss of, 527
    vegetated, 515
Wet year, and flow strategies, 314
White sturgeon, 155–157, 155f
Wilderness, and habitat loss, 234
Wildfire, and riverine habitat, 199
Wild fish, 497f
    effects of hatcheries on, 444
    genetic effects of hatchery fish on
        direct, 446–448
        indirect, 448–449
    interactions with hatchery-origin fish of,
        444–446
    producing, 446
Wild salmon
    in harvest management, 467
    protection of, 486–490
    recovery of, 496
Willamette Basin, discharge statistics for,
        175t
Willamette Falls, 290
Willamette River, 279, 281, 284, 287–288,
        293
Windermere Lake, 126
Wind River, 653f
Wind River Canyon, 653f
Wood
    and riverine habitat, 188–190, 189f
    submerged, 224
Working group, on predation, 376, 378
World view, 4. *See also* Conceptual
        foundation

Y

Yakima Basin
    connected wetlands of, 213
    discharge statistics for, 175t
Yakima River, 254, 389, 473, 585f–586f, 643f
    juveniles in, 69, 81t
    timing for smolts in, 214–215

Yakima Tribe, 288
Yearling Chinook, 279–280
  in estuary, 287
  and flow structure, 281–285, 282f–284f
  in mid-Columbia Reach, 293
  in reservoirs, 285–287
  travel time of, 298
  and water transit time, 296–297, 299f
Yearlings
  diurnal periodicity of, 285–286
  flushing of, 280–281
  spiraling migration of, 261f
Yearling steelhead, 292
Yellowstone cutthroat trout, 148t
  life history of, 70
  propagation efforts for, 150

Yellowstone River, cutthroat trout in, 148
Yield management, basis of, 499
Young's bay fishery, 131

Z

Zig-zagging, 271
Zig-zag migrations, 261f
Zooplankton, 227, 530–531
  and Aleutian Low, 554
  in California Current, 534
  and climatic shifts, 545
  distribution of, 537, 537f
  off southern California, 536
  patterns of abundance of, 532
  and thermocline, 541

Printed and bound by CPI Group (UK) Ltd, Croydon, CR0 4YY

03/10/2024

01040412-0015